GATE PREP SERIES

GATE

Graduate Aptitude Test in Engineering

2022

22 Years

Chapter-Wise Solved Papers

(2000-2021)

Physics

G. K. PUBLICATIONS (P) LTD.

Title : GATE 2022 - Physics - 22 Years' Chapter-wise Solved Papers (2000-2021)

Language : English

Editor's Name : Vinit Garg

Typeset & Published by :

Career Launcher Infrastructure (P) Ltd.

A-45, Mohan Cooperative Industrial Area, Near Mohan Estate Metro Station, New Delhi - 110044

Marketed by :

G.K. Publications (P) Ltd.

Plot No. 9A, Sector-27A, Mathura Road, Faridabad, Haryana-121003

ISBN : **978-93-91061-26-5**

Contents

Technical Section

GATE

Graduate Aptitude Test in Engineering

IIT Institutes

GATE 2021 conducted by
Indian Institute of Technology, Mumbai

GATE 2022 will be conducted by
Indian Institute of Technology, Kharagpur

Indian Institute of Technology, Delhi

Indian Institute of Technology, Chennai

Indian Institute of Technology, Guwahati

Indian Institute of Technology, Roorkee

Indian Institute of Technology, Kanpur

Preface

The Graduate Aptitude Test in Engineering (GATE) is an online exam conducted by the IITs for admissions to PG courses in IITs, IISc Bangalore, NITs and many state run universities as well as private universities. Also there are more than 37 PSUs that use GATE score for recruitments. A large number of corporates are also using GATE score as a tool to screen students for placements.

GK Publications is well known as the "publisher of choice" to students preparing for GATE and other technical examinations in the country. We published the first set of books in 1994 when GATE exam, both objective and conventional, was conducted in the paper and pencil environment, and used as a check point for entry to post graduate courses in IITs and IISCs. At that time, students had little access to technology and relied mainly on instructor led learning followed by practice with books available for these examinations.

A lot has changed since then!

Today, GATE is conducted in an online only mode with multiple choice and numerical based questions. The score is valid for three years and is used not only for post graduate courses but is also used by major PSUs for recruitment. Today's students have easy access to technology and the concept of a monologue within the classroom has changed to dialogue where students come prepared with concepts and then discuss topics. They learn a lot of things on the go with their mobile devices and practice for mock tests online.

We, as a leading publisher of GATE books, have also embraced change. Today, our books are no more guides and papers only but come with a fully supported mobile app and a web portal. The mobile app provides access to video lectures, short tests and regular updates about the exam. The web portal in additional to what is available on the app provides full length mock tests to mimic the actual exam and help you gauge your level of preparedness. The combination of practice content in print, video lectures, and short and full length tests on mobile and web makes this product a complete courseware for GATE preparation.

This book includes previous years GATE questions along with detailed solution of each question for better understanding. It will help the GATE aspirants to know an idea about the pattern of questions asked in GATE examination.

We also know that improvement is a never ending process and hence we welcome your suggestions and feedback or spelling and technical errors if any. Please write to us at gkp@gkpublications.com

We hope that our small effort will help you prepare well for the examination.

We wish you all the best!

GK Publications Pvt. Ltd.

About GATE

The Graduate Aptitude Test in Engineering (GATE) conducted by IISc and IITs has emerged as one of the bench mark tests for engineering and science aptitude in facilitating admissions for higher education (M.Tech./Ph.D.) in IITs, IISc and various other Institutes/Universities/Laboratories in India. With the standard and high quality of the GATE examination in 27 disciplines of engineering and science, Humanities and Social Sciences subjects, it identifies the candidate's understanding of a subject and aptitude and eligibility for higher studies. During the last few years, GATE score is also being used as one of the criteria for recruitment in Government Organizations such as Cabinet Secretariat, and National/State Public Sector Undertakings in India. Because of the importance of the GATE examination, the number of candidates taking up GATE exams has increased tremendously. GATE exams are conducted by the IITs and IISc as a computer based test having multiple choice questions and numerical answer type questions. The questions are mostly fundamental, concept based and thought provoking. From 2017 onwards GATE Exam is being held in Bangladesh, Ethiopia, Nepal, Singapore, Sri Lanka and United Arab Emirates. An Institute with various nationalities in its campus widens the horizons of an academic environment. A foreign student brings with him/her a great diversity, culture and wisdom to share. Many GATE qualified candidates are paid scholarships/assistantship, especially funded by Ministry of Human Resources Development, Government of India and by other Ministries. IIT, Kharagpur is the Organizing Institute for GATE 2022.

Why GATE?

Admission to Post Graduate and Doctoral Programmes

Admission to postgraduate programmes with MHRD and some other government scholarships/ assistantships in engineering colleges/institutes is open to those who qualify through GATE. GATE qualified candidates with Bachelor's degree in Engineering/Technology/Architecture or Master's degree in any branch of Science/Mathematics/Statistics/Computer Applications are eligible for admission to Master/Doctoral programmes in Engineering/Technology/Architecture as well as for Doctoral programmes in relevant branches of Science with MHRD or other government scholarships/ assistantships. Candidates with Master's degree in Engineering/Technology/Architecture may seek admission to relevant Ph.D programmes with scholarship/assistantship without appearing in the GATE examination.

Financial Assistance

A valid GATE score is essential for obtaining financial assistance during Master's programs and direct Doctoral programs in Engineering/Technology/Architecture, and Doctoral programs in relevant branches of Science in Institutes supported by the MHRD or other Government agencies. As per the directives of the MHRD, the following procedure is to be adopted for admission to the post-graduate programs (Master's and Doctoral) with MHRD scholarship/assistantship. Depending upon the norms adopted by a specific institute or department of the Institute, a candidate may be admitted directly into a course based on his/her performance in GATE only **or** based on his/her performance in GATE **and** an admission test/interview conducted by the department to which he/she has applied **and/or** the candidate's academic record. If the candidate is to be selected through test/interview for post-graduate programs, a minimum of 70% weightage will be given to the performance in GATE and the remaining 30% weightage will be given to the candidate's performance in test/interview and/or academic record, as per MHRD guidelines. The admitting institutes could however prescribe a minimum passing percentage of marks in the test/interview. Some colleges/institutes specify GATE qualification as the mandatory requirement even for admission without MHRD scholarship/assistantship.

To avail of the financial assistance (scholarship), the candidate must first secure admission to a program in these Institutes, by a procedure that could vary from institute to institute. Qualification in GATE is also a minimum requirement to apply for various fellowships awarded by many Government organizations. Candidates are advised to seek complete details of admission procedures and availability of MHRD scholarship/assistantship from the concerned admitting institution. The criteria for postgraduate admission with scholarship/assistantship could be different for different institutions. The management of the post-graduate scholarship/assistantship is also the responsibility of the admitting institution. Similarly, reservation of seats under different categories is as per the policies and norms prevailing at the admitting institution and Government of India rules. *GATE offices will usually not entertain any enquiry about admission, reservation of seats and / or award of scholarship / assistantship.*

PSU Recruitments

As many as 37 PSUs are using GATE score for recruitment. It is likely that more number of PSUs may start doing so by next year. Below is the list of PSUs:

MDL, BPCL, GAIL, NLC LTD, CEL, Indian Oil, HPCL, NBPC, NECC, BHEL, WBSEDCL, NTPC, ONGC, Oil India, Power Grid, Cabinet Secretariat, Govt. of India, BAARC, NFL, IPR, PSPCL, PSTCL, DRDO, OPGC Ltd., THDC India Ltd., BBNL, RITES, IRCON, GHECL, NHAI, KRIBHCO, Mumbai Railway Vikas Corporation Ltd. (MRVC Ltd.), National Textile Corporation, Coal India Ltd., BNPM, AAI, NALCO, EdCIL India.

Important :

1. Admissions in IITs/IISc or other Institutes for M.Tech./Ph.D. through GATE scores shall be advertised separately by the Institutes and GATE does not take the responsibility of admissions.
2. Cabinet Secretariat has decided to recruit officers for the post of Senior Field Officer (Tele) (From GATE papers of EC, CS, PH), Senior Research Officer (Crypto) (From GATE papers of EC, CS, MA), Senior Research Officer (S&T) (From GATE papers EC, CS, CY, PH, AE, BT) in the Telecommunication Cadre, Cryptographic Cadre and Science & Technology Unit respectively of Cabinet Secretariat. The details of the scheme of recruitment shall be published in National Newspaper/Employment News by the concerned authority.
3. Some PSUs in India have expressed their interest to utilize GATE scores for their recruitment purpose. The Organizations who intend to utilize GATE scores shall make separate advertisement for this purpose in Newspapers etc.

Who Can Appear for GATE?

Eligibility for GATE

Before starting the application process, the candidate must ensure that he/she meets the eligibility criteria of GATE given in Table.

Eligibility Criteria for GATE 2022

Degree/Program	Qualifying Degree/Examination	Description of Eligible Candidates	Expected Year of Completion
B.E. / B.Tech. / B.Pharm.	Bachelor's degree in Engineering / Technology (4 years after 10+2 or 3 years after B.Sc. / Diploma in Engineering / Technology)	Currently in the 3rd year or higher grade or already completed	2022
B. Arch.	Bachelor's degree of Architecture (5- year course) / Naval Architecture (4- year course) / Planning (4- year course)	Currently in the 3rd year or higher grade or already completed	2023 (for 5-year program), 2022 (for 4-year program)
B.Sc. (Research) / B.S.	Bachelor's degree in Science (Post-Diploma/4 years after 10+2)	Currently in the 3rd year or higher grade or already completed	2022
Pharm. D. (after 10+2)	6 years degree program, consisting of internship or residency training, during third year onwards	Currently in the 3rd/4th/5th/6th year or already completed	2024
M.B.B.S.	Degree holders of M.B.B.S. and those who are in the 5th/6th/7th semester or higher semester of such programme.	5th, 6th, 7th or higher semester or already completed	2022
M. Sc. / M.A. / MCA or equivalent	Master's degree in any branch of Arts/Science/Mathematics/Statistics/ Computer Applications or equivalent	Currently in the first year or higher or already Completed	2022
Int. M.E./ M.Tech. (Post-B.Sc.)	Post-B.Sc Integrated Master's degree programs in Engineering/ Technology (4-year program)	Currently in the 1st / 2nd/3rd/4th year or already completed	2024
Int. M.E./ M.Tech. or Dual Degree (after Diploma or 10+2)	Integrated Master's degree program or Dual Degree program in Engineering/Technology (5-year program)	Currently in the 3rd /4th/5th year or alreadycompleted	2023
B.Sc. / B.A. / B.Com.	Bachelor degree in any branch of Science / Arts / Commerce (3 years program)	Currently in the 3rd year or already completed	2021
Int. M.Sc. / Int. B.S. / M.S.	Integrated M.Sc. or 5-year integrated B.S.-M.S. program	Currently in the 3rd year or higher or already completed	2022
Professional Society Examinations (equivalent to B.E. / B.Tech. / B.Arch.)	B.E./B.Tech./B.Arch. equivalent examinations of Professional Societies, recognized by MoE/UPSC/AICTE (e.g. AMIE by Institution of Engineers-India, AMICE by the Institute of Civil Engineers-India and so on)	Completed Section A or equivalent of such professional courses	NA

In case a candidate has passed one of the qualifying examinations as mentioned above in 2020 or earlier, the candidate has to submit the degree certificate / provisional certificate / course completion certificate / professional certificate / membership certificate issued by the society or institute. In case, the candidate is expected to complete one of the qualifying criteria in 2022 or later as mentioned above, he/she has to submit a certificate from Principal or a copy of marks card for section A of AMIE.

Certificate From Principal

Candidates who have to submit a certificate from their college Principal have to obtain one from his/her institution beforehand and upload the same during the online submission of the application form.

Candidates With Backlogs

Candidates, who have appeared in the final semester/year exam in 2022, but with a backlog (arrears/failed subjects) in any of the papers in their qualifying degree should upload a copy of any one of the mark sheets of the final year,

OR

obtain a declaration from their Principal along with the signature and seal beforehand and upload the same during the online submission of the application form.

GATE Structure

GATE 2022 Papers

GATE 2022 will be conducted on 29 subjects (papers). Table below shows the list of papers and paper codes for GATE 2022. A candidate is allowed to appear in ANY ONE or UP TO TWO papers of the GATE examination. However, note that the combination of TWO papers in which a candidate can appear MUST be from the pre-defined list as given in Table. Also note that for a paper running in multiple sessions, a candidate will be mapped to appear for the examination in one of the sessions ONLY.

List of GATE 2022 Papers and Corresponding Codes

GATE Paper	Code	GATE Paper	Code
Aerospace Engineering	AE	Instrumentation Engineering	IN
Agricultural Engineering	AG	Mathematics	MA
Architecture and Planning	AR	Mechanical Engineering	ME
Bio-medical Engineering	BM	Mining Engineering	MN
Biotechnology	BT	Metallurgical Engineering	MT
Civil Engineering	CE	Naval Architecture and Marine Engineering	NM
Chemical Engineering	CH	Petroleum Engineering	PE
Computer Science and Information Technology	CS	Physics	PH
Chemistry	CY	Production and Industrial Engineering	PI
Electronics and Communication Engineering	EC	Statistics	ST
Electrical Engineering	EE	Textile Engineering and Fibre Science	TF
Environmental Science & Engineering	ES	Engineering Sciences	XE*
Ecology and Evolution	EY	Humanities & Social Sciences	XH**
Geomatics Engineering	GE	Life Sciences	XL***
Geology and Geophysics	GG		

Note: *"Geomatics Engineering (GE)" and "Naval Architecture and Marine Engineering (NM)" are two new papers introduced in GATE 2022.*

*XE Paper Sections	Code	**XH Paper Sections	Code	***XL Paper Sections	Code
Engineering Mathematics (Compulsory) (15 marks)	A	Reasoning and Comprehension (Compulsory) (25 marks)	B1	Chemistry (Compulsory) (25 marks)	P
Any TWO optional Sections (2x35 = 70 marks)		Any ONE optional Section (60 marks)		Any TWO optional Sections (2x30 = 60 marks)	
Fluid Mechanics	B	Economics	C1	Biochemistry	Q
Materials Science	C	English	C2	Botany	R
Solid Mechanics	D	Linguistics	C3	Microbiology	S
Thermodynamics	E	Philosophy	C4	Zoology	T
Polymer Science and Engineering	F	Psychology	C5	Food Technology	U
Food Technology	G	Sociology	C6		
Atmospheric and Oceanic Sciences	H				

*XE (Engineering Sciences), **XH (Humanities & Social Sciences), ***XL (Life Sciences), papers are of general nature and will be comprised of Sections listed in the above table

Note: *Each subject/paper is of total 100 marks. General Aptitude (GA) section of 15 marks is common for all papers. Hence remaining 85 marks are for the respective subject/paper code.*

Combination of Two Papers Allowed to Appear in GATE 2022 (subject to availability of infrastructure and schedule)

Code of The First (Primary) Paper	Codes of Papers Allowed as The Second Paper
AE	XE
AG	ES
AR	CE
BM	BT / XL
BT	BM / XL
CE	AR / ES
CH	CY / PE / XE
CS	MA
CY	CH / XL
EC	IN / PH
EE	IN
ES	AG / CE
EY	XL
GG	MN / PE / PH
IN	EC / EE / PH
MA	CS / PH / ST
ME	XE
MN	GG / XE
MT	PH / XE
PE	CH / GG / XE
PH	EC / GG / IN / MA / MT / ST
PI	XE
ST	MA / PH
TF	XE
XE	AE / CH / ME / MN / MT / PE / PI / TF
XH	----
XL	BM / BT / CY / EY

General Aptitude Questions

All the papers will have a few questions that test the General Aptitude (Language and Analytical Skills), apart from the core subject of the paper.

Duration and Examination Type

All the papers of the GATE 2022 examination will be for 3 hours duration and they consist of 65 questions for a total of 100 marks. Since the examination is an ONLINE computer based test (CBT), at the end of the stipulated time (3-hours), the computer screen will automatically close the examination inhibiting any further action.

Candidates will be permitted to occupy their allotted seats 40 minutes before the scheduled start of the examination. Candidates can login and start reading the instructions 20 minutes before the start of examination. The late login time (if any) recorded by the computer system MUST NOT be beyond 30 minutes from the actual starting time of the examination. Under NO circumstances, will a candidate be permitted to login after 30 minutes from the actual examination starting time. Candidates will NOT be permitted to leave the examination hall before the end of the examination.

Pattern of Question Papers

GATE 2022 may contain questions of THREE different types in all the papers:

(i) **Multiple Choice Questions (MCQ)** carrying 1 or 2 marks each, in all the papers and sections. These questions are objective in nature, and each will have choice of four answers, out of which ONLY ONE choice is correct.

Negative Marking for Wrong Answers: For a wrong answer chosen in a MCQ, there will be negative marking. For 1-mark MCQ, 1/3 mark will be deducted for a wrong answer. Likewise, for 2-mark MCQ, 2/3 mark will be deducted for a wrong answer.

(ii) **Multiple Select Questions (MSQ)** carrying 1 or 2 marks each in all the papers and sections. These questions are objective in nature, and each will have choice of four answers, out of which ONE or MORE than ONE choice(s) is / are correct.

Note: There is NO negative marking for a wrong answer in MSQ questions. However, there is NO partial credit for choosing partially correct combinations of choices or any single wrong choice.

(iii) **Numerical Answer Type (NAT)** Questions carrying 1 or 2 marks each in most of the papers and sections. For these questions, the answer is a signed real number, which needs to be entered by the candidate using the virtual numeric keypad on the monitor (keyboard of the computer will be disabled). No choices will be shown for these types of questions. The answer can be a number such as 10 or –10 (an integer only). The answer may be in decimals as well, for example, 10.1 (one decimal) or 10.01 (two decimals) or –10.001 (three decimals). These questions will be mentioned with, up to which decimal places, the candidates need to present the answer. Also, for some NAT type problems an appropriate range will be considered while evaluating these questions so that the candidate is not unduly penalized due to the usual round-off errors. Candidates are advised to do the rounding off at the end of the calculation (not in between steps). Wherever required and possible, it is better to give NAT answer up to a maximum of three decimal places.

Note: *There is NO negative marking for a wrong answer in NAT questions.*
Also, there is NO partial credit in NAT questions.

Marking Scheme – Distribution of Marks and Questions

General Aptitude (GA) Questions

In all papers, GA questions carry a total of 15 marks. The GA section includes 5 questions carrying 1-mark each (sub-total 5 marks) and 5 questions carrying 2-marks each (sub-total 10 marks).

Question Papers other than GG, XE, XH and XL

These papers would contain 25 questions carrying 1-mark each (sub-total 25 marks) and 30 questions carrying 2-marks each (sub-total 60 marks) consisting of some MCQ type questions, while the remaining may be MSQ and / or NAT questions.

GG (Geology and Geophysics) Paper

Apart from the General Aptitude (GA) section, the GG question paper consists of two parts: Part A and Part B. Part A is compulsory for all the candidates. Part B contains two sections: Section 1 (Geology) and Section 2 (Geophysics). Candidates will have to attempt questions in Part A and questions in either Section 1 or Section 2 of Part B. The choice of Section 1 OR Section 2 of Part B has to be made at the time of online registration in GOAPS. At the examination hall, candidate cannot request for change of section.

Part A consists of 25 questions carrying 1-mark each (sub-total 25 marks and some of these will be MSQ and/or numerical answer type questions while remaining questions will be MCQ type). Either section of Part B (Section 1 and Section 2) consists of 30 questions carrying 2-marks each (sub-total 60 marks and some of these will be MSQ and/or numerical answer type questions while remaining questions will be MCQ type).

XE Paper (Engineering Sciences)

A candidate appearing in the XE paper has to answer the following:

- GA – General Aptitude carrying a total of 15 marks.
- Section A – Engineering Mathematics (Compulsory): This section contains 11 questions carrying a total of 15 marks: 7 questions carrying 1-mark each (sub-total 7 marks), and 4 questions carrying 2-marks each (sub-total 8 marks). Some questions will be of numerical answer type while remaining questions will be MCQ type.
- Any two of XE Sections B to H: The choice of two sections from B to H can be made during the examination after viewing the questions. Only TWO optional sections can be answered at a time. A candidate wishing to change midway of the examination to another optional section must first choose to deselect one of the previously chosen optional sections (B to H). Each of the optional sections of the XE paper (Sections B through H) contains 22 questions carrying a total of 35 marks: 9 questions carrying 1-mark each (sub-total 9 marks) and 13 questions carrying 2-marks each (sub-total 26 marks). Some questions will be of MSQ and/or numerical answer type while remaining questions will be MCQ type.

XH Paper (Humanities and Social Sciences)

A candidate appearing in the XH paper has to answer the following:

- **GA – General Aptitude** carrying a total of 15 marks.
- **Section B1 – Reasoning and Comprehension (Compulsory):** This section contains 15 questions carrying a total of 25 marks: 5 questions carrying 1-mark each (sub-total 5 marks) and 10 questions carrying 2-marks each (sub-total 20 marks). Some questions will be of MSQ and/or numerical answer type while remaining questions will be MCQ type.
- **Any ONE of XH Sections C1 to C6:** The ONE choice of section from C1 to C6 has to be made at the time of online registration in GOAPS. At the examination hall, candidate cannot request for change of section. Each of the optional sections of the XH paper (Sections C1 through C6) contains 40 questions carrying a total of 60 marks: 20 questions carrying 1-mark each (sub-total 20 marks) and 20 questions carrying 2-marks each (sub-total 40 marks). Some questions will be of MSQ and/or numerical answer type while remaining questions will be MCQ type.

XL Paper (Life Sciences)

A candidate appearing in the XL paper has to answer the following:

- **GA – General Aptitude** carrying a total of 15 marks.
- **Section P–Chemistry (Compulsory):** This section contains 15 questions carrying a total of 25 marks: 5 questions carrying 1-mark each (sub-total 5 marks) and 10 questions carrying 2-marks each (sub-total 20 marks). Some questions will be of MSQ and/or numerical answer type while remaining questions will be MCQ type.
- **Any two of XL Sections Q to U:** The choice of two sections from Q to U can be made during the examination after viewing the questions. Only TWO optional sections can be answered at a time. A candidate wishing to change midway of the examination to another optional section must first choose to deselect one of the previously chosen optional sections (Q to U). Each of the optional sections of the XL paper (Sections Q through U) contains 20 questions carrying a total of 30 marks: 10 questions carrying 1-mark each (sub-total 10 marks) and 10 questions carrying 2-marks each (sub-total 20 marks). Some questions will be of MSQ and/or numerical answer type while remaining questions will be MCQ type.

GATE Score

After the evaluation of the answers, the actual (raw) marks obtained by a candidate will be considered for computing the GATE Score. For multi-session papers (subjects), raw marks obtained by the candidates in different sessions will be converted to Normalized marks for that particular subject. Thus, raw marks (for single session papers) or normalized marks (for multi-session papers) will be used for computing the GATE Score, based on the qualifying marks.

Calculation of Normalized Marks for Multi-Session Papers

In GATE 2022 examination, some papers may be conducted in multi-sessions. Hence, for these papers, a suitable normalization is applied to take into account any variation in the difficulty levels of the question papers across different sessions. The normalization is done based on the fundamental assumption that "in all multi-session GATE papers, the distribution of abilities of candidates is the same across all the sessions". This assumption is justified since the number of candidates appearing in multi-session papers in GATE 2022 is large and the procedure for allocation of session to candidates is random. Further, it is also ensured that for the same multi-session paper, the number of candidates allotted in each session is of the same order of magnitude.

Based on the above, and considering various normalization methods, the committee arrived at the following formula for calculating the normalized marks for the multi-session papers.

Nsormalization mark of jth candidate in the ith session $\widehat{M_{ij}}$ is given by

$$\widehat{M_{ij}} = \frac{\bar{M}_t^g - M_q^g}{\bar{M}_{ti} - M_{iq}}(M_{ij} - M_{iq}) + M_q^g$$

where

M_{ij} : is the actual marks obtained by the jth candidate in i^{th} session

$\bar{M}_t^g$: is the average marks of the top 0.1% of the candidates considering all sessions

M_q^g : is the sum of mean and standard deviation marks of the candidates in the paper considering all sessions

$\bar{M}_{ti}$: is the average marks of the top 0.1% of the candidates in the ith session

M_{iq} : is the sum of the mean marks and standard deviation of the ith session

Calculation of GATE Score for All Papers

For all papers for which there is only one session, actual marks obtained by the candidates will be used for calculating the GATE 2022 Score. For papers in multi-sessions, normalized marks will be calculated corresponding to the raw marks obtained by a candidate and the GATE 2022 Score will be calculated based on the normalized marks.

The GATE 2022 score will be computed using the formula given below.

$$\text{GATE Score} = S_q + (S_t - S_q)\frac{(M - M_q)}{(\bar{M}_t - M_q)}$$

where

- M : marks obtained by the candidate (actual marks for single session papers and normalized marks for multi-session papers)
- M_q : is the qualifying marks for general category candidate in the paper
- $\bar{M}_t$: is the mean of marks of top 0.1% or top 10 (whichever is larger) of the candidates who appeared in the paper (in case of multi-session papers including all sessions)
- S_q : 350, is the score assigned to M_q
- S_t : 900, is the score assigned to $\bar{M}_t$

In the GATE 2022 the qualifying marks (M_q) for general category student in each subject will be 25 marks (out of 100) or $\mu + \sigma$, whichever is larger. Here μ is the mean and σ is the standard deviation of marks of all the candidates who appeared in the paper.

After the declaration of results, GATE Scorecards can be downloaded by the GATE qualified candidates ONLY.

The GATE 2022 Committee has the authority to decide the qualifying mark/score for each GATE paper. In case of any claim or dispute with respect to GATE 2022 examination or score, the Courts and Tribunals in Mumbai alone will have the exclusive jurisdiction to entertain and settle them.

GATE Syllabus

GENERAL APTITUDE

Verbal Aptitude

Basic English Grammar: tenses, articles, adjectives, prepositions, conjunctions, verb-noun agreement, and other parts of speech.

Basic Vocabulary: words, idioms, and phrases in context, Reading and comprehension, Narrative sequencing.

Quantitative Aptitude

Data Interpretation: data graphs (bar graphs, pie charts, and other graphs representing data), 2-and 3-dimensional plots, maps, and tables.

Numerical Computation and Estimation: ratios, percentages, powers, exponents and logarithms, permutations and combinations, and series, Mensuration and geometry, Elementary statistics and probability.

Analytical Aptitude

Logic: Deduction and induction, Analogy, Numerical relations and reasoning.

Spatial Aptitude

Transformation of Shapes: translation, rotation, scaling, mirroring, assembling, and grouping Paper folding, cutting, and patterns in 2 and 3 dimensions.

Section 1: Mathematical Physics

Vector calculus: linear vector space: basis, orthogonality and completeness; matrices; similarity transformations, diagonalization, eigenvalues and eigenvectors; linear differential equations: second order linear differential equations and solutions involving special functions; complex analysis: Cauchy-Riemann conditions, Cauchy's theorem, singularities, residue theorem and applications; Laplace transform, Fourier analysis; elementary ideas about tensors: covariant and contravariant tensors.

Section 2: Classical Mechanics

Lagrangian formulation: D'Alembert's principle, Euler-Lagrange equation, Hamilton's principle, calculus of variations; symmetry and conservation laws; central force motion: Kepler problem and Rutherford scattering; small oscillations: coupled oscillations and normal modes; rigid body dynamics: interia tensor, orthogonal transformations, Euler angles, Torque free motion of a symmetric top; Hamiltonian and Hamilton's equations of motion; Liouville's theorem; canonical transformations: action-angle variables, Poisson brackets, Hamilton-Jacobi equation.

Special theory of relativity: Lorentz transformations, relativistic kinematics, mass-energy equivalence.

Section 3: Electromagnetic Theory

Solutions of electrostatic and magnetostatic problems including boundary value problems; method of images; separation of variables; dielectrics and conductors; magnetic materials; multipole expansion; Maxwell's equations; scalar and vector potentials; Coulomb and Lorentz gauges; electromagnetic waves in free space, non-conducting and conducting media; reflection and transmission at normal and oblique incidences; polarization of electromagnetic waves; Poynting vector, Poynting theorem, energy and momentum of electromagnetic waves; radiation from a moving charge.

Section 4: Quantum Mechanics

Postulates of quantum mechanics; uncertainty principle; Schrodinger equation; Dirac Bra-Ket notation, linear vectors and operators in Hilbert space; one dimensional potentials: step potential, finite rectangular well, tunneling from a potential barrier, particle in a box, harmonic oscillator; two and three dimensional systems: concept of degeneracy; hydrogen atom; angular momentum and spin; addition of angular momenta; variational method and WKB approximation, time independent perturbation theory; elementary scattering theory, Born approximation; symmetries in quantum mechanical systems.

Section 5: Thermodynamics and Statistical Physics

Laws of thermodynamics; macrostates and microstates; phase space; ensembles; partition function, free energy, calculation of thermodynamic quantities; classical and quantum statistics; degenerate Fermi gas; black body radiation and Planck's distribution law; Bose-Einstein condensation; first and second order phase transitions, phase equilibria, critical point.

Section 6: Atomic and Molecular Physics

Spectra of one-and many-electron atoms; spin-orbit interaction: LS and jj couplings; fine and hyperfine structures; Zeeman and Stark effects; electric dipole transitions and selection rules; rotational and vibrational spectra of diatomic molecules; electronic transitions in diatomic molecules, Franck-Condon principle; Raman effect; EPR, NMR, ESR, X-ray spectra; lasers: Einstein coefficients, population inversion, two and three level systems.

Section 7: Solid State Physics

Elements of crystallography; diffraction methods for structure determination; bonding in solids; lattice vibrations and thermal properties of solids; free electron theory; band theory of solids: nearly free electron and tight binding models; metals, semiconductors and insulators; conductivity, mobility and effective mass; Optical properties of solids; Kramer's-Kronig relation, intra- and inter-band transitions; dielectric properties of solid; dielectric function, polarizability, ferroelectricity; magnetic properties of solids; dia, para, ferro, antiferro and ferri-magnetism, domains and magnetic anisotropy; superconductivity: Type-I and Type II superconductors, Meissner effect, London equation, BCS Theory, flux quantization.

Section 8: Electronics

Semiconductors in equilibrium: electron and hole statistics in intrinsic and extrinsic semiconductors; metal-semiconductor junctions; Ohmic and rectifying contacts; PN diodes, bipolar junction transistors, field effect transistors; negative and positive feedback circuits; oscillators, operational amplifiers, active filters; basics of digital logic circuits, combinational and sequential circuits, flip-flops, timers, counters, registers, A/D and D/A conversion.

Section 9: Nuclear and Particle Physics

Nuclear radii and charge distributions, nuclear binding energy, electric and magnetic moments; semi-empirical mass formula; nuclear models; liquid drop model, nuclear shell model; nuclear force and two nucleon problem; alpha decay, beta-decay, electromagnetic transitions in nuclei; Rutherford scattering, nuclear reactions, conservation laws; fission and fusion; particle accelerators and detectors; elementary particles; photons, baryons, mesons and leptons; quark model; conservation laws, isospin symmetry, charge conjugation, parity and time-reversal invariance.

Verbal Ability

1

CHAPTER

English Grammar

MCQ TYPE QUESTIONS

2021

1. (i) Arun and Aparna are here.

(ii) Arun and Aparna is here.

(iii) Arun's families is here.

(iv) Arun's family is here.

Which of the above sentences are grammatically CORRECT?

(*a*) (i) and (ii) (*b*) (i) and (iv)

(*c*) (ii) and (iv) (*d*) (iii) and (iv)

2020

2. I do not think you know the case well enough to have opinions. Having said that, I agree with your other point.

What does the phrase "having said that" mean in the given text?

(*a*) as opposed to what I have said

(*b*) despite what I have said

(*c*) in addition to what I have said

(*d*) contrary to what I have said

2016

3. Computers were invented for performing only high-end useful computations. However, it is no understatement that they have taken over our world today. The internet, for example, is ubiquitous. Many believe that the internet itself is an unintended consequence of the original invention. With the advent of mobile computing on our phones, a whole new dimension is now enabled. One is left wondering if all these developments are good or, more importantly, required.

Which of the statement(s) below is/are logically valid and can be inferred from the above paragraph?

(*i*) The author believes that computers are not good for us.

(*ii*) Mobile computers and the internet are both intended inventions

(*a*) (*i*) only (*b*) (*ii*) only

(*c*) both (*i*) and (*ii*) (*d*) neither (*i*) nor (*ii*)

4. All hill-stations have a lake. Ooty has two lakes.

Which of the statement(s) below is/are logically valid and can be inferred from the above sentences?

(*i*) Ooty is not a hill-station.

(*ii*) No hill-station can have more than one lake.

(*a*) (*i*) only (*b*) (*ii*) only

(*c*) both (*i*) and (*ii*) (*d*) neither (*i*) nor (*ii*)

5. Identify the correct spelling out of the given options :

(*a*) Managable (*b*) Manageable

(*c*) Mangaeble (*d*) Managible

6. Choose the statement(s) where the underlined word is used correctly :

(*i*) A prone is a dried plum.

(*ii*) He was lying prone on the floor.

(*iii*) People who eat a lot of fat are prone to heart disease.

(*a*) (*i*) and (*iii*) only (*b*) (*iii*) only

(*c*) (*i*) and (*ii*) only (*d*) (*ii*) and (*iii*) only

7. The policeman asked the victim of a theft, "What did you ?"

(*a*) loose (*b*) lose

(*c*) loss (*d*) louse

8. In a world filled with uncertainty, he was glad to have many good friends. He had always assisted them in times of need and was confident that they would reciprocate. However, the events of the last week proved him wrong.

Which of the following inference(s) is/are logically valid and can be inferred from the above passage?

(*i*) His friends were always asking him to help them.

(*ii*) He felt that when in need of help, his friends would let him down.

(*iii*) He was sure that his friends would help him when in need.

(*iv*) His friends did not help him last week.

(*a*) (*i*) and (*ii*) (*b*) (*iii*) and (*iv*)

(*c*) (*iii*) only (*d*) (*iv*) only

9. An apple costs Rs. 10. An onion costs Rs. 8.
Select the most suitable sentence with respect to grammar and usage.
(*a*) The price of an apple is greater than an onion.
(*b*) The price of an apple is more than onion.
(*c*) The price of an apple is greater than that of an onion.
(*d*) Apples are more costlier than onions.

10. He turned a deaf ear to my request. What does the underlined phrasal verb mean?
(*a*) ignored (*b*) appreciated
(*c*) twisted (*d*) returned

2015

11. Choose the statement where underlined word is used correctly
(*a*) The minister insured the victims that everything would be all right.
(*b*) He ensured that the company will not have to bear any loss.
(*c*) The actor got himself ensured against any accident.
(*d*) The teacher insured students of good results

12. The following question presents a sentence, part of which is underlined. Beneath the sentence you find four ways of phrasing the underline part. Following the requirements of the standard written English, select the answer that produces the most effective sentence.
Tuberculosis, together with its effects, ranks one of the leading causes of death in India.
(*a*) ranks as one of the leading causes of death
(*b*) rank as one of the leading causes of death
(*c*) has the rank of one of the leading causes of death
(*d*) are one of the leading causes of death

13. Choose the statement where underlined word is used correctly.
(*a*) When the teacher eludes to different authors, he is being elusive
(*b*) When the thief keeps eluding the police, he is being elusive
(*c*) Matters that are difficult to understand, identify or remember are allusive
(*d*) Mirages can be allusive, but a better way to express them is illusory

14. Fill in the blank with the correct idiom/phrase.
That boy from the town was a __________ in the sleepy village.
(*a*) Dog out of herd (*b*) Sheep from the heap
(*c*) Fish out of water (*d*) Bird from the flock

15. Select the appropriate option in place of underlined part of the sentence.
Increased productivity necessary reflects greater efforts made by the employees.
(*a*) Increase in productivity necessary
(*b*) Increase productivity is necessary
(*c*) Increase in productivity necessarily
(*d*) No improvement required

16. Ram and Shyam shared a secret and promised to each other that it would remain between them. Ram expressed himself in one of the following ways as given in the choices below. Identify the correct way as per standard English.
(*a*) It would remain between you and me
(*b*) It would remain between me and you
(*c*) It would remain between you and I
(*d*) It would remain with me

17. What is the adverb for the given word below? Misogynous
(*a*) Misogynousness (*b*) Misogynity
(*c*) Misogynously (*d*) Misogynous

18. In the following sentence certain parts are underlined and marked P, Q and R. One of the parts may contain certain error or may not be acceptable in standard written communication. Select the part containing an error. Choose D as your Answer: if there is no error.

The student corrected all the errors (P) that the instructor marked (Q) on the answer book (R)

(*a*) P (*b*) Q
(*c*) R (*d*) No Error

19. Choose the statement where underlined word is used correctly.
(*a*) The industrialist had a personnel jet.
(*b*) I write my experience in my personnel diary.
(*c*) All personnel are being given the day off.
(*d*) Being religious is a personnel aspect.

20. Out of the following four sentences, select the most suitable sentence with respect to grammer and usage:
(*a*) Since the report lacked needed information, it was of no use to them.
(*b*) The report was useless to them because there were no needed information in it.
(*c*) Since the report did not contain the needed information, it was not real useful to them.
(*d*) Since the report lacked needed information, it would not had been useful to them.

21. Select the alternative meaning of the underlined part of the sentence.

The chain snatchers <u>took to their heels</u> when the police party arrived.

(*a*) took shelter in a thick jungle
(*b*) open indiscriminate fire
(*c*) took to flight
(*d*) unconditionally surrendered

2014

22. Which of the following options is the closest in meaning to the word underlined in the sentence below?

In a democracy, everybody has the freedom to <u>disagree</u> with the government.

(*a*) dissent (*b*) descent
(*c*) decent (*d*) decadent

23. 'Advice' is ________.

(*a*) a verb (*b*) a noun
(*c*) an adjective (*d*) both a verb and a noun

24. Which of the options given below best completes the following sentence?

She will feel much better if she ________.

(*a*) will get some rest
(*b*) gets some rest
(*c*) will be getting some rest
(*d*) is getting some rest

25. <u>While trying to collect</u> (I) an envelope <u>from under the table,</u> (II) <u>Mr. X fell down</u> (III) and <u>was losing consciousness.</u> (IV)

Which one of the above underlined parts of the sentence is NOT appropriate?

(*a*) I (*b*) II
(*c*) III (*d*) IV

26. Which of the following options is the closest in meaning to the phrase underlined in the sentence below?

It is fascinating to see life forms <u>cope with</u> varied environmental conditions.

(*a*) Adopt to (*b*) Adapt to
(*c*) Adept in (*d*) Accept with

27. In a press meet on the recent scam, the minister said, "The buck stops here". What did the minister convey by the statement?

(*a*) He wants all the money
(*b*) He will return the money
(*c*) He will assume final responsibility
(*d*) He will resist all enquiries

2013

28. <u>The professor</u> (I) <u>ordered to</u> (II) <u>the students to go</u> (III) <u>out of the class.</u> (IV)

Which of the above underlined parts of the sentence is grammatically incorrect?

(*a*) I (*b*) II
(*c*) III (*d*) IV

29. Choose the grammatically INCORRECT sentence:

(*a*) He is of Asian origin.
(*b*) They belonged to Africa.
(*c*) She is an European.
(*d*) They migrated from India to Australia.

30. Choose the grammatically CORRECT sentence:

(*a*) Two and two add four.
(*b*) Two and two become four.
(*c*) Two and two are four.
(*d*) Two and two make four.

31. <u>All engineering students</u> (I) <u>should learn mechanics,</u> (II) <u>mathematics and</u> (III) <u>how to do computation.</u> (IV)

Which of the above underlined parts of the sentence is not appropriate?

(*a*) I (*b*) II
(*c*) III (*d*) IV

2012

32. Choose the most appropriate alternative from the options given below to complete the following sentence :

Suresh's dog is the one ______ was hurt in the stampede.

(*a*) that (*b*) which
(*c*) who (*d*) whom

33. Choose the grammatically **INCORRECT** sentence :

(*a*) They gave us the money back less the service charges of Three Hundred rupees.

(*b*) This country's expenditure is not less than that of Bangladesh.

(*c*) The committee initially asked for a funding of Fifty Lakh rupeesr, but later settled for a lesser sum.

(*d*) This country's expenditure on educational reforms is very less.

34. One of the parts (A, B, C, D) in the sentence given below contains an ERROR. Which one of the following is **INCORRECT** ?

I requested that he should be given the driving test today instead of tomorrow.

(*a*) requested that

(*b*) should be given

(*c*) the driving test

(*d*) instead of tomorrow

ANSWERS

MCQ Type Questions

1. (*b*)	**2.** (*b*)	**3.** (*d*)	**4.** (*d*)	**5.** (*b*)	**6.** (*d*)	**7.** (*b*)	**8.** (*b*)	**9.** (*c*)	**10.** (*a*)
11. (*b*)	**12.** (*a*)	**13.** (*b*)	**14.** (*c*)	**15.** (*c*)	**16.** (*a*)	**17.** (*c*)	**18.** (*b*)	**19.** (*c*)	**20.** (*a*)
21. (*c*)	**22.** (*a*)	**23.** (*b*)	**24.** (*b*)	**25.** (*d*)	**26.** (*b*)	**27.** (*c*)	**28.** (*b*)	**29.** (*c*)	**30.** (*d*)
31. (*d*)	**32.** (*a*)	**33.** (*d*)	**34.** (*b*)						

EXPLANATIONS

MCQ TYPE QUESTIONS

1. 'is' use for singular
and 'are' use for plural
So (i) and (iv) are correct.
Hence, option (b) is correct answer.

2. having said that : despite what one just said.

6. 'lying prone' means lie down flat. 'Prone to' means vulnerable to.

7. Lose (verb)

11. *Insured*–the person, group, or organization whose life or property is covered by an insurance policy.

Ensured–to secure or guarantee

13. Elusive: Difficult to answer.

14. From the statement, it appears that boy found it tough to adapt to a very different situation.

18. The is not required in 'Q'.

32. The correct answer is 'that'. In the given sentence 'was hurt in the stampede' determines which dog belongs to Suresh.

■■

2
CHAPTER

Sentence Completion

MCQ TYPE QUESTIONS

2021

1. The author said, "Musicians rehearse before their concerts. Actors rehearse their roles before the opening of a new paly. On the other hand, I find it strange that many public speakers think they can just walk on to the stage and start speaking. In my opinion, it is no less important for public speakers to rehearse their talks."

Based on the above passage, which one of the following is TRUE?

(a) The author is of the opinion that rehearsing is important for musicians, actors and public speakers.

(b) The author is of the opinion that rehearsal is less important for public speakers than for musicians and actors.

(c) The author is of the opinion that rehearsing is more important only for musicians than public speakers.

(d) The author is of the opinion that rehearsal is more important for actors than musicians.

2020

2. He is known for his unscrupulous ways. He always sheds _____ tears to deceive people.

(*a*) fox's (*b*) crocodile's
(c) crocodile (*d*) fox

3. Jofra Archer, the England fast bowler, is ______ than accurate.

(*a*) more fast (*b*) faster
(*c*) less fast (*d*) more faster

2019

4. Until Iran come along. India had never been ______ in kabaddi.

(*a*) defeated (*b*) defeating
(*c*) defeat (*d*) defeatist

5. The fishermen ,_______the flood victims owed their lives, were rewarded by the government

(*a*) whom (*b*) to which
(*c*) to whom (*d*) that

2018

6. "When she fell down the _______, she received many _______ but little help."

The words that best fill the blanks in the above sentence are

(*a*) stairs, stares (*b*) stairs, stairs
(*c*) stares, stairs (*d*) stares, stares

7. "In spite of being warned repeatedly, he failed to correct his ________ behaviour."

The word that best fills the blank in the above sentence is

(*a*) rational (*b*) reasonable
(*c*) errant (*d*) good

2017

8. It is _______ to read this year's textbook ______ the last year's.

(*a*) easier, than
(*b*) most easy, than
(*c*) easier, from
(*d*) easiest, from

2016

9. The man who is now Municipal Commissioner worked as __________________.

(*a*) the security guard at a university
(*b*) a security guard at the university
(*c*) a security guard at university
(*d*) the security guard at the university

10. Nobody knows how the Indian cricket team is going to <u>cope with</u> the difficult and seamer-friendly wickets in Australia.

Choose the option which is closest in meaning to the underlined phrase in the above sentence.

(*a*) put up with (*b*) put in with
(*c*) put down to (*d*) put up against

11. The chairman requested the aggrieved shareholders to ________ him.

(*a*) bare with (*b*) bore with
(*c*) bear with (*d*) bare

12. The unruly crowd demanded that the accused be ____________ without trial.

(*a*) hanged (*b*) hanging

(*c*) hankering (*d*) hung

13. Which of the following is CORRECT with respect to grammar and usage?

Mount Everest is ___________.

(*a*) the highest peak in the world

(*b*) highest peak in the world

(*c*) one of highest peak in the world

(*d*) one of the highest peak in the world

14. Despite the new medicine's ____________ in treating diabetes, it is not ____________widely.

(*a*) effectiveness --- prescribed

(*b*) availability --- used

(*c*) prescription --- available

(*d*) acceptance --- proscribed

15. The students _____ the teacher on teachers' day for twenty years of dedicated teaching.

(*a*) facilitated (*b*) felicitated

(*c*) fantasized (*d*) facillitated

16. After India's cricket world cup victory in 1985, Shrotria who was playing both tennis and cricket till then, decided to concentrate only on cricket. And <u>the rest is history</u>.

What does the underlined phrase mean in this context?

(*a*) history will rest in peace

(*b*) rest is recorded in history books

(*c*) rest is well known

(*d*) rest is archaic

17. If I were you, I _________ that laptop. It's much too expensive.

(*a*) won't buy (*b*) shan't buy

(*c*) wouldn't buy (*d*) would buy

18. Choose the most appropriate set of words from the options given below to complete the following sentence.

_____ ________ is a will, ________ is a way.

(*a*) Wear, there, their (*b*) Were, their, there

(*c*) Where, there, there (*d*) Where, their, their

2015

19. Choose the most appropriate word from the options given below to complete the following sentence. The official answered ________ that the complaints of the citizen would be looked into.

(*a*) respectably (*b*) respectfully

(*c*) reputably (*d*) respectively

20. Choose the most appropriate word from the options given below to complete the following sentence.

The principal presented the chief guest with a ___________, as token of appreciation.

(*a*) momento (*b*) memento

(*c*) momentum (*d*) moment

21. Choose the appropriate word/phrase, out of the four options given below, to complete the following sentence : Frogs ________.

(*a*) Croak (*b*) Roar

(*c*) Hiss (*d*) Patter

22. Extreme focus on syllabus and studying for tests has become such a dominant concern of Indian students that they close their minds to anything _____ to the requirements of the exam

(*a*) related (*b*) extraneous

(*c*) outside (*d*) useful

23. The Tamil version of _____ John Abraham-starrer Madras café _____ cleared by the Censor Board with no cuts last week but the film's distributors _____ no takers among the exhibitors for a release in Tamil Nadu _____ this Friday.

(*a*) Mr., was, found, on (*b*) a, was, found, at

(*c*) the, was, found, on (*d*) a, being, find at

24. Choose the appropriate word/phase, out of the four options given below, to complete the following sentence:

Apparent lifelessness ______ dormant life.

(*a*) harbours (*b*) lead to

(*c*) supports (*d*) affects

25. Choose the correct verb to fill in the blank below :

Let us ____________.

(*a*) Introvert (*b*) Alternate

(*c*) Atheist (*d*) Altruist

26. Choose the most appropriate word from the options given below to complete the following sentence :

If the Athlete had wanted to come first in the race, he __________ several hours everyday

(*a*) should practice (*b*) should have practiced

(*c*) practiced (*d*) should be practicing

27. Choose the appropriate word/phrase, out of the four options given below, to complete the following sentence:

Dhoni, as well as the other team members of Indian team, _____ present on the occasion.

(*a*) were (*b*) was

(*c*) has (*d*) have

28. We _______ our friend's birthday and we _______ how to make it up to him.

(*a*) completely forgot — don't just known

(*b*) forget completely — don't just know

(*c*) completely forget —just don't know

(*d*) forgot completely —just don't know

29. Which of the following options is the closest in meaning to the sentence below?

She enjoyed herself immensely at the party.

(*a*) She had a terrible time at the party

(*b*) She had a horrible time at the party

(*c*) She had a terrific time at the party

(*d*) She had a terrifying time at the party

30. Didn't you buy ______ when you went shopping?

(*a*) any paper

(*b*) much paper

(*c*) no paper

(*d*) a few paper

2014

31. Choose the most appropriate word from the options given below to complete the following sentence:

A person suffering from Alzheimer's disease ________ short-term memory loss.

(*a*) experienced (*b*) has experienced

(*c*) is experiencing (*d*) experiences

32. Choose the most appropriate word from the options given below to complete the following sentence:

________ is the key to their happiness; they are satisfied with what they have.

(*a*) Contentment (*b*) Ambition

(*c*) Perseverance (*d*) Hunger

33. Choose the most appropriate word from the options given below to complete the following sentence:

One of his biggest _____ was his ability to forgive.

(*a*) vice (*b*) virtues

(*c*) choices (*d*) strength

34. After the discussion, Tom said to me, 'Please revert!' He expects me to ________.

(*a*) retract (*b*) get back to him

(*c*) move in reverse (*d*) retreat

35. The value of one U.S. Dollar is 65 Indian Rupees today, compared to 60 last year. The Indian Rupee has ________.

(*a*) Depressed (*b*) Depreciated

(*c*) Appreciated (*d*) Stabilized

36. Choose the most appropriate word from the options given below to complete the following sentence:

Communication and interpersonal skills are ________ important in their own ways.

(*a*) each (*b*) both

(*c*) all (*d*) either

37. Choose the most appropriate pair of words from the options given below to complete the following sentence:

She could not _______ the thought of _______ the election to her bitter rival.

(*a*) bear, loosing (*b*) bare, loosing

(*c*) bear, losing (*d*) bare, losing

38. Choose the most appropriate phrase from the options given below to complete the following sentence:

The aircraft ________ take off as soon as its flight plan was filed.

(*a*) is allowed to (*b*) will be allowed to

(*c*) was allowed to (*d*) has been allowed to

39. Choose the most appropriate word from the options given below to complete the following sentence:

Many ancient cultures attributed disease to supernatural causes. However, modern science has largely helped _____ such notions.

(*a*) impel (*b*) dispel

(*c*) propel (*d*) repel

40. If she __________ how to calibrate the instrument, she ________ done the experiment.

(*a*) knows, will have

(*b*) knew, had

(*c*) had known, could have

(*d*) should have known, would have

41. Who _______ was coming to see us this evening ?

(*a*) you said (*b*) did you say

(*c*) did you say that (*d*) had you said

42. Choose the most appropriate word from the options given below to complete the following sentence:

He could not understand the judges awarding her the first prize, because he thought that her performance was quite _________________.

(*a*) Superb (*b*) Medium

(*c*) Mediocre (*d*) Exhilarating

2013

43. Friendship, no matter how _______ it is, has its limitations.

(*a*) cordial (*b*) intimate

(*c*) secret (*d*) pleasant

44. Were you a bird, you_____ in the sky.
(*a*) would fly (*b*) shall fly
(*c*) should fly (*d*) shall have flown

45. Complete the sentence:
Dare __________ mistakes.
(*a*) commit (*b*) to commit
(*c*) committed (*d*) committing

46. The Headmaster ______ to speak to you.
Which of the following options is incorrect to complete the above sentence?
(*a*) is wanting (*b*) wants
(*c*) want (*d*) was wanting

2012

47. Choose the most appropriate alternative from the options given below to complete the following sentence :
Despite several ______ the mission succeeded in its attempt to resolve the conflict.
(*a*) attempts (*b*) setbacks
(*c*) meetings (*d*) delegations

48. Choose the most appropriate alternative from the options given below to complete the following sentence :
If the tired soldier wanted to lie down, he ___ the mattress out on the balcony.
(*a*) should take (*b*) shall take
(*c*) should have taken (*d*) will have taken

49. Choose the most appropriate word from the options given below to complete the following sentence :
Given the seriousness of the situation that he had to face, his ___ was impressive.
(*a*) beggary (*b*) nomenclature
(*c*) jealousy (*d*) nonchalance

2011

50. Choose the most appropriate word from the options given below to complete the following sentence:
If you are trying to make a strong impression on your audience, you cannot do so by being understated, tentative or _______ .
(*a*) hyperbolic (*b*) restrained
(*c*) argumentative (*d*) indifferent

51. Choose the most appropriate word(s) from the options given below to complete the following sentence.
I contemplated ______ Singapore for my vacation but decided against it.
(*a*) to visit (*b*) having to visit
(*c*) visiting (*d*) for a visit

52. Choose the most appropriate word from the options given below to complete the following sentence:
Under ethical guidelines recently adopted by the Indian Medical Association, human genes are to be manipulated only to correct diseases for which_______treatments are unsatisfactory.
(*a*) similar (*b*) most
(*c*) uncommon (*d*) available

53. Choose the most appropriate word from the options given below to complete the following sentence:
It was her view that the country's problems had been_________by foreign technocrats, so that to invite them to come back would be counter-productive.
(*a*) identified (*b*) ascertained
(*c*) exacerbated (*d*) analysed

2010

54. Choose the most appropriate word from the options given below to complete the following sentence:
If we manage to _________ our natural resources, we would leave a better planet for our children.
(*a*) uphold (*b*) restrain
(*c*) cherish (*d*) conserve

55. Choose the most appropriate word from the options given below to complete the following sentence:
His rather casual remarks on politics ________his lack of seriousness about the subject.
(*a*) masked (*b*) belied
(*c*) betrayed (*d*) suppressed

ANSWERS

MCQ Type Questions

1. (a)	**2.** (c)	**3.** (a)	**4.** (*a*)	**5.** (*c*)	**6.** (*a*)	**7.** (*c*)	**8.** (*a*)	**9.** (*b*)	**10.** (*a*)
11. (*c*)	**12.** (*a*)	**13.** (*a*)	**14.** (*a*)	**15.** (*b*)	**16.** (*c*)	**17.** (*c*)	**18.** (*c*)	**19.** (*b*)	**20.** (*b*)
21. (*a*)	**22.** (*b*)	**23.** (*c*)	**24.** (*a*)	**25.** (*b*)	**26.** (*b*)	**27.** (*b*)	**28.** (*c*)	**29.** (*c*)	**30.** (*a*)
31. (*d*)	**32.** (*a*)	**33.** (*b*)	**34.** (*b*)	**35.** (*b*)	**36.** (*b*)	**37.** (*c*)	**38.** (*c*)	**39.** (*b*)	**40.** (*c*)
41. (*b*)	**42.** (*c*)	**43.** (*b*)	**44.** (*a*)	**45.** (*b*)	**46.** (*c*)	**47.** (*b*)	**48.** (*a*)	**49.** (*d*)	**50.** (*b*)
51. (*c*)	**52.** (*d*)	**53.** (*c*)	**54.** (*a*)	**55.** (*c*)					

EXPLANATIONS

MCQ TYPE QUESTIONS

1. From given passage author said in last line "In my opinion, it is no less important for public speakers to reheasse thiner talks".

So, option (a) is correct.

2. Crocodile tears: A false, insincere display of emotion.

3. Jofra Archar, the England fast bowler, is more fast than accurate.

6. Stares - To look at someone for a long time

Stairs - A construction designed to bridge a large vertical distance by dividing it into smaller vertical distances, called steps

7. Errant means misbehaving, exhibiting inappropriate behaviour / offending conduct.

8. easier, than

10. From the given underlined phrase <u>copewith</u> means put up with.

12. 'hanged' means death by hanging 'hung' is used only with things and not with people.

14. Here 'effectiveness' is noun and 'prescribed' is verb. So these words are apt and befitting with the given word 'medicine'.

20. The principal presented the chief guest with a memento, as token of appreciation.

21. Frogs make 'croak' sound.

22. Extraneous -irrelevant or unrelated to the subject being dealt with.

23. John-Abraham starrer Madras Cafe talks about the movie not the person, so Mr. is ruled out.

'Find no takers' is not the correct phrase. At this Friday is incorrect.

So, option (*c*) is correct.

24. Apparent: looks like; dormant: hidden

Harbour: give shelter;
Effect (verb): results in

26. For condition regarding something which already happened, should have practiced is the correct choice.

31. The correct word to fill in the blank is experiences. This is because the sentence is in present tense and the sense is in present indefinite.

32. Contentment is the right word which is appropriate to the meaning of the sentence. Contentment is the sign of true happiness as it gives a sense of fulfillment.

47. The word despite tells us that mission succeeded even though there were problems. Hence the correct answer is Setbacks.

50. Tone of the sentence clearly indicates a word that is similar to understated is needed for the blank. Alternatively, word should be antonym of strong (fail to make strong impression). Therefore, best choice is restrained which means controlled / reserved / timid.

51. Contemplate is a transitive verb and hence is followed by a gerund Hence the correct usage of contemplate is verb + ing form.

52. The context seeks to take a deviation only when existing / present / current / alternative treatments are unsatisfactory. So word for the blank should be a close synonym of existing / present / current / alternative. Available is the closest of all.

53. Clues: foreign technocrats did something negatively to the problems – so it is counter-productive to invite them. All other options are non-negative. Best choice is exacerbated which means aggravated or worsened.

55. Betrayed, **means** 'showed' or revealed.

3 CHAPTER
Synonyms

MCQ TYPE QUESTIONS

2016

1. The Buddha said, "Holding on to anger is like <u>grasping</u> a hot coal with the intent of throwing it at someone else; you are the one who gets burnt."

Select the word below which is closest in meaning to the word underlined above.

(*a*) burning (*b*) igniting
(*c*) clutching (*d*) flinging

2015

2. The word similar in meaning to 'dreary' is

(*a*) cheerful (*b*) dreamy
(*c*) hard (*d*) dismal

3. Choose the word most similar in meaning to the given word :

Educe

(*a*) Exert (*b*) Educate
(*c*) Extract (*d*) Extend

4. Choose the most suitable one word substitute for the following expression

Connotation of a road or way

(*a*) Pertinacious (*b*) Viaticum
(*c*) Clandestine (*d*) Ravenous

5. Choose the word most similar in meaning to the given word:

Awkward

(*a*) Inept (*b*) Graceful
(*c*) Suitable (*d*) Dreadful

6. A generic term that includes various items of clothing such as a skirt, a pair of trousers and a shirt as

(*a*) fabric
(*b*) textile
(*c*) fibre
(*d*) apparel

7. Which of the following combinations is incorrect?

(*a*) Acquiescence – Submission
(*b*) Wheedle – Roundabout
(*c*) Flippancy – Lightness
(*d*) Profligate – Extravagant

2014

8. A student is required to demonstrate a high level of <u>comprehension</u> of the subject, especially in the social sciences.

The word closest in meaning to <u>comprehension</u> is

(*a*) understanding (*b*) meaning
(*c*) concentration (*d*) stability

9. While receiving the award, the scientist said, "I feel vindicated". Which of the following is closest in meaning to the word 'vindicated'?

(*a*) punished (*b*) substantiated
(*c*) appreciated (*d*) chastened

10. Choose the word that is opposite in meaning to the word "coherent".

(*a*) sticky (*b*) well-connected
(*c*) rambling (*d*) friendly

11. Match the columns:

Column-1	**Column-2**
1. eradicate	P. misrepresent
2. distort	Q. soak completely
3. saturate	R. use
4. utilize	S. destroy utterly

(*a*) 1 : S, 2 : P, 3 : Q, 4 : R
(*b*) 1 : P, 2 : Q, 3 : R, 4 : S
(*c*) 1 : Q, 2 : R, 3 : S, 4 : P
(*d*) 1 : S, 2 : P, 3 : R, 4 : Q

2013

12. Which of the following options is the closest in meaning to the word given below:

Primeval

(*a*) Modern (*b*) Historic
(*c*) Primitive (*d*) Antique

13. Which one of the following options is the closest in meaning to the word given below?

Nadir

(*a*) Highest (*b*) Lowest
(*c*) Medium (*d*) Integration

14. They were requested not to **quarrel** with others.

Which one of the following options is the closest in meaning to the word **quarrel**?

(*a*) make out (*b*) call out
(*c*) dig out (*d*) fall out

2012

15. Which one of the following options is the closest in meaning to the word given below?

Mitigate

(*a*) Diminish (*b*) Divulge

(*c*) Dedicate (*d*) Denote

16. Which one of the following options is the closest in meaning to the word given below ?

Latitude

(*a*) Eligibility (*b*) Freedom

(*c*) Coercion (*d*) Meticulousness

2011

17. Which of the following options is the closest in the meaning to the word below :

Inexplicable

(*a*) Incomprehensible (*b*) Indelible

(*c*) Inextricable (*d*) Infallible

2010

18. Which of the following options is the closest in meaning to the word below:

Circuitous

(*a*) cyclic (*b*) indirect

(*c*) confusing (*d*) crooked

ANSWERS

MCQ Type Questions

1. (*c*)	**2.** (*d*)	**3.** (*c*)	**4.** (*b*)	**5.** (*a*)	**6.** (*d*)	**7.** (*b*)	**8.** (*a*)	**9.** (*b*)	**10.** (*c*)
11. (*a*)	**12.** (*c*)	**13.** (*b*)	**14.** (*b*)	**15.** (*a*)	**16.** (*b*)	**17.** (*a*)	**18.** (*b*)		

EXPLANATIONS

MCQ TYPE QUESTIONS

1. The meaning of underlined word grasping means clutching (or holding something tightly).

2. *dreary* - depressingly dull and bleak or repetitive.

3. The word similar in meaning to Educe is Extract.

13. Given: A word "Nadir"

To find: Closest meaning word to 'Nadir' in the given options.

Analysis: The dictionary meaning of 'Nadir' is 'Lowest' which is potion (*b*)

Source: Google

Given: A sequence 44, 42, 40,......

To find: Maximum sum of the sequence

Analysis: To find the maximum sum, we have to ignore all the negative terms (they will only decrease the sum). Hence, we terminate the sequence at 0.

Now, the sequence is 44, 42,......,2, 0.

This sequence is an Arithmetic progression with common difference –2. Now, the problem reduces to finding the sum of this A.P.

To find the sum of A.P., we have to find the number of terms in A.P.

Let n be the number of terms

a be the first term

l be the last term

d be the common difference.

$n = ?$

$a = 44, \quad l = 0, \quad d = -2$

Now, $l = a + (n - 1)d$

15. 'Mitigate' means to make something less harmful, serious, intense etc. Hence correct answer is 'Diminis'

17. Inexplicable means not explicable; that cannot be explained, understood, or accounted for. So the best synonym here is incomprehensible.

18. Circuitous means round about or not direct

So circuitous : indirect.

4 CHAPTER Antonyms

MCQ TYPE QUESTIONS

2015

1. Which word is not a synonym for the word *vernacular*?

(*a*) regional

(*b*) indigeneous

(*c*) indigent

(*d*) colloquial

2011

2. Choose the word from the options given below that is most nearly opposite in meaning to the given word:

Amalgamate

(*a*) merge

(*b*) split

(*c*) collect

(*d*) separate

3. Choose the word from the options given below that is most nearly **opposite** in meaning to the given word :

Frequency

(*a*) periodicity

(*b*) rarity

(*c*) gradualness

(*d*) persistency

ANSWERS

MCQ Type Questions

1. (*c*) **2.** (*b*) **3.** (*b*)

EXPLANATIONS

MCQ TYPE QUESTIONS

1. *Vernacular* - expressed or written in the native language of a place

   Indigent - deficient in what is requisite

2. Amalgamate means combine or unite to form one organization or structure. So the best option here is split. Separate on the other hand, although a close synonym, it is too general to be the best antonym in the given question while Merge is the synonym; Collect is not related.

3. Best antonym is rarity which means shortage or scarcity.

5 CHAPTER

Reasoning Ability

MCQ TYPE QUESTIONS

2021

1.

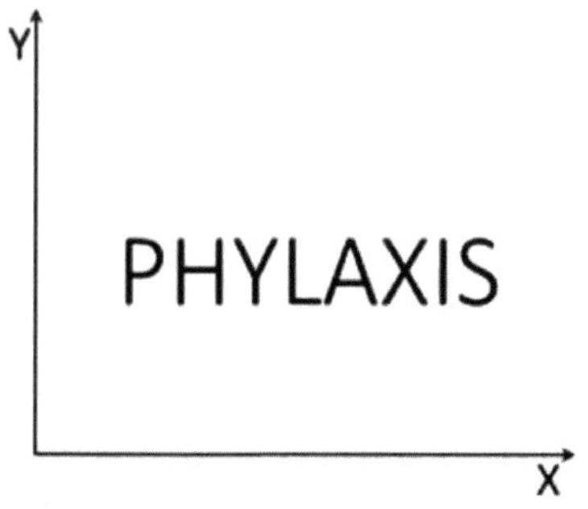

The mirror image of the above text about the x-axis is

(a)  (b) 

(c)  (d) 

2. Four persons P, Q, R and S are to be seated in a row. R should not be seated at the second position from the left end of the row. The number of distinct seating arrangements possible is:

(a) 6 (b) 9

(c) 18 (d) 24

3. On a planar field, you travelled 3 units East from a point O. Next you travelled 4 units South to arrive at point P. Then you travelled from P in the North-East direction such that you arrive at a point that is 6 units East of point O. Next, you travelled in the North-West direction, so that you arrive at point Q that is 8 units North of point P.

The distance of point Q to point O, in the same units, should be _______

(a) 3 (b) 4

(c) 5 (d) 6

4. 1. Some football players play cricket.

2. All cricket players play hockey.

Among the options given below, the statement that logically follows from the two statements 1 and 2 above, is:

(a) No football player plays hockey.

(b) Some football players play hockey.

(c) All football players play hockey.

(d) All hockey players play football.

2020

5. Select the word that fits the analogy:

Build : Building : : Grow : _____

(*a*) Grown (*b*) Grew

(*c*) Growth (*d*) Growed

6. Crowd funding deals with mobilisation of funds for a project from a large number of people, who would be willing to invest smaller amounts through web-based platform in the project.

Based on the above paragraph, which of the following is correct about crowd funding?

(*a*) Funds raised through unwilling contributions on web-based platforms.

(*b*) Funds raised through large contributions on web-based platform.

(*c*) Funds raised through coerced contributions on web-based platforms.

(*d*) Funds raised through voluntary contributions on web-based platforms.

7. P, Q, R and S are to be uniquely coded using α and β. If P is coded as $\alpha\alpha$ and Q as $\alpha\beta$, then R and S, respectively, can be coded as _____.

(*a*) $\beta\alpha$ and $\alpha\beta$ (*b*) $\beta\beta$ and $\alpha\alpha$

(*c*) $\alpha\beta$ and $\beta\beta$ (*d*) $\beta\alpha$ and $\beta\beta$

2019

8. Five numbers 10, 7, 5, 4, 2 are arranged in a sequence from left to right following the directions given below:

(1) No two odd or even numbers are next to each other.

(2) The second number from left is exactly half of the left -most number.

(3) The middle number is exactly twice the right most number.

Which is the second number from the right ?

(*a*) 2 (*b*) 4

(*c*) 7 (*d*) 10

9. "Some students were not involved in the strike". If the above statement is true, which of the following conclusions is/are logically necessary ?

1. Some who were involved in strike were students.
2. No student was involved in the strike.
3. At least one student was involved in the strike.
4. Some who were not involved in the strike were students.

(*a*) 1 and 2 (*b*) 3
(*c*) 4 (*d*) 2 and 3

10. "I read somewhere that in ancient times the prestige of a kingdom depended upon the number of taxes that it was able to levy on its people. It was very much like the prestige of a head-hunter in his own community".

Based on the paragraph above, the prestige of a head-hunter depended upon ____

(*a*) the prestige of the kingdom
(*b*) the prestige of the heads
(*c*) the number of taxes he could levy
(*d*) the number of head she could gather

11. The nomenclature of Hindustani music has changed over centuries. Since the medieval period *dhrupad* styles were identified as *baanis*. Terms like *gayaki* and *baaj* were used to refer to vocal and instrumental styles, respectively. With institutionalization of music education the term *gharana* became acceptable. *Gharana* originally referred to hereditary musicians from a particular lineage, including disciples and ground disciples. Which one of the following pairings is NOT Correct ?

(*a*) Dhrupad, baani (*b*) Gayaki, Vocal
(*c*) Baaj, institution (*d*) Gharana, lineage

12. Since the last one year, after a 125 basis point reduction in repo rate by the Reserve Bank of India, banking institutions have been making a demand to reduce interest rates on small savings schemes. Finally, the government announced yesterday a reduction in interest rates on small saving schemes to bring them on par with fixed deposit interest rates.

Which one of the following statements can be inferred from the given passage ?

(*a*) Whenever the Reserve Bank of India reduces the repo rate, the interest rates on small saving schemes are also reduced.
(*b*) Interest rates on small saving schemes are always maintained on par with fixed deposit interest rates.
(*c*) The government sometimes takes into consideration the demands of banking institutions before reducing the interest rates on small saving schemes.
(*d*) A reduction in interest rates on small saving schemes follow only after a reduction in repo rate by the Reserve Bank of India.

2018

13. Arrange the following three-dimensional objects in the descending order of their volumes:

(*i*) A cuboid with dimensions 10 cm, 8 cm and 6 cm
(*ii*) A cube of side 8 cm
(*iii*) A cylinder with base radius 7 cm and height 7 cm
(*iv*) A sphere of radius 7 cm

(*a*) (*i*), (*ii*), (*iii*), (*iv*)
(*b*) (*ii*), (*i*), (*iv*), (*iii*)
(*c*) (*iii*), (*ii*), (*i*), (*iv*)
(*d*) (*iv*), (*iii*), (*ii*), (*i*)

14. A set of 4 parallel lines intersect with another set of 5 parallel lines. How many parallelograms are formed?

(*a*) 20 (*b*) 48
(*c*) 60 (*d*) 72

15. If $x^2 + x - 1 = 0$ what is the value of $x^4 + \frac{1}{x^4}$?

(*a*) 1 (*b*) 5
(*c*) 7 (*d*) 9

2017

16. The ninth and the tenth of this month are Monday and Tuesday ________.

(*a*) figuratively
(*b*) retrospectively
(*c*) respectively
(*d*) rightfully

17. A rule states that in order to drink beer, one must be over 18 years old. In a bar, there are 4 people. P is 16 years old, Q is 25 years old, R is drinking milkshake and S is drinking a beer. What must be checked to ensure that the rule is being followed?

(*a*) Only P's drink
(*b*) Only P's drink and S's age
(*c*) Only S's age
(*d*) Only P's drink, Q's drink and S's age

18. Fatima starts from point P, goes North for 3 km, and then East for 4 km to reach point Q. She then turns to face point P and goes 15 km in that direction. She then goes North for 6 km. How far is she from point P, and in which direction should she go to reach point P?

(*a*) 8 km, East

(*b*) 12 km, North

(*c*) 6 km, East

(*d*) 10 km, North

19. "If you are looking for a history of India, or for an account of the rise and fall of the British Raj, or for the reason of the cleaving of the subcontinent into two mutually antagonistic parts and the effects this mutilation will have in the respective sections, and ultimately on Asia, you will not find it in these pages; for though I have spent a lifetime in the country. I lived too near the seat of events, and was too intimately associated with the actors, to get the perspective needed for the impartial recording of these matters."

Which of the following statements best reflects the author's opinion?

(*a*) An intimate association does not allow for the necessary perspective.

(*b*) Matters are recorded with an impartial perspective.

(*c*) An intimate association offers an impartial perspective.

(*d*) Actors are typically associated with the impartial recording of matters.

20. Each of P, Q, R, S, W, X, Y and Z has been married at most once. X and Y are married and have two children P and Q. Z is the grandfather of the daughter S of P. Further, Z and W are married and are parents of R. Which one of the following must necessarily be FALSE?

(*a*) X is the mother-in-law of R

(*b*) P and Q are not married to each other

(*c*) P is a son of X and Y

(*d*) Q cannot be married to R

2016

21. Find the odd one in the following group of words.

mock, deride, praise, jeer

(*a*) mock (*b*) deride

(*c*) praise (*d*) jeer

22. Pick the odd one from the following options.

(*a*) CADBE (*b*) JHKIL

(*c*) XVYWZ (*d*) ONPMQ

23. Pick the odd one out in the following :

13, 23, 33, 43, 53

(*a*) 23 (*b*) 33

(*c*) 43 (*d*) 53

24. R2D2 is a robot. R2D2 can repair aeroplanes. No other robot can repair aeroplanes. Which of the following can be logically inferred from the above statements?

(*a*) R2D2 is a robot which can only repair aeroplanes.

(*b*) R2D2 is the only robot which can repair aeroplanes.

(*c*) R2D2 is a robot which can repair only aeroplanes.

(*d*) Only R2D2 is a robot.

25. A poll of students appearing for masters in engineering indicated that 60% of the students believed that mechanical engineering is a profession unsuitable for women. A research study on women with masters or higher degrees in mechanical engineering found that 99% of such women were successful in their professions.

Which of the following can be logically inferred from the above paragraph?

(*a*) Many students have misconceptions regarding various engineering disciplines.

(*b*) Men with advanced degrees in mechanical engineering believe women are well suited to be mechanical engineers.

(*c*) Mechanical engineering is a profession well suited for women with masters or higher degrees in mechanical engineering.

(*d*) The number of women pursuing higher degrees in mechanical engineering is small.

26. Sourya committee had proposed the establishment of Sourya Institutes of Technology (SITs) in line with Indian Institutes of Technology (IITs) to cater to the technological and industrial needs of a developing country.

Which of the following can be logically inferred from the above sentence?

Based on the proposal,

(*i*) In the initial years, SIT students will get degrees from IIT.

(*ii*) SITs will have a distinct national objective.

(*iii*) SIT like institutions can only be established in consulation with IIT.

(*iv*) SITs will serve technological needs of a developing country.

(*a*) (*iii*) and (*iv*) only (*b*) (*i*) and (*iv*) only

(*c*) (*ii*) and (*iv*) only (*d*) (*ii*) and (*iii*) only

27. Fact : If it rains, then the field is wet.

Read the following statements:

(*i*) It rains (*ii*) The field is not wet

(*iii*) The field is wet (*iv*) It did not rain

Which one of the options given below is NOT logically possible, based on the given fact?

(*a*) If (*iii*), then (*iv*) (*b*) If (*i*), then (*iii*)

(*c*) If (*i*), then (*ii*) (*d*) If (*ii*), then (*iv*)

28. Students taking an exam are divided into two groups, **P** and **Q** such that each group has the same number of students. The performance of each of the students in a test was evaluated out of 200 marks. It was observed that the mean of group **P** was 105, while that of group **Q** was 85. The standard deviation of group **P** was 25, while that of group **Q** was 5. Assuming that the marks were distributed on a normal distribution, which of the following statements will have the highest probability of being **TRUE**?

(*a*) No student in group **Q** scored less marks than any student in group **P**.

(*b*) No student in group **P** scored less marks than any student in group **Q**.

(*c*) Most students of group **Q** scored marks in a narrower range than students in group **P**.

(*d*) The median of the marks of group **P** is 100.

29. A smart city integrates all modes of transport, uses clean energy and promotes sustainable use of resources. It also uses technology to ensure safety and security of the city, something which critics argue, will lead to a surveillance state.

Which of the following can be logically inferred from the above paragraph?

(*i*) All smart cities encourage the formation of surveillance states.

(*ii*) Surveillance is an integral part of a smart city.

(*iii*) Sustainability and surveillance go hand in hand in a smart city.

(*iv*) There is a perception that smart cities promote surveillance.

(*a*) (*i*) and (*iv*) only (*b*) (*ii*) and (*iii*) only

(*c*) (*iv*) only (*d*) (*i*) only

30. Find the missing sequence in the letter series.

B, FH, LNP, _ _ _ _.

(*a*) SUWY (*b*) TUVW

(*c*) TVXZ (*d*) TWXZ

31. Michael lives 10 km away from where I live. Ahmed lives 5 km away and Susan lives 7 km away from where I live. Arun is farther away than Ahmed but closer than Susan from where I live. From the information provided here, what is one possible distance (in km) at which I live from Arun's place?

(*a*) 3.00 (*b*) 4.99

(*c*) 6.02 (*d*) 7.01

32. Leela is older than her cousin Pavithra. Pavithra's brother Shiva is older than Leela. When Pavithra and Shiva are visiting Leela, all three like to play chess. Pavithra wins more often than Leela does.

Which one of the following statements must be TRUE based on the above?

(*a*) When Shiva plays chess with Leela and Pavithra, he often loses.

(*b*) Leela is the oldest of the three.

(*c*) Shiva is a better chess player than Pavithra.

(*d*) Pavithra is the youngest of the three.

33. Based on the given statements, select the appropriate option with respect to grammar and usage.

Statements :

I. The height of Mr. X is 6 feet.

II. The height of Mr. Y is 5 feet.

(*a*) Mr. X is longer than Mr. Y.

(*b*) Mr. X is more elongated than Mr. Y

(*c*) Mr. X is taller than Mr. Y

(*d*) Mr. X is lengthier than Mr. Y

34. Social science disciplines were in existence in an amorphous form until the colonial period when they were institutionalized. In varying degrees, they were intended to further the colonial interest. In the time of globalization and the economic rise of postcolonial countries like India, conventional ways of knowledge production have become obsolete.

Which of the following can be logically inferred from the above statements?

I. Social science disciplines have become obsolete.

II. Social science disciplines had a pre-colonial origin.

III. Social science disciplines always promote colonialism.

IV. Social science must maintain disciplinary boundaries.

(*a*) II only (*b*) I and III only

(*c*) II and IV only (*d*) III and IV only

35. Two and a quarter hours back, when seen in a mirror, the reflection of a wall clock without number markings seemed to show 1 : 30. What is the actual current time shown by the clock?

(*a*) 8:15 (*b*) 11:15
(*c*) 12:15 (*d*) 12:45

36. M and N start from the same location. M travels 10 km East and then 10 km North-East. N travels 5 km South and then 4 km South-East. What is the shortest distance (in km) between M and N at the end of their travel?

(*a*) 18.60 (*b*) 22.50
(*c*) 20.61 (*d*) 25.00

37. M has a son Q and a daughter R. He has no other children. E is the mother of P and daughter-inlaw of M. How is P related to M?

(*a*) P is the son-in-law of M.
(*b*) P is the grandchild of M.
(*c*) P is the daughter-in-law of M.
(*d*) P is the grandfather of M.

38. The number that least fits this set: (324, 441, 97 and 64) is ________.

(*a*) 324 (*b*) 441
(*c*) 97 (*d*) 64

39. The overwhelming number of people infected with rabies in India has been flagged by the World Health Organization as a source of concern. It is estimated that inoculating 70% of pets and stray dogs against rabies can lead to a significant reduction in the number of people infected with rabies.

Which of the following can be logically inferred from the above sentences ?

(*a*) The number of people in India infected with rabies is high
(*b*) The number of people in other parts of the world who are infected with rabies is low
(*c*) Rabies can be eradicated in India by vaccinating 70% of stray dogs
(d) Stray dogs are the main source of rabies worldwide

40. A flat is shared by four first year undergraduate students. They agreed to allow the oldest of them to enjoy some extra space in the flat. Manu is two months older than Sravan, who is three months younger than Trideep. Pavan is one month older than Sravan. Who should occupy the extra space in the flat ?

(*a*) Manu (*b*) Sravan
(*c*) Trideep (*d*) Pavan

41. Today, we consider Ashoka as a great ruler because of the copious evidence he left behind in the form of stone carved edicts. Historians tend to correlate greatness of a king at his time with the availability of evidence today. Which of the following can be logically inferred from the above sentences?

(*a*) Emperors who do not leave significant sculpted evidence are completely forgotten.
(*b*) Ashoka produced stone carved edicts to ensure that later historians will respect him.
(*c*) Statues of kings are a reminder of their greatness.
(*d*) A king's greatness, as we know him today, is interpreted by historians.

42. **Fact 1 :** Humans are mammals.

Fact 2 : Some humans are engineers.

Fact 3 : Engineers build houses.

If the above statements are facts, which of the following can be logically inferred ?

I. All mammals build houses.
II. Engineers are mammals.
III. Some humans are not engineers.

(*a*) II only (*b*) III only
(*c*) I, II and III (*d*) I only

2015

43. Mr. Vivek walks 6 meters North-East, then turns and walks 6 meters South- East, both at 60 degrees to East. He further moves 2 meters South and 4 meters West. What is the straight distance in meters between the point he started from and the point he finally reached?

(*a*) $2\sqrt{2}$ (*b*) 2
(*c*) $\sqrt{2}$ (*d*) $1/\sqrt{2}$

44. There are 16 teachers who can teach Thermodynamics (TD), 11 who can teach Electrical Sciences (ES), and 5 who can teach both TD and Engineering Mechanics (EM). There are a total of 40 teachers, 6 cannot teach any of the three subjects, i.e. EM, ES or TD. 6 can teach only ES. 4 can teach all three subjects, i.e. EM, ES and TD. 4 can teach ES and TD. How many can teach both ES and EM but not TD?

(*a*) 1 (*b*) 2
(*c*) 3 (*d*) 4

45. The given question is followed by two statements: select the most appropriate option that solves the question

Capacity of a solution tank A is 70% of the capacity of tank B. How many gallons of solution are in tank A and tank B?

Statements:

I. Tank A is 80% full and tank B is 40% full

II. Tank A if full contains 14,000 gallons of solution

(*a*) Statement I alone is sufficient

(*b*) Statement II alone is sufficient

(*c*) Either statement I or II alone is sufficient

(*d*) Both the statements I and II together are sufficient

46. Operators □, ◇ and → are defined by

$a \,\square\, b = \dfrac{a-b}{a+b}$; $a \diamond b = \dfrac{a+b}{a-b}$; $a \rightarrow b = ab$.

Find the value $(66 \square 6) \rightarrow (66 \diamond 6)$

(*a*) –2 (*b*) –1

(*c*) 1 (*d*) 2

47. Humpty Dumpty sits on a wall every day while having lunch. The wall sometimes breaks. A person sitting on the wall falls if the wall breaks.

Which one of the statements below is logically valid and can be inferred from the above sentences?

(*a*) Humpty Dumpty always falls while having lunch

(*b*) Humpty Dumpty does not fall sometimes while having lunch

(*c*) Humpty Dumpty never falls during dinner

(*d*) When Humpty Dumpty does not sit on the wall, the wall does not break

48. If ROAD is written as URDG, then SWAN should be written as :

(*a*) VXDQ (*b*) VZDQ

(*c*) VZDP (*d*) UXDQ

49. Select the pair that best expresses a relationship similar to that expressed in the pair :

Children : Pediatrician

(*a*) Adult : Orthopaedist

(*b*) Females : Gynaecologist

(*c*) Kidney : Nephrologist

(*d*) Skin : Dermatologist

50. The head of a newly formed government desires to appoint five of the six selected members P,Q,R,S,T and U to portfolios of Home, Power, Defense, Telecom, and Finance. U does not want any portfolio if S gets one of the five. R wants either Home or Finance or no portfolio. Q says that if S gets either Power or Telecom, then she must get the other one. T insists on a portfolio if P gets one.

Which is the valid distribution of portfolios?

(*a*) P-Home, Q-Power, R-Defense, S-Telecom, T-Finance

(*b*) R-Home, S-Power, P-Defense, Q-Telecom, T-Finance

(*c*) P-Home, Q-Power, T-Defense, S-Telecom, U-Finance

(*d*) Q-Home, U-Power, T-Defense, R-Telecom, P-Finance

51. Most experts feel that in spite of possessing all the technical skills required to be a batsman of the highest order, he is unlikely to be so due to lack of requisite temperament. He was guilty of throwing away his wicket several times after working hard to lay a strong foundation. His critics pointed out that until he addressed this problem success at the highest level will continue to elude him.

Which of the statement (s) below is/are logically valid and can be inferred from the above passage?

(*i*) He was already a successful batsman at the highest level

(*ii*) He has to improve his temperament in order to become a great batsman

(*iii*) He failed to make many of his good starts count

(*iv*) Improving his technical skills will guarantee success

(*a*) (*iii*) and (*iv*) (*b*) (*ii*) and (*iii*)

(*c*) (*i*), (*ii*) and (*iii*) (*d*) (*ii*) only

52. Alexander turned his attention towards India, since he had conquered Persia.

Which one of the statements below is logically valid and can inferred from the above sentence?

(*a*) Alexander would not have turned his attention towards India had he not conquered Persia.

(*b*) Alexander was not ready to rest on his laurels, and wanted to march to India

(*c*) Alexander was completely in control of his army and could command it to move towards India

(*d*) Since Alexander's kingdom extended to Indian borders after the conquest of Persia, he was keen to move further

53. Tanya is older than Enc.

Cliff is older than Tanya.

Eric is older than Cliff.

If the first two statements are true, then the third statement is

(*a*) True (*b*) False

(*c*) Uncertain (*d*) Data insufficient

54. Given below are two statements followed by two conclusions. Assuming these statements to be true, decide which one logically follows:

Statements:

I. No manager is a leader.

II. All leaders are executive.

Conclusions:

I. No manager is a executive.

II. All executive is a manager.

(*a*) Only conclusion I follows.

(*b*) Only conclusion II follows.

(*c*) Neither conclusion I nor II follows.

(*d*) Both conclusion I and II follow.

55. Find the missing sequence in the letter series below :

A, CD, GHI, _?___, UVWXY

(*a*) LMN (*b*) MNO

(*c*) MNOP (*d*) NOPQ

56. Ms. X will be in Bagdogra from 1/5/14 to 20/5/14 and from 22/5/14 to 31/5/14. On the morning of 21/5/14, she will reach Kochi via Mumbai.

Which one of the statements below is logically valid and can be inferred from the above sentences?

(*a*) Ms. X will be in Kochi for only one day only in May

(*b*) Ms. X will be in Kochi for only one day in May

(*c*) Ms. X will be only in Kochi for one day in May

(*d*) Only Ms. X will be in Kochi for one day in May

57. Given below are two statements followed by two conclusions. Assuming these statements to be true, decide which one logically follows.

Statement:

I. All film stars are playback singers.

II. All film directors are film stars.

Conclusions:

I. All film directors are playback singers.

II. Some film stars are film directors.

(*a*) Only conclusion I follows

(*b*) Only conclusion II follows

(*c*) Neither conclusion I nor II follows

(*d*) Both conclusions I and II follow

58. Lamenting the gradual sidelining of the arts in school curricula, a group of prominent artists wrote to the Chief Minister last year, asking him to allocate more funds to support arts education in schools. However, no such increase has been announced in this year's Budget. The artists expressed their deep anguish at their request not being approved, but many of them remain optimistic about funding in the future.

Which of the statement(*s*) below is/are logically valid and can be inferred from the above statements?

i. The artists expected funding for the arts to increase this year.

ii. The Chief Minister was receptive to the idea of increasing funding for the arts.

iii. The Chief Minister is a prominent artist.

iv. Schools are giving less importance to arts education nowadays.

(*a*) *iii* and *iv* (*b*) *i* and *iv*

(*c*) *i*, *ii* and *iv* (*d*) *i* and *iii*

59. Four branches of a company are located at M, N, O, and P. M is north of N at a distance of 4 km; P is south of O at a distance of 2 km; N is southeast of O by 1 km. What is the distance between M and P in km?

(*a*) 5.34 (*b*) 6.74

(*c*) 28.5 (*d*) 45.49

60. The given statement is followed by some courses of action. Assuming the statement to be true, decide the correct option.

Statement:

There has been a significant drop in the water level in the lakes supplying water to the city.

Course of action:

I. The water supply authority should impose a partial cut in supply to tackle the situation.

II. The government should appeal to all the residents through mass media for minimal use of water.

III. The government should ban the water supply in lower areas.

(*a*) Statements I and II follow.

(*b*) Statements I and III follow.

(*c*) Statements II and III follow.

(*d*) All statements follow.

2014

61. In a group of four children, Som is younger to Riaz. Shiv is elder to Ansu. Ansu is youngest in the group. Which of the following statements is/are required to find the eldest child in the group?

Statements

1. Shiv is younger to Riaz.
2. Shiv is elder to Som.

(*a*) Statement 1 by itself determines the eldest child.

(*b*) Statement 2 by itself determines the eldest child.

(*c*) Statements 1 and 2 are both required to determine the eldest child.

(*d*) Statements 1 and 2 are not sufficient to determine the eldest child.

62. X is 1 km northeast of Y. Y is 1 km southeast of Z. W is 1 km west of Z. P is 1 km south of W. Q is 1 km east of P. What is the distance between X and Q in km?

(*a*) 1 (*b*) $\sqrt{2}$

(*c*) $\sqrt{3}$ (*d*) 2

63. Rajan was not happy that Sajan decided to do the project on his own. On observing his unhappiness, Sajan explained to Rajan that he preferred to work independently.

Which one of the statements below is logically valid and can be inferred from the above sentences?

(*a*) Rajan has decided to work only in a group.

(*b*) Rajan and Sajan were formed into a group against their wishes.

(*c*) Sajan had decided to give in to Rajan's request to work with him.

(*d*) Rajan had believed that Sajan and he would be working together.

64. Find the odd one in the following group: ALRVX, EPVZB, ITZDF, OYEIK

(*a*) ALRVX (*b*) EPVZB

(*c*) ITZDF (*d*) OYEIK

65. Anuj, Bhola, Chandan, Dilip, Eswar and Faisal live on different floors in a six-storeyed building (the ground floor is numbered 1, the floor above it 2, and so on). Anuj lives on an even-numbered floor. Bhola does not live on an odd numbered floor. Chandan does not live on any of the floors below Faisal's floor. Dilip does not live on floor number 2. Eswar does not live on a floor immediately above or immediately below Bhola. Faisal lives three floors above Dilip. Which of the following floor-person combinations is correct?

	Anuj	*Bhola*	*Chandan*	*Dilip*	*Eswar*	*Faisal*
(*a*)	6	2	5	1	3	4
(*b*)	2	6	5	1	3	4
(*c*)	4	2	6	3	1	5
(*d*)	2	4	6	1	3	5

66. Find the next term in the sequence: 13M, 17Q, 19S, ________

(*a*) 21W (*b*) 21V

(*c*) 23W (*d*) 23V

67. If 'KCLFTSB' stands for 'best of luck' and 'SHSWDG' stands for 'good wishes', which of the following indicates 'ace the exam'?

(*a*) MCHTX (*b*) MXHTC

(*c*) XMHCT (*d*) XMHTC

68. "India is a country of rich heritage and cultural diversity."

Which one of the following facts best supports the claim made in the above sentence?

(*a*) India is a union of 28 states and 7 union territories.

(*b*) India has a population of over 1.1 billion.

(*c*) India is home to 22 official languages and thousands of dialects.

(*d*) The Indian cricket team draws players from over ten states.

69. In which of the following options will the expression $P < M$ be definitely true?

(*a*) $M < R > P > S$

(*b*) $M > S < P < F$

(*c*) $Q < M < F = P$

(*d*) $P = A < R < M$

70. Find the next term in the sequence:

7G, 11K, 13M, ________

(*a*) 15Q (*b*) 17Q

(*c*) 15P (*d*) 17P

71. The multi-level hierarchical pie chart shows the population of animals in a reserve forest. The correct conclusions from this information are:

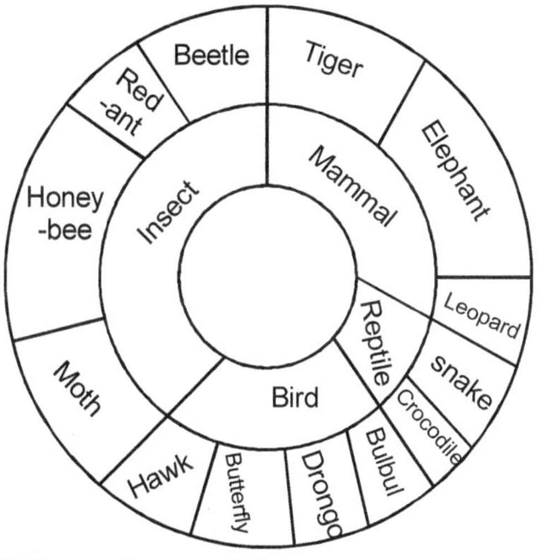

(*i*) Butterflies are birds

(*ii*) There are more tigers in this forest than red ants

(*iii*) All reptiles in this forest are either snakes or crocodiles

(*iv*) Elephants are the largest mammals in this forest

(*a*) (*i*) and (*ii*) only

(*b*) (*i*), (*ii*), (*iii*) and (*iv*)

(*c*) (*i*), (*iii*) and (*iv*) only

(*d*) (*i*), (*ii*) and (*iii*) only

72. Find the odd one in the following group

Q,W,Z,B B,H,K,M W,C,G,J M,S,V,X

(*a*) Q, W, Z, B (*b*) B, H, K, M

(*c*) W, C, G, J (*d*) M, S, V, X

73. Lights of four colors (red, blue, green, yellow) are hung on a ladder. On every step of the ladder there are two lights. If one of the lights is red, the other light on that step will always be blue. If one of the lights on a step is green, the other light on that step will always be yellow. Which of the following statements is not necessarily correct?

(*a*) The number of red lights is equal to the number of blue lights

(*b*) The number of green lights is equal to the number of yellow lights

(*c*) The sum of the red and green lights is equal to the sum of the yellow and blue lights

(*d*) The sum of the red and blue lights is equal to the sum of the green and yellow lights

74. Read the statements:

All women are entrepreneurs.

Some women are doctors.

Which of the following conclusions can be logically inferred from the above statements?

(*a*) All women are doctors

(*b*) All doctors are entrepreneurs

(*c*) All entrepreneurs are women

(*d*) Some entrepreneurs are doctors

75. Find the odd one from the following group:

(*a*) W,E,K,O (*b*) I,Q,W,A

(*c*) F,N,T,X, (*d*) N,V,B,D

76. For submitting tax returns, all resident males with annual income below ₹ 10 lakh should fill up Form P and all resident females with income below ₹ 8 lakh should fill up Form Q. All people with incomes above ₹ 10 lakh should fill up Form R, except non residents with income above ₹ 15 lakhs, who should fill up Form S. All others should fill Form T. An example of a person who should fill Form T is

(*a*) a resident male with annual income ₹ 9 lakh

(*b*) a resident female with annual income ₹ 9 lakh

(*c*) a non-resident male with annual income ₹ 16 lakh

(*d*) a non-resident female with annual income ₹ 16 lakh

77. Which number does not belong in the series below?

2, 5, 10, 17, 26, 37, 50, 64

(*a*) 17 (*b*) 37

(*c*) 64 (*d*) 26

78. A dance programme is scheduled for 10.00 a.m. Some students are participating in the programme and they need to come an hour earlier than the start of the event. These students should be accompanied by a parent. Other students and parents should come in time for the programme. The instruction you think that is appropriate for this is

(*a*) Students should come at 9.00 a.m. and parents should come at 10.00 a.m.

(*b*) Participating students should come at 9.00 a.m. accompanied by a parent, and other parents and students should come by 10.00 a.m.

(*c*) Students who are not participating should come by 10.00 a.m. and they should not bring their parents. Participating students should come at 9.00 a.m.

(*d*) Participating students should come before 9.00 a.m. Parents who accompany them should come at 9.00 a.m. All others should come at 10.00 a.m.

79. The number of people diagnosed with dengue fever (contracted from the bite of a mosquito) in north India is twice the number diagnosed last year. Municipal authorities have concluded that measures to control the mosquito population have failed in this region.

Which one of the following statements, if true, does not contradict this conclusion?

(*a*) A high proportion of the affected population has returned from neighbouring countries where dengue is prevalent

(*b*) More cases of dengue are now reported because of an increase in the Municipal Office's administrative efficiency

(*c*) Many more cases of dengue are being diagnosed this year since the introduction of a new and effective diagnostic test

(*d*) The number of people with malarial fever (also contracted from mosquito bites) has increased this year

80. Geneticists say that they are very close to confirming the genetic roots of psychiatric illnesses such as depression and schizophrenia, and consequently, that doctors will be able to eradicate these diseases through early identification and gene therapy.

On which of the following assumptions does the statement above rely?

(*a*) Strategies are now available for eliminating psychiatric illnesses

(*b*) Certain psychiatric illnesses have a genetic basis

(*c*) All human diseases can be traced back to genes and how they are expressed

(*d*) In the future, genetics will become the only relevant field for identifying psychiatric illnesses

2013

81. Select the pair that best expresses a relationship similar to that expressed in the pair:

Medicine: Health

(*a*) Science : Experiment

(*b*) Wealth : Peace

(*c*) Education : Knowledge

(*d*) Money : Happiness

82. Abhishek is elder to Savar.

Savar is younger to Anshul.

Which of the given conclusions is logically valid and is inferred from the above statements?

(*a*) Abhishek is elder to Anshul

(*b*) Anshul is elder to Abhishek

(*c*) Abhishek and Anshul are of the same age

(*d*) No conclusion follows

83. Complete the sentence:

Universalism is to particularism as diffuseness is to _____

(*a*) specificity

(*b*) neutrality

(*c*) generality

(*d*) adaptation

84. After several defeats in wars, Robert Bruce went in exile and wanted to commit suicide. Just before committing suicide, he came across a spider attempting tirelessly to have its net. Time and again, the spider failed but that did not deter it to retrain from making attempts. Such attempts by the spider made Bruce curious. Thus, Bruce started observing the near-impossible goal of the spider to have the net. Ultimately, the spider succeeded in having its net despite several failures. Such act of the spider encouraged Bruce not to commit suicide. And then, Bruce went back again and won many a battle, and the rest is history.

Which one of the following assertions is best supported by the above information?

(*a*) Failure is the pillar of success.

(*b*) Honesty is the best policy.

(*c*) Life begins and ends with adventures.

(*d*) No adversity justifies giving up hope.

85. Statement: You can always give me a ring whenever you need.

Which one of the following is the best inference from the above statement?

(*a*) Because I have a nice caller tune.

(*b*) Because I have a better telephone facility.

(*c*) Because a friend in need is a friend indeed.

(*d*) Because you need not pay towards the telephone bills when you give me a ring.

86. Statement: There were different streams of freedom movements in colonial India carried out by the moderates, liberals, radicals, socialists, and so on.

Which one of the following is the best inference from the above statement?

(*a*) The emergence of nationalism in colonial India led to our Independence.

(*b*) Nationalism in India emerged in the context of colonialism.

(*c*) Nationalism in India is homogeneous.

(*d*) Nationalism in India is heterogeneous.

87. Select the pair that best expresses a relationship similar to that expressed in the pair :

water : pipe : :

(*a*) cart : road

(*b*) electricity : wire

(*c*) sea : beach

(*d*) music: instrument

88. All professors are researchers

Some scientists are professors

Which of the given conclusions is logically valid and is inferred from the above arguments :

(*a*) All scientists are researchers

(*b*) All professors are scientists

(*c*) Some researchers are scientists

(*d*) No conclusion follows

2012

89. Wanted Temporary, Part-time persons for the post of Field Interviewer to conduct personal interviews to collect and collate economic data. Requirements: High School-pass, must be available for Day, Evening and Saturday work. Transportation paid, expenses reimbursed.

Which one of the following is the best inference from the above advertisement?

(*a*) Gender-discriminatory

(*b*) Xenophobic

(*c*) Not designed to make the post attractive

(*d*) Not gender-discriminatory

90. Given the sequence of terms, AD CG FK JP, the next term is

(*a*) OV (*b*) OW

(*c*) PV (*d*) PW

2011

91. A transporter receives the same number of orders each day. Currently, he has some pending orders (backlog) to be shipped. If he uses 7 trucks, then at the end of the 4th day he can clear all the orders. Alternatively, if he uses only 3 trucks, then all the orders are cleared at the end of the 10th day. What is the minimum number of trucks required so that there will be no pending order at the end of the 5th day?

(*a*) 4 (*b*) 5

(*c*) 6 (*d*) 7

92. Few school curricula include a unit on how to deal with bereavement and grief, and yet all students at some point in their lives suffer from losses through death and parting.

Based on the above passage which topic would not be included in a unit on bereavement?

(*a*) How to write a letter of condolence

(*b*) What emotional stages are passed through in the healing process

(*c*) What the leading causes of death are

(*d*) How to give support to a grieving friend

93. The question below consists of a pair of related words followed by four pairs of words. Select the pair that best expresses the relation in the original pair :

Gladiator : Arena

(*a*) dancer : stage

(*b*) commuter : train

(*c*) teacher : classroom

(*d*) lawyer : courtroom

94. The horse has played a little known but very important role in the field of medicine. Horses were injected with toxins of diseases until their blood built up immunities. Then a serum was made from their blood. Serums to fight with diphtheria and tetanus were developed this way.

It can be inferred from the passage, that horses were

(*a*) given immunity of diseases

(*b*) generally quite immune to diseases

(*c*) given medicines to fight toxins

(*d*) given diphtheria and tetanus serums

2010

95. The question below consists of a pair of related words followed by four pairs of words. Select the pair that best expresses the relation in the original pair.

Unemployed : Worker

(*a*) fallow : land (*b*) unaware : sleeper

(*c*) wit : jester (*d*) renovated : house

96. Hari (H), Gita (G), Irfan (I) and Saira (S) are siblings (i.e. brothers and sisters). All were born on 1st January. The age difference between any two successive siblings (that is born one after another) is less than 3 years. Given the following facts:

I. Hari's age + Gita age > Irfan's age + Saira's age

II. The age difference between Gita and Saira is 1 year. However, Gita is not the oldest and Saira is not the youngest.

III. There are no twins.

In what order were they born (oldest first)?

(*a*) HSIG (*b*) SGHI

(*c*) IGSH (*d*) IHSG

97. Modern warfare has changed from large scale clashes of armies to suppression of civilian populations. Chemical agents that do their work silently appear to be suited to such warfare; and regretfully, there exist people in military establishments who think that chemical agents are useful tools for their cause.

Which of the following statements best sums up the meaning of the above passage:

(*a*) Modern warfare has resulted in civil strife

(*b*) Chemical agents are useful in modern warfare

(*c*) Use of chemical agents in warfare would be undesirable

(*d*) People in military establishments like to use chemical agents in war

98. If 137 + 276 = 435 how much is 731 + 672?

(*a*) 534 (*b*) 1403

(*c*) 1623 (*d*) 1513

NUMERICAL TYPE QUESTIONS

2015

1. Fill in the missing value

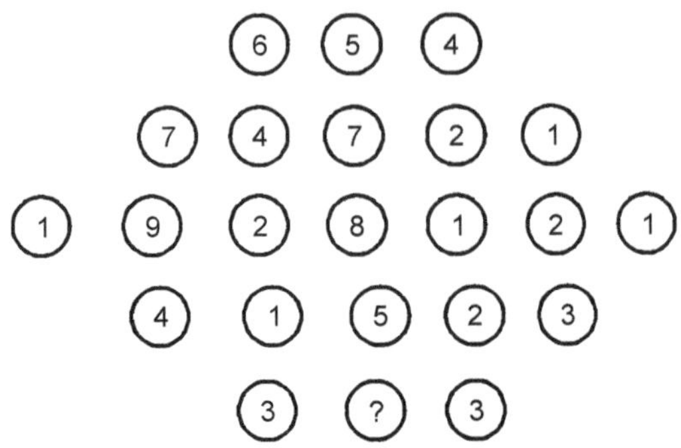

2014

2. The next term in the series 81, 54, 36, 24,.... is ________

3. Fill in the missing number in the series.

2 3 6 15 ___ 157.5 630

4. What is the next number in the series?

12 35 81 173 357 ____

ANSWERS

MCQ Type Questions

1. (b)	**2.** (c)	**3.** (c)	**4.** (b)	**5.** (c)	**6.** (d)	**7.** (d)	**8.** (*c*)	**9.** (*c*)	**10.** (*d*)
11. (*c*)	**12.** (*c*)	**13.** (*d*)	**14.** (*c*)	**15.** (*c*)	**16.** (*c*)	**17.** (*b*)	**18.** (*a*)	**19.** (*a*)	**20.** (*d*)
21. (*c*)	**22.** (*d*)	**23.** (*b*)	**24.** (*b*)	**25.** (*a*)	**26.** (*c*)	**27.** (*c*)	**28.** (*c*)	**29.** (*b*)	**30.** (*c*)
31. (*c*)	**32.** (*d*)	**33.** (*c*)	**34.** (*a*)	**35.** (*d*)	**36.** (*c*)	**37.** (*b*)	**38.** (*c*)	**39.** (*a*)	**40.** (*c*)
41. (*d*)	**42.** (*b*)	**43.** (*a*)	**44.** (*a*)	**45.** (*d*)	**46.** (*c*)	**47.** (*b*)	**48.** (*b*)	**49.** (*b*)	**50.** (*b*)
51. (*b*)	**52.** (*a*)	**53.** (*b*)	**54.** (*c*)	**55.** (*c*)	**56.** (*b*)	**57.** (*d*)	**58.** (*b*)	**59.** (*a*)	**60.** (*a*)
61. (*a*)	**62.** (*c*)	**63.** (*d*)	**64.** (*d*)	**65.** (*b*)	**66.** (*c*)	**67.** (*b*)	**68.** (*c*)	**69.** (*d*)	**70.** (*b*)
71. (*d*)	**72.** (*c*)	**73.** (*d*)	**74.** (*d*)	**75.** (*d*)	**76.** (*b*)	**77.** (*c*)	**78.** (*b*)	**79.** (*d*)	**80.** (*b*)
81. (*c*)	**82.** (*d*)	**83.** (*a*)	**84.** (*d*)	**85.** (*c*)	**86.** (*d*)	**87.** (*b*)	**88.** (*c*)	**89.** (*d*)	**90.** (*a*)
91. (*c*)	**92.** (*c*)	**93.** (*d*)	**94.** (*b*)	**95.** (*d*)	**96.** (*b*)	**97.** (*c*)	**98.** (*c*)		

Numerical Type Questions

1. (3) **2.** (16) **3.** (45) **4.** (725 to 725)

EXPLANATIONS

MCQ TYPE QUESTIONS

1. Mirror image of given text is

Hence, option (b) is correct answer.

2.

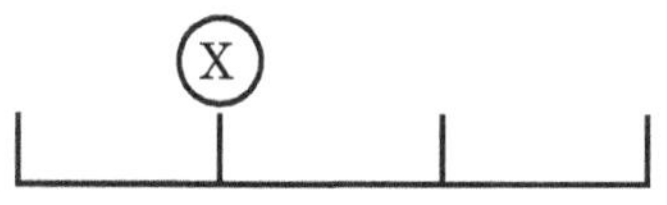

No. of vacant positions for R = 3

The Number of distinct seating arrangements possible is

$= 3 \times 3!$

$= 3 \times 3 \times 2 \times 1$

$= 18$

Hence, option (c) is the correct answer.

3.

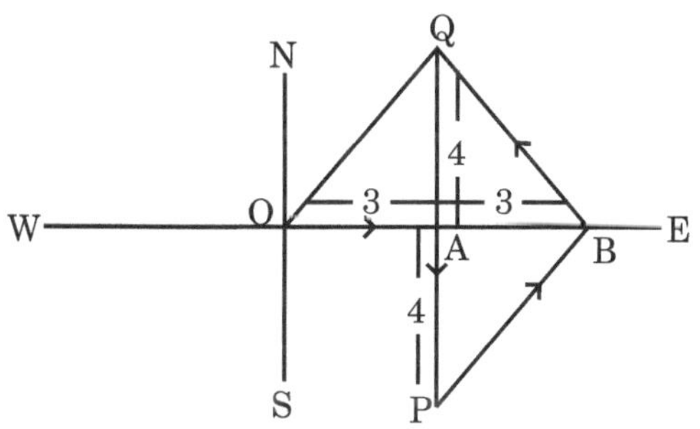

Given: OA = 3 unit, PQ = 8 unit, QA = 4 unit, OB = 6unit

In ΔOQA

$OQ^2 = OA^2 + AQ^2$

$OQ = \sqrt{3^2 + 4^2}$

$= \sqrt{9+16} = \sqrt{25} = 5$ unit

So, option (c) is correct.

4.

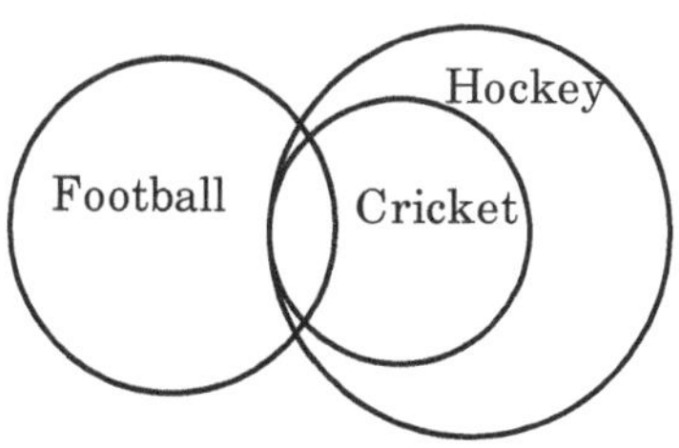

From above diagram;

Some football player play Hockey option (b) is correct answer.

5. Build : Building (Noun)

so, Grow : Growth (Noun)

6. The correct statement about crowd funding is to fund raised through voluntary contributions on web-based platforms.

7.

P,	Q,	R,	S
$\alpha\alpha$	$\alpha\beta$	$\boxed{\beta\alpha}$	$\boxed{\beta\beta}$

8. Numbers are 2, 4, 5, 7, and 10

The correct order of arrangement 10, 5, 4, 7, and 2

Thus an arrangement follows given three conditions

Second number from right = 7

13. (i) Volume of cuboid = $(8 \times 10 \times 6)\ cm^3 = 480\ cm^3$

(ii) Volume of cube = $(8 \times 8 \times 8)\ cm^3 = 512\ cm^3$

(iii) Volume of cylinder = $\pi r^2 h$

$$\left(\frac{22}{7} \times 7 \times 7 \times 7\right) cm^3 = 1078\ cm^3$$

(iv) Volume of Sphere $= \frac{4}{3}\pi r^3 = \left(\frac{4}{3}\pi \times (7)^2\right) cm^3$

$= 1436.75\ cm^3$

Hence, the descending order of their volumes is (iv), (iii), (ii), (i).

14. Number of parallelogram = ${}^4C_2 \times {}^5C_2$

$$= \frac{4!}{2! \times 2!} \times \frac{5!}{3! \times 2!}$$

$$= 6 \times \frac{5 \times 4}{2} = 60$$

15.

$$x^2 - 1 = -x$$

$$x - \frac{1}{x} = -1$$

$$x^2 + \frac{1}{x^2} = 0.2$$

Now, $$x^4 + \frac{1}{x^4} = \left(x^2 + \frac{1}{x^2}\right)^2 - 2$$

$$= (3)^2 - 2$$

$$= 9 - 2 = 7$$

16. respectively

17. P = 16 years

Q = 25 years

R = milkshake

S = Bear

With above known values, for the rule to be followed we have to check. P's drink and Sage because he is drinking beer.

18.

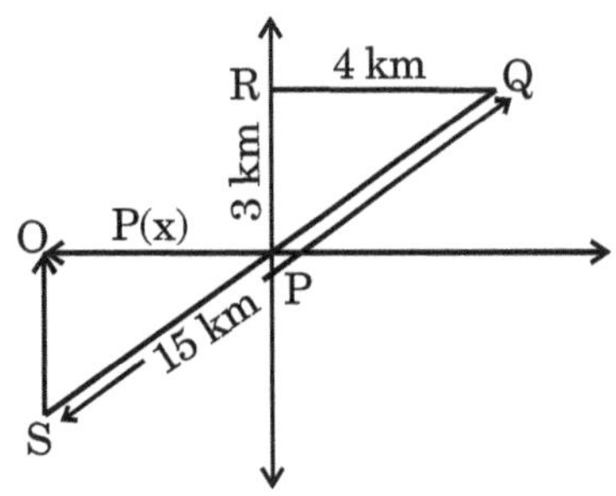

In ΔPQR

By pythagorus

$(4)^2 + (3)^2 = PQ^2$

PQ = 5 km

In Now PS = QS – PQ

= 15 – 5 = 10 km.

In ΔPOS

$x^2 = (10)^2 - (6)^2 = 64$

= 8 km. (East)

19. An intimate association does not allow for the necessary perspective.

20. With diagram:

X and Y are married and have 2 children P and Q

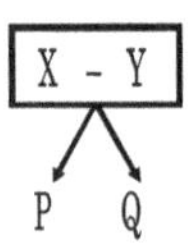

S is daughter of P and Z is S' grandfather

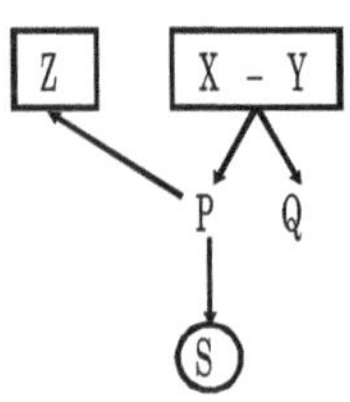

Z is married to W

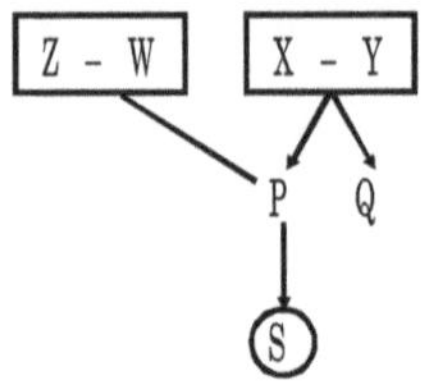

W and Z are parents of R

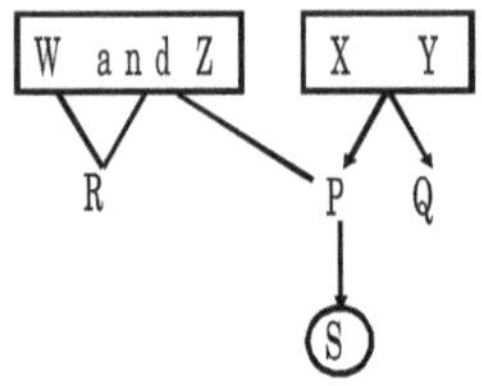

We can simply

Z and P are married and are parents of S. Hence above conditions prove d is the answer.

21. Given words 'mock, deride and jeer' are synonyms which means mockery. So, the given word praise is odd one.

22.

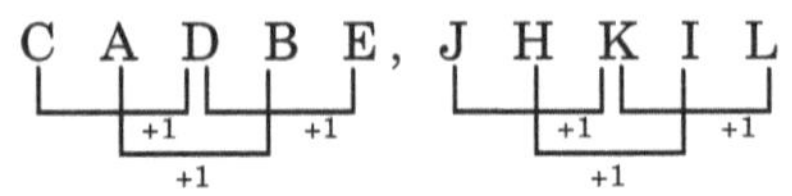

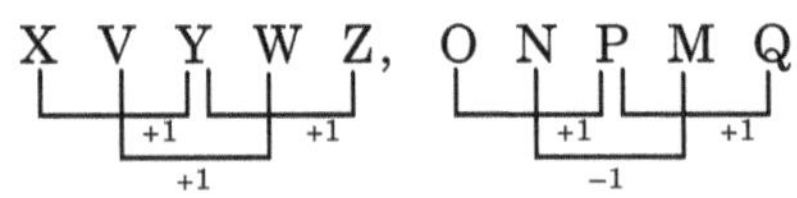

So, option (d) is odd one from the given group.

23. The given number 33 is odd one out, because the remaining numbers are prime number.

27. Statements i and ii are not logically possible based on the given fact.

28. Group 'P'

Mean (μ) = 105

Standard deviation (σ_1) = 25

$Pr(\mu - \sigma \le x \le \mu + \sigma) \approx 0.6827$

∴ 68% within one standard deviation

$\mu_1 - \sigma_1 = 105 - 25 = 80$

$\mu_1 + \sigma_1 = 105 + 25 = 130$

∴ range = 80 to 130

Distribution of P:

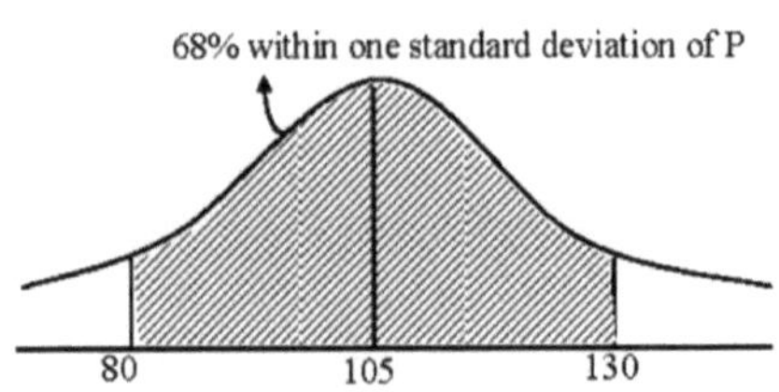

Group 'Q'

Mean $(\mu_2) = 85$

Standard deviation $(\sigma_2) = 5$

$\Pr(\mu - \sigma \le x \le \mu + \sigma) \approx 0.6827$

$\therefore$ 68% within one standard deviation

$$\mu_2 - \sigma_2 = 85 - 5 = 80$$

$$\mu_2 + \sigma_2 = 85 + 5 = 90$$

$\therefore$ Range of Q in one standard deviation is 80 to 90

Distribution of 'Q'

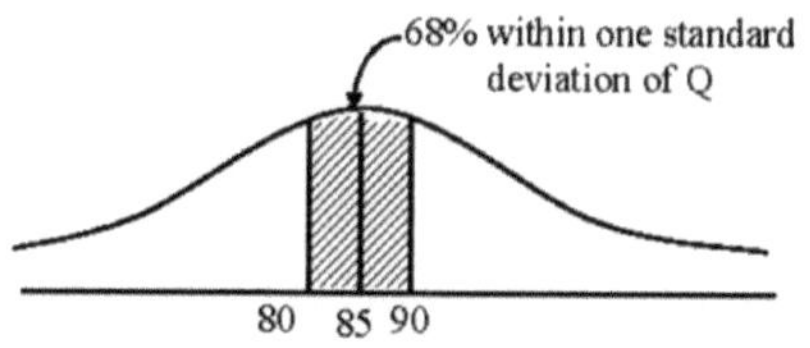

68% within one standard deviation of Q is narrower

$\therefore$ 68% within one standard deviation of Q means most students of group Q.

$\therefore$ Most students of group 'Q' scored marks in a narrower range than students in group 'P'

30. The following letter series is in the order of even letters series

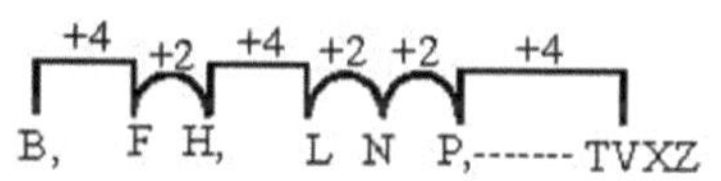

31. From given data, the following diagram can be possible.

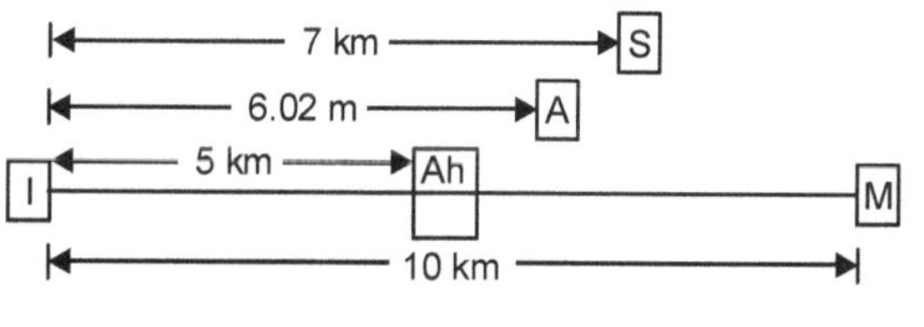

Here S → Susan

A → Arun

Ah → Ahmed

M → Michael

From the above diagram, Arun lives farthest away than Ahmed means more than 5 km but closer than Susan means less than 7 km, from the given alternatives.

So, only option 'c' is possible.

32. From the given question two statements will be followed.

For statment I

Arrange the given data according to their ages

For statement II

Arrange the given data according to their winning

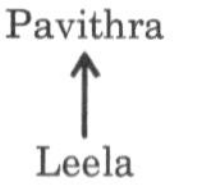

So, from statement I and II, it is clear that only option (*d*) is possible (i.e., statement I).

33.

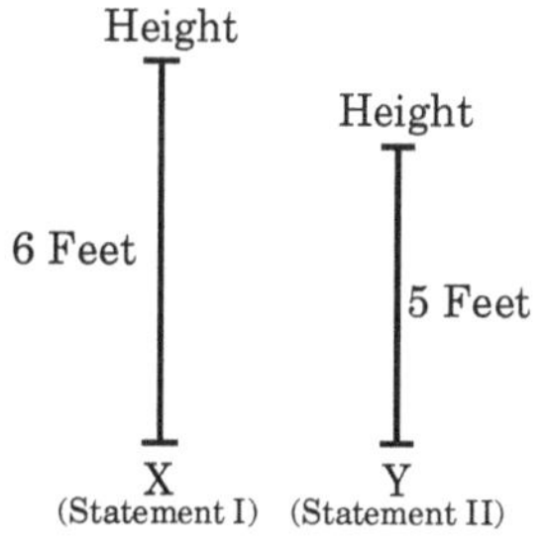

Hence from the given figure Mr. X is taller than Mr. Y by 1 foot.

34. Social science disciplines had a pre-colonial origin.

35.

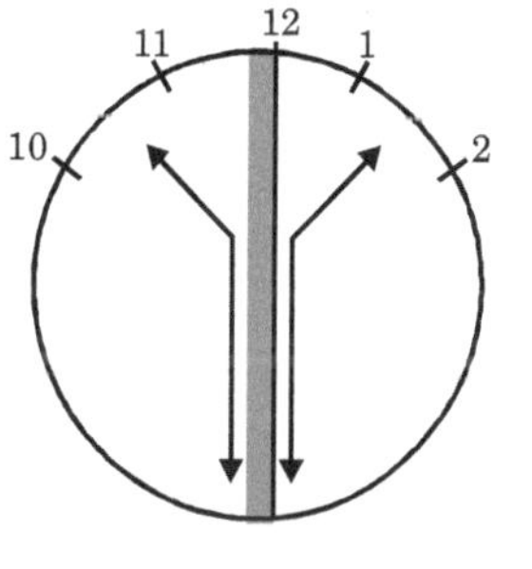

Mirror image of 1 : 20 is 10 : 30

10 : 30 was the time two and quarter hour back so time now will be 12 : 45

36.

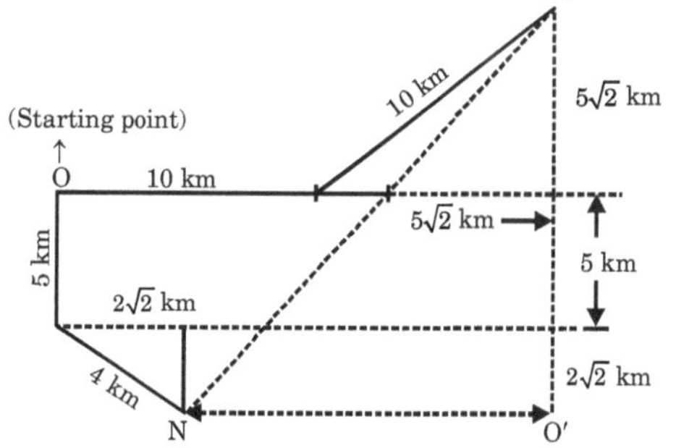

From the given figure

$$MN = \sqrt{(O'M)^2 + (O'N)^2}$$

$$O'M = 5\sqrt{2} + 5 + 2\sqrt{2} = 5 + 7\sqrt{2}$$

$$O'N = 10 + 5\sqrt{2} - 2\sqrt{2} = 10 + 3\sqrt{2}$$

$$MN = \sqrt{(5+7\sqrt{2})^2 + (10+3\sqrt{2})^2}$$

$$= \sqrt{25+98+70\sqrt{2}+100+18+60\sqrt{2})}$$

≈ 20.61 km

37. 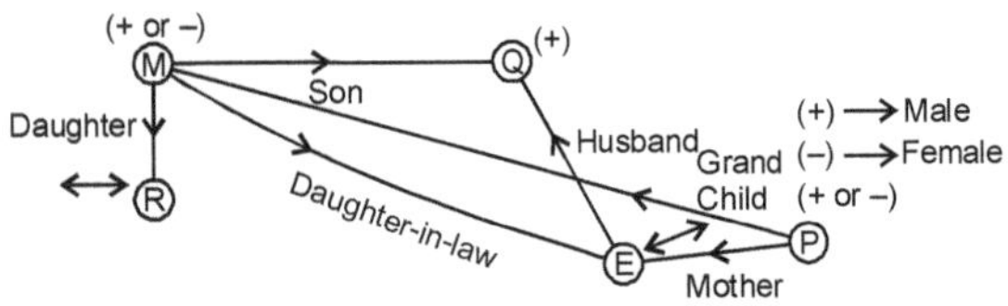

So, from the above relation diagram it is clear that P is the grandchild of M.

40. Manu age = sravan age + 2 months

Manu age = Trideep age – 3 months

Pavan age = Sravan's age + 1 month

From the above statement

Trideep age > Man > Pavan > Sravan

Hence, Trideep can occupy the extra space in the flat.

44. 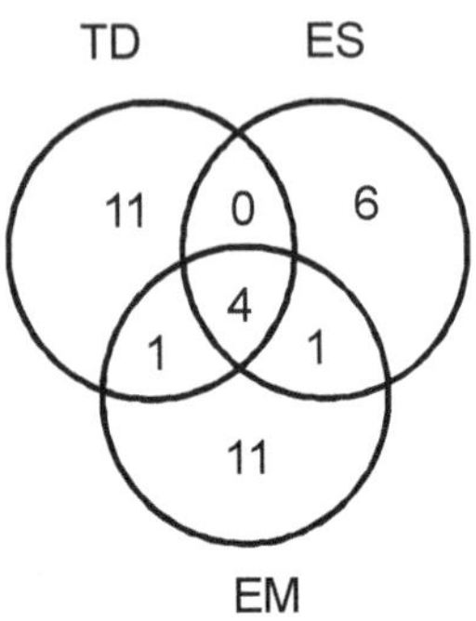

45. Statement-I can be used to solve the question if capacity of both tanks is already known

Statement-II can be used if it is known what quantity of each tank is full/empty.

Therefore, by using both statements

Let capacity of tank B is x

$$\frac{70}{100}x = 14000$$

$$\Rightarrow \quad x = 20000 \text{ gallons}$$

$$\text{Solution in tank A} = \frac{80}{100} \times 14000 = 11200 \text{ gallons}$$

$$\text{Solution in tank B} = \frac{40}{100} \times 20000 = 8000 \text{ gallons}$$

$\therefore$ Total solution = 11200 + 8000 = 19200 gallons

46. $66\square 6 = \dfrac{66-6}{66+6} = \dfrac{60}{72} = \dfrac{5}{6}$

$66\diamond 6 = \dfrac{66+6}{66-6} = \dfrac{72}{60} = \dfrac{6}{5}$

$(66\square 6) \rightarrow (66\diamond 6) = \dfrac{5}{6} \times \dfrac{6}{5} = 1$

48. R + 3 = U, O + 3 = R,
A + 3 = D, D + 3 = G;
S + 3 = V, W + 3 = Z,
A + 3 = D, N + 3 = Q

49. Community of people : Doctor

50. Since U does not want any portfolio, (*c*) and (*d*) are ruled out. R wants Home, or Finance or No portfolio, (*a*) is not valid.

Hence option (*b*) is correct.

54. S + 1: S + 2 :

Therefore concluding diagram can be

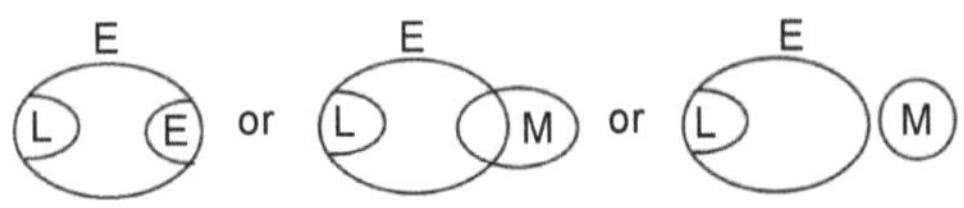

55. 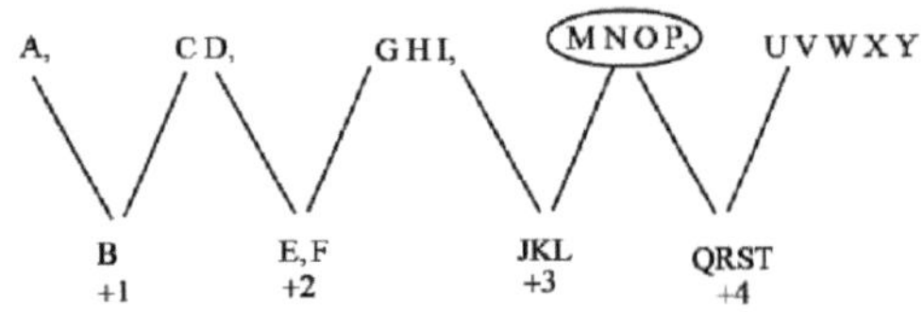

56. Second sentence says that Ms. X reaches Kochi on 21/05/2014. Also she has to be in Bagdogora on 22/05/2014.

$\therefore$ She stays in Kochi for only one day in may.

59. 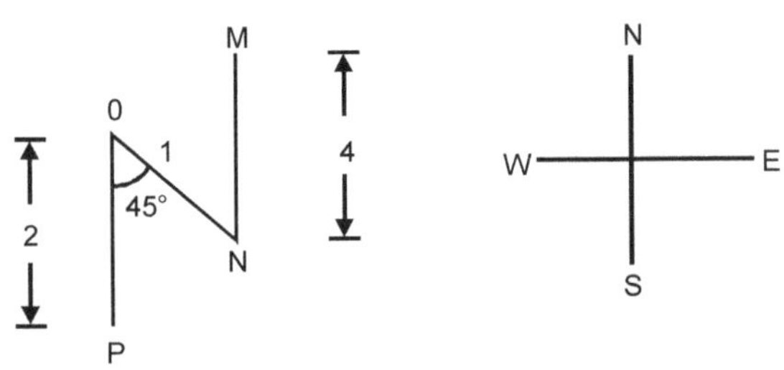

61. Riaz > Som ...(*i*)

Shiv > Ansu ...(*ii*)

Ansu is youngest in the group

∴ From equations (*i*) and (*ii*)

∴ Riaz > Som > Shiv > Ansu

62.

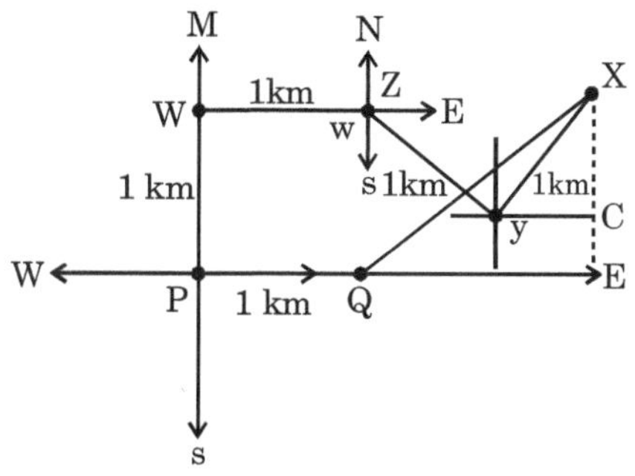

Here XE = 1 km

$$QE = \frac{2}{\sqrt{2}} = \sqrt{2}$$

∴ using Pythagoras theorem,

$$XQ = \sqrt{(QE)^2 + (XE)^2}$$

$$= \sqrt{2+1} = \sqrt{3} \text{ km}$$

∴ Statement I itself determines the eldest child which statement 2 is incorrect as Ansu is youngest in the group is not satisfied by statement 2

Hence option (*a*) is correct

64. ALRVX → only one vowel

EPVZB → only one vowel

ITZDF → only one vowel

OYEIK → three vowels

65. (*a*) Anuj: Even numbered floor (2, 4, 6)

(*b*) Bhola: Even numbered floor (2, 4, 6)

(*c*) Chandan lives on the floor above that of Faisal.

(*d*) Dilip: not on 2nd floor.

(*e*) Eswar: does not live immediately above or immediately below Bhola

From the options its clear, that only option (*b*) satisfies condition (*e*).

So, correct Ans is (*b*).

67. KCLFTSB: BST-Best, F-Of, LCK-Luck (Reverse order)

SHSWDG: GD-Good, WSHS-Wishes (Reverse order)

Similarly "ace the Exam'- C-Ace, T-The, XM-Exam

68. Diversity is shown in terms of difference language.

71. It is not mentioned that elephant is the largest animal.

72.

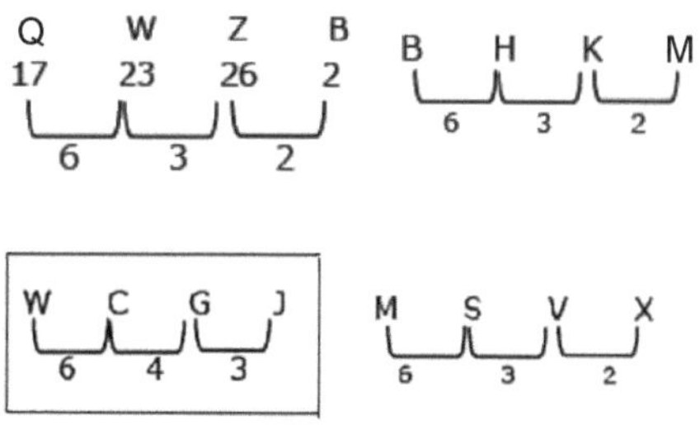

75.

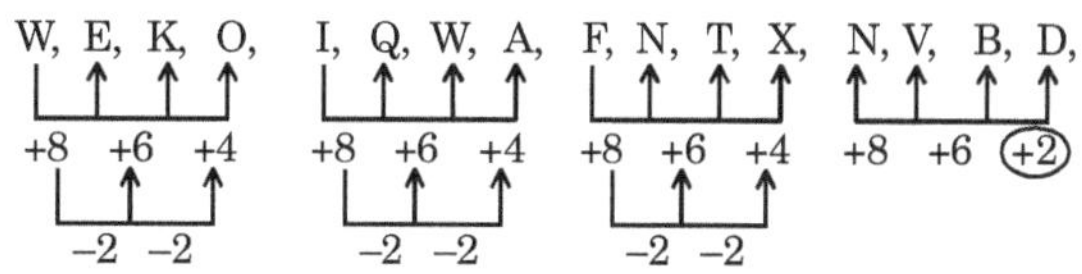

Hence the odd one from the following group is N, V, B, D.

76. By reading the instructions, it is clear that person who fills form T is a resident female with annual income of ₹ 9 Lakh.

79. Among given option, option (*d*) *i.e.* the number of people with malarial fever has increased this year do not contradic the conclusion.

90.

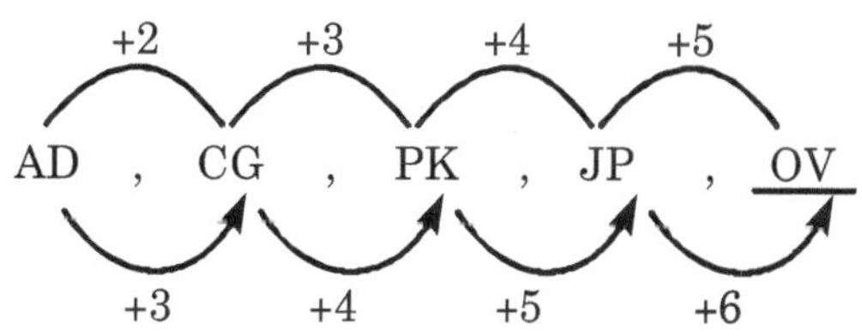

91. Let each truck carry 100 units.

$$2800 = 4n + e$$

where n = normal

$$3000 = 10n + e$$

where e = excess/pending

$$\therefore \quad n = \frac{100}{3}, \quad e = \frac{8000}{3}$$

$$5 \text{ days} \Rightarrow 500\,x = \frac{5100}{3} + \frac{8000}{3}$$

$$\Rightarrow \quad 500x = \frac{8500}{3} = 17$$

$$\Rightarrow \quad x > 5$$

Minimum possible = 6

92. The given passage clearly deals with how to deal with bereavement and grief and so after the tragedy occurs and not about precautions. Therefore, irrespective of the causes of death, a school student rarely gets into details of causes—which is beyond the scope of the context. Rest all are important in dealing with grief.

93. Given relationship is worker: workplace. A gladiator is

(*i*) a person, usually a professional combatant trained to entertain the public by engaging in mortal combat with another person or a wild.

(*ii*) A person engaged in a controversy or debate, especially in public.

94. From the passage it cannot be inferred that horses are given immunity as in option (*a*), since the aim is to develop medicine and in turn immunize humans. Option (*b*) is correct since it is given that horses develop immunity after some time. Refer "until their blood built up immunities". Even option(*c*) is invalid since medicine is not built till immunity is developed in the horses. Option (*d*) is incorrect since specific examples are cited to illustrate and this cannot capture the essence.

96. $H + G > I + S$ ...(1)

and $G - S = 1$ ...(2)

G is not oldest, S is not youngest

$\therefore$ $H + 1 > I$

Irfan older than Hari

Gita older than Sarita

From given option SGHI

97. Use of chemical agents in warfare would be undesirable

98.

$$\begin{array}{rl} 137 - & \overline{001}\ \ \overline{011}\ \ \overline{111} \\ +276 - & 010\ \ 111\ \ 110 \\ \hline & 100\ \ 011\ \ 101 = 435 \end{array}$$

Octal number system

$$\therefore \quad \begin{array}{rl} 731 - & 111\ \ 011\ \ 001 \\ 672 - & 110\ \ 111\ \ 110 \\ \hline & \underline{100}\ \ 010\ \ 011 \end{array}$$

Hence 1623

NUMERICAL TYPE QUESTIONS

1. Middle number is the average of number on both sides

$\therefore$ Average of 3 and 3 is $\frac{3+3}{2} = \frac{6}{2} = 3$

2. $81 - 54 = 27;\ 27 \times \frac{2}{3} = 18$

$54 - 36 = 18;\ 18 \times \frac{2}{3} = 12$

$36 - 24 = 12;\ 12 \times \frac{2}{3} = 8$

$\therefore$ $24 - 8 = 16$

3.

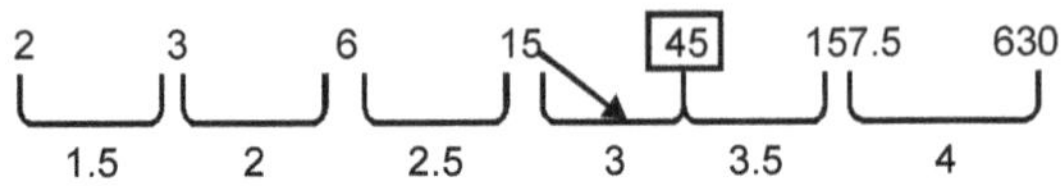

$\frac{\text{2nd number}}{\text{1st number}}$ is in increasing order as shown above

4. The given series is

12, 35, 81, 173, 357, ...

The given series follows the following pattern

$12 \times 2 + 11 = 35$

$35 \times 2 + 11 = 81$

$81 \times 2 + 11 = 173$

$173 \times 2 + 11 = 357$

$357 \times 2 + 11 = 725$

Numerical Ability

1 CHAPTER Numbers and Algebra

MCQ TYPE QUESTIONS

2021

1. $\oplus$ and $\odot$ are two operators on numbers p and q such that $p \odot q = p - q$, and $p \oplus q = p \times q$

Then, $(9 \odot (6 \oplus 7)) \odot (7 \oplus (6 \odot 5)) =$

(a) 40 (b) × 26

(c) – 33 (d) – 40

2020

2. The sum of the first n terms in the sequence 8, 88, 888, 8888, ... is ____

(*a*) $\frac{81}{80}(10^n - 1) + \frac{9}{8}n$ (*b*) $\frac{81}{80}(10^n - 1) - \frac{9}{8}n$

(*c*) $\frac{80}{81}(10^n - 1) + \frac{8}{9}n$ (*d*) $\frac{80}{81}(10^n - 1) - \frac{8}{9}n$

2017

3. The number of 3-digit numbers such that the digit 1 is never to the immediate right of 2 is

(*a*) 781 (*b*) 791

(*c*) 881 (*d*) 891

2016

4. In a quadratic function, the value of the product of the roots (α, β) is 4. Find the value of

$$\frac{\alpha^n + \beta^n}{\alpha^{-n} + \beta^{-n}}$$

(*a*) n^4 (*b*) 4^n

(*c*) 2^{2n-1} (*d*) 4^{n-1}

5. Among 150 faculty members in an institute, 55 are connected with each other through Facebook® and 85 are connected through WhatsApp®. 30 faculty members do not have Facebook® or WhatsApp® accounts. The number of faculty members connected only through Facebook® accounts is ____________.

(*a*) 35 (*b*) 45

(*c*) 65 (*d*) 90

6. If $|9y - 6| = 3$, then $y^2 - \frac{4y}{3}$ is ________.

(*a*) 0 (*b*) $+\frac{1}{3}$

(*c*) $-\frac{1}{3}$ (*d*) Undefined

7. If $q^{-a} = \frac{1}{r}$ and $r^{-b} = \frac{1}{s}$ and $s^{-c} = \frac{1}{q}$, the value of abc is .

(*a*) $(rqs)^{-1}$ (*b*) 0

(*c*) 1 (*d*) r^+q^+s

8. The sum of the digits of a two digit number is 12. If the new number formed by reversing the digits is greater than the original number by 54, find the original number.

(*a*) 39 (*b*) 57

(*c*) 66 (*d*) 93

9. Two finance companies, P and Q, declared fixed annual rates of interest on the amounts invested with them. The rates of interest offered by these companies may differ from year to year. Year-wise annual rates of interest offered by these companies are shown by the line graph provided below.

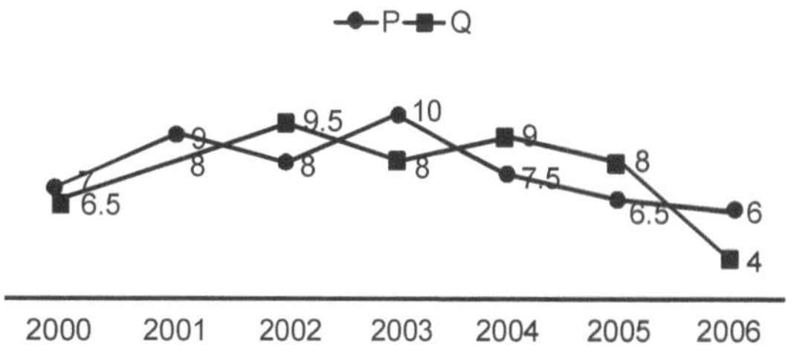

If the amounts invested in the companies, P and Q, in 2006 are in the ratio 8 : 9, then the amounts received after one year as interests from companies P and Q would be in the ratio:

(*a*) 2 : 3 (*b*) 3 : 4

(*c*) 6 : 7 (*d*) 4 : 3

2015

10. Read the following table giving sales data of five types of batteries for years 2006 to 2012

Year	Type I	Type I	Type I	Type I	Type I
2006	75	144	114	102	108
2007	90	126	102	84	126
2008	96	114	75	105	135
2009	105	90	150	90	75
2010	90	75	135	75	90
2011	105	60	165	45	120
2012	115	85	160	100	145

Out of the following, which type of battery achieved highest growth between the years 2006 and 2012?

(*a*) Type V (*b*) Type III

(*c*) Type II (*d*) Type I

11. If $a^2 + b^2 + c^2 = 1$, then $ab + bc + ac$ lies in the interval

(*a*) $\left[1, \frac{2}{3}\right]$ (*b*) $\left[\frac{-1}{2}, 1\right]$

(*c*) $\left[-1, \frac{1}{2}\right]$ (*d*) [2,–4]

12. If the list of letters, P, R, S, T, U is an arithmetic sequence, which of the following are also in arithmetic sequence?

I. 2P, 2R, 2S, 2T, 2U

II. P–3, R–3, S–3, T–3, U–3

III. P^2, R^2, S^2, T^2, U^2

(*a*) I only (*b*) I and II

(*c*) II and III (*d*) I and III

13. The number of students in a class who have answered correctly, wrongly, or not attempted each question in an exam, are listed in the table below. The marks for each question are also listed. There is no negative or partial marking.

Q. No.	Marks	Answered Correctly	Answered Wrongly	Not Attempted
1	2	21	17	6
2	3	15	27	2
3	1	11	29	4
4	2	23	18	3
5	5	31	12	1

What is the average of the marks obtained by the class in the examination?

(*a*) 2.290 (*b*) 2.970

(*c*) 6.795 (*d*) 8.795

2014

14. The monthly rainfall chart on 50 years of rainfall in Agra is shown in the following figure. Which of the following are true? (k percentile is the value such that k percent of the data fall below that value)

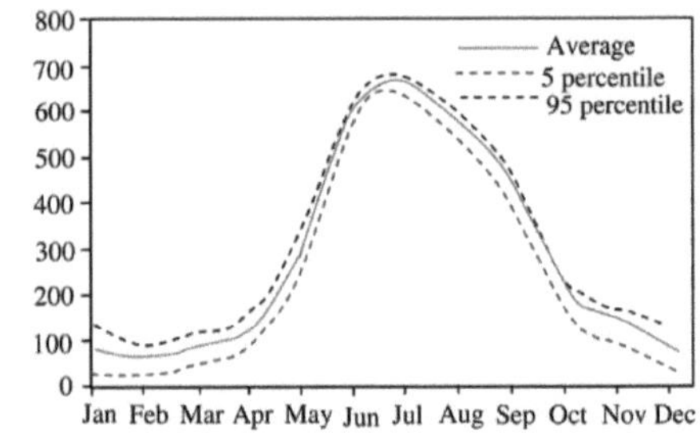

(*i*) On average, it rains more in July than in December

(*ii*) Every year, the amount of rainfall in August is more than that in January

(*iii*) July rainfall can be estimated with better confidence than February rainfall

(*iv*) In August, there is at least 500 mm of rainfall

(*a*) (*i*) and (*ii*) (*b*) (*i*) and (*iii*)

(*c*) (*ii*) and (*iii*) (*d*) (*iii*) and (*iv*)

15. The total exports and revenues from the exports of a country are given in the two charts shown below. The pie chart for exports shows the quantity of each item exported as a percentage of the total quantity of exports. The pie chart for the revenues shows the percentage of the total revenue generated through export of each item. The total quantity of exports of all the items is 500 thousand tonnes and the total revenues are 250 crore rupees. Which item among the following has generated the maximum revenue per kg?

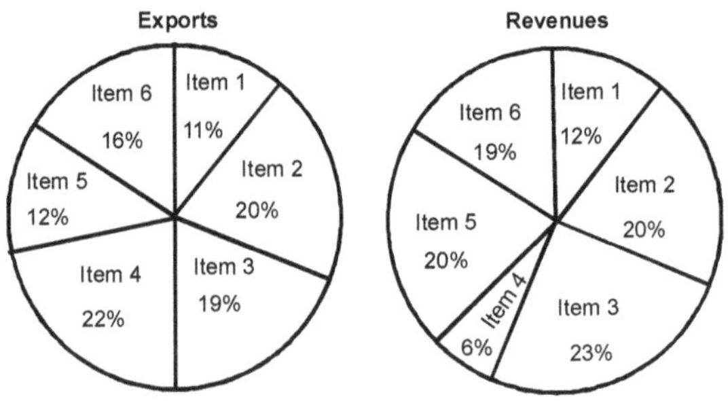

(*a*) Item 2 (*b*) Item 3

(*c*) Item 6 (*d*) Item 5

16. The statistics of runs scored in a series by four batsmen are provided in the following table. Who is the most consistent batsman of these four?

Batsman	Average	Standard deviation
K	31.2	5.21
L	46.0	6.35
M	54.4	6.22
N	17.9	5.90

(*a*) K (*b*) L
(*c*) M (*d*) N

17. The exports and imports (in crores of ₹) of a country from 2000 to 2007 are given in the following bar chart. If the trade deficit is defined as excess of imports over exports, in which year is the trade deficit 1/5th of the exports?

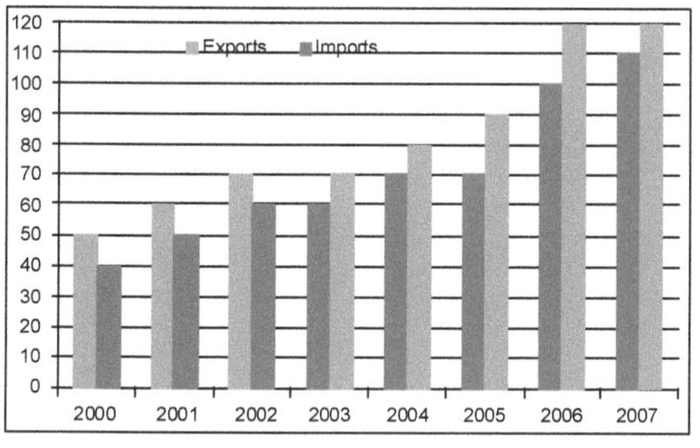

(*a*) 2005
(*b*) 2004
(*c*) 2007
(*d*) 2006

18. The table below has question-wise data on the performance of students in an examination. The marks for each question are also listed. There is no negative or partial marking in the examination.

Q.No.	Marks	Answered Correctly	Answered Wrongly	Not Attempted
1	2	21	17	6
2	3	15	27	2
3	2	23	18	3

What is the average of the marks obtained by the class in the examination?

(*a*) 1.34 (*b*) 1.74
(*c*) 3.02 (*d*) 3.91

19. The ratio of male to female students in a college for five years is plotted in the following line graph. If the number of female students in 2011 and 2012 is equal, what is the ratio of male students in 2012 to male students in 2011?

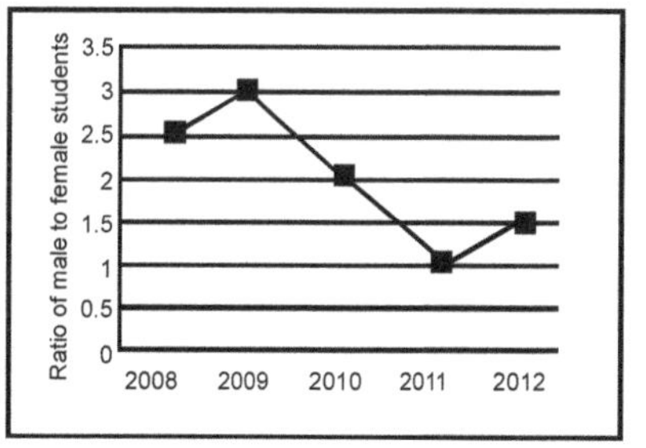

(*a*) 1 : 1 (*b*) 2 : 1
(*c*) 1.5 : 1 (*d*) 2.5 : 1

20. What is the average of all multiples of 10 from 2 to 198?

(*a*) 90 (*b*) 100
(*c*) 110 (*d*) 120

21. The value of $\sqrt{12+\sqrt{12+\sqrt{12+...}}}$ is

(*a*) 3.464 (*b*) 3.932
(*c*) 4.000 (*d*) 4.444

22. If x is real and $|x^2 - 2x + 3| = 11$, then possible values of $|-x^3 + x^2 - x|$ include

(*a*) 2, 4 (*b*) 2, 14
(*c*) 4, 52 (*d*) 14, 52

2013

23. A number is as much greater than 75 as it is smaller than 117. The number is

(*a*) 91
(*b*) 93
(*c*) 89
(*d*) 96

24. Following table provides figures (in rupees) on annual expenditure of a firm for two years - 2010 and 2011.

Category	2010	2011
Raw material	5200	6240
Power & fuel	7000	9450
Salary & wages	9000	12600
Plant & machinery	20000	25000
Advertising	15000	19500
Research & Development	22000	26400

In 2011, which of the following two categories have registered increase by same percentage?

(*a*) Raw material and Salary & wages
(*b*) Salary & wages and Advertising
(*c*) Power & fuel and Advertising
(*d*) Raw material and Research & Development

25. What will be the maximum sum of 44, 42, 40,... ?

(*a*) 502 (*b*) 504
(*c*) 506 (*d*) 500

26. The current erection cost of a structure is ₹ 13,200. If the labour wages per day increase by 1/5 of the current wages and the working hours decrease by 1/24 of the current period, then the new cost of erection in ₹ is

(*a*) 16,500 (*b*) 15,180
(*c*) 11,000 (*d*) 10,120

27. Find the sum of the expression

$$\frac{1}{\sqrt{1}+\sqrt{2}}+\frac{1}{\sqrt{2}+\sqrt{3}}+\frac{1}{\sqrt{3}+\sqrt{4}}+.....+\frac{1}{\sqrt{80}+\sqrt{81}}$$

(*a*) 7 (*b*) 8
(*c*) 9 (*d*) 10

28. In the summer of 2012, in New Delhi, the mean temperature of Monday to Wednesday was 41°C and of Tuesday to Thursday was 43°C. If the temperature on Thursday was 15% higher than that of Monday, then the temperature in °C on Thursday was

(*a*) 40 (*b*) 43
(*c*) 46 (*d*) 49

29. The set of values of p for which the roots of the equation $3x^2 + 2x + p(p - 1) = 0$ are of opposite sign is

(*a*) $(-\infty, 0)$ (*b*) $(0, 1)$
(*c*) $(1, \infty)$ (*d*) $(0, \infty)$

30. Find the sum to n terms of the series 10 + 84 + 734 + ...

(*a*) $\frac{9(9^n+1)}{10}+1$ (*b*) $\frac{9(9^n-1)}{8}+1$
(*c*) $\frac{9(9^n-1)}{8}+n$ (*d*) $\frac{9(9^n-1)}{8}+n^2$

31. If $3 \le X \le 5$ and $8 \le Y \le 11$ then which of the following options is TRUE?

(*a*) $\frac{3}{5}\le\frac{X}{Y}\le\frac{8}{5}$ (*b*) $\frac{3}{11}\le\frac{X}{Y}\le\frac{5}{8}$
(*c*) $\frac{3}{11}\le\frac{X}{Y}\le\frac{8}{5}$ (*d*) $\frac{3}{5}\le\frac{X}{Y}\le\frac{8}{11}$

32. Following table gives data on tourists from different countries visiting India in the year 2011.

Country	Number of Tourists
USA	2000
England	3500
Germany	1200
Italy	1100
Japan	2400
Australia	2300
France	1000

Which two countries contributed to the one third of the total number of tourists who visited India in 2011?

(*a*) USA and Japan
(*b*) USA and Australia
(*c*) England and France
(*d*) Japan and Australia

2012

33. If $(1.001)^{1259} = 3.52$ and $(1.001)^{2062} = 7.85$, then $(1.001)^{3321} =$

(*a*) 2.23 (*b*) 4.33
(*c*) 11.37 (*d*) 27.64

34. Raju has 14 currency notes in his pocket consisting of only ₹ 20 notes and ₹ 10 notes. The total money value of the notes is ₹ 230. The number of ₹ 10 notes that Raju has is

(*a*) 5 (*b*) 6
(*c*) 9 (*d*) 10

35. The data given in the following table summarizes the monthly budget of an average household.

Category	Amount (Rs.)
Food	4000
Clothing	1200
Rent	2000
Savings	1500
Other expenses	1800

The approximate percentage of the monthly budget **NOT** spent on savings is

(*a*) 10% (*b*) 14%
(*c*) 81% (*d*) 86%

2011

36. P, Q, R and S are four types of dangerous microbes recently found in a human habitat. The area of each circle with its diameter printed in brackets represents the growth of a single microbe surviving human immunity system within 24 hours of entering the body. The danger to human beings varies proportionately with the toxicity, potency and growth attributed to a microbe shown in the figure below :

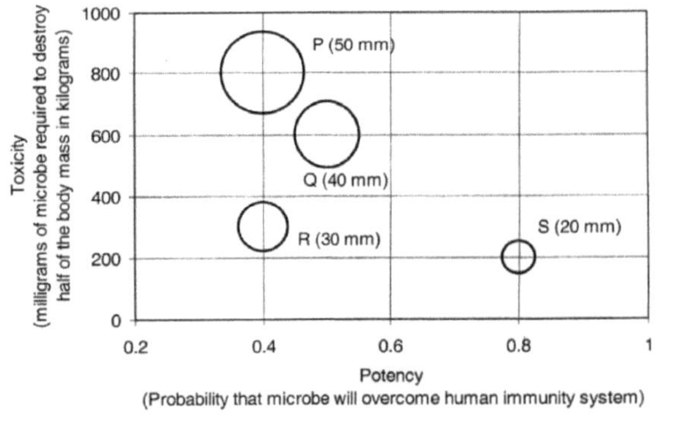

A pharmaceutical company is contemplating the development of a vaccine against the most dangerous microbe. Which microbe should the company target in its first attempt?

(*a*) P (*b*) Q

(*c*) R (*d*) S

37. The sum of n terms of the series 4 + 44 + 444 + ... is

(*a*) (4/81) $[10^{n+1} - 9n - 1]$

(*b*) (4/81) $[10^{n-1} - 9n - 1]$

(*c*) (4/81) $[10^{n+1} - 9n - 10]$

(*d*) (4/81) $[10^{n} - 9n - 10]$

38. Given that $f(y) = |y|/y$, and q is any non-zero real number, the value of $|f(q) - f(-q)|$ is

(*a*) 0 (*b*) –1

(*c*) 1 (*d*) 2

NUMERICAL TYPE QUESTIONS

2016

1. The following graph represents the installed capacity for cement production (in tonnes) and the actual production (in tonnes) of nine cement plants of a cement company. Capacity utilization of a plant is defined as ratio of actual production of cement to installed capacity. A plant with installed capacity of at least 200 tonnes is called a large plant and a plant with lesser capacity is called a small plant. The difference between total production of large plants and small plants, in tonnes is ________.

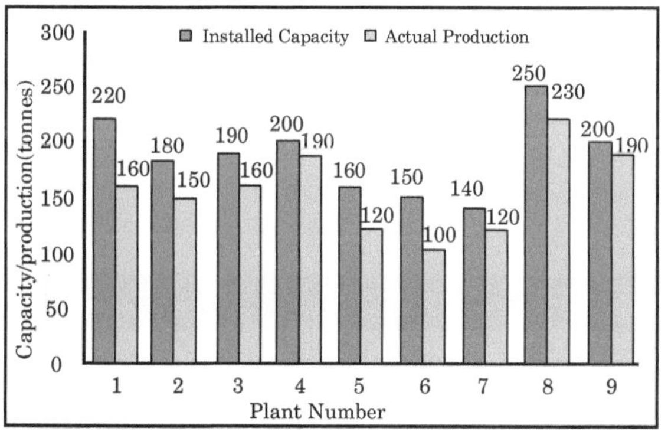

2. The numeral in the units position of

$211^{870} + 146^{127} \times 3^{424}$ is ________.

2015

3. The exports and imports (in crores of Rs.) of a country from the year 2000 to 2007 are given in the following bar chart. In which year is the combined percentage increase in imports and exports the highest ?

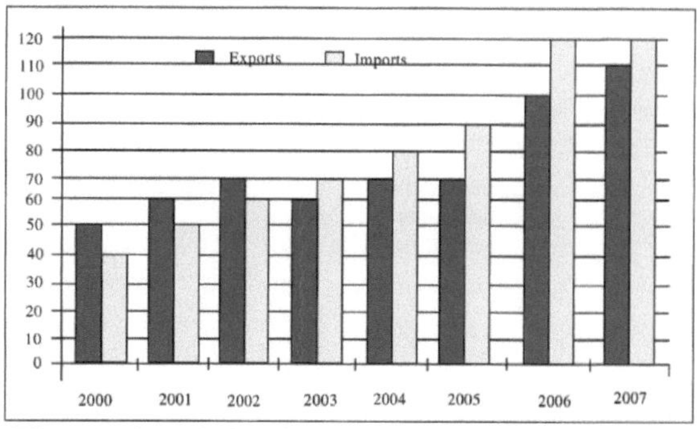

4. The pie chart below has the breakup of the number of students from different departments in an engineering college for the year 2012. The proportion of male to female students in each department is 5 : 4. There are 40 males in Electrical Engineering. What is the difference between numbers of female students in the Civil department and the female students in the Mechanical department?

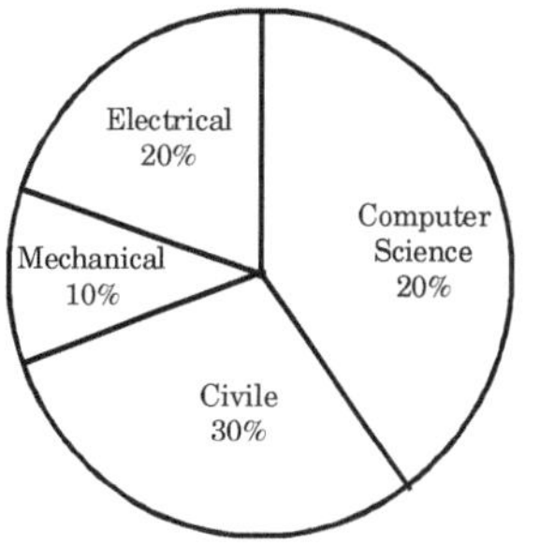

2014

5. A foundry has a fixed daily cost of ₹ 50,000 whenever it operates and variable cost of ₹ 800Q, where Q is the daily production in tonnes. What is the cost of production in ₹ per tonne for a daily production of 100 tonnes?

6. The smallest angle of a triangle is equal to two thirds of the smallest angle of a quadrilateral. The ratio between the angles of the quadrilateral is 3 : 4 : 5 : 6. The largest angle of the triangle is twice its smallest angle. What is the sum, in degrees, of the second largest angle of the triangle and the largest angle of the quadrilateral?

7. In a sequence of 12 consecutive odd numbers, the sum of the first 5 numbers is 425. What is the sum of the last 5 numbers in the sequence?

8. A firm producing air purifiers sold 200 units in 2012. The following pie chart presents the share of raw material, labour, energy, plant & machinery, and transportation costs in the total manufacturing cost of the firm in 2012. The expenditure on labour in 2012 is ₹ 4,50,000. In 2013, the raw material expenses increased by 30% and all other expenses increased by 20%. What is the percentage increase in total cost for the company in 2013?

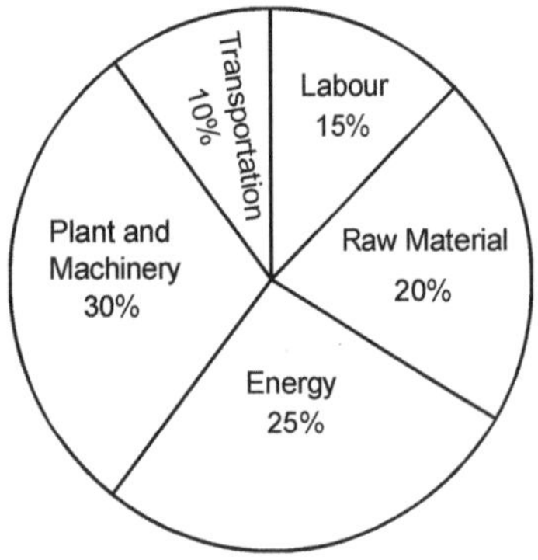

9. A firm producing air purifiers sold 200 units in 2012. The following pie chart presents the share of raw material, labour, energy, plant & machinery, and transportation costs in the total manufacturing cost of the firm in 2012. The expenditure on labour in 2012 is ₹ 4,50,000. In 2013, the raw material expenses increased by 30% and all other expenses increased by 20%. If the company registered a profit of ₹ 10 lakhs in 2012, at what price (in ₹) was each air purifier sold?

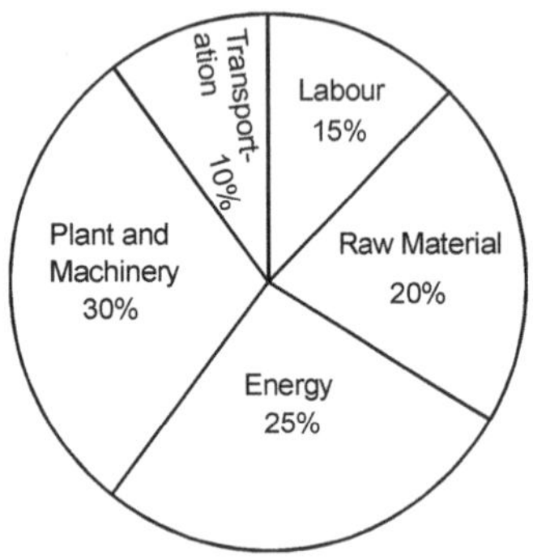

10. The sum of eight consecutive odd numbers is 656. The average of four consecutive even numbers is 87. What is the sum of the smallest odd number and second largest even number?

11. The ratio of male to female students in a college for five years is plotted in the following line graph. If the number of female students doubled in 2009, by what percent did the number of male students increase in 2009?

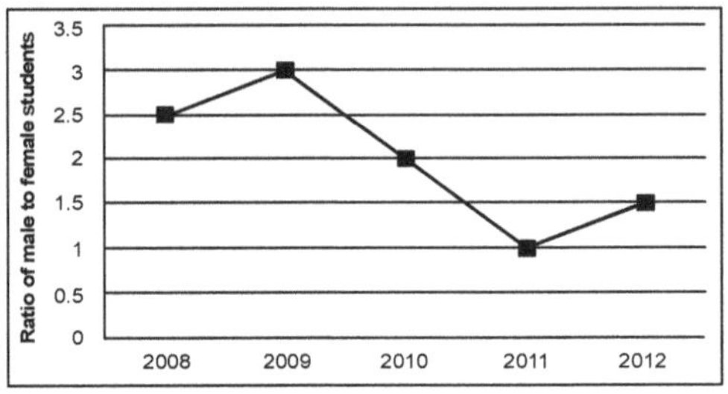

12. If $(z + 1/z)^2 = 98$, compute $(z^2 + 1/z^2)$

13. In a survey, 300 respondents were asked whether they own a vehicle or not. If yes, they were further asked to mention whether they own a car or scooter or both. Their responses are tabulated below. What percent of respondents do not own a scooter?

		Men	Women
Own vehicle	Car	40	34
	Scooter	30	20
	Both	60	46
Do not own vehicle		20	50

14. Round–trip tickets to a tourist destination are eligible for a discount of 10% on the total fare. In addition, groups of 4 or more get a discount of 5% on the total fare. If the one way single person fare is ₹ 100, a group of 5 tourists purchasing round–trip tickets will be charged ₹ _______.

ANSWERS

MCQ Type Questions

1. (*d*)	**2.** (*d*)	**3.** (*c*)	**4.** (*b*)	**5.** (*a*)	**6.** (*c*)	**7.** (*c*)	**8.** (*a*)	**9.** (*d*)	**10.** (*d*)
11. (*b*)	**12.** (*b*)	**13.** (*c*)	**14.** (*b*)	**15.** (*d*)	**16.** (*a*)	**17.** (*d*)	**18.** (*c*)	**19.** (*c*)	**20.** (*b*)
21. (*c*)	**22.** (*d*)	**23.** (*d*)	**24.** (*d*)	**25.** (*c*)	**26.** (*a*)	**27.** (*b*)	**28.** (*c*)	**29.** (*b*)	**30.** (*d*)
31. (*b*)	**32.** (*c*)	**33.** (*d*)	**34.** (*a*)	**35.** (*d*)	**36.** (*d*)	**37.** (*c*)	**38.** (*d*)		

Numerical Type Questions

1. (120) **2.** (7) **3.** (2006) **4.** (32) **5.** (1300 to 1300) **6.** (180 to 180) **7.** (495) **8.** (22)
9. (20,000) **10.** (163) **11.** (140 to 140) **12.** (96) **13.** (48) **14.** (850)

EXPLANATIONS

MCQ TYPE QUESTIONS

1. $\odot \rightarrow -,\ \oplus \rightarrow \times$

$$(9 \odot (6 \oplus 7)) \odot (7 \oplus (6 \odot 5))$$

$= (9 - (6 \times 7)) - (7 \times (6 - 5))$

$= (9 - 42) - (7 \times 1)$

$= -33 - 7$

$= -40$

Hence, Option (d) is the correct answer.

2. $S = 8 + 88 + 888 + \ldots n$

$$= \frac{8}{9}(9 + 99 + 999 + \ldots .. n)$$

$$= \frac{8}{9}\left[(10^1 - 1) + (100^2 - 1) + \ldots + (10^n - 1)\right]$$

$$= \frac{8}{9}\left[(10^1 + 100^2 + \ldots 100^n) - n\right]$$

$$= \frac{8}{9}\left[\frac{10 \times (10^n - 1)}{10 - 1}\right] - \frac{8}{9}n$$

$$= \frac{80}{81}(10^n - 1) - \frac{8}{9}n$$

3. 3 digit number

can be filled from number 0 to 9 hence

–	–	–

in 10 ways.

Hundredth place – can't have 0 and 1, hence can be filled in 8 ways.

Ten's place – can't have 1 or 2, hence can be filled in 8 ways.

One's place – It can only be filled in 1 way.

So, number becomes 881.

4. $$\frac{\alpha^n + \beta^n}{\alpha^{-n} + \beta^{-n}} = \frac{\alpha^n + \beta^n}{\frac{1}{\alpha^n} + \frac{1}{\beta^n}} = \frac{\alpha^n + \beta^n}{\frac{\beta^n + \alpha^n}{\alpha^n \times \beta^n}} = \alpha^n \times \beta^n$$

$\alpha^n \times \beta^n = (\alpha \times \beta)^n = (4)^n$

So, the value of $\frac{\alpha^n + \beta^n}{\alpha^{-n} + \beta^{-n}}$ is 4^n.

5. From question

Total number of faculty members = 150

The number of faculty members having facebook account = (FB) = 55

The number of faculty members having whatsapp = (W) = 85

The number of faculty members do not have face book or Whats App accounts = 30

The number of faculty members having any account = 150 – 30 = 120

The number of faculty members having both the accounts = (FB + W) – 120 = (55 + 85) – 120 = 20

∴ The number of faculty members connected only through Facebook accounts = 55 – 20 = 35

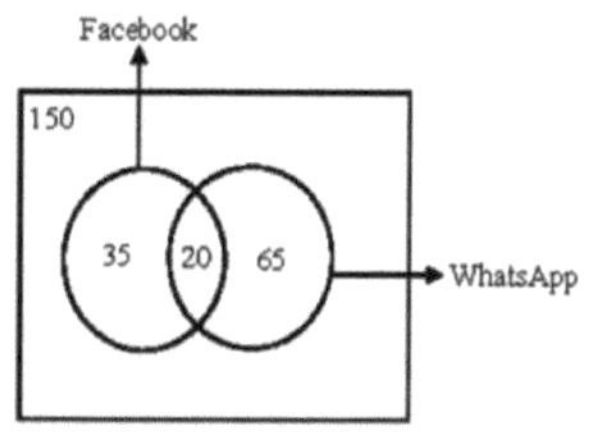

So the number of faculty members connected only through facebook accounts is 35.

6. Given, $|9y - 6| = 3$

It has two values either

$$(9y - 6) = 3 \quad \ldots(i)$$

Or $$-(9y - 6) = 3$$

$$9y - 6 = -3 \quad \ldots(ii)$$

From equation (*i*),

$$9y = 9$$

∴ $$y = 1$$

Or from equation (*ii*),

$$9y = 3$$

∴ $$y = \frac{1}{3}$$

Now $$y^2 = \frac{4y}{3}$$

put $y = 1$,

we get $$= 1 - \frac{4}{3} = -\frac{1}{3}$$

and also put $y = \frac{1}{3}$

we get, $$\frac{1}{9} - \frac{4}{9} = \frac{-3}{9} = \frac{-1}{3}$$

So, the required value is $\frac{-1}{3}$.

7. Given $q^{-a} = \frac{1}{r}$

$\Rightarrow \quad \frac{1}{q^a} = \frac{1}{r}$

$\Rightarrow \quad q^a = r \quad ...(i)$

$r^{-b} = \frac{1}{s}$

$\Rightarrow \quad \frac{1}{r^b} = \frac{1}{s}$

$\Rightarrow \quad s = r^b \quad ...(ii)$

$s^{-c} = \frac{1}{q}$

$\Rightarrow \quad \frac{1}{s^c} = \frac{1}{q}$

$\Rightarrow \quad s^c = q \quad ...(iii)$

From equation (*i*),

$q^a = r$

$(s^c)^a = r$ (from eq. (*iii*))

$S^{ac} = r$

$s^{ac} = (s)^{\frac{1}{b}}$ (from eq. (*ii*))

$s^{abc} = s = s^1$

$\therefore \quad abc = 1$

So the value of *abc* is 1

8. Let the two digit number be xy

Now sum of the digit of a two digit number is 12

$\therefore \quad x + y = 12 \quad ...(i)$

Now according to question,

$(10y + x) - (10x + y) = 54$

$\therefore \quad 9y - 9x = 54$

$\therefore \quad y - x = 6 \quad ...(ii)$

Now solving equation (*i*) and (*ii*) we get,

$x = 3$ and $y = 9$

Hence the original number is 39.

$\therefore$ Option (*a*) is correct.

9. Amount invested by Company P in 2006 = ₹ 8x

and amount invested by Company Q in 2006 = ₹ 9*x*

Interest from Company P

$$= \frac{\frac{₹\, 8x \times 6}{100}}{\frac{₹\, 9x \times 4}{100}} = \frac{4}{3}$$

$\therefore$ Option (*d*) is correct.

10. Type-I achieved a growth of 53% in the period which is higher than any other type of battery

13. Total number of students in class = 44

Average of marks,

$$= \frac{(2\times21)+(3\times15)+(1\times11)+(2\times23)+(5\times31)}{42}$$

$$= \frac{42+45+11+46+155}{44} = \frac{299}{44} = 6.795$$

$\therefore$ Option (*c*) is correct.

14. In the question the monthly average rainfall chart for 50 years has been given.

Let us check the options.

(*i*) On average, it rains more in July than in December ⇒ correct.

(*ii*) Every year, the amount of rainfall in August is more than that in January.

⇒ may not be correct because average rainfall is given in the question.

(*iii*) July rainfall can be estimated with better confidence than February rainfall.

⇒ From chart it is clear the gap between 5 percentile and 95 percentile from average is higher in February than that in July ⇒ correct.

(*iv*) In August at least 500 mm rainfall

⇒ May not be correct, because its 50 year average.

So correct option (*b*) (*i*) and (*iii*).

15. Item : 2

$$\frac{\frac{20}{100}\times 250\times 10^7}{\frac{20}{100}\times 500\times 10^3}$$

$0.5\times 10^4 = 5\times 10^3 \boxed{1} = \text{Item } 2$

Item : 3

$$\frac{23\times 250\times 10^7}{19\times 500\times 10^3}$$

1.2 = Item 3

Item : 6

$\frac{19}{16} = 1.18 = \text{Item } 6$

Item : 5

$\frac{20}{12} = \frac{5}{3} = 1.6 \Rightarrow \boxed{1.6 = \text{Item } 5}$

17. For 2005,

trade deficit = (90 – 70) crores = 20 crores

Now $\frac{1}{5}^{th}$ of export = $\frac{1}{5}$(70) crores

= 14 crores ≠ 20 crores.

Hence option (*a*) is wrong.

For 2004,

Trade deficit =(80 – 70) crores = 10 crores

Now $\frac{1}{5}^{th}$ of export = $\frac{1}{5}$(70) crores

= 14 crores ≠ 20 crores.

Hence option (*b*) is wrong.

For 2007

Trade deficit = (120 – 110) crores

= 10 crores

Now $\frac{1}{5}^{th}$ of export = $\frac{1}{5}$(110) crores

= 22 crores ≠ 10 crores.

Hence option (*c*) is wrong.

For 2006,

Trade deficit = (120 – 100) crores

= 20 crores

Now $\frac{1}{5}^{th}$ of export = $\frac{1}{5}$(100) crores

= 20 crores = Trade defects

Hence option (*d*) is correct.

18. Total question

$44 \times 2 = 88$

$44 \times 3 = 132$

$\frac{144}{132} = \frac{88}{308}$

Total marks obtained = (21×2) + (15×3) + (23×2)

= 133

Total Number of students = 44

Average = $\frac{133}{44}$ = 3.02

19. Take number of female students in 2011 = 100

∴ Number of male in 2011 = 100

No. of female in 2012 = 100

No. of male in 2012 = 150

Ratio = $\frac{150}{100}$

20. All multiples of 10 from 2 to 198

= 10, 20,............190

Now $a_n = a + (n-1)d$

$190 = 10 + (n-1) \times 10$

$180 = (n-1) \times 10$

$18 = n - 1$

∴ $n = 19$

Hence total number of multiples are 19.

Now $S_n = \frac{n}{2}[2a + (n-1)d]$

$= \frac{19}{2}[20 + (18)(10)]$

$= \frac{19}{2}[20 + 180] = 1900$

Now average of all multiples of 10 from 2 to198

$= \frac{10 + 20 + ... + 190}{19}$

∴ $= \frac{1900}{19} = 100$

21. $\sqrt{12 + \sqrt{12 + \sqrt{12 + ...}}}$

For a particular type of question

i.e. $\sqrt{a + \sqrt{a + \sqrt{a + ...}}}$ we find a two consecutive number whose product is *a*. The greater number among the two consecutive number is the answer.

Here $a = 12$

i.e. $12 = 3 \times 4$
↓
greater number

Hence $\sqrt{12 + \sqrt{12 + \sqrt{12 + ...}}} = 4$

22. Here it is given that *x* is real

Now $|x^2 - 2x + 3| = 11$

⇒ $x^2 - 2x + 3 = 11$ or $x^2 - 2x + 3 = -11$

For $x^2 - 2x + 3 = 11 \Rightarrow x^2 - 2x - 8 = 0$

$b^2 - 4ac = 4 - 4(1)(-8) = 36 > 0$

Hence roots are real.

and for $x^2 - 2x + 14 = 0$

$b^2 - 4ac = 4 - 4(1)(14) < 0$

Hence roots are imaginary

∴ The value of *x* is the root of $x^2 - 2x - 8 = 0$

Now $x = \frac{+2 \pm \sqrt{4 - 4(1)(-8)}}{2}$

$$= \frac{2 \pm \sqrt{36}}{2}$$

$$= \frac{2 \pm 6}{2}$$

$\therefore \quad x = 4, -2.$

Now for $x = 4$,

$$|-x^3 + x^2 - x| = |-64 + 16 - 4| = 52.$$

and for $x = -2$

$$|-x^3 + x^2 - x| = |+8 + 4 + 2| = 14$$

Hence possible values of $|-x^3 + x^2 - x| = 14, 52$

25. $0 = 44 + (n - 1)(-2)$

$$n - 1 = 22$$

$$n = 23$$

$$\text{Sum} = \frac{n}{2}(a + l)$$

$$= \frac{23}{2}(44 + 0)$$

$$= 23 \times 22 = 506$$

Hence (c) is the answer

26. Let 'd' be the # of days required

Let 'h' be the daily working hours.

$\therefore$ dh can be taken as a measure of the total work be done in order to erect the structure.

Now working hours per day $h' = \frac{23}{24}h$

$$\therefore \text{ \# of days} \quad d' = \frac{dh}{h'} = \frac{dh}{\left(\frac{23}{24}h\right)} = \frac{24}{23}d$$

Let $\quad c$ = daily wages earlier

$$\therefore \quad c = \frac{13,200}{d}$$

Now daily wages $c' = \frac{6}{5}c$

$\therefore$ Total new cost $= c' \times d'$

$$= \frac{6}{5} \times \frac{13200}{d} \times \frac{24}{23}d \approx 16,500$$

27. Given: The expression

$$S = \frac{1}{\sqrt{1}+\sqrt{2}} + \frac{1}{\sqrt{2}+\sqrt{3}} + \ldots\ldots + \frac{1}{\sqrt{80}+\sqrt{81}}$$

To find: Sum of the expression.

Analysis: Each term of the expression is having irrational denominator. One can also see, that both the elements in the denominator of a term differ by only 1.

This gives an idea that on rationalising the denominator, each denominator will become 1, so the sum will be easier than earlier.

Rationalising the denominator,

$$S = \frac{\sqrt{2}-\sqrt{1}}{(\sqrt{2})^2 - (\sqrt{1})^2} + \frac{\sqrt{3}-\sqrt{2}}{(\sqrt{3})^2 - (\sqrt{2})^2} + \ldots + \frac{\sqrt{81}-\sqrt{80}}{(\sqrt{81})^2 - (\sqrt{80})^2}$$

$$= \frac{\sqrt{2}-\sqrt{1}}{1} + \frac{\sqrt{3}-\sqrt{2}}{3-2} + \ldots. + \frac{\sqrt{81}-\sqrt{80}}{81-80}$$

$$= \sqrt{2} - \sqrt{1} + \sqrt{3} - \sqrt{2} + \ldots. + 81 - \sqrt{80}$$

Now, one can see that the terms will cancel out.

Each term except $\sqrt{1}$ & $\sqrt{81}$

Occurs twice, once as positive and once a negative. So, after cancelling out, we are left with the following:

$$S = \sqrt{81} - \sqrt{1} = 9 - 1 = 8$$

So, the answer is (b).

28. $\frac{\text{Mon + Tues + Wed.}}{3} = 41$

Mon + Tues + Wed. = 123 ...(1)

$\frac{\text{Tues + Wed + Thurs.}}{3} = 43°$

Tue + Wed + Thu. = 129° ...(2)

(2) – (1)

Tues + Wed + Thu – (Mon + Tues. + Wed.)

$$= 129 - 123 = 6°$$

Thu. – Mon. = 6° $\Rightarrow \frac{115x}{100} - x = 6°$

Thus. = Mon $\times \frac{115}{100} = \frac{15x}{100} = 6°$

Mon = $x \quad x = 40°$

Thurs $= \frac{115x}{100}$

$$\therefore \quad \text{Thurs} = \frac{115 \times 40}{100} = 46°$$

33. Let $\quad 1.001 = x$

$x^{1259} = 3.52$, and $x^{2062} = 7.85$

$$\therefore \quad x^{3321} = x^{1259} \cdot x^{2062}$$

$$= 3.52 \times 7.85$$

$$= 27.64$$

34. Let number of ₹ 20 notes be x and ₹ 10 notes be y

$\therefore \quad 20x + 10y = 230$ ...(1)

and $\quad x + y = 14$ ...(2)

Solving equations (1) and (2), we have

$$x = 9 \text{ and } y = 5$$

Hence numbers of 10 rupee notes are 5.

35. Total budget = 10,500

Expenditure other than savings = 9000

Hence approximate percentage of monthly budget

$$= \frac{9000}{10500} = 86\%$$

36. By observation of the table, we can say S

	P	Q	R	S
Requirement	800	600	300	200
Potency	0.4	0.5	0.4	0.8

37. Let S = 4 (1 + 11 + 111 + ...)

$$= \frac{4}{9}(9 + 99 + 999 + ...)$$

$$= \frac{4}{9}\{(10-1) + (10^2-1) + (10^3-1) + ...\}$$

$$= \frac{4}{9}\{(10 + 10^2 + ...+ 10^n) - n\}$$

$$= \frac{4}{9}\left\{10\frac{(10^n-1)}{9} - n\right\}$$

$$= \frac{4}{81}\{10^{n+1} - 9n - 10\}$$

38. Given : $f(y) = \frac{|y|}{y}$

$$\Rightarrow \quad f(q) = \frac{|q|}{q}$$

$$f(-q) = \frac{|-q|}{-q}$$

$$= \frac{-|q|}{q}$$

$$|f(q) - f(q)| = \frac{|q|}{q} + \frac{|q|}{q}$$

$$= \frac{2|q|}{q} = 2$$

NUMERICAL TYPE QUESTIONS

1. From the given graph large plants which are having installed capacity of at least 200 tonnes are 1, 4, 8 and 9.

So, the total production of large plants

= 160 + 190 + 230 + 190

= 770

Now the remaining plants with installed capacity is less then 200 tonnes are 2, 3, 5, 6 and 7.

So, total production of small plants

= 150 + 160 + 120 + 100 + 120

= 650

Then difference = 770 – 650

= 120

2. Unit digit of $211^{870} + 146^{127} \times 3^{424}$ is $1 + 6 \times 1 = 7$

3. Increase in exports in 2006

$$= \frac{100-70}{70} = 42.8\%$$

Increase in imports in 2006

$$= \frac{120-90}{90} = 33.3\%$$

which is more than any other year.

4. According to question,

There are 40 males in Electrical Engg.

∴ Total number of females in Electrical Engg.

= 72 – 40 = 32

∴ Total number of students in Electrical Engg. = 72

⇒ 20% = 72

∴ 30% = Total number of students in Civil Engg.

∴ Total number of students in Civil Engg.

$$= \frac{30 \times 72}{20} = 36 \times 3 = 108$$

∴ Number of female students in Civil Engg.

$$= \frac{4 \times 108}{9} = 4 \times 12 = 48$$

Now, total number of students in Mechanical Engg.

$$= \frac{10 \times 72}{20} = 36$$

∴ Number of female students in Mechanical Engg.

$$= \frac{4 \times 36}{9} = 16$$

∴ Difference between female students in Civil and Mechanical departments = 48 – 16 = 32

5. Fixed cost = ₹ 50,000

Variable cost = ₹ 800Q

Q = daily production in tones

For 100 tonnes of production daily, total cost of production = 50,000 + 800 × 100 = 130,000

So, cost of production per tonne of daily production

$$= \frac{130,000}{100} = ₹\ 1300.$$

6. Let the angles of quadrilateral are $3x$, $4x$, $5x$, $6x$

So, $3x + 4x + 5x + 6x = 360$

$$x = 20$$

Smallest angle of quadrilateral = $3 \times 20 = 60°$

Smallest angle of triangle = $\frac{2}{3} \times 60° = 40°$

Largest angle of triangle = $2 \times 40° = 60°$

Three angles of triangle are 40°, 60°, 80°

Largest angle of quadrilateral is 120°

Sum (2nd largest angle of triangle + largest angle of quadrilateral) = 60° + 120° = 180°.

7. 8th observation is $7 \times 2 = 14$ more than 1st observation

9th observation is 14 more than 2nd observation

10th observation is 14 more than 3rd observation

11th observation is 14 more than 4th observation

12th observation is 14 more than 5th observation

Total $14 \times 5 = 70$

Sum of the first five numbers = 425

Sum of last five numbers = 495

8.

	2012	2013
Transport (10%)	300,000	360,000
Labour (15%)	450,000	540,000
Raw material (20%)	750,000	780,000
Energy (25%)	750,000	900,000
Plant and Machinery (30%)	900,000	1,080,000
Total	**3,000,000**	**3,660,000**

Percentage increase in total cost = 22%

9. Total expenditure = $\frac{15}{100} x = 4,50,000$

$$x = 3 \times 10^6$$

Profit = 10 lakhs

So, total selling price = 40,00,000 ...(1)

Total purifies = 200 ...(2)

S.P. of each purifier = (1)/(2) = 20,000

10. Eight consecutive odd number = 656

$a - 6, a - 1, a - 2, a, a + 2, a + 4, a + 6$

$$a + 8 = 656$$

$$a = 81$$

Smallest $m = 75$ ...(1)

Average consecutive even numbers

$$\Rightarrow \frac{a-2+a+a+2+a+4}{4} = 87$$

$\Rightarrow a = 86$

Second largest number = 88 ...(2)

Adding equation (1) and (2),

$$75 + 88 = 163$$

11. Ratio of male to female student in 2008 = 5 : 2

In 2009, let the male student be $5x$ and female student be $2x$

Now if number of female students double in 2009

$\therefore$ Number of female student = $4x$

$\therefore$ Number of male student = $12x$

Gender	2008	2009
Male	$5x$	$12x$
Female	$2x$	$4x$

% of male student increase in 2009

$$= \left[\left(\frac{12x - 5x}{5x}\right) \times 100\right]\%$$

$$= \left[\frac{7}{5} \times 100\right]\%$$

$$= 140\%$$

12. Expanding

$$\left(z + \frac{1}{z}\right)^2 = 98$$

$$\therefore \quad z^2 + \frac{1}{z^2} + 2(z)\left(\frac{1}{z}\right) = 98$$

$$\therefore \quad z^2 + \frac{1}{z^2} = 96$$

13. Total respondents = 300

Now men respondents who do not have scooter

$$= 40 + 20 = 60$$

And women respondents who do not have scooter

$$= 34 + 50 = 84$$

$\therefore$ Total respondents who do not have scooter

$$= 60 + 84 = 144$$

$\therefore$ Required percent $= \left(\frac{144}{300} \times 100\right)\% = 48\%$

14. One way single person fare = ₹ 100

$\therefore$ Two way fare for single person = ₹ 200

$\therefore$ For 5 persons two way fare = ₹ 1000

Now, total discount = (10 + 5)% = 15%

Discount amount = ₹ $\left[\frac{15}{100} \times 1000\right]$ = ₹ 150

Amount to be paid = ₹(1000 – 150) = ₹ 850

2
CHAPTER

Percentage and Its Applications

MCQ TYPE QUESTIONS

2019

1. The radius as well as the height of a circular cone is increased by 10%. The percentage increase in its volume is ___________.
   (a) 17.1 (b) 21.0
   (c) 33.1 (d) 72.8
2. In a country of 1400 million population, 70% own mobile phones. Among the mobile phone owners only 294 million access the internet. Among these Internet users, only half buy goods from e-commerce portals. What is the percentage of these buyers in the country ?
   (a) 10.50 (b) 14.70
   (c) 15.00 (d) 50.00

2016

3. In a huge pile of apples and oranges, both ripe and unripe mixed together, 15% are unripe fruits. Of the unripe fruits, 45% are apples. Of the ripe ones, 66% are oranges. If the pile contains a total of 5692000 fruits, how many of them are apples?
   (a) 2029198 (b) 2467482
   (c) 2789080 (d) 3577422
4. A person moving through a tuberculosis prone zone has a 50% probability of becoming infected. However, only 30% of infected people develop the disease. What percentage of people moving through a tuberculosis prone zone remains infected but does not show symptoms of disease?
   (a) 15 (b) 33
   (c) 35 (d) 37
5. (x % of y) + (y % of x) is equivalent to _______.
   (a) 2 % of xy (b) 2 % of (xy/100)
   (c) xy % of 100 (d) 100 % of xy

2014

6. The population of a new city is 5 million and is growing at 20% annually. How many years would it take to double at this growth rate?
   (a) 3 – 4 years (b) 4 – 5 years
   (c) 5 – 6 years (d) 6 – 7 years
7. One percent of the people of country X are taller than 6 ft. Two percent of the people of country Y are taller than 6 ft. There are thrice as many people in country X as in country Y. Taking both countries together, what is the percentage of people taller than 6 ft?
   (a) 3.0 (b) 2.5
   (c) 1.5 (d) 1.25
8. Industrial consumption of power doubled from 2000-2001 to 2010-2011. Find the annual rate of increase in percent assuming it to be uniform over the years.
   (a) 5.6 (b) 7.2
   (c) 10.0 (d) 12.2
9. The Gross Domestic Product (GDP) in Rupees grew at 7% during 2012-2013. For international comparison, the GDP is compared in US Dollars (USD) after conversion based on the market exchange rate. During the period 2012-2013 the exchange rate for the USD increased from ₹ 50/ USD to ₹ 60/ USD. India's GDP in USD during the period 2012-2013
   (a) increased by 5% (b) decreased by 13%
   (c) decreased by 20% (d) decreased by 11%

2013

10. A firm is selling its product at ₹ 60 per unit. The total cost of production is ₹ 100 and firm is earning total profit of ₹ 500. Later, the total cost increased by 30%. By what percentage the price should be increased to maintained the same profit level.
    (a) 5 (b) 10
    (c) 15 (d) 30

2012

11. An automobile plant contracted to buy shock absorbers from two suppliers X and Y. X supplies 60% and Y supplies 40% of the shock absorbers. All shock absorbers are subjected to a quality test. The ones that pass the quality test are considered reliable. Of X's shock absorbers, 96% are reliable. Of Y's shock absorbers, 72% are reliable.
    The probability that a randomly chosen shock absorber, which is found to be reliable, is made by Y is
    (a) 0.288 (b) 0.334
    (c) 0.667 (d) 0.720

2011

12. There are two candidates P and Q in an election. During the campaign, 40% of the voters promised to vote for P, and rest for Q. However, on the day of election 15% of the voters went back on their promise to vote for P and instead voted for Q. 25% of the voters went back on their promise to vote for Q and instead voted for P. Suppose, P lost by 2 votes, then what was the total number of voters?

(*a*) 100 (*b*) 110

(*c*) 90 (*d*) 95

NUMERICAL TYPE QUESTION

2015

1. From a circular sheet of paper of radius 30 cm, a sector of 10% area is removed. If the remaining part is used to make a conical surface, then the ratio of the radius and height of the cone is ____.

ANSWERS

MCQ Type Questions

1. (*c*) 2. (*a*) 3. (*a*) 4. (*c*) 5. (*a*) 6. (*a*) 7. (*d*) 8. (*b*) 9. (*d*) 10. (*a*)

11. (*b*) 12. (*a*)

Numerical Type Question

1. (2.06)

EXPLANATIONS

MCQ TYPE QUESTIONS

1. Initial Stage — After increasing 10% for radius & height

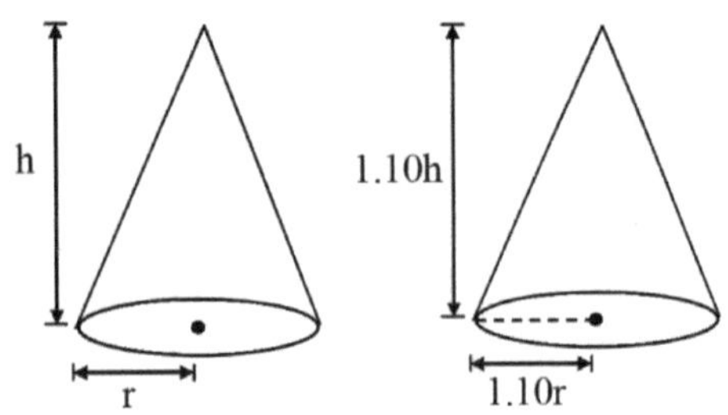

Volume of circular cone, $V_i = \frac{1}{3}\pi r^2 h$

Volume, $V_f = \frac{1}{3}\pi(1.1r)^2(1.1h)$

$= \frac{1.331}{3}\pi r^2 h$

Percentage increase

$= \frac{V_f - V_i}{V_i} \times 100$

$= \frac{\frac{1.331}{3}\pi r^2 h - \frac{1}{3}\pi r^2 h}{\frac{1}{3}\pi r^2 h} \times 100$

$= \frac{1.331 - 1}{1} \times 100$

$= 33.1\%$

2. Total population = 1400 million

Number of people whose having own mobile phones

= 70% of 1400

= 0.7 × 1400 = 980 million

Number of people whose accesses the internet = 294 million

Number of people who buy goods from e-commercial portals

= Half of internet users

$= \frac{294}{2} = 147$ million

Percentage buyers $= \frac{147 \text{ million}}{1400 \text{ million}} \times 100 = 10.5\%$

3. Given,

Total no. of fruits = 5692000

Unripe type of apples = 45% of 15% of 5692000

$= \frac{45}{100} \times \frac{15}{100} \times 5692000$

= 384210

Ripe type of apples = (100 – 66)% × (100 – 15)% × 5692000

$= \frac{34}{100} \times \frac{85}{100} \times 5692000$

= 1644988

∴ Total no. of apples = 384210 + 1644988

= 2029198

4. The percentage of people moving through a tuberculosis prone zone remains infected but does not show symptoms of disease.

$$= (100-30)\% \times 50\% = \frac{70}{100} \times \frac{50}{100} = \frac{35}{100} = 35\%$$

5. (x% of y) + (y% of x)

$$\frac{x}{100} \times y + \frac{y}{100} \times x = \frac{2xy}{100} = 2\% \text{ of } xy$$

$\therefore$ Option (a) is correct.

6.
$$A = P\left[1+\frac{r}{100}\right]^t$$

$$10 \text{ million} = 5 \text{ million}\left[1+\frac{20}{100}\right]^t$$

$$2 = \left[1+\frac{1}{5}\right]^t$$

$$2 = \left(\frac{6}{5}\right)^t = \log 2 = t \log\left(\frac{6}{5}\right)$$

$$0.301 = t\, 0.0791$$

$$t = \frac{0.301}{0.0791} = 3.8 \text{ years}$$

$\therefore$ Hence around 3–4 years would it take to double the growth rate

7. Let number of people in country y = 100

So, number of people in country x = 300

Total number of people taller than 6ft in both the countries $= 300 \times \frac{1}{100} + 100 \times \frac{2}{100} = 5$

% of people taller than 6ft in both the countries

$$= \frac{5}{400} \times 100 = 1.25\%.$$

9. Per ₹ 100 final value ₹ 107

$\Rightarrow$ per $\frac{100}{50}$ Dollars final value $\frac{107}{60}$

for 100 dollars _____?

$$= \frac{100 \times 50}{100} \times \frac{107}{60} = 89.16$$

Descreased by 11%

11.

	X	Y
Supply	60%	40%
Reliable	96%	72%
Overall	0.576	0.288

$$\therefore \quad P(x) = \frac{0.288}{0.576+0.288} = 0.334$$

12.

P	Q
40%	60%
–6%	+6%
+15%	–15%
49%	51%

$\therefore$ 2% = 2 $\Rightarrow$ 100% = 100

NUMERICAL TYPE QUESTION

1. 90% of area of sheet = Cross sectional area of cone

$\Rightarrow 0.9 \times \pi \times 30 \times 30 = \pi \times r_1 \times 30$

($\because$ Slant height of cone (l) = 30)

$\Rightarrow$ 27 cm = r_1

$\therefore$ Height of the cone = $\sqrt{30^2 - 27^2}$ = 13.08 cm

Hence ratio of radius and height is 2.06.

3 CHAPTER

Time and Work

MCQ TYPE QUESTIONS

2019

1. Two trains started at 7 AM from the same point. The first train travelled towards north at a speed of 80 km/h and the second train travelled south at a speed of 100 km/h. The time at which they were 540 km apart is ______ AM.

(*a*) 9 (*b*) 10

(*c*) 11 (*d*) 11:30

2018

2. An automobile travels from city A to city B and returns to city A by the same route. The speed of the vehicle during the onward and return journeys were constant at 60 km/h and 90 km/h, respectively. What is the average speed in km/h for the entire journey?

(*a*) 72 (*b*) 73

(*c*) 74 (*d*) 75

2017

3. 1200 men and 500 women can build a bridge in 2 weeks. 900 men and 250 women will take 3 weeks to build the same bridge. How many men will be needed to build the bridge in one week?

(*a*) 3000 (*b*) 3300

(*c*) 3600 (*d*) 3900

2016

4. P, Q, R and S are working on a project. Q can finish the task in 25 days, working alone for 12 hours a day. R can finish the task in 50 days, working alone for 12 hours per day. Q worked 12 hours a day but took sick leave in the beginning for two days. R worked 18 hours a day on all days. What is the ratio of work done by Q and R after 7 days from the start of the project?

(*a*) 10:11 (*b*) 11:10

(*c*) 20:21 (*d*) 21:20

5. S, M, E and F are working in shifts in a team to finish a project. M works with twice the efficiency of others but for half as many days as E worked. S and M have 6 hour shifts in a day whereas E and F have 12 hours shifts. What is the ratio of contribution of M to contribution of E in the project?

(*a*) 1 : 1 (*b*) 1 : 2

(*c*) 1 : 4 (*d*) 2 : 1

6. It takes 10 s and 15 s, respectively, for two trains travelling at different constant speeds to completely pass a telegraph post. The length of the first train is 120 m and that of the second train is 150 m. The magnitude of the difference in the speeds of the two trains (in m/s) is ______.

(*a*) 2.0 (*b*) 10.0

(*c*) 12.0 (*d*) 22.0

7. The velocity V of a vehicle along a straight line is measured in m/s and plotted as shown with respect to time in seconds. At the end of the 7 seconds, how much will the odometer reading increase by (in m) ?

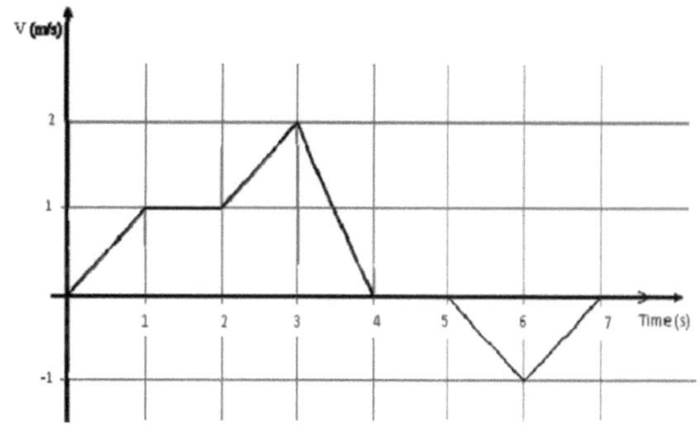

(*a*) 0 (*b*) 3

(*c*) 4 (*d*) 5

8. Ananth takes 6 hours and Bharath takes 4 hours to read a book. Both started reading copies of the book at the same time. After how many hours is the number of pages **to be** read by Ananth, twice that **to be** read by Bharath? Assume Ananth and Bharath read all the pages with constant pace.

(*a*) 1 (*b*) 2

(*c*) 3 (*d*) 4

2015

9. An electric bus has onboard instruments that report the total electricity consumed since the start of the trip as well as the total distance covered. During a single day of operation, the bus travels on stretches M, N, O and P, in that order. The cumulative distances travelled and the corresponding electricity consumption are shown in the table below.

Stretch	Comulative distance(km)	Electricity used (kWh)
M	20	12
N	45	25
O	75	45
P	100	57

The stretch where the electricity consumption per km is minimum is

(*a*) M (*b*) N
(*c*) O (*d*) P

2014

10. It takes 30 minutes to empty a half-full tank by draining it at a constant rate. It is decided to simultaneously pump water into the half-full tank while draining it. What is the rate at which water has to be pumped in so that it gets fully filled in 10 minutes?

(*a*) 4 times the draining rate
(*b*) 3 times the draining rate
(*c*) 2.5 times the draining rate
(*d*) 2 times the draining rate

2013

11. A tourist covers half of his journey by train at 60 km/h, half of the remainder by bus at 30 km/h and the rest by cycle at 10 km/h. The average speed of the tourist in km/h during his entire journey is

(*a*) 36 (*b*) 30
(*c*) 24 (*d*) 18

12. A car travels 8 km in the first quarter of an hour, 6 km in the second quarter and 16 km in the third quarter. The average speed of the car in km per hour over the entire journey is

(*a*) 30 (*b*) 36
(*c*) 40 (*d*) 24

2011

13. The fuel consumed by a motorcycle during a journey while traveling at various speeds is indicated in the graph below.

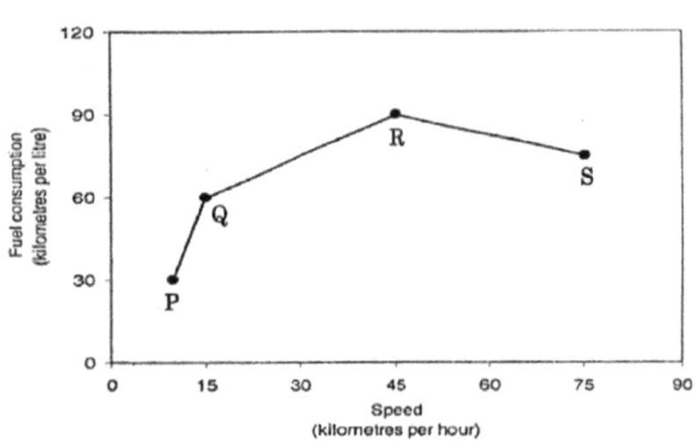

The distances covered during four laps of the journey are listed in the table below

Lap	Distance (kilometers)	Average speed (kilometers per hour)
P	15	15
Q	75	45
R	40	75
S	10	10

From the given data, we can conclude that the fuel consumed per kilometre was least during the lap

(*a*) P (*b*) Q
(*c*) R (*d*) S

2010

14. 5 skilled workers can build a wall in 20 days; 8 semi-skilled workers can build a wall in 25 days; 10 unskilled workers can build a wall in 30 days. If a team has 2 skilled, 6 semi-skilled and 5 unskilled workers, how long will it take to build the wall?

(*a*) 20 days (*b*) 18 days
(*c*) 16 days (*d*) 15 days

NUMERICAL TYPE QUESTIONS

2015

1. A tiger is 50 leaps of its own behind a deer. The tiger takes 5 leaps per minute to the deer's 4. If the tiger and the deer cover 8 metre and 5 metre per leap respectively, what distance in meters will the tiger have a run before it catches the deer?

2014

2. A man can row at 8 km per hour in still water. If it takes him thrice as long to row upstream, as to row downstream, then find the stream velocity in km per hour.

3. A train that is 280 metres long, travelling at a uniform speed, crosses a platform in 60 seconds and passes a man standing on the platform in 20 seconds. What is the length of the platform in metres?

ANSWERS

MCQ Type Questions

1. (*b*) **2.** (*a*) **3.** (*c*) **4.** (*c*) **5.** (*b*) **6.** (*a*) **7.** (*d*) **8.** (*c*) **9.** (*b*) **10.** (*a*)
11. (*c*) **12.** (*c*) **13.** (*a*) **14.** (*d*)

Numerical Type Questions

1. (800) **2.** (4) **3.** (560 to 560)

EXPLANATIONS

MCQ TYPE QUESTIONS

1.

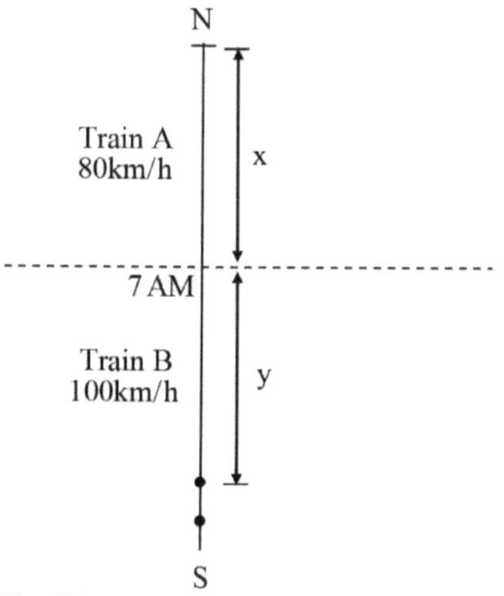

For X

Time taken = t

Distance x = velocity × time

$$x = 80t \Rightarrow t = \frac{x}{80} \quad ...(1)$$

For y time taken = t

Distance y = 100t

$$\Rightarrow t = \frac{y}{100} \quad ...(2)$$

$$x + y = 540 \text{ km} \quad ...(3)$$

From (1) and (2)

$$t = \frac{x}{80} = \frac{y}{100} \Rightarrow x = 0.8y$$

$x + y = 540 \Rightarrow x = 0.8y + y = 540$

$1.8y = 540 \Rightarrow y = 300\text{km}$

$$\text{Time taken } = \frac{y}{100} = \frac{300}{100} = 3\text{hrs}$$

Time at which these trains

= 7.00 AM + 3hrs = 10.00 AM

2. Average speed $= \dfrac{2(S_1 \times S_2)}{S_1 + S_2}$

$$= \frac{2(60 \times 90)}{(60 + 90)} \text{km/h}$$

$$= \frac{2 \times 60 \times 90}{150} = 72 \text{ km/h}$$

3. Turn week to reciprocals to make into 1

$$1200 \text{ M} + 500 \text{ W} = \frac{1}{2}$$

$$900 \text{ M} + 250 \text{ W} = \frac{1}{3}$$

Now 2(1200 M + 500 W) = 1

3(900 M + 250 W) = 1

Which becomes

2400 M + 1000 W = 1

2700 M + 750 W = 1

equating 2700 M – 2400 M

= 100 W – 750 W = 1000 W – 750 W

$$300 \text{ M} - 250 \text{ W} \Rightarrow \text{W} = \frac{300}{250} = \frac{6}{5}\text{M}.$$

Put $W = \frac{6}{5}M$ in equation equal to 1 week

$$2400M + 1000 \times \frac{6}{5}M = 1$$

M = 3600

4. Q can finish 1 day's 1 hour's work

$$= \frac{1}{25 \times 12} = \frac{1}{300}$$

R can finish 1 day's 1 hour's work

$$= \frac{1}{50 \times 12}$$

$$= \frac{1}{600}$$

Now Q working hours

$= (7 - 2) \times 12$

$= 60$ hr

$\therefore$ R working hours = $7 \times 18 = 126$

After 7 days, the ratio of work done by Q & R is

Q : R

$\frac{60}{300} : \frac{126}{600}$

$Q : R = 120 : 126$

$= 20 : 21$

5. M is twice as efficient as E but worked for half as many days. So in this case they will do equal work if their shifts had same timings. But M's shift is for 6 hours, while E's shift for 12 hours. Hence, E will do twice the work as M.

Ratio of contribution of M : E in work is 1 : 2.

6. Here the speed of first train $(S_1) = \frac{120}{10} = 12$ m/s

and the speed of second train $(S_2) = \frac{150}{15} = 10$ m/s

$\therefore$ The magnitude of difference in the speeds of two trains = $(S_1 - S_2) = (12 - 10)$ m/s = 2 m/s

7. From the given figure.

The odometer reading increases from starting point to end point.

Odometer reading = Area of the given diagram

Area of the velocity and time graph per second

1st sec $\Rightarrow$ triangle

$= \frac{1}{2} \times 1 \times 1 = \frac{1}{2}$

2nd sec $\Rightarrow$ square

$= 1 \times 1 = 1$

3rd sec $\Rightarrow$ square + triangle

$= 1 \times 1 + \frac{1}{2} \times 1 \times 1 = 1\frac{1}{2}$

4th sec $\Rightarrow$ triangle

$= \frac{1}{2} \times 1 \times 2 = 1$

5th sec $\Rightarrow$ straight line

$= 0$

6th sec $\Rightarrow$ triangle

$= \frac{1}{2} \times 1 \times 1 = \frac{1}{2}$

7th sec $\Rightarrow$ triangle

$= \frac{1}{2} \times 1 \times 1 = \frac{1}{2}$

Total Odometer reading at 7 seconds

$= \left(\frac{1}{2} + 1 + 1\frac{1}{2} + 1 + 0 + \frac{1}{2} + \frac{1}{2}\right)$m

$= 5$ m

So, the odometer reading is increased by 5 m.

9. For $M \Rightarrow \frac{12}{20} = 0.6$

$N \Rightarrow \frac{25}{45} = 0.55$

$O \Rightarrow \frac{45}{75} = 0.6$

$P \Rightarrow \frac{57}{100} = 0.57$

10. $V_{half} = 30(s)$ drawing rate = s

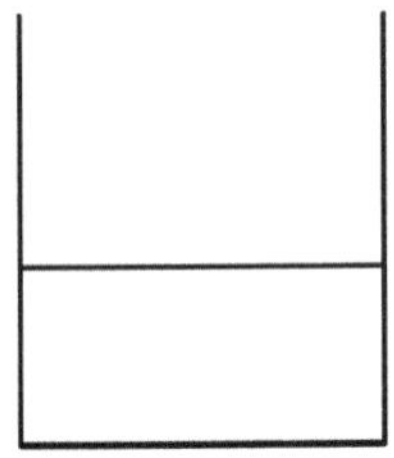

Total volume = 60 S tank

$(s^1)(10) - (s)10 = 30s$

$s^1(s) - s = 3s$

$s1 = 4s$

$s^1 = 4$ drawing rate

11. Given: A tourist covers half the journey at 60 km/h, one fourth at 30 km/h and the remaining one fourth at 10 km/h

To find: Avearge speed of the whole journey.

Analysis: One approach to this is to find individual time consumed in each of the three parts and use that value to get total time, and find average speed.

So,

Lets assume total length of journey = x km

$\therefore$ First phase at 60 km/h

Time $t_1 = \frac{\text{Distance}}{\text{Speed}} = \frac{x}{2}\frac{1}{60} = \frac{x}{120} h$

Second phase at 30 km/h

Distance = $\frac{x}{4}$ km

Time $t_2 = \frac{x}{4.30} = \frac{x}{120}$

Third phase at 10 km/h

Distance = $\frac{x}{4}$ km

Time $t_3 = \frac{x}{4.10} = \frac{x}{40} h$

$\therefore$ Total time $= \frac{x}{120} + \frac{x}{120} + \frac{x}{40} = \frac{x}{40}\left(\frac{1}{3} + \frac{1}{3} + 1\right)$

$$= \frac{x}{40}\frac{(1+1+3)}{3} = \frac{5x}{120}$$

$\therefore$ Average speed $= \frac{\text{Total Distance}}{\text{Total time}}$

$$= \frac{x \cdot 120}{5x} = 24 \text{ km/}h$$

Hence, the answer is (*c*)

12. Average Speed $= \frac{\text{Total distance}}{\text{total time taken}}$

Total Distance $= \frac{8+6+16}{15+15+15} = \frac{30 \times 60}{45}$

= 40km/h.

13.

	Fuel consumption	Actual
P	60 km/l	$\frac{15}{60} = \frac{1}{4}l$
Q	90 km/l	$\frac{75}{90} = \frac{5}{6}l$
R	75 km/l	$\frac{40}{75} = \frac{8}{15}l$
S	30 km/l	$\frac{10}{30} = \frac{1}{3}l$

14. 5 skilled workers build wall in 20 days

1 skilled worker build wall in 20 × 5 days

Hence in 1 day, part of work done by skilled work

$$= \frac{1}{100}$$

Similarly in 1 day part of work done by semi-skilled workers

$$= \frac{1}{25 \times 8}$$

and in 1 day part of work done by un-skilled worker

$$= \frac{1}{30 \times 10}$$

So part of work done in 1 day by 2 skilled, 6 semi-skilled and 5 unskilled

$$= \frac{2}{100} + \frac{6}{200} + \frac{5}{300} = \frac{1}{15}$$

So work done by given workers in days = 15

NUMERICAL TYPE QUESTIONS

1. Tiger – 1 leap ⇒ 8 meter

Speed = 5 leap/hr = 40m/min

Deer → 1 leap = 5 meter

speed = 4hr = 20m/min

Let at time '*t*' the tiger catches the deer.

$\therefore$ Distance travelled by deer + initial distance between them

50 × 8 ⇒ 400m = distance covered by tiger.

$\Rightarrow$ $40 \times t = 400 + 20t$

$\Rightarrow$ $t = \frac{400}{200} = 20$ min

$\Rightarrow$ total distance ⇒ $400 + 20 \times t = 800$ m

2. Speed of man = 8;

Left distance = d

Time taken $= \frac{d}{8}$

Upstream: Speed of stream = s

$\Rightarrow$ speed upstream = S' = (8 – s)

$$t' = \left(\frac{d}{8-s}\right)$$

Downstream:

Given speed downstream $= t'' = \frac{d}{8+s}$

$\Rightarrow 3t' = t'' \Rightarrow \frac{3d}{8-s} = \frac{d}{8+s}$

$\Rightarrow \frac{3d}{8-s} = \frac{d}{8+s}$

$\Rightarrow$ s = 4 km/hr

3. Let the length of platform = x.m

and train length = 280 m (given)

According to Question

$$\frac{x+280}{60} = \frac{280}{20}$$

$\therefore$ $x = 560$ m.

4

CHAPTER

Ratio, Proportion and Mixtures

MCQ TYPE QUESTIONS

2015

1. A cube of side 3 units is formed using a set of smaller cubes of side 1 unit. Find the proportion of the number of faces of the smaller cubes visible to those which are NOT visible.

   (*a*) 1 : 4

   (*b*) 1 : 3

   (*c*) 1 : 2

   (*d*) 2 : 3

2011

2. A container originally contains 10 litres of pure spirit. From this container 1 litre of spirit is replaced with 1 litre of water. Subsequently, 1 litre of the mixture is again replaced with 1 litre of water and this process is repeated one more time. How much spirit is now left in the container?

   (*a*) 7.58 litres

   (*b*) 7.84 litres

   (*c*) 7 litres

   (*d*) 7.29 litres

ANSWERS

MCQ Type Questions

1. (*c*) **2.** (*d*)

EXPLANATIONS

MCQ TYPE QUESTIONS

1.

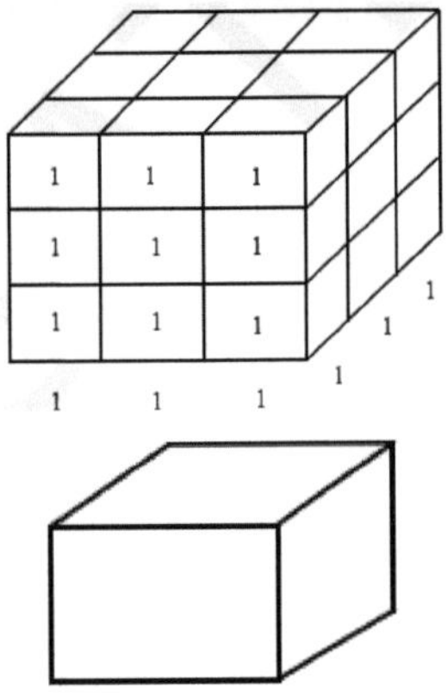

Number of faces per cube = 6

Total number of cubes = $9 \times 3 = 27$

$\therefore$ Total number of faces = $27 \times 6 = 162$

$\therefore$ Total number of non visible faces = $162 - 54 = 108$

$$\therefore \frac{\text{Number of visible faces}}{\text{Number of non visible faces}} = \frac{54}{108} = \frac{1}{2}$$

2. $10\left(\frac{10-1}{10}\right)^3 = 10\left(\frac{9}{10}\right)^3 = \frac{729}{1000}$

$$\therefore \frac{729}{1000} \times 1 = 7.29 \text{ litres}$$

5 CHAPTER Permutations and Combinations & Probability

MCQ TYPE QUESTIONS

2021

1. Two identical cube shaped dice each with faces numbered 1 to 6 are rolled simultaneously. The probability that an even number is rolled out on each dice is:

(a) $\frac{1}{36}$ (b) $\frac{1}{12}$

(c) $\frac{1}{8}$ (d) $\frac{1}{4}$

2.

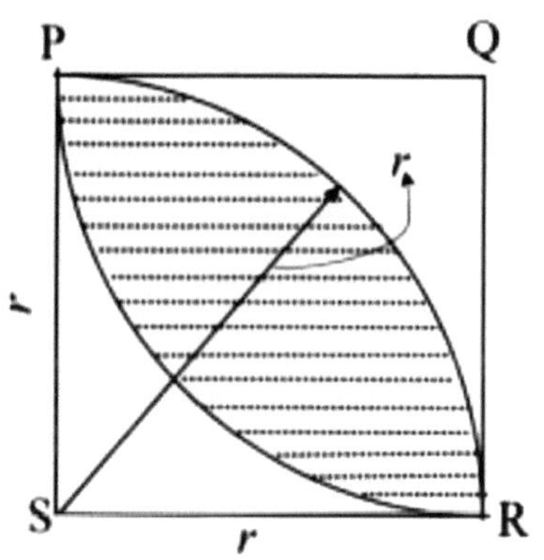

In the figure shown above, PQRS is a square. The shaded portion is formed by the intersection of sectors of circles with radius equal to the side of the square and centers at S and Q.

The probability that any point picked randomly within the square falls in the shaded area is _______

(a) $4-\frac{\pi}{2}$ (b) $\frac{1}{2}$

(c) $\frac{\pi}{2}-1$ (d) $\frac{\pi}{4}$

2018

3. To pass a test, a candidate needs to answer at least 2 out of 3 questions correctly. A total of 6,30,000 candidates appeared for the test. Question A was correctly answered by 3,30,000 candidates. Question B was answered correctly by 2,50,000 candidates. Question C was answered correctly by 2,60,000 candidates. Both questions A and B were answered correctly by 1,00,000 candidates. Both questions B and C were answered correctly by 90,000 candidates. Both questions A and C were answered correctly by 80,000 candidates. If the number of students answering all questions correctly is the same as the number answering none, how many candidates failed to clear the test?

(*a*) 30,000 (*b*) 2,70,000

(*c*) 3,90,000 (*d*) 4,20,000

2016

4. Shaquille O' Neal is a 60% career free throw shooter, meaning that he successfully makes 60 free throws out of 100 attempts on average. What is the probability that he will successfully make exactly 6 free throws in 10 attempts?

(*a*) 0.2508 (*b*) 0.2816

(*c*) 0.2934 (*d*) 0.6000

2015

5. Four cards are randomly selected from a pack of 52 cards. If the first two cards are kings, what is the probability that the third card is a king?

(*a*) 4/52 (*b*) 2/50

(*c*) 1/52 × (1/52) (*d*) 1/52 × (1/52) × (1/50)

6. Five teams have to compete in a league, with every team playing every other team exactly once, before going to the next round. How many matches will have to be held complete the league round of matches?

(*a*) 20 (*b*) 10

(*c*) 8 (*d*) 5

7. Right triangle PQR is to be constructed in the xy – plane so that the right angle is at P and line PR is parallel to the-axis. The x and y coordinates of P, Q, and R are to be integers that satisfy the inequalities: $-4 \leq x \leq 5$ and $6 \leq y \leq 16$. How many different triangles could be constructed with these properties?

(*a*) 110 (*b*) 1,100

(*c*) 9,900 (*d*) 10,000

8. A coin is tossed thrice. Let X be the event that head occurs in each of the first two tosses. Let Y be the event that a tail occurs on the third toss. Let Z be the event that two tails occurs in three tosses. Based on the above information, which one of the following statements is TRUE?

(*a*) X and Y are not independent

(*b*) Y and Z are dependent

(*c*) Y and Z are independent

(*d*) X and Z independent

9. Ram and Ramesh appeared in an interview for two vacancies in the same department. The probability of Ram's selection is 1/6 and that of Ramesh is 1/8. What is the probability that only one of them will be selected?

(*a*) $\frac{47}{48}$ (*b*) $\frac{1}{4}$

(*c*) $\frac{13}{48}$ (*d*) $\frac{35}{48}$

10. Given set A = {2, 3, 4, 5} and Set B = {11, 12, 13, 14, 15}, two numbers are randomly selected, one from each set. What is probability that the sum of the two numbers equals 16 ?

(*a*) 0.20 (*b*) 0.25

(*c*) 0.30 (*d*) 0.33

11. The probabilities that a student passes in Mathematics, Physics and Chemistry are m, p and c respectively. Of these subjects, the student has 75% chance of passing in at least one, a 50% chance of passing in at least two and a 40% chance of passing in exactly two. Following relations are drawn in $m, p.c$:

I. $p + m + c = \frac{27}{20}$ II. $p + m + c = \frac{13}{20}$

III. $(p) \times (m) \times (c) = \frac{1}{10}$

(*a*) Only relation I is true.

(*b*) Only relation II is true.

(*c*) Relations II and III are true.

(*d*) Relations I and III are true.

2014

12. A five digit number is formed using the digits 1,3,5,7 and 9 without repeating any of them. What is the sum of all such possible five digit numbers?

(*a*) 6666660 (*b*) 6666600

(*c*) 6666666 (*d*) 6666606

13. You are given three coins: one has heads on both faces, the second has tails on both faces, and the third has a head on one face and a tail on the other. You choose a coin at random and toss it, and it comes up heads. The probability that the other face is tails is

(*a*) 1/4 (*b*) 1/3

(*c*) 1/2 (*d*) 2/3

2013

14. Out of all the 2-digit integers between 1 and 100, a 2-digit number has to be selected at random. What is the probability that the selected number is not divisible by 7?

(*a*) 13/90 (*b*) 12/90

(*c*) 78/90 (*d*) 77/90

15. What is the chance that a leap year, selected at random, will contain 53 Saturdays?

(*a*) 2/7 (*b*) 3/7

(*c*) 1/7 (*d*) 5/7

16. In a factory, two machines M1 and M2 manufacture 60% and 40% of the autocomponents respectively. Out of the total production, 2% of M1 and 3% of M2 are found to be defective. If a randomly drawn autocomponent from the combined lot is found defective, what is the probability that it was manufactured by M2?

(*a*) 0.35 (*b*) 0.45

(*c*) 0.5 (*d*) 0.4

2012

17. A and B are friends. They decide to meet between 1 PM and 2 PM on a given day. There is a condition that whoever arrives first will not wait for the other for more than 15 minutes. The probability that they will meet on that day is

(*a*) $\frac{1}{4}$ (*b*) $\frac{1}{16}$

(*c*) $\frac{7}{16}$ (*d*) $\frac{9}{16}$

2010

18. Given digits 2, 2, 3, 3, 3, 4, 4, 4, 4 how many distinct 4 digit numbers greater than 3000 can be formed?

(*a*) 50 (*b*) 51

(*c*) 52 (*d*) 54

NUMERICAL TYPE QUESTIONS

2015

1. How many four digit numbers can be formed with the 10 digits 0, 1, 2, ... 9 if no number can start with 0 and if repetitions are not allowed?

2014

2. In any given year, the probability of an earthquake greater than Magnitude 6 occurring in the Garhwal Himalayas is 0.04. The average time between successive occurrences of such earthquakes is _____ years.

3. 10% of the population in a town is HIV^+. A new diagnostic kit for HIV detection is available; this kit correctly identifies HIV^+ individuals 95% of the time, and HIV^- individuals 89% of the time. A particular patient is tested using this kit and is found to be positive. The probability that the individual is actually positive is ______

4. A batch of one hundred bulbs is inspected by testing four randomly chosen bulbs. The batch is rejected if even one of the bulbs is defective. A batch typically has five defective bulbs. The probability that the current batch is accepted is ________

ANSWERS

MCQ Type Questions

1. (d) **2.** (c) **3.** (*d*) **4.** (*a*) **5.** (*b*) **6.** (*b*) **7.** (*c*) **8.** (*d*) **9.** (*b*) **10.** (*a*)

11. (*d*) **12.** (*b*) **13.** (*b*) **14.** (*d*) **15.** (*a*) **16.** (*c*) **17.** (*c*) **18.** (*b*)

Numerical Type Questions

1. 4536 **2.** (25 to 25) **3.** (0.47 to 0.48) **4.** (0.8145)

EXPLANATIONS

MCQ TYPE QUESTIONS

1. Possible combinations are 9.

(2, 2), (4, 4), (6, 6)

(2, 4), (4, 2), (6, 4)

(2, 6), (4, 6), (6, 2)

Total Number of Combinations are 36.

$$\text{Probability} = \frac{9}{36}$$

$$= \frac{1}{4}$$

Hence, option (d) is the correct answer.

2.

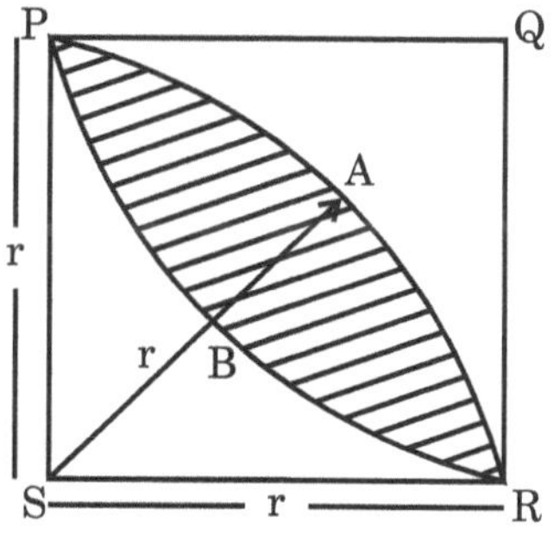

Side = r

Area of shaded region = Area of PQRS

– 2(Area of PQRS – Area of Quadrant PARSP)

$$= r^2 - 2\left(r^2 - \frac{\pi r^2}{4}\right)$$

$$= = r^2 - 2r^2 + \frac{\pi r^2}{2}$$

$$\Rightarrow \frac{\pi r^2}{2} - r^2 \Rightarrow r^2\left(\frac{\pi}{2} - 1\right) \text{ sq. unit}$$

$$\text{Probability} = \frac{\text{Area of shaded Region}}{\text{Area of Square}}$$

$$\Rightarrow \frac{r^2\left(\frac{\pi}{2} - 1\right)}{r^2}$$

$$= \left(\frac{\pi}{2} - 1\right), \text{ So option (c).}$$

3.

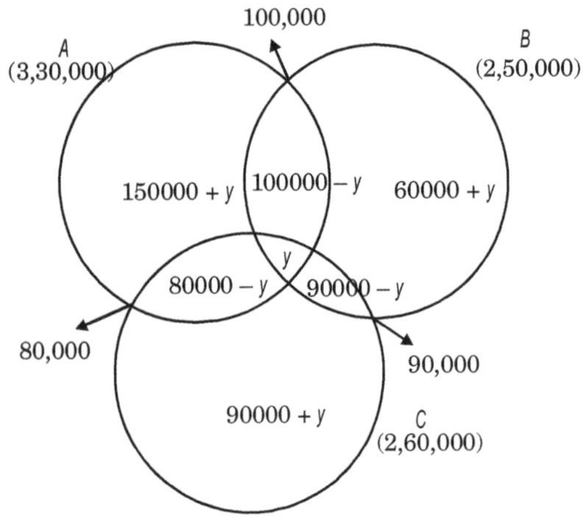

From above diagram,

$\therefore$ 6,30,000 = 2y + 1,50,000 + 100000 + 80000 + 60000 + 90000 + 90000

$\therefore$ 63000 = 2y + 57000

$\therefore$ 6,30,000 – 5,70,000 = 2y

$\therefore$ 60000 = 2y

$\therefore$ y = 30000

Students who failed to clear the test

= 150000 + 60000 + 90000 + 4y

= 300000 + 4 × 30000 ($\because$ y = 30000)

= 420000

Students who failed to clear the test = 420000

4. The question clearly explains that shaquille makes 60 free throws out of 100

Hence, Probability of free throw = $\frac{60}{100} = 0.6$

And probability of NOT free throw

= 1 – 0.6 = 0.4

So required probability of exactly 6 throws in 10 attempts will be given by

$^{10}C_6\ 0.6^6 \times 0.4^4 = 0.2508$

5. There are 4 kings in a pack of 52 cards.

If 2 cards are selected and both are kings, remaining cards will be 50 out of which 2 will be kings.

6. For a match to be played, we need 2 teams

No. of matches = No. of ways of selections 2 teams out of 5

$= {}^5C_2 = 10$

8. x = {HHT, HHH}

y depends on x

z = {TTH, TTT}

$\therefore$ 'd' is the correct choice.

9. P (Ram) = 1/6; p(Ramesh) = 1/8

p(only at) = p (Ram) × p (not ramesh) + p(Ramesh) × p($n_0 \times R_{am}$)

$= \frac{1}{6} + \frac{7}{8} + \frac{1}{8} \times \frac{5}{6}$

$\Rightarrow \frac{12}{40} = \frac{1}{4}$

10. Here the given sets are

A = {2, 3, 4, 5}

and B = {11, 12, 13, 14, 15}

desired outcome = {(2, 14), (3, 13), (4, 12), (5, 11)}

$\therefore$ n(E) = 4

and total outcome = 4 × 5 = 20

$\therefore$ $n(s) = 20$

$\therefore$ Probability of an event

$= \frac{n(E)}{n(s)}$

$= \frac{4}{20} = \frac{1}{5} = 0.2$

12. The digit in unit place is selected in 4! Ways

The digit in tens place is selected in 4! Ways

The digit in hundreds place is selected in 4! Ways

The digit in thousands place is selected in 4! Ways

The digit in ten thousands place is selected in 4! Ways

Sum of all values for 1

$4! \times 1 \times (10^0 + 10^1 + 10^2 + 10^3 + 10^4)$

$= 4! \times 11111 \times 1$

Similarly for '3' 4! × (11111) × 3

Similarly for '5' 4! × (11111) × 5

Similarly for '7' 4! × (11111) × 7

Similarly for '9' 4! × (11111) × 9

$\therefore$ sum of all such numbers

4! ×(11111)× (1+ 3 + 5 + 7 + 9)

= 24 × (11111) × 25

= 6666600

13. The probability that the other face is tail is $\frac{1}{3}$.

14. **Given:** All the 2-digit numbers

To find: Probability of selecting a number at random not divisible by 7

Analysis: Total number of 2-digit numbers

10, 11, ..., 99 → 90 numbers

Unfavourable numbers *i.e.* numbers divisible by 7 : 14, 21, ..., 98

Number of unfavourable numbers = 13

This can be hand counted or we can use the following properties:

Last term = First time + (no. – 1) Diff

98 = 14 + (no. – 1) 7

$\Rightarrow$ no. = 13

$\Rightarrow$ Probability = $\frac{\text{Total} - \text{unfavourable}}{\text{Total}}$

$= \frac{90-13}{90} = 77/90$

Hence, the solution is (*d*)

15. at a leap year

52 weeks, and 2 extra day they are (Mon, Tues) (Tues, Wed) (Wed Thu.) (Thus. Fri) (Fri. Sat) (Sat. Sun) (Sun Mond.)

$n(s) = 7$

$n(\text{E}) = 2 \quad \text{P(E)} = \frac{2}{7}$

17.

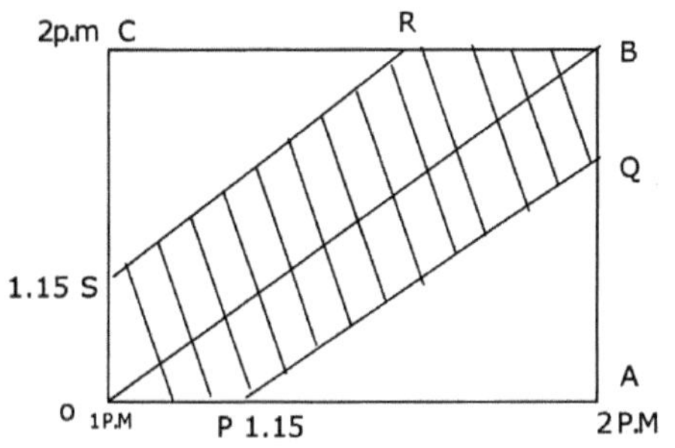

OB is the line when both A and B arrive at same time.

Total sample sapce = 60 × 60 = 3600

Favourable cases = Area of OABC – 2(Area of SRC)

$= 3600 - 2 \times \left(\frac{1}{2} \times 45 \times 45\right)$

$= 1575$

$\therefore$ Required probability $= \frac{1575}{3600} = \frac{7}{16}$

18. 2, 2, 3, 3, 4, 4, 4, 4 $\quad x > 3000$

322 — 3, 4
323 — 2, 3, 4
324 — 2, 3, 4
344 — 2, 3, 4
422 — 3, 4
423 — 2, 3, 4
332 — 2, 3, 4
333 — 2, 4
334 — 2, 3, 4
342 — 2, 3, 4
343 — 2, 3, 4
424 — 2, 3, 4
432 — 2, 3, 4
433 — 2, 3, 4
434 — 2, 3, 4
442 — 2, 3, 4
443 — 2, 3, 4
444 — 2, 3, 4

NUMERICAL TYPE QUESTIONS

1. In thousands place, 9 digits except 0 can be placed

In hundreds place, 9 digits can be placed (including 0, excluding the one used in thousands place)

In tens place, 8 digits can be placed (excluding the ones used in thousands and hundreds place)

In ones place, 7 digits can be placed (excluding the one used in thousands, hundreds and tens place)

Total number of combinations

$= 9 \times 9 \times 8 \times 7 = 4536$

2. Given, in one year, probability of an earthquake > 6 magnitude = 0.04

So, average time between successive earthquakes = 1/0.04 = 25 years

3. Given, 10% of the population is HIV^+, so probability = 0.1 for HIV^+ and 0.9 for HIV^-

The diagnostic kit identifies $\rightarrow$ 0.95 HIV^+ and 0.89 HIV^- correctly

So, the required probability that the person is actually HIV^+ = 0.95/2 = 0.475

4. Probability for one bulb to be non-defective is

$\frac{95}{100}$

$\therefore$ Probabilities that none of the bulbs is defectives $\left(\frac{95}{100}\right)^4 = 0.8145$

■■

6 CHAPTER Miscellaneous

MCQ TYPE QUESTIONS

2021

1. In an equilateral triangle PQR, side PQ is divided into four equal parts, side QR is divided into six equal parts and side PR is divided into eight equal parts. The length of each subdivided part in cm is an integer.

The minimum area of the triangle PQR possible, in cm^2, is

(a) 18 (b) 24

(c) $48\sqrt{3}$ (d) $144\sqrt{3}$

2020

2. Define $[x]$ as the greatest integer less than or equal to x, for each $x \in (-\infty, \infty)$. If $y = [x]$, then area under y for $x \in [1, 4]$ is ______

(*a*) 1 (*b*) 3

(*c*) 4 (*d*) 6

3. Select the graph that schematically represents BOTH $y = x^m$ and $y^{1/m}$ properly in the interval $0 \le x \le 1$, for integer values of m, where $m > 1$.

(*a*)

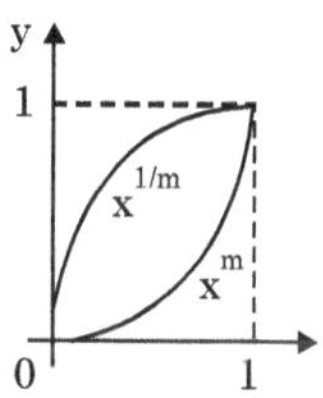

(*b*)

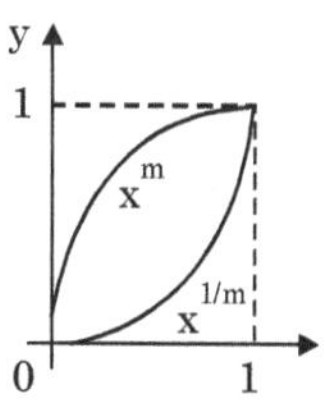

(*c*)

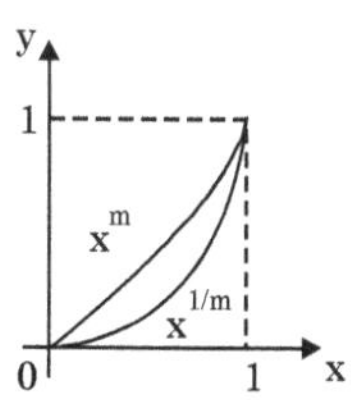

(*d*)

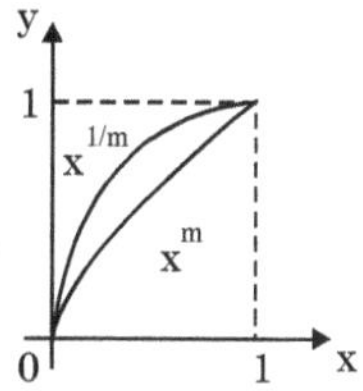

4. The bar graph shows the data of the students who appeared and passed in an examination for four schools P, Q, R and S. The average of success rates (in percentage) of these four schools is _____.

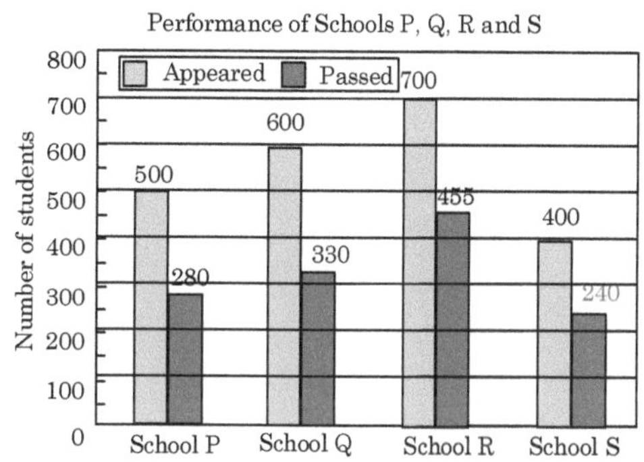

(*a*) 58.5 % (*b*) 58.8 %

(*c*) 59.0 % (*d*) 59.3%

2018

5. For $0 \le x \le 2\pi$, sin x and cos x are both decreasing functions in the interval ________.

(*a*) $\left(0, \dfrac{\pi}{2}\right)$

(*b*) $\left(\dfrac{\pi}{2}, \pi\right)$

(*c*) $\left(\pi, \dfrac{3\pi}{2}\right)$

(*d*) $\left(\dfrac{3\pi}{2}, 2\pi\right)$

6. The area of an equilateral triangle is $\sqrt{3}$. What is the perimeter of the triangle?

(*a*) 2 (*b*) 4

(*c*) 6 (*d*) 8

7. In a detailed study of annual crow births in India, it was found that there was relatively no growth during the period 2002 to 2004 and a sudden spike from 2004 to 2005. In another unrelated study, it was found that the revenue from cracker sales in India which remained fairly flat from 2002 to 2004, saw a sudden spike in 2005 before declining again in 2006. The solid line in the graph below refers to annual sale of crackers and the dashed line refers to the annual crow births in India. Choose the most appropriate inference from the above data.

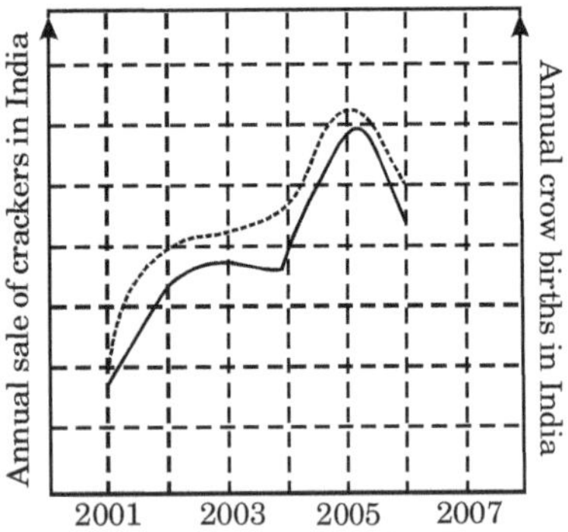

(*a*) There is a strong correlation between crow birth and cracker sales.

(*b*) Cracker usage increases crow birth rate.

(*c*) If cracker sale declines, crow birth will decline.

(*d*) Increased birth rate of crows will cause an increase in the sale of crackers.

2017

8. 500 students are taking one or more courses out of Chemistry, Physics, and Mathematics. Registration records indicate course enrolment as follows : Chemistry (329), Physics (186), Mathematics (295), Chemistry and Physics (83), Chemistry and Mathematics (217) and Physics and Mathematics (63). How many students are taking all 3 subjects?

(*a*) 37 (*b*) 43

(*c*) 47 (*d*) 53

9. A contour line joins locations having the same height above the mean sea level. The following is a contour plot of a geographical region. Contour lines are shown at 25 m intervals in this plot.

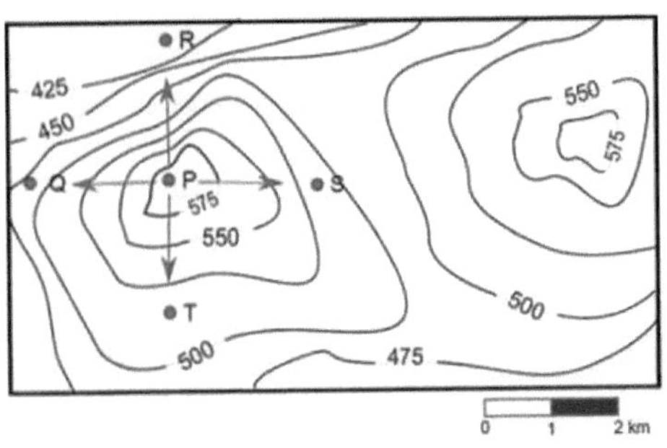

Which of the following is the steepest path leaving from P?

(*a*) P to Q (*b*) P to R

(*c*) P to S (*d*) P to T

2016

10. In a 2×4 rectangle grid shown below, each cell is a rectangle. How many rectangles can be observed in the grid?

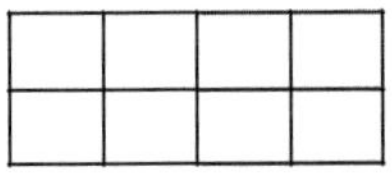

(*a*) 21 (*b*) 27

(*c*) 30 (*d*) 36

11.

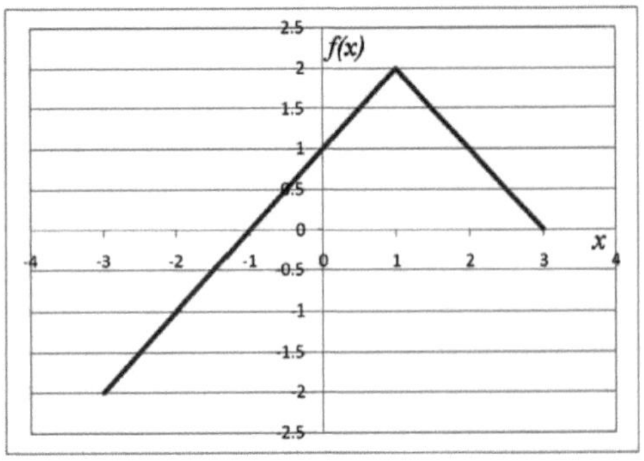

Choose the correct expression for $f(x)$ given in the graph.

(*a*) $f(x) = 1 - |x - 1|$

(*b*) $f(x) = 1 + |x - 1|$

(*c*) $f(x) = 2 - |x - 1|$

(*d*) $f(x) = 2 + |x - 1|$

12. The volume of a sphere of diameter 1 unit is ________ than the volume of a cube of side 1 unit.

(*a*) least (*b*) less

(*c*) lesser (*d*) low

13. A window is made up of a square portion and an equilateral triangle portion above it. The base of the triangular portion coincides with the upper side of the square. If the perimeter of the window is 6 m, the area of the window in m^2 is ________.

(*a*) 1.43 (*b*) 2.06

(*c*) 2.68 (*d*) 2.88

14. The binary operation $\square$ is defined as $a \square b = ab + (a + b)$, where a and b are any two real numbers. The value of the identity element of this operation, defined as the number x such that $a \square x = a$, for any a, is ________.

(*a*) 0 (*b*) 1

(*c*) 2 (*d*) 10

15. Which of the following curves represents the function $y = \ln\left(\left|e^{\left[\left|\sin(|x|)\right|\right]}\right|\right)$ for $|x| < 2\pi$?

Here, x represents the abscissa and y represents the ordinate.

(*a*)

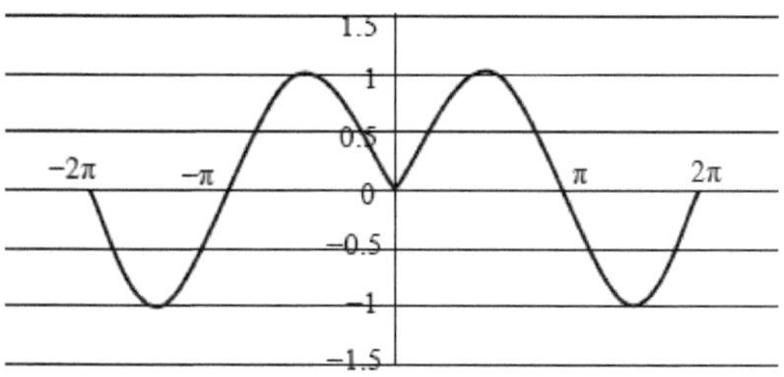

(*b*)

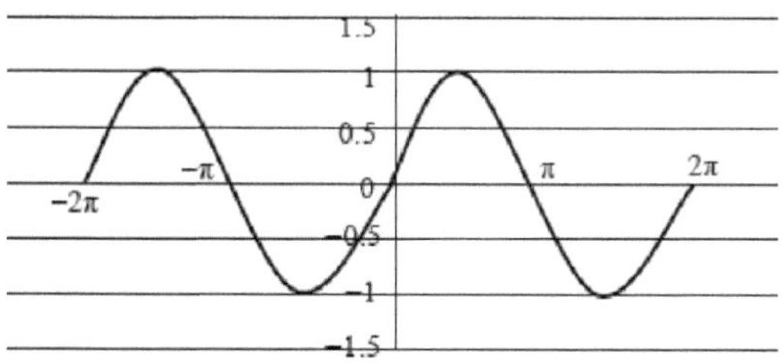

(*c*)

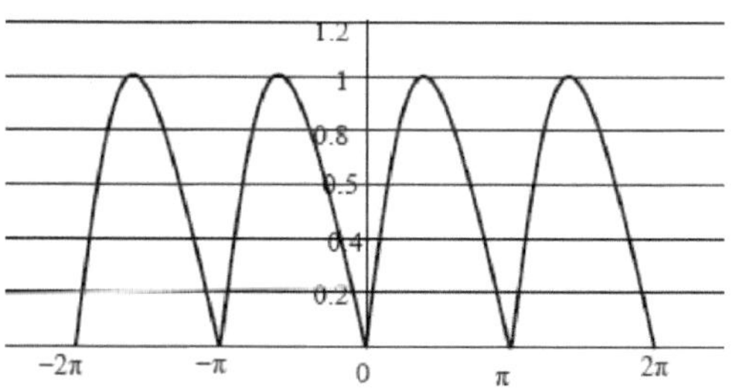

(*d*)

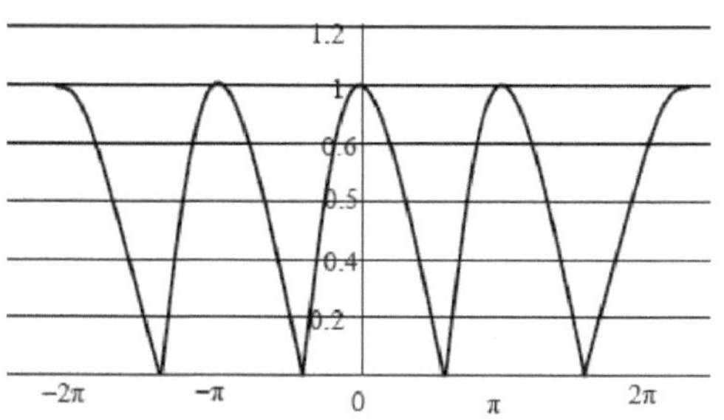

16. Given $(9 \text{ inches})^{1/2} = (0.25 \text{ yards})^{1/2}$, which one of the following statements is TRUE?

(*a*) 3 inches = 0.5 yards

(*b*) 9 inches = 1.5 yards

(*c*) 9 inches = 0.25 yards

(*d*) 81 inches = 0.0625 yards

17. The Venn diagram shows the preference of the student population for leisure activities.

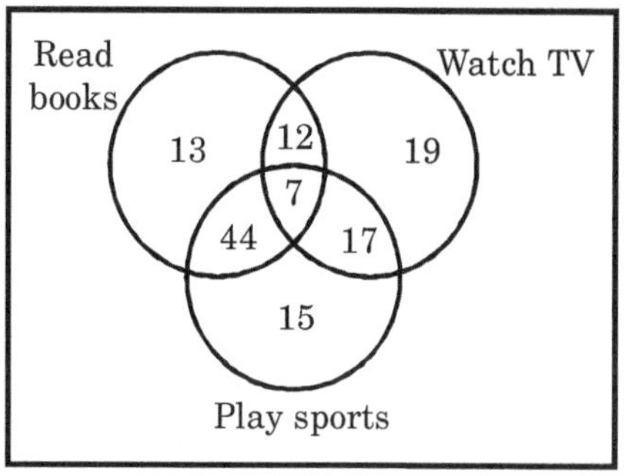

From the data given, the number of students who like to read books or play sports is ____.

(*a*) 44 (*b*) 51 (*c*) 79 (*d*) 108

18. A wire of length 340 mm is to be cut into two parts. One of the parts is to be made into a square and the other into a rectangle where sides are in the ratio of 1:2. What is the length of the side of the square (in mm) such that the combined area of the square and the rectangle is a MINIMUM?

(*a*) 30 (*b*) 40 (*c*) 120 (*d*) 180

19. Find the area bounded by the lines $3x + 2y = 14$, $2x - 3y = 5$ in the first quadrant

(*a*) 14.95 (*b*) 15.25 (*c*) 15.70 (*d*) 20.35

20. A straight line is fit to a data set $(\ln x, y)$. This line intercepts the abscissa at $\ln x = 0.1$ and has a slope of -0.02. What is the value of y at $x = 5$ from the fit?

(*a*) –0.030 (*b*) –0.014 (*c*) 0.014 (*d*) 0.030

21. A square pyramid has a base perimeter x, and the slant height is half of the perimeter. What is the lateral surface area of the pyramid?

(*a*) x^2 (*b*) $0.75\,x^2$ (*c*) $0.50\,x^2$ (*d*) $0.25\,x^2$

2015

22. If $\log_x\left(\frac{5}{7}\right) = -\frac{1}{3}$, then the value of x is

(*a*) $\frac{343}{125}$ (*b*) $\frac{125}{343}$ (*c*) $-\frac{25}{49}$ (*d*) $-\frac{49}{25}$

23. A function $f(x)$ is linear and has a value of 29 at $x = -2$ and 39 at $x = 3$. Find its value at $x = 5$.

(*a*) 59 (*b*) 45 (*c*) 43 (*d*) 35

24. If $x > y > $ I, which of the following must be true?

(*i*) $\ln x > \ln y$ (*ii*) $e^x > e^y$

(*iii*) $y^x > x^y$ (*iv*) $\cos x > \cos y$

(*a*) *i* and *ii* (*b*) *i* and *iii*

(*c*) *iii* and *iv* (*d*) *ii* and *iv*

25. log tan1° + log tan2° +-------+ log tan89° is ____.

(*a*) 1 (*b*) $1/\sqrt{2}$ (*c*) 0 (*d*) –1

26. Based on the given statements, select the most appropriate option to solve the given question.

What will be the total weight of 10 poles each of same weight?

Statements:

(I) One fourth of the weight of a pole is 5 kg

(II) The total weight of these poles is 160 kg more than the total weight of two poles.

(*a*) Statement II alone is not sufficient

(*b*) Statement II alone is not sufficient

(*c*) Either I or II alone is sufficient

(*d*) Both statements I and II together are not sufficient.

27. Consider a function $f(x) = 1 - |x|$ on $-1 \leq x \leq 1$. The value of x at which the function attains a maximum, and the maximum value of function are:

(*a*) 0, –1 (*b*) –1, 0

(*c*) 0, 1 (*d*) –1, 2

28. In a triangle PQR, PS is the angle bisector of $\angle$QPR and $\angle$QPS = 60°. What is the length of PS?

(*a*) $\frac{(q+r)}{qr}$

(*b*) $\frac{qr}{(q+r)}$

(*c*) $\sqrt{(q^2+r^2)}$

(*d*) $\frac{(q+r)^2}{qr}$

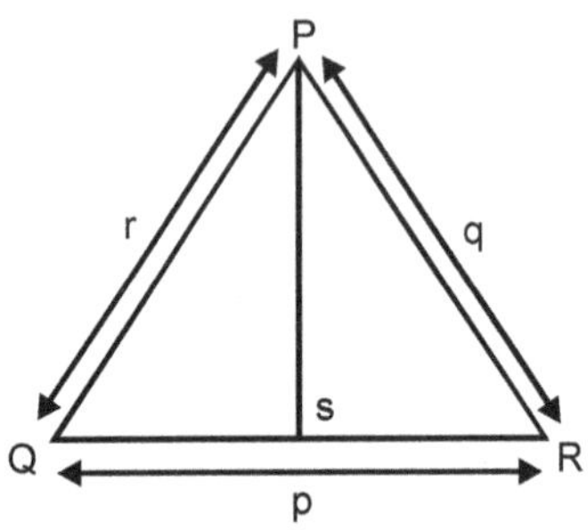

29. Based on the given statements, select the most appropriate option to solve the given question.

If two floors in a certain building are 9 feet apart, how many steps are there in a set of stairs that extends from the first floor to the second floor of the building?

Statements:

I. Each step is ¾ foot high.

II. Each step is 1 foot wide.

(*a*) Statement I alone is sufficient, but statement II alone is not sufficient.

(*b*) Statement II alone is sufficient, but statementI alone is not sufficient.

(*c*) Both statements together are sufficient, but neither statement alone is sufficient.

(*d*) Statement I and II together are not sufficient.

2014

30. If $y = 5x^2 + 3$, then the tangent at $x = 0, y = 3$

(*a*) passes through $x = 0, y = 0$

(*b*) has a slope of +1

(*c*) is parallel to the x-axis

(*d*) has a slope of –1

31. Let $f(x, y) = x^n y^m = \text{P}$. If x is doubled and y is halved, the new value of f is

(*a*) 2^{n-m} P (*b*) 2^{m-n} P

(*c*) $2(n - m)$ P (*d*) $2(m - n)$ P

32. A regular die has six sides with numbers 1 to 6 marked on its sides. If a very large number of throws show the following frequencies of occurrence: 1 → 0.167; 2 → 0.167; 3 → 0.152; 4 → 0.166; 5 → 0.168; 6 → 0.180. We call this die

(*a*) irregular (*b*) biased

(*c*) Gaussian (*d*) insufficient

33. Consider the equation : $(7526)_8 - (Y)_8 = (4364)_8$, where $(X)_N$ stands for X to the base N. Find Y.

(*a*) 1634 (*b*) 1737 (*c*) 3142 (*d*) 3162

34. At what time between 6 *a.m.* and 7 *a.m.* will the minute hand and hour hand of a clock make an angle closest to 60°?

(*a*) 6 : 22 *a.m.* (*b*) 6 : 27 *a.m.*

(*c*) 6 : 38 *a.m.* (*d*) 6 : 45 *a.m.*

35. The roots of $ax^2 + bx + c = 0$ are real and positive a, b and c are real. Then $ax^2 + b\,|x| + c = 0$ has

(*a*) No roots (*b*) 2 real roots

(*c*) 3 real roots (*d*) 4 real roots

2012

36. The cost function for a product in a firm is given by $5q^2$, where q is the amount of production. The firm can sell the product at a market price of ₹ 50 per unit. The number of units to be produced by the firm such that the profit is maximized is

(*a*) 5 (*b*) 10 (*c*) 15 (*d*) 25

37. Which of the following assertions are **CORRECT ?**

P : Adding 7 to each entry in a list adds 7 to the mean of the list

Q : Adding 7 to each entry in a list adds 7 to the standard deviation of the list

R : Doubling each entry in a list doubles the mean of the list

S : Doubling each entry in a list leaves the standard deviation of the list unchanged

(*a*) P, Q (*b*) Q, R (*c*) P, R (*d*) R, S

38. A political party orders an arch for the entrance to the ground in which the annual convention is being held. The profile of the arch follows the equation $y = 2x - 0.1x^2$ where y is the height of the arch in meters. The maximum possible height of the arch is

(*a*) 8 meters (*b*) 10 meters

(*c*) 12 meters (*d*) 14 meters

39. There are eight bags of rice looking alike, seven of which have equal weight and one is slightly heavier. The weighing balance is of unlimited capacity. Using this balance, the minimum number of weighings required to identify the heavier bag is

(*a*) 2 (*b*) 3 (*c*) 4 (*d*) 8

2011

40. If log (P) = (1/2) Log (Q) = (1/3) Log (R), then which of the following options is TRUE?

(*a*) $P^2 = Q^3 R^2$ (*b*) $Q^2 = PR$

(*c*) $Q^2 = R^3 P$ (*d*) $R = P^2 Q^2$

41. The variable cost (V) of manufacturing a product varies according to the equation V = 4q, where q is the quanity produced. The fixed cost (F) of production of same product reduces with *q* according to the equation F = 100/*q*. How many units should be produced to minimize the total cost (V + F)?

(*a*) 5 (*b*) 4 (*c*) 7 (*d*) 6

42. Three friends R, S and T shared toffee from a bowl. R took 1/3rd of the toffees, but returned four to the bowl. S took 1/4th of what was left but returned three toffees to the bowl. T took half of the remainder but returned two back into the bowl. If the bowl had 17 toffees left, how many toffees were originally there in the bowl?

(*a*) 38 (*b*) 31 (*c*) 48 (*d*) 41

NUMERICAL TYPE QUESTIONS

2015

1. In the given figure angle Q is a right angle, PS : QS = 3:1, RT : QT = 5:2 and PU : UR = 1:1. If area of triangle QTS is 20 cm², then the area of triangle PQR in cm² is ______.

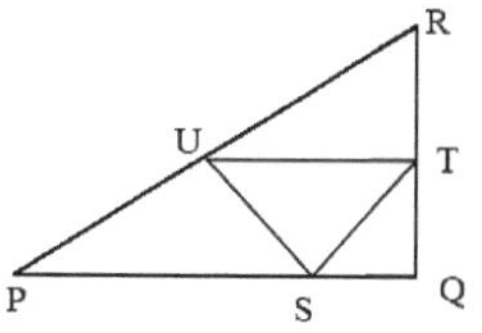

2. If *p, q, r, s* are distinct integers such that:

$f(p, q, r, s) = \max(p, q, r, s)$

$g(p, q, r, s) = \min(p, q, r, s)$

$h(p, q, r, s)$ = remainder of $(p \times q)/(r \times s)$ if $(p \times q) > (r \times s)$ or remainder of $(r \times s)/(p \times q)$ if $(r \times s) > (p \times q)$

Also a function $fgh(p, q, r, s) = f(p, q, r, s) \times g(p, q, r, s) \times h(p, q, r, s)$

Also the same operations are valid with two variable function of the form $f(p, q)$.

What is the value of $fg(h(2, 5, 7, 3), 4, 6, 8$

2014

3. When a point inside of a tetrahedron (a solid with four triangular surfaces) is connected by straight lines to its corners, how many (new) internal planes are created with these lines? ________.

ANSWERS

MCQ Type Questions

1. (d)	**2.** (d)	**3.** (a)	**4.** (c)	**5.** (*b*)	**6.** (*c*)	**7.** (*a*)	**8.** (*d*)	**9.** (*b*)	**10.** (*c*)
11. (*c*)	**12.** (*c*)	**13.** (*b*)	**14.** (*a*)	**15.** (*c*)	**16.** (*c*)	**17.** (*d*)	**18.** (*b*)	**19.** (*b*)	**20.** (*a*)
21. (*d*)	**22.** (*a*)	**23.** (*c*)	**24.** (*a*)	**25.** (*c*)	**26.** (*c*)	**27.** (*c*)	**28.** (*b*)	**29.** (*a*)	**30.** (*c*)
31. (*a*)	**32.** (*b*)	**33.** (*c*)	**34.** (*a*)	**35.** (*d*)	**36.** (*a*)	**37.** (*c*)	**38.** (*b*)	**39.** (*a*)	**40.** (*b*)
41. (*a*)	**42.** (*c*)								

Numerical Type Questions

1. 280 **2.** (8) **3.** (6)

EXPLANATIONS

MCQ TYPE QUESTIONS

1. Since we divide equilateral triangle

The sides of triangle PQR must be divisible by 4, 6 and 8, therefore

For option (a),

Area = 18

$$\frac{\sqrt{3}}{4}(\text{side})^2 = 18$$

side ≠ integer (wrong)

For option (b)

Area = 24

$$\frac{\sqrt{3}}{4}(\text{side})^2 = 24$$

side ≠ integer (wrong)

For option (c)

$$\frac{\sqrt{3}}{4}(\text{side})^2 = 48\sqrt{3}$$

$$\text{Side} = \sqrt{48 \times 4}$$

side ≠ integer (wrong)

For option (d)

$$\frac{\sqrt{3}}{4}(\text{side})^2 = 144\sqrt{3}$$

$$\text{Side} = \sqrt{144 \times 4}$$

side = 12 × 2

$$\boxed{\text{side} = 24}$$

Which is divisible by 4, 6 and 8.

Hence, option (d) is correct.

2. Graph of y = [x] for x ∈ [1, 4]

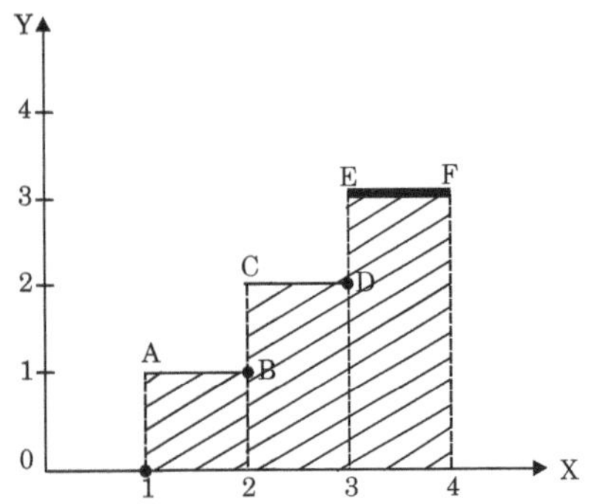

The total area, A = (1 × 1) + (2 × 1) + (3 × 1)

= 1 + 2 + 3 = 6

3. Graph of $y = x^m$ and $y = x^{1/m}$ for m > 1

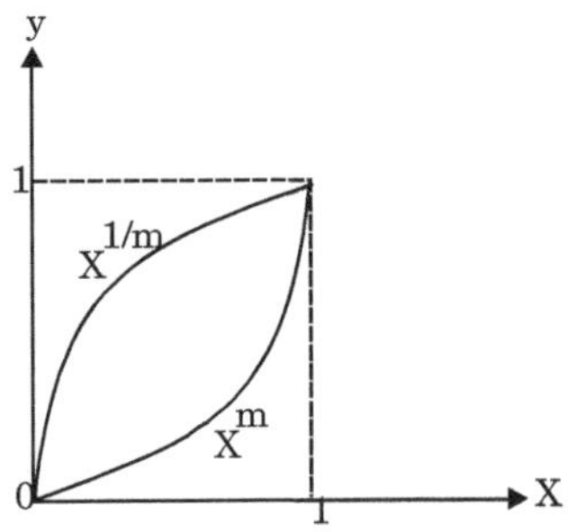

Correct option is (a).

4. Success rate = $\dfrac{\text{Number of candidates who passed the exam}}{\text{Number of appeared candidates}}$

So, Average success rate,

$$= \frac{\frac{280}{500} + \frac{330}{600} + \frac{455}{700} + \frac{240}{400}}{4}$$

= 0.59 or 59%

Correct option is (c).

5.

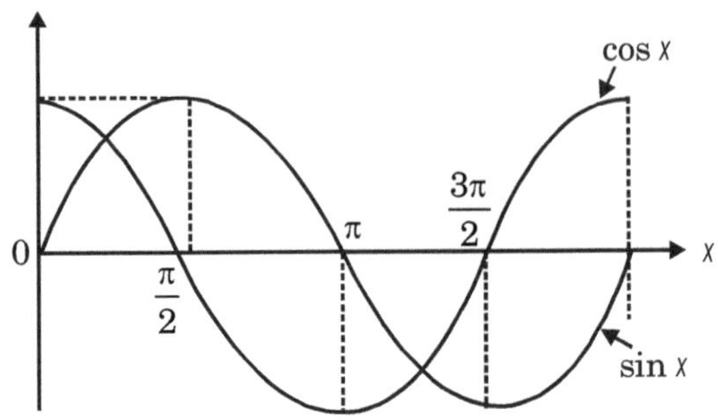

From the curve it is clear that sin x and cos x both are decreasing in the interval $\left(\frac{\pi}{2}, \pi\right)$

6. Area of equilateral triangle = $\dfrac{\sqrt{3}}{4}a^2$

$$\therefore \quad \sqrt{3} = \frac{\sqrt{3}}{4}a^2$$

$$\therefore \quad a^2 = 4$$

$$\therefore \quad a = 2$$

Now perimeter of equilateral triangle

= 3a = 3 × 2 = 6

8. $P(A \cap B \cap C) = P((A) + P(B) + P(C)$

$$-[P(A \cap B) + P(B \cap C) + P(A \cap C)]$$

$$-P(A \cap B \cap C)$$

$$500 = 329 + 186 + 295 - [83 + 217 + 63]$$

$$-P(A \cap B \cap C)$$

$P(A \cap B \cap C) = 53.$

9. P to R is the steepest path because the path from P to R erasses the most number of contour lines which are 25 metres away.

10. In the given 2×4 rectangle grid, the following type of rectangles are present.

One figured rectangles = 8

Two figured rectangles = 10

Three figured rectangles = 4

Four figured rectangles = 5

Six figured rectangles = 2

Eight figured rectangles = 1

∴ Total number of rectangles

$$= 8 + 10 + 4 + 5 + 2 + 1 = 30$$

So the number of rectangles observed in the given grid is 30.

11. From the given graph function $f(x)$ must be equals to zero at $x = 3$.

From Option A :

$f(x) = 1 - |x - 1|$

at $x = 3$

$f(x) = 1 - |3 - 1| = 1 - 2 = -1$

So, it is false

From Option B :

$f(x) = 1 + |x - 1|$

at $x = 3$

$f(x) = 1 + |3 - 1| = 1 + 2 = 3$

So, it is false

From Option C :

$f(x) = 2 - |x - 1|$

at $x = 3$

$f(x) = 2 - |3 - 1| = 2 - 2 = 0$

So, it is true.

From Option D :

$f(x) = 2 + |x - 1|$

at $x = 3$

$f(x) = 2 - |3 - 1| = 2 + 2 = 4$

So, it is false.

Hence the correct expression for $f(x)$ given in the graph is $2 - |x - 1|$.

12. 'Lesser than' is apt because the sentence should be in comparative degree.

13. The possible window will be as follows :

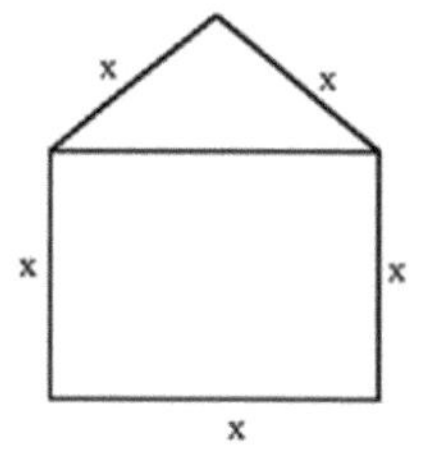

Perimeter of window = $5x = 6$m

$x = 1.2$ m

So area of window = Area of triangle + Area of square

$$= \frac{\sqrt{3}}{4}x^2 + x^2$$

$$= \frac{\sqrt{3}}{4}(1.2)^2 + (1.2)^2 = 2.06 \text{ m}^2$$

14. The binary operation □ is defined

⇒ $a \square b = ab + (a + b)$

$a \square x = a$

∴ From the equation 'b' is the variable

Option A: $x = 0$

$a \square o = a \times 0 + (a + 0) = 0 + a = a$

Option B: $x = 1$

$a \square 1 = a \times 1 + (a + 1) = a + a + 1 = 2a + 1$

Option C: $x = 2$

$a \square 2 \Rightarrow a \times 2 + (a + 2) = 2a + a + 2 = 3a + 2$

Option D: $x = 10$

$a \square 10 \Rightarrow a \times 10 + (a + 10) = 10a + a + 10$

$= 11a + 10$

∴ Option 'A' only True.

16. $(9 \text{ inches})^{1/2} = (0.25 \text{ yards})^{1/2}$,

Solving we get 9 inch = 0.25 yards

(since 1 inch = 0.028 yard)

17. Given Venn diagram is

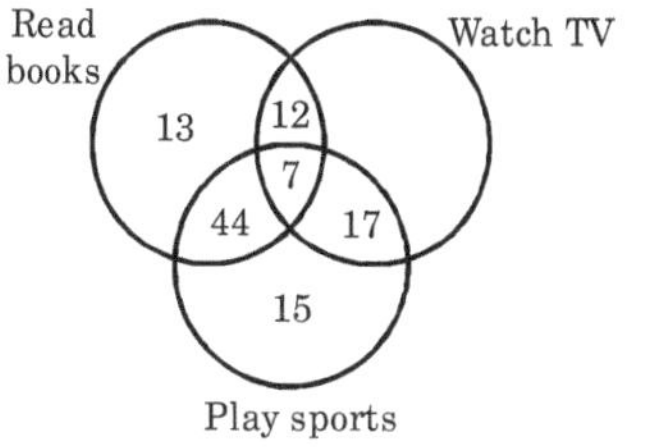

The number of students who like to read books or play sports

$$= 13 + 12 + 44 + 7 + 15 + 17 = 108$$

19. $A = \left[\frac{14}{3}, 0\right]$

$B = [0, 7]$,

$C = \left[\frac{5}{2}, 0\right]$

$D = \left[0, \frac{-5}{3}\right]$

$E = [4, 1]$,

$F = [0, 1]$

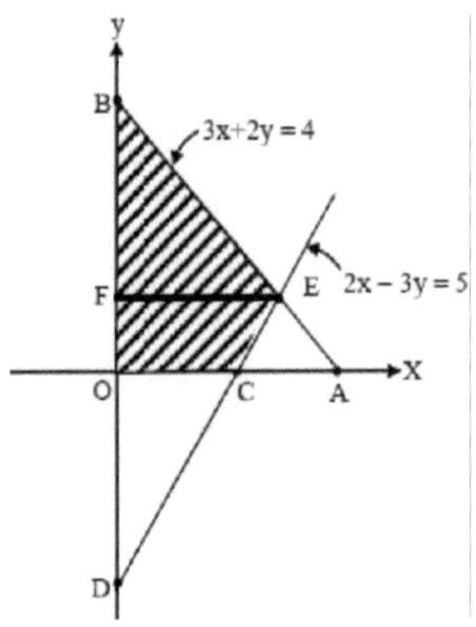

Required area = Area of Δ BFE + Area of quadrilateral FEOC

$$= \frac{1}{2} \times 4 \times 6 + \frac{1}{2} \times (4 + 2.5) \times 1$$

$$= 2 \times 6 + \frac{1}{2} \times 6.5$$

$= 12 + 3.25 = 15.25$ sq. units

So, the area bounded by the given lines is 15.25 sq. units.

20. The equation of a straight line is

$y = mx + c$

m = slope = – 0.02

set (log x, y)

Now, the line intercepts the abscissa at

$\log x = 0.1$

So, at abscissa $y = 0$

$\therefore \quad y = mx + c$

$0 = -0.02 \times 0.1 + c$

$\therefore \quad c = 0.002$

$y = mx + c$

$y = -0.02 \times \log x + c$

At $\quad x = 5$

$\therefore \quad y = -0.02 \times \log 5 + 0.002 = -0.030$

So, the value of y is – 0.030.

22. $\frac{5}{7} = x^{-1/3} \Rightarrow \frac{7}{5} = x^{1/3} \Rightarrow \left(\frac{7}{5}\right)^3 = x \Rightarrow x = \frac{343}{125}$

23. $f(x) = 2x + 33 = 2(5) + 33 = 43$

24. For whole numbers, greater the value greater will be its log.

Same logic for power of e.

25. $\log \tan 1° + \log \tan 89° = \log(\tan 1° \times \tan 89°)$

$= \log(\tan 1° \times \cot 1°)$

$= \log 1 = 0$

Using the same logic total sum is '0'.

27. The given function is

$f(x) = 1 - |x|$ on $-1 \le x \le 1$

In order to find maximum and minimum value of function in interval [–1, 1]

At $x = -1$, $\quad f(-1) = 1 - |-1| = 1 - 1 = 0$

At $x = -0.5$, $\quad f(-0.5) = 1 - |-0.5| = 1 - 0.5 = 0.5$

At $x = 0$, $\quad f(0) = 1 - |0| = 1$

At $x = 0.5$, $\quad f(0.5) = 1 - |0.5| = 0.5$

At $x = 1$, $\quad f(1) = 1 - |1| = 1 - 1 = 0$

∴ Hence, maximum value occurs at $x = 0$ and its maximum value is 1.

∴ Option (c) is correct.

30. $y = 5x^2 + 3$, $\frac{dy}{dx} = 10x$

Slope of tangent $= \left(\frac{dy}{dx}\right)_{x=0,\, y=3} = 10 \times 0 = 0$

Slope = 0

⇒ tangent is parallel to x-axis.

31. $$P' = 2^n X^n \left(\frac{1}{2}\right)^m y^m$$

$$= 2^{n-m} X^n Y^m = 2^{n-m} P$$

32. For a very large number of throws, the frequency should be same for unbiased throw. As it not same, then the die is baised.

33. $(7526)_8 - (y)_8 = (4364)_8$

$\Rightarrow y_8 = (7526)_8 - (4364)_8$

	4	(8 + 2=10)	
7	~~5~~	2	6
4	3	6	4
3	1	4	2

When we have base 8, we borrow 8 instead of 10 as done in normal subtraction.

34.

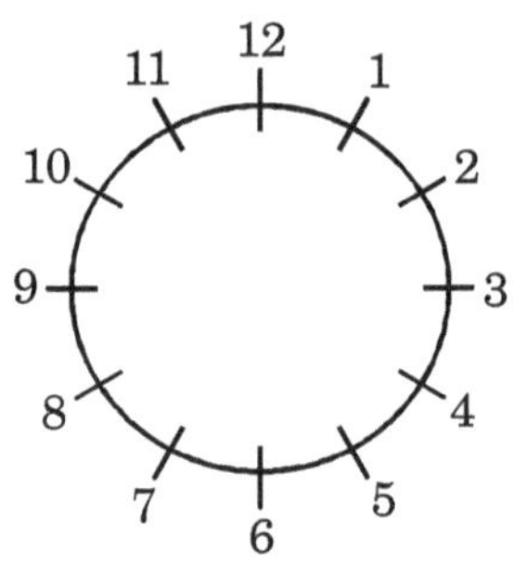

From option

At 6 a.m.

Both hand – 180° aparts,

At 6 : 20 a.m. – both hand apart

$\Rightarrow \quad 180° + 10° - 120° = 70°$

At 6 : 22 a.m. – both hand apart

$\Rightarrow \quad 180° + 12° - 132° = 60°$ (closest)

35. $ax^2 + bx + c = 0$

Roots are real and positive

$\therefore \quad b^2 - 4ac > 0$

By the above condition we get two positive real roots

Now, $ax^2 + b|x| + c = 0$

This can be written as;

$ax^2 + bx + c$

$\Delta = b^2 - 4ac > 0$

Now for $ax^2 - bx + c$

$(-b)^2 - 4ac$

$\Rightarrow \quad b^2 - 4ac$

Again $b^2 - 4ac > 0$

So again we get real roots

Thus we have 4 real roots of $ax^2 + b|x| + c = 0$

36. $P = 50q - 5q^2$

$$\frac{dp}{dq} = 50 - 10q; \ \frac{d^2p}{dq^2} < 0$$

$\therefore$ p is maximum at $50 - 10q = 0$ or, $q = 5$

Else check with options

37. P and R always holds true

Else consider a sample set {1, 2, 3, 4} and check accordingly

38. $y = 2x - 0.1x^2; \ \dfrac{dy}{dx} = 2 - 0.2x; \ \dfrac{d^2y}{dx^2} < 0$

$\therefore$ y maximises at $2 - 0.2x = 0$

$\Rightarrow x = 10$

$\therefore$ $y = 20 - 10 = 10$m

39. Let us categorize bags in three groups as

$A_1 A_2 A_3 \quad B_1 B_2 B_3 \quad C_1 C_2$

1st weighing A vs B

Case-1	**Case-2**
$A_1 A_2 A_3 = B_1 B_2 B_3$	$A_1 A_2 A_3 \neq B_1 B_2 B_3$
Then either C_1 or C_2 is heavier	Either A or B would be heavier (Say A > B)
2nd weighing	
C_1 vs C_2	A_1 vs A_2
If $C_1 > C_2$, then C_1	If $A_1 = A_2$, then A_3
If $C_1 < C_2$, then C_2	If $A_1 > A_2$, then A_1
If $A_1 < A_2$, then A_2	

40. $\log P = \dfrac{1}{2} \log Q$

$= \dfrac{1}{3} \log (R) = k$

$\therefore \quad P = b^k, Q = b^{2k}, R = b^{3k}$

Hence, $Q^2 = b^{4k} = b^{3k} b^k = PR$

41. Checking with all options in formula: $(4q+100/q)$ i.e. (V+F). Option (*a*) gives the minimum cost.

42. Let total number of toffees in bowl be x

R took $\frac{1}{3}$ of toffees and returned 4 to the bowl

$\therefore$ Number of toffees with R $= \dfrac{1}{3}x - 4$

Remaining of toffees in bowl $= \dfrac{2}{3}x + 4$

Number of toffees with S $= \dfrac{1}{4}\left(\dfrac{2}{3}x + 4\right) - 3$

Remaining toffees in bowl $= \dfrac{3}{4}\left(\dfrac{2}{3}x + 4\right) + 4$

Number of toffees with T $= \dfrac{1}{2}\left(\dfrac{3}{4}\left(\dfrac{2}{3}x + 4\right) + 4\right) + 2$

Remaining toffees in bowl

$$= \frac{1}{2}\left[\frac{3}{4}\left(\frac{2}{3}x+4\right) + 4\right] + 2$$

Given : $\dfrac{1}{2}\left[\dfrac{3}{4}\left(\dfrac{2}{3}x+4\right) + 4\right] + 2 = 17$

$\Rightarrow \quad \dfrac{3}{4}\left(\dfrac{2}{3}x+4\right) = 27$

$\Rightarrow \quad x = 48$

NUMERICAL TYPE QUESTIONS

1. Let area of triangle PQR be 'A'

$$\frac{SQ}{PQ} = \frac{1}{1+3} = \frac{1}{4}$$

$$\frac{QT}{QR} = \frac{2}{2+5} = \frac{2}{7}$$

$\therefore$ Area of $\Delta QTS = \dfrac{1}{2} \times SQ \times QT$

$$= \frac{1}{2} \times \left(\frac{1}{4}PQ\right) \times \left(\frac{2}{7}QR\right)$$

$$= \frac{1}{4} \times \frac{2}{7} \times \left(\frac{1}{2} \times PQ \times QR\right)$$

$$= \frac{1}{14} \times \text{Area of } \Delta PQR$$

Given $20 \text{ cm}^2 = \dfrac{1}{14} \times A$

$\therefore \quad A = 14 \times 20 = 280 \text{ cm}^2$

2. $fg(h(2, 5, 7, 3), 4, 6, 8) = fg(1, 4, 6, 8)$

$= f(1, 4, 6, 8) \times g(1, 4, 6, 8)$

$= 8 \times 1 = 8$

Technical Section

1 CHAPTER Mathematical Physics

2021

1. If $y_n(x)$ is a solution of the differential equation

$$y'' - 2xy' + 2ny = 0$$

where n is an integer and the prime ($'$) denotes differentiation with respect to x, then acceptable plot(s) of $\psi_n(x) = e^{-x^2/2} y_n(x)$, is (are)

(a)

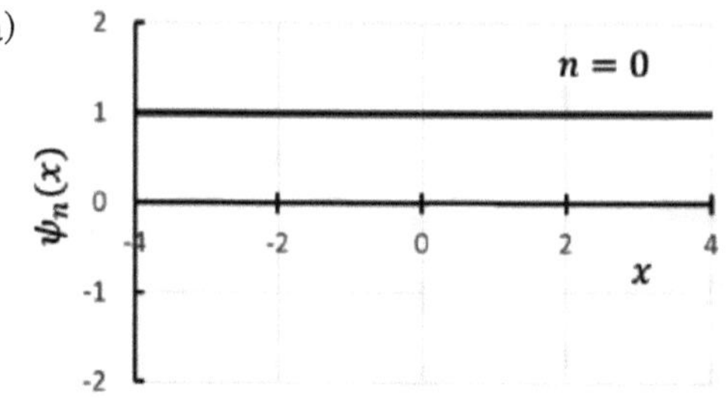

(b)

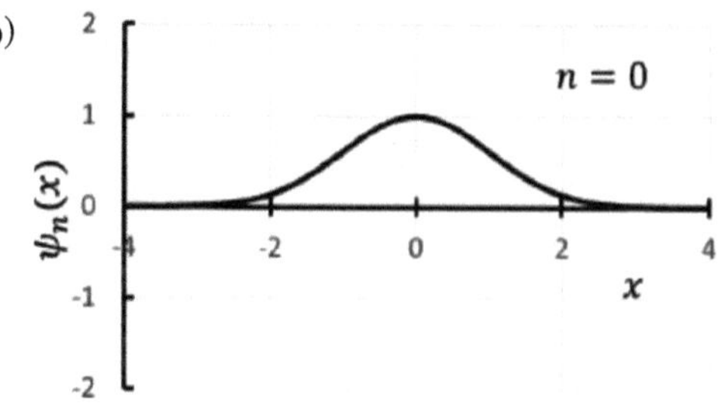

(c)

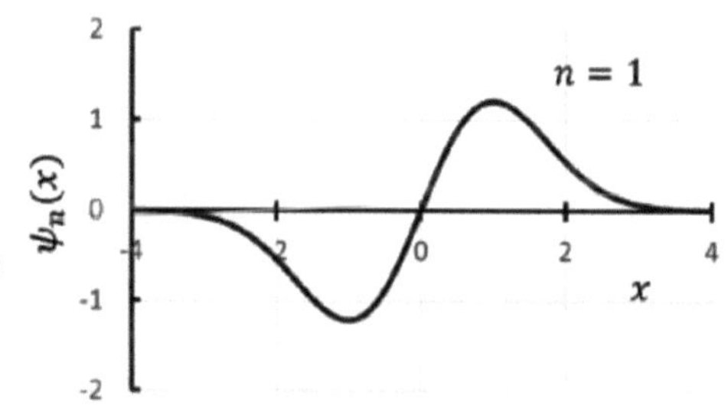

(d)

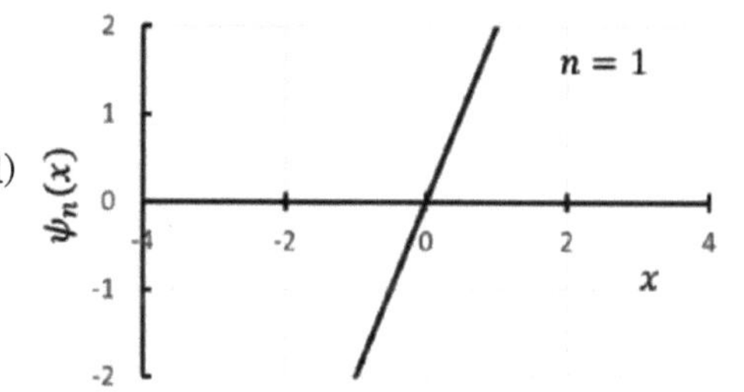

2. A function $f(t)$ is defined only for $t \geq 0$. The Laplace transform of $f(t)$ is

$$L(f;s) = \int_0^{\infty} e^{-st} f(t)dt$$

whereas the Fourier transform of $f(t)$ is

$$\tilde{f}(\omega) = \int_0^{\infty} f(t)e^{-i\omega t}dt.$$

The correct statement(s) is(are)

(a) The variable s is always real.

(b) The variable s can be complex.

(c) $L(f; s)$ and $\tilde{f}(\omega)$ can never be made connected.

(d) $L(f; s)$ and $\tilde{f}(\omega)$ can be made connected.

3. A contour intergral is defined as

$$I_n = \oint_c \frac{dz}{(z-n)^2 + \pi^2}$$

where n is a positive integer and C is the closed contour, as shown in the figure, consisting of the line from −100 to 100 and the semicircle traversed in the counter-clockwise sense.

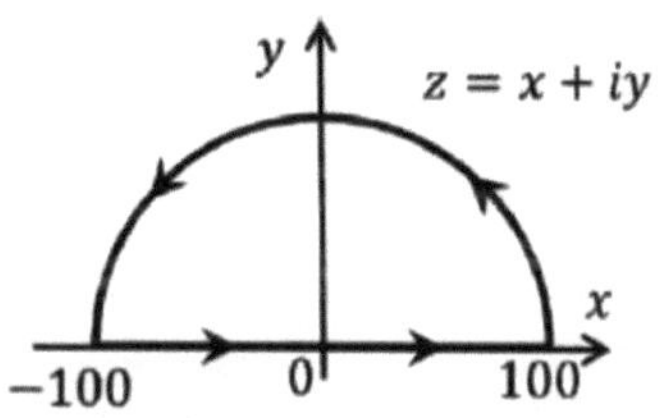

The value of $\sum_{n=1}^{5} I_n$ (in integer) is ______.

2020

4. Which one of the following is a solution of $\frac{d^2u(x)}{dx^2} = k^2u(x)$, for k real?

(a) e^{-kx} (b) $\sin kx$

(c) $\cos kx$ (d) $\sinh x$

5. For a complex variable z and the contour $c: |z| = 1$ taken in the counter clockwise direction, $\frac{1}{2\pi i}\oint_C \left(z - \frac{2}{z} + \frac{3}{z^2}\right)dz =$ _____.

6. The product of eigenvalues of $\begin{pmatrix} 0 & 0 & 1 \\ 0 & 1 & 0 \\ 1 & 0 & 0 \end{pmatrix}$ is

(a) –1 (b) 1

(c) 0 (d) 2

7. Let $|e_1\rangle \equiv \begin{pmatrix} 1 \\ 0 \\ 0 \end{pmatrix}, |e_2\rangle \equiv \begin{pmatrix} 1 \\ 1 \\ 0 \end{pmatrix}$ and $|e_3\rangle \equiv \begin{pmatrix} 1 \\ 1 \\ 1 \end{pmatrix}$, Let $S = \{|e_1\rangle, |e_2\rangle, |e_3\rangle\}$. Let $\mathbb{R}^3$ denote the three-dimensional real vector space. Which one of the following is correct?

(*a*) S is an orthonormal set

(*b*) S is a linearly dependent set

(c) S is a basis for $\mathbb{R}^3$

(*d*) $\sum_{i=1}^{3} |e_i\rangle\langle e_i| = \begin{pmatrix} 1 & 0 & 0 \\ 0 & 1 & 0 \\ 0 & 0 & 1 \end{pmatrix}$

8. If $x = \sum_{k=1}^{\infty} a_k \sin kx$, for $-\pi \le x \le \pi$, the value of a_2 is _______.

9. Let $f_x(x) = \begin{cases} 0, & x < -\frac{1}{2n} \\ n, & -\frac{1}{2n} < x < \frac{1}{2n} \\ 0, & \frac{1}{2n} < x. \end{cases}$

The value of $\lim_{x\to\infty} \int_{-\infty}^{\infty} f_n(x) \sin x \, dx$ is ______.

2019

10. For the differential equation

$$\frac{d^2y}{dx^2} - n(n+1)\frac{y}{x^2} = 0,$$

where n is a constant, the product of its two independent solutions is

(*a*) $\frac{1}{x}$ (*b*) x

(*c*) x^n (*d*) $\frac{1}{x^{n+1}}$

11. During a rotation, vectors along the axis of raotation remain unchanged. For the rotation matrix $\begin{pmatrix} 0 & 1 & 0 \\ 0 & 0 & -1 \\ -1 & 0 & 0 \end{pmatrix}$, the unit verctor along the axis of rotation is

(*a*) $\frac{1}{3}(2\hat{i} - \hat{j} + 2\hat{k})$ (*b*) $\frac{1}{\sqrt{3}}(\hat{i} + \hat{j} - \hat{k})$

(*c*) $\frac{1}{\sqrt{3}}(\hat{i} - \hat{j} - \hat{k})$ (*d*) $\frac{1}{3}(2\hat{i} + 2\hat{j} - \hat{k})$

12. The pole of the function f (Z) = cot z at z = 0 is

(*a*) a removable singularity

(*b*) an essential singularity

(*c*) a simple pole

(*d*) a second order pole

13. The value of the integral $\int_{-\infty}^{\infty} \frac{\cos(kx)}{x^2 + a^2} dx$, where k > 0 and a > 0, is

(*a*) $\frac{\pi}{\alpha}e^{-k\alpha}$ (*b*) $\frac{2\pi}{\alpha}e^{-k\alpha}$

(*c*) $\frac{\pi}{2\alpha}e^{-k\alpha}$ (*d*) $\frac{3\pi}{2\alpha}e^{-k\alpha}$

14. Let q be a variable in the range $-\pi \le \theta < \pi$. Now consider a function

$\psi(\theta) = 1$ for $-\pi/2 \le \theta < \pi/2$

$= 0$ otherwise.

If its Fourier-series is written as $\psi(\theta) = \sum_{m}^{\infty} = -Cme^{-im\theta}$, then the value of $|C_3|^2$ (rounded off to three decimal places) is_____________

2018

15. Given $\vec{V}_1 = \hat{\imath} - \hat{\jmath}$ and $\vec{V}_2 = -2\hat{\imath} + 3\hat{\jmath} + 2\hat{k}$, which one of the following $\vec{V}_3$ makes $(\vec{V}_1, \vec{V}_2, \vec{V}_3)$ a complete set for a three dimensional real linear vector space?

(*a*) $\vec{V}_3 = \hat{\imath} + \hat{\jmath} + 4\hat{k}$ (*b*) $\vec{V}_3 = 2\hat{\imath} - \hat{\jmath} + 2\hat{k}$

(*c*) $\vec{V}_3 = \hat{\imath} + 2\hat{\jmath} + 6\hat{k}$ (*d*) $\vec{V}_3 = 2\hat{\imath} + \hat{\jmath} + 4\hat{k}$

16. Given $\frac{d^2f(x)}{dx^2} - 2\frac{df(x)}{dx} + f(x) = 0$, and boundary conditions $f(0) = 1$ and $f(1) = 0$, the value of $f(0.5)$ is ________ (up to two decimal places).

17. The absolute value of the integral $\int \frac{5z^3+3z^2}{z^2-4}dz,$ over the circle $|z-1.5| = 1$ in complex plane, is ______ (up to two decimal places).

2017

18. The contour integral $\oint \frac{dz}{1+z^2}$ evaluated along a contour going from $-\infty$ to $+\infty$ along the real axis and closed in the lower half-plane by a half circle is equal to ______. (up to two decimal places).

19. The coefficient of e^{ikx} in the Fourier expansion of $u(x) = A\sin^2(\alpha x)$ for $k = -2\alpha$ is

(a) $\frac{A}{4}$ (b) $-\frac{A}{4}$

(c) $\frac{A}{2}$ (d) $-\frac{A}{2}$

20. The integral $\int_0^\infty x^2 e^{-x^2} dx$ is equal to ______. (up to two decimal places).

21. The imaginary part of an analytic complex function is $v(x, y) = 2xy + 3y$. The real part of the function is zero at the origin. The value of the real part of the function at $1 + i$ is ______. (up to two decimal places).

22. Consider the differential equation $\frac{dy}{dx} + y\tan(x) = \cos(x)$. If $y(0) = 0$, $y\left(\frac{\pi}{3}\right)$ is ______. (up to two decimal places).

2016

23. Consider the linear differential equation $\frac{dy}{dx} = xy$. If y = 2 at x = 0, then the value of y at x = 2 is given by

(a) e^{-2} (b) $2e^{-2}$

(c) e^2 (d) $2e^2$

24. Which of the following is an analytic function of z everywhere in the complex plane?

(a) z^2 (b) $(z^*)^2$

(c) $|z|^2$ (d) $\sqrt{z}$

25. Which of the following operators is Hermitian?

(a) $\frac{d}{dx}$ (b) $\frac{d^2}{dx^2}$

(c) $i\frac{d^2}{dx^2}$ (d) $\frac{d^3}{dx^3}$

26. The direction of $\vec{\nabla}f$ for a scalar field $f(x, y, z) = \frac{1}{2}x^2 - xy + \frac{1}{2}z^2$ at the point P(1, 1, 2) is

(a) $\frac{(-\hat{j}-2\hat{k})}{\sqrt{5}}$ (b) $\frac{(-\hat{j}+2\hat{k})}{\sqrt{5}}$

(c) $\frac{(\hat{j}-2\hat{k})}{\sqrt{5}}$ (d) $\frac{(\hat{j}+2\hat{k})}{\sqrt{5}}$

27. σ_x, σ_y and σ_z are the Pauli matrices. The expression $2\sigma_x\sigma_y + \sigma_y\sigma_x$ is equal to

(a) $-3i\sigma_z$ (b) $-i\sigma_z$

(c) $i\sigma_z$ (d) $3i\sigma_z$

28. A periodic function f(x) of period 2π is defined in the interval $(-\pi < x < \pi)$ as:

$$f(x) = \begin{cases} -1, -\pi < x < 0 \\ 1, 0 < x < \pi \end{cases}$$

The appropriate Fourier series expansion for f(x) is

(a) $f(x) = (4/\pi)$ [sin x + (sin 3x) / 3 + (sin 5x) / 5 + ...]

(b) $f(x) = (4/\pi)$ [sin x − (sin 3x) / 3 + (sin 5x) / 5 − ....]

(c) $f(x) = (4/\pi)$ [cos x + (sin 3x) / 3 + (cos 5x) / 5 + ...]

(d) $f(x) = (4/\pi)$ [cos x − (cos 3x) / 3 + (cos 5x) / 5 − ....]

2015

29. Consider a complex function

$$f(z) = \frac{1}{z\left(z+\frac{1}{2}\right)\cos(z\pi)}.$$

Which one of the following statements is correct?

(a) $f(z)$ has simple poles at $z = 0$ and $z = -\frac{1}{2}$

(b) $f(z)$ has a second order pole at $z = -\frac{1}{2}$

(c) $f(z)$ has infinite number of second order poles

(d) $f(z)$ has all simple poles

30. The value of $\int_0^3 t^2\delta(3t-6)dt$ is ______ (upto one decimal place)

31. If $f(x) = e^{-x^2}$ and $g(x) = |x|e^{-x^2}$, then

(a) f and g are differentiable everywhere

(b) f is differentiable everywhere but g is not

(c) g is differentiable everywhere but f is not

(d) g is discontinuous at $x = 0$

32. Consider $w = f(z) = u(x, y) + iv(x, y)$ to be an analytic function in a domain D. Which one of the following options is NOT correct?

(*a*) $u(x, y)$ satisfies Laplace equation in D

(*b*) $v(x, y)$ satisfies Laplace equation in D

(*c*) $\int_{z_1}^{z_2} f(z)dz$ is dependent on the choice of the contour between z_1 and z_2 in D

(*d*) $f(z)$ can be Taylor expanded in D

33. A function $y(z)$ satisfies the ordinary differential equation $y'' + \frac{1}{z}y' - \frac{m^2}{z^2}y = 0$, where $m = 0, 1, 2, 3,\ldots..$ Consider the four statement P, Q, R, S as given below.

P : z^m and z^{-m} are linearly independent solutions for all values of m

Q : z^m and z^{-m} are linearly independent solutions for all values of $m>0$

R : In z and 1 are linearly independent solutions for $m = 0$

S : z^m and z are linearly independent solutions for all values of m

The correct option for the combination of valid statements is

(*a*) P, R and S only

(*b*) P and R only

(*c*) Q and R only

(*d*) R and S only

34. The Heaviside function is defined as $H(t) = \begin{cases} +1 & \text{for } t > 0 \\ -1 & \text{for } t < 0 \end{cases}$ and its Fourier transform is given $-2i/\omega$. The Fourier transform of $\frac{1}{2}\left[H\left(\frac{t+1}{2}\right) - H\left(\frac{t-1}{2}\right)\right]$ is

(*a*) $\dfrac{\sin\left(\frac{\omega}{2}\right)}{\frac{\omega}{2}}$ (*b*) $\dfrac{\cos\left(\frac{\omega}{2}\right)}{\frac{\omega}{2}}$

(*c*) $\sin\left(\frac{\omega}{2}\right)$ (*d*) 0

2014

35. The unit vector perpendicular to the surface $x^2 + y^2 + z^2 = 3$ at the point (1,1,1) is

(*a*) $\dfrac{\hat{x}+\hat{y}-\hat{z}}{\sqrt{3}}$ (*b*) $\dfrac{\hat{x}-\hat{y}-\hat{z}}{\sqrt{3}}$

(*c*) $\dfrac{\hat{x}-\hat{y}+\hat{z}}{\sqrt{3}}$ (*d*) $\dfrac{\hat{x}+\hat{y}+\hat{z}}{\sqrt{3}}$

36. The matrix

$$A = \frac{1}{\sqrt{3}}\begin{bmatrix} 1 & 1+i \\ 1-i & -1 \end{bmatrix} \text{ is}$$

(*a*) orthogonal (*b*) symmetric

(*c*) anti-symmetric (*d*) unitary

37. The length element ds of an arc is given by, $(ds)^2 = 2(dx^1)^2 + (dx^2)^2 + \sqrt{3}\, dx^1 dx^2$. The metric tensor g_{ij} is

(*a*) $\begin{pmatrix} 2 & \sqrt{3} \\ \sqrt{3} & 1 \end{pmatrix}$ (*b*) $\begin{pmatrix} 2 & \sqrt{\frac{3}{2}} \\ \sqrt{\frac{3}{2}} & 1 \end{pmatrix}$

(*c*) $\begin{pmatrix} 2 & 1 \\ \sqrt{\frac{3}{2}} & \sqrt{\frac{3}{2}} \end{pmatrix}$ (*d*) $\begin{pmatrix} 1 & \sqrt{\frac{3}{2}} \\ \sqrt{\frac{3}{2}} & 2 \end{pmatrix}$

38. The value of the integral

$$\oint_C \frac{z^2}{e^z + 1} dz,$$

where C is the circle $|z| = 4$, is

(*a*) $2\pi i$ (*b*) $2\pi^2 i$

(*c*) $4\pi^3 i$ (*d*) $4\pi^2 i$

39. The solution of the differential equation

$$\frac{d^2y}{dt^2} - y = 0,$$

subject to the boundary conditions $y(0) = 1$ and $y(\infty) = 0$, is

(*a*) $\cos t + \sin t$ (*b*) $\cosh t + \sinh t$

(*c*) $\cos t - \sin t$ (*d*) $\cosh t - \sinh t$

2013

40. $f(x)$ is a symmetric periodic function of x i.e. $f(x) = f(-x)$. Then, in general, the Fourier series of the function $f(x)$ will be of the form

(*a*) $f(x) = \sum_{n=1}^{\infty} (a_n \cos(nkx) + b_n \sin(nkx))$

(*b*) $f(x) = a_0 + \sum_{n=1}^{\infty} (a_n \cos(nkx))$

(*c*) $f(x) = \sum_{n=1}^{\infty} (b_n \sin(nkx))$

(*d*) $f(x) = a_0 + \sum_{n=1}^{\infty} (b_n \sin(nkx))$

41. In the most general case, which one of the following quantities is NOT a second order tensor?

(*a*) Stress (*b*) Strain

(*c*) Moment of inertia (*d*) Pressure

42. For a scalar function φ satisfying the Laplace equation, $\nabla\varphi$ has

(*a*) zero curl and non-zero divergence

(*b*) non-zero curl and zero divergence

(*c*) zero curl and zero divergence

(*d*) non-zero curl and non-zero divergence

43. Which of the following pairs of the given function $F(t)$ and its Laplace transform $f(s)$ is NOT CORRECT?

(*a*) $F(t) = \delta(t)$, $f(s) = 1$,(Singularity at +0)

(*b*) $F(t) = 1$, $f(s) = \dfrac{1}{s}$, $(s > 0)$

(*c*) $F(t) = \sin kt$, $f(s) = \dfrac{s}{s^2+k^2}$, $(s > 0)$

(*d*) $F(t) = te^{kt}$, $f(s) = \dfrac{1}{(s-k)^2}$, $(s > k, s > 0)$

44. If $\vec{A}$ and $\vec{B}$ are constant vectors, then $\nabla(\vec{A}\cdot\vec{B}\times\vec{r})$ is

(*a*) $\vec{A}\cdot\vec{B}$ (*b*) $\vec{A}\times\vec{B}$

(*c*) $\vec{r}$ (*d*) Zero

45. $\Gamma\left(n+\dfrac{1}{2}\right)$ is equal to [Given $\Gamma(n+1) = n\,\Gamma(n)$ and $\Gamma(1/2) = \sqrt{2}$]

(*a*) $\dfrac{n!}{2^n}\sqrt{\pi}$ (*b*) $\dfrac{2n!}{n!2^n}\sqrt{\pi}$

(*c*) $\dfrac{2n!}{n!2^{2n}}\sqrt{\pi}$ (*d*) $\dfrac{n!}{2^{2n}}\sqrt{\pi}$

46. For the function $f(z) = \dfrac{16z}{(z+3)(z-1)^2}$, the residue at the pole $z = 1$ is (your answer should be an integer)__________.

2012

47. Identify the CORRECT statement for the following vectors $\vec{a} = 3\hat{i} + 2\hat{j}$ and $\vec{b} = \hat{i} + 2\hat{j}$

(*a*) The vectors $\vec{a}$ and $\vec{b}$ are linearly independent

(*b*) The vectors $\vec{a}$ and $\vec{b}$ are linearly dependent

(*c*) The vectors $\vec{a}$ and $\vec{b}$ are orthogonal

(*d*) The vectors $\vec{a}$ and $\vec{b}$ are normalized

48. The number of independent components of the symmetric tensor A_{ij} with indices $i, j = 1, 2, 3$ is

(*a*) 1 (*b*) 3

(*c*) 6 (*d*) 9

49. The eigenvalues of the matrix $\begin{pmatrix} 0 & 1 & 0 \\ 1 & 0 & 1 \\ 0 & 1 & 0 \end{pmatrix}$ are

(*a*) 0, 1, 1 (*b*) $0, -\sqrt{2}, \sqrt{2}$

(*c*) $\dfrac{1}{\sqrt{2}}, \dfrac{1}{\sqrt{2}}, 0$ (*d*) $\sqrt{2}, \sqrt{2}, 0$

50. Given $\vec{F} = \vec{r}\times\vec{B}$, where $\vec{B} = B_0(\hat{i}+\hat{j}+\hat{k})$ is a constant vector and $\vec{r}$ is the position vector. The value of $\oint_C \vec{F}\cdot d\vec{r}$, where C is a circle of unit radius centered at origin is,

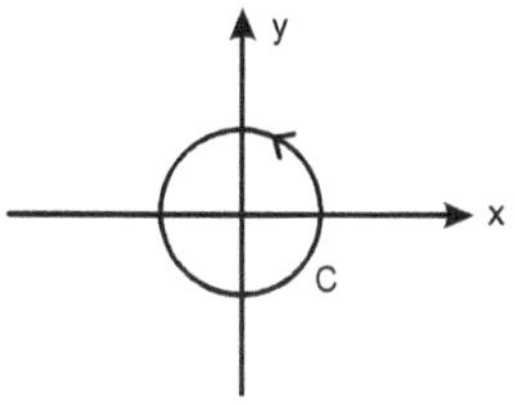

(*a*) 0 (*b*) $2\pi B_0$

(*c*) $-2\pi B_0$ (*d*) 1

51. The value of the integral $\oint_C e^{1/z}dz$, using the contour C of circle with unit radius $|z| = 1$ is

(*a*) 0 (*b*) $1 - 2\pi i$

(*c*) $1 + 2\pi i$ (*d*) $2\pi i$

2011

52. If a force $\vec{F}$ is derivable from a potential function $V(r)$, where r is the distance from the origin of the coordinate system, it follows that

(*a*) $\vec{\nabla}\times\vec{F} = 0$ (*b*) $\vec{\nabla}\cdot\vec{F} = 0$

(*c*) $\vec{\nabla}V = 0$ (*d*) $\nabla^2 V = 0$

53. The unit vector normal to the surface $x^2 + y^2 - z = 1$ at the point P(1, 1, 1) is

(*a*) $\dfrac{\hat{i}+\hat{j}-\hat{k}}{\sqrt{3}}$ (*b*) $\dfrac{2\hat{i}+\hat{j}-\hat{k}}{\sqrt{6}}$

(*c*) $\dfrac{\hat{i}+2\hat{j}-\hat{k}}{\sqrt{6}}$ (*d*) $\dfrac{2\hat{i}+2\hat{j}-\hat{k}}{3}$

54. Consider a cylinder of height h and radius a, closed at both ends, centered at the origin. Let $\hat{i}x + \hat{j}y + \hat{k}z$ be the position vector and $\hat{n}$ a unit vector normal to the surface. The surface integral $\int_S \vec{r}\cdot\hat{n}\,ds$ over the closed surface of the cylinder is

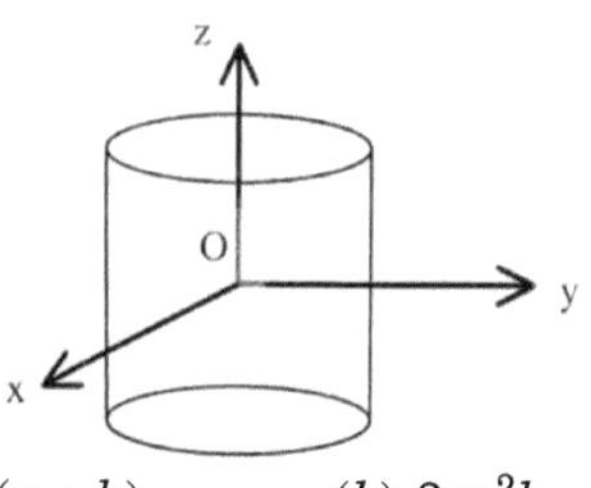

(*a*) $2\pi a^2(a + h)$ (*b*) $3\pi a^2 h$

(*c*) $2\pi a^2 h$ (*d*) zero

55. The solutions to the differential equation

$$\frac{dy}{dx} = -\frac{x}{y+1}$$

are a family of

(*a*) circles with different radii

(*b*) circles with different centres

(*c*) straight lines with different slopes

(*d*) straight lines with different intercepts on the *y*-axis

Common Data for Questions 56 and 57:

Consider a function $f(z) = \frac{z\sin z}{(z-\pi)^2}$ of a complex variable z.

56. Which of the following statements is **TRUE** for the function $f(z)$?

(*a*) $f(z)$ is analytic everywhere in the complex plane

(*b*) $f(z)$ has a zero at $z = \pi$

(*c*) $f(z)$ has a pole of order 2 at $z = \pi$

(*d*) $f(z)$ has a simple pole at $z = \pi$

57. Consider a counterclockwise circular contour $|z| = 1$ about the origin. The integral $\oint f(z)dz$ over this contour is

(*a*) $-i\pi$ (*b*) zero

(*c*) $i\pi$ (*d*) $2i\pi$

2010

Some useful physical constants

Speed of light	$c = 3 \times 10^8$ ms^{-1}
Plank's constant	$h = 6.63 \times 10^{-34}$ Js
Boltzmann constant	$k_B = 1.38 \times 10^{-23}$ JK^{-1}
Charge of electron	$e = 1.6 \times 10^{-19}$C

58. Consider an anti-symmetric tensor P_{ij} with the indices i and j running from 1 to 5. The number of independent components of the tensor is

(*a*) 3 (*b*) 10

(*c*) 9 (*d*) 6

59. The value of the integral $\oint_C \frac{e^z \sin(z)}{z^2} dz$, where the contour C is the unit circle: $|z-2| = 1$, is

(*a*) $2\pi i$ (*b*) $4\pi i$

(*c*) πi (*d*) 0

60. If $f(x)\begin{cases} 0 & \text{for } x < 3, \\ x-3 & \text{for } x \geq 3, \end{cases}$ then the Laplace transform of $f(x)$ is

(*a*) $s^{-2}e^{3s}$ (*b*) s^2e^{-3s}

(*c*) s^{-2} (*d*) $s^{-2}e^{-3s}$

61. For the complex function, $f(z) = \frac{e^{\sqrt{z}} - e^{-\sqrt{z}}}{\sin(\sqrt{z})}$, which of the following statements is correct?

(*a*) $z = 0$ is a branch point

(*b*) $z = 0$ is a pole of order one

(*c*) $z = 0$ is a removable singularity

(*d*) $z = 0$ is an essential singularity

62. The solution of the differential equation for $y(t)$: $\frac{d^2y}{dt^2} - y = 2\cosh(t)$, subject to the initial conditions $y(0) = 0$ and $\left.\frac{dy}{dt}\right|_{t=0} = 0$, is

(*a*) $\frac{1}{2}\cosh(t) + t\sinh(t)$ (*b*) $-\sinh(t) + t\cosh(t)$

(*c*) $t\cosh(t)$ (*d*) $t\sinh(t)$

63. Given the recurrence relation for the Legendre polynomials

$(2n+1)\,x\,P_n(x) = (n+1)\,P_{n+1}(x) + n\,P_{n-1}(x)$,

which of the following integrals has a non-zero value?

(*a*) $\int_{-1}^{+1} x^2 P_n(x)P_{n+1}(x)\,dx$ (*b*) $\int_{-1}^{+1} x\,P_n(x)P_{n+2}(x)\,dx$

(*c*) $\int_{-1}^{+1} x\,[P_n(x)]^2\,dx$ (*d*) $\int_{-1}^{+1} x^2\,P_n(x)P_{n+2}(x)\,dx$

2009

64. The value of the contour integral, $\left|\int_C \vec{r} \times d\vec{\theta}\right|$, for a circle C of radius r with centre at the origin is

(*a*) $2\pi r$ (*b*) $\frac{r^2}{2}$

(*c*) πr^2 (*d*) r

65. Which one of the following curves gives the solution of the differential equation $k_1\frac{dx}{dt} + k_2x = k_3$, where k_1, k_2 and k_3 are positive constants with initial conditions $x = 0$ at $t = 0$?

(*a*)

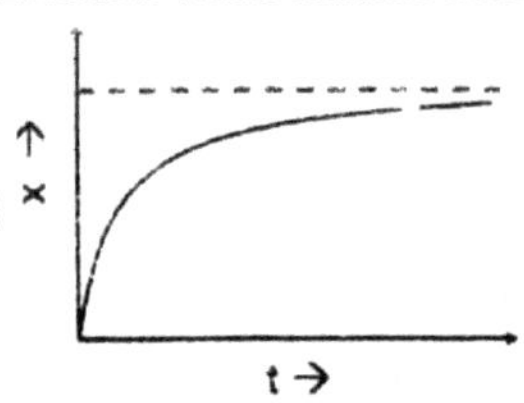

(*b*)

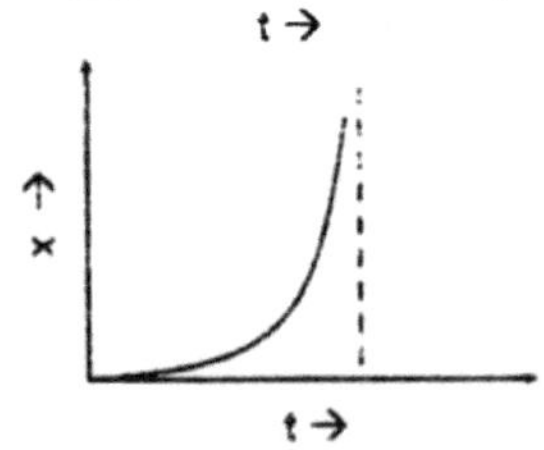

(c) 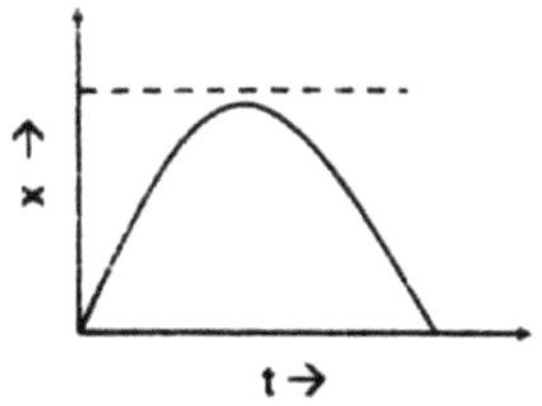

(d) 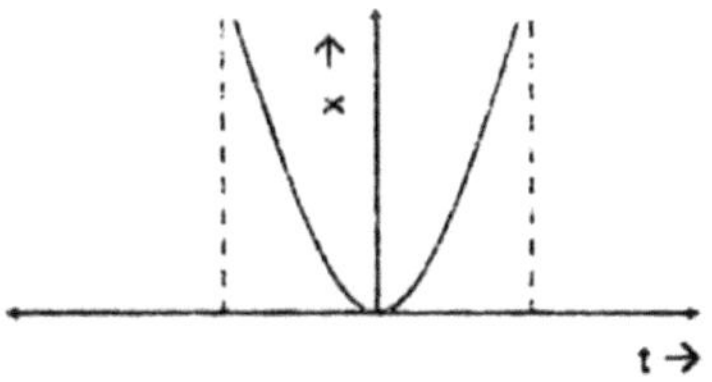

66. The value of the integral $\int_C \frac{e^z}{z^2 - 3z + 2} dz$, where the contour C is the circle $|z| = \frac{3}{2}$ is

(a) $2\pi ie$ (b) πie

(c) $-2\pi ie$ (d) $-\pi ie$

67. Let $T_{ij} = \sum_k \varepsilon_{ijk} a_k$ and $\beta_k = \sum_{i,j} \varepsilon_{ijk} T_{ij}$, where ε_{ijk} is the Levi-Civita density, defined to be zero if two of the indices coincide and +1 and –1 depending on whether *ijk* is even or odd permutation of 1, 2, 3. Then β_3 is equal to

(a) $2a_3$ (b) $-2a_3$

(c) a_3 (d) $-a_3$

68. Consider the set of vectors in three-dimensional real vector space

$\mathbb{R}^3$, S = {(1, 1, 1), (1, – 1, 1), (1, 1, —1)}. Which one of the following statements is true?

(a) S is not a linearly independent set.

(b) S is a basis for $\mathbb{R}^3$.

(c) The vectors in S are orthogonal.

(d) An orthogonal set of vectors cannot be generated from S.

2008

69. An unitary matrix $\begin{pmatrix} ae^{i\alpha} & b \\ ce^{i\beta} & d \end{pmatrix}$ is given, where *a*, *b*, *c*, *d* α ad β are real. The inverse of the matrix is

(a) $\begin{pmatrix} ae^{i\alpha} & -ce^{i\beta} \\ b & d \end{pmatrix}$ (b) $\begin{pmatrix} ae^{i\alpha} & ce^{i\beta} \\ b & d \end{pmatrix}$

(c) $\begin{pmatrix} ae^{-i\alpha} & b \\ ce^{-i\beta} & d \end{pmatrix}$ (d) $\begin{pmatrix} ae^{-i\alpha} & ce^{-i\beta} \\ b & d \end{pmatrix}$

70. The curl of a vector field $\vec{F}$ is $2\hat{x}$. Identify the appropriate vector field $\vec{F}$ from the choices given below.

(a) $\vec{F} = 2z\hat{x} + 3z\hat{y} + 5y\hat{z}$ (b) $\vec{F} = 3z\hat{y} + 5y\hat{z}$

(c) $\vec{F} = 3x\hat{y} + 5y\hat{z}$ (d) $\vec{F} = 2\hat{x} + 5y\hat{z}$

71. The eigenvalues of the matrix $\begin{pmatrix} \cos\theta & -\sin\theta \\ \sin\theta & \cos\theta \end{pmatrix}$ are

(a) $\frac{1}{2}(\sqrt{3} \pm i)$ when θ = 45°

(b) $\frac{1}{2}(\sqrt{3} \pm i)$ when θ = 30°

(c) ±1 since the matrix is unitary

(d) $\frac{1}{\sqrt{2}}(1 \pm i)$ when θ = 30°

72. If the Fourier transform $F[\delta(x - a)] = \exp(-i2\pi v a)$, the $F^{-1}(\cos 2\pi av)$ will correspond to

(a) $\delta(x - a) - \delta(x + a)$

(b) a constant

(c) $\frac{1}{2}[\delta(x-a) + i\delta(x+a)]$

(d) $\frac{1}{2}[\delta(x-a) + \delta(x+a)]$

73. If $I \equiv \oint dz \text{Ln}(z)$, where C is the unit circle taken anticlockwise and Ln(*z*) is the principal branch of the Logarithm function, which one of the following is correct?

(a) I = 0 by residue theorem.

(b) I is not defined since Ln(*z*) has a branch cut.

(c) I ≠ 0

(d) $\oint dz \text{Ln}(z^2) = 2I$

74. The value of $\int_{-i}^{i} \pi(z+1) dz$ is

(a) 0 (b) $2\pi i$

(c) $-2\pi i$ (d) $(-1 + 2i)\pi$

75. Consider the Bessel equation (*v* = 0), $\frac{d^2y}{dz^2} + \frac{1}{z}\frac{dy}{dz} + y = 0$. Which one of the following statements is correct?

(a) Equation has regular singular points at *z* = 0 and *z* = ∞

(b) Equation has 2 linearly independent solutions that are entire.

(c) Equaton has an entire solution and a second linearly independent solution singular at *z* = 0.

(d) Limit *z* → ∞, taken along *x* axis, exists for botgh the linearly independent solutions.

76. Under a certain rotation of coordinate axes, a rank-1 tensor $v_a (a = 1, 2, 3)$ transforms according to the orthogonal transformation defined by the relations

$$v_1 = \frac{1}{\sqrt{2}}(v_1 + v_2); v_2 = \frac{1}{\sqrt{2}}(-v_1 + v_2); v_3 = v_3.$$

Under the same rotation a rank- tensor $T_{a,b}$ would transform such that

(a) $T_{1,1} = T_{1,1}T_{1,2}$

(b) $T_{1,1} = T_{1,1}$

(c) $T_{1,1} = T_{1,1} + T_{2,2} - T_{2,1}$

(d) $T_{1,1} = \frac{1}{2}(T_{1,1} + T_{2,2} + T_{1,2} + T_{2,1})$

Statement for Linked Answer Questions 77 and 78:

The set V of all polynomials of a real variable x of degree two or less and with real coefficients, constitutes a real linear vector space $V \equiv \{c_0 + c_1x + c_2x^2 : c_0, c_1, c_2 \in R\}$.

77. $f(x) = a_0 + a_1x + a_2x^2 \in V$ and

$g(x) = b_0 + b_1x + b_2x^2 \in V$,

which one of the following constitutes an accpetable scalar product?

(a) $(f, g) = a_0^2b_0 + a_1^2b_1 + a_2^2b_2$

(b) $(f, g) = a_0^2b_0^2 + a_1^2b_1^2 + a_2^2b_2^2$

(c) $(f, g) = a_0b_0 - a_1b_1 + a_2b_2$

(d) $(f, g) = a_0b_0 + \frac{a_1b_1}{2} + \frac{a_2b_2}{3}$

78. Using the scalar product obtained in the above question, identify the subspace of V that is orthogonal to $(1 + x)$:

(a) $\{f(x) : b(1-x) + cx^2; b, c \in R\}$

(b) $\{f(x) = b(1-2x) + cx^2; b, c \in R\}$

(c) $\{f(x) : b + cx^2; b, c \in R\}$

(d) $\{f(x) : bx + cx^2; b, c \in R\}$

2007

79. The eigenvalues of a matrxi are i, $2i$ and $3i$. The matrix is

(a) unitary (b) anti-unitary.

(c) Hermitian. (d) anti-Hermitian.

80. A space station moving in a circular orbit around the Earth goes into a new bound orbit by firing its engine radially outwards. This orbit is

(a) a larger circle. (b) a smaller circle.

(c) an ellipse. (d) a parabola.

81. The eigenvalues and eignvectors of the matrix $\begin{bmatrix} 5 & 4 \\ 1 & 2 \end{bmatrix}$ are

(a) 6, 1 and $\begin{bmatrix} 4 \\ 1 \end{bmatrix}, \begin{bmatrix} 1 \\ -1 \end{bmatrix}$

(b) 2, 5 and $\begin{bmatrix} 4 \\ 1 \end{bmatrix}, \begin{bmatrix} 1 \\ -1 \end{bmatrix}$

(c) 6, 1, and $\begin{bmatrix} 1 \\ 4 \end{bmatrix}, \begin{bmatrix} 1 \\ -1 \end{bmatrix}$

(d) 2, 5 and $\begin{bmatrix} 1 \\ 4 \end{bmatrix}, \begin{bmatrix} 1 \\ -1 \end{bmatrix}$

82. The contour integral $\oint \frac{dz}{z^2 + a^2}$ is to be evaluated on a circle of radius $2a$ centered at the origin. It will have contributions only from the points.

(a) $\frac{1+i}{\sqrt{2}}a$ and $-\frac{1+i}{\sqrt{2}}a$

(b) ia and $-ia$

(c) $ia, -ia, \frac{1+i}{\sqrt{2}}a$ and $-\frac{1-i}{\sqrt{2}}a$

(d) $\frac{1+i}{\sqrt{2}}a, -\frac{1+i}{\sqrt{2}}a, \frac{1-i}{\sqrt{2}}a$ and $-\frac{1-i}{\sqrt{2}}a$

83. The points, where the series solution of the Legendre differential equation

$$(1-x^2)\frac{d^2y}{dx^2} - 2x\frac{dy}{dx} + \frac{3}{2}\left(\frac{3}{2}+1\right)y = 0$$ will diverge, are located at

(a) 0 and 1 (b) 0 and –1

(c) –1 and 1 (d) $\frac{3}{2}$ and $\frac{5}{2}$

84. Solution of the differential equation $x\frac{dy}{dx} + y = x^4$, with the boundary condition that $y = 1$, at $x = 1$, is

(a) $y = 5x^4 - 4$ (b) $y = \frac{x^4}{5} + \frac{4x}{5}$

(c) $y = \frac{4x^4}{5} + \frac{1}{5x}$ (d) $y = \frac{x^4}{5} + \frac{4}{5x}$

2006

85. The trace of a 3 × 3 matrix is 2. Two of its eigenvalues are 1 and 2. The third eigenvalue is

(a) – 1 (b) 0

(c) 1 (d) 2

86. The value of $\oint \vec{A}.\vec{dl}$ along a square loop of side L in a uniform field $\vec{A}$ is

(a) 0 (b) 2LA

(c) 4LA (d) L^2 A

87. A linear transformation T, defined as $T\begin{pmatrix}x_1\\x_2\\x_3\end{pmatrix}=\begin{pmatrix}x_1+x_2\\x_2-x_3\end{pmatrix}$, transforms a vector $\vec{x}$ for as three dimensional real space to a two-dimensional real space. The transformaiton matrix T is

(*a*) $\begin{pmatrix}1 & 1 & 0\\0 & 1 & -1\end{pmatrix}$ (*b*) $\begin{pmatrix}1 & 0 & 0\\0 & 1 & 0\end{pmatrix}$

(*c*) $\begin{pmatrix}1 & 1 & 1\\-1 & 1 & 1\end{pmatrix}$ (*d*) $\begin{pmatrix}1 & 0 & 0\\0 & 0 & 1\end{pmatrix}$

88. The value of $\oint_S \frac{\vec{r}.d\vec{S}}{r^3}$, where $\vec{r}$ is the position vector and S is a closed surface enclosing the origin, is

(*a*) 0 (*b*) π

(*c*) 4π (*d*) 8π

89. The value of $\oint_C \frac{e^{2z}}{(z+1)^4}dz$, where C is a circle defined by $|z| = 3$, is

(*a*) $\frac{8\pi i}{3}e^{-2}$ (*b*) $\frac{8\pi i}{3}e^{-1}$

(*c*) $\frac{8\pi i}{3}e$ (*d*) $\frac{8\pi i}{3}e^{2}$

90. The kth Fourier component of $f(x) = \delta(x)$ is

(*a*) 1 (*b*) 0

(*c*) $(2\pi)^{-1/2}$ (*d*) $(2\pi)^{-3/2}$

91. An atom with net magnetic moment $\vec{\mu}$ and net angular momentum $\vec{L}(\vec{\mu}=\gamma\vec{L})$ is kept in a uniform magnetic induction $\vec{B}=B_0\hat{k}$. The magnetic moment $\vec{\mu}(=\mu_x)$ is

(*a*) $\frac{d^2\mu_x}{dt^2}+\gamma B_0\mu_x=0$

(*b*) $\frac{d^2\mu_x}{dt^2}+\gamma B_0\frac{d\mu_x}{dt}+\mu_x=0$

(*c*) $\frac{d^2\mu_x}{dt^2}+\gamma^2 B_0^2\,\mu_x=0$

(*d*) $\frac{d^2\mu_x}{dt^2}+2\gamma B_0\mu_x=0$

Common Data for Questions 92, 93:

One of the eigenvalues of the matrix $\begin{pmatrix}2 & 3 & 0\\3 & 2 & 0\\0 & 0 & 1\end{pmatrix}$ is 5.

92. The other two eigenvalues are

(*a*) 0 and 0 (*b*) 1 and 1

(*c*) 1 and –1 (*d*) –1 and –1

93. The normalized eigenvector corresponding to the eigenvalue 5 is

(*a*) $\frac{1}{\sqrt{2}}\begin{pmatrix}0\\-1\\1\end{pmatrix}$ (*b*) $\frac{1}{\sqrt{2}}\begin{pmatrix}-1\\1\\0\end{pmatrix}$

(*c*) $\frac{1}{\sqrt{2}}\begin{pmatrix}1\\0\\-1\end{pmatrix}$ (*d*) $\frac{1}{\sqrt{2}}\begin{pmatrix}1\\1\\0\end{pmatrix}$

2005

94. The average value of the function $f(x) = 4x^3$ in the interval 1 to 3 is

(*a*) 15 (*b*) 20

(*c*) 40 (*d*) 80

95. The unit normal to the curve $x^3y^2 + xy = 17$ at the point (2, 0) is

(*a*) $(\hat{i}+\hat{j})/\sqrt{2}$ (*b*) $-\hat{i}$

(*c*) $-\hat{j}$ (*d*) $\hat{j}$

96. The value of the integral $\int_c \frac{dz}{z+3}$ where C is a circle (anticlockwise) with $|z| = 4$, is

(*a*) 0 (*b*) πi

(*c*) $2\pi i$ (*d*) $4\pi i$

97. The determinant of a 3 × 3 real symmetric matrix is 36. If two of its eigen values are 2 and 3 then the third eigenvalue is

(*a*) 4 (*b*) 6

(*c*) 8 (*d*) 9

98. If a vector field $\vec{F} = x\hat{i}+2y\hat{j}+3z\hat{k}$, then $\vec{\nabla}\times(\vec{\nabla}\times\vec{F})$ is

(*a*) 0 (*b*) $\hat{i}$

(*c*) $2\hat{j}$ (*d*) $3\hat{k}$

99. All solutions of the equation $e^z = -3$ are

(*a*) $z = i\,n\,\pi\ln 3$, $n = \pm1,\pm2,\ldots\ldots.$

(*b*) $z = \ln 3 + i\,(2n+1)\,\pi$, $n = 0, \pm1,\pm2,\ldots\ldots.$

(*c*) $z = \ln 3 + i\,2n\pi$, $n = 0,\pm1\pm2,\ldots\ldots.$

(*d*) $z = i\,3n\pi$, $n = \pm1\pm2,\ldots\ldots.$

100. If $\bar{f}(s)$ is the Laplace transform of $f(t)$ the Laplace transform of $f(at)$, where a is a constant, is

(*a*) $\frac{1}{a}\bar{f}(s)$ (*b*) $\frac{1}{a}\bar{f}(s/a)$

(*c*) $\bar{f}(s)$ (*d*) $\bar{f}(s/a)$

101. Given the four vectors

$$u_1 = \begin{pmatrix} 1 \\ 2 \\ 1 \end{pmatrix}, u_2 = \begin{pmatrix} 3 \\ -5 \\ 1 \end{pmatrix}, u_3 = \begin{pmatrix} 2 \\ 4 \\ -8 \end{pmatrix}, u_4 = \begin{pmatrix} 3 \\ 6 \\ -12 \end{pmatrix},$$

the inearly dependent pair is

(*a*) u_1, u_2 (*b*) u_1, u_3

(*c*) u_1, u_4 (*d*) u_3, u_4

102. Which of the following functions of the complex variable z is NOT analytic everywhere ?

(*a*) e^z (*b*) $\sin z/z$

(*c*) z^3 (*d*) $|z|^3$

103. Eigen values of the matrix

$$\begin{pmatrix} 0 & 1 & 0 & 0 \\ 1 & 0 & 0 & 0 \\ 0 & 0 & 0 & -2i \\ 0 & 0 & 2i & 0 \end{pmatrix}$$

are

(*a*) –2, –1, 1, 2 (*b*) –1, 1, 0, 2

(*c*) 1, 0, 2, 3 (*d*) –1, 1, 0, 3

Statement for Linked Answer Questions 104 & 105 :

For the differential equation

$$\frac{d^2y}{dx^2} - 2\frac{dy}{dx} + y = 0$$

104. One of the solutions is

(*a*) e^x (*b*) $\ln x$

(*c*) e^{-x^2} (*d*) e^{x^2}

105. The second linearly independent solution is

(*a*) e^{-x} (*b*) xe^x

(*c*) x^2e^x (*d*) x^2e^{-x}

2004

106. For the function $\phi = x^2 y + xy$, the value of $|\vec{\nabla}\phi|$ at $= x = y = 1$ is

(*a*) 5 (*b*) $\sqrt{5}$

(*c*) 13 (*d*) $\sqrt{13}$

107. The average of the function $f(x) = \sin x$ in the interval $(0, \pi)$ is

(*a*) $\frac{1}{2}$ (*b*) $\frac{2}{\pi}$

(*d*) $\frac{1}{\pi}$ (*d*) $\frac{4}{\pi}$

108. Identify the points of unstable equilibrium for the potential shown in the figure.

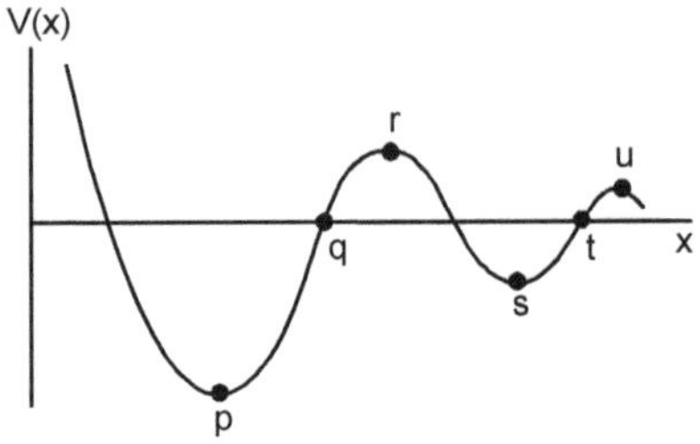

(*a*) *p* and *s* (*b*) *q* and *t*

(*c*) *r* and *u* (*d*) *r* and *s*

109. A real traceless 4 × 4 unitary matrix has two eigenvalues – 1 and + 1. The other eigenvalues are

(*a*) zero and +2 (*b*) – 1 and + 1

(*c*) zero and + 1 (*d*) + 1 and + 1

110. The eigenvalues of the matrix $\begin{pmatrix} 1 & i \\ -i & 1 \end{pmatrix}$ are

(*a*) + 1 and + 1 (*b*) zero and + 1

(*c*) zero and + 2 (*d*) – 1 and + 1

111. The inverse of the complex number $\frac{3+4i}{3-4i}$ is

(*a*) $\frac{7}{25} + i\frac{24}{25}$ (*b*) $-\frac{7}{25} + i\frac{24}{25}$

(*c*) $\frac{7}{25} - i\frac{24}{25}$ (*d*) $-\frac{7}{25} - i\frac{25}{25}$

112. The value of $\int_C \frac{dz}{(z^2+a^2)}$, where C is a unit circle (anti clockwise) centered at the origin in the complex *z*-plane is

(*a*) π for $a = 2$ (*b*) zero for $a = \frac{1}{2}$

(*c*) 4π for $a = 2$ (*d*) $\frac{\pi}{2}$ for $a = \frac{1}{2}$

113. The Laplace transform of $f(t) = \sin \pi t$ is $F(s) = \frac{\pi}{(s^2+\pi^2)}$, $s > 0$. Therefore, the Laplace transform of $t\sin\pi t$ is

(*a*) $\frac{\pi}{s^2(s^2+\pi^2)}$

(*b*) $\frac{2\pi}{s^2(s^2+\pi^2)^2}$

(*c*) $\frac{2\pi s}{(s^2+\pi^2)^2}$

(*d*) $\frac{2\pi}{(s^2+\pi^2)^2}$

114. A periodic function $f(x) = x$ for $-\pi < x < +\pi$ has the Fourier series representation $f(x) = \sum_{n=1}^{\infty}\left(-\frac{2}{n}\right)(-1)^n \sin nx$. Using this, one finds the sum $\sum_{n=1}^{\infty} n^{-2}$ to be

(*a*) 2 ln 2 (*b*) $\frac{\pi^2}{3}$

(*c*) $\frac{\pi^2}{6}$ (*d*) $\pi \ln 2$

115. The Fourier transform F(*k*) of a function $f(x)$ is defined as $F(k) = \int_{-\infty}^{\infty} dx\, f(x) \exp(ikx)$. Then F(*k*) for $f(x) = \exp(-x^2)$ is

[Given : $\int_{-\infty}^{\infty} \exp(-x^2)\, dx = \sqrt{\pi}$]

(*a*) $\pi \exp(-k)$

(*b*) $\sqrt{\pi} \exp\left(\frac{-k^2}{4}\right)$

(*c*) $\frac{\sqrt{\pi}}{2} \exp\left(\frac{-k^2}{2}\right)$

(*d*) $\sqrt{2\pi} \exp(-k^2)$

Data for Q. 116-117:

Consider the differential equation

$y'' + p(x)y' + q(x)\, y(x) = 0.$

116. If $xp(x)$ and $x^2q(x)$ have the Taylor series expansions

$xp(x) = 4 + x + x^2 + \ldots\ldots\ldots$

$x^2q(x) = 2 + 3x + 5x^2 + \ldots\ldots\ldots.$

then the roots of the indicial equation are

(*a*) –1, 0 (*b*) – 1, – 2

(*c*) – 1, 1 (*d*) – 1, 2

117. If $p(x) = 0$ with the Wronskian at $x = 0$ as $W(x = 0) = 1$ and one of the solutions is x, then the other linearly indpendent solution which vanishes at $x = 1/2$ is

(*a*) 1 (*b*) $1 - 4x^2$

(*c*) x (*d*) $-1 + 2x$

2003

118. The two vectors P = i, q = (i + j)/$\sqrt{2}$ are

(*a*) related by a rotation

(*b*) related by a reflection through the *xy*-plane

(*c*) related by an inversion

(*d*) not linearly independent

119. A 3 × 3 matrix has eigenvalues 0, 2 + i and 2 – i. Which one of the following statements is correct ?

(*a*) The matrix is Hermitian

(*b*) The matrix is unitary

(*c*) The inverse of the matrix exists

(*d*) The determinant of the matrix is zero

120. The value of the integral $\int_C dz/z^2$, where z is a complex variable and C is the unit circle with the origin as its centre, is

(*a*) 0 (*b*) $2\pi i$

(*c*) $4\pi i$ (*d*) $-4\pi i$

121. A particle with an initial velocity v_0 i enters a region with an electric field E_0 j and a magnetic field B_0 j. The trajectory of the particle will

(*a*) be an ellipse

(*b*) be a cycloid

(*c*) be a helix with constant pitch

(*d*) not be confined to any plane

122. An object of mass m rests on a surface with coefficient of static friction μ. Which of the following is NOT correct ?

(*a*) The force of friction is exactly $\mu m g$

(*b*) The maximum force of friction is $\mu m g$

(*c*) The force of friction is along the surface

(*d*) The force of friction opposes any effort to move the object

123. The curl of the vector A = z i + x j + y k is given by

(*a*) i + j + k (*b*) i – j + k

(*c*) i + j – k (*d*) –i – j – k

124. Consider the differential equation $d^2x/dt^2 + 2\, dx/dt + x = 0$. At time t = 0, it is given that x = 1 and dx/dt = 0. At t = 1, the value of x is given by

(*a*) 1/*e* (*b*) 2/*e*

(*c*) 1 (*d*) 3/*e*

125. S_{ij} and A_{ij} represent a symmetric and an antisymmetric real-valued tensor respectively in three dimensions. The number of independent components of S_{ij} and A_{ij} are

(*a*) 3 and 6 respectively

(*b*) 6 and 3 respectively

(*c*) 6 and 6 respectively

(*d*) 9 and 6 respectively

126. Consider the four statements given below about the function $f(x) = x^4 - x^2$ in the range $-\infty < x < +\infty$. Which one of the following statements is correct?

P the plot of f(x) versus x has two maxima and two minima

Q the plot of f(x) versus x cuts the x axis at four points

R the plot of f(x) versus x has three extrema

S no part of the plot f(x) versus x lies in the fourth quadrant

Pick the right combination of correct choices from those given below

(a) P and R (b) R only

(c) R and S (d) P and Q

127. The Fourier transform of the function f(x) is F(k) $= \int e^{ikx}$ f(x) dx. The Fourier transform of df(x)/dx is

(a) $dF(k)/dk$ (b) $\int F(k)/dk$

(c) $-ikF(k)$ (d) $ikF(k)$

Data for Q. 128 – 129 :

Consider the vector $V = r/r^3$

128. The surface integral of this vector over the surface of a cube of size a and centered at the origin

(a) 0 (b) 2π

(c) $2\pi a^3$ (d) 4π

129. Which one of the following is NOT correct?

(a) Value of the line integral of this vector around any closed curve is zero

(b) This vector can be written as the gradient of some scalar function

(c) The line integral of this vector from point P to point Q is independent of the path taken

(d) This vector can represent the magnetic field of some current distribution

2002

130. If two matrices A and B can be diagonalized simultaneously, which of the following is true?

(a) $A^2B = B^2A$

(b) $A^2B^2 = B^2A$

(c) $AB = BA$

(d) $AB^2 AB = BABA^2$

131. Which one of the following matrices is the inverse of the matrix $\begin{pmatrix} 1 & -1 \\ 0 & 1 \end{pmatrix}$?

(a) $\begin{pmatrix} 1 & 1 \\ -1 & 1 \end{pmatrix}$ (b) $\begin{pmatrix} 1 & 0 \\ 1 & 1 \end{pmatrix}$

(c) $\begin{pmatrix} 1 & 1 \\ 0 & 1 \end{pmatrix}$ (d) $\begin{pmatrix} -1 & 1 \\ 0 & -1 \end{pmatrix}$

132. Which of the following vectors is orthogonal to the vector $(a\hat{i} + b\hat{j})$, where a and b ($a \neq b$) are constants, and $\hat{i}$ and $\hat{j}$ are unit orthogonal vectors?

(a) $-b\hat{i} + a\hat{j}$ (b) $-a\hat{i} + b\hat{j}$

(c) $-a\hat{i} - b\hat{j}$ (d) $-b\hat{i} - a\hat{j}$

133. Fourier transform of which of the following functions does not exist?

(a) $e^{-|x|}$ (b) xe^{-x^2}

(c) e^{x^2} (d) e^{-x^2}

134. The unit vector normal to the surface $3x^2 + 4y = z$ at the point (1, 1, 7) is

(a) $(-6\hat{i} + 4\hat{j} + \hat{k})/\sqrt{53}$

(b) $(4\hat{i} + 6\hat{j} - \hat{k})/\sqrt{53}$

(c) $(6\hat{i} + 4\hat{j} - \hat{k})/\sqrt{53}$

(d) $(4\hat{i} + 6\hat{j} + \hat{k})/\sqrt{53}$

135. The solution of the differential equation

$$(1 + x)\frac{d^2y(x)}{dx^2} + x\frac{dy(x)}{dx} - y(x) = 0 \text{ is}$$

(a) $Ax^2 + B$ (b) $Ax + Be^{-x}$

(c) $Ax + Be^{x}$ (d) $Ax + Bx^2$

where A and B are constants.

2001

139. If S is the closed surface enclosing a volume V and $\hat{n}$ is the unit normal vector to the surface and $\vec{r}$ is the positive vector, then the value of the following integral $\iint_S \hat{n} dS$ is

(a) V (b) 2V

(c) 0 (d) 3V

140. The value of the integral $\int_C z^{10} dz$, where C is the unit circle with the origin as the center is

(a) 0 (b) $z^{11}/11$

(c) $2\pi i z^{11}/11$ (d) 1/11

141. Consider the set of vectors $\frac{1}{\sqrt{2}}$ (1, 1, 0), $\frac{1}{\sqrt{2}}$ (0, 1, 1) and $\frac{1}{\sqrt{2}}$ (1, 0, 1).

(*a*) The three vectors are orthonormal.

(*b*) Three vectors are linearly independent.

(*c*) The three vectors cannot form a basis in a three-dimensional real vector space.

(*d*) $\frac{1}{\sqrt{2}}$(1, 1, 0) can be written as a linear combination of $\frac{1}{\sqrt{2}}$ (0, 1, 1) and $\frac{1}{\sqrt{2}}$ (1, 0, 1).

142. If $A = x\hat{e}_x + y\hat{e}_y + z\hat{e}_z$, then $\nabla^2 \vec{A}$ equals

(*a*) 1 (*b*) 3

(*c*) 0 (*d*) – 3

143. The value of the residue of $\frac{\sin z}{z^6}$ is

(*a*) $-\frac{1}{5!}$ (*b*) $\frac{1}{5!}$

(*c*) $\frac{2\pi i}{5!}$ (*d*) $-\frac{2\pi i}{5!}$

144. If $F[f(x)] = \frac{1}{\sqrt{2\pi}} \int_{-\infty}^{\infty} f(x)e^{-ikx}dx$, then $\hat{F}^2\,[f(x)]$ is equal to

(*a*) $f(x)$ (*b*) $-f(x)$

(*c*) $f(-x)$ (*d*) $[f(x) + f(-x)]/2$

145. The solution of the system of differential equations

$$\frac{dy}{dx} = y - z \text{ and } \frac{dz}{dx} = -4y + z$$

is given by (for A and B are arbitrary constants)

(*a*) $y(x) = Ae^{3x} + Be^{-x}; z(x) = -2Ae^{3x} + 2Be^{-x}$

(*b*) $y(x) = Ae^{3x} + Be^{-x}; z(x) = 2Ae^{3x} + 2Be^{-x}$

(*c*) $y(x) = Ae^{3x} + Be^{-x}; z(x) = 2Ae^{3x} - 2Be^{-x}$

(*d*) $y(x) = Ae^{3x} + Be^{-x}; z(x) = -2Ae^{3x} - 2Be^{-x}$

2000

149. A square matrix a is unitary if

(*a*) $A^+ = A$

(*b*) $A^+ = A^{-1}$

(*c*) $Tr(A) = 1$

(*d*) dct (A) = 1

150. If Z_1 and Z_2 are complex numbers, which of the following is always true?

(*a*) $Z_1^*Z_2^* + Z_2^*Z_1^* = 2Z^*Z_2$

(*b*) $Z_1^*Z_2 + Z_2^*Z_1 = 2Z_1Z_2$

(*c*) $Z_1^*Z_2 + Z_2^*Z_1 \le 2|Z_1||Z_2|$

(*d*) $Z_1^*Z_2 + Z_2^*Z_1 \ge 2|Z_1||Z_2|$

151. The eigenvalues of the matrix $\begin{pmatrix} 1 & 1 \\ 1 & 1 \end{pmatrix}$ are

(*a*) 1, 0 (*b*) 1, 1

(*c*) 1, 2 (*d*) 0, 2

152. The Dirac delta function $\delta(x)$ can be represented by,

(*a*) $\frac{1}{\pi} \lim_{N \to 0} \frac{\sin(Nx)}{x}$

(*b*) $\frac{1}{\pi} \lim_{\varepsilon \to 0+} \frac{\varepsilon}{x^2 + \varepsilon^2}$

(*c*) $\frac{1}{\sqrt{x}} \lim_{\varepsilon \to +\infty} \frac{1}{\sqrt{3}} \exp(-x^2/\varepsilon)$

(*d*) $\frac{1}{2\pi} \int_{-\infty}^{+\infty} \exp\left(i|x - x^4|u\right)du$

ANSWERS

1. (b,c) **2.** (b,d) **3.** (5 to 5) **4.** (*a*) **5.** (–2 to –2) **6.** (*a*) **7.** (*c*) **8.** (–1 to –1)
9. (0 to 0) **10.** (*b*) **11.** (*b*) **12.** (*c*) **13.** (*a*) **14.** (0.010 to 0.013)
15. (*d*) **16.** (0.81 to 0.84) **17.** (81.60 to 81.80) **18.** 3.13 to 3.15 **19.** (*b*) **20.** 0.43 to 0.45
21. 2.90 to 3.10 **22.** 0.51 to 0.53 **23.** (*d*) **24.** (*a*) **25.** (*b*) **26.** (*b*) **27.** (*c*) **28.** (*a*)
29. (*b*) **30.** (1.3) **31.** (*b*) **32.** (*c*) **33.** (*c*) **34.** (*a*) **35.** (*d*) **36.** (*d*) **37.** (*b*) **38.** (*c*)
39. (*d*) **40.** (*b*) **41.** (*d*) **42.** (*c*) **43.** (*c*) **44.** (*b*) **45.** (*c*) **46.** (3) **47.** (*a*) **48.** (*c*)
49. (*b*) **50.** (*a*) **51.** (*d*) **52.** (*a*) **53.** (*d*) **54.** (*b*) **55.** (*a*) **56.** (*c*) **57.** (*b*) **58.** (*b*)
59. (*d*) **60.** (*d*) **61.** (*c*) **62.** (*d*) **63.** (*a*) **64.** (*a*) **65.** (*b*) **66.** (*a*) **67.** (*a*) **68.** (*a*)
69. (*d*) **70.** (*b*) **71.** (*b*) **72.** (*d*) **73.** (*a*) **74.** (*d*) **75.** (*a*) **76.** (*c*) **77.** (*d*) **78.** (*b*)
79. (*d*) **80.** (*c*) **81.** (*a*) **82.** (*b*) **83.** (*c*) **84.** (*d*) **85.** (*a*) **86.** (*a*) **87.** (*a*) **88.** (*a*)
89. (*a*) **90.** (*c*) **91.** (*c*) **92.** (*c*) **93.** (*d*) **94.** (*c*) **95.** (*d*) **96.** (*c*) **97.** (*b*) **98.** (*a*)
99. (*b*) **100.** (*b*) **101.** (*d*) **102.** (*b*) **103.** (*a*) **104.** (*a*) **105.** (*b*) **106.** (*d*) **107.** (*b*) **108.** (*c*)
109. (*b*) **110.** (*c*) **111.** (*c*) **112.** (*b*) **113.** (*c*) **114.** (*c*) **115.** (*b*) **116.** (*b*) **117.** (*d*) **118.** (*a*)
119. (*a*) **120.** (*a*) **121.** (*d*) **122.** (*a*) **123.** (*a*) **124.** (*b*) **125.** (*b*) **126.** (*c*) **127.** (*c*) **128.** (*a*)
129. (*d*) **130.** (*c*) **131.** (*c*) **132.** (*a*) **133.** (*b*) **134.** (*c*) **135.** (*b*) **139.** (*d*) **140.** (*a*) **141.** (*b*)
142. (*c*) **143.** (*a*) **144.** (*a*) **145.** (*a*) **149.** (*b*) **150.** (*c*) **151.** (*d*) **152.** (*d*)

EXPLANATIONS

1. $y'' - 2xy' + 2ny = 0$ is the Hermite's differential equation

$y_n(x) = H_n(x)$

Now, $\psi_n(x) = H_n(x)\, e^{-\frac{x^2}{2}}$

Wave equation for Harmonic oscillator option (d).

2. The Laplace transformation evaluated at $s = i\omega$ is equal to the fourier transformation is its region of convergence contains the imaginary axis

So, $L(f;s)$ and $\hat{f}(\omega)$ can be made connected The variable s can be complex.

Hence, both are right.

3. In $\oint_c \frac{dz}{(z-n)^2 + \pi^2}$

Here $f(z) = \frac{1}{(z-n)^2 + \pi^2}$

$$f(z) = \frac{1}{[z-(n+i\pi)][z-(n-i\pi)]}$$

$z = n + i\pi,\ n - i\pi$

For positive upper half part

$z = n + i\pi$

Residue at $z = n + ip$

$$= \lim_{z \to n+i\pi} [z-(n+i\pi)] \frac{1}{[z-(n+i\pi)][z-(n-i\pi)]}$$

$$= \frac{1}{n + i\pi - n + i\pi}$$

Residue $= \frac{1}{2\pi i}$

Now $\oint \frac{dz}{(z-n)^2 + \pi^2} = 2\pi i \sum(\text{Residue})$

$$= 2\pi i \times \frac{1}{2\pi i} = 1$$

$$\sum_{n=1}^{5} In = I_1 + I_2 + I_3 + I_4 + I_5$$

$= 1 + 1 + 1 + 1 + 1 = 5$

4.
$$\frac{d^2u(x)}{dx^2} = k^2u(x)$$
$$(D^2 - k^2) = 0$$
$$D^2 = k^2$$
$$(D - k)(D + k) = 0$$
$$C.F = C_1e^{-kx} + C_2e^{kx}$$

5.
$$b_1 = \frac{1}{2\pi i}\oint f(z)dz$$
= sum of residue

Factor of $\frac{1}{2} = -2$

$$\frac{1}{2\pi i}\oint f(z)dz = \frac{2\pi i}{2\pi i}(-2)$$
$$= -2$$

6. $A = \begin{pmatrix} 0 & 0 & 1 \\ 0 & 1 & 0 \\ 1 & 0 & 0 \end{pmatrix}$

Product of eigenvalues

$= \lambda_1, \lambda_2, \lambda_3$

= Determinant of A

$= 1(0 - 1)$

$= -1$

7.
$M \to \lambda_j = 1, 2, 3$

$|e_j\rangle$ = Eigen state

$$M^{-1}|e_j\rangle = \frac{1}{\lambda}|e_j\rangle$$

$M \to \lambda_1, \lambda_2, \lambda_3$

$$M^{-1} \to \frac{1}{\lambda_1}, \frac{1}{\lambda_2}, \frac{1}{\lambda_3}$$

M and M^{-1} has same eigen state.

8. $f(x) = x$

a_o and $a_n = 0$, for odd wave functions

$$b_n = \frac{2}{T}\int_d^{d+T} f(x)\sin x\, dx$$

$$a_k = \frac{2}{2\pi}\int_{-\pi}^{-\pi} x\sin x\, dx$$

$$= \frac{-2}{n}(-1)^n$$

for n = 2

$$a_2 = \frac{-2}{2}(-1)^2 = -1$$

9. $f(x) = \begin{cases} 0 & x \le -\frac{1}{2n} \\ n & -\frac{1}{2n} < x < \frac{1}{2n} \\ 0 & x \ge \frac{1}{2n} \end{cases}$

$$I = \lim_{n\to\infty}\left[\int_{-\infty}^{-\frac{1}{2n}} 0 + \int_{-\frac{1}{2n}}^{\frac{1}{2n}} n\sin x\, dx + \int_{\frac{1}{2n}}^{\infty} 0\right]$$

$= 0$

10. $x = e^t$, $t = \ln x$

$$x^2\frac{d^2y}{dx^2} = \frac{d^2y}{dt^2} - \frac{dy}{dt}$$

$$x^2\frac{d^2y}{dx^2}n(n+1)y = 0$$

$$\frac{d^2y}{dt^2} - \frac{dy}{dt} - n(n+1)y = 0$$

$D^2y - Dy - n(n + 1)y = 0$

$D^2 - D - n(n + 1) = 0$

$$m_1, m_2 = \frac{1 \pm \sqrt{1+4n(n+1)}}{2}$$

$$y_1 = e^{\frac{1+\sqrt{1+4n(n+1)}}{2}}, y_2 = e^{\frac{1-\sqrt{1+4n(n+1)}}{2}}$$

$y_1.y_2 = e^t.$

$y_1.\ y_2 = x$

11.
$$\frac{1}{3}\begin{pmatrix} 0 & 1 & 0 \\ 0 & 0 & -1 \\ -1 & 0 & 0 \end{pmatrix}\begin{pmatrix} 2 \\ -1 \\ 2 \end{pmatrix} = \frac{1}{3}\begin{pmatrix} -1 \\ -2 \\ -2 \end{pmatrix}$$

$$\frac{1}{3}\begin{pmatrix} 0 & 1 & 0 \\ 0 & 0 & -1 \\ -1 & 0 & 0 \end{pmatrix}\begin{pmatrix} 1 \\ 1 \\ -1 \end{pmatrix} = \frac{1}{\sqrt{3}}\begin{pmatrix} -1 \\ 1 \\ -1 \end{pmatrix}$$

12. $f(z) = \cot z = \frac{\cos z}{\sin z}$

$\sin z = 0$

$z = n\pi$

at $z = 0$, $\cot(z) \to \infty$

$z = 0$ is a simple pole.

13. $f(x) = \frac{\cos(kx)}{x^2a^2}$

$$\int_{-\infty}^{\infty}\frac{\cos(kx)}{x^2+a^2}dx = \text{R.P.of}\int_{-\infty}^{\infty}\frac{e^{-ikz}}{z^2+a^2}, \quad x^2 + a^2 = 0$$

$x = \pm ia$

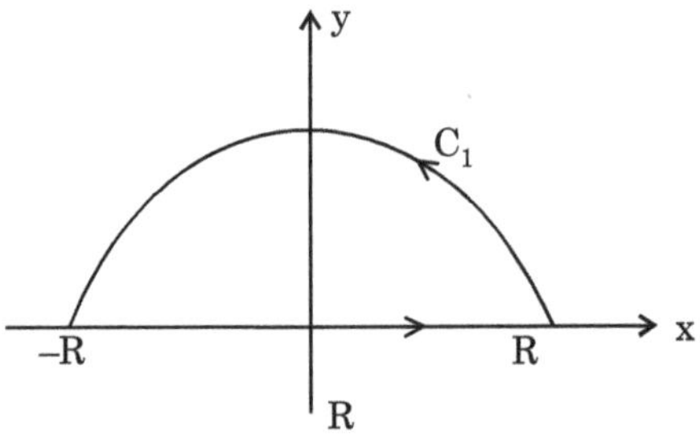

$$\oint \frac{e^{ikx}}{x^2+a^2}dx = \int_{-R}^{R} f(x)\,dx + \int_{c_1} f(x)\,dx$$

when $R \to \infty$

$$\oint \frac{e^{ikx}}{x^2+a^2}dz = \int_{-\infty}^{\infty} f(x)\,dx = 2\pi i(\Sigma \text{Res})$$

$$= 2\pi i\left[\lim_{x\to ia}(x-ia)\frac{e^{ikx}}{x^2+a^2}\right]$$

$$= \frac{\pi}{\alpha}e^{ka}$$

14. $C_m = \dfrac{1}{b-a}\displaystyle\int_a^b \psi(\theta)\, e^{im\theta}.d\theta$

$$C_3 = \frac{1}{2\pi}\int_{\frac{-\pi}{2}}^{\frac{\pi}{2}} \psi(\theta)e^{im\theta}.d\theta$$

$$C_3 = \frac{1}{2\pi}\int_{\frac{-\pi}{2}}^{\frac{+\pi}{2}} 1.e^{im\theta}.d\theta$$

$$C_3 = \frac{1}{2\pi}\int_{\frac{-\pi}{2}}^{\frac{+\pi}{2}} e^{i3\theta}.d\theta$$

$$C_3 = \frac{1}{2\pi}\left(\frac{-1}{3i}\right)\left[e^{3i\theta}\right]_{\frac{-\pi}{2}}^{\frac{\pi}{2}}$$

$$C_3 = \frac{-1}{3\pi}$$

$$|C_3|^2 = .011$$

15. Given $\vec{V_1} = \hat{i} - \hat{j}$

and $\vec{V_2} = -2\hat{i} + 3\hat{j} + 2\hat{k}$

The vector $\vec{V_2}$ can be written as a linear combination of the vectors in $\vec{V_3}$.

Then $\vec{V_3} = 2\hat{i} + (-2+3)\hat{j} + 2(2\hat{k})$

$= 2\hat{i} + \hat{j} + 4\hat{k}$

16. The given differential equation is

$$\frac{d^2f(x)}{dx^2} - 2\frac{df(x)}{dx} + f(x) = 0$$

with boundary condition

$f(0) = 1$

$f(1) = 0$

The auxiliary equation is

$\lambda^2 - 2\lambda + 1 = 0$

$\Rightarrow \lambda = 1, 1$

Therefore, Complementary function is

$f(x) = (c_1 + c_2 x)e^x$

On applying boundary conditions ,we get

$c_1 = 1$

$c_2 = -1$

Hence $f(x) = (1 - x)e^x$

So $f(0.5) = (1 - 0.5)e^{0.5}$

$= 0.5e^{0.5} = 0.82$

17. The given circle

$|z - 1.5| = 1$

includes only $z = 2$ but not include $z = -2$. Hence the given integral becomes

$$\int \frac{\left(\frac{5z^3+3z^2}{z+2}\right)}{z-2}dz$$

$$= 2\pi i \times \frac{5\times 2^3 + 3\times 2^2}{2+2}$$

$= 2\pi i \times 13$

Therefore, absolute value of given integral is $2\pi \times 13 = 81.68$

18. The contour integral $\oint \dfrac{dz}{1+z^2}$

$$f(z) = \frac{1}{1+z^2}$$

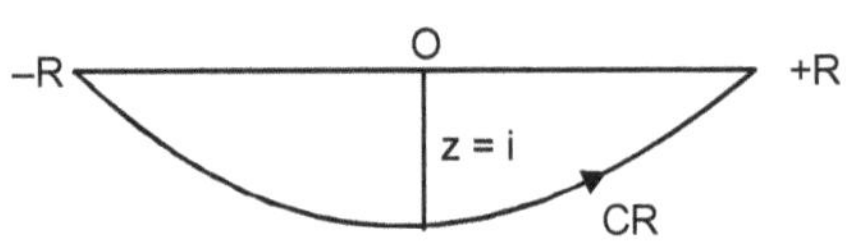

Poles of $f(z)$ are $1 + z^2 = 0 \Rightarrow z^2 = -1$

$\Rightarrow z = \pm i$.

The only pole which lie within contour is at $z = i$.

The Residue of $f(z)$ at $z = i$

$$\underset{z\to a}{Lt}\; z - a\, f(z) \;\Rightarrow\; \underset{z\to i}{Lt}\frac{(z-i)}{1+z^2}$$

$$= \underset{z\to i}{Lt}\frac{1}{i+z} = \frac{1}{2i}$$

By cauchy's residue theorem we have

$$\int_c f(z)dz = 2\pi i \times \text{ sum of residues.}$$

$$\int_c f(z)dz = 2\pi i \times \frac{1}{2i} = \pi$$

$$\Rightarrow \int_{-R}^{R}\frac{1}{1+z^2}dz = \pi$$

$$\Rightarrow \int_{-R}^{R}\frac{1}{1+x^2}dx = \pi$$

Equating real parts

$$\int_{-\infty}^{+\infty}\frac{1}{1+x^2}dx = \pi$$

$$\int_{-\infty}^{+\infty}\frac{1}{1+x^2}dx = 3.14$$

19. $f(x) = \dfrac{a_0}{2} + a_1\cos x + a_2\cos 2x + ...a_n\cos nx$

$$+ b_1 \sin x \;......\; b_n \sin x$$

$$a_0 = \frac{1}{\pi}\int_0^{2\pi} A\sin^2(\alpha x)dx$$

$$= \frac{A}{\pi}\int_0^{2\pi}\sin^2(\alpha x)dx = A$$

$$a_n = \frac{1}{\pi}\int_0^{2\pi} A\sin^2(\alpha x)\cos nx\; dx = 0$$

$$b_n = \frac{1}{\pi}\int_0^{2\pi} A\sin^2(\alpha x)\sin nx\; dx = 0$$

Hence coefficient of e^{ikx} in the A $\sin^2(\alpha dx) = \dfrac{-A}{4}$.

20. $I(n) = \sum\limits_0^{\infty} x^{n-1}e^{-x}dx$

$$I = \int_0^{\infty} x^2 e^{-x^2} dx \quad 2xdx = dy$$

$$\left(\text{Put } x^2 = y \;\; dx = \frac{dy}{2\sqrt{y}}\right)$$

$$I = \int_0^{\infty} ye^{-y}\frac{dy}{2\sqrt{y}} \qquad \left(I\left(\frac{1}{2}\right) = \pi\right)$$

$$= \frac{1}{2}\int_0^{\infty} y^{-1/2}e^{-y}dy = \frac{1}{2}I\left(\frac{3}{2}\right) = \frac{1}{2}.\frac{1}{2}.I\left(\frac{1}{2}\right)$$

$$= \frac{\sqrt{\pi}}{4} = 0.44$$

21. We know that

$f(z) = u(x, y) + i\, v(x, y)$

$\quad = u(x, y) + i\,(2xy + 3y)$

$u(x, y) = 0$ at $(0, 0)$

$f(z) = 0$ at $(0, 0)$

By Mi ne – Thompson method

$v(x, y) = 2xy + 3y$

$$\frac{\delta v}{\delta x} = 2y,\; \frac{\delta v}{\delta y} = 2x + 3$$

$$\phi(x, y) = 2y, \psi(x, y) = 2x + 3.$$

Put $x = z$, $y = 0$

$\phi(z, 0) = 0,\ \psi(z, 0) = 2z + 3$

$f'(z) = 2z + 3$

$f(z) = z^2 + 3z$

$\therefore z^2 + 3z = u(x, y) + i\,(2xy + 3y)$

$(x + iy)^2 + 3(x + iy) = u(x, y) + i\,(2x + 3y)$

$x^2 - y^2 + 2ixy + 3x + 3iy$

$= u(x, y) + i(2xy + 3y)$

$u(x, y) = i^2 - y^2 + 3x$

$u(1, 1) = 1 - 1 + 3 = 3.$

22. $\dfrac{dy}{dx} + y\tan x = \cos x$

$$IF = e^{\int \tan x dx} = e^{\log\sec x} = \sec x$$

$$y\cdot\sec x = \int\cos x\cdot\frac{1}{\cos x}dx + C$$

$$y\sec x = x + C$$

$$y(0) = 0 \Rightarrow 0 + C = 0 \Rightarrow C = 0$$

$$\therefore y = \frac{x}{\sec x} \Rightarrow y\left(\frac{\pi}{3}\right) = \frac{\pi}{3}\cos\frac{\pi}{3}$$

$$= \frac{\pi}{3}\cdot\frac{1}{2} = \frac{3.14}{6} = 0.52$$

23. $$\frac{dy}{dx} = xy$$

$$\Rightarrow \quad \frac{1}{y}dy = x\,dx$$

$$\Rightarrow \quad \ln y = \frac{x^2}{2} + \ln C$$

$$\Rightarrow \quad y = Ce^{x^2/2}$$

If $y = 2$, at $x = 0$, C = 2

$$\Rightarrow \quad y = 2e^{x^2/2}$$

Value of y at $x = 2$ is $y = 2e^2$

24. $z^2 = (x + iy)^2 = x^2 - y^2 + i(2xy)$

$\Rightarrow u = x^2 - y^2, \quad v = 2xy$

$$\frac{\partial u}{\partial x} = \frac{\partial v}{\partial y} = 2x$$

$$\frac{\partial v}{\partial x} = \frac{-\partial u}{\partial y} = 2y \quad \text{satisfies}$$

25. Consider

$$\int_{-\infty}^{\infty} \varphi_m^* \frac{d^2\varphi_n}{dx^2}.dx = \left[\varphi_m^* \frac{d\psi_n}{dx}\right]_{-\infty}^{+\infty} - \int_{-\infty}^{+\infty} \frac{d\varphi_n}{dx} \times \frac{d\varphi_m^*}{dx}.dx$$

$$= \left[-\frac{d\varphi_m^*}{dx}\psi_n\right]_{-\infty}^{+\infty} + \int_{-\infty}^{+\infty} \varphi_n \frac{d^2\varphi_m^*}{dx^2}dx$$

$$= \int_{-\infty}^{+\infty} \frac{d^2\varphi_m^*}{dx^2}.\varphi_n\,dx$$

Two arbitary w.f. $\varphi_m(x)\ \varphi_n(x)$

Operator A is hermition if

$$\int_{-\infty}^{+\infty} \varphi_m^*\ A\ \varphi_n\,dx = \int_{-\infty}^{+\infty} (A\varphi_m)^*\ \varphi_n\,dx$$

$$= \left[\int_{-\infty}^{+\infty} \varphi_n^*\ A\ \varphi_m\,dx\right]^*$$

26. $$f(x, y, z) = \frac{x^2}{2} - xy + \frac{z^2}{2}$$

$$\nabla f = \hat{i}\frac{\partial f}{\partial x} + \hat{j}\frac{\partial f}{\partial y} + \hat{k}\frac{\partial f}{\partial z}$$

$$= \hat{i}(x - y) + \hat{j}(-x) + \hat{k}(z)$$

At (1, 1, 2)

$$\nabla f = -\hat{j} + 2\hat{k}$$

$$\text{Direction} = \frac{-\hat{j}+2\hat{k}}{\sqrt{1^2+2^2}} = \frac{-\hat{j}+2\hat{k}}{\sqrt{5}}$$

27. $$\sigma_x = \begin{bmatrix} 0 & 1 \\ 1 & 0 \end{bmatrix}$$

$$\sigma_y = \begin{bmatrix} 0 & -i \\ i & 0 \end{bmatrix}$$

$$\sigma_z = \begin{bmatrix} 1 & 0 \\ 0 & -1 \end{bmatrix}$$

$2\sigma_x\sigma_y + \sigma_y\,\sigma_x$

$$= 2\begin{bmatrix} 0 & 1 \\ 1 & 0 \end{bmatrix}\begin{bmatrix} 0 & -i \\ i & 0 \end{bmatrix} + \begin{bmatrix} 0 & -i \\ i & 0 \end{bmatrix}\begin{bmatrix} 0 & 1 \\ 1 & 0 \end{bmatrix}$$

$$= 2\begin{bmatrix} i & 0 \\ 0 & -i \end{bmatrix} + \begin{bmatrix} -i & 0 \\ 0 & i \end{bmatrix}$$

$$= 2\begin{bmatrix} i & 0 \\ 0 & -i \end{bmatrix} - \begin{bmatrix} i & 0 \\ 0 & -i \end{bmatrix}$$

$$= 2i\begin{bmatrix} 1 & 0 \\ 0 & -1 \end{bmatrix} - i\begin{bmatrix} 1 & 0 \\ 0 & -1 \end{bmatrix}$$

$$= 2i\sigma_z - i\sigma_z = i\sigma_z$$

28. $$f(x) = \begin{cases} -1, & -\pi < x < 0 \\ 1, & 0 < x < \pi \end{cases}$$

$$f(x) = a_0 + \sum_{n=1}^{\infty}(a_n \cos nx + b_n \sin nx)$$

$$\therefore \quad a_0 = \frac{1}{2\pi}\int_{-\pi}^{+\pi} f(x)dx$$

$$= \frac{1}{2\pi}\left[\int_{\pi}^{0}(-1)dx + \int_{0}^{\pi}(1)dx\right]$$

$$= 0$$

and $$a_n = \frac{1}{\pi}\int_{-\pi}^{\pi} f(x)\cos nx\,dx = 0$$

and $$b_n = \frac{1}{\pi}\int_{-\pi}^{+\pi} f(x)\sin nx\,dx$$

$$= \frac{1}{\pi}\left[\left\{\frac{\cos nx}{n}\right\}_{-\pi}^{0} - \left\{\frac{\cos nx}{n}\right\}_{0}^{\pi}\right]$$

$$= \frac{1}{\pi}\left[\frac{1}{n} - \frac{(-1)^n}{n} - \frac{(-1)^n}{n} + \frac{1}{n}\right]$$

$$= \frac{1}{\pi}\left[\frac{2}{n} - \frac{2(-1)^n}{n}\right]$$

If n is even $b_n = 0$

and if n is odd $b_n = \dfrac{4}{n\pi}$

Thus Fourier Series is,

$$f(x) = \frac{4}{\pi}\left[\sin x + \frac{1}{3}\sin 3x + \frac{1}{5}\sin 5x + ...\right]$$

30. $\int_0^3 t^2\,\delta(3t-6)dt$

As we know shifting properties of dirac delta function is defined as

$$\int_{-\infty}^{\infty} f(x)\delta(x-a)dx = f(a)$$

And also

$$\delta(kx) = \frac{1}{|k|}\delta(x)$$

Therefore $= \int_0^3 t^2\,\delta(3(t-2))dt$

$$= \frac{1}{|3|}\int_0^3 t^2\,\delta(t-2)\,dt$$

$$= \frac{1}{3}.4 = \frac{4}{3} = 1.333$$

$$\int_0^3 t^2\,\delta(3t-6)\,dt = 1.333$$

31. $f(x) = e^{-x^2}$ is differentiable but $g(x) = |x|e^{-x^2}$ is not differentiable.

$$g(x) = \begin{cases} -xe^{-x^2}; & x<0 \\ xe^{-x^2}; & x>0 \end{cases}$$

Left hand limit, $\lim\limits_{n\to 0} g(x-n) = -[x-n]e^{-(x-n)^2}$

Right hand limit, $\lim\limits_{n\to 0} g(x+n) = (x+n)e^{-(x+n)^2}$

$\Rightarrow \lim\limits_{n\to 0} g(x-n) \neq \lim\limits_{n\to 0} g(x+n)$

32. w = f(z) = u(x, y) + iv(x, y) to be an analytic function in a Domain D,

$\left|_{z_1}^{z_2} f(z)dz\right.$ is independent of the choice of the contour between z_1 and z_1 and z_2 in D.

33. $y^{11} + \frac{1}{z}y' - \frac{m^2}{z^2}y = 0$

$\Rightarrow z^2 y^{11} + zy^1 - m^2 y = 0$

m = 0, 1, 2, 3...,

$z = e^x$, $D = \frac{d}{dx}$

If m = 0; $z^2y^n + zy^1 = 0$, $[D(D-1) + D]y = 0$

$\Rightarrow [D^2 - D + D]\,y = 0$

$D^2y = 0 \Rightarrow y = c_1 + c_2\,x \Rightarrow y = c_1 + c_2\,\text{I n z}$

(R is correct)

And if m ≠ 0, m > 0, then m ≠ 0,

then $(D^2 - m^2)\,y = 0$

$\Rightarrow D = \pm\, m$

$Y = c_1\,e^{mx} + c^2\,e^{-mx}$

$= c_1\,e^{mlogz} + c_2\,e^{-\,mlogz}$

$c_1\,z^m + c_2\,z^{-m}$

or if m ≠ 0, m > 0, then

$y = c_1 \cos n\,(m\log(z)) + ic_2 \sin n\,(m\log(x)),\ m > 0$

34. $H(f) = \int_{-\infty}^{+\infty} H(t)e^{-2\pi ft}dt$, for a function H(t),

$$H(f) = -\frac{2i}{\omega}$$

For $H(t-t_0)$, Fourier transform is $e^{-i2\pi ft0}H(f)$

Shifting Theorem,

For $\frac{1}{2}\left[H\left(t+\frac{1}{2}\right) - H\left(t-\frac{1}{2}\right)\right]$

$$= \frac{1}{2}\left[e^{\frac{i\omega}{2}} - e^{\frac{-i\omega}{2}}\right]\frac{-2i}{\omega}$$

$$= \frac{1}{2i}\left[e\frac{i\omega}{2} - e^{\frac{-i\omega}{2}}\right]\frac{-2i}{\omega}\times 2$$

For Fourier Transform of,

$$\frac{1}{2}\left[H\left(t+\frac{1}{2}\right) - H\left(t-\frac{1}{2}\right)\right]$$

$$= \frac{\sin\left(\frac{\omega}{2}\right)}{\frac{\omega}{2}}$$

35. The given surface $\phi = x^2 + y^2 + z^2 - 3$

Now $\nabla\phi = \left(\frac{\delta}{\delta x}\hat{x} + \frac{\delta}{\delta y}\hat{y} + \frac{\delta}{\delta z}\hat{z}\right)\cdot(x^2+y^2+z^2-3)$

$$= 2x\,\hat{x} + 2y\,\hat{y} + 2z\,\hat{z}$$

Now $|\nabla\phi| = \sqrt{(2x)^2 + (2y)^2 + (2z)^2}$

$$= \sqrt{4x^2 + 4y^2 + 4z^2}$$

$$= \sqrt{4(x^2+y^2+z^2)}$$

Now unit vector $\perp^r$ to surface ϕ is given by

$$\hat{n} = \frac{\nabla\phi}{|\nabla\phi|} = \frac{2x\,\hat{x} + 2y\,\hat{y} + 2z\,\hat{z}}{\sqrt{4\,(x^2+y^2+z^2)}}$$

$$= \frac{x.\hat{x} + y.\hat{y} + z.\hat{z}}{\sqrt{(x^2 + y^2 + z^2)}}$$

Now $\hat{n}$ at (1, 1, 1)

$$\therefore \hat{n}(1, 1, 1) = \left(\frac{\nabla\phi}{|\nabla\phi|}\right)_{(1,1,1)} = \frac{\hat{x} + \hat{y} + \hat{z}}{\sqrt{3}}$$

36. $A = \frac{1}{\sqrt{3}}\begin{bmatrix} 1 & 1+i \\ 1-i & -1 \end{bmatrix}$

$AA^* = I$ for unitary transformation

$$AA^* = \left(\frac{1}{\sqrt{3}}\right)^2 \begin{bmatrix} 1 & 1+i \\ 1-i & -1 \end{bmatrix}\begin{bmatrix} 1 & 1+i \\ 1-i & -1 \end{bmatrix}$$

$$= \frac{1}{3}\begin{bmatrix} 1+1-i^2 & 1+i-1-i \\ 1-i-1+i & 1-i^2+1 \end{bmatrix}$$

$$\Rightarrow \frac{1}{3}\begin{bmatrix} 3 & 0 \\ 0 & 3 \end{bmatrix} = \frac{3}{3}\begin{bmatrix} 1 & 0 \\ 0 & 1 \end{bmatrix} = I$$

37. $(ds)^2 = 2(dx^1)^2 + (dx^2)^2 + \sqrt{3}\ dx^1dx^2$ ...(*i*)

Now as we have relation of tensor (g_{ij}) with ds^2 *i.e.*

$$ds^2 = g_{ij}\, dx^i dx^j$$

$$\therefore \quad ds^2 = g_{11}\, dx^1dx^1 + g_{12}\, dx^1dx^2 + g_{21}\, dx^2dx^1 + g_{22}\, dx^2dx^2$$

$$= g_{11}\,(dx^1)^2 + g_{22}(dx^2)^2 + \frac{\sqrt{3}}{2}\, dx^1dx^2 + \frac{\sqrt{3}}{2}\, dx^2dx^1 \quad ...(ii)$$

Comparing equation (*ii*) with equation (*i*) we get

$$g_{11} = 2, g_{12} = \frac{\sqrt{3}}{2}, g_{21} = \frac{\sqrt{3}}{2}, g_{22} = 1$$

$$\therefore \text{ tensor } g_{ij} = \begin{pmatrix} g_{11} & g_{12} \\ g_{21} & g_{22} \end{pmatrix} = \begin{pmatrix} 2 & \sqrt{3}/2 \\ \sqrt{3}/2 & 1 \end{pmatrix}$$

38. Pole, $e^z = -1$

$e^z = e^{i(2m+1)\pi}$,

where, m = 0, 1, 2, 3

$$\text{For } z = i\pi,\ \text{Res} = \lim_{z\to i\pi}\frac{\phi(z)}{\phi'(z)} = \frac{-\pi^2}{e^{i\pi}} = \pi^2$$

Similarly, for $z = -i\pi$, Res $= \pi^2$

$\therefore\ I = 2\pi i(\pi^2 + \pi^2) = 4\pi^3 i$

39. $$\frac{d^2y}{dt^2} - y = 0$$

$(D^2 - 1)\, y = 0$

The auxilary equation is $m^2 = 1$

$\therefore\ \boxed{m = \pm 1}$

Hence solution for differential equation is

$$y = C_1 e^t + C_2 e^{-t}$$

Put at $t = 0$, $y(0) = 1$

$$1 = C_1 + C_2 \quad ...(i)$$

at $t = \infty$, $y(\infty) = 0$

$$y = C_1 e^t + C_2 e^{-t}$$

$$\therefore \quad 0 = C_1 e + C_2 e^{-\infty}$$

$$a = 0 \quad [\because e^{-\infty} = 0]$$

Putting the value of C_1 in equation (*i*) we get $C_2 = 1$

$$\therefore \quad y = C_2 e^{-t} = e^{-t}$$

$$y = \cosh t - \sinh t$$

$[\because e^{-t} = \cosh t - \sinh t]$

40. f(x) = f(–x)

The f(x) will be the even function.

So, by fourier series,

$$f(x) = a_0 + \sum_{n_7}^{-\infty}[a_n \cos(nkx) + b_n \sin(nkx)]$$

For even function, $b_n = 0$

$$\text{Then, } f(x) = a_0 + \sum_{n=1}^{\infty}(a_n \cos(nkx))$$

41. Pressure is a scalar quantity not a vector quantity. It has a magnitude but not direction associated with it.

42. The Curl of gradient is always zero.

Thus, $\vec{v} = \nabla\phi$

Taking of the curl both sides, we have

$$\nabla \times \vec{v} = \underbrace{\nabla}_{\text{curl}} \times \underbrace{\nabla\phi}_{\text{gradient}}$$

$$\nabla \times \vec{v} = 0,$$

$$\nabla^2\phi = 0$$

$$\nabla.\nabla\phi = 0,$$

43. f(t) = sinKT

$$L[f(t)] = \frac{K}{S^2 + K^2},\ S > 0\ \left\{L(\sin at) = \frac{a}{S^2 + a^2}\right\}$$

44. Let $\vec{A} = A_0(\hat{x} + \hat{y} + \hat{z})$, $\vec{B} = B_0(\hat{x} + \hat{y} + \hat{z})$

and $\vec{r} = x\hat{x} + y\hat{y} + z\hat{z}$

$$\vec{B} \times \vec{r} = \hat{x}(z - y)B_0 - \hat{y}(z - x)B_0 + \hat{z}(y - x)B_0$$

$$\Rightarrow \vec{\nabla}(\vec{A}.(\vec{B} \times \vec{r})) = 0$$

46. At z = 1, Pole of order 2.

So Residue is $\dfrac{1}{\lfloor 2-1}\dfrac{d^{2-1}}{dz^{2-1}}\left[\dfrac{(z-1)^2\,16\,z}{(z+3)(z-1)^2}\right]_{z=1}$ ...(1)

From general formulaes of order n,

$$\text{Res } (z = a) = \frac{1}{(n-1)_1^1}\left\{\frac{d^{n-1}}{dz^{n-1}}\left[(z-a)^n f(z)\right]\right\}$$

On solving equation (1), we have

$$= \frac{1}{1_1^1}\frac{d}{dz}\left[\frac{16z}{z+3}\right]_{z=1} = \left\{\frac{(z+3).16-16.z\times 1}{(z+3)^2}\right\}_{z=1}$$

$$= \frac{48}{16} = 3$$

47. (*a*) If $\vec{a} = 3\hat{i}+2\hat{j}, \vec{b} = \hat{i}+2\hat{j}$ are linearly dependent

$\vec{a}+m\vec{b} = 0$ for some values of m

but 3 + m = 0 and 2 + 2m = 0 do not have any solution so they are linearly independent.

$\vec{a}\cdot\vec{b} \neq 0$ (Not orthogonal)

$(\vec{a}\times\vec{b} \neq 0)$ (Not normalized)

48. For symmetric tensor

$$A_{ij} = \begin{bmatrix} A_{11} & A_{12} & A_{13} \\ A_{21} & A_{22} & A_{23} \\ A_{31} & A_{32} & A_{33} \end{bmatrix}$$

$\because$ $A_{12} = A_{21}$, $A_{23} = A_{32} = A_{13} = A_{31}$ are 3 components and A_{11}, A_{22}, A_{33} are 3 components. Hence, there are 6 independent components

49. $$A = \begin{vmatrix} 0 & 1 & 0 \\ 1 & 0 & 1 \\ 0 & 1 & 0 \end{vmatrix}$$

Diagonalising the matrix,

$$|A - \lambda| = 0$$

$$\begin{vmatrix} 0 & 1 & 0 \\ 1 & 0 & 1 \\ 0 & 1 & 0 \end{vmatrix} - \lambda\begin{vmatrix} 1 & 0 & 0 \\ 0 & 1 & 0 \\ 0 & 0 & 1 \end{vmatrix} = 0$$

$$= \begin{vmatrix} -\lambda & 1 & 0 \\ 1 & -\lambda & 1 \\ 0 & 1 & -\lambda \end{vmatrix} = 0$$

$$= -\lambda(\lambda^2 - 1) + \lambda = 0$$

$$\lambda(\lambda^2 - 2) = 0$$

$$\lambda = 0, +\sqrt{2}, -\sqrt{2}$$

50. Given $\vec{r} = r(\cos\phi i + \sin\phi j)$

$$\vec{F} = \vec{r}\times\vec{B}$$

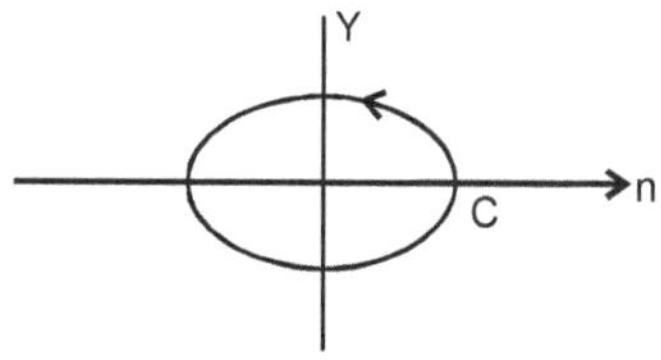

$$= r\left(\cos\phi\hat{i}+\sin\phi\hat{j}\right)\times B_0\left(\hat{i}+j+\hat{k}\right)$$

$$= B_0 r\left(\cos\phi\hat{k}-\cos\theta\hat{j}-\sin\phi\hat{k}+\sin\phi\hat{j}\right)$$

$$d\vec{i} = rd\phi\left(-\cos\phi\hat{i}+\sin\phi\hat{i}\right)$$

Now,

$$\oint\vec{F}\cdot\vec{d} = B_0 r^2\int_0^{2\pi}\left[(\cos\phi-\sin\phi)\hat{k}-\cos\phi\hat{j}+\sin\phi\hat{i}\right]\cdot\left(-\cos\phi i+\sin\phi\hat{j}\right)d\phi$$

$$= B_0 r^2\int_0^{2\pi}(-2\sin\phi\cos\phi)d\phi = 0$$

51. Let $= \oint_c e^{1/2}dz$

$$= \oint_c\left\{1+\frac{1}{2}+\frac{1}{2}z^2+\ldots\right\}dz$$

Here singularity of $e^{1/2}$ is z = 0 is lie in the contour $|z| = 1$

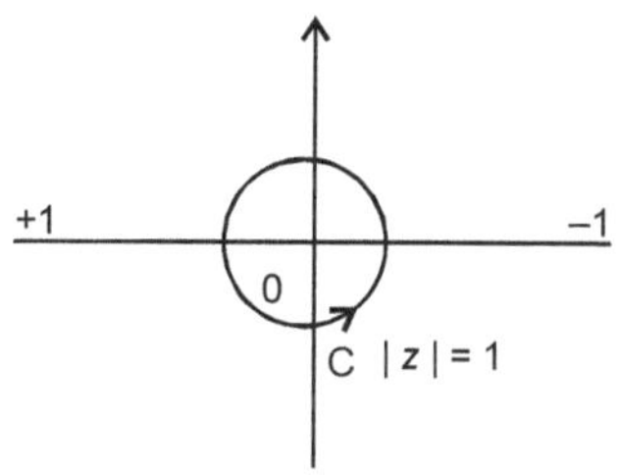

Now, by Cauchy residue theorem

$$\oint_e e^{1/2}dz = 2\pi i\Sigma R$$

$$= 2\pi i\left(\lim_{z\to 0}[i]\right)$$

$$= 2\pi i \times 1$$

$$= 2\pi i$$

52. Since, F is derivable from potential $v(r)$

Hence, $\vec{F} = -\vec{\nabla}\, v(r)$

Now, $\vec{\nabla}\times\vec{F} = -\vec{\nabla}\times\vec{\nabla}[v(r)] = 0$

$\because$ $-\vec{\nabla}$ and $\vec{\nabla}[v(r)]$ is parallel is each other

53. The equation of the surface is

$$f(x, y, z) \equiv (x^2 + y^2 - Z - 1) = 0$$

$$\vec{\nabla}f = \left(\frac{\partial}{\partial x}\hat{i}+\frac{\partial}{\partial y}\hat{j}+\frac{\partial}{\partial z}\hat{k}\right)(x^2+y^2-z-1)$$

$$= \left[\frac{\partial}{\partial x}(x^2+y^2-z-1)i+\frac{\partial}{\partial y}(x^2+y^2-z-1)\hat{j}+\frac{\partial}{\partial z}(x^2+y^2-z-1)\hat{k}\right]$$

$$= 2\pi\hat{i} + 2y\hat{j} - \hat{k} \text{ value of } \vec{\nabla} f \text{ point P(1, ,1, 1)} ki$$

$$(\vec{\nabla} f)(1,1,1) = (2\hat{i} + 2\hat{j} - \hat{k})$$

$$|(\vec{\nabla} f)(1,1,1)| = \sqrt{2^2 + 2^2 + 1^2} = 3$$

Hence unit normal vector to the given surface of the point P(1, 1, 1) is

$$\frac{(\vec{\nabla} f)_{(1,1,1)}}{|(\vec{\nabla} f)_{(1,1,1)}|} = \frac{(2i + 2\hat{j} - \hat{k})}{3}$$

54. For calculation of surface integral, we need the surface of the cyclinder shown in adjacent figure

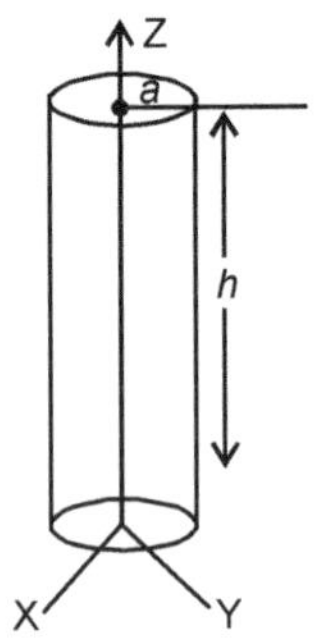

$$\hat{x} = \hat{k} \text{ and } \vec{r} \cdot \hat{n} = \vec{r} \cdot \hat{k} = z$$

Where $\vec{r} = x\hat{i} + yj + z\hat{k}$ is the given

Position vector. Thus

$$\iint \vec{r} \cdot \overline{ds} = \iint \vec{r} \cdot \vec{n}\, ds$$

Top surface

$$n \iint ds = nha^2 \text{ (on top surface)}$$

On the bottom surface of the cyclinder

$$n = -\hat{k}$$

$$\vec{r} \cdot \hat{n} = -\vec{r} \cdot \hat{n} = -z = 0$$

($\because$ for the bottom surface z = 0)

Thus, $\iint \vec{r} \cdot \overline{ds} = \iint \vec{r} \cdot \hat{n}\, ds = 0$

Top bottom surface

$$n \iint ds = \pi ha^2 \text{ (on top surface)}$$

Vector $x\hat{i} + y\hat{j}$ is normal to the current surface of the cyclinder and therefore for this surface we have

$$\hat{x} = \frac{x\hat{i} + y\hat{j}}{\sqrt{x^2 + y^2}} = \frac{x\hat{i} + yj}{a}$$

$$\vec{r} \cdot \hat{x} = \frac{x^2 + y^2}{a} = a$$

Then, $\iint \vec{r} \cdot \hat{x}\, ds = a \iint ds = a$

Current surface (area of curved surface)

$$= a \cdot 2\pi ah = 2\pi a^2 h$$

The value of $\oiint \vec{r} \cdot \hat{n}\, ds$ are the whole surface of cylinder is then

$$\pi a^2 h + 2\pi a^2 h = 3\pi a^2 h$$

55. $$\frac{dy}{dx} = -\frac{x}{y+1}$$

$$\Rightarrow \quad x\,dx + y\,dy + dy = 0$$

$$\Rightarrow \quad \frac{x^2}{2} + \frac{y^2}{2} + y = C_1$$

$$\Rightarrow \quad x^2 + y^2 + 2y = 2C_1$$

$$\Rightarrow \quad (x-0)^2 + (y+1)^2 = 2C_1 + C$$

Which is family of circles with differente radii.

56. $f(z) = \dfrac{z \log z}{(z - \pi)}$ has a pole of order z at

$$z = \pi$$

57. $$\oint f(z)dz = 0$$

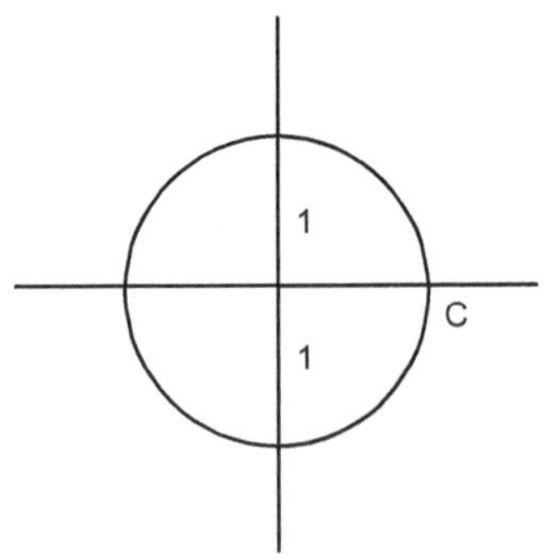

58. The number of independent components of the tensor

$$\frac{1}{2}(N^2 - N) = \frac{1}{2}(25 - 5) = 10$$

59. $z^2 = 0 \quad \Rightarrow \quad z = 0$

It means $f(z)$ has poles at $z = 0$ of the order of two, which lie outside the circle $(z - 2) = 1$

$$\therefore \quad \oint_c = \frac{e^z \sin z}{z^2} dz = 2\pi i(0) = 0$$

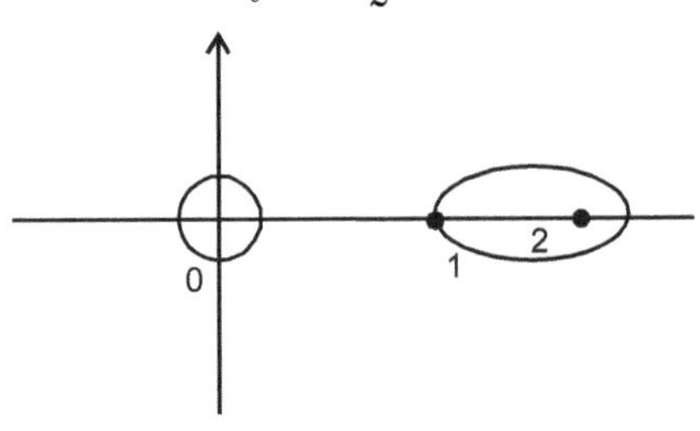

60. $4(f(x)) = \int_0^{\infty} e^{-sx} f(x)\,dx$

$$= \int_0^{3} e^{-sx} f(x)\,dx + \int_3^{\infty} e^{-sx} f(x)\,dx$$

$$= \int_0^{3} e^{-sx}\cdot 0dx + \int_3^{\infty} e^{-sx}(x-3)\,dx$$

$$= 0 + (x-3)\frac{e^{-sx}}{-5}\int_0^{\infty} - \int_0^{\infty} 1\cdot\left(\frac{e^{-sx}}{-5}\right)dx$$

$$= 0 - 0 - \left(\frac{e^{sx}}{s^2}\right)_3^{\infty}$$

$$= -\left(0 - \frac{e^{35}}{s^2}\right) = s^{-2}e^{-35}$$

62. $(D^2 - 1)y = 2\cos ht = e^t + e^t$

Auxillary equation of the above differential equation is

$$m^2 - 1 = 0$$

$$m = \pm 1$$

Complementry factor = $C_1e^t + C_2e^{-t}$

A cos ht + B sin t

Particular integral of

$$e^t = \frac{1}{b^2-1}e^t = t\frac{1}{20}e^t = \frac{t}{2}e^t$$

Particular integral of

$$e^t = \frac{1}{b^2-1}e^{-t} = t\cdot\frac{1}{2_D}e^{-t}$$

$$\frac{t}{2}e^t$$

$$\rightarrow \quad e^{-t} = \frac{1}{b^2-1}e^{-t} = t\cdot\frac{1}{2_D}e^{-t}$$

$$\Rightarrow \quad \frac{t}{2}(e^{-t})$$

64. $\left|\int \vec{r}\times d\vec{\theta}\right|$

For a circle C, $\vec{r}$ is perpendicular to $d\vec{\theta}$

$$\left|\int(r\,d\,\theta\sin 90°)\right| = r\int_C d\theta = 2\pi r$$

68. The set of vectors are (1, 1, 1), (1, –1, 1) and (1, 1, –1)

The volume of the vector space is

$$V = \begin{vmatrix} 1 & 1 & 1 \\ 1 & -1 & 1 \\ 1 & 1 & -1 \end{vmatrix}$$

$$= 1(1-1) - 1(-1-1) + (1+1)$$

$$\Rightarrow \quad V = 0 + 2 + 2 = 4 \neq 0$$

Hence, S is a linearly dependent set.

69. We know that

The unitary matrix is

$$AA^{\theta} = l \qquad \ldots (i)$$

When Å = conjugate transpose

Given that = $\begin{bmatrix} ae^{i\alpha} & b \\ ce^{i\beta} & d \end{bmatrix}$

From eq. (i)

$$A^{-1}(AA^{\theta}) = l.A^{-1}$$

$$A^{\theta} = A^{-1} \qquad \ldots (ii)$$

$$\bar{A} = \begin{vmatrix} ae^{-i\alpha} & b \\ ce^{-i\beta} & d \end{vmatrix}$$

$$|\bar{A}|^{-1}\begin{vmatrix} ae^{-i\alpha} & ce^{-i\beta} \\ b & d \end{vmatrix}$$

70. The general equation for a wave propogating in space is

$$\left|\nabla^2 - \frac{1}{V^2}\frac{\partial^2}{\partial t^2}\right|\psi(\pi, t) = 0$$

∵ Wave is propogating along + x-axis without change in its shape

$$\frac{\partial^2}{\partial y^2} = \frac{\partial^2}{\partial z^2} = 0$$

Thus,

$$\because \quad \nabla^2 = \frac{\partial^2}{\partial x^2} + \frac{\partial^2}{\partial y^2} + \frac{\partial^2}{\partial z^2}$$

$$\left(\frac{\partial^2}{\partial x^2} - \frac{1}{V^2}\frac{\partial^2}{\partial t^2}\right)\psi(r, t) = 0$$

71. Le t $A = \begin{bmatrix} \cos\theta & -\sin\theta \\ \sin\theta & \cos\theta \end{bmatrix}$

$$[A - \lambda I] = 0$$

$$\begin{vmatrix} \cos\theta - \lambda & -\sin\theta \\ \sin\theta & \cos\theta - \lambda \end{vmatrix} = 0$$

$$\cos^2\theta - 2\lambda\cos\theta + \lambda^2 + \sin^2\theta = 0$$

$$\lambda^2 - 2\lambda\cos\theta + 1 = 0$$

$$\lambda = \frac{2\cos \pm \sqrt{4\cos^2\theta - 4}}{2}$$

$$= \cos\theta \pm i\sin\theta$$

$$= \frac{\sqrt{3}}{2} \pm \frac{i}{2} \qquad \theta = 30°$$

$$= \frac{1}{2}(\sqrt{3} \pm i) \qquad \theta = 30°$$

72. $F[\delta(x-a)] = \exp(-i2\pi va)$

$= \cos 2\pi va - i \sin 2\pi va$

$\Rightarrow F[\delta(y+a)] = \exp(i\, 2\pi va)$

$= \cos 2\pi va + i \sin 2\pi va$

$\Rightarrow F[\delta(x-a)] + \delta(x+a) = 2\cos 2\pi va$

$$\Rightarrow F[\cos(2\pi va)] = \frac{1}{2}[\delta(x-a)+\delta(x+a)]$$

74.
$$\int_{-i}^{i} \pi(z+1)dz = \pi\int_{-i}^{+i}(z+1)dz$$
$$= \pi\left|\frac{z^2}{2}+z\right|_{-i}^{i}$$
$$= \pi\left[\frac{(i)^2-(-i)^2}{z}+i-(-i)\right]$$
$$= \pi\left[\frac{-1-(+1)}{z}+2i\right]$$
$$= \pi(-1+2i)$$

77. $f(x) = a_0 + a_1x + a_2x^2$

$g(x) = b_0 + b_1x + b_2x^2$

$(f(x)\, g(x)) = a_0b_0 + a_1b_1(x/x) + a_2b_2(x^2/x^2)$

$$= a_0b_0 + \frac{a_1b_1}{2} + \frac{a_2b_2}{3}$$

79. The given eigen values are $i_1 - 2i$ and 3_i. Since all eigen values are imaginary. Hence matrix is anti-hermitian.

80. The orbit will be an ellipse.

81. The characterstic equations is

$$|A - \lambda I| = 0$$
$$\begin{vmatrix} 5-\lambda & 4 \\ 1 & 2-\lambda \end{vmatrix} = 0$$
$$10 - 5\lambda - 2\lambda + \lambda^2 - 4 = 0$$
$$\lambda^2 - \lambda\lambda + 6 = 0$$
$$\lambda = 1, 6$$

To find out the eigen vectors for the corresponding eigen values, we will consider the matrix. equation.

$|A - \lambda I|X = 0$

$$\begin{bmatrix} 5-\lambda & 4 \\ 1 & 2-\lambda \end{bmatrix}\begin{bmatrix} X \\ Y \end{bmatrix} = \begin{bmatrix} 0 \\ 0 \end{bmatrix} \quad ...(i)$$

(1) Eigen vector corresponding to the eigen value $\lambda = 1$. By putting $\lambda = 1$. the matrix eq. (i) will become.

$$\begin{bmatrix} 4 & 4 \\ 1 & 1 \end{bmatrix}\begin{bmatrix} X \\ Y \end{bmatrix} = \begin{bmatrix} 0 \\ 0 \end{bmatrix}$$

$$\begin{bmatrix} 4X+4Y=0 \\ X+\ \ Y=0 \end{bmatrix}$$

$X = K, Y = -K$

Eigen vector X_1 is $\begin{bmatrix} K \\ -K \end{bmatrix}$ or $K\begin{bmatrix} 1 \\ -1 \end{bmatrix}$

(2) Eigen vector corresponding to the eigen value $\lambda = 6$.

By putting $\lambda = 6$, the matrix Eqation(i) will become.

$$\begin{bmatrix} -1 & 4 \\ 1 & -4 \end{bmatrix}\begin{bmatrix} X \\ Y \end{bmatrix} = \begin{bmatrix} 0 \\ 0 \end{bmatrix}$$

$$\Rightarrow \begin{bmatrix} -X+4Y=0 \\ X+4Y=0 \end{bmatrix}$$

Let $X = 4K, Y = K$

eigen vector X_2 is $\begin{bmatrix} 4K \\ K \end{bmatrix}$ or $K\begin{bmatrix} 4 \\ 1 \end{bmatrix}$

82. $Z^2 + a^2 = 0$

$Z = \pm ia$

83. The given equation is

$$(1-x^2)\frac{d^2y}{dx^2} - 2x\frac{dy}{dx} + \frac{3}{2}\left(\frac{3}{2}+1\right) = 0$$

will diverge if $(1 - x^2) = 0 \Rightarrow x = \pm 1$

84.
$$x\frac{dy}{dx} + y = x^4$$
$$\frac{dy}{dx} + \frac{y}{x} = x^3 \quad ...(i)$$

The Eq. (i) is a linear differential equation

$$IF = e^{\int \frac{1}{x}dx}$$

Solution of Equation (i) is

$$y \cdot x = \int x \cdot x^3 dx + c = \int x^4 dx + c$$
$$y \cdot x = \frac{x^5}{5} + c \quad ...(ii)$$

Given $y = 1$ at $x = 1$, then from Eq. (ii), we have.

$$1 \cdot 1 = \frac{1}{5} + c \Rightarrow c = 1 - \frac{1}{5} = \frac{4}{5}$$

Hence $y \cdot x = \dfrac{x^5}{5} + \dfrac{4}{5}$

$$y = \frac{x^4}{5} + \frac{4}{5x}$$

85. Trace = Sum of diagonal elements

= Sum of eigen values

$= \lambda_1 + \lambda_2 + \lambda_3$

or $2 = \lambda_1 + \lambda_2 + \lambda_3 = 1 + 2 + \lambda_3$

or $\lambda_3 = -1$

86. $$\mathbf{I} = \oint_c \bar{A} \cdot d\bar{l} = \iint_s (\bar{\nabla} \times \bar{A}).d\bar{s} = \iint_s 0.d\bar{s} = 0$$

$$\left[\text{As } \bar{A} \text{ is uniform, } \bar{\nabla} \times \bar{A} = 0\right]$$

87. $$\begin{pmatrix} 1 & 1 & 0 \\ 0 & 1 & -1 \end{pmatrix} \begin{pmatrix} n_1 \\ n_2 \\ n_3 \end{pmatrix} = \begin{pmatrix} n_1 + n_2 \\ n_2 - n_3 \end{pmatrix};$$

$$T = \begin{pmatrix} 1 & 1 & 0 \\ 0 & 1 & -1 \end{pmatrix}$$

88. $$I = \oint_s \frac{\bar{r}.\bar{ds}}{r^3} = \oint_s \left(\frac{\bar{r}}{r^3}\right).\bar{ds} = \int_v \bar{\nabla}\left(\frac{\bar{r}}{r^3}\right) dV$$

(by divergence theorem)

Now, $$\bar{\nabla}.\left(\frac{\bar{r}}{r^3}\right) = \sum_1^3 \frac{\partial}{\partial x}\left(\frac{n}{r^3}\right)$$

$$= \sum_1^3 \left[\frac{1}{r^3} - \frac{3n}{r^4}.\frac{n}{r}\right]; \text{ as } \frac{\partial r}{\partial n} = \frac{n}{r}$$

$$= \frac{3}{r^3} - \frac{3}{r^3} = 0$$

$\therefore \quad I = 0$

89. $$I = \oint_C \frac{e^{2z}}{(z+1)^4} dz. \text{ for } C, |z| = 3$$

Integrant has a pole of degree 4 at $z = -1$

$$I = 2\pi i \text{ Res (I)}|_{z=-1}$$

$$= 2\pi i.\frac{4}{3}e^{-2} = \frac{8\pi i}{3}e^{-2}$$

$$I = \frac{e^{2z}}{(z+1)^4}, \qquad \text{Res}(I)|_{z=-1}$$

$$= \frac{1}{3i}\left[\frac{\partial^3}{\partial z^3}(e^{2z})\right]\Bigg|_{z=-1}$$

90. $$f(k) = \frac{1}{\sqrt{2\pi}} \int e^{-ikn} \delta(x) dx$$

$$\simeq \frac{1}{\sqrt{2\pi}}$$

91. Torque on a moment $\bar{\mu}$ in magnetic field $\bar{B}$,

$$\bar{\tau} = \bar{\mu} \times \bar{B} = \frac{d\bar{L}}{dt}$$

(= rate of change of angular momentum)

Now, $\bar{\mu} = \gamma \bar{L}$

$$\frac{d\bar{L}}{dt} = \frac{1}{\gamma}\frac{d\bar{\mu}}{dt} = (\bar{\mu} \times \bar{B})$$

$$\frac{d\bar{\mu}}{dt} = \gamma(\bar{\mu} \times \bar{B}), \qquad \bar{\mu} = \mu_x \hat{i}$$

$$= -\gamma \mu_n B_0 \hat{j} \qquad \bar{B} = B_0 \hat{k}$$

$$\frac{d^2\bar{\mu}}{dt^2} = \gamma\left(\frac{d\bar{\mu}}{dt} \times \bar{B}\right) = \gamma^2\left[(\bar{\mu} \times \bar{B}) \times \bar{B}\right]$$

$$= -\gamma^2 \mu_x (B_0 \hat{j} \times B_0 \hat{k})$$

$$= -\gamma^2 \mu_x B_0^2 \hat{i}$$

$$\frac{d^2\mu}{dt^2} = -\gamma^2 \mu_x B_0^2$$

$$\therefore \frac{d^2\mu_x}{dt^2} + \gamma^2 B_0^2 \mu_x = 0$$

92. det A = – 5, Tr A = 5.

$\lambda_1 = 5; \lambda_2 = 1; \lambda_3 = -1$

93. $$\begin{pmatrix} 2-5 & 3 & 0 \\ 3 & 2-5 & 0 \\ 0 & 0 & 1-5 \end{pmatrix} \begin{pmatrix} x_1 \\ x_2 \\ x_3 \end{pmatrix} = 0$$

$$\left.\begin{aligned} -3x_1 + 3x_2 &= 0 \\ 3x_1 - 3x_2 &= 0 \end{aligned}\right\} x_1 = x_2$$

$$3x_1 - 3x_2 = 0$$

$$-4x_3 = 0, \; x_3 = 0$$

$$X_1 = \frac{1}{\sqrt{2}} \begin{pmatrix} 1 \\ 1 \\ 0 \end{pmatrix}$$

94. $$f(x) = 4x^3$$

$$\text{Average value} = \frac{\int_1^3 f(x) dx}{\int_1^3 dx}$$

$$= \frac{\int_1^3 4x^3 dx}{2}$$

$$= \frac{1}{2}\left[x^4\right]_1^3$$

$$= \frac{80}{2}$$

$$= 40$$

95. $$f(x) = x^3y^2 + xy = 17$$

Let $\vec{N}$ be normal to the curve

$$\therefore \quad \vec{N} = \nabla f = \left(i\frac{\partial}{\partial x} + i\frac{\partial}{\partial y}\right)(x^3y^2 + xy)$$

$$= (3x^2y^2 + y)\hat{i} + (2x^3y + x)\hat{j}$$

$$\vec{N}_{(2,0)} = 0\hat{i} + 2\hat{j} = 2\hat{j}$$

$$\therefore \text{ Unit normal} = \frac{\vec{N}}{|\vec{N}|} = \frac{2\hat{j}}{2} = \hat{j}$$

96. $$\int_c \frac{dz}{z+3} = \int_0^{2\pi} \frac{1}{re^{i\theta}} ire^{i\theta} d\theta$$

$$= i\int_0^{2\pi} d\theta = 2\pi i$$

97. $D = 36$

Now for real symmetric 3×3 matrix,

$$D = \lambda_1 \lambda_2 \lambda_3$$

where $\lambda_1, \lambda_2, \lambda_3$ are eigen values

Given: $\lambda_1 = 2, \lambda_2 = 3$

$\therefore \quad 36 = 2 \times 3 \times \lambda_3$

$\therefore \quad \lambda_3 = 6$

98. Hence, $\vec{F} = x\hat{i} + 2y\hat{j} + 32\vec{k}$

$$\Rightarrow \nabla \times \left(\nabla \times \vec{F}\right) = (\nabla . F)\nabla - (\nabla . \nabla)F$$

$$= 0$$

99. Here, $z = (ln3 + i(2n+1)\pi$

Now, $e^z = e^{(ln3 + i(2n+1)\pi}$

$= 3. e^{i(2n+1)\pi}$

$= 3[\cos(2n+1)\pi + i\sin(2n+1)\pi]$

$= -3$

101. $$\frac{3}{2}U_3 - U_4 = \frac{3}{2}\begin{bmatrix} 2 \\ 4 \\ -8 \end{bmatrix} - \begin{bmatrix} 3 \\ 6 \\ -12 \end{bmatrix}$$

$$= \begin{bmatrix} 3 \\ 6 \\ -12 \end{bmatrix} - \begin{bmatrix} 3 \\ 6 \\ -12 \end{bmatrix} = 0$$

Hence, u_3, u_4 linearly dependent.

102. Because derivative $f'(z)$ does not exist at all points z of a region r. Also it does not satisfy Cauchy-Riemann equation.

103. Solving for eigen values, we get

$$\lambda^4 - 5\lambda^2 + 4 = 0$$

$$\Rightarrow (\lambda^2 - 1)(\lambda^2 - 4) = 0$$

$$\Rightarrow \lambda = \pm 1, \pm 2$$

104. $y = e^x$

$$\therefore \quad \frac{dy}{dx} = e^x, \text{ and } \frac{d^2y}{dx^2} = e^x$$

$$\therefore \quad \frac{d^2y}{dx^2} - \frac{2dy}{dx} + y = e^x$$

$$e^x - 2e^x + e^x = 0$$

105. $y = xe^x$

$$\therefore \quad \frac{dy}{dx} = e^x + xe^x$$

$$\text{and} \quad \frac{d^2y}{dx^2} = 2e^x + xe^x$$

106. $$\nabla\phi = \left(\hat{i}\frac{\partial}{\partial x} + \hat{j}\frac{\partial}{\partial y} + \hat{k}\frac{\partial}{\partial z}\right)\phi$$

$$\Rightarrow \quad \nabla\phi = 3\hat{i} + 2\hat{j}$$

$$\therefore \quad |\nabla\phi| = \sqrt{3^2 + 2^2} = \sqrt{9+4} = \sqrt{13}.$$

107. $f(x) = \sin x$,

$$\text{average value } f(x) = \frac{1}{\pi}\int_0^{\pi} \sin x\, dx$$

$$f(x) = \frac{1}{\pi}[-\cos x]_0^{\pi}$$

$$= \frac{2}{\pi}.$$

108. Here q and t are equilibrium point where $V(x) = 0$.

For stable point $\frac{\partial v}{\partial r} = -ve$ so that p and s are stable but for point s and u, $\frac{\partial v}{\partial r} = +ve$, so r and u are unstable points.

109. 4×4 unitary matrix is trace less

Trace (diagonal element adding) = 0

Here suppose eigen values are $\lambda_1, \lambda_2, \lambda_3$ and λ_4

Trace (A) $= 0 = \lambda_1 + \lambda_2 + \lambda_3 + \lambda_4$

If $\lambda_1 = -1$ and $\lambda_2 = +1$, then

$$-1 + 1 + \lambda_3 + \lambda_4 = 0$$

$$\Rightarrow \quad \lambda_3 = -\lambda_4$$

So only possibility is, $\lambda_3 = 1, \quad \lambda_4 = -1$.

110. $A = \begin{pmatrix} 1 & i \\ -i & 1 \end{pmatrix}$

Since $|A - Ix| = 0$

$\Rightarrow \begin{pmatrix} 1-x & i \\ -i & 1-x \end{pmatrix} = 0$

$\Rightarrow (1-x)^2 + i^2 = 0$

$\Rightarrow (1-x)^2 - 1 = 0$

$\Rightarrow 1 - 2x + x^2 - 1 = 0$

$\Rightarrow x^2 - 2x = 0$

$\Rightarrow x(x-2) = 0$

$\Rightarrow x = 0$ and $(x-2) = 0$

$\Rightarrow x = 0, +2.$

111. SinceZ $Z^{-1} = 1$

$\Rightarrow Z^{-1} = \frac{1}{Z}$

Since $Z = \left(\frac{3+4i}{3-4i}\right)$

$\Rightarrow Z^{-1} = \frac{1}{\left(\frac{3+4i}{3-4i}\right)} = \left(\frac{3-4i}{3+4i}\right)$

$= \left(\frac{3-4i}{3+4i}\right) \times \left(\frac{3-4i}{3-4i}\right)$

$= \frac{(3-4i)^2}{(9-16i^2)} = \frac{9-16-24i}{(9+16)}$

$= \frac{(-7-24i)}{25} = \frac{-7}{25} - i\frac{24}{25}.$

112. $\int_C \frac{dz}{(z^2+a^2)}$ $\quad [\because f(z) = \frac{1}{(z^2+a^2)}]$

poles $(z^2 + a^2) = 0$

$(z - ia)(z + ia) = 0$

$\Rightarrow z = \pm ia$

$\therefore \int f(z)\, dz = 2\pi i\, \Sigma R_+$

$\therefore \Sigma R_+ = R_1 + R_2$

The residue of $f(z)$ at $z = ia$ is

$R_1 = \underset{z \to ia}{\text{Lt}} (z - ia)\frac{1}{(z-ia)(z+ia)} = \frac{1}{2ia}$

The residue of $f(z)$ at $z = -ia$ is

$R_2 = \underset{z \to -ia}{\text{Lt}} (z + ia)\frac{1}{(z-ia)(z+ia)} = -\frac{1}{2ia}$

So the pole at $a = \frac{1}{2}$ is $z = \frac{+i}{2}, -\frac{i}{2}$

inside the unit circle.

$\Sigma R_+ = \frac{1}{2i\left(\frac{1}{2}\right)} - \frac{1}{2i\left(\frac{1}{2}\right)} = \frac{1}{i} - \frac{1}{i} = 0$

$\Rightarrow \int_C \frac{dz}{(z^2+a^2)} = 2\pi i \times 0 = 0$ at $a = \frac{1}{2}.$

113. $F(s) = \int_0^\infty e^{-st} F(t)\, dt$

$\Rightarrow \int_0^\infty e^{-st} \sin \pi t\, dt = \frac{\pi}{(s^2+\pi^2)}$

Since $F(s) = \int_0^\infty e^{-st} t \sin \pi t\, dt = \frac{2\pi s}{(s^2+\pi^2)^2}.$

114. $\because \sum_{n=1}^{\infty} n^{-2} = \sum_{n=1}^{\infty} \frac{1}{n^2}$

$= \left(1 + \frac{1}{2^2} + \frac{1}{3^2} + \frac{1}{4^2} + \ldots\right)$

$= \frac{\pi^2}{6}.$

115. $F(k) = \int_{-\infty}^{\infty} dx\, f(x) \exp(i\, kx)$

$\because f(x) = \exp(-x^2)$

$F(k) = \int_{-\infty}^{\infty} \exp(-x^2) \exp(ikx)\, dx$

$= \int_{-\infty}^{\infty} \exp(-x^2 + ikx)\, dx$

$= \int_{-\infty}^{\infty} \exp\left[-x^2 + ikx - \frac{i^2k^2}{4} + \frac{i^2k^2}{4}\right] dx$

$= \int_{-\infty}^{\infty} \exp\left(\frac{i^2k^2}{4}\right) \exp\left(-\left\{x^2 - ikx + \frac{i^2k^2}{4}\right\}\right) dx$

$= \exp\left(\frac{i^2k^2}{4}\right) \int_{-\infty}^{\infty} \exp\left\{-\left(x - \frac{ik}{4}\right)^2\right\} dx$

Let $x - \frac{ik}{4} = z \Rightarrow dx = dz$

$\therefore F(k) = \exp\left(\frac{-k^2}{4}\right) \int_{-\infty}^{\infty} \exp(-z^2)\, dz$

Since, here given $\int_{-\infty}^{\infty} \exp(-x^2)\, dx = \sqrt{\pi}$

$F(k) = \sqrt{\pi} \exp\left(-\frac{k^2}{4}\right).$

116. $y'' + p(x)\,y' + q(x)\,y(x) = 0$

Frobenier series power by using

$$y = \sum_{m=0}^{\infty} C_m\, x^{k+m}$$

Indicial equation, $k^2 + (p_0 - 1)\,k + q_0 = 0$

By taylor series expansion $p_0 = 4$, $q_0 = 2$

$$k^2 + (4-1)\,k + 2 = 0$$

$\Rightarrow$ $k^2 + 3k + 2 = 0$

$\Rightarrow$ $(k+2)(k+1) = 0$

$\Rightarrow$ $k = -1, -2.$

117. Wronskian $W = \begin{vmatrix} y_1 & y_2 \\ y'_1 & y'_2 \end{vmatrix} \neq 0$ for linearly independent

$$\begin{vmatrix} x & y \\ 1 & \frac{dy}{dx} \end{vmatrix} = W$$

$[\because$ $W \neq 0$ Condition $\left.\begin{matrix} W(x=0) = 1 \\ y = 0 \text{ at } x = \frac{1}{2} \end{matrix}\right\}$

$\Rightarrow$ $x\dfrac{dy}{dx} - y = W$

$\Rightarrow$ at $x = \dfrac{1}{2}$, $y = 0$

or $\dfrac{1}{2}\dfrac{dy}{dx} = W \Rightarrow \dfrac{dy}{dx} = 2W$ or $y = 2Wx + C$

If $x = \dfrac{1}{2}$, then $0 = W + C, \Rightarrow C = -W$

$y = 2Wx - W = (2x - 1)\,W$

so second solution in term of W multiple is,

$$y = (2x-1).$$

122. $N = mg$

$$\frac{F_{limit}}{N} = \mu$$

$\Rightarrow$ $F_{limit} = \mu\, mg\ (> N = mg)$

$$F_{max} = \mu\, mg$$

123. $\begin{vmatrix} i & j & k \\ \partial/\partial x & \partial/\partial y & \partial/\partial z \\ z & x & y \end{vmatrix} = i + j + k$

124. $x = (A + Bt)\,e^{-t}$

For $t = 0$, $\dfrac{dx}{dt} = 0$

$\Rightarrow$ $x = 0$

$$x = (1+t)\,e^{-t}$$

$$x/_{t=1} = \frac{2}{e}$$

129. It is a conservative field.

131. $A^{-1} = \dfrac{1}{|A|}\,\text{Adj } A$

$\text{Adj } A = \begin{bmatrix} 1 & 1 \\ 0 & 1 \end{bmatrix}$ and $|A| = 1$

133. Dirichlet's conditions are not satisfied.

134. Let $f = 3x^2 + 4y - z$

Unit normal vector to the surface

$f = 0$ is $\dfrac{\nabla f}{|\nabla f|}$

Now $\nabla f = 6x\hat{i} + 4\hat{j} - \hat{k}$

∇f at $(1, 1, 7) = 6\hat{i} + 4\hat{j} - \hat{k}$

$\therefore$ Required unit vector

$$= \frac{1}{\sqrt{53}}(6\hat{i} + 4\hat{j} - \hat{k})$$

139. By Gauss's divergence theorem

$$\iint_S \hat{r}\cdot\hat{n}\,ds = \iiint_V \nabla\cdot\hat{r}\,dV = 3\iiint_V dV = 3V$$

140. $f(z) = z^{10}$ is analytic inside.

143. Let $f(z) = \dfrac{\sin z}{z^6}$

$f(z)$ has a pole of order 6 at $z = 0$

Let $\phi(z) = \sin z$

$\therefore$ Residue $= \dfrac{\phi^5(0)}{\underline{|5}} = -\dfrac{1}{\underline{|5}}$

151. Characteristics equation for the given matrix

$A = \begin{bmatrix} 1 & 1 \\ 1 & 1 \end{bmatrix}$ is

$$|A - \lambda I| = 0$$

$$\begin{vmatrix} 1-\lambda & 1 \\ 1 & 1-\lambda \end{vmatrix} = (1-\lambda)^2 - 1 = 0$$

$\Rightarrow$ $1 + \lambda^2 - 2\lambda - 1 = \lambda(\lambda - 2) = 0$

$\Rightarrow$ $\lambda = 0$ or $\lambda = 2$

■■

CHAPTER 2
Classical Mechanics

2021

1. Two observers O and O' observe two events P and Q. The observers have a constant relative speed of 0.5c. In the units, where the speed of light, c, is taken as unity, the observer O obtained the following coordinates:

   Event P: $x = 5, y = 3, z = 5, t = 3$

   Event Q: $x = 5, y = 1, z = 3, t = 5$

   The length of the space-time interval between these two events, as measured by O', is L. The value of $|L|$ (in integer) is ______.

2. A hoop of mass M and radius R rolls without slipping along a straight line on a horizontal surface as shown in the figure. A point mass m slides without friction along the inner surface of the hoop, performing small oscillations about the mean position. The number of degrees of freedom of the system (in integer) is ______.

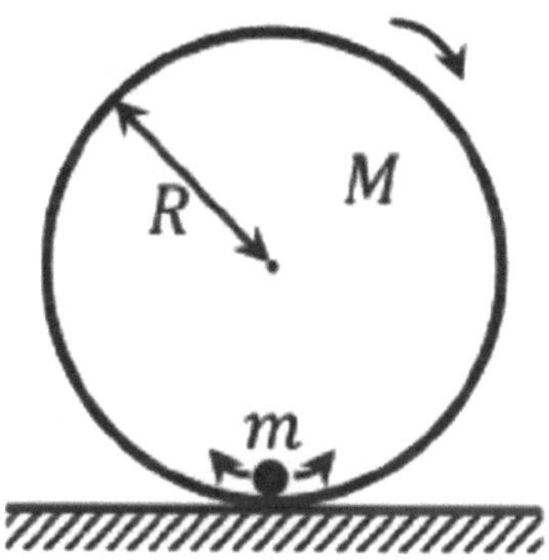

3. Consider a point charge $+Q$ of mass m suspended by a massless, inextensible string of length l in free space (permittivity ε_0) as shown in the figure. It is placed at a height d ($d > l$) over an infinitely large, grounded conducting plane. The gravitational potential energy is assumed to be zero at the position of the conducting plane and is positive above the plane.

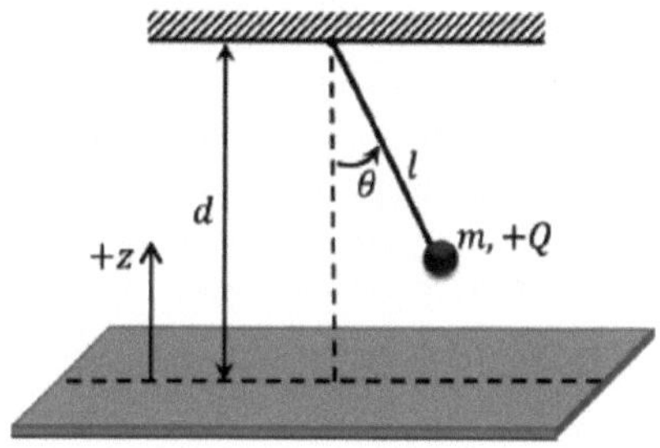

If θ represents the angular position and p_θ its corresponding canonical momentum, then the correct Hamiltonian of the system is

(a) $\dfrac{p_\theta^2}{2ml^2} - \dfrac{Q^2}{16\pi\varepsilon_0(d - l\cos\theta)} - mg(d - l\cos\theta)$

(b) $\dfrac{p_\theta^2}{2ml^2} - \dfrac{Q^2}{8\pi\varepsilon_0(d - l\cos\theta)} + mg(d - l\cos\theta)$

(c) $\dfrac{p_\theta^2}{2ml^2} - \dfrac{Q^2}{8\pi\varepsilon_0(d - l\cos\theta)} - mg(d - l\cos\theta)$

(d) $\dfrac{p_\theta^2}{2ml^2} - \dfrac{Q^2}{16\pi\varepsilon_0(d - l\cos\theta)} + mg(d - l\cos\theta)$

4. Consider a spherical galaxy of total mass M and radius R, having a uniform matter distribution. In this idealized situation, the orbital speed v of a star of mass m ($m << M$) as a function of the distance r from the galactic centre is best described by

   (*G is the universal gravitational constant*)

(a) 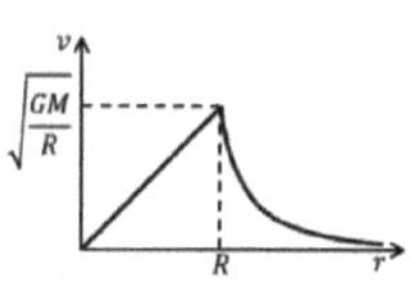

(b) 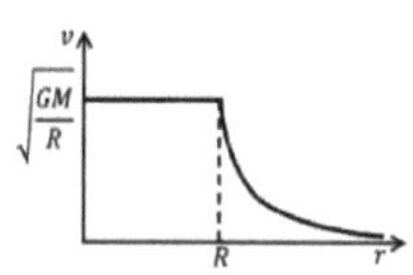

(c) 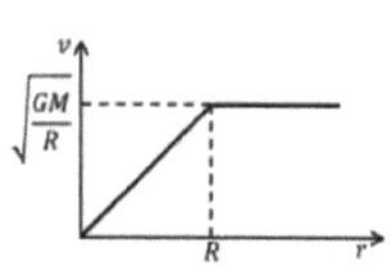

(d) 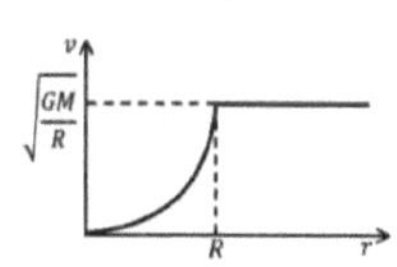

5. A uniform block of mass M slides on a smooth horizontal bar. Another mass m is connected to it by an inextensible string of length l of negligible mass, and is constrained to oscillate in the X-Y plane only. Neglect the sizes of the masses. The number of degrees of freedom of the system is two and the generalized coordinates are chosen as x and θ, as shown in the figure.

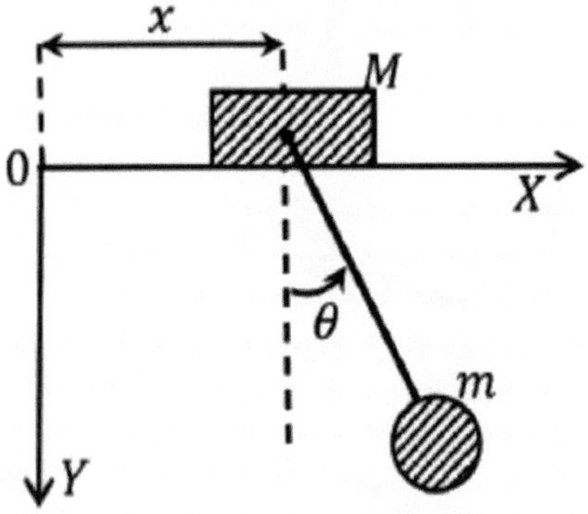

If p_x and p_θ are the generalized momenta corresponding to x and θ, respectively, then the correct option(s) is(are)

(a) $p_x = (m + M)\dot{x} + ml \cos\theta\,\dot{\theta}$

(b) $p_\theta = ml^2\dot{\theta} - ml\cos\theta\,\dot{x}$

(c) p_x is conserved

(d) p_θ is conserved

6. The time derivative of a differentiable function $g(q_i, t)$ is added to a Lagrangian $L(q_i, \dot{q}_i, t)$ such that

$$L' = L\left(q_i, \dot{q}_i, t\right) + \frac{d}{dt}g(q_i, t)$$

where q_i, $\dot{q}_i$, t are the generalized coordinates, generalized velocities and time, respectively. Let p_i be the generalized momentum and H the Hamiltonian associated with $L(q_i, \dot{q}_i, t)$. If p'_t and $\dot{H}$ are those associated with L', then the correct option(s) is(are)

(a) Both L and $\dot{L}$ satisfy the Euler-Lagrange's equations of motion

(b) $\dot{p}_i = p_i + \frac{\partial}{\partial q_i}g(q_i, t)$

(c) If p_i is conserved, then $\dot{p}_i$ is necessarily conserved

(d) $\dot{H} = H + \frac{d}{dt}(q_i, t)$

2020

7. A particle is moving in a central force field given by $\vec{F} = -\frac{k}{r^3}\hat{r}$, where $\hat{r}$ is the unit vector pointing away from the center of the field. The potential energy of the particle is given by

(a) $\frac{k}{r^2}$

(b) $\frac{k}{2r^2}$

(c) $-\frac{k}{r^2}$

(d) $-\frac{k}{2r^2}$

8. A small disc is suspended by a fiber such that it is free to rotate about the fiber axis (see figure). For small angular deflections, the Hamiltonian for the disc is given by

$$H = \frac{p_\theta^2}{2I} + \frac{1}{2}\alpha\theta^2,$$

where I is the moment of inertia and α is the restoring torque per unit deflection. The disc is subjected to angular deflections (θ) due to thermal collisions from the surrounding gas at temperature T and p_θ is the momentum conjugate to θ. The average and the root-mean-square angular deflection, θ_{avg} and θ_{rms}, respectively are

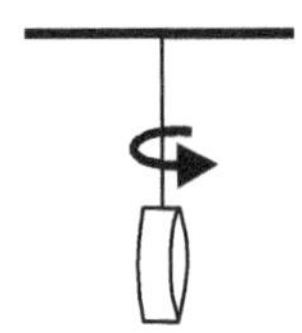

(a) $\theta_{avg} = 0$ and $\theta_{rms} = \left(\frac{k_B T}{\alpha}\right)^{\frac{3}{2}}$

(b) $\theta_{avg} = 0$ and $\theta_{rms} = \left(\frac{k_B T}{\alpha}\right)^{\frac{1}{2}}$

(c) $\theta_{avg} \neq 0$ and $\theta_{rms} = \left(\frac{k_B T}{\alpha}\right)^{\frac{1}{2}}$

(d) $\theta_{avg} \neq 0$ and $\theta_{rms} = \left(\frac{k_B T}{\alpha}\right)^{\frac{3}{2}}$

9. Let p be the momentum conjugate to the generalized coordinate q. If the transformation.

$$Q = \sqrt{2}q^m \cos p$$
$$P = \sqrt{2}q^m \sin p$$

is canonical, then m = _____.

10. If a particle is moving along a sinusoidal curve, the number of degrees of freedom of the particle is_____.

11. Consider the Lagrangian

$$L = a\left(\frac{dx}{dt}\right)^2 + b\left(\frac{dy}{dt}\right)^2 + cxy,$$

where a, b and c are constants. If p_x and p_y are the momenta conjugate to the coordinates x and y respectively, then the Hamiltonian is

(*a*) $\frac{p_x^2}{4a} + \frac{p_y^2}{4b} - cxy$ (*b*) $\frac{p_x^2}{2a} + \frac{p_y^2}{2b} - cxy$

(*c*) $\frac{p_x^2}{2a} + \frac{p_y^2}{2b} + cxy$ (*d*) $\frac{p_x^2}{a} + \frac{p_y^2}{b} + cxy$

12. The potential energy of a particle of mass m is given by

$$U(x) = a \sin\left(k^2 x - \frac{\pi}{2}\right), \; a > 0, k^2 > 0.$$

The angular frequency of small oscillations of the particle about x = 0 is

(a) $k^2\sqrt{\frac{2a}{m}}$ (b) $k^2\sqrt{\frac{a}{m}}$

(c) $k^2\sqrt{\frac{a}{2m}}$ (d) $2k^2\sqrt{\frac{a}{m}}$

2019

13. Consider a transformation from one set of generalized coordinate and momentum (q, p) to another set (Q, P) denoted by,

$Q = pq^s$; $P = q^r$

where s and r are constants. The transformation is canonical if

(*a*) s = 0 and r = 1 (*b*) s = 2 and r = –1

(*c*) S = 0 and r = –1 (*d*) s = 2 and r = 1

14. The Hamiltonian for a particle of mass m is $H = \frac{p^2}{2m} + kqt$ where q and p are the generalized coordinate and momentum, respectively, t is time and k is a constant. For the initial condition, q = 0 and p = 0 at t = 0, $q(t) \propto t\alpha$. The value of α is______

15. A ball bouncing off a rigid floor is described by the potential energy function

$$V(x) = mgx \quad \text{for } x > 0$$
$$= \infty \quad \text{for } x \le 0$$

Which of the following schematic diagrams best represents the phase space plot of the ball?

(*a*)

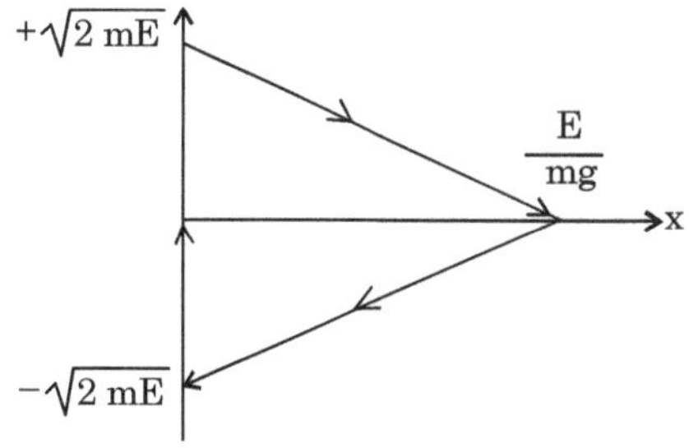

(*b*)

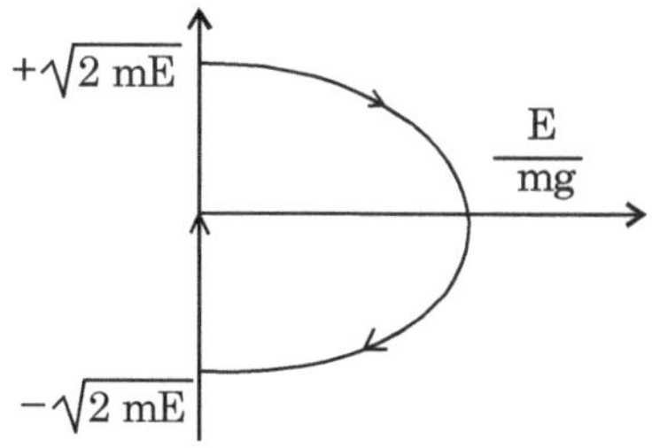

(*c*)

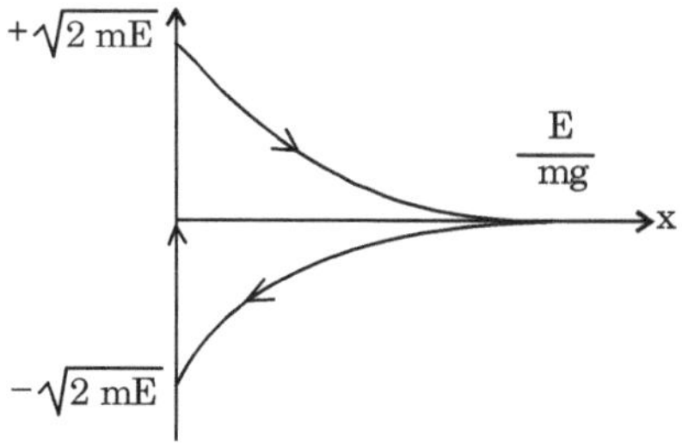

(*d*)

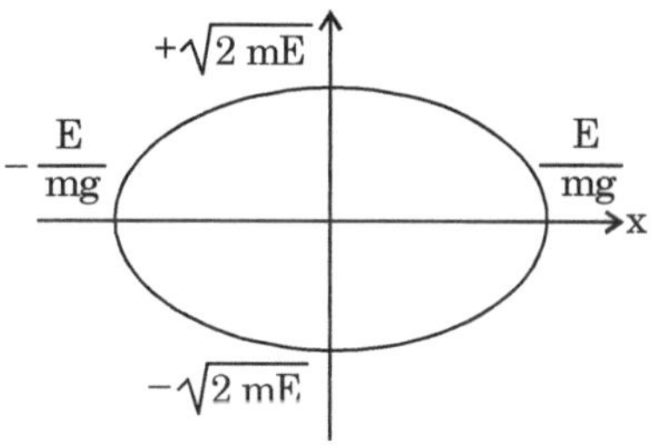

16. Consider the Hamiltonian $H(q,p) = \frac{\alpha p^2 q^4}{2} + \frac{\beta}{q^2}$, where α and β are parameters with appropriate dimensions, and q and p are the generalized coordinate and momentum, respectively. The corresponding Lagrangian L (q,q) is

(*a*) $\frac{1}{2\alpha}\frac{q^2}{q^4} - \frac{\beta}{q^2}$ (*b*) $\frac{2}{\alpha}\frac{q^2}{q^4} + \frac{\beta}{q^2}$

(*c*) $\frac{1}{\alpha}\frac{q^2}{q^4} + \frac{\beta}{q^2}$ (*d*) $-\frac{1}{2\alpha}\frac{q^2}{q^4} + \frac{\beta}{q^2}$

17. A projeetile of mass 1 kg is launched at an angle of 30° form the horizontal direction at t = 0 and takes time T before hitting the ground. If its initial speed is 10 ms^{-1}, the value of the action integral for the entire flight in the units of kg m^2s^{-1}, the value of the action integral for the entire flight in the units fo kg m^2s^{-1} (rounded off to one decimal place) is_____

[Take g = 10 ms^{-2}]

18. Two spaceships A and B, each of the same rest length L, are moving in the same direction with speeds $\frac{4c}{5}$ and $\frac{3c}{5}$ respectively, where c is is the speed of light. As measured by B, the time taken By A to completely overtake B [see figure below] in units of L/c (to the nearst integer) is

(i)

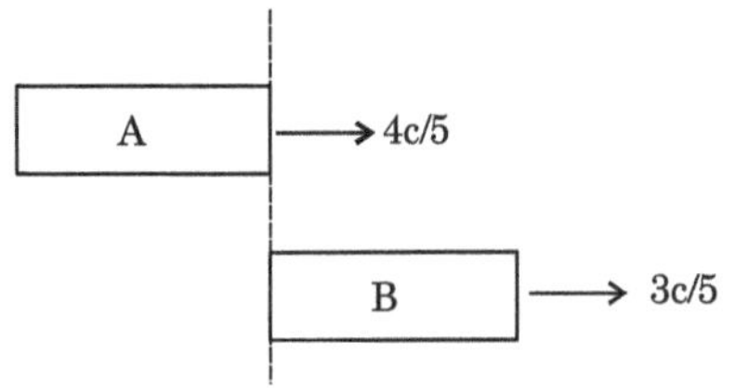

(ii)

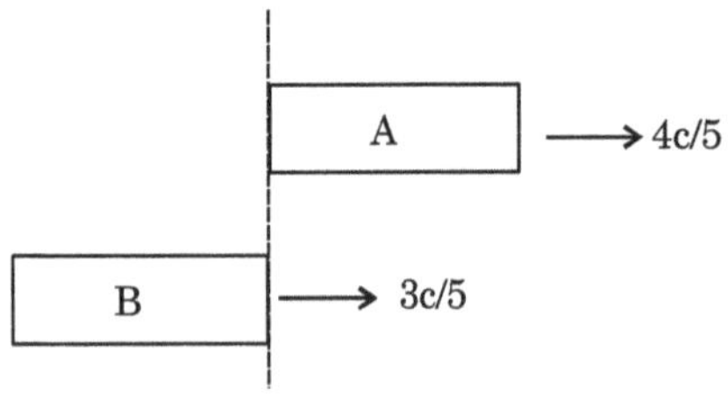

19. Two events, one on the earth and the other one on the Sun, occur simultaneously in the earth's frame. The time difference between the two events as seen by an observer in a spaceship moving with velocity 0.5c in the earth's frame along the line joining the earth to the Sun is Δt, where c is the speed of light. Given that light travels from the Sun to the earth in 8.3 minutes in the earth's frame, the value of $|\Delta t|$ in minutes (rounded off to two decimal places) is ________

(Take the earth's frame to be inertial and neglect the relative motion between the earth and the sun)

2018

20. The scale factors corresponding to the covariant metric tensor g_{ij} in spherical polar coordinates are

(*a*) $1, r^2, r^2 \sin^2 \theta$ (*b*) $1, r^2, \sin^2 \theta$

(*c*) 1, 1, 1 (*d*) $1, r, r \sin \theta$

21. In the context of small oscillations, which one of the following does NOT apply to the normal coordinates?

(*a*) Each normal coordinate has an eigen-frequency associated with it

(*b*) The normal coordinates are orthogonal to one another

(*c*) The normal coordinates are all independent

(*d*) The potential energy of the system is a sum of squares of the normal coordinates with constant coefficients

22. A spaceship is travelling with a velocity of $0.7c$ away from a space station. The spaceship ejects a probe with a velocity $0.59c$ opposite to its own velocity. A person in the space station would see the probe moving at a speed Xc, where the value of X is ________ (up to three decimal places).

23. An interstellar object has speed v at the point of its shortest distance R from a star of much larger mass M. Given $v^2 = 2GM / R$, the trajectory of the object is

(*a*) circle (*b*) ellipse

(*c*) parabola (*d*) hyperbola

24. A particle moves in one dimension under a potential $V(x) = \alpha |x|$ with some non-zero total energy. Which one of the following best describes the particle trajectory in the phase space?

(*a*)

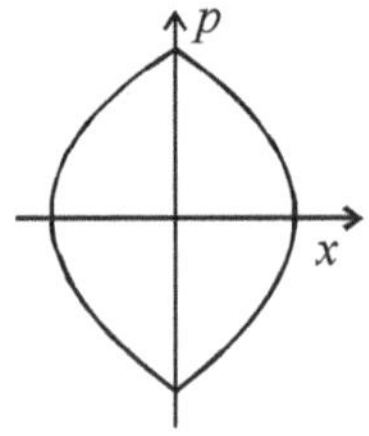

(*b*)

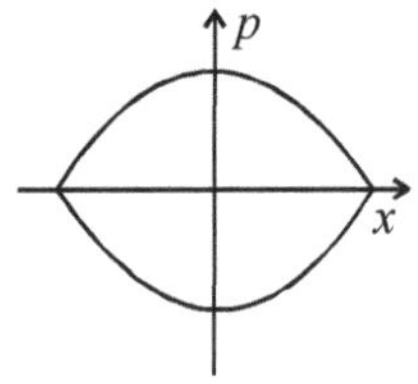

(*c*)

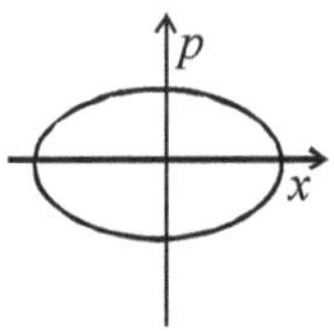

(*d*)

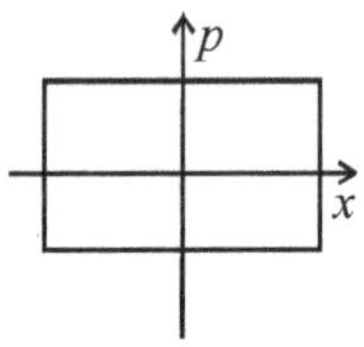

25. A uniform circular disc of mass m and radius R is rotating with angular speed ω about an axis passing through its center and making an angle $\theta = 30°$ with the axis of the disc. If the kinetic energy of the disc is $\alpha m\omega^2 R^2$, the value of α is ________ (up to 2 decimal places).

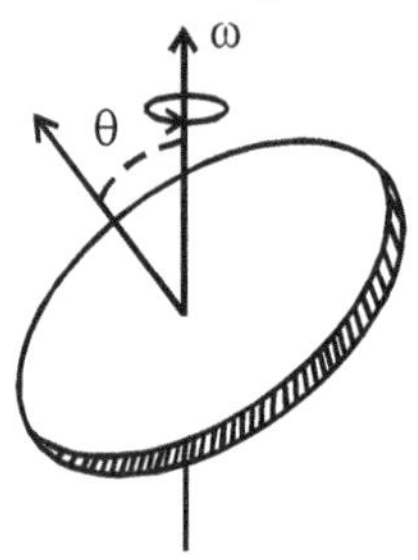

2017

26. If the Lagrangian $L_0 = \frac{1}{2}m\left(\frac{dq}{dt}\right)^2 - \frac{1}{2}m\omega^2 q^2$ is modified to $L = L_0 + \alpha q\left(\frac{dq}{dt}\right)$, which one of the following is TRUE?

(*a*) Both the canonical momentum and equation of motion do not change (*b*) Canonical momentum changes, equation of motion does not change

(*c*) Canonical momentum does not change, equation of motion changes

(*d*) Both the canonical momentum and equation of motion change.

27. Two identical masses of 10 gm each are connected by a massless spring of spring constant 1 N/m. The non-zero angular eigenfrequency of the system is ______ rad/s. (up to two decimal places).

28. The phase space trajectory of an otherwise free particle bouncing between two hard walls elastically in one dimension is a

(*a*) straight line (*b*) parabola

(*c*) rectangle (*d*) circle

29. The Poisson bracket $[x, xp_y + yp_x]$ is equal to

(*a*) $-x$ (*b*) y

(*c*) $2p_x$ (*d*) p_y

30. An object travels along the x-direction with velocity $c/2$ in a frame O. An observer in a frame O′ sees the same object travelling with velocity $c/4$. The relative velocity of O′ with respect to O in units of c is ______. (up to two decimal places).

31. A uniform solid cylinder is released on a horizontal surface with speed 5 m/s without any rotation (slipping without rolling). The cylinder eventually starts rolling without slipping. If the mass and radius of the cylinder are 10 gm and 1 cm respectively, the final linear velocity of the cylinder is ______ m/s. (up to two decimal places).

32. A person weighs w_p at Earth's north pole and w_e at the equator. Treating the Earth as a perfect sphere of radius 6400 km, the value $100 \times (w_p - w_e)/w_p$ is ______. (up to two decimal places). (Take $g = 10\ \text{ms}^{-2}$).

2016

33. Two blocks are connected by a spring of spring constant k. One block has mass m and the other block has mass 2m. If the ratio $k/m = 4\ \text{s}^{-2}$, the angular frequency of vibration ω of the two block spring system in s^{-1} is ______. (Give your answer upto two decimal places)

34. A particle of mass m = 0.1 kg is initially at rest at origin. It starts moving with a uniform acceleration $\vec{a} = 10\hat{i}\ \text{ms}^{-2}$ at t = 0. The action s of the particle, in units of J-s, at t = 2 s is ______. (Give your answer upto two decimal places)

35. In an inertial frame of reference S, an observer finds two events occurring at the same time at coordinates $x_1 = 0$ and $x_2 = d$. A different inertial frame S′ moves with velocity v with respect to S along the positive x-axis. An observer in S′ also notices these two events and finds them to occur at times t'_1 and t'_2 and at positions x'_1 and x'_2 respectively.

If $\Delta t' = t'_2 - t'_1$, $\Delta x' = x'_2 - x'_1$ & $\gamma = \dfrac{1}{\sqrt{1-\dfrac{v^2}{c^2}}}$

which of the following statements is true?

(*a*) $\Delta t' = 0, \Delta x' = \gamma d$

(*b*) $\Delta t' = 0, \Delta x' = d/\gamma$

(*c*) $\Delta t' = -\gamma vd/c^2, \Delta x' = \gamma d$

(*d*) $\Delta t' = -\gamma vd/c^2, \Delta x' = d/\gamma$

36. Consider a system having three energy levels with energies 0, 2ε and , 3ε with respective degeneracies of 2, 2 and 3. Four bosons of spin zero have to be accommodated in these levels such that the total energy of the system is 10ε. The number of ways in which it can be done is ____.

37. The Lagrangian of a system is given by

$L = \frac{1}{2}ml^2\left[\dot{\theta}^2 + \sin^2\theta\,\dot{\varphi}^2\right] - mgl\cos\theta$, where m, l and g are constants.

Which of the following is conserved?

(*a*) $\dot{\varphi}\sin^2\theta$ (*b*) $\dot{\varphi}\sin\theta$

(*c*) $\dfrac{\dot{\varphi}}{\sin\theta}$ (*d*) $\dfrac{\dot{\varphi}}{\sin^2\theta}$

2015

38. A satellite is moving in a circular orbit around the Earth. If T, V and E are its average kinetic, average potential and total energies, respectively, then which one of the following options is correct?

(*a*) V = −2T ; E = −T (*b*) V = −T ; E = 0

(*c*) $V = \dfrac{-T}{2}; E = \dfrac{T}{2}$ (*d*) $V = -\dfrac{3T}{2}; E = -\dfrac{T}{2}$

39. Four forces are given below in Cartesian and spherical polar coordinates.

(*i*) $\vec{F}_1 = K\exp(-r^2/R^2)\hat{r}$

(*ii*) $\vec{F}_2 = K(x^3\hat{y} - y^3\hat{z})$

(*iii*) $\vec{F}_3 = K(x^3\hat{x} + y^3\hat{y})$

(*iv*) $\vec{F}_4 = K(\hat{\phi}/r)$

where K is a constant. Identify the correct option.

(*a*) (*iii*) and (*iv*) are conservative but (*i*) and (*ii*) are not

(*b*) (*i*) and (*ii*) are conservative but (*iii*) and (*iv*) are not

(c) (ii) and (iii) are conservative but (i) and (iv) are not

(d) (i) and (iii) are conservative but (ii) and (iv) are not

40. An operator for a spin-½ particle is given by $\hat{A} = \lambda\vec{\sigma}.\vec{B}$, where $\vec{B} = \frac{B}{\sqrt{2}}(\hat{x}+\hat{y}), \vec{\sigma}$ denotes Pauli matrices and λ is a constant. The eigen values of $\hat{A}$ are

(a) $\frac{\pm\lambda B}{\sqrt{2}}$ (b) $\pm\lambda B$

(c) 0, λ B (d) 0, $-\lambda$ B

41. In an inertial frame S, two events A and B take place at $(ct_A = 0, \vec{r}_A = 0)$ and $(ct_B = 0, \vec{r}_B = 2)$ respectively. The times at which these events take place in a frame S′ moving with a velocity $0.6c\,\hat{y}$ with respect to S are given by

(a) $ct'_A = 0; ct'_B = -3/2$ (b) $ct'_A = 0; ct'_B = 0$

(c) $ct'_A = 0; ct'_B = 3/2$ (d) $ct'_A = 0; ct'_B = 1/2$

42. The Hamiltonian for a system of two particles of masses m_1 and m_2 at $\vec{r}_1$ and $\vec{r}_2$ having velocities $\vec{v}_1$ and $\vec{v}_2$ is given by

$$H = \frac{1}{2}m_1v_1^2 + \frac{1}{2}m_2v_2^2 + \frac{C}{(\vec{r}_1 - \vec{r}_2)^2}\hat{z}.(\vec{r}_1 \times \vec{r}_2),$$

where C is a constant. Which one of the following statements is correct?

(a) The total energy and total momentum are conserved

(b) Only the total energy is conserved

(c) The total energy and the z – component of the total angular momentum are conserved

(d) The total energy and total angular momentum are conserved

43. A particle with rest mass M is at rest and decays into two particles of equal rest masses $\frac{3}{10}$M which move along the z axis. Their velocities are given by

(a) $\vec{v}_1 = \vec{v}_2 = (0.8c)\hat{z}$

(b) $\vec{v}_1 = -\vec{v}_2 = (0.8c)\hat{z}$

(c) $\vec{v}_1 = -\vec{v}_2 = (0.6c)\hat{z}$

(d) $\vec{v}_1 = (0.6c)\hat{z};\ \vec{v}_2 = (-0.8c)\hat{z}$

44. The Lagrangian for a particle of mass m at a position $\vec{r}$ moving with a velocity $\vec{v}$ is given by $L = \frac{m}{2}\vec{v}^2 + C\vec{r}.\vec{v} - V(r)$, where V($r$) is a potential and C is constant. If $\vec{p}_c$ is the canonical momentum, then its Hamiltonian is given by

(a) $\frac{1}{2m}(\vec{p}_c + C\vec{r})^2 + V(r)$

(b) $\frac{1}{2m}(\vec{p}_c - C\vec{r})^2 + V(r)$

(c) $\frac{p_c^2}{2m} + V(r)$

(d) $\frac{1}{2m}p_c^2 + C^2r^2 + V(r)$

45. A charge $-q$ is distributed uniformly over a sphere, with a positive charge q at its center in Also in (ii), a charge $-q$ is distributed uniformly over an ellipsoid with a positive charge q at center. With respect to the origin of the coordinate system, which one of the following statement is correct ?

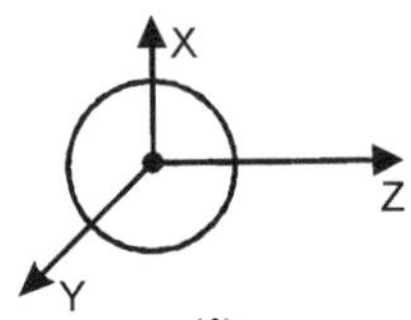

(i)

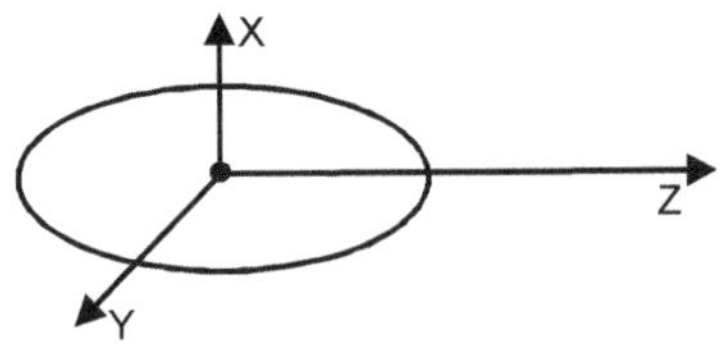

(ii)

(a) The dipole moment is zero in both (i) and (ii)

(b) The dipole moment is non-zero in (i) but zero in (ii)

(c) The dipole moment is zero in (i) but non-zero in (ii)

(d) The dipole moment is non-zero in both (i) and (ii)

46. Consider the motion of the Sun with respect to the rotation of the Earth about its axis. If $\vec{F}_c$ and $\vec{F}_{Co}$ denote the centrifugal and the Coriolis forces, respectively, acting on the Sun, then

(a) $\vec{F}_c$ is radially outward and $\vec{F}_{Co} = \vec{F}_c$

(b) $\vec{F}_c$ is radially inward and $\vec{F}_{Co} = -2\vec{F}_c$

(c) $\vec{F}_c$ is radially outward and $\vec{F}_{Co} = -2\vec{F}_c$

(d) $\vec{F}_c$ is radially outward and $\vec{F}_{Co} = 2\vec{F}_c$

47. A particle of mass 0.01 kg falls freely in the earth's gravitational field with an initial velocity $v(0) = 10\ ms^{-1}$. If the air exerts a frictional force of the form, $f = -kv$, then for $k = 0.05\ Nm^{-1}s$, the velocity (in ms^{-1}) at time $t = 0.2\ s$ is ________ (upto two decimal places) (use $g = 10\ ms^{-2}$ and $e = 2.72$)

2014

48. Which one of the following quantities is invariant under Lorentz transformation?

(*a*) Charge density (*b*) Charge

(*c*) Current (*d*) Electric field

49. If the half-life of an elementary particle moving with speed 0.9*c* in the laboratory frame is $5 \times 10^{-8} s$, then the proper half-life is ______ $\times 10^{-8} s$. ($c = 3 \times 10^8\ m/s$)

50. Two masses *m* and 3*m* are attached to the two ends of a massless spring with force constant K. If $m = 100\ g$ and $K = 0.3\ N/m$, then the natural angular frequency of oscillation is ________ *Hz*.

51. The Hamilton's canonical equations of motion in terms of Poisson Brackets are

(*a*) $\dot{q} = \{q, H\}; \dot{p} = \{p, H\}$

(*b*) $\dot{q} = \{H, q\}; \dot{p} = \{H, p\}$

(*c*) $\dot{q} = \{H, p\}; \dot{p} = \{H, q\}$

(*d*) $\dot{q} = \{p, H\}; \dot{p} = \{q, H\}$

52. A bead of mass *m* can slide without friction along a massless rod kept at 45^0 with the vertical as shown in the figure. The rod is rotating about the vertical axis with a constant angular speed ω. At any instant, *r* is the distance of the bead from the origin. The momentum conjugate to *r* is

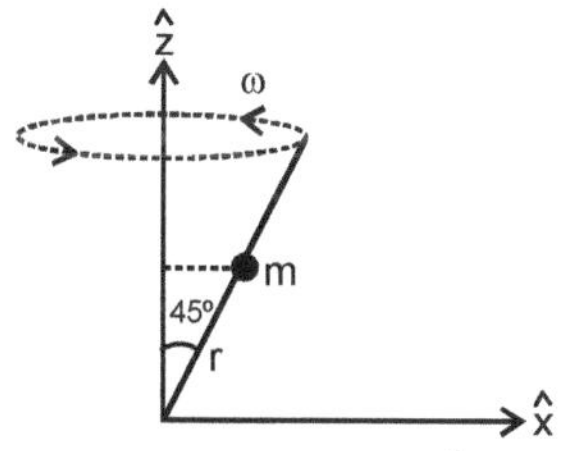

(*a*) $m\dot{r}$ (*b*) $\frac{1}{\sqrt{2}} m\dot{r}$

(*c*) $\frac{1}{2} m\dot{r}$ (*d*) $\sqrt{2} m\dot{r}$

53. A particle of mass *m* is in a potential given by

$$V(r) = -\frac{a}{r} + \frac{a r_0^2}{3r^3},$$

where *a* and r_0 are positive constants. When disturbed slightly from its stable equilibrium position, it undergoes a simple harmonic oscillation. The time period of oscillation is

(*a*) $2\pi\sqrt{\frac{m r_0^3}{2a}}$ (*b*) $2\pi\sqrt{\frac{m r_0^3}{a}}$

(*c*) $2\pi\sqrt{\frac{2m r_0^3}{a}}$ (*d*) $4\pi\sqrt{\frac{m r_0^3}{a}}$

54. A planet of mass *m* moves in a circular orbit of radius r_0 in the gravitational potential $V(r) = -\frac{k}{r}$, where *k* is a positive constant. The orbital angular momentum of the planet is

(*a*) $2r_0 km$ (*b*) $\sqrt{2r_0 km}$

(*c*) $r_0 km$ (*d*) $\sqrt{r_0 km}$

55. Given that the linear transformation of a generalized coordinate *q* and the corresponding momentum *p*,

$Q = q + 4ap$

$P = q + 2p$

is canonical, the value of the constant *a* is ______.

56. The value of the magnetic field required to maintain non-relativistic protons of energy 1 MeV in a circular orbit of radius 100 mm is ________ Tesla.

(Given: $m_p = 1.67 \times 10^{-27}$ kg, $e = 1.6 \times 10^{-19}$ C)

2013

57. An electron is moving with a velocity of 0.85*c* in the same direction as that of a moving photon. The relative velocity of the electron with respect to photon is

(*a*) *c* (*b*) $-c$

(*c*) 0.15*c* (*d*) $-0.15c$

58. The Lagrangian of a system with one degree of freedom *q* is given by $L = \alpha\dot{q}^2 + \beta q^2$, where α and β are non-zero constants. If p_q denotes the canonical momentum conjugate to *q* then which one of the following statements is CORRECT?

(*a*) $p_q = 2\beta q$ and it is a conserved quantity.

(*b*) $p_q = 2\beta q$ and it is not a conserved quantity.

(*c*) $p_q = 2\alpha\dot{q}$ and it is a conserved quantity.

(*d*) $p_q = 2\alpha\dot{q}$ and it is not a conserved quantity.

59. The relativistic form of Newton's second law of motion is

(*a*) $F = \frac{mc}{\sqrt{c^2 - v^2}} \frac{dv}{dt}$ (*b*) $F = \frac{m\sqrt{c^2 - v^2}}{c} \frac{dv}{dt}$

(*c*) $F = \frac{mc^2}{c^2 - v^2} \frac{dv}{dt}$ (*d*) $F = m\frac{c^2 - v^2}{c^2} \frac{dv}{dt}$

60. A uniform circular disk of radius R and mass M is rotating with angular speed ω about an axis, passing through its center and inclined at an angle 60 degrees with respect to its symmetry axis. The magnitude of the angular momentum of the disk is,

(*a*) $\frac{\sqrt{3}}{4}\omega MR^2$ (*b*) $\frac{\sqrt{3}}{8}\omega MR^2$

(*c*) $\frac{\sqrt{7}}{8}\omega MR^2$ (*d*) $\frac{\sqrt{7}}{4}\omega MR^2$

61. Consider two small blocks, each of mass M, attached to two identical springs. One of the springs is attached to the wall, as shown in the figure. The spring constant of each spring is k. The masses slide along the surface and the friction is negligible. The frequency of one of the normal modes of the system is,

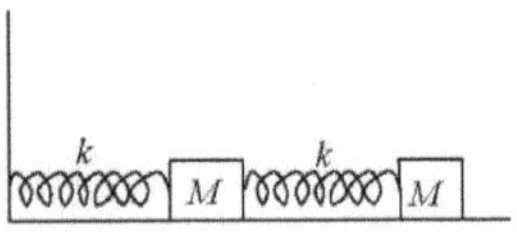

(*a*) $\sqrt{\frac{3+\sqrt{2}}{2}}\sqrt{\frac{k}{M}}$ (*b*) $\sqrt{\frac{3+\sqrt{3}}{2}}\sqrt{\frac{k}{M}}$

(*c*) $\sqrt{\frac{3+\sqrt{5}}{2}}\sqrt{\frac{k}{M}}$ (*d*) $\sqrt{\frac{3+\sqrt{6}}{2}}\sqrt{\frac{k}{M}}$

2012

62. Two uniform thin rods of equal length, L, and masses M_1 and M_2 are joined together along the length. The moment of inertia of the combined rod of length 2L about an axis passing through the mid-point and perpendicular to the length of the rod is,

(*a*) $(M_1+M_2)\frac{L^2}{12}$ (*b*) $(M_1+M_2)\frac{L^2}{6}$

(*c*) $(M_1+M_2)\frac{L^2}{3}$ (*d*) $(M_1+M_2)\frac{L^2}{2}$

63. In a central force field, the trajectory of a particle of mass m and angular momentum L in plane polar coordinates is given by,

$$\frac{1}{r}=\frac{m}{L^2}(1+\varepsilon\cos\theta)$$

where, ε is the eccentricity of the particle's motion. Which one of the following choices for ε gives rise to a parabolic trajectory?

(*a*) $\varepsilon = 0$ (*b*) $\varepsilon = 1$

(*c*) $0 < \varepsilon < 1$ (*d*) $\varepsilon > 1$

64. A particle of unit mass moves along the x-axis under the influence of a potential, $V(x) = x(x-2)^2$. The particle is found to be in stable equilibrium at the point $x = 2$. The time period of oscillation of the particle is

(*a*) $\frac{\pi}{2}$ (*b*) π

(*c*) $\frac{3\pi}{2}$ (*d*) 2π

65. A rod of proper length l_0 oriented parallel to the x-axis moves with speed $2c/3$ along the x-axis in the S-frame, where c is the speed of the light in free space. The observer is also moving along the x-axis with speed $c/2$ with respect to the S-frame. The length of the rod as measured by the observer is

(*a*) $0.35\ l_0$ (*b*) $0.48\ l_0$

(*c*) $0.87\ l_0$ (*d*) $0.97\ l_0$

66. A particle of mass m is attached to a fixed point O by a weightless inextensible string of length a. It is rotating under the gravity as shown in the figure. The Lagrangian of the particle is

$$L(\theta,\phi)=\frac{1}{2}ma^2(\dot\theta^2+\sin^2\theta\dot\phi^2)-mga\cos\theta$$

where θ and ϕ are the polar angles.

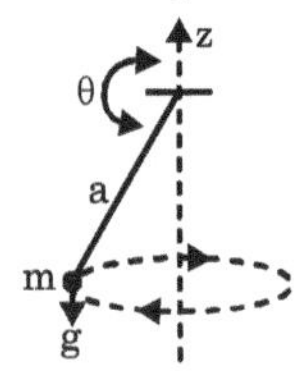

The Hamiltonian of the particle is

(*a*) $H=\frac{1}{2ma^2}\left(p_\theta^2+\frac{p_\phi^2}{\sin^2\theta}\right)-mga\cos\theta$

(*b*) $H=\frac{1}{2ma^2}\left(p_\theta^2+\frac{p_\phi^2}{\sin^2\theta}\right)+mga\cos\theta$

(*c*) $H=\frac{1}{2ma^2}\left(p_\theta^2+p_\phi^2\right)-mga\cos\theta$

(*d*) $H=\frac{1}{2ma^2}\left(p_\theta^2+p_\phi^2\right)+mga\cos\theta$

Linked Answer Q. 67 and 68 :

A particle of mass m slides under the gravity without friction along the parabolic path $y = ax^2$ as shown in the figure. Here a is a constant.

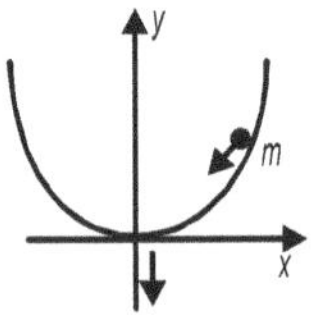

67. The Lagrangian for this particle is given by,

(*a*) $L = \frac{1}{2}m\dot{x}^2 - mgax^2$

(*b*) $L = \frac{1}{2}m(1+4a^2x^2)\dot{x}^2 - mgax^2$

(*c*) $L = \frac{1}{2}m\dot{x}^2 + mgax^2$

(*d*) $L = \frac{1}{2}m(1+4a^2x^2)\dot{x}^2 + mgax^2$

68. The Lagrange's equation of motion of the particle is

(*a*) $\ddot{x} = 2gax$

(*b*) $m(1+4a^2x^2)\ddot{x} = -2mgax - 4ma^2x\dot{x}^2$

(*c*) $m(1+4a^2x^2)\ddot{x} = 2mgax + 4ma^2x\dot{x}^2$

(*d*) $\ddot{x} = -2gax$

2011

69. A heavy symmetrical top is rotating about its own axis of symmetry (the z-axis). If I_1, I_2 and I_3 are the principal moments of inertia along x, y and z axes respectively, then

(*a*) $I_2 = I_3$; $I_1 \neq I_2$ (*b*) $I_1 = I_3$; $I_1 \neq I_2$

(*c*) $I_1 = I_2$; $I_1 \neq I_3$ (*d*) $I_1 \neq I_2 \neq I_3$

70. A particle is moving under the action of a generalized potential

$$V(q, \dot{q}) = \frac{(1+\dot{q})}{q^2}$$

The magnitude of the generalized force is

(*a*) $\frac{2(1+\dot{q})}{q^3}$ (*b*) $\frac{2(1-\dot{q})}{q^3}$

(*c*) $\frac{2}{q^3}$ (*d*) $\frac{q}{q^3}$

71. Two bodies of mass m and $2m$ are connected by spring constant k. The frequency of the normal mode is

(*a*) $\sqrt{3k/2m}$ (*b*) $\sqrt{k/m}$

(*c*) $\sqrt{2k/3m}$ (*d*) $\sqrt{k/2m}$

72. Let (p, q) and (P, Q) be two pairs of canonical variables. The transformation

$$Q = q^{\alpha} \cos(\beta p)$$
$$P = q^{\alpha} \sin(\beta p)$$

is canonical for

(*a*) $\alpha = 2, \beta = 1/2$ (*b*) $\alpha = 2, \beta = 2$

(*c*) $\alpha = 1, \beta = 1$ (*d*) $\alpha = 1/2, \beta = 2$

73. Two particles, each of rest mass m collide head-on and stick together. Before collision, the speed of each mass was 0.6 times the speed of light in free space. The mass of the final entity is

(*a*) $5m/4$

(*b*) $2m$

(*c*) $5m/2$

(*d*) $25m/8$

2010

74. A system of N non-interacting classical point particles is constrained to move on the two-dimensional surface of a sphere. The internal energy of the system is

(*a*) $\frac{3}{2}Nk_BT$ (*b*) $\frac{1}{2}Nk_BT$

(*c*) Nk_BT (*d*) $\frac{5}{2}Nk_BT$

75. For the set of all Lorentz transformations with velocities along the x-axis, consider the two statements given below:

P : If L is a Lorentz transformation them, L^{-1}, is also a Lorentz transformation.

Q : If L_1 and L_2 are Lorentz transforma-tions then, L_1L_2 is necessarily a Lorentz transformation.

Choose the correct option.

(*a*) P is true and Q is false

(*b*) Both P and Q are true

(*c*) Both P and Q are false

(*d*) P is false and Q is true

76. A particle is placed in a region with the potential $V(x) = \frac{1}{2}kx^2 - \frac{\lambda}{3}x^3$, where $k, l > 0$. Then,

(*a*) $x = 0$ and $x = \frac{k}{\lambda}$ are points of stable equilibrium

(*b*) $x = 0$ is a point of stable equilibrium and $x = \frac{k}{\lambda}$ is a point of unstable equilibrium

(*c*) $x = 0$ and $x = \frac{k}{\lambda}$ are points of unstable equilibrium

(*d*) There are no points of stable or unstable equilibrium

Statement for Linked Answer Questions 77 and 78:

The Lagrangian for a simple pendulum is given by:

$$L = \frac{1}{2} ml^2 \dot{\theta}^2 - mgl(1 - \cos\theta)$$

77. Hamilton's equations are then given by

(a) $\dot{p}_\theta = -mgl\sin\theta;\ \dot{\theta} = \frac{P_\theta}{ml^2}$

(b) $\dot{p}_\theta = mgl\sin\theta;\ \dot{\theta} = \frac{P_\theta}{ml^2}$

(c) $\dot{p}_\theta = -m\ddot{\theta};\ \dot{\theta} = \frac{P_\theta}{m}$

(d) $\dot{p}_\theta = -\left(\frac{g}{l}\right)\theta;\ \dot{\theta} = \frac{P_\theta}{ml}$

78. The Poisson bracket between θ and $\dot{\theta}$ is

(a) $\{\theta, \dot{\theta}\} = 1$ (b) $\{\theta, \dot{\theta}\} = \frac{1}{ml^2}$

(c) $\{\theta, \dot{\theta}\} = \frac{1}{m}$ (d) $\{\theta, \dot{\theta}\} = \frac{g}{l}$

2009

79. The Lagrangian of a free particle in spherical polar co-ordinates is given by

$$L = \frac{1}{2} m\,(\dot{r}^2 + r^2\dot{\theta}^2 + r^2\dot{\phi}^2\sin^2\theta).$$

The quantity that is conserved is

(a) $\frac{\partial L}{\partial \dot{r}}$ (b) $\frac{\partial L}{\partial \dot{\theta}}$

(c) $\frac{\partial L}{\partial \dot{\phi}}$ (d) $\frac{\partial L}{\partial \dot{\phi}} + \dot{r}\dot{\theta}$

80. If p and q are the position and momentum variables, which one of the following is NOT a canonical transformation?

(a) $Q = \alpha q$ and $P = \frac{1}{\alpha}p$, for $\alpha \neq 0$

(b) $Q = \alpha q + \beta p$ and $P = \beta q + \alpha p$ for α, β real and $\alpha^2 - \beta^2 = 1$

(c) $Q = p$ and $P = q$

(d) $Q = p$ and $P = -q$

81. Assuming the mean life time of a muon (in its rest frame) to be 2×10^{-6} s, its life time in the laboratory frame, when it is moving with a velocity $0.95c$ is

(a) 6.4×10^{-6} s (b) 0.62×10^{-6} s

(c) 2.16×10^{-6} s (d) 0.19×10^{-6} s

82. A classical particle is moving in an external potential field $V(x, y, z)$ which is invariant under the following infinitesimal transformations

$$x \to x' = x + \delta x,$$
$$y \to y' = y + \delta y,$$
$$\begin{pmatrix} x \\ y \end{pmatrix} \to \begin{pmatrix} x' \\ y' \end{pmatrix} = R_z \begin{pmatrix} x \\ y \end{pmatrix},$$

where R_z is the matrix corresponding to rotation about the z axis. The conserved quantities are (the symbols have their usual meaning)

(a) p_x, p_z, L_z (b) p_x, p_y, L_z, E

(c) p_y, L_z, E (d) p_y, p_z, L_x, E

83. The Lagrangian of a diatomic molecule is given by $L = \frac{m}{2}(\dot{x}_1^2 + \dot{x}_2^2) - \frac{k}{2}x_1x_2$, where m is the mass of each of the atoms and x_1 and x_2 are the displacements of atoms measured from the equilibrium position and $k > 0$. The normal frequencies are

(a) $\pm\left(\frac{k}{m}\right)^{1/2}$ (b) $\pm\left(\frac{k}{m}\right)^{1/4}$

(c) $\pm\left(\frac{k}{2m}\right)^{1/4}$ (d) $\pm\left(\frac{k}{2m}\right)^{1/2}$

84. The Lagrangian of a particle of mass m moving in one dimension is $L = \exp(\alpha t)\left[\frac{m\dot{x}^2}{2} - \frac{kx^2}{2}\right]$, where α and k are positive constants. The equation of motion of the particle is

(a) $\ddot{x} + \alpha\dot{x} = 0$ (b) $\ddot{x} + \frac{k}{m}x = 0$

(c) $\ddot{x} - \alpha\dot{x} + \frac{k}{m}x = 0$ (d) $\ddot{x} + \alpha\dot{x} + \frac{k}{m}x = 0$

2008

85. A rigid body is rotating about its centre of mass, fixed at the origin, with an angular velocity $\vec{\omega}$ and angular acceleration $\vec{\alpha}$. If the torque acting on it $\vec{\tau}$ and its angular momentum is $\vec{L}$, the rate of change of its kinetic energy is

(a) $\frac{1}{2}\vec{\tau}.\vec{\omega}$ (b) $\frac{1}{2}\vec{L}.\vec{\omega}$

(c) $\frac{1}{2}(\vec{\tau}.\vec{\omega} + \vec{L}.\vec{\alpha})$ (d) $\frac{1}{2}\vec{L}.\vec{\alpha}$

86. A cylinder of mass M and radius R is rolling down without slipping on an inclined plane of angle of inclination θ. The number of generalized coordinates required to describe the motion of this system is

(a) 1 (b) 2

(c) 4 (d) 6

87. A finite wave train, of an unspecified nature, propagates along the positive x axis with a constant speed v and without any change of shape. The differential equation among the four listed below, whose solution it must be, is

(a) $\left(\frac{\partial^2}{\partial x^2} - \frac{1}{V^2}\frac{\partial^2}{\partial t^2}\right)\Psi(x, t) = 0$

(b) $\left(\nabla^2 - \frac{1}{V^2}\frac{\partial^2}{\partial t^2}\right)\Psi(\vec{r}, t) = 0$

(c) $\left(-\frac{\hbar}{2m}\frac{\partial^2}{\partial x^2} - i\hbar\frac{\partial^2}{\partial t}\right)\Psi(x, t) = 0$

(d) $\left(\nabla^2 - a\frac{\partial}{\partial t}\right)\Psi(\vec{r}, t) = 0$

88. The Lagrangian of a system is given by $L = \frac{1}{2}\dot{q}^2 + q\dot{q} - \frac{1}{2}q^2$. It describes the motion of

(a) a harmonic oscillator.

(b) a damped harmonic oscillator

(c) an anharmonic oscillator.

(d) a system with unbounded motion.

89. The moment of inertia tensor of a rigid body is given by $I = \begin{pmatrix} 8 & 0 & -4 \\ 0 & 4 & 0 \\ -4 & 0 & 8 \end{pmatrix}$. The magnitude of the moment of inertia about an axis $\hat{n} = \left(\frac{1}{2}, \frac{\sqrt{3}}{2}, 0\right)$ is

(a) 6 (b) 5

(c) 2 (d) $\frac{8}{3}$

90. A hoop of radiius R is pivoted at a point on the circumference. The period of small oscillations in the plane of the hoop is

(a) $2\pi\sqrt{\frac{2R}{g}}$ (b) $2\pi\sqrt{\frac{R}{4g}}$

(c) $2\pi\sqrt{\frac{R}{g}}$ (d) $2\pi\sqrt{\frac{9R}{7g}}$

91. A mass m is constrained to move on a horizontal frictionless suurface. It is set in circular motion with radius r_0 and angular speed ω_0 by an applied force $\vec{F}$ communicated through an inextensible thread that passes through a hole on the surface as shown in the figure. This force is then suddenly doubled. The magnitude of the radial velocity of the mass

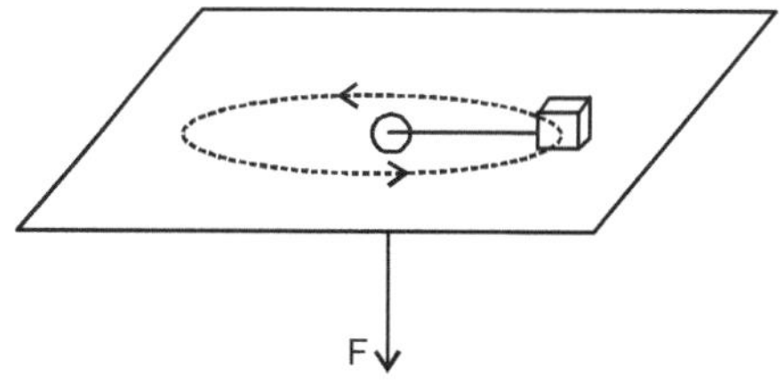

(a) increases till the mass falls into the hole.

(b) decreases till the mass falls into the hole.

(c) remains costant.

(d) becomes zero at a radius r_1 where $0 < r_1 < r_0$.

92. For a simple harmonic oscillator the Largrangian is given by $L = \frac{1}{2}\dot{q}^2 - \frac{1}{2}q^2$. If $A(p, d) = \frac{p + iq}{\sqrt{2}}$ and $H(p, q)$ is the Hamiltoniam of the system, the Poisson bracket $\{A(p, q), H(p, q)\}$ is given by

(a) $iA(p, q)$ (b) $A^*(p, q)$

(c) $-iA^*(p, q)$ (d) $-iA(p, q)$

93. An inelastic ball of mass m has been thrown vertically upwards from the ground at $z = 0$. The initial kinetic energy of the ball is E. The phase trajectory of the ball after successive bouncing on the ground is

(a)

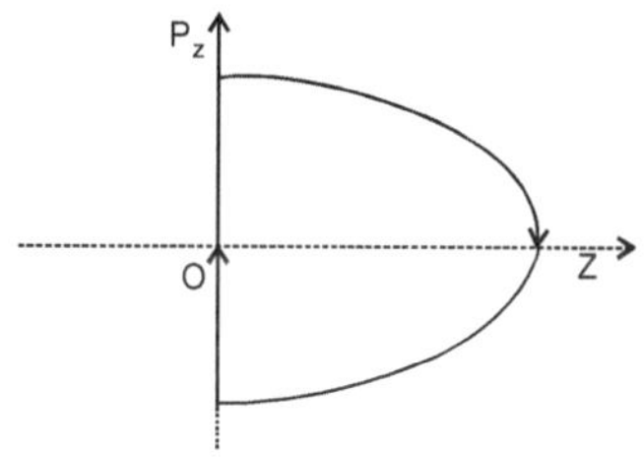

(b)

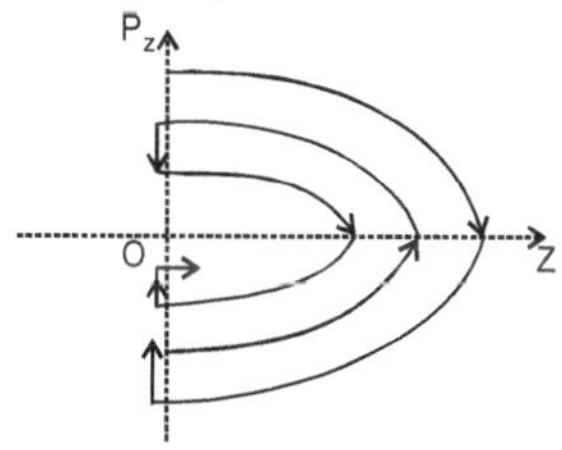

(c)

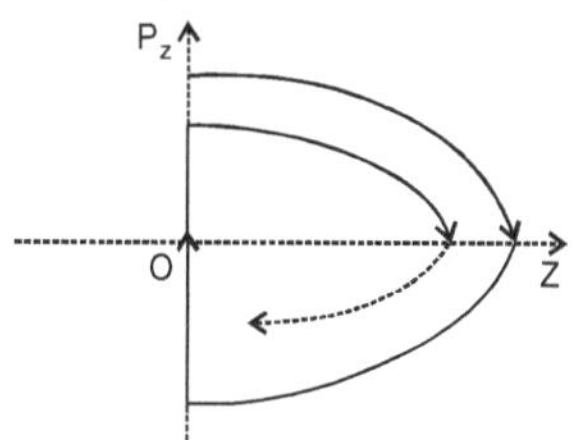

(d)

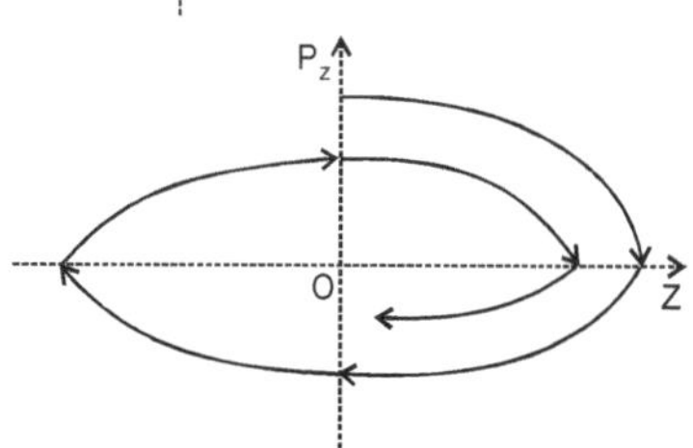

94. A system containing N non-interacting localized particles of spine $\frac{1}{2}$ and magnetic moment μ each is kept in constant external magnetic field B and in normal equilibrium at temperature T. The magnetization of the system is,

(a) $N\mu \coth\left(\frac{\mu B}{k_B T}\right)$ (b) $N\mu \tanh\left(\frac{\mu B}{k_B T}\right)$

(c) $N\mu \sinh\left(\frac{\mu B}{k_B T}\right)$ (d) $N\mu \cosh\left(\frac{\mu B}{k_B T}\right)$

Statement for Linked Answer Questions 95 and 96:

The rate of a clock in a spaceship "Suryashakti" is observed from earth to be $\frac{3}{5}$ of the rate of the clocks on earth.

95. The speed of the spaceship "Suryashakti" relative to earth is

(a) $\frac{4}{5}c$ (b) $\frac{3}{5}c$

(c) $\frac{9}{10}c$ (d) $\frac{2}{5}c$

96. The rate of a clock in a spaceship "Aakashganga" is observed from earth to be $\frac{5}{13}$ of the rate of the clocks on earth. If both Aakashganga and Suryashakti are moving in the same direction relative to someone on earth, then the speed of Aakashganga relative to Suryashakti is

(a) $\frac{12}{13}c$ (b) $\frac{4}{5}c$

(c) $\frac{8}{17}c$ (d) $\frac{5}{6}c$

2007

97. Inverse Laplace transform of $\frac{s+1}{s^2-4}$ is

(a) $\cos 2x + \frac{1}{2}\sin 2x$ (b) $\cos x + \frac{1}{2}\sin x$

(c) $\cosh x + \frac{1}{2}\sinh x$ (d) $\cosh 2x + \frac{1}{2}\sinh 2x$

98. Match the following

P : rest mass	1. timelike vector
Q : charge	2. Lorentz invariant
R : four-momentum	3. temnsor of rank 2
S : electromagnetic field	4. conserved and Lorentz invariant

(a)	(b)	(c)	(d)
P-2	P-4	P-2	P-4
Q-4	Q-2	Q-4	Q-2
R-3	R-1	R-1	R-3
S-1	S-3	S-3	S-1

99. The moment of inertia of a uniform sphere of radius r about an axis passing through its centre is given by $\frac{2}{5}\left(\frac{4\pi}{3}r^5\rho\right)$. A rigid sphere of uniform mass density ρ and radius R has two smaller sphere of radius R / 2 hollowed out of it, as shown in the figure. The moment of intertia of the resuulting body about the Y axis is

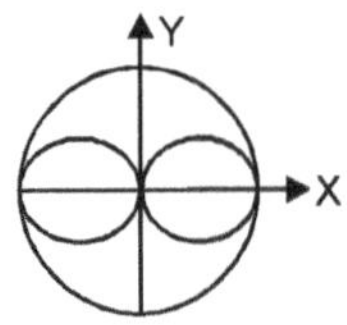

(a) $\frac{\pi\rho R^5}{4}$ (b) $\frac{5\pi\rho R^5}{12}$

(c) $\frac{7\pi\rho R^5}{12}$ (d) $\frac{3\pi\rho R^5}{4}$

100. The Lagrangian of a particle of mass m is

$$L = \frac{m}{2}\left[\left(\frac{dx}{dt}\right)^2 + \left(\frac{dy}{dt}\right)^2 + \left(\frac{dz}{dt}\right)^2\right] - \frac{V}{2}(x^2 + y^2) + W$$

sin ωt, where V, W and ω are constants. The conserved quantitites are

(a) energy and z-component of leaner momentum only.

(b) energy and z-component of angular momentum only.

(c) z-components of both linear and angular momenta only.

(d) energy and z-components of both linear and angular momenta.

101. Three particles of mass m each situated at $x_1(t)$, $x_2(t)$, and $x_3(t)$ respectively are connected by two springs of springs constant k and un-stretched lenght l. The system is free to oscillate only in one dimension along the straighnt line joining all the three particles. The lagragian of the system is

(a) $L = \frac{m}{2}\left[\left(\frac{dx_1}{dt}\right)^2 + \left(\frac{dx_2}{dt}\right)^2 + \left(\frac{dx_3}{dt}\right)^2\right]$

$$-\frac{k}{2}(x_1 + x_2 - l)^2 + \frac{k}{2}(x_3 - x_2 - l)^2$$

(b) $L = \frac{m}{2}\left[\left(\frac{dx_1}{dt}\right)^2 + \left(\frac{dx_2}{dt}\right)^2 + \left(\frac{dx_3}{dt}\right)^2\right]$
$- \frac{k}{2}(x_1 - x_3 - l)^2 + \frac{k}{2}(x_3 - x_2 - l)^2$

(c) $L = \frac{m}{2}\left[\left(\frac{dx_1}{dt}\right)^2 + \left(\frac{dx_2}{dt}\right)^2 + \left(\frac{dx_3}{dt}\right)^2\right]$
$- \frac{k}{2}(x_1 - x_2 - l)^2 + \frac{k}{2}(x_3 - x_2 + l)^2$

(d) $L = \frac{m}{2}\left[\left(\frac{dx_1}{dt}\right)^2 + \left(\frac{dx_2}{dt}\right)^2 + \left(\frac{dx_3}{dt}\right)^2\right]$
$- \frac{k}{2}(x_1 + x_2 - l)^2 - \frac{k}{2}(x_3 - x_2 - l)^2$

102. The Hamiltonion of a particle is $H = \frac{P^2}{2m} + pq$, where q is the generalized coordinate and p is the corresponding cononical momentum. The Lagrangian is

(a) $\frac{m}{2}\left(\frac{dq}{dt} + q\right)^2$

(b) $\frac{m}{2}\left(\frac{dq}{dt} - q\right)^2$

(c) $\frac{m}{2}\left[\left(\frac{dq}{dt}\right)^2 + q\frac{dq}{dt} - q^2\right]$

(d) $\frac{m}{2}\left[\left(\frac{dq}{dt}\right)^2 - q\frac{dq}{dt} + q^2\right]$

Statement for Linked Answcr Questions 103 & 104:

In the laboratory frame, a particle P of rest mass m_0 is moving in the positive x direction with a speed of $\frac{5c}{19}$. It approaches an identical particle Q, moving in the negative x direction with a speed of $\frac{2c}{5}$.

103. The speed of the particle P in the rest frame of the particle Q is

(a) $\frac{7c}{95}$ (b) $\frac{13c}{85}$

(c) $\frac{3c}{5}$ (d) $\frac{63c}{95}$

104. The energy of the particle P in the rest frame of the particle Q is

(a) $\frac{1}{2}m_0c^2$ (b) $\frac{5}{4}m_0c^2$

(c) $\frac{19}{13}m_0c^2$ (d) $\frac{11}{9}m_0c^2$

2006

105. A particle of charge q, mass m and linear momentum $\vec{p}$ enters an electromagnetic field of vector potential $\vec{A}$ and scalar potential ϕ. The Hamiltonian of the particle is

(a) $\frac{p^2}{2m} + q\theta + \frac{A^2}{2m}$

(b) $\frac{1}{2m}\left(\vec{p} - \frac{q}{c}\vec{A}\right)^2 + q\phi$

(c) $\frac{1}{2m}\left(\vec{p} - \frac{q}{c}\vec{A}\right)^2 + \vec{p}.\vec{A}$

(d) $\frac{p^2}{2m} q\phi - \vec{p}.\vec{A}$

106. a particle is moving in an inverse square force field. If the total energy of the particle is positive, then trajectory of the particle is

(a) circular (b) elliptical

(c) parabolic (d) hyperbolic

107. A free particle is moving in +x direction with a linear momentum p. The wavefunction of the particle normalised in a length L is

(a) $\frac{1}{\sqrt{L}}\sin\frac{p}{\eta}x$ (b) $\frac{1}{\sqrt{L}}\cos\frac{p}{\hbar}x$

(c) $\frac{1}{\sqrt{L}}e^{-i\frac{p}{\hbar}x}$ (d) $\frac{1}{\sqrt{L}}e^{i\frac{p}{\hbar}x}$

108. For the given transformations (i) $Q = p, P = -q$ and (ii) $Q = p, P = q$, where p and q are canonically conjugate variables, which one of the following statement is true?

(a) Both (i) and (ii) are canonical

(b) Only (i) is canonical

(c) Only (ii) is canonical

(d) Neither (i) nor (ii) is canonical

109. The mass m of a moving particale is $\frac{2m_0}{\sqrt{3}}$, where m_0 is its rest mass. The linear momentum of the particle is

(a) $2m_0c$ (b) $\frac{2m_0}{\sqrt{3}}$

(c) m_0c (d) $\frac{m_0c}{\sqrt{3}}$

110. Three point charges q, q and $-2q$ are located at $(0, -a, a)$, $(0, a, a)$ and $(0, 0, -a)$, respectively. The net dipole moment of this charge distribution is

(a) $4qa\hat{k}$ (b) $2qa\hat{k}$

(c) $-4qa\hat{i}$ (d) $-2qa\hat{j}$

111. The commutator $[L_x, y]$, where L_x is the x-component of the angular momentum operator and y is the y–component of the position operator, is equal to

(*a*) 0 (*b*) $0\hbar x$

(*c*) $i\hbar y$ (*b*) $i\hbar z$

Common Data for Questions 112, 113, 114:

An unperturbed two-level system has energy eigenvalues E_1 and E_2, and eigen functions $\begin{pmatrix}1\\0\end{pmatrix}$ and $\begin{pmatrix}0\\1\end{pmatrix}$. When perturbed, its Hamiltonian is represented by $\begin{pmatrix}E_1 & A\\A^* & E_2\end{pmatrix}$

112. The first-order correction to E_1 is

(*a*) 4 A (*b*) 2 A

(*c*) A (*d*) 0

113. The second-order correction to E_1 is

(*a*) 0 (*b*) A

(*c*) $\dfrac{A^2}{E_2 - E_1}$ (*d*) $\dfrac{A^2}{E_1 - E_2}$

114. The first-order correction to the eigenfunction $\begin{pmatrix}1\\0\end{pmatrix}$ is

(*a*) $\begin{pmatrix}0\\A^*/(E_1 - E_2)\end{pmatrix}$ (*b*) $\begin{pmatrix}0\\1\end{pmatrix}$

(*c*) $\begin{pmatrix}A^*/(E_1 - E_2)\\0\end{pmatrix}$ (*d*) $\begin{pmatrix}1\\1\end{pmatrix}$

Statement for Linked Answer Questions 115 and 116:

A particle of mass m is constrained to move in a vertical plane along a trajectory given by $x = A\cos\theta$, $y = A\sin\theta$, where A is a constant.

115. The Lagrangian of the particle is

(*a*) $\frac{1}{2}mA^2\theta^2 - mg\,A\cos\theta$

(*b*) $\frac{1}{2}mA^2\theta^2 - mg\,A\sin\theta$

(*c*) $\frac{1}{2}mA^2\theta^2$

(*d*) $\frac{1}{2}mA^2\theta^2 + mgA\cos\theta$

116. The equation of motion of the particle is

(*a*) $\ddot{\theta} - \frac{g}{A}\cos\theta = 0$ (*b*) $\ddot{\theta} + \frac{g}{A}\sin\theta = 0$

(*c*) $\ddot{\theta} = 0$ (*d*) $\ddot{\theta} - \frac{g}{A}\sin\theta = 0$

2005

117. For a particle moving in a central field

(*a*) the kinetic energy is a constant of motion

(*b*) the potential energy is velocity dependent

(*c*) the motion is confined in a plane

(*d*) the total energy is not conserved

118. A bead of mass m slides along a straight frictionless rigid wire rotating in a horizontal plane with a constant angular speed ω. The axis of rotation is perpendicular to the wire and passes through one end of the wire. If r is the distance of the mass from the axis of rotation and v is its speed then the magnitude of the Coriolis force is

(*a*) $\dfrac{mv^2}{r}$ (*b*) $\dfrac{2mv^2}{r}$

(*c*) $mv\omega$ (*d*) $2mv\omega$

119. If for a system of N particles of different masses $m_1, m_2, ..., m_N$ with position vectors $\vec{r}_1, \vec{r}_2, ..., \vec{r}_N$ and corresponding velocities $\vec{v}_1, \vec{v}_2, ..., \vec{v}_N$, respectively, such that $\sum_i \vec{v}_i = 0$, then

(*a*) the total momentum MUST be zero

(*b*) the total angular momentum MUST be independent of the choice of the origin

(*c*) the total force on the system MUST be zero

(*d*) the total torque on the system MUST be zero

120. If a particle moves outward in a plane along a curved trajectory described by $r = a\theta$, $0 = \omega t$, where a and ω are constants, then its

(*a*) kinetic energy is conserved

(*b*) angular momentum is conserved

(*c*) total momentum is conserved

(*d*) radial momentum is conserved

121. A circular hoop of mass M and radius a rolls without slipping with constant angular speed ω along the horizontal x-axis in the x-y plane. When the centre of the hoop is at a distance $d = \sqrt{2}a$ from the origin, the magnitude of the total angular moment 4 m of the hoop about the origin is

(*a*) $Ma^2\omega$ (*b*) $\sqrt{2}\,Ma^2\omega$

(*c*) $2Ma^2\,\omega$ (*d*) $3Ma^2\omega$

122. Two solid spheres of radius R and mass M each are connected by a thin rigid rod of negligible mass. The distance between the centres is 4R. The moment of inertia about an axis passing through the centre of symmetry and perpendicular to the line joining the spheres is

(a) $\frac{11}{5}MR^2$ (b) $\frac{22}{5}MR^2$

(c) $\frac{44}{5}MR^2$ (d) $\frac{88}{5}MR^2$

123. A car is moving with constant linear acceleration a along horizontal x-axis. A solid sphere of mass M and radius R is found rolling without slipping on the horizontal floor of the car in the same direction as seen from an inertial frame outside the car. The acceleration of the sphere in the inertial frame is

(a) $\frac{a}{7}$ (b) $\frac{2a}{7}$

(c) $\frac{3a}{7}$ (d) $\frac{5a}{7}$

124. A rod of length l_0 makes an angle θ_0 with the y-axis in its rest frame, while the rest frame moves to the right along the x-axis with relativistic speed v with respect to the lab frame. If $\gamma = (1 - v^2/c^2)^{-1/2}$, the angle θ in the lab frame is

(a) $\theta = \tan^{-1}(\gamma \tan \theta_0)$

(b) $\theta = \tan^{-1}(\gamma \cot \theta_0)$

(c) $\theta = \tan^{-1}\left(\frac{1}{\gamma}\tan \theta_0\right)$

(d) $\theta = \tan^{-1}\left(\frac{1}{\gamma}\cot \theta_0\right)$

125. A particle of mass m moves in a potential V(x) = $\frac{1}{2}m\omega^2x^2 + \frac{1}{2}m\mu v^2$ where x is the position coordinate, v is the speed, and **ω** and μ are constants. The canonical (conjugate) momentum of the particle is

(a) $p = m(1 + \mu)v$ (b) $p = mv$

(c) $p = m\mu v$ (d) $p = m(1 - \mu)v$

Statement for Linked Answer Questions 126 & 127 :

The Lagrangian of two coupled oscillators of mass m each is

$$L = \frac{1}{2}m\left(\dot{x}_1^2 + \dot{x}_2^2\right) - \frac{1}{2}m\omega_0^2\left(x_1^2 + x_2^2\right) + m\omega_0^2\mu x_1 x_2$$

126. the equations of motion are

(a) $\ddot{x}_1 + \omega_0^2 x_1 = \omega_0^2\mu x_1, \ddot{x}_2 + \omega_0^2 x_2 = \omega_0^2\mu x_2$

(b) $\ddot{x}_1 + \omega_0^2 x_1 = \omega_0^2\mu x_2, \ddot{x}_2 + \omega_0^2 x_2 = \omega_0^2\mu x_1$

(c) $\ddot{x}_1 + \omega_0^2 x_1 = \omega_0^2\mu x_1, \ddot{x}_2 + \omega_0^2 x_2 = -\omega_0^2\mu x_2$

(d) $\ddot{x}_1 + \omega_0^2 x_1 = \omega_0^2\mu x_1, \ddot{x}_2 + \omega_0^2 x_2 = \omega_0^2\mu x_1$

127. The normal modes of the system are

(a) $\omega_0\sqrt{\mu^2 - 1}, \omega_0\sqrt{\mu^2 + 1}$

(b) $\omega_0\sqrt{1 - \mu^2}, \omega_0\sqrt{1 + \mu^2}$

(c) $\omega_0\sqrt{\mu - 1}, \omega_0\sqrt{\mu + 1}$

(d) $\omega_0\sqrt{1 - \mu}, \omega_0\sqrt{1 + \mu}$

2004

128. Which one of the following remains invariant under Lorentz transformations ?

(a) $\frac{\partial}{\partial x} + \frac{\partial}{\partial y} + \frac{\partial}{\partial z} - \frac{1}{c^2}\frac{\partial}{\partial t}$

(b) $\frac{\partial^2}{\partial x^2} + \frac{\partial^2}{\partial y^2} + \frac{\partial^2}{\partial z^2} + \frac{1}{c^2}\frac{\partial^2}{\partial t^2}$

(c) $\frac{\partial^2}{\partial x^2} + \frac{\partial^2}{\partial y^2} + \frac{\partial^2}{\partial z^2} - \frac{1}{c^2}\frac{\partial^2}{\partial t^2}$

(d) $\frac{\partial^2}{\partial x^2} + \frac{\partial^2}{\partial y^2} + \frac{\partial^2}{\partial z^2}$

129. The Hamiltonian of a particle is given by H = $\frac{p^2}{2m} + V(|\vec{r}|) + \phi(+|\vec{r}|)\vec{L}\cdot\vec{S}$ where $\vec{S}$ is the spin, $V(|\vec{r}|)$ and $\phi(|\vec{r}|)$ are potential functions and $\vec{L}(= \vec{r} \times \vec{p})$ is the angular momentum. The Hamiltonian does NOT commute with

(a) $\vec{L} + \vec{S}$ (b) $\vec{S}^2$

(c) L_z (d) $\vec{L}^2$

130. The Lagrangian of a particle moving in a plane under the influence of a central potential is given by $L = \frac{1}{2}m(\dot{r}^2 + r^2\dot{\theta}^2) - V(r)$. The generalized momenta corresponding to r and θ are given by

(a) $m\dot{r}$ and $mr^2\dot{\theta}$ (b) $m\dot{r}$ and $mr\dot{\theta}$

(c) $m\dot{r}^2$ and $mr^2\dot{\theta}$ (d) $m\dot{r}^2$ and $mr^2\dot{\theta}^2$

131. A particle of mass m is attached to a thin uniform rod of length a and mass 4 m. The distance of the particle from the center of mass of the rod is $a/4$. The moment of inertia of the combination about an axis passing through O normal to the rod is

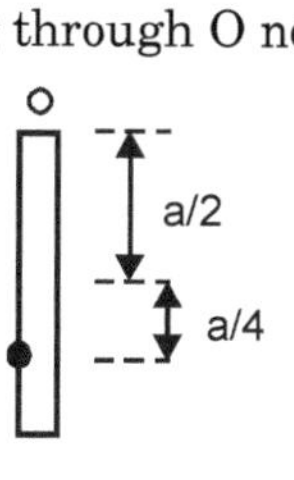

(a) $\frac{64}{48}ma^2$ (b) $\frac{91}{48}ma^2$

(c) $\frac{27}{48}ma^2$ (d) $\frac{51}{48}ma^2$

132. The Hamiltonian corresponding to the Langrangian L = $a\dot{x}^2 + b\dot{y}^2 - kxy$ is

(a) $\frac{p_x^2}{2a} + \frac{p_y^2}{2b} + kxy$

(b) $\frac{p_x^2}{4a} + \frac{p_y^2}{4b} - kxy$

(c) $\frac{p_x^2}{4a} + \frac{p_y^2}{4b} + kxy$

(d) $\frac{p_x^2 + p_y^2}{4ab} + kxy$

133. The value of the Poisson bracket $[\vec{a}\cdot\vec{r}, \vec{b}\cdot\vec{p}]$, where $\vec{a}$ and $\vec{b}$ are constant vectors, is

(a) $\vec{a}\vec{b}$ (b) $\vec{a} - \vec{b}$

(c) $\vec{a} + \vec{b}$ (d) $\vec{a}\cdot\vec{b}$

134. A mass m is connected on either side with a spring each of spring constants k_1 and k_2. The free ends of springs are tied to rigid supports. The displacement of the mass is x from equilibrium position. Which one of the following is TRUE ?

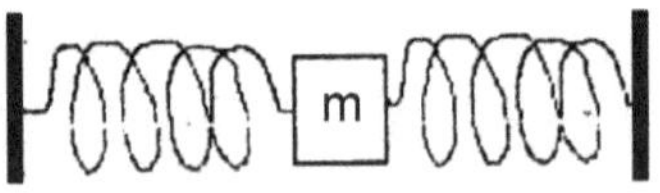

(a) The force acting on the mass is $-(k_1 k_2)^{1/2} x$.

(b) The angular momentum of the mass is zero about the equilibrium point and its Lagrangian is $\frac{1}{2} m\dot{x}^2 - \frac{1}{2}(k_1 + k_2) x^2$.

(c) The total energy of the system is $\frac{1}{2} m\dot{x}^2$

(d) The angular momentum of the mass is $mx\dot{x}$ and the Lagrangian of the system is $\frac{m}{2}\dot{x}^2 + \frac{1}{2}(k_1 + k_2) x^2$.

2003

135. The Lagrangian of a particle of mass m moving in a plane is given by L = (½) m[$v_x^2 + v_y^2$] + a(x v_y − y v_x), where v_x and v_y are velocity components and a is a constant. The canonical momenta of the particle are given by

(a) $p_x = m v_x$ and $p_y = m v_y$

(b) $p_x = m v_x + a y$ and $p_y = m v_y + a x$

(c) $p_x = m v_x - a y$ and $p_y = m v_y + a x$

(d) $p_x = m v_x - a y$ and $p_y = m v_y + a x$

136. Two events are separated by a distance of 6 × 10^5 km and the first event occurs 1's before the second event. The interval between the two events

(a) is time-like

(b) is light-like (null)

(c) is space-like

(d) cannot be determined from the information given

137. A particle of mass m is moving in a potential of the form V(x, y, z) = (1/2) mω^2(3x^2 + 3y^2 + 2z^2 + 2xy). The oscillation frequencies of the three normal modes of the particle are given by

(a) ω, √3 ω and √3 ω

(b) √2 ω, √3 ω and √3 ω

(c) √2 ω, √2 ω and 2 ω

(d) √2 ω, 2 ω and 2 ω

Data for Q. 138 – 139 :

A particle of mass m moving with speed v collides with a stationary particle of equal mass. After the collision, both the particles move. Let θ be the angle between the two velocity vectors

138. If the collision is elastic, then

(a) θ is always less than 90°

(b) θ is always equal to 90°

(c) θ is always greater than 90°

(d) θ cannot be deduced from the given data

139. If the collision is inelastic, then

(a) θ is always less than 90°

(b) θ is always equal to 90°

(c) θ is always greater than 90°

(d) θ could assume any value in the range 0° to 180°

2002

140. If a function $f(z) = u(x, y) + iv(x, y)$ of the complex variable $z = x + iy$, where x, y, u and v are real, is analytic in a domain D of z, then which of the following is true ?

(a) $\frac{\partial u}{\partial x} = \frac{\partial v}{\partial x}$

(b) $\frac{\partial u}{\partial x} = \frac{\partial v}{\partial y}$ and $\frac{\partial u}{\partial y} = -\frac{\partial v}{\partial x}$

(c) $\frac{\partial u}{\partial x} = \frac{\partial v}{\partial x}$ and $\frac{\partial u}{\partial y} = \frac{\partial v}{\partial y}$

(d) $\frac{\partial^2 u}{\partial x \partial y} = \frac{\partial^2 v}{\partial x \partial y}$

141. The homogeneity of time leads to the law of conservation of

(a) linear momentum

(b) angular momentum

(c) energy

(d) parity

142. Hamilton canonical equations of motion for a conservative system are

(a) $-\frac{dq_i}{dt} = \frac{\partial H}{\partial p_i}$ and $-\frac{dp_i}{dt} = \frac{\partial H}{\partial q_i}$

(b) $\frac{dp_i}{dt} = \frac{\partial H}{\partial p_i}$ and $\frac{dp_i}{dt} = \frac{\partial H}{\partial q_i}$

(c) $-\frac{dq_i}{dt} = \frac{\partial H}{\partial p_i}$ and $\frac{dp_i}{dt} = \frac{\partial H}{\partial q_i}$

(d) $\frac{dq_i}{dt} = \frac{\partial H}{\partial p_i}$ and $-\frac{dp_i}{dt} = \frac{\partial H}{\partial q_i}$

where q_i, p_i and H are generalized coordinate, generalized momentum and Hamiltonian respectively.

143. A particle of mass M moving in a straight line with speed v collides with a stationary particle of the same mass. In the center of mass coordinate system, the first particle is deflected by 90º. The speed of the second particle, after collision, in the laboratory system will be

(a) $v/\sqrt{2}$ (b) $\sqrt{2}\,v$

(c) v (d) $v/2$

144. Two particles of equal mass are connected by an inextensible string of length L. One of the masses is constrained to move on the surface of a horizontal table. The string passes through a small hole in the table and the other mass is hanging below the table. The only constraint is that the first mass moves on the surface of the table. The number of degrees of freedom of the masses-string system is

(a) five (b) four

(c) two (d) one

145. An electron is accelerated from rest by 10.2 million volts. The percent increase in its mass is

(a) 20,000 (b) 2,000

(c) 200 (d) 20

2001

146. For any operator A, $i(A^* - A)$ is

(a) Hermitian

(b) anti-Hermitian

(c) unitary

(d) orthogonal

147. The Lagrangian for the Kepler problem is given by

$$L = \frac{1}{2}\left(\dot{r}^2 + r^2\dot{\theta}^2\right) + \frac{\mu}{r}, (\mu > 0),$$

where (r, θ)denote the polar coordinates and the mass of the particle is unity. Then

(a) $p_\theta = 2r^2\dot{\theta}$

(b) $p_r = 2\dot{r}$

(c) the angular momentum of the particle about the centre of attraction is a constant

(d) the total energy of the particle is time dependent

148. Which of the following equations is relativistically invariant ? (α, β, γ and δ are constants of suitable dimensions).

(a) $\frac{\partial \phi(x,t)}{\partial t} = \alpha \frac{\partial^2 \phi(x,t)}{\partial x^2}$

(b) $\frac{\partial^2 \phi(x,t)}{\partial t^2} = \beta \frac{\partial^2 \phi(x,t)}{\partial x^2}$

(c) $\frac{\partial^2 \phi(x,t)}{\partial t^2} = \gamma \frac{\partial \phi(x,t)}{\partial x}$

(d) $\frac{\partial \phi(x,t)}{\partial t} = \delta \frac{\partial^3 \phi(x,t)}{\partial x^3}$

149. The Lagrangian for a three particle system is given by :

$$L = \frac{1}{2}\left(\dot{\eta}_1^2 + \dot{\eta}_2^2 + \dot{\eta}_3^2\right) - a^2\left(\eta_1^2 + \eta_2^2 + \eta_3^2 - \eta_1\eta_3\right),$$

where a is real. Then one of the normal coordinates has a frequency ω given by

(a) $\omega^2 = a^2$ (b) $\omega^2 = a^2/2$

(c) $\omega^2 = 2a^2$ (d) $\omega^2 = \sqrt{2}\,a^2$

150. A particle of mass m is constrained to move on the plane curve $xy = C$ (C > 0) under gravity (y axis vertical). The Lagrangian of the particle is given by

(a) $\frac{1}{2}m\dot{x}^2\left(1 + \frac{C^2}{x^4}\right) + \frac{mgC}{x}$

(b) $\frac{1}{2}m\dot{x}^2\left(1 + \frac{C^2}{x^4}\right) - \frac{mgC}{x}$

(c) $\frac{1}{2} m\dot{x}^2 \left(1+\frac{C}{x^2}\right) + \frac{mgC}{x}$

(d) $\frac{1}{2} m\dot{x}^2 \left(1+\frac{C}{x^2}\right) - \frac{mgC}{x}$

2000

151. A planet moves around the Sun in an elliptical orbit with semi-major axis a and time period T.T is proportional to

(a) a^2 (b) $a^{1/2}$

(c) $a^{3/2}$ (d) a^3

152. A particle moves in central force field $\bar{f} = -kr^n\hat{r}$, where k is a constant, r, the distance of the particle from the origin and $\hat{r}$ is the unit vector in the direction of position vector $\bar{r}$. Closed stable orbits are possible for

(a) $n = 1$ and $n = 2$

(b) $n = 1$ and $n = -1$

(c) $n = 2$ and $n = -2$

(d) $n = 1$ and $n = -2$

153. A particle constrained to move along the x-axis in a potential $V = kx^2$, is subjected to an external time-dependent force $\bar{f}(t)$.

Here k is a constant, x the distance from the origin, and t the time. At some time T, when the particle has zero velocity at $x = 0$, the external force is removed. The particle will then,

(a) execute simple harmonic motion

(b) move along $+x$ direction

(c) move along $-$ x direction

(d) remain at rest

154. Consider two particles with position vectors $\bar{r}_1$ and $\bar{r}_2$. The force exerted by particle 2 on particle 1 is,

$\bar{f}(\bar{r}_1,\bar{r}_2,) = (\dot{r}_2 - \dot{r}_1)(\bar{r}_2 - \bar{r}_1)$.

The force is

(a) central and consesrvative

(b) non-central and conservative

(c) central and non-conservative

(d) non-central and non-conservative

ANSWERS

1. (2 to 2) **2.** (2 to 2) **3.** (d) **4.** (a) **5.** (a,c) **6.** (a,b) **7.** (d) **8.** (b)
9. (0.5 to 0.5) **10.** (1 to 1) **11.** (a) **12.** (b) **13.** (b) **14.** (3 to 3) **15.** (b)
16. (a) **17.** (33.2 to 33.4) **18.** (5 to 5) **19.** (4.77 to 4.81) **20.** (d)
21. (b) **22.** (0.185 to 0.189) **23.** (c) **24.** (a) **25.** (0.21 to 0.23) **26.** (b) **27.** 14.10 to 14.20
28. (c) **29.** (b) **30.** 0.27 to 0.31 2 **31.** 3.30 to 3.35 **32.** 0.32 to 0.36 **33.** (2.41 : 2.49)
34. (26.65 : 26.68) **35.** (c) **36.** (18 : 18) **37.** (a) **38.** (a) **39.** (d) **40.** (b) **41.** (a)
42. (c) **43.** (b) **44.** (b) **45.** (a) **46.** (c) **47.** (4.93 to 4.98) **48.** (b) **49.** (2.1 to 2.3)
50. (1.99 to 2.01) **51.** (a) **52.** (a) **53.** (a) **54.** (d) **55.** (0.24 to 0.26) **56.** (1.41 to 1.47)
57. (b) **58.** (d) **59.** (c) **60.** (c) **61.** (c) **62.** (c) **63.** (b) **64.** (c) **65.** (d) **66.** (b)
67. (b) **68.** (c) **69.** (a) **70.** (a) **71.** (a) **72.** (d) **73.** (a) **74.** (c) **75.** (a) **76.** (b)
77. (a) **78.** (a) **79.** (c) **80.** (c) **81.** (d) **82.** (b) **83.** (d) **84.** (d) **85.** (c) **86.** (b)
87. (a) **88.** (a) **89.** (b) **90.** (a) **91.** (d) **92.** (a) **93.** (b) **94.** (b) **95.** (a) **96.** (c)
97. (d) **98.** (c) **99.** (b) **100.** (a) **101.** (a) **102.** (b) **103.** (c) **104.** (b) **105.** (b) **106.** (d)
107. (c) **108.** (b) **109.** (d) **110.** (a) **111.** (d) **112.** (d) **113.** (d) **114.** (c) **115.** (c) **116.** (c)
117. (c) **118.** (d) **119.** (c) **120.** (b) **121.** (c) **122.** (c) **123.** (d) **124.** (a) **125.** (d) **126.** (b)
127. (b) **128.** (c) **129.** (c) **130.** (a) **131.** (b) **132.** (c) **133.** (a) **134.** (b) **135.** (c) **136.** (d)
137. (c) **138.** (b) **139.** (a) **140.** (b) **141.** (c) **142.** (d) **143.** (a) **144.** (c) **145.** (b) **146.** (a)
147. (c) **148.** (b) **149.** (c) **150.** (b) **151.** (c) **152.** (d) **153.** (d) **154.** (d)

EXPLANATIONS

2021

1. As me know that space-time interval is invariant quantity then

$$|L| = \sqrt{(x_2 - x_1)^2 + (y_2 - y_1)^2 + (z_2 - z_1)^2 + (t_2 - t_1)^2 (i)^2}$$

(Since C = 1)

$$= \sqrt{(5-5)^2 + (1-3)^2 + (3-5)^2 - (5-3)^2}$$

$$= \sqrt{0 + 4 + 4 - 4}$$

$$\boxed{|L| = 2}$$

2. A hoop of mass M

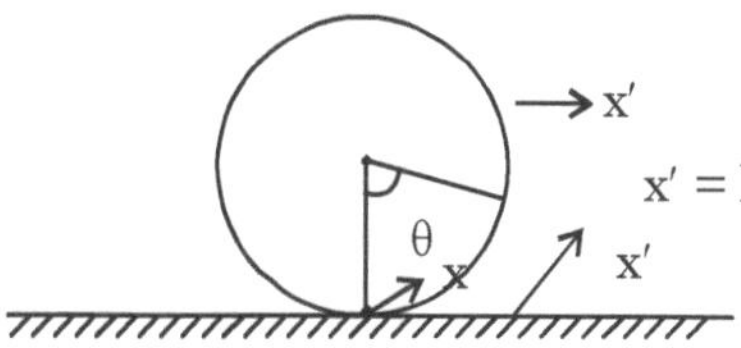

Here generalized

Co-ordinate (x, θ)

So Degree of freedom = 2

3. Consider a point charge + Q

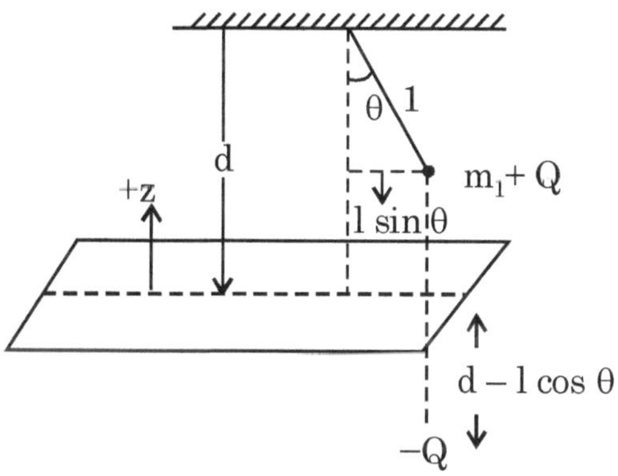

Here θ is generalised coordinate

$$T = \frac{P_\theta^2}{2ml^2}$$

$V = V_1$ (due to mage charge) + V_2 (due to mass m)

and $V_1 = \frac{-kq^2}{4d^1}$ (From method of mage charge)

$\because$ $q = Q$ and $d^1 = d - l\cos\theta$

$$V_1 = -\frac{Q^2}{16\pi \epsilon_0 (d - l\cos\theta)}$$

and $V_1 = mg(d - l\cos\theta)$

Now Hamiltonian is given by

$H = T + V$

$$H = \frac{P_\theta^2}{2ml^2} - \frac{Q^2}{16\pi \epsilon_0 (d - l\cos\theta)} + mg(d - l\cos\theta)$$

Hence, option (b) is the correct answer.

4. For r > R

$$\frac{mv^2}{r} = \frac{GMm}{r}$$

$$v = \sqrt{\frac{GMm}{r}}$$

$$v \propto \sqrt{\frac{1}{r}} \qquad ...(1)$$

Let mass density is ρ

$$M = \rho \times \frac{4}{3}\pi R^3$$

at distance r < R

$$M' = \rho \times \frac{4}{3}\pi r^3$$

$$M' = \frac{M}{\frac{4}{3}\pi R^3} \times \frac{4}{3}\pi r^3$$

$$M' = \frac{Mr^3}{R^3} \qquad ...(2)$$

Now, For r < R

$$\frac{mv^2}{\pi} = \frac{GM'm}{r^2}$$

From (2)

$$\frac{mv^2}{\pi} = \frac{GMmr^3}{r^2 R^3}$$

$$v = \sqrt{\frac{GM}{R^3}}\, r$$

$$v \propto r \qquad ...(3)$$

Hence, option (c) is correct answer.

5.

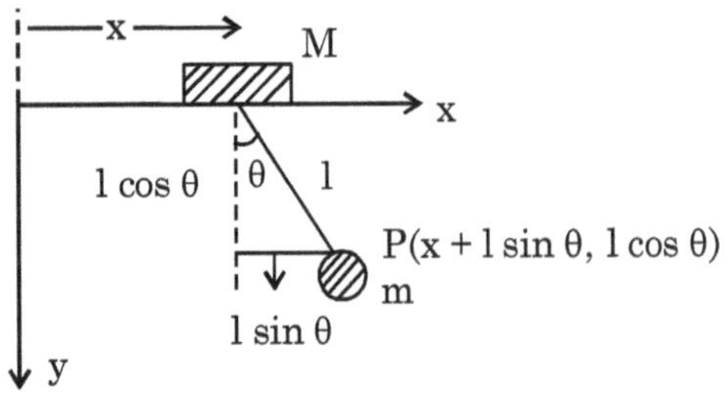

$$T = K.E = \frac{1}{2}M\dot{x}^2 + \frac{1}{2}m\left(\dot{x}_1^2 + \dot{y}_1^2\right)$$

$$\frac{1}{2}M\dot{x}^2 + \frac{1}{2}m\left[\left(\dot{x} + l\cos\theta\,\dot{\theta}\right)^2 + l^2\sin^2\theta\,\dot{\theta}^2\right]$$

$$T = \frac{1}{2}M\dot{x}^2 + \frac{1}{2}m\left(\dot{x}^2 + l^2\dot{\theta}^2 + 2\dot{x}l\cos\theta\,\dot{\theta}\right)$$

$V = P.E = -mgl\cos\theta$

$L = T - V$

$$L = \frac{1}{2}(m + M)\dot{x}^2 + \frac{1}{2}m\left(l^2\dot{\theta}^2 + 2\dot{x}l\cos\theta\,\dot{\theta}\right)$$

Here x is cyclic coordinate then P_x is conserved.

$$P_x = \frac{\partial L}{\partial \dot{x}}$$

$$P_x = (m + M)\dot{x} + m\,l\cos\theta\dot{\theta}$$

So option 1

6. $L' = L(q_1, \dot{q}_i, t) + \frac{d}{dt}g(q_i, t)$

$$L' = L + \frac{\partial q}{\partial q}\dot{q}_1 + \frac{\partial g}{\partial t}$$

$$\text{Now, } P_i' = \frac{\partial L'}{\partial \dot{q}i}$$

$$P = \frac{\partial L}{\partial \dot{q}i} + \frac{\partial g}{\partial g_i} \times \frac{\partial \dot{g}_i}{\partial \dot{g}_1} + 0$$

$$P_i' = P_i + \frac{\partial g}{\partial \dot{q}i}$$

Now P_i is conserved then P'_i is not conserved because

$$\frac{dP_i'}{dt} = \frac{dP_i}{dt} + \frac{d}{dt}\left(\frac{\partial g}{\partial q_i}\right)$$

$$\frac{dP_i'}{dt} + \frac{d}{dt}\left(\frac{\partial g}{\partial q_i}\right)$$

$$\text{Now } \frac{d}{dt}\left(\frac{\partial L}{\partial \dot{q}_i}\right) - \frac{\partial L}{\partial q_i} = 0$$

$$\frac{dP_i}{dt} - \frac{\partial L}{\partial q_i} = 0 \qquad \text{... (i)}$$

Similarly

$$\frac{d}{dt}\left(\frac{\partial L'}{\partial \dot{q}_1}\right) - \frac{\partial L'}{\partial q_i} = 0$$

$$\frac{d}{dt}\left(P_i\frac{\partial}{\partial \dot{q}_1}\right) - \frac{\partial L}{\partial q_i}\left(L + \frac{\partial g}{\partial q_i}\right) = 0$$

$$\frac{dP_i}{dt} + \frac{d}{dt}\left(\frac{\partial g}{\partial \dot{q}_1}\right) - \frac{\partial L}{\partial q_i} - \frac{d}{dt}\left(\frac{\partial g}{\partial q_i}\right) - 0$$

$$\frac{dP_i}{dt} - \frac{\partial L}{\partial q_i} = 0$$

From equation (i) and (ii) we can say that both L and L′ satisfy the Eular-Lagrange's equation of motion . Now, Hamiltonian

$H = P_i\dot{q}_1 - L$ ... (iii)

$H' = P_i'\dot{q}_1 - L'$

$$= P_i'\dot{q}_1 + \frac{\partial g}{\partial q_i}\dot{q} - L - \frac{\partial g}{\partial q_i}\dot{q}_i - \frac{\partial g}{dt}$$

$$H' = P_i\dot{q}_i - L - \frac{\partial g}{dt}$$

$$H' = H - \frac{\partial g}{dt}$$

2020

7.

$$F = -\frac{k}{r^3}\hat{r}$$

$$-\frac{\partial u}{\partial r} = -\frac{k}{r^3}$$

$$U = \int \frac{k}{r^3}dr$$

$$U = -\frac{k}{2r^2}$$

9. $Q = \sqrt{2}q^m\cos p$

$P = \sqrt{2}q^m\sin p$

$m = ?$

$$\frac{\partial Q}{\partial q}\cdot\frac{\partial P}{\partial P} - \frac{\partial Q}{\partial P}\cdot\frac{\partial Q}{\partial q} = 1$$

$$m = \frac{1}{2}$$

10. Degree of freedom = 2M – K

$= 2 \times 1 - 1$

$= 2 - 1$

$= 1$

11. $$L = a\left(\frac{dx}{dt}\right)^2 + b\left(\frac{dy}{dt}\right)^2 + cxy$$

H = ?

$$p_x = \frac{\partial L}{\partial \dot{x}} = 2a\dot{x}$$

$$p_y = \frac{\partial L}{\partial \dot{y}} = 2a\dot{y}$$

$$\dot{y} = \frac{p_y}{2b}$$

$$\dot{x} = \frac{p_x}{2a}$$

$$H = \Sigma p_j q_j - L$$

$$= (2a\dot{x})\dot{x} + (2b\dot{y})\dot{y} - L$$

$$H = \frac{p_x^2}{4a} + \frac{p_y^2}{4b} - cxy$$

12. $$V = a\sin\left(k^2x - \frac{\pi}{2}\right)$$

$$\omega = \sqrt{\frac{k}{m}}$$

$$k = \left.\frac{\partial^2 v}{\partial x^2}\right|_{x=0}$$

$$k = k^4 a$$

$$\omega = k^2\sqrt{\frac{a}{m}}$$

2019

13. $\{Q, P\}_{p,q} = 1$

$$\left[\frac{\partial Q}{\partial q}\cdot\frac{\partial P}{\partial p} - \frac{\partial Q}{\partial p}\cdot\frac{\partial P}{\partial q}\right] = 1 \quad \text{...(A)}$$

$$\frac{\partial Q}{\partial q} = spq^{s-1}, \frac{\partial P}{\partial P} = 0; \frac{\partial Q}{\partial P} = q^2, \frac{\partial P}{\partial q} = rq^{r-1}$$

By putting above values in (A), we get

$$-q^s rq^{r-1} = 1 \Rightarrow -rq^{s+r-1} = 1 = 1.q^0$$

$$s + r - 1 = 0$$

$$s - 1 - 1 = 0 \Rightarrow s = 2$$

17.

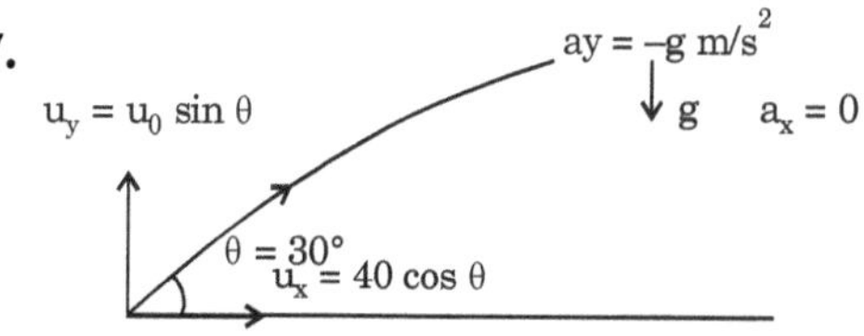

$u_0 = 10$ m/s

m = 1 kg

$$T = \frac{2u_0\sin\theta}{g}$$

$$= \frac{2 \times 10 \times \sin 30°}{10} = 1\text{ sec}$$

$V_x = u_x = u_0\cos\theta$.

$$V_y^2 = u_y^2 + 2(-g)y$$

$$V_y^2 = u_y^2 - 2gy$$

$$L = T - V$$

$$V = mgy$$

$$L = \frac{1}{2}mV_x^2 + \frac{1}{2}mV_y^2 - mgy$$

$$L = \frac{1}{2}m\left(u_x^2 + u_y^2 - 2gy\right) - mgy$$

$$L = \frac{1}{2}m(u_0\cos\theta)^2 + \frac{1}{2}m(u_0\sin\theta)^2 - mgy - mgy$$

$$L = \frac{1}{2}mu_0^2 - 2mgy$$

$$I = \int L\,dt$$

$$I = \frac{1}{2}mu_0^2[t]_0^1 - \int_0^1\left(2mgu_0\sin\theta t + mg^2t^2\right)dt$$

$$I = 33.33$$

18.

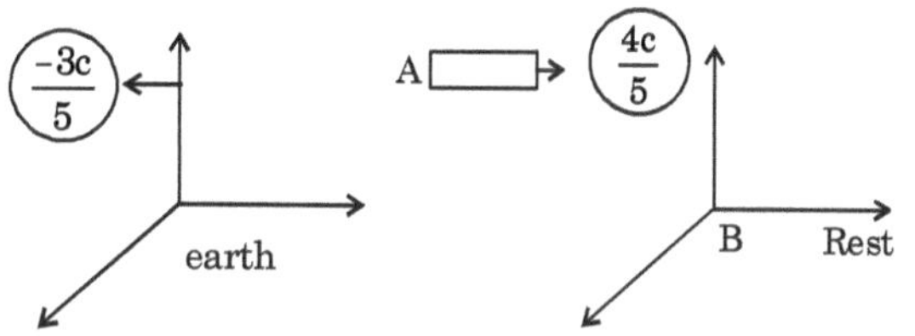

$$V_{AB} = \frac{(V_{A\,earth} + V)}{1 + V_{A\,earth} \times \frac{v}{c^2}}$$

$$V = \frac{-3c}{5}$$

$$V_{A\,earth} = \frac{4c}{5}\uparrow$$

$$V_{AB} = \frac{\left(\frac{4c}{5} - \frac{3c}{5}\right)}{1 - \frac{12}{25}} = \frac{5c}{13}$$

$$\Delta x = L + L\sqrt{1 - \frac{V_{AB}^2}{c^2}} = L\left(1 + \sqrt{1 - \frac{V_{AB}^2}{c^2}}\right)$$

$$= L\left(1 + \sqrt{\frac{144}{169}}\right) = L\left(1 + \frac{12}{13}\right)$$

$$\Delta x = \frac{25L}{13}$$

$$\Delta t = \frac{\Delta x}{v_{AB}} = \dfrac{\frac{2SL}{13}}{\frac{Sc}{\substack{1\\13}}} = 6\left(\frac{L}{C}\right)$$

$$\Delta t = 5\left(\frac{L}{C}\right)$$

19.

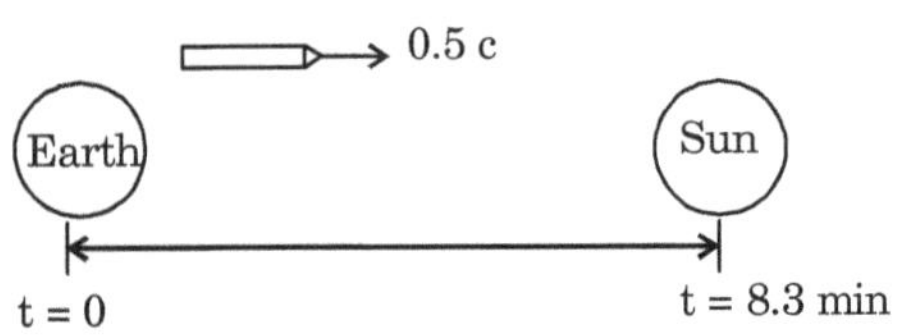

$\Delta t = 0$

$\Delta t' = ?$

$$\Delta t' = \frac{\left(\Delta t - \frac{v\Delta x}{c^2}\right)}{\sqrt{1 - \frac{v^2}{c^2}}}$$

$$\Delta t' = \frac{0 - \frac{0.5c}{c} \times \left(\frac{\Delta x}{c}\right)}{\sqrt{1 - 0.5^2}}$$

$$\Delta t' = \frac{-0.5 \times 8.3}{\sqrt{0.75}}$$

$\Delta t' = 4.80$ minutes

2018

20. The shape of coordinate lines of a constant radial distance, azimuthal as well as polar angle.

The component of matric tensor

$$g_{ij} = \begin{bmatrix} 1 & 0 & 0 \\ 0 & r & 0 \\ 0 & 0 & r\sin\theta \end{bmatrix}$$

So, the scale factors corresponding to the convenient metric tensor in spherical polar coordinates $g_{ij} = 1, r, r\sin\theta$.

21. In the context of small oscillations the normal coordinates are orthogonal to one another does not apply to the normal coordinates.

22. The value of $X = \frac{0.7c - 0.59c}{0.59c}$

$$= \frac{0.11c}{0.59c} = 0.186$$

23. If a object be projected from a point has a speed v in any direction, then

$$v_e^2 = \frac{2GM}{R}$$

or $$v_e = \sqrt{\frac{2GM}{R}}$$

Here $v = v_e$, the object escapes from the gravitational pull of the star and path is parabola.

24. For potential $V(x) = \alpha|x|$

$$\frac{dV(x)}{dx} = \alpha \text{ (with non-zero energy)}$$

Then best describes the particle trajectory in the phase space in option (*a*)

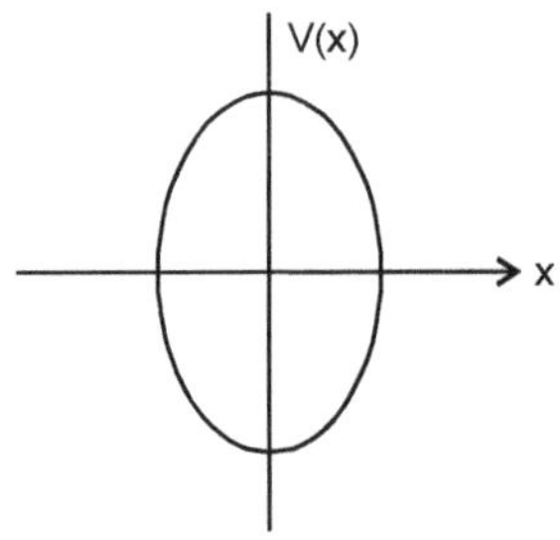

25. (0.21 to 0.23)

2017

26. If Lagrangian $L_0 = \frac{1}{2}m\left(\frac{dq}{dt}\right)^2 - \frac{1}{2}m\omega^2 q^2$ is modified to $L = L_0 = \alpha q\left(\frac{dq}{dt}\right)$ then canonical momentum changes and equation of motion does not change.

27. The non-zero angular eigen frequency of the systems is given by

$$W_n = \sqrt{\frac{k(m_1 + m_2)}{m_1 m_2}}$$

$$= \sqrt{\frac{1(10 \times 10^{-3}) + (10 + 10^{-3})}{(10 \times 10^{-3})(10 \times 10^{-3})}}$$

$$= \sqrt{\frac{20 \times 10^{-3}}{100 \times 10^{-6}}}$$

$$= \sqrt{0.2 \times 10^3}$$

$$= \sqrt{200}$$

= 14.14 rad/s.

28. The phase space trajectory of free particles bouncing between two hard walls elastically in one dimension is rectangle.

29. The poisson bracket $[x, xp_y + yp_x]$ is equal

$= [x, xp_y] + [x, yp_x]$

$= x\,[x, p_y] + y[x, p_x]$

$$= 0 + y^{(1)} = y \qquad \left(\begin{aligned} [q_i p_j]_{q,p} &= \Sigma \delta_{ik} \delta_{jk} \\ &= \delta_{ij} \\ &= 0 \; if \; i \neq j \\ &= 1 \; if \; i = j \end{aligned} \right)$$

30. The relative velocity of O′ with respect to O is given by:-

$$W = \frac{u - v}{1 - \left(\frac{uv}{c^2}\right)} = \frac{\frac{c}{2} - \frac{c}{4}}{1 - \frac{\frac{c^2}{8}}{c^2}} = \frac{2c/8}{\frac{7}{8}} = \frac{2}{7}c$$

= 0.28 c.

31. $\frac{1}{2} m V_1^2 = \frac{1}{2} I w^2$

$$V_1^2 = \frac{V_2^2}{2} \Rightarrow V_1 = \frac{V_2}{2} = \frac{5}{2} = 2.5 m/s.$$

32. $\Delta g = g_p - g_c = Rw^2$

$$= \left(6.4 \times 10^6 \times 9.3 \times 10^{-5}\right)^2$$

$$= 3.41 \times 10^{-2} \text{ m/s}^2$$

$$100 \times \left(\frac{w_p - w_e}{w_p}\right) = 100 \times \left(\frac{mg_p - mg_e}{mg_p}\right)$$

$$= 100 \times \left(\frac{g_p - g_e}{g_p}\right) = 100 \times 3.41$$

$$= 34.1 \times 10^{-2} = 0.34.$$

33. $$f = \frac{1}{2\pi}\sqrt{\frac{k}{\mu}}$$

where $$\mu = \frac{m_1 m_2}{m_1 + m_2}$$

$$\mu = \frac{m \times 2m}{m + 2m} = \frac{2m}{3}$$

$$\therefore \quad f = \frac{1}{2\pi}\sqrt{\frac{3k}{2m}} \qquad ...(i)$$

Now $\omega = 2\pi f$

$$\therefore \quad f = \frac{\omega}{2\pi} \qquad ...(ii)$$

Comparing equation (i) and (ii) we get

$$\therefore \quad \omega = \sqrt{\frac{3k}{2m}}$$

Given $\frac{k}{m} = 4$

$$\therefore \quad \omega = \sqrt{\frac{3}{2} \times 4} = 2.44$$

34. Given data is insufficient.

35. $$t_2' - t_1' = \frac{t_2 - \frac{vx_2}{c^2}}{\sqrt{1 - \frac{v^2}{c^2}}} - \frac{t_1 - \frac{vx_1}{c^2}}{\sqrt{1 - \frac{v^2}{c^2}}}$$

$$= \frac{t_2 - t_1}{\sqrt{1 - \frac{v^2}{c^2}}} - \frac{\frac{v}{c^2}(x_2 - x_1)}{\sqrt{1 - \frac{v^2}{c^2}}}$$

or $\Delta t' = \gamma\, \Delta t - \frac{v}{c^2} \gamma\, \Delta x$

Given $\Delta t = 0, \quad \Delta x = d$

$$\therefore \quad \Delta t' = -\frac{v}{c^2} \gamma\, d$$

$$x_2' - x_1' = \frac{x_2 - vt_2}{\sqrt{1 - \frac{v^2}{c^2}}} - \frac{x_1 - vt_1}{\sqrt{1 - \frac{v^2}{c^2}}}$$

$$= \frac{x_2 - x_1}{\sqrt{1 - v^2/c^2}} - \frac{v(t_2 - t_1)}{\sqrt{1 - v^2/c^2}}$$

$$\therefore \quad \Delta x' = \gamma\, \Delta x - v\gamma\, \Delta t$$

Given $\Delta t = 0, \quad \Delta x = d$

$\therefore \quad \Delta x' = \gamma d$

$\therefore \quad \Delta t' = \frac{-\gamma v d}{c^2},$

$\therefore \quad \Delta x' = \gamma d$

36. $W = \frac{\pi(n_i + g_i - 1)!}{(n_i)!(g_i - 1)!}$

$n_1 = 2, g_1 = 2, n_2 = 2, g_2 = 3$

$$\therefore \quad W = \frac{(2+2-1)!}{2!(2-1)!} \times \frac{(2+3-1)!}{2!(3-1)!}$$

$$= \frac{3!}{2!\,1!} \times \frac{4!}{2!\,2!}$$

$$= \frac{6}{2} \times \frac{4\times3\times2}{2\times2}$$

$$= 3 \times 6$$

$$= 18$$

37. $L = \frac{ml^2}{2}[\dot\theta^2 + \sin^2\theta\,\dot\phi^2] - mgl\cos\theta$

$$h = \frac{\partial L}{\partial \dot\phi} = ml^2\,\dot\phi\,\sin^2\theta \text{ is const.}$$

$$\therefore \quad \frac{d}{dt}(ml^2\,\dot\phi\,\sin^2\theta) = 0$$

or $ml^2\,\dot\phi\,\sin^2\theta = \text{const.}$

or $\dot\phi\,\sin^2\theta = \text{const.}$

38. Virial theorem for the potential

$V(r) = Kr^n$ is

$$(\text{K.E.})\ T = \frac{nE}{n+2}$$

and $(\text{P.E.})\ V = \frac{2E}{n+2}$

for circular orbit $V(r) = \frac{1}{r}$

i.e. $\quad n = -1$

$$\frac{T}{V} = \frac{n}{2}$$

$$\frac{T}{V} = -\frac{1}{2}$$

$$\boxed{2T = -V}$$

and $\quad T = \frac{(-1)E}{-1+2}$

$$T = -E$$

or $\quad \boxed{E = -T}$

Therefore answer (*a*) is correct answer.

41. Velocity of S with respect to S is V = 0.6 c

$$t'_A = \frac{t_A - \frac{v}{c^2}y}{\sqrt{1-\frac{v^2}{c^2}}}$$

For event A, $t_A = 0, y = 0$.

So $ct'_A = 0$

$$t'_B = \frac{t_B - \frac{v}{c^2}y}{\sqrt{1-\frac{v^2}{c^2}}}$$

For event B, $t_B = 0, y = 2$.

So, $ct'_B = \frac{-3}{2}$

42. Lagrangian is not a function of time. So energy is converted and component of $\left(\vec{r_1} \times \vec{r_2}\right)$ is only in $\hat{z}$ direction means potential is symmetric under ϕ, So L_z is conserved.

43. $M \rightarrow \frac{3}{10}M + \frac{3}{10}M$

For momentum conservation

$O = \vec{P_1} + \vec{P_2}$

$\Rightarrow \vec{P_1} = -\vec{P_2}$

$\Rightarrow |P_1| = |P_2|$

From energy conservation.

$E = E_1 + E_2$

$$\Rightarrow Mc^2 = \frac{3}{10}\frac{Mc^2}{\sqrt{1-\frac{v^2}{c^2}}} + \frac{3}{10}\frac{Mc^2}{\sqrt{1-\frac{v^2}{c^2}}}$$

$$\Rightarrow Mc^2 = \frac{3}{5}\frac{Mc^2}{\sqrt{1-\frac{v^2}{c^2}}}$$

$$\left(1-\frac{v^2}{c^2}\right)=\frac{9}{15}$$

$$\Rightarrow \frac{v^2}{c^2}=\frac{16}{25}$$

$$\Rightarrow \boxed{v = 0.8\,c}$$

44. $\lambda = \frac{m}{2}\vec{v}^2 + \left(\therefore \vec{r}.v - v(r)\right),$

where $v =$

$$H = \sum \dot{r}\, P_c - L$$

$$\dot{r} P_c - L$$

$$\Rightarrow \frac{\partial L}{\partial \dot{r}} = P_c = \left(m\dot{r} + cr\right)$$

$$\Rightarrow \dot{r} = \frac{P_c - Cr}{m}$$

$$\Rightarrow H = \left(\frac{P_c - Cr}{m}\right)P_c - \frac{m}{2}\left(\frac{P_c - Cr}{m}\right)^2 - Cr\left(\frac{P_c - Cr}{m}\right) + v(r)$$

$$\Rightarrow H = \left(\frac{P_c - Cr}{m}\right)(P_c - Cr) - \frac{m}{2}\left(\frac{P_c - Cr}{m}\right)^2 + v(r),$$

$$H = \frac{(P_c - Cr)^2}{m} - \frac{(P_c - Cr)^2}{2m} + v(r)$$

$$\Rightarrow H = \frac{1}{2m}(P_c - Cr)^2 + v(r)$$

48. According to Noether charge theorem

$Q = \int J^{\circ}\, d^3x$ shows the scalar invariant.

Hence charge is invariant under Lorentz transformation.

49. Proper half-life time is given by

$$\tau = \left(\frac{1}{r}\right)t$$

where 't' is half-life of elementary particle

Now
$$r = \frac{1}{\sqrt{1-\frac{v^2}{c^2}}}$$

$$= \frac{1}{\sqrt{1-\frac{(0.9)^2c^2}{c^2}}}$$

$$= \frac{1}{\sqrt{1-(0.9)^2}}$$

$$= \frac{1}{\sqrt{1-0.81}}$$

$$= \frac{1}{\sqrt{0.19}}$$

$$\therefore \quad \tau = \left(\frac{1}{r}\right)t$$

$$= \left(\sqrt{0.19}\right)t$$

$$= \sqrt{0.19}\times 5\times 10^{-8}$$

$$= 2.1794\times 10^{-8}\ \text{sec}$$

50.

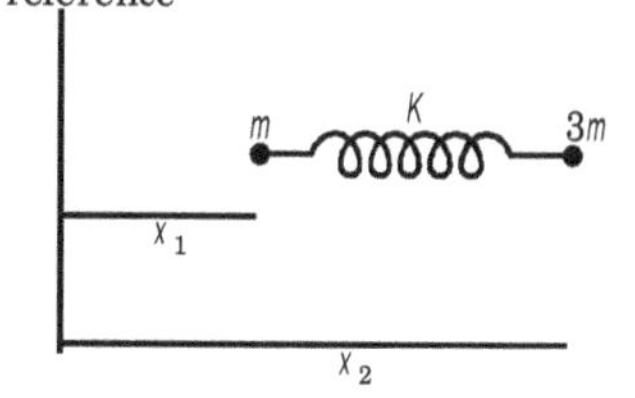

$$T = \frac{1}{2}m\dot{x}_1^2 + \frac{1}{2}(3m)\dot{x}_2^2$$

$$= \frac{1}{2}m\dot{x}_1\dot{x}_1 + \frac{1}{2}(3m)\dot{x}_2\dot{x}_2$$

$$V = \frac{1}{2}K(x_2 - x_1)^2$$

$$= \frac{1}{2}K(x_2x_2 + x_1x_1 - x_1x_2 - x_2x_1)$$

The equation of motion

$|V - \omega^2 T| = 0$

$$\begin{bmatrix} V_{11} & V_{12} \\ V_{21} & V_{22} \end{bmatrix} - \omega^2 \begin{bmatrix} T_{11} & T_{12} \\ T_{21} & T_{22} \end{bmatrix} = 0$$

$$\begin{vmatrix} K - m\omega^2 & -K \\ -K & K - 3m\omega^2 \end{vmatrix} = 0$$

$\therefore\ (K - m\omega^2)(K - 3m\omega^2) - K^2 = 0$

$K^2 - 3mK\omega^2 - mK\omega^2 + 3m^2\omega^4 - K^2 = 0$

$m\omega^2[-4K + 3m\omega^2] = 0$

$\therefore\ \omega = 0$

or $3m\omega^2 - 4K = 0$

$$\therefore\ \omega_1 = 0 \text{ or } \omega = \sqrt{\frac{4K}{3m}}$$

$$\omega = \sqrt{\frac{4\times 0.3}{3\times 0.1}} = \sqrt{4} \quad \therefore\ \omega = 2\text{Hz}$$

$\omega = 2$Hz

51. The equation of motion in terms of poisson bracket

If t explicite then $\frac{dA}{dt} = \{A, H\} + \frac{\partial A}{\partial t}$

If t is not explicit then $\frac{\partial A}{\partial t} = 0$

$\therefore \frac{dA}{dt} = \dot{A} = \{A, H\}$

$\dot{q} = \{q, H\},\ \dot{p} = \{p, H\}$

52.

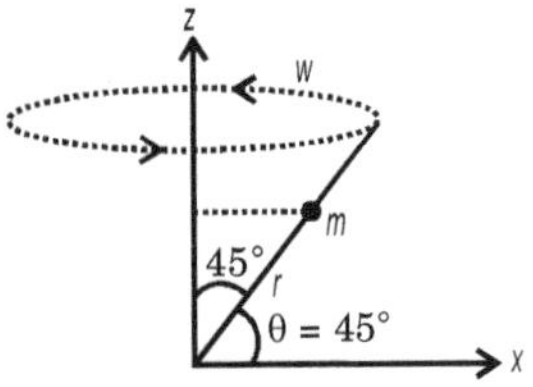

$T = \frac{1}{2}m(\dot{r}^2 + r^2\dot{\theta}^2)$ and $V = mgr\cos\theta$

Now, $L = T - V = \frac{1}{2}m(\dot{r}^2 + r^2\dot{\theta}^2) - mgr\cos\theta$

The momentum conjugate $P = \frac{\partial L}{\partial \dot{r}} = m\dot{r}$

53. In case of simple harmonic oscillation,

We have $V = \frac{1}{2}kx^2$

Differentiating V w.r.t. to x we get

$$\frac{dV}{dx} = \frac{1}{2}k(2x)$$

$$\frac{dV}{dx} = kx$$

Again differentiating, we get

$$\boxed{\frac{d^2V}{dx^2} = k} \quad ...(i)$$

Here $V(r) = \frac{-a}{r} + \frac{ar_0^2}{3r^3}$ (given)

Differentiating V(r) w.r.t. to r

$$\therefore \quad \frac{dV(r)}{dr} = +\frac{a}{r^2} - \frac{ar_0^2}{r^4}$$

Again differentiating,

$$\frac{d^2V(r)}{dr^2} = \frac{-2a}{r^3} + \frac{4ar_0^2}{r^5}$$

$$\text{Now } \left.\frac{d^2V(r)}{dr^2}\right|_{r=r_0} = \frac{-2a}{r_0^3} + \frac{4ar_0^2}{r_0^5}$$

$$= \frac{-2a}{r_0^3} + \frac{4a}{r_0^3}$$

$$= \frac{2a}{r_0^3} = k \quad \text{(from equation } (i))$$

$$\therefore \quad k = \frac{2a}{r_0^3}$$

Now $w = \frac{2\pi}{T}$

$\therefore T = \frac{2\pi}{w}$ and $w = \sqrt{\frac{k}{m}}$

$$\therefore T = \frac{2\pi\sqrt{m}}{\sqrt{k}}$$

$$= 2\pi\sqrt{\frac{m}{k}}$$

$$\boxed{T = 2\pi\sqrt{\frac{mr_0^3}{2a}}}$$

54.

$$F = -\frac{dv}{dr} = -\frac{k}{r^2} \quad ...(i)$$

The centripetal force $= -\frac{mv^2}{r}$ ...(ii)

$$\therefore \quad -\frac{mv^2}{r_0} = -\frac{k}{r_0^2}$$

$$v^2 = \frac{k}{mr_0}$$

The orbital angular momentum $l = mr_0V$

$$l = mr_0\sqrt{\frac{k}{mr_0}} = \sqrt{kmr_0}$$

$$= \sqrt{mr_0k}$$

55.

$$Q = q + 4ap \quad ...(i)$$

$$P = q + 2p \quad ...(ii)$$

For canonical condition,

$[Q\ P] = 1$

i.e., $\frac{\partial Q}{\partial q}\frac{\partial q}{\partial p} - \frac{\partial Q}{\partial P}\frac{\partial p}{\partial q} = 1$ ...(iii)

Now from equations (i) and (ii),

$$\frac{\partial Q}{\partial q} = 1;\ \frac{\partial Q}{\partial P} = 4a;\ \frac{\partial p}{\partial q} = 1;\ \frac{\partial q}{\partial p} = 2$$

Putting the value in equation (iii),

$\therefore 1 \times 2 - 4a \times 1 = 1$

$4a = 2 - 1$

$\therefore a = \frac{1}{4}$

$\therefore a = 0.25$

56. $B = \frac{1}{R}\sqrt{\frac{2VM_P}{e}}$

$$B = \frac{1}{0.1}\sqrt{\frac{2\times10^{6}\times1.67\times10^{-27}}{1.6\times10^{-19}}}$$

$$= \frac{1}{0.1}\sqrt{\frac{2\times1.67}{1.6}\times10^{-2}}$$

$$= \frac{10^{-1}}{0.1}\sqrt{\frac{2\times1.67}{1.6}}$$

$$= \left(\frac{0.1}{0.1}\right)\sqrt{\frac{2\times1.67}{1.6}}$$

B = 1.4448 Tesla.

58. As $\frac{\partial L}{\partial \dot{q}} = P_q$

But, $\frac{\partial L}{\partial q} \neq 0$.

Thus it is not conserved quantity.

59. $P = \frac{mv}{\sqrt{1-\frac{v^2}{c^2}}}$

$$\Rightarrow F = \frac{dP}{dt}$$

$$\Rightarrow m = \frac{dv}{dt}\cdot\frac{1}{\sqrt{1-\frac{v^2}{c^2}}} + mv\left(-\frac{1}{2}\right)\cdot\frac{1}{\left(1-\frac{v^2}{c^2}\right)^{\frac{3}{2}}}\cdot\frac{-2v}{c^2}\frac{dv}{dt}$$

$$\Rightarrow f = m\frac{dv}{dt}\cdot\frac{1}{\sqrt{1-\frac{v^2}{c^2}}}\left[1+\frac{1}{2}\frac{\frac{v^2}{c^2}}{\left(1-\frac{v^2}{c^2}\right)}\right]$$

$$= m\frac{dv}{dt}\left(\frac{1-\frac{v^2}{2c^2}}{\left(1-\frac{v^2}{c^2}\right)^{\frac{3}{2}}}\right)$$

$$\Rightarrow m\frac{dv}{dt}\left[\frac{\left(1-\frac{v^2}{c^2}\right)^{\frac{1}{2}}}{\left(1-\frac{v^2}{c^2}\right)\left(1-\frac{v^2}{c^2}\right)^{\frac{1}{2}}}\right]$$

$$= \frac{mc^2}{\left(c^2-v^2\right)}\cdot\frac{dv}{dt}$$

61. $T = \frac{1}{2}m\dot{x}_1^{\,2} + \frac{1}{2}m\dot{x}_2^{\,2}$,

$$v = \frac{1}{2}Kx_1^{\,2} + \frac{1}{2}k\left(x_2 - x_1\right)^2$$

$$= \frac{1}{2}Kx_1^{\,2} + \frac{1}{2}K\left[x_2^{\,2} + x_1^{\,2} - 2x_2x_1\right]$$

$$= \frac{1}{2}K\left[2x_1^{\,2} + x_2^{\,2} - 2x_2x_1\right]$$

By secular equation $\left|V - \omega^2 T\right| = 0$

$$T = \begin{pmatrix} m & 0 \\ 0 & m \end{pmatrix};\ V = \begin{pmatrix} 2K & K \\ -K & K \end{pmatrix}$$

$$\therefore \begin{vmatrix} 2K-\omega^2 m & -K \\ -K & K-\omega^2 m \end{vmatrix} = 0$$

$$\Rightarrow \left(2K-\omega^2 m\right)\left(K-\omega^2 m\right) - K^2 = 0$$

$$\Rightarrow \omega = \sqrt{\frac{3+\sqrt{5}}{2}}\sqrt{\frac{K}{m}}$$

62. Moment inertia of uniform rod about an axis passing through the end and perpendicular to the length of rod if

$$l = \frac{ML^2}{3}$$

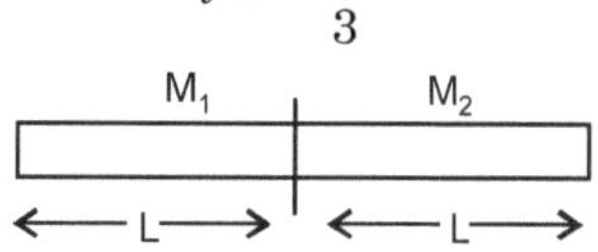

$l = l_1 + l_2$

Hence, $l = l_1 + l_2$

$$= \frac{M_1L^2}{3} + \frac{M_2L^2}{3} = \frac{1}{3}(M_1+M_2)L^2$$

63.

$$\varepsilon = 0 - \text{Circle}$$
$$\varepsilon = 1 - \text{Parabola}$$
$$0 < \varepsilon < 1 = \text{Ellipse}$$
$$0 < \varepsilon > 1 = \text{Hyperbola}$$

64. $v(x) = x(x-2)^2$

$$\frac{\partial v}{\partial x} = (x-2)^2 \cdot 1 + 2x[x-2] = 0$$

$$\Rightarrow x = 2,\ x = \frac{2}{3}$$

$$\frac{\partial^2 v}{\partial x^2} = 2(x-2) + 2(x-2) + 2x$$

$$\left.\frac{\partial^2 v}{\partial x^2}\right|_{x=2} = 2 \times 2 = 4$$

$$\Rightarrow w = \sqrt{\left.\frac{\partial^2 v}{\partial x^2}\right|_{x=2}}$$

$$\Rightarrow w = \frac{2\pi}{T} = 2$$

$$\Rightarrow T = \pi$$

65. Relative velocity

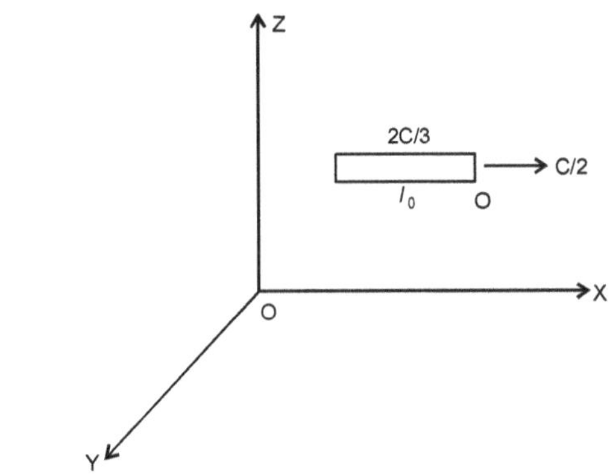

$$V = \frac{\frac{2C}{3} - \frac{C}{2}}{1 - \left[\frac{\left(\frac{2C}{3}\right)\left(\frac{C}{2}\right)}{C_2}\right]} \Rightarrow \frac{C}{4}$$

The length of rod with respect to observer

$$= \sqrt[0]{1 - \frac{V^2}{C^2}} = \sqrt[0]{1 - \left(\frac{1}{4}\right)^2}$$

$$= \frac{\sqrt[0]{15}}{4} \cong 0.97$$

66. Lagrangian

$$L(\theta, \phi) = \frac{1}{2}ma^2(\dot{\theta}^2 + \sin^2\theta\dot{\phi}^2) - mga\cos\theta$$

$$P_0 = \frac{\partial L}{\partial \theta} = ma^2\sin^2\theta$$

$$P\phi = \frac{\partial L}{\partial \theta} = ma^2\sin^2\theta\phi$$

Now, $H = \sum p\dot{q} - L$

$$= p\dot{\theta} + p\phi\dot{\phi} - L$$

$$= ma^2\theta^2 + ma^2\sin^2\theta\dot{\phi}^2 - \frac{1}{2}ma^2(\dot{\theta}^2 + \sin^2\theta\dot{\phi}^2) + mga\cos\theta$$

$$\Rightarrow \frac{1}{2}ma^2\dot{\theta}^2 + \frac{1}{2}ma^2\sin^2\theta\dot{\phi}^2 + mga\cos\theta$$

Putting $\dot{\theta} = \frac{p\theta}{ma^2}$ and $\dot{\phi} = \frac{p\phi}{ma^2\sin^2\theta}$

$$H = \frac{1}{2ma^2}\left(p\theta^2 + \frac{p\phi^2}{\sin^2\theta}\right) + mga\cos\theta$$

67. Lagrangian for the particle is $L = T - V$

Know, Kinetic energy of particle

$$T = \frac{1}{2}m(\dot{x}^2 + \dot{y}^2)$$

For parabola $y = ax^2$

$$\dot{y} = 2a\dot{x}\,x$$

$$\dot{y}^2 = 4a^2x^2\,\dot{x}^2$$

$$T = \frac{1}{2}m(1 + 4a^2x^2)\dot{x}^2$$

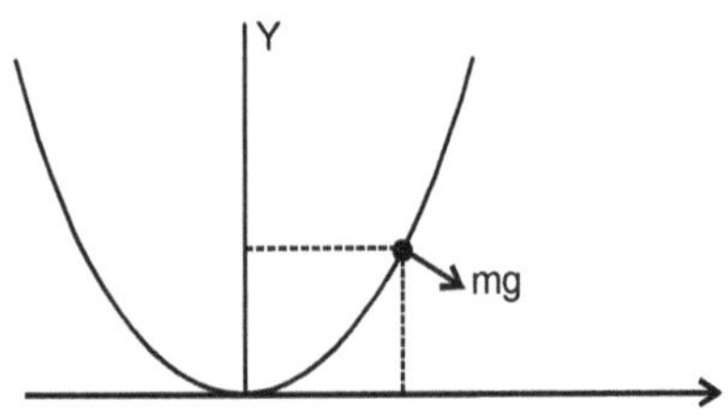

Potential energy of particle,

$$V = mg\,y = mg.ax^2$$

So, $L = T - V$

$$= \frac{1}{2}m(1 + 4a^2x^2)\dot{x}^2 - mgax^2$$

68. Lagrangian equation of motion

$$\frac{d}{dt}\left(\frac{\partial L}{\partial x}\right) - \frac{\partial L}{\partial x} = 0$$

$$\frac{\partial L}{\partial \dot{x}} = m(1 + 4a^2x^2)\dot{x}$$

$$\frac{d}{dt}\left(\frac{\partial L}{\partial m}\right) = m[(1 + 4a^2x)]$$

69. By symmetry $I_\pi = I_y$

$$I_1 = T_2$$

By perpendicular axis π theorme

$$lz = lx + ly$$

$$\Rightarrow \quad l_3 = 2l_1$$

Hence $\quad l_1 = l_2$ and $l_3 \neq l_1$

70. $$v(q,\dot{q}) = \frac{1+q^0}{q^2}$$

Generalised force

$$= \frac{\partial v}{\partial q} = \frac{-(1+\dot{q})2q}{q^4}$$

Magnitude of generalised force

$$= \left|\frac{\partial v}{\partial q}\right| = \frac{2(1+\dot{q})}{q^3}$$

71. The reduced mass of the system is

$$\mu = \frac{m \times 2m}{m+2m} = \frac{2m}{3m}$$

Lagrangian $L = \frac{1}{2}\mu\dot{x}^2 - \frac{1}{2}kx^2$

$$\Rightarrow \quad L = \frac{1}{2}\times\frac{2m}{3}\dot{x}^2 - \frac{1}{2}kx^2$$

$$\frac{\partial L}{\partial \dot{x}} = \frac{2m}{3}\dot{x} \text{ and } \frac{\partial L}{\partial x} = -vx$$

$$\frac{d}{dt}\left(\frac{\partial L}{\partial x}\right) - \frac{\partial L}{\partial \dot{x}} = 0$$

$$\Rightarrow \quad \frac{2m}{3}\ddot{x} + kx = 0$$

$$\Rightarrow \quad \ddot{x} + \frac{3x}{2m}x = 0$$

$$\Rightarrow \quad \ddot{x} + w^2x = 0$$

$$w = \sqrt{\frac{3x}{2m}}$$

72. Here $\quad Q = q^\alpha \cos \beta P$

$$\frac{\partial Q}{\partial P} = \alpha q^{\alpha-1}\cos\beta\, P$$

$$\frac{\partial Q}{\partial P} = -\beta q^\alpha \sin\beta P$$

$$P = q^\alpha \sin\beta P$$

$$\frac{\partial P}{\partial q} = \alpha q^{\alpha-1}\sin\beta q$$

$$\frac{\partial P}{\partial P} = \beta q^\alpha \cos\beta P$$

Now, $\left(\frac{\partial P}{\partial q}, \frac{\partial P}{\partial P} - \frac{\partial Q}{\partial q}\frac{\partial Q}{\partial P}\right)$

$$= \left(\alpha\beta q^{2\alpha-1}\sin\beta_P \cos\beta_P + \alpha\beta q^{2\alpha-1}\sin\beta_P\cos\beta_P\right)$$

$$= \left(\alpha\beta\, q^{2\alpha-1} \times \sin 2\beta P\right)$$

If becomes a perfect differential.

73. Here $\psi_n(x) = \sqrt{\frac{2}{a}}\sin\frac{n\pi x}{a}$

For ground state $\quad n = 1$

$$\psi_1(x) = \sqrt{\frac{2}{a}}\sin\frac{\pi x}{a}$$

The shift in ground state energy

$$E = \int_0^{a/2}\psi_1^*(x)v'(x)\psi_1(x)dx$$

$$= \frac{2}{a}v_0\int_0^{a/2}\sin^2\frac{\pi x}{a}\cos\frac{\pi x}{a}dx$$

Let $\sin\frac{\pi x}{a} = t$

$$\Rightarrow \quad \frac{\pi}{a}\cos\frac{\pi x}{a}dx = dt$$

$$\Rightarrow \quad \cos\frac{\pi x}{a}dx = \frac{a}{n}dt$$

Now when $x = 0, t = 0$

and when $x = \frac{a}{2},\ t = \sin\frac{\pi}{2} = 1$

$$E = \frac{2v_0}{a}\times\frac{a}{\pi}\int_0^1 t^2dt$$

$$\Rightarrow \quad E = \frac{2v_0}{3\pi}$$

74. Since particles is moving on two dimensional surface its degree of freedom = 2

Hence for N–non interacting particle total degree of freedom = 2N

The energy for degree of freedom = $\frac{1}{2}k_BT$

Hence total energy $\quad v = 2N \times \frac{1}{2}k_BT$

$$v = Nk_BT$$

77. $$L = \frac{1}{2}ml^2\theta^2 - mgl(1-\cos\theta)$$

$$P_\theta = \frac{\partial L}{\partial \theta} = ml^2\theta$$

$$\Rightarrow \quad \dot{\theta} = \frac{P_\theta}{ml^2}$$

$$\frac{\partial L}{\partial \theta} = mgl\sin\theta$$

$$\Rightarrow \quad \dot{P}_0 = -\frac{\partial L}{\partial \theta} = -mgl\sin\theta$$

78. $\dot{\theta}$ is a constant of motion

$$[\theta, \dot{\theta}] = 1$$

80. $PdQ - qdP$ is an exact differential for

$$Q = P \quad P = q$$

83. $$L = \frac{m}{2}(\dot{x}_1^2 + \dot{x}_2^2) - \frac{k}{2}x_1x_2$$

$$\frac{\partial L}{\partial \dot{x}_1} = m\dot{x}_1, \quad \frac{\partial L}{\partial x_1} = \frac{-k}{2}x_2$$

$$\frac{\partial L}{\partial \dot{x}_2} = m\dot{x}_2, \quad \frac{\partial L}{\partial x_2} = \frac{-k}{2}x_1$$

Now $\frac{d}{dt}\left(\frac{\partial L}{\partial \dot{x}_1}\right) - \frac{\partial L}{\partial x_1} = 0$

$$m\ddot{x}_1 + \frac{k}{2}x_2 = 0 \qquad \text{... } (i)$$

and $\frac{d}{dt}\left(\frac{\partial L}{\partial \dot{x}_2}\right) - \frac{\partial L}{\partial x_2} = 0$

$$m\ddot{x}_2 + \frac{k}{2}x_1 = 0 \qquad \text{... } (ii)$$

Subtracting equation (i) from (ii)

$$m(\ddot{x}_2 - \ddot{x}_1) - \frac{k}{2}(x_1 - x_2) = 0$$

Let $x_2 - x_1 = z$

$$\Rightarrow \quad \ddot{x}_2 - \ddot{x}_1 = \ddot{z}$$

$$\Rightarrow \quad \ddot{z} = \frac{k}{2m}z$$

$$w = \pm\sqrt{\frac{k}{2m}}$$

84. $$L = \exp\left[(\alpha t)\left(\frac{m\dot{x}^2}{2} - kx^2\right)\right]$$

$$\frac{\partial L}{\partial \dot{x}} = \exp\left[(\alpha t)\left(\frac{m\dot{x}^2}{2} - kx^2\right)\right] \times m\dot{x}$$

$$\frac{\partial L}{\partial x} = \exp\left[(\alpha t)\left(\frac{m\dot{x}^2}{2} - kx^2\right)\right] \times (-kx)$$

$$\frac{d}{dt}\left(\frac{\partial L}{\partial \dot{x}}\right) = m\dot{x}\alpha\left[(\alpha t)\left(\frac{m\dot{x}^2}{2} - kx\right)\right] + \exp\left[(\alpha t)\left(\frac{m\dot{x}^2}{2} - kx\right)\right] + m\ddot{x}\exp\left[(\alpha t)\left(\frac{mx^2}{2} - kx\right)\right]$$

Now, $\frac{d}{dt}\left(\frac{\partial L}{\partial \dot{x}}\right) - \frac{\partial L}{\partial x} = 0$

Putting the above values, we get

$$\ddot{x} + \alpha\dot{x} + \frac{k}{m}x = 0$$

85. $$kE = \frac{1}{2}(lw^2 + mv^2)$$

$$\frac{d}{dt}(kE) = \frac{1}{2}\left(l2w\frac{dw}{dt} + m\,^2v\frac{dv}{dt}\right)$$

$$\Rightarrow \quad \frac{dk}{dt} = \frac{1}{2}(\vec{\tau}, \vec{w}, +\vec{L}, \vec{\alpha})$$

86. θ is fixed only R and θ are required to describe the motion, hence generalised coordinates are 2.

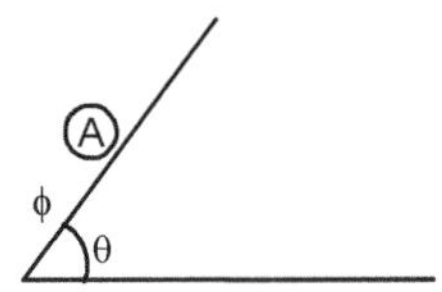

88. $$L = \frac{1}{2}\dot{q}^2 + 4\dot{q} - \frac{1}{2}q^2$$

$$\frac{\partial L}{\partial \dot{q}} = \dot{q} + q, \frac{\partial L}{\partial q} = \dot{q} - q$$

$$\frac{d}{dt}\left(\frac{\partial L}{\partial \dot{q}}\right) - \frac{\partial L}{\partial q} = 0$$

$$\dot{q} + \dot{q} - \dot{q} + q = 0$$

$\ddot{q} + q = 0$ equation of SHM.

89. $$I = \begin{vmatrix} 8 & 0 & -4 \\ 0 & 4 & 0 \\ -4 & 0 & 8 \end{vmatrix}$$

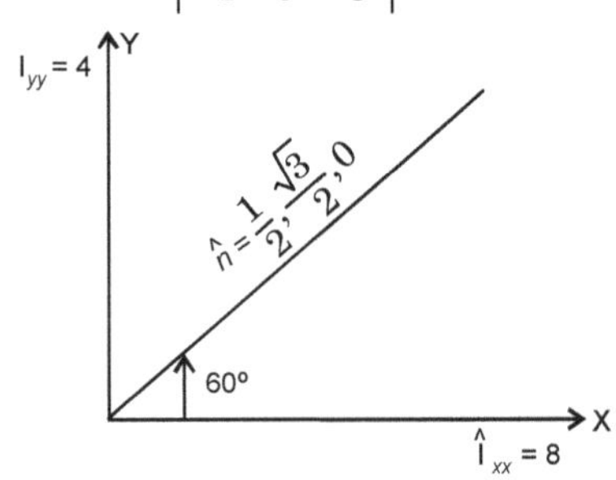

$$lav = \frac{l_{xx} + l_{yy} + l_{zz}}{2}$$

$$= \frac{8+4+8}{2} = 10$$

Along the line $l = 10 \times \frac{1}{2} = 5$

90. 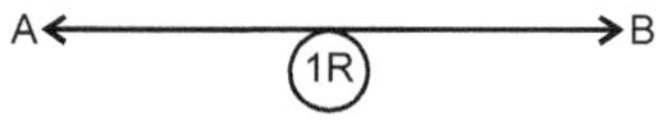

$$l_{AB} = MR^2 + MR^2 = 2MR^2$$

Torque $\tau = MR^2$

$$\tau = l\alpha$$

$$\Rightarrow \alpha = \frac{1}{2}$$

$$T = 2\pi\sqrt{\frac{R}{g\alpha}} = 2\pi\sqrt{\frac{R}{g \times \frac{1}{2}}} = 2\pi\sqrt{\frac{2R}{g}}$$

91. Let T is the tension in string then

$$F + T = mrw_0^{\,2}$$

and $$F = mV$$

if F becomes doubled then

$$2F + T = mr_0 w_0^{\,2}$$

$$2mV + mrw_0^{\,2} = mr_0\, w_0^{\,2}$$

$$V = \frac{rw_0^2 - r_0 w_0^2}{2}$$

$$= \left(\frac{r - r_0}{2}\right) w_0^2$$

$$\Rightarrow V \to C;\ 0 < r_1 < r_0$$

92.

$$L = \frac{1}{2}\dot{q}^2 - \frac{1}{2}q^2$$

$$Pq = \frac{\partial L}{\partial \dot{q}} = \dot{q}, \frac{\partial L}{\partial \dot{q}} = -q$$

$$H = Pq\,\dot{q} - L = \dot{q}^2 - \frac{1}{2}\dot{q}^2 + \frac{1}{2}q^2$$

$$= \frac{q^2}{2} + \frac{1}{2}q^2, H = iA(P, q)$$

94. The magnetization can be written as

$$Ms = \mu N\ \tan h \frac{(\mu_0\ \mu\lambda\ Ms)}{k_B T}$$

But $B = \mu_0 \times Ms$

$$Ms = \mu N \tan h\left(\frac{\mu B}{k_B T}\right)$$

95.

$$DE = \frac{\Delta t \sin P}{\sqrt{1 - \frac{V^2}{C^2}}}$$

$$= \frac{\Delta t \sin P}{\sqrt{1 - \frac{V^2}{C^2}}}$$

$$= \frac{V_1^2}{C^2} = 1 - \frac{25}{169}$$

$$V_1 = \frac{12}{13}C$$

$$\text{Relative speed} = \frac{\frac{12}{13}C - \frac{4}{5C}}{1 - \frac{48}{13 \times 5}}$$

$$= \frac{8}{17}C$$

$$\Rightarrow \qquad = \frac{8}{17}C$$

$$V_0 = -\left(\frac{R}{3R}V_a + \frac{R}{3R}V_b + \frac{R}{3R}V_c\right)$$

$$= -\frac{1}{3}(V_a + V_b + V_c)$$

$$V_{01} = \frac{-R}{R}V_0 = -\left[-\frac{1}{3}(V_a + V_b + V_c)\right]$$

$$= -\frac{1}{3}(V_a + V_b + V_c)$$

97. $\alpha^{-1}\left(\frac{s+1}{s^2-4}\right) = \alpha^{-1}\left\{\frac{s}{s^2-4}\right\} + \alpha^{-1}\left\{\frac{s}{s^2-4}\right\}$

$$= \cosh 2x + \frac{1}{2}\sinh 2x.$$

99. Let I is the moment of inertia of bigger sphere.

hence $I = \frac{2}{5}\left(\frac{4\pi}{3}\right)R^5\rho$ ...(*i*)

for small sphare of $\frac{R}{2}$ about its centre.

$$I_1 = \frac{2}{5}\left(\frac{4\pi}{3}\right)\left(\frac{R}{2}\right)^5 \rho$$

Moment of inertia for sphere about *y*-axis

$$I^1 = \frac{2}{5}\left(\frac{4\pi}{3}\right)\left(\frac{R}{2}\right)^5 \rho \times \left(\frac{R}{2}\right)^2$$

$$= \frac{4\pi}{3}\rho\left[\frac{2}{5}\left(\frac{R}{5}\right)^5 + \left(\frac{R}{2}\right)^5\right]$$

$$= \frac{4\pi}{3}\rho\left(\frac{R}{2}\right)^5 \times \frac{7}{5}$$

For second sphere, $I'' = I' = \frac{4\pi}{3}\rho\left(\frac{R}{2}\right)^5 \times \frac{7}{5}$

Total $I_t = 2I' = \frac{4\pi}{3}\rho\left(\frac{R}{2}\right)^5 \times \frac{14}{5}$

Now, $I_r = I_1 - I_t = \frac{4\pi}{3}\rho R^5\left(\frac{2}{5} - \frac{7}{5 \times 32}\right)$

$$I_R = \frac{5\pi\rho R^5}{12}$$

100. $L = \frac{m}{2}\left[\left(\frac{dx}{dt}\right)^2 + \left(\frac{dy}{dt}\right)^2 + \left(\frac{dz}{dt}\right)^2\right] - \frac{V}{2}\left(x^2 + y^2\right) + w$

$= \frac{m}{2}\left[\dot{x}^2 + \dot{y}^2 + \dot{Z}^2\right] - \frac{V}{2}\left(x^2 + y^2\right) + w$

∵ Potential is independent of velocity, So energy is conserved.

Now, $\frac{\partial L}{\partial \dot{Z}} = m\dot{Z}$ and $\frac{\partial L}{\partial Z} = 0$

$$\frac{d}{dt}\left(\frac{\partial L}{\partial \dot{Z}}\right) - \frac{\partial L}{\partial Z} = 0 \Rightarrow m\ddot{Z} = 0$$

$\Rightarrow \quad F_z = 0$

$\Rightarrow \quad \frac{dP_z}{dt} = 0 \Rightarrow P_z = \text{constant}$

Hence, Z-component of linear momentum is constant.

101.

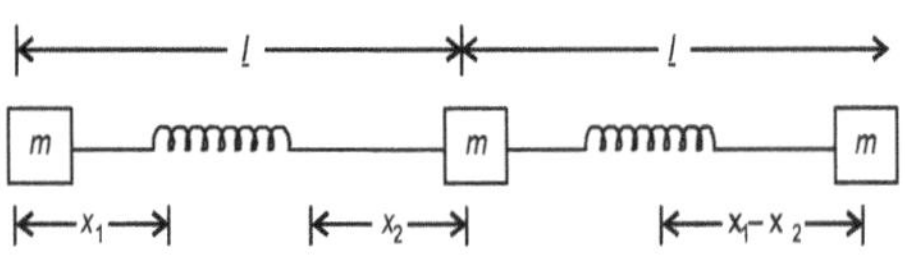

$$T = \frac{1}{2}m\left(\dot{x}_1^2 + \dot{x}_2^2 + \dot{x}_3^2\right)$$

$$V = \frac{k}{2}\left[\left(x_1 - x_2 - l\right)^2 + \left(x_3 - x_2 - l\right)^2\right]$$

$$L = T - V$$

$$= \frac{m}{2}\left[\left(\frac{dx_1}{dt}\right)^2 + \left(\frac{dx_2}{dt}\right)^2 + \left(\frac{dx_3}{dt}\right)^2\right] - \frac{k}{2}\left[\left(x_1 - x_2 - l\right)^2 + \left(x_3 - x_2 - l\right)^2\right]$$

102. $H = \frac{p^2}{2m} + pq$

Now $\ddot{q} = \frac{p}{m}$ and $H = p\dot{q} - L$

$$\frac{p^2}{2m} + pq = \frac{p^2}{m} - L$$

$$\Rightarrow L = \frac{p^2}{2m} - pq = \frac{m}{2}\left[\dot{q}^2 + q^2 - q\dot{q}\right] = \frac{m}{2}\left[\frac{dq}{dt} - q\right]^2$$

103. Speed of particle P in rest frame of Q is

$$V = \frac{v_1 + v_2}{1 + \frac{v_1 v_2}{c}} = \frac{\frac{5c}{19} + \frac{2c}{5}}{1 + \frac{5c \times 2c}{95c^2}} = \frac{3}{5}c$$

104. $E = mc^2 = \frac{m_0}{\sqrt{1 - v^2/c^2}} = \frac{m_0 c^2}{\sqrt{1 - \frac{9}{25}}} = \frac{5}{4}m_0 c^2$

105. K.E. of the particle, $E_K = \frac{1}{2}mv^2$

P.E. of the particle = Electric field contribution + Magnetic field contribution

$$= q\phi - \frac{q}{c}\left(\vec{v} \cdot \vec{A}\right)$$

∴ Lagrangian

$$L = \frac{1}{2}m\,\vec{v} \cdot \vec{v} - q\phi + \frac{q}{c}\left(\vec{v} \cdot \vec{A}\right)$$

Generalised momenutm,

$$\vec{p} = \frac{\partial L}{\partial \bar{v}} = \left(m\bar{v} + \frac{q\bar{A}}{c}\right)$$

$$\Rightarrow \quad m\bar{v} = \left(\bar{p} - \frac{q}{c}\vec{A}\right)$$

Hamiltonian

$$H = \vec{p} \cdot \vec{v} - L$$

$$= \left(m\vec{v} + \frac{q\bar{A}}{c}\right) \cdot \vec{v} - \frac{1}{2}m\,\vec{v} \cdot \vec{v} + q\phi - \frac{q}{c}\left(\vec{v} \cdot \vec{A}\right)$$

$$= \frac{1}{2}m(\vec{v} \cdot \vec{v}) + q\phi$$

$$H = \frac{1}{2m}\left(p - \frac{q\vec{A}}{c}\right)^2 + q\phi$$

106. Eccentricity of a trajectory,

$$\in = \left[1 + \frac{2El^2}{mk^2}\right]^{1/2}$$

where, E = energy

e = angular moment per unit mass, and

k = force constant. $\left[F_{\text{central}} = \frac{-k}{r^2}\right]$

As $E > 0$, $\in > 1$ and trajectory is hyperbola.

107. For force particler moving in $+ve$ x–direction

$$\Psi(n) \simeq \frac{1}{\sqrt{L}}e^{-\frac{ipn}{h}}$$

108. For (i) transformation,

$$Q = p,\ P = -q$$

Poisson bracket $[Q, P]_{q,p}$

$$= \frac{\partial Q}{\partial q}\frac{\partial P}{\partial p} - \frac{\partial Q}{\partial p}\frac{\partial P}{\partial q}$$

$$= -1.(-1) = 1$$

(*ii*) transformation

$$Q = p, \quad P = q,$$

poisson bracket $[Q, P]_{q,p} = -1$

Thus only (*i*) transformation is canonical.

109. Relativistic mass correction,

$$m = \frac{m_0}{\sqrt{1-\beta^2}} = \frac{2m_0}{\sqrt{3}}$$

$$\Rightarrow \sqrt{1-\beta^2} = \frac{\sqrt{3}}{2}$$

$$1-\beta^2 = \frac{3}{4}$$

$$\beta^2 = \frac{1}{4},\ v = \frac{c}{2}$$

∴ Linear momentum

$$= mv = \frac{2m_0}{\sqrt{3}}.\frac{c}{2} = \frac{m_0 c}{\sqrt{3}}$$

110. $\mu_1 = \sqrt{5}\ aq\ \hat{n}_1$(along θ_1)

$\mu_2 = \sqrt{5}\ aq\ \hat{n}_2$(along θ_2)

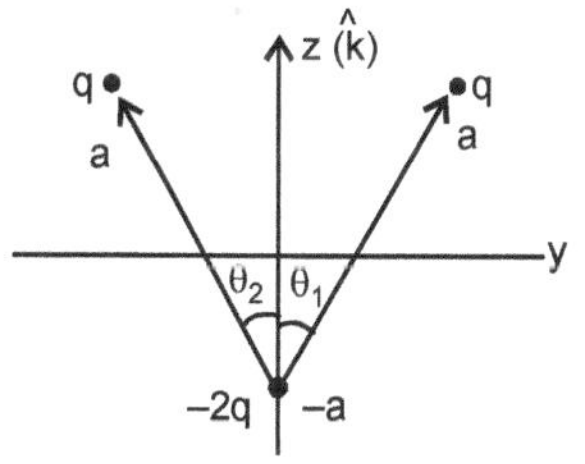

Net *d.p.* along $\hat{k}$

$$\mu = \mu_1 \cos\theta_1 + \mu_2 \cos\theta_2$$

Here, $\tan\theta_1 = \tan\theta_2 = \frac{a}{2a} = \frac{1}{2}$

$$\cos\theta_1 = \cos\theta_2 = \frac{2}{\sqrt{5}}$$

$$\mu = \sqrt{5}aq.\frac{2}{\sqrt{5}} + \sqrt{5}aq.\frac{2}{\sqrt{5}} = 4aq$$

$$\bar{\mu} = 4qa\hat{k}$$

111. $L_X = (\bar{r}\times\bar{p})_x = [yp_z - zp_y]$

$$[L_X, y] = [yp_z, y] - [zp_y, y]$$

$$= 0 - [p_y - y]z$$

$$= -(-i\hbar)z = i\hbar z$$

112. $|\psi_1\rangle^0 = \begin{pmatrix}1\\0\end{pmatrix}; |\psi_2\rangle^0 = \begin{pmatrix}0\\1\end{pmatrix} H_p = \begin{pmatrix}E_1 & A\\ A^* & E_2\end{pmatrix}$

1st order, correction :

$$E_1^{(i)}\ \langle\psi_1|H_p|\psi_1\rangle = (10)\begin{pmatrix}E_1 & A\\ A^* & E_2\end{pmatrix}\begin{pmatrix}1\\0\end{pmatrix} = (10)\begin{pmatrix}E_1\\A^*\end{pmatrix} = E_1$$

$$E_1^{(i)} = E_1$$

∴ 1st order correction = 0

113. 1st order correction to eigenfunction. $|\psi_1\rangle$

$$|\psi_n\rangle^{(1)} = \sum_{\mu\neq n}\frac{\langle\psi_\mu|H_p|\psi_n\rangle}{E_n^0 - E_\mu^0}|\psi_\mu\rangle^{(0)}\ ;\ |\psi_1\rangle^{(1)} = \sum_{\mu\neq 1}\langle\psi_\mu|H_p|\psi_1\rangle|\psi_\mu\rangle^0$$

$$|\psi_1\rangle^{(1)} = \frac{\langle\psi_2|H_p|\psi_1\rangle}{E_1^0 - E_2^0}|\psi_2\rangle = \frac{(01)\begin{pmatrix}E_1 & A\\ A^* & E_2\end{pmatrix}\begin{pmatrix}1\\0\end{pmatrix}}{E_1 - E_2}|\psi_2\rangle$$

$$= \frac{A^*}{E_1 - E_2}|\psi_2\rangle = \frac{A^*}{E_1 - E_2}\begin{pmatrix}0\\1\end{pmatrix} = \begin{pmatrix}A^*/(E_1 - E_2)\\0\end{pmatrix}$$

2nd order energy correction :

$$E_1^{(2)} = \langle\psi_1^{(0)}|H_p|\psi_1^{(1)}\rangle$$

$$|\psi_1\rangle^{(1)} = \left(\frac{A^*}{E_1 - E_2}\right);$$

$$E_1^{(2)} = (10)\begin{pmatrix}E_1 & A\\ A^* & E_2\end{pmatrix}\begin{pmatrix}A^*/(E_1 - E_2)\\0\end{pmatrix}$$

$$E_1^{(2)} = (1\ 0)\begin{pmatrix}E_1A^*/(E_1 - E_2)\\A^*A^*/(E_1 - E_2)\end{pmatrix} = \frac{A^*}{E_1 - E_2}$$

2nd order correction to energy

$$E_1 = \sum_{n\neq 1}\frac{\langle 1|H|n\rangle}{(E_1 - E_n)} = \frac{A^2}{E_1 - E_2}$$

114. 1st order correction to

$|\psi_1\rangle$ is $\begin{pmatrix}A^*/(E_1 - E_2)\\0\end{pmatrix}$

115. $L = \frac{1}{2}m(\dot{x}^2 + \dot{y}^2)$

$$= \frac{1}{2}m\left[A^2(-\sin\theta\theta)^2 + A^2(\cos\theta\theta)^2\right]$$

$$= \frac{1}{2}mA^2\dot{\theta}^2$$

116. $\frac{\partial}{\partial t}\left(\frac{\partial L}{\partial\theta}\right) - \left(\frac{\partial L}{\partial\theta}\right) = 0$

$$\frac{\partial L}{\partial\theta} = 0$$

$$\Rightarrow \frac{d}{dt}\left(\frac{\partial L}{\partial \theta}\right) = 0;\ \frac{d}{dt}\left(\frac{1}{2}mA^2\dot{\theta}\times 2\right) = 0$$

or $\quad mA^2\ddot{\theta} = 0$

or $\quad \ddot{\theta} = 0$

117. Orbit always lie in a plane which is perpendicular to the fixed direction of angular motion.

118. Coriolis force $= 2m\vec{\omega}\times\vec{v} = 2\omega.mv.\sin\theta$

Here, $\quad \theta = 90°$

$\therefore \quad F = 2mv\omega$

121.

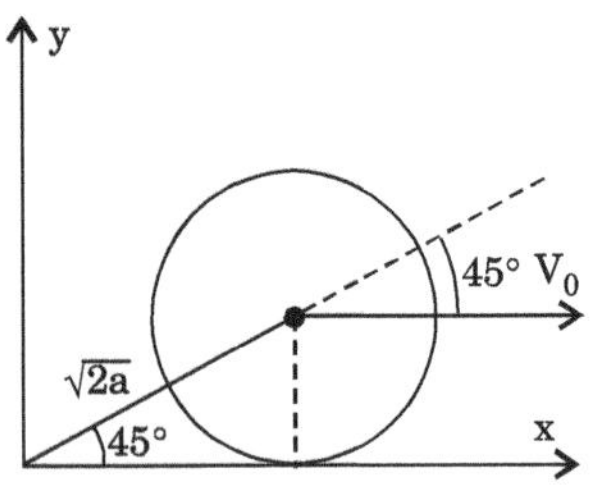

$$\vec{L} = \vec{L}_{cm} + M\vec{r_0}\times\vec{V_0}$$
$$= I\omega + Mr_0V_0\sin 45°$$
$$= Ma^2\omega + M\sqrt{2}a\omega a\frac{1}{\sqrt{2}}$$
$$= 2M\omega a^2$$

122.
$$I = I_1 + I_2$$
$$= \left[\frac{2}{5}MR^2 + M\times(2R)^2\right] + \left[\frac{2}{5}MR^2 + M\times(2R)^2\right]$$
$$= \frac{44}{5}MR^2$$

124. We know, $l_x = l_0\cos\theta_0$,

and $\quad l_y = l_0\sin\theta_0$

$$\therefore \quad l'_x = l_x\sqrt{1-\frac{v^2}{c^2}}$$
$$= l_0\cos\theta°\sqrt{1-\frac{v^2}{c^2}}$$

and $\quad l'_y = l_y = l_0\sin\theta°$

$$\therefore \quad \tan\theta = \frac{l'_y}{l'_x} = \frac{l_0\sin\theta_0}{\frac{l_0\cos\theta_0}{\gamma}}$$
$$= \tan\theta_0\gamma$$
$$\therefore \quad \theta = \tan^{-1}(\nu\tan\theta_0)$$

126. Equation of motion :
$$\frac{d}{dx_1}\left[\frac{\partial L}{\partial x_1}\right] - \frac{\partial L}{\partial x_1} = 0$$
and
$$\frac{d}{dx_2}\left[\frac{\partial L}{\partial x_2}\right] - \frac{\partial L}{\partial x_2} = 0$$

127.
$$\oint B.dt = \mu_0 I$$
$$B\times 2\pi R = \mu_0\frac{\sigma dA}{dt}$$
$$\left[\because\ I = \frac{d\phi}{dt} = \frac{d(\sigma A)}{dt} = \frac{\sigma dA}{dt}\right]$$

$\therefore$ Surface current,
$$J_s = M = \frac{B}{\mu_0} = \frac{\sigma}{2\pi R}\frac{dA}{dt}$$

Now, $A = \pi R^2$
$$\therefore \quad \frac{dA}{dt} = 2\pi R\frac{dR}{dt} = 2\pi R 2\omega$$
$$\therefore \quad J_s = M = \frac{\sigma}{2\pi R}\times 2\pi R^2\omega = \sigma R^2\omega$$

128. Since $\Box = \hat{i}\frac{\partial}{\partial x_1} + \hat{j}\frac{\partial}{\partial x_2} + \hat{k}\frac{\partial}{\partial x_3} + \hat{x}_4\frac{\partial}{\partial x_4}$

$$\Rightarrow \Box\cdot\Box = \left(\frac{\partial^2}{\partial x_1^2} + \frac{\partial^2}{\partial x_2^2} + \frac{\partial^2}{\partial x_3^2} + \frac{\partial^2}{\partial x_4^2}\right)$$

= Lorentz invariant

But $x_1 = x,\ x_2 = y,\ x_3 = z$ and $x_4 = ict$

$$\therefore \quad \Box^2 = \left(\frac{\partial^2}{\partial x^2} + \frac{\partial^2}{\partial y^2} + \frac{\partial^2}{\partial z^2} - \frac{1}{c^2}\frac{\partial^2}{\partial t^2}\right)$$

= invariant under Lorentz transformation.

129. Because $\quad [H, L] \neq 0$

So, $\quad [H, Lz]$ is not commute also.

130.
$$L = \frac{1}{2}m(\dot{r}^2 + r^2\dot{\theta}^2) - v(r)$$

Since, Generalized momenta, $p = \left(\frac{\partial L}{\partial \dot{q}_i}\right)$

$$p = \left(\frac{\partial L}{\partial \dot{r}}\right) \text{ and } \left(\frac{\partial L}{\partial \dot{\theta}}\right)$$

$$\Rightarrow \quad p = m\dot{r} \text{ and } mr^2\dot{\theta}.$$

131. Moment of inertia $= m\left(\frac{3a}{4}\right)^2 + m_1\frac{a^2}{3}$

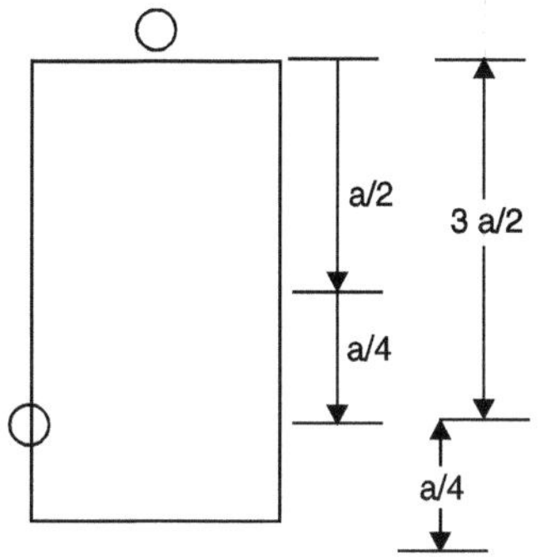

For the center of rod

$$\left(\frac{m_1a^2}{12} + \frac{m_1a^2}{4}\right) = \frac{m_1a^2}{3}$$

$$\therefore \quad m_1 = 4m$$

$$\text{Total I} = m\left(\frac{3a}{4}\right)^2 + \frac{4ma^2}{3}$$

$$= \frac{9ma^2}{16} + \frac{4\,ma^2}{3}$$

$$= \frac{(27+64)}{48}ma^2 = \frac{91}{48}ma^2.$$

132. $\quad L = a\dot{x}^2 + b\dot{y}^2 - kxy$

Since $\quad H = \sum_i p_i\dot{q}_i - L = \Sigma(p_x\,\dot{x} + p_y\,\dot{y}) - L$

$$\therefore \quad p_x = \frac{\partial L}{\partial \dot{x}} = 2a\,\dot{x} \quad \Rightarrow \quad \dot{x} = \frac{p_x}{2a}$$

and $\quad p_y = \frac{\partial L}{\partial \dot{y}} = 2b\,\dot{y} \quad \Rightarrow \quad \dot{y} = \frac{p_y}{2b}$

$$H = \left(\frac{p_x^2}{2a} + \frac{p_y^2}{2b}\right) - a\,\dot{x}^2 - b\dot{y}^2 + kxy$$

$$= \left(\frac{p_x^2}{2a} + \frac{p_y^2}{2b}\right)\frac{-p_x^2}{4a}\frac{-p_y^2}{4b} + kxy$$

$$= \boldsymbol{\frac{p_x^2}{4a} + \frac{p_x^2}{4b} + kxy}$$

133. $\quad [u, v] = \sum\left(\frac{\partial u}{\partial q}\cdot\frac{\partial v}{\partial p} - \frac{\partial u}{\partial p}\frac{\partial v}{\partial q}\right)$

Here use $u = \vec{a}\,.\,\vec{r}$ and $v = \vec{b}\,.\,\vec{q}$

$$\left[\vec{a}\,.\,\vec{r}\,,\,\vec{b}\,.\,\vec{p}\right] = \left(a\frac{\partial r}{\partial r}.\,b\frac{\partial p}{\partial p} - a\frac{\partial r}{\partial p}b\frac{\partial p}{\partial r}\right)$$

$$= (a \times 1\,.\,b \times 1 - a \times 0.b \times 0)$$

$$= (ab - 0)$$

$$= \boldsymbol{ab.}$$

134.

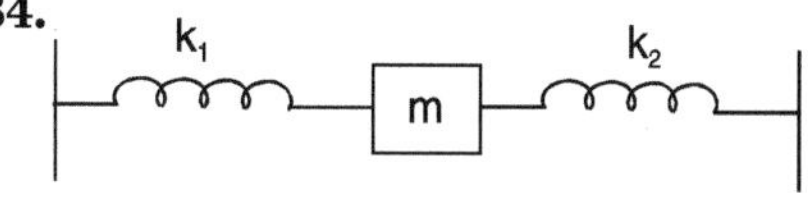

$$L = T - V$$

where, $\quad T = \frac{1}{2}m\,\dot{x}^2$

$$V = \frac{1}{2}k_1x^2 + \frac{1}{2}k_2x^2 = \frac{1}{2}(k_1 + k_2)x^2$$

$$\therefore \quad L = \frac{1}{2}m\,\dot{x}^2 - \frac{1}{2}(k_1 + k_2)x^2.$$

143. Velocity oif centre of mass $= \frac{MV}{M+M} = \frac{V}{2}$

In centre of mass frame, total momentum before and after collision must be zero.

Therefore second particle must move in just opposite direction to the first particle with same mangitude of momentum after collison, in COM frame.

$\therefore$ Velocity of second particle after collision

$= \frac{V}{2}$ along perpendicular direction to the velocity of COM

$\therefore$ Velocity of second particle in laboratory frame

$$= \sqrt{\frac{V^2}{4} + \frac{V^2}{4}} = \frac{V}{\sqrt{2}}$$

145. Mass of electron after being accelerated by 10.2 M volt is equivalent to

= 10.2 MeV + rest mass energy

= 10.2 MeV + 0.5 MeV

$\therefore$ Percent increase in mass

$$= \frac{10.2}{0.5} \times 100 = 2000$$

146. An operator X is said to b Hermition if

$$X^* = X$$

Now $\quad [i\,(A^+ - A)]^* = -i\,[(A^+)^* - A^*]$

$$= i\,(A^* - A)$$

150. $\quad L = T - U$

$$T = \frac{1}{2}m\dot{x}^2 + \frac{1}{2}m\dot{y}^2$$

Now, $\quad xy = c$

$$\Rightarrow \quad y = \frac{c}{x}$$

$$\Rightarrow \qquad \dot{y} = \frac{-c}{x^2}\dot{x}$$

$$\therefore \qquad \dot{y}^2 = \frac{-c}{x^4}\dot{x}^2$$

$$U = mgy = \frac{mgc}{x}$$

$$\therefore \qquad L = T - U$$

$$= \frac{1}{2}m\dot{x}^2\left(1+\frac{x^2}{x^4}\right) - \frac{mgc}{x}$$

153. The potential cure is shown below

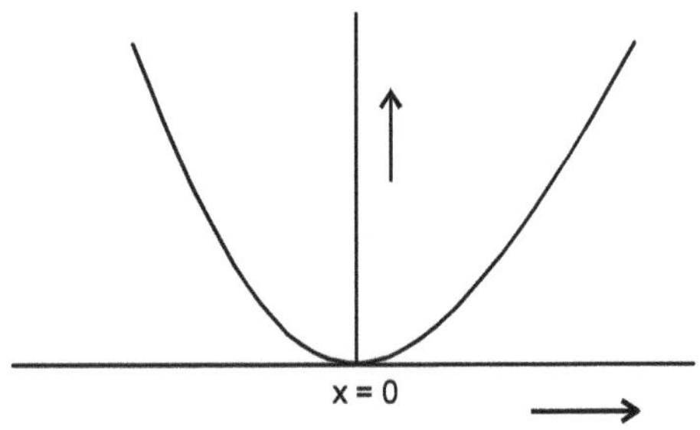

Velocity at $x = 0$, is zero.

Therefore K.E. at $x = 0$ is also zero. From the potential curve it is evident that particle must have some energy to move from $x = 0$.

■ ■

3 CHAPTER Electromagnetic Theory

2021

1. Consider a tiny current loop driven by a sinusoidal alternating current. If the surface integral of its time-averaged Poynting vector is constant, then the magnitude of the time-averaged magnetic field intensity, at any arbitrary position, $\vec{r}$, is proportional to

(a) $\frac{1}{r^3}$ (b) $\frac{1}{r^2}$

(c) $\frac{1}{r}$ (d) r

2. Consider a solenoid of length L and radius R, where $R << L$. A steady-current flows through the solenoid. The magnetic field is uniform inside the solenoid and zero outside.

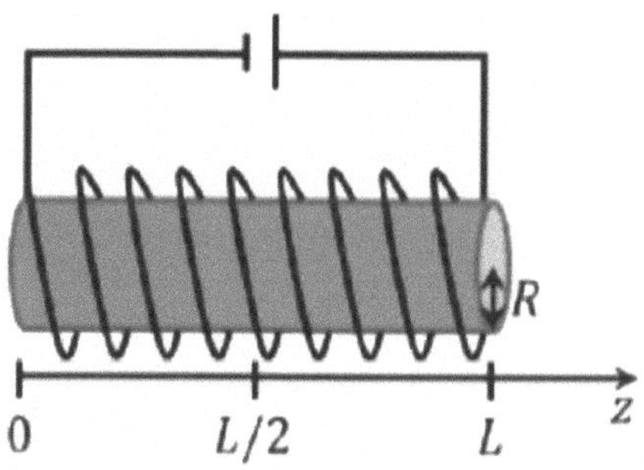

Among the given options, choose the one that best represents the variation in the magnitude of the vector potential, $(0, A_\varphi, 0)$ at $z = L/2$, as a function of the radial distance (r) in cylindrical coordinates.

Useful information: The curl of a vector $\vec{F}$, in cylindrical coordinates is

$$\vec{\Delta} \times \vec{F}(r,\varphi,z) = \hat{r}\left[\frac{1}{r}\frac{\partial F_z}{\partial \varphi} - \frac{\partial F_\varphi}{\partial z}\right] + \hat{\varphi}\left[\frac{\partial F_r}{\partial z} - \frac{\partial F_z}{\partial r}\right] + \hat{z}\frac{1}{r}\left[\frac{\partial (rF_\varphi)}{\partial r} - \frac{\partial F_r}{\partial \varphi}\right]$$

(a)

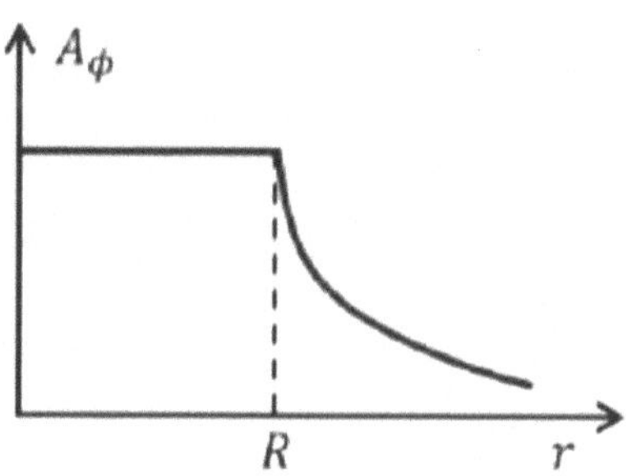

(b)

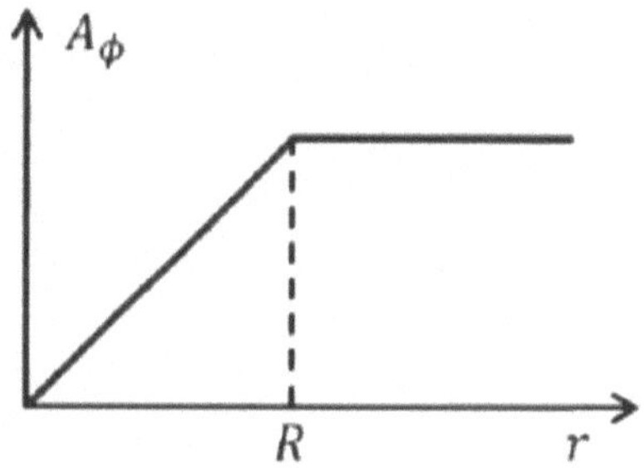

(c)

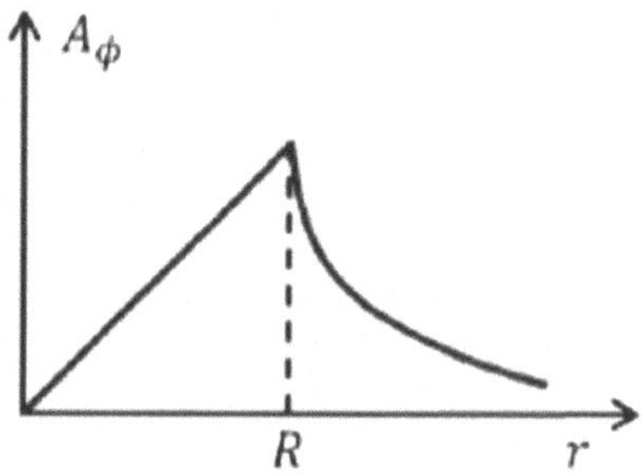

(d)

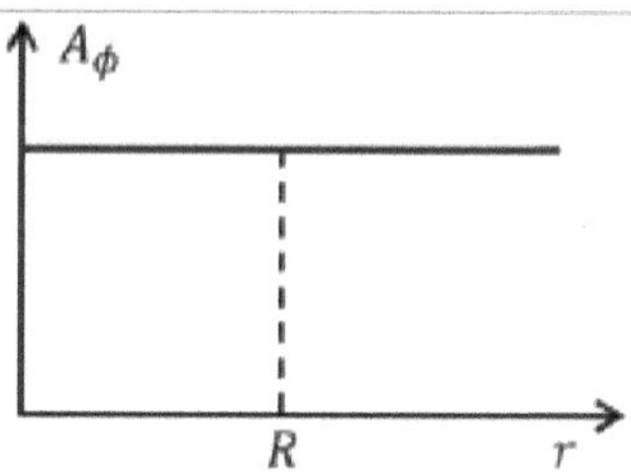

3. As shown in the figure, an electromagnetic wave with intensity I_I is incident at the interface of two media having refractive indices $n_1 = 1$ and $n_2 = \sqrt{3}$. The wave is reflected with intensity I_R and transmitted with intensity I_T. Permeability of each medium is the same. (Reflection coefficient $R = I_R/I_I$ and Transmission coefficient $T = I_T/I_I$).

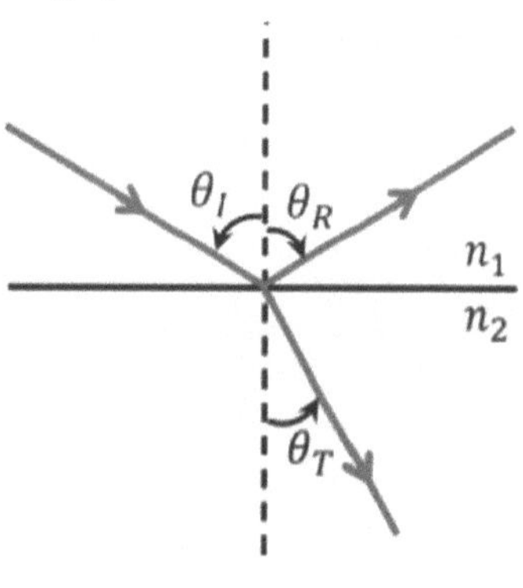

Choose the correct statement(s).

(a) R = 0 if θ_I = 0° and polarization of incident light is parallel to the plane of incidence.

(b) T = 1 if θ_I = 60° and polarization of incident light is parallel to the plane of incidence.

(c) R = 0 if θ_I = 60° and polarization of incident light is perpendicular to the plane of incidence.

(d) T = 1 if θ_I = 60° and polarization of incident light is perpendicular to the plane of incidence.

4. A material is placed in a magnetic field intensity H. As a result, bound current density J_b is induced and magnetization of the material is M. The magnetic flux density is B. Choose the correct option(s) valid at the *surface* of the material.

(a) $\nabla.M = 0$ (b) $\nabla.B = 0$

(c) $\nabla.H = 0$ (d) $\nabla.J_b = 0$

5. An electromagnetic wave having electric field $E = 8\cos(kz - \omega t)\,\hat{y}$ Vcm^{-1} is incident at 90° (normal incidence) on a square slab from vacuum (with refractive index n_0 = 1.0) as shown in the figure. The slab is composed of two different materials with refractive indices n_1 and n_2. Assume that the permeability of each medium is the same. After passing through the slab for the first time, the electric field amplitude, in V cm^{-1}, of the electromagnetic wave, which emerges from the slab in region 2, is closest to

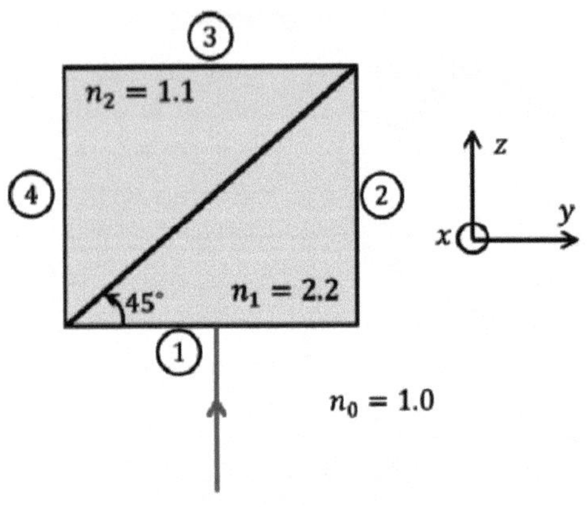

(a) $\dfrac{11}{1.6}$ (b) $\dfrac{11}{3.2}$

(c) $\dfrac{11}{13.8}$ (d) $\dfrac{11}{25.6}$

6. Consider two concentric conducting spherical shells as shown in the figure. The inner shell has a radius a and carries a charge $+Q$. The outer shell has a radius b and carries a charge $-Q$. The empty space between them is half- filled by a hemispherical shell of a dielectric having permittivity ε_1. The remaining space between the shells is filled with air having the permittivity ε_0.

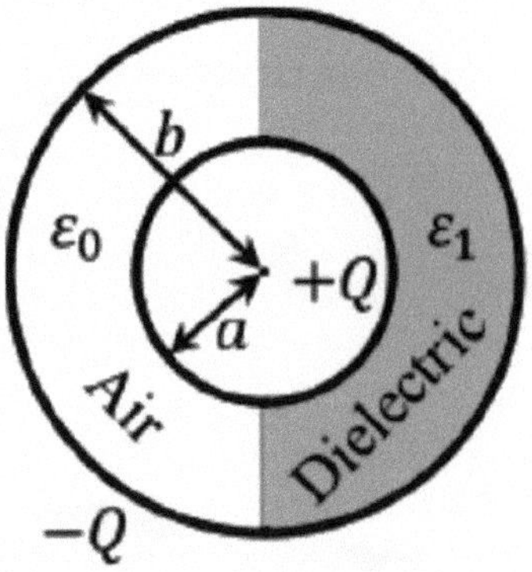

The electric field at a radial distance r from the center and between the shells ($a < r < b$) is

(a) $\dfrac{Q}{2\pi(\varepsilon_0 + \varepsilon_1)}\dfrac{\hat{r}}{r^2}$ everywhere

(b) $\dfrac{Q}{4\pi\varepsilon_0}\dfrac{\hat{r}}{r^2}$ on the air side and $\dfrac{Q}{4\pi\varepsilon_1}\dfrac{\hat{r}}{r^2}$ on the dielectric side

(c) $\dfrac{Q}{2\pi\varepsilon_0}\dfrac{\hat{r}}{r^2}$ on the air side and $\dfrac{Q}{2\pi\varepsilon_1}\dfrac{\hat{r}}{r^2}$ on the dielectric side

(d) $\dfrac{Q}{4\pi(\varepsilon_0 + \varepsilon_1)}\dfrac{\hat{r}}{r^2}$ everywhere

7. The Coulomb energy component in the binding energy of a nucleus is 18.432 MeV. If the radius of the uniform and spherical charge distribution in the nucleus is 3 fm, the corresponding atomic number (rounded off to the nearest integer) is ________.

(Given: $\dfrac{e^2}{4\pi\varepsilon_0}$ =1.44 MeV fm)

8. Consider a cross-section of an electromagnet having an air-gap of 5 cm as shown in the figure. It consists of a magnetic material ($\mu = 20000\mu_0$) and is driven by a coil having $NI = 10^4$A, where N is the number of turns and I is the current in Ampere.

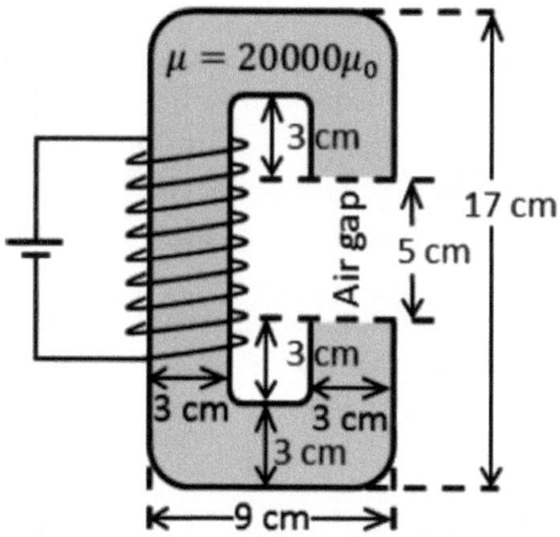

Ignoring the fringe fields, the magnitude of the magnetic field $\vec{B}$ (in Tesla, rounded off to two decimal places) in the air-gap between the magnetic poles is ______.

2020

9. A medium ($\varepsilon_r > 1, \mu_r = 1, \sigma > 0$) is semi-transparent to an electromagnetic wave when

(a) Conduction current >> Displacement current

(b) Conduction current << Displacement current

(c) Conduction current = Displacement current

(d) Both conduction current and Displacement current are zero

10. Which one of the following relations determines the manner in which the electric field lines are refracted across the interface between two dielectric media having dielectric constants ε_1 and ε_2 (see figure)?

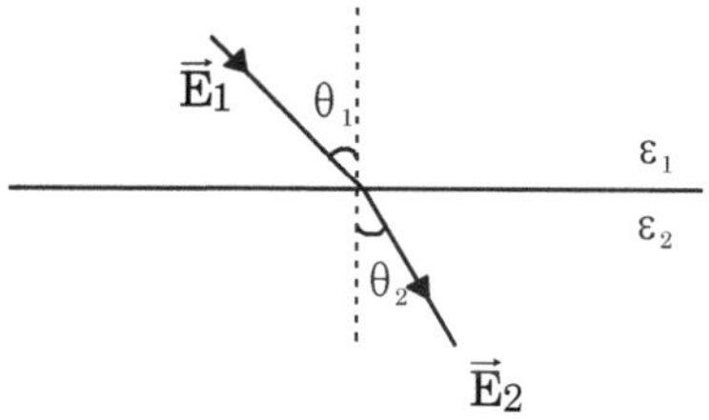

(a) $\varepsilon_1 \sin\theta_1 = \varepsilon_2 \sin\theta_2$ (b) $\varepsilon_1 \cos\theta_1 = \varepsilon_2 \cos\theta_2$

(c) $\varepsilon_1 \tan\theta_1 = \varepsilon_2 \tan\theta_2$ (d) $\varepsilon_1 \cot\theta_1 = \varepsilon_2 \cot\theta_2$

11. Far from the Earth, the Earth's magnetic field can be approximated as due to a bar magnet of magnetic pole strength 4×10^{14} Am. Assume this magnetic field is generated by a current carrying loop encircling the magnetic equator. The current required to do so is about 4×10^n A, where n is an integer. The value of n is _____.

(Earth's circumference: 4×10^7 m)

12. A conducting sphere of radius 1 m is placed in air. The maximum number of electrons that can be put on the sphere to avoid electrical breakdown is about 7×10^n, where n is an integer. The value of n is ______.

Assume:

Breakdown electric field strength in air is

$$|\vec{E}| = 3 \times 10^6 \text{ V/m}$$

Permittivity of free space ε_0

$$= 8.85 \times 10^{-12} \text{ F/m}$$

Electron charge $e = 1.60 \times 10^{-19}$ C

13. Which one of the following matrices does NOT represent a proper rotation in a plane?

(a) $\begin{pmatrix} -\sin\theta & \cos\theta \\ -\cos\theta & -\sin\theta \end{pmatrix}$ (b) $\begin{pmatrix} \cos\theta & \sin\theta \\ -\sin\theta & \cos\theta \end{pmatrix}$

(c) $\begin{pmatrix} \sin\theta & \cos\theta \\ -\cos\theta & \sin\theta \end{pmatrix}$ (d) $\begin{pmatrix} -\sin\theta & \cos\theta \\ -\cos\theta & \sin\theta \end{pmatrix}$

14. A uniform magnetic field $\vec{B} = B_0\hat{y}$ exists in an inertial frame K. A perfect conducting sphere moves with a constant velocity $\vec{v} = v_0\hat{x}$ with respect to this inertial frame. The rest frame of the sphere is K' (see figure). The electric and magnetic fields in K and K' are related as

$$\left.\begin{array}{ll} \vec{E}'_{\parallel} = \vec{E}_{\parallel} & \vec{E}'_{\perp} = \gamma\left(\vec{E}_{\perp} + \vec{v}\times\vec{B}\right) \\ \vec{E}'_{\parallel} = \vec{E}_{\parallel} & \vec{E}'_{\perp} = \gamma\left(\vec{B}_{\perp} + \dfrac{\vec{v}}{c^2}\times\vec{E}\right) \end{array}\right\},$$

$$\gamma = \frac{1}{\sqrt{1-(v/c)^2}}.$$

The induced surface charge density on the sphere (to the lowest order in v/c) in the frame K′ is

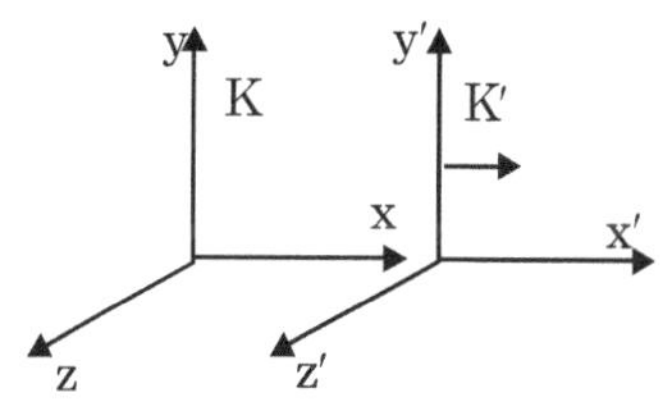

(a) maximum along z′

(b) maximum along y′

(c) maximum along x′

(d) uniform over the sphere

15. A charge q moving with uniform speed enters a cylindrical region in free space at t = 0 and exits the region at t= τ (see figure). Which one of the following options best describes the time dependence of the total electric flux φ(t), through the entire surface of the cylinder?

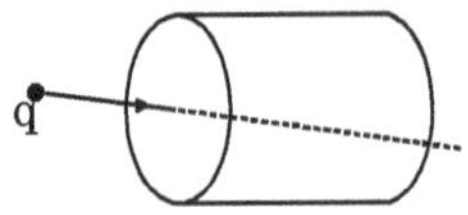

(a)
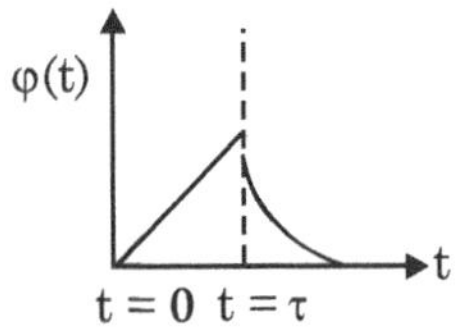

(b)
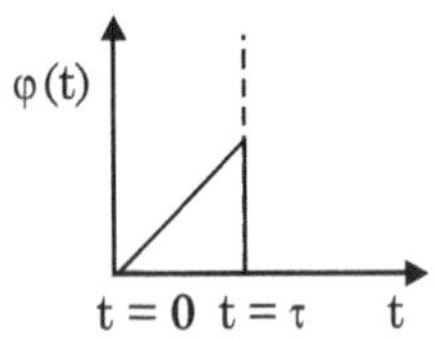

(c)
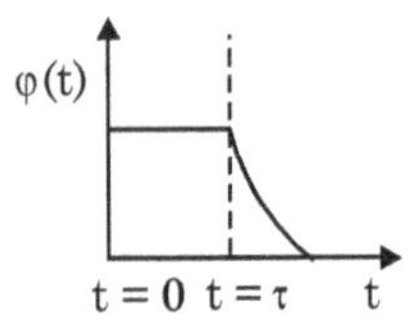

(d)
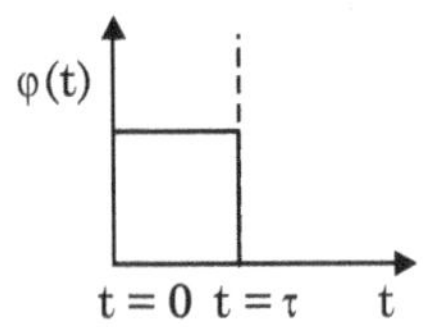

16. Let u^μ denote the 4-velocity of a relativistic particle whose square $u^\mu u_\mu = 1$. If $\varepsilon_{\mu\nu\rho\sigma}$ is the Levi-Civita tensor then the value of $\varepsilon_{\mu\nu\rho\sigma} u^\mu u^\nu u^\rho u^\sigma$ is _______.

17. A plane electromagnetic wave of wavelength λ is incident on a circular loop of conducting wire. The loop radius is a (a << λ). The angle (in degrees), made by the poynting vector with the normal to the plane of the loop to generate a maximum induced electrical signal, is ______.

18. A sinusoidal voltage of the form $V(t) = V_0 \cos(\omega t)$ is applied across a parallel plate capacitor placed in vacuum. Ignoring the edge effects, the induced emf within the region between the capacitor plates can be expressed as a power series in ω. The lowest non-vanishing exponent in ω is _____.

2019

19. The electric field of an electromagnetic wave is given by $\vec{E} = 3\sin(kz - \omega t)\hat{x} + 4\cos(kz - \omega t)\hat{y}$ .the wave is

(a) Linearly polarized at an angle $\tan^{-1}\left(\frac{4}{3}\right)$ from the x–axis

(b) Linearly polarized at an angle $\tan^{-1}\left(\frac{3}{4}\right)$ from the x–axis

(c) Elliptically polarized in clockwise direction when seen travelling towards the observer

(d) Elliptically polarized in counter-clockwise direction when seen travelling towards the observer

20. An infinitely long thin cylindrical shell has is axis coinciding with the z–axis. It carries a surface charge density $\sigma_0 \cos\phi$, where φ is the polar angle and σ_0 is a constant. The

(a) 0 (b) $\frac{\sigma_0}{2\epsilon_0}$

(c) $\frac{\sigma_0}{3\epsilon_0}$ (d) $\frac{\sigma_0}{4\epsilon_0}$

21. A circular loop made of a thin wire has radius 2 cm and resistance 2 Ω. It is placed perpendicular to a uniform magnetic field of magnitude $|\vec{B}_0|$ = 0.01 Tesla. At time t = 0 the field starts decaying as $\vec{B} = \vec{B}_0 e^{-t/t_0}$, where t_0 = 1s. The total charge that passes through a cross section of the wire during the decay is Q. The Value of Q in mC (rounded off to two decimal places) is__________

22. The electric field of an electromagnetic wave in vacuum is given by

$$\vec{E} = E_0 \cos\left(3y + 4z - 1.5\times 10^9 t\right)\hat{x}$$

The wave is reflected form the z = 0 surface. If the pressure exerted on the surface is $\alpha\, \epsilon_0 E_0^2$, the value of α (rounded off to one decimal place) is__________

23. A solid cylinder of radius R has total charge Q distributed uniformly over its volume. It is rotating about its axis with angular speed ω. The magnitude of the total magnetic moment of the cylinder is

(a) $QR^2\omega$ (b) $\frac{1}{2}QR^2\omega$

(c) $\frac{1}{4}QR^2\omega$ (d) $\frac{1}{8}QR^2\omega$

24. Ina set of N successive polarizers, the mth polarizer makes an angle $\left(\frac{m\pi}{2N}\right)$ with the vertical. A vertically polarized light beam of intensity I0 is incident on two such sets with $N = N_1$ and $N = N_2$, where $N_2 > N_1$. Let the intensity of light beams coming out be $I(N_1)$ and $I(N_2)$, respectively. Which of the following statement is correct about the two outgoing beams?

(*a*) $I(N_2) > (N_1)$; the polarization in each case is vertical

(*b*) $I(N_2) < I(N_1)$; the polarization in each case is vertical

(*c*) $I(N_2) > I(N_1)$; the polarization in each case is horizontal

(*d*) $I(N_2) < I(N_1)$; the polarization in each case is horizontal

25. An infinitely long wire parallel to the x-axis is kept at z = d and carries a current I in the positive x direction above a superconductor filling the region z ≤ =0 (see figure). The magnetic field $\vec{B}$ inside the superconductor is zero so that the field just outside the superconductor is parallel to its surface. The magnetic field due to this configuration at a point (x,y, z >0) is

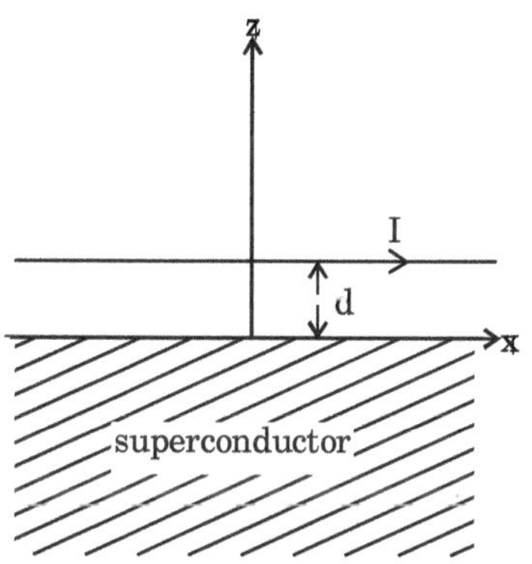

(*a*) $\left(\frac{\mu_0 I}{2\pi}\right)\frac{-(z-d)\hat{j}+y\hat{k}}{\left[y^2+(z-d)^2\right]}$

(*b*) $\left(\frac{\mu_0 I}{2\pi}\right)\left[\frac{-(z-d)\hat{j}+y\hat{k}}{y^2+(z-d)^2}+\frac{-(z-d)\hat{j}+y\hat{k}}{y^2+(z+d)^2}\right]$

(*c*) $\left(\frac{\mu_0 I}{2\pi}\right)\left[\frac{-(z-d)\hat{j}+y\hat{k}}{y^2+(z-d)^2}-\frac{-(z+d)\hat{j}-y\hat{k}}{y^2+(z+d)^2}\right]$

(*d*) $\left(\frac{\mu_0 I}{2\pi}\right)\left[\frac{y\hat{j}+(z-d)\hat{k}}{y^2+(z-d)^2}-\frac{y\hat{j}-(z+d)\hat{k}}{y^2+(z+d)^2}\right]$

26. The vector potential inside a long solenoid with n turns per unit length and carrying current I, written in cylindrical coordinates is $\vec{A}(s,\phi,z)=\frac{\mu_0 ni}{2}s\hat{\phi}$. If the term $\frac{\mu_0 nl}{2}s(\alpha\cos\phi\hat{\phi}+\beta\sin\phi\hat{s})$, where $\alpha\neq 0, \beta\neq 0$, is added to $\vec{A}(s,\phi,z)$, the magnetic field remains the same if

(*a*) $\alpha = \beta$ (*b*) $\alpha = -\beta$

(*c*) $\alpha = 2\beta$ (*d*) $\alpha = \frac{\beta}{2}$

useful formula $\vec{\nabla}t=\frac{\partial t}{\partial s}\hat{s}+\frac{1}{s}\frac{\partial r}{\partial \phi}\hat{\phi}+\frac{\partial t}{\partial z}\hat{z}$;

$$\left[\vec{\nabla}\times\vec{v}=\left(\frac{1}{s}\frac{\partial v_z}{\partial\phi}-\frac{\partial v_\phi}{\partial z}\right)\hat{s}+\left(\frac{\partial v_s}{\partial z}-\frac{\partial v_z}{\partial s}\right)\hat{\phi}+\frac{1}{s}\left(\frac{\partial(sv_\phi)}{\partial s}-\frac{\partial v_s}{\partial\phi}\right)\hat{z}\right]$$

27. Consider a system of three charges as shown in the figure below:

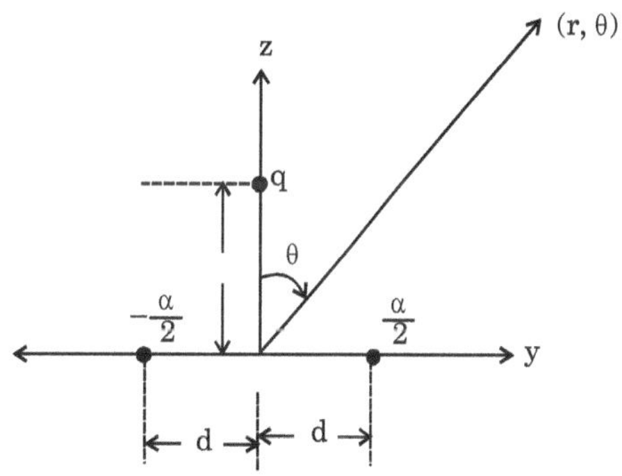

For r = 10 m; q = 60° q = 10^{-6} Coulomb, and d = 10^{-3} m, the electric dipole potential in volts (rounded off to three decimal places) at a point (r, q) is

$$\left[\text{Use}: \frac{1}{4\pi\epsilon_0}=9\times10^9\frac{Nm^2}{C^2}\right]$$

2018

28. An infinitely long straight wire is carrying a steady current *I*. The ratio of magnetic energy density at distance r_1 to that at $r_2(=2r_1)$ from the wire is ____.

29. A light beam of intensity I_0 is falling normally on a surface. The surface absorbs 20% of the intensity and the rest is reflected. The radiation pressure on the surface is given by XI_0/c, where X is ________ (up to one decimal place). Here *c* is the speed of light.

30. The number of independent components of a general electromagnetic field tensor is________.

31. Consider an infinitely long solenoid with N turns per unit length, radius R and carrying a current $I(t) = \alpha \cos \omega t$, where α is a constant and ω is the angular frequency. The magnitude of electric field at the surface of the solenoid is

(*a*) $\frac{1}{2}\mu_0 NR\omega\alpha \sin \omega t$ (*b*) $\frac{1}{2}\mu_0 \omega NR \cos \omega t$

(*c*) $\mu_0 NR\omega\alpha \sin \omega t$ (*d*) $\mu_0 \omega NR \cos \omega t$

32. A constant and uniform magnetic field $\vec{B} = B_0\hat{k}$ pervades all space. Which one of the following is the correct choice for the vector potential in Coulomb gauge?

(*a*) $-B_0(x+y)\hat{\imath}$ (*b*) $B_0(x+y)\hat{\jmath}$

(*c*) $B_0 x\hat{\jmath}$ (*d*) $-\frac{1}{2}B_0(x\hat{\imath} - y\hat{\jmath})$

33. A long straight wire, having radius a and resistance per unit length r, carries a current I. The magnitude and direction of the Poynting vector on the surface of the wire is

(*a*) $I^2r/2\pi a$, perpendicular to axis of the wire and pointing inwards

(*b*) $I^2r/2\pi a$, perpendicular to axis of the wire and pointing outwards

(*c*) $I^2r/\pi a$, perpendicular to axis of the wire and pointing inwards

(*d*) $I^2r/\pi a$, perpendicular to axis of the wire and pointing outwards

34. A quarter wave plate introduces a path difference of $\lambda/4$ between the two components of polarization parallel and perpendicular to the optic axis. An electromagnetic wave with $\vec{E} = (\hat{x}+\hat{y})E_0\, e^{i(kz-\omega t)}$ is incident normally on a quarter wave plate which has its optic axis making an angle $135°$ with the x-axis as shown.

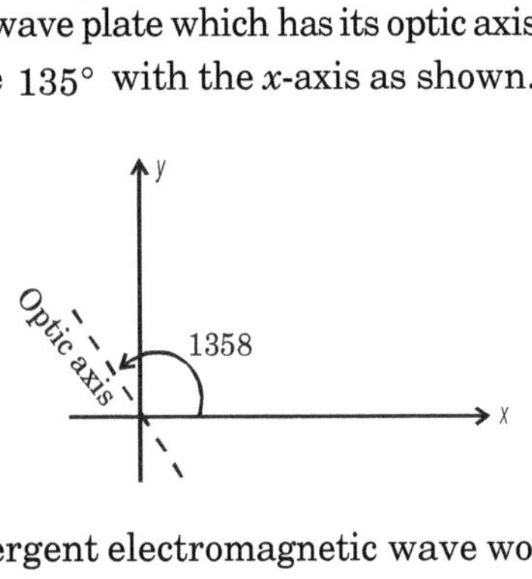

The emergent electromagnetic wave would be

(*a*) elliptically polarized

(*b*) circularly polarized

(*c*) linearly polarized with polarization as that of incident wave

(*d*) linearly polarized but with polarization at $90°$ to that of the incident wave

35. An electromagnetic plane wave is propagating with an intensity $I = 1.0\times10^5$ Wm^{-2} in a medium with $\varepsilon = 3\varepsilon_0$ and $\mu = \mu_0$.
The amplitude of the electric field inside the medium is ______ $\times10^3$ Vm^{-1} (up to one decimal place).

$(\varepsilon_0 = 8.85\times10^{-12}\text{C}^2\text{N}^{-1}\text{m}^{-2},$
$\mu_0 = 4\pi\times10^{-7}\text{NA}^{-2}, c = 3\times10^8\text{ms}^{-1})$

2017

36. Identical charges q are placed at five vertices of a regular hexagon of side a. The magnitude of the electric field and the electrostatic potential at the centre of the hexagon are respectively

(*a*) 0, 0 (*b*) $\frac{q}{4\pi\varepsilon_0 a^2}, \frac{q}{4\pi\varepsilon_0 a}$

(*c*) $\frac{q}{4\pi\varepsilon_0 a^2}, \frac{5q}{4\pi\varepsilon_0 a}$ (*d*) $\frac{\sqrt{5}q}{4\pi\varepsilon_0 a^2}, \frac{\sqrt{5}q}{4\pi\varepsilon_0 a}$

37. A parallel plate capacitor with square plates of side 1 m separated by 1 micro meter is filled with a medium of dielectric constant of 10. If the charges on the two plates are 1 C and – 1 C, the voltage across the capacitor is _______ kV. (up to two decimal places). ($\varepsilon_0 = 8.854\times10^{-12}$ F/m)

38. Light is incident from a medium of refractive index $n = 1.5$ onto vacuum. The smallest angle of incidence for which the light is not transmitted into vacuum is _______ degrees. (up to two decimal places).

39. A monochromatic plane wave in free space with electric field amplitude of 1 V/m is normally incident on a fully reflecting mirror. The pressure exerted on the mirror is ________ $\times10^{-12}$ Pa. (up to two decimal places) ($\varepsilon_0 = 8.854\times10^{-12}$ F/m).

40. Three charges (2 C, – 1 C, – 1 C) are placed at the vertices of an equilateral triangle of side 1 m as shown in the figure. The component of the electric dipole moment about the marked origin along the $\hat{y}$ direction is _______ C m.

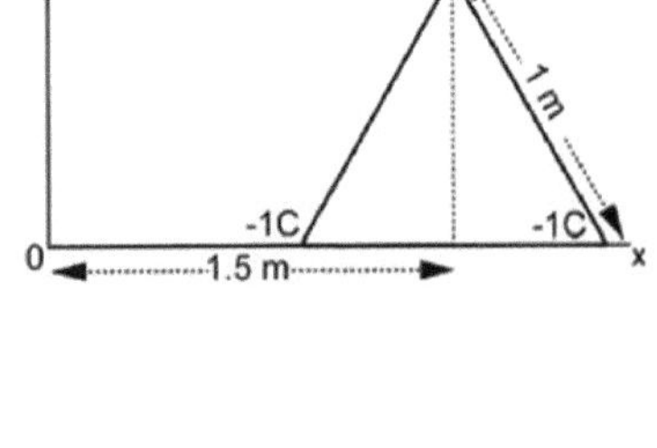

41. An infinite solenoid carries a time varying current I(t) = At^2, with A ¹ 0. The axis of the solenoid is along the $\hat{z}$ direction. $\hat{r}$ and $\hat{\theta}$ are the usual radial and polar directions in cylindrical polar coordinates. $\vec{B} = B_r\hat{r} + B_\theta\hat{\theta} + B_z\hat{z}$ is the magnetic field at a point outside the solenoid. Which one of the following statements is true?

(a) $B_r = 0$, $B_\theta = 0$, $B_z = 0$
(b) $B_r \neq 0$, $B_\theta \neq 0$, $B_z = 0$
(c) $B_r \neq 0$, $B_\theta \neq 0$, $B_z \neq 0$
(d) $B_r = 0$, $B_\theta = 0$, $B_z \neq 0$

42. A uniform volume charge density is placed inside a conductor (with resistivity 10^{-2} Ωm). The charge density becomes 1/(2.718) of its original value after time _______ femto seconds. (up to two decimal places) ($\varepsilon_0 = 8.854 \times 10^{-12}$ F/m)

2016

43. An infinite, conducting slab kept in a horizontal plane carries a uniform charge density σ. Another infinite slab of thickness t, made of a linear dielectric material of dielectric constant k, is kept above the conducting slab. The bound charge density on the upper surface of the dielectric slab is

(a) $\frac{\sigma}{2k}$ (b) $\frac{\sigma}{k}$

(c) $\frac{\sigma(k-2)}{2k}$ (d) $\frac{\sigma(k-1)}{k}$

44. The electric field component of a plane electromagnetic wave travelling in vacuum is given by $\vec{E}(z,t) = E_0\cos(kz - \omega t)\hat{i}$. The Poynting vector for the wave is

(a) $(c\varepsilon_0/2)E_0^2\cos^2(kz-\omega t)\hat{j}$

(b) $(c\varepsilon_0/2)E_0^2\cos^2(kz-\omega t)\hat{k}$

(c) $c\varepsilon_0 E_0^2\cos^2(kz-\omega t)\hat{j}$

(d) $c\varepsilon_0 E_0^2\cos^2(kz-\omega t)\hat{k}$

45. The x-y plane is the boundary between free space and a magnetic material with relative permeability μ_r. The magnetic field in the free space is $B_x\hat{i} + B_z\hat{k}$. The magnetic field in the magnetic material is

(a) $B_x\hat{i} + B_z\hat{k}$

(b) $B_x\hat{i} + \mu_r B_z\hat{k}$

(c) $\frac{1}{\mu}B_x\hat{i} + B_z\hat{k}$

(d) $\mu_r B_x\hat{i} + B_z\hat{k}$

46. The state of a system is given by

$$|\psi\rangle = |\phi_1\rangle + 2|\phi_2\rangle + 3|\phi_3\rangle$$

where $|\phi_1\rangle$, $|\phi_2\rangle$ and $|\phi_3\rangle$ from an orthonormal set. The probability of finding the system in the state $|\phi_2\rangle$ is ________. (Give your answer upto two decimal places)

47. A particle of mass m and energy E, moving in the positive x direction, is incident on a step potential at x = 0, as indicated in the figure. The height of the potential is V_0, where $V_0 > E$. At x = x_0 where x_0, 0, the probability of finding the electron is 1/e times the probability of finding it at x = 0.

If $\alpha = \sqrt{\frac{2m(V_0 - E)}{\hbar^2}}$, the value of x_0 is

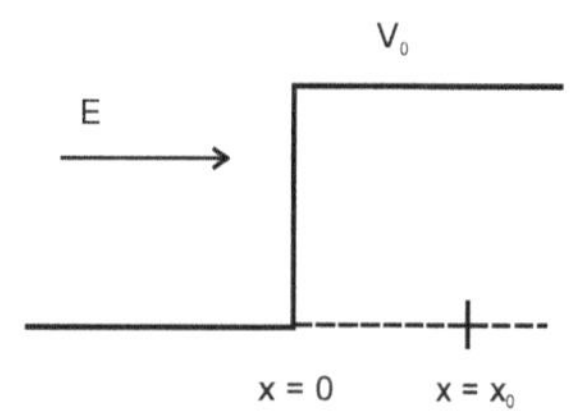

(a) $\frac{2}{\alpha}$ (b) $\frac{1}{\alpha}$

(c) $\frac{1}{2\alpha}$ (d) $\frac{1}{4\alpha}$

2015

48. Given that the magnetic flux through the closed loop PQRSP is ϕ. If $\int_P^R \vec{A}.d\vec{l} = \phi_1$ along PQR, value of $\int_P^R \vec{A}.d\vec{l}$ along PSR is

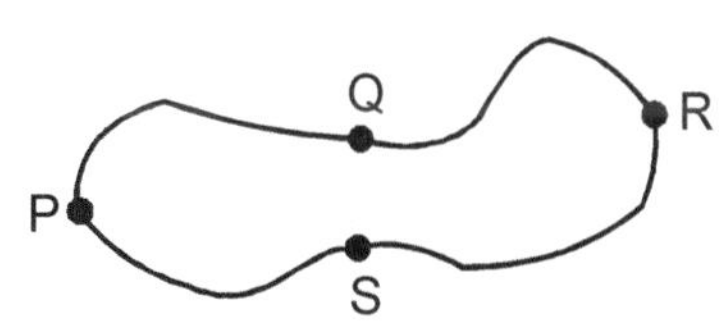

(a) $\phi - \phi_1$ (b) $\phi_1 - \phi$
(c) $-\phi_1$ (d) ϕ_1

49. The space between two plates of a capacitor carrying +Q and –Q is filled with two different dielectric materials, as shown in the figure. Across the interface of the two dielectric materials, which one of the following statements is correct?

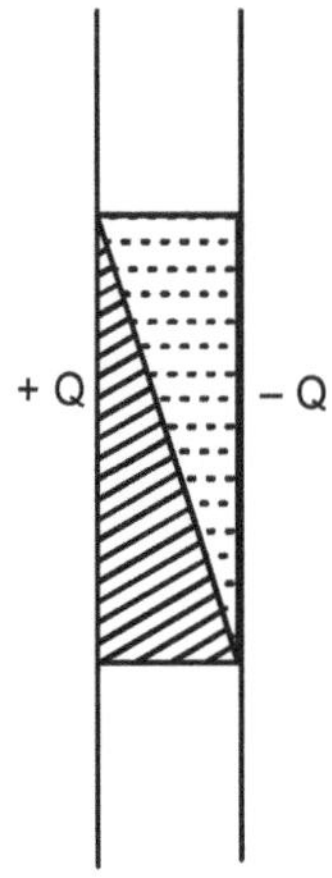

(a) $\vec{E}$ and $\vec{D}$ are continuous

(b) $\vec{E}$ is continuous and $\vec{D}$ is discontinuous

(c) $\vec{D}$ is continuous and $\vec{E}$ is discontinuous

(d) $\vec{E}$ and $\vec{D}$ are discontinuous

50. A point charge is placed between two semi-infinite conducting plates which are inclined at an angle of 30° with respect to each other. The number of image charges is ________

51. A monochromatic plane wave (wave length = 600 nm) $E_0 \exp[i(kz - \omega t)]$ is incident normally of a diffraction grating giving rise to a plane wave $E_1 \exp[i(\vec{k}_1 \cdot \vec{r} - \omega t)]$ in the first order of diffraction.

Here $E_1 < E_0$ and $\vec{k}_1 = |\vec{k}_1| \left[\frac{1}{2}\hat{x} + \frac{\sqrt{3}}{2}\hat{z}\right]$.

The period (in μm) of the diffraction grating is _______ (upto one decimal place).

52. A plane wave $(\hat{x} + i\hat{y})E_0 \exp[i(kz - \omega t)]$ after passing through an optical element emerges as $(\hat{x} - i\hat{y})E_0 \exp[i(kz - \omega t)]$, where k and ω are the wavevector and the angular frequency respectively. The optical element is a

(a) quarter wave plate

(b) half wave plate

(c) polarizer

(d) Faraday rotator

53. A long solenoid is embedded in a conducting medium and is insulated from the medium. If the current through the solenoid is increased at a constant rate, the induced current in the medium as function of the radial distance r from the axis of the solenoid is proportional to

(a) r^2 inside the solenoid and $\frac{1}{r}$ outside

(b) r inside the solenoid and $\frac{1}{r^2}$ outside

(c) r^2 inside the solenoid and $\frac{1}{r^2}$ outside

(d) r inside the solenoid and $\frac{1}{r}$ outside

2014

54. An unpolarized light wave is incident from air on a glass surface at the Brewster angle. The angle between the reflected and the refracted wave is

(a) 0° (b) 45°

(c) 90° (d) 120°

55. The electric field of a uniform plane wave propagating in a dielectric, non-conducting medium is given by,

$$\vec{E} = \hat{x}10\cos(6\pi \times 10^7 t - 0.4\pi z)\text{V}/m.$$

The phase velocity of the wave is __________ × $10^8 m/s$.

56. If the vector potential

$$\vec{A} = \alpha x\hat{x} + 2y\hat{y} - 3z\hat{z},$$

satisfies the Coulomb gauge, the value of the constant α is ____________.

57. A ray of light inside Region 1 in the xy- plane is incident at the semicircular boundary that carries no free charges. The electric field at the point $P(r_0, \pi/4)$ in plane polar coordinates is $\vec{E}_1 = 7\hat{e}_r - 3\hat{e}_\varphi$, where $\hat{e}_r$ and $\hat{e}_\varphi$ are the unit vectors. The emerging ray in Region 2 has the electric field $\vec{E}_2$ parallel to x-axis. If ε_1 and ε_2 are the dielectric constants of Region 1 and Region 2 respectively, then $\frac{\varepsilon_2}{\varepsilon_1}$ is ___

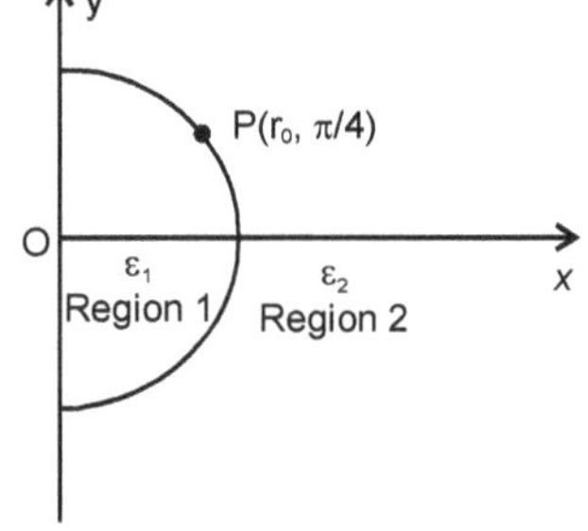

58. In an interference pattern formed by two coherent sources, the maximum and the minimum of the intensities are $9I_0$ and I_0, respectively. The intensities of the individual waves are

(a) $3I_0$ and I_0 (b) $4I_0$ and I_0
(c) $5I_0$ and $4I_0$ (d) $9I_0$ and I_0

2013

59. At a surface current, which one of the magnetostatic boundary condition is NOT CORRECT?

(a) Normal component of the magnetic field is continuous.

(b) Normal component of the magnetic vector potential is continuous.

(c) Tangential component of the magnetic vector potential is continuous.

(d) Tangential component of the magnetic vector potential is not continuous.

60. Interference fringes are seen at an observation plane $z = 0$, by the superposition of two plane waves $A_1 exp\left[i\left(\vec{k}_1.\vec{r}-\omega t\right)\right]$ and $A_2 exp\left[i\left(\vec{k}_2.\vec{r}-\omega t\right)\right]$, where A_1 and A_2 are real amplitudes. The condition for interference maximum is

(a) $(\vec{k}_1 - \vec{k}_2).\vec{r} = (2m+1)\pi$

(b) $(\vec{k}_1 - \vec{k}_2).\vec{r} = 2m\pi$

(c) $(\vec{k}_1 + \vec{k}_2).\vec{r} = (2m+1)\pi$

(d) $(\vec{k}_1 + \vec{k}_2).\vec{r} = 2m\pi$

61. A circularly polarized monochromatic plane wave is incident on a dielectric interface at Brewster angle. Which one of the following statements is CORRECT?

(a) The reflected light is plane polarized in the plane of incidence and the transmitted light is circularly polarized.

(b) The reflected light is plane polarized perpendicular to the plane of incidence and the transmitted light is plane polarized in the plane of incidence.

(c) The reflected light is plane polarized perpendicular to the plane of incidence and the transmitted light is elliptically polarized.

(d) There will be no reflected light and the transmitted light is circularly polarized.

62. A charge distribution has the charge density given by $\rho = Q\{\delta(x - x_0) - \delta(x - x_0)\}$. For this charge distribution the electric field at $(2x_0, 0, 0)$

(a) $\dfrac{2Q\hat{x}}{9\pi\varepsilon_0 x_0^2}$ (b) $\dfrac{Q\hat{x}}{4\pi\varepsilon_0 x_0^3}$

(c) $\dfrac{Q\hat{x}}{4\pi\varepsilon_0 x_0^2}$ (d) $\dfrac{Q\hat{x}}{16\pi\varepsilon_0 x_0^2}$

63. A monochromatic plane wave at oblique incidence undergoes reflection at a dielectric interface. If $\hat{k}_i, \hat{k}_r$ and $\hat{n}$ are the unit vectors in the directions of incident wave, reflected wave and the normal to the surface respectively, which one of the following expressions is correct?

(a) $(\hat{k}_i - \hat{k}_r) \times \hat{n} \neq 0$ (b) $(\hat{k}_i - \hat{k}_r) \cdot \hat{n} = 0$

(c) $(\hat{k}_i \times \hat{n}) \cdot \hat{k}_r = 0$ (d) $(\hat{k}_i \times \hat{n}) \cdot \hat{k}_r \neq 0$

64. In a constant magnetic field of 0.6 Tesla along the z direction, find the value of the path integral $\oint \vec{A}.\vec{dl}$ in the units of (Tesla m²) on a square loop of side length $(1/\sqrt{2})$ meters. The normal to the loop makes an angle of 60° to the z-axis, as shown in the figure. The answer should be up to two decimal places. ________

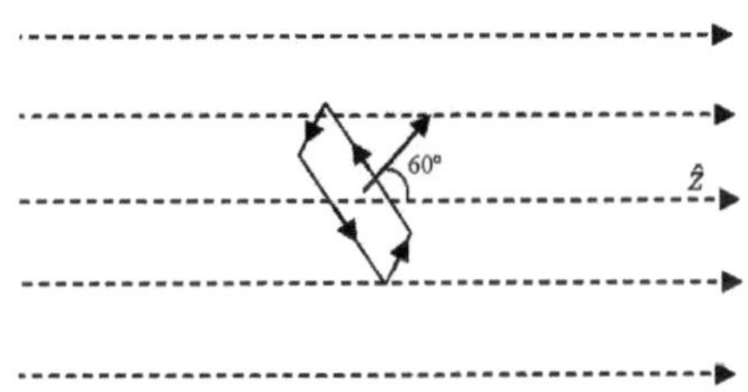

2012

65. The space-time dependence of the electric field of a linearly polarized light in free space is given by $\hat{x}E_0 \cos(\omega t - kz)$ where E_0, ω and k are the amplitude, the angular frequency and the wavevector, respectively. The time averaged energy density associated with the electric field is

(a) $\dfrac{1}{4}\varepsilon_0 E_0^2$ (b) $\dfrac{1}{2}\varepsilon_0 E_0^2$

(c) $\varepsilon_0 E_0^2$ (d) $2\varepsilon_0 E_0^2$

66. A plane electromagnetic wave traveling in free space is incident normally on a glass plate of refractive ndex 3/2. If there is no absorption by the glass, its reflectivity is

(a) 4% (b) 16%
(c) 20% (d) 50%

67. The electric and magnetic fields $\vec{E}(z,t)$ and $\vec{B}(z,t)$, respectively corresponding to the scalar potential $\phi(z, t) = 0$ and vector potential $\vec{A}(z,t) = \hat{i}tz$ are

(*a*) $\vec{E} = \hat{i}z$ and $\vec{B} = -\hat{j}t$

(*b*) $\vec{E} = \hat{i}z$ and $\vec{B} = \hat{j}t$

(*c*) $\vec{E} = -\hat{i}z$ and $\vec{B} = -\hat{j}t$

(*d*) $\vec{E} = -\hat{i}z$ and $\vec{B} = \hat{j}t$

68. A plane polarized electromagnetic wave in free space at time $t = 0$ is given by $\vec{E}(x,z) = 10\hat{j}\exp[i(6x+8z)]$. The magnetic field $\vec{B}(x,z,t)$ is given by

(*a*) $\vec{B}(x,z,t) = \frac{1}{c}(6\hat{k}-8\hat{i})\exp[i(6x+8z-10ct)]$

(*b*) $\vec{B}(x,z,t) = \frac{1}{c}(6\hat{k}+8\hat{i})\exp[i(6x+8z-10ct)]$

(*c*) $\vec{B}(x,z,t) = \frac{1}{c}(6\hat{k}-8\hat{i})\exp[i(6x+8z-ct)]$

(*d*) $\vec{B}(x,z,t) = \frac{1}{c}(6\hat{k}+8\hat{i})\exp[i(6x+8z+ct)]$

69. Two infinitely extended homogeneous isotropic dielectric media (medium-1 and medium-2 with dielectric constants $\frac{\varepsilon_1}{\varepsilon_0} = 2$ and $\frac{\varepsilon_2}{\varepsilon_0} = 5$, respectively) meet at the $z = 0$ plane as shown in the figure. A uniform electric field exists everywhere. For $z \geq 0$, the electric field is given by $\vec{E} = 2\hat{i} - 3\hat{j} + 5\hat{k}$. The interface separating the two media is charge free. The electric displacement vector in the medium-2 is given by

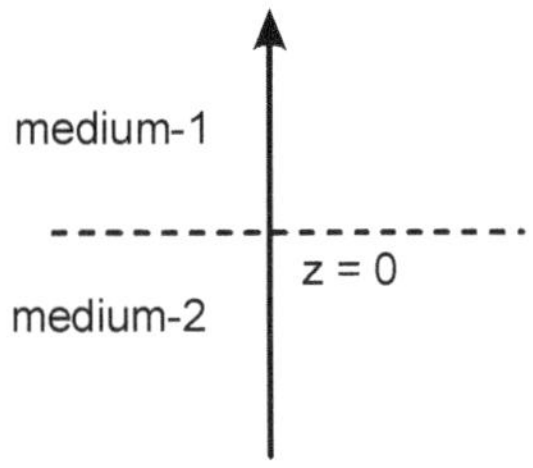

(*a*) $\vec{D}_2 = \varepsilon_0[10\hat{i}+15\hat{j}+10\hat{k}]$

(*b*) $\vec{D}_2 = \varepsilon_0[10\hat{i}-15\hat{j}+10\hat{k}]$

(*c*) $\vec{D}_2 = \varepsilon_0[4\hat{i}-6\hat{j}+10\hat{k}]$

(*d*) $\vec{D}_2 = \varepsilon_0[4\hat{i}+6\hat{j}+10\hat{k}]$

Linked Answer Q. 70 and 71 :

In a hydrogen atom, consider that the electronic charge is uniformly distributed in a spherical volume of radius a(= 0.5×10^{-10}m) around the proton. The atom is placed in a uniform electric field E = 30×10^5 V/m. Assume that the spherical distribution of the negative charge remains undistorted under the electric field.

70. In the equilibrium condition, the separation between the positive and the negative charge centers is

(*a*) 8.66×10^{-16} m (*b*) 2.60×10^{-15} m

(*c*) 2.60×10^{-16} m (*d*) 8.66×10^{-15} m

71. The polarizability of the hydrogen atom in unit of (C^2m/N) is

(*a*) 2.0×10^{-40} (*b*) 1.4×10^{-41}

(*c*) 1.4×10^{-40} (*d*) 2.0×10^{-39}

2011

72. Two charges q and $2q$ are placed along the x-axis in front of a grounded, infinite conducting plane, as shown in the figure. They are located respectively at a distance of 0.5 m and 1.5 m from the plane. The force acting on the charge q is

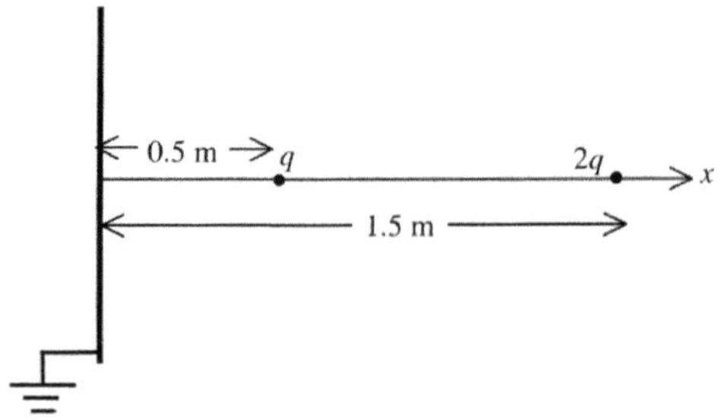

(*a*) $\frac{1}{4\pi\varepsilon_0}\frac{7q^2}{2}$ (*b*) $\frac{1}{4\pi\varepsilon_0}2q^2$

(*c*) $\frac{1}{4\pi\varepsilon_0}q^2$ (*d*) $\frac{1}{4\pi\varepsilon_0}\frac{q^2}{2}$

73. A uniform surface current is flowing in the positive y-direction over an infinite sheet lying in x-y plane. The direction of the magnetic field is

(*a*) along $\hat{i}$ for $z > 0$ and along $-\hat{i}$ for $z < 0$

(*b*) along $\hat{k}$ for $z > 0$ and along $-\hat{k}$ for $z < 0$

(*c*) along $-\hat{i}$ for $z > 0$ and along $\hat{i}$ for $z < 0$

(*d*) along $-\hat{k}$ for $z > 0$ and along $\hat{k}$ for $z < 0$

74. A magnetic dipole of dipole moment $\vec{m}$ is placed in a non-uniform magnetic field $\vec{B}$. If the position vector of the dipole is $\vec{r}$, the torque acting on the dipole about the origin is

(a) $\vec{r} \times (\vec{m} \times \vec{B})$

(b) $\vec{r} \times \vec{\nabla}(\vec{m} . \vec{B})$

(c) $\vec{m} \times \vec{B}$

(d) $\vec{m} \times \vec{B} + \vec{r} \times \vec{\nabla}(\vec{m} . \vec{B})$

75. Which of the following expressions for a vector potential $\vec{A}$ DOES NOT represent a uniform magnetic field of magnitude B_0 along the z-direction?

(a) $\vec{A} = (0, B_0 x, 0)$

(b) $\vec{A} = (-B_0 y, 0, 0)$

(c) $\vec{A} = \left(\frac{B_0 x}{2}, \frac{B_0 y}{2}, 0\right)$

(d) $\vec{A} = \left(-\frac{B_0 y}{2}, \frac{B_0 x}{2}, 0\right)$

76. A spherical conductor of radius a is placed in a uniform electric field $\vec{E} = E_0 \hat{k}$. The potential at a point $P(r, \theta)$ for $r > a$, is given by

$$\varphi(r, \theta) = \text{constant} - E_0 r \cos\theta + \frac{E_0 a^3}{r^2} \cos\theta$$

where r is the distance of P from the centre O of the sphere and θ is the angle OP makes with the z-axis.

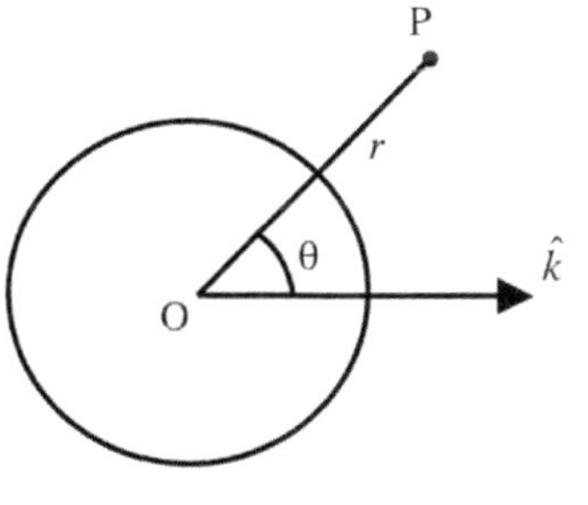

The charge density on the sphere at $\theta = 30°$ is

(a) $3\sqrt{3}\varepsilon_0 E_0/2$ (b) $3\varepsilon_0 E_0/2$

(c) $\sqrt{3}\varepsilon_0 E_0/2$ (d) $\varepsilon_0 E_0/2$

Statement for Linked Answer Questions 77 and 78:

A plane electromagnetic wave has the magnetic field given by

$$\vec{B}(x, y, z, t) = B_0 \sin\left((x+y)\frac{k}{\sqrt{2}} + \omega t\right)\hat{k}$$

where k is the wave number and $\hat{i}$, $\hat{j}$ and $\hat{k}$ are the cartesian unit vectors in x, y and z directions, respectively.

77. The electric field $\vec{E}(x, y, z, t)$ corresponding to the above wave is given by

(a) $cB_0 \sin\left[(x+y)\frac{k}{\sqrt{2}} + \omega t\right]\frac{(\hat{i}-\hat{j})}{\sqrt{2}}$

(b) $cB_0 \sin\left[(x+y)\frac{k}{\sqrt{2}} + \omega t\right]\frac{(\hat{i}+\hat{j})}{\sqrt{2}}$

(c) $cB_0 \sin\left[(x+y)\frac{k}{\sqrt{2}} + \omega t\right]\hat{i}$

(d) $cB_0 \sin\left[(x+y)\frac{k}{\sqrt{2}} + \omega t\right]\hat{j}$

78. The average Poynting vector is given by

(a) $\frac{cB_0^2}{2\mu_0}\frac{(\hat{i}-\hat{j})}{\sqrt{2}}$ (b) $-\frac{cB_0^2}{2\mu_0}\frac{(\hat{i}-\hat{j})}{\sqrt{2}}$

(c) $\frac{cB_0^2}{2\mu_0}\frac{(\hat{i}+\hat{j})}{\sqrt{2}}$ (d) $-\frac{cB_0^2}{2\mu_0}\frac{(\hat{i}+\hat{j})}{\sqrt{2}}$

2010

79. An insulating sphere of radius a carries a charge density $\rho(\vec{r}) = \rho_0 (a^2 - r^2)\cos\theta$; $r < a$. The leading order term for the electric field at a distance d, for away from the charge distribution, is proportional to

(a) d^{-1} (b) d^{-2}

(c) d^{-3} (d) d^{-4}

80. Two magnetic dipoles of magnitude m each are placed in a plane as shown.

The energy of interaction is given by

(a) Zero (b) $\frac{\mu_0}{4\pi}\frac{m^2}{d^3}$

(c) $\frac{3\mu_0}{2\pi}\frac{m^2}{d^3}$ (d) $-\frac{3\mu_0}{8\pi}\frac{m^2}{d^3}$

Statement for Linked Answer Questions 81 and 82:

Consider the propagation of electromagnetic waves in a linear, homogeneous and isotropic material medium with electric permittivity ε, and magnetic permeability μ.

81. For a plane wave of angular frequency ω and propagation vector $\vec{k}$ propagating in the medium Maxwell's equations reduce to

(a) $\vec{k}\cdot\vec{E} = 0$; $\vec{k}\cdot\vec{H} = 0$; $\vec{k}\times\vec{E} = \omega\varepsilon\vec{H}$; $\vec{k}\times\vec{H} = -\omega\mu\vec{E}$

(b) $\vec{k}\cdot\vec{E}=0$; $\vec{k}\cdot\vec{H}=0$; $\vec{k}\times\vec{E}=-\omega\varepsilon\vec{H}$; $\vec{k}\times\vec{H}=\omega\mu\vec{E}$

(c) $\vec{k}\cdot\vec{E} = 0$; $\vec{k}\cdot\vec{H} = 0$; $\vec{k}\times\vec{E} = -\omega\mu\vec{H}$; $\vec{k}\times\vec{H} = \omega\varepsilon\vec{E}$

(d) $\vec{k}\cdot\vec{E} = 0$, $\vec{k}\cdot\vec{H} = 0$; $\vec{k}\times\vec{E} = \omega\mu\vec{H}$; $\vec{k}\times\vec{H} = -\omega\varepsilon\vec{E}$

82. If ε and μ assume negative values in a certain frequency range, then the directions of the propagation vector $\vec{k}$ and the Poynting vector $\vec{S}$ in that frequency range are related as

(a) $\vec{k}$ and $\vec{S}$ are parallel

(b) $\vec{k}$ and $\vec{S}$ are anti-parallel

(c) $\vec{k}$ and $\vec{S}$ are perpendicular to each other

(d) $\vec{k}$ and $\vec{S}$ make an angle that depends on the magnitude of $|\varepsilon|$ and $|\mu|$

2009

83. A conducting loop L of surface area S is moving with a velocity $\vec{v}$ in a magnetic field $\vec{B}(\vec{r},t) = \vec{B}_0 t^2$, B_0 is a positive constant of suitable dimensions. The emf induced, V_{emf}, in the loop is given by

(a) $-\int_S \frac{\partial\vec{B}}{\partial t}\cdot d\vec{S}$

(b) $\oint_L (\vec{v}\times\vec{B})\cdot d\vec{L}$

(c) $-\int_S \frac{\partial\vec{B}}{\partial t}\cdot d\vec{S} - \oint_L (\vec{v}\times\vec{B})\cdot d\vec{L}$

(d) $-\int_S \frac{\partial\vec{B}}{\partial t}\cdot d\vec{S} + \oint_L (\vec{v}\times\vec{B})\cdot d\vec{L}$

84. The magnetic field (in A m^{-1}) inside a long solid cylindrical conductor of radius $a = 0.1\ m$ is,

$$\vec{H} = \frac{10^4}{r}\left[\frac{1}{\alpha^2}\sin(\alpha r) - \frac{r}{\alpha}\cos(\alpha r)\right]\hat{\phi},$$

where $\alpha = \frac{\pi}{2a}$. What is the total current (in A) in the conductor ?

(a) $\frac{\pi}{2a}$ (b) $\frac{800}{\pi}$

(c) $\frac{400}{\pi}$ (d) $\frac{300}{\pi}$

85. Which one of the following current densities, $\vec{J}$, can generate the magnetic vector potential $\vec{A} = (y^2\hat{i} + x^2\hat{j})$?

(a) $\frac{2}{\mu_0}(x\hat{i} + y\hat{j})$ (b) $-\frac{2}{\mu_0}(\hat{i} + \hat{j})$

(c) $-\frac{2}{\mu_0}(\hat{i} - \hat{j})$ (d) $\frac{2}{\mu_0}(x\hat{i} - y\hat{j})$

86. In a non-conducting medium characterized by $\varepsilon = \varepsilon_0$, $\mu = \mu_0$ and conductivity $\sigma = 0$, the electric field (in Vm^{-1}) is given by

$$\vec{E} = 20\sin[10^8 t - kz]\ \hat{j}.$$

The magnetic field, $\vec{H}$ (in A m^{-1}), is given by

(a) $20k\cos[10^8 t - kz]\ \hat{i}$

(b) $\frac{20k}{10^8\mu_0}\sin[10^8 t - kz]\ \hat{j}$

(c) $-\frac{20k}{10^8\mu_0}\sin[10^8 t - kz]\ \hat{i}$

(d) $-20k\cos[10^8 t - kz]\ \hat{j}$

87. A cylindrical rod of length L and radius r, made of an inhomogeneous dielectric, is placed with its axis along the z direction with one end at the origin as shown below.

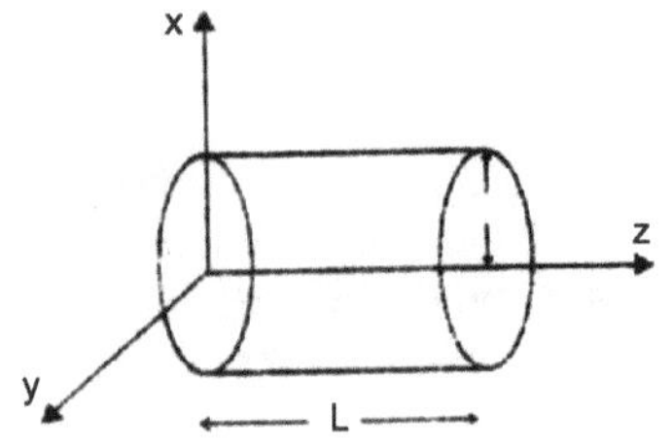

If the rod carries a polarization, $\vec{P} = (5z^2 + 7)\hat{k}$, the volume bound charge inside the dielectric is

(a) Zero (b) $10\pi r^2 L$

(c) $-5\pi r^2 L$ (d) $-5\pi r^2 L^2$

88. Two monochromatic waves having frequencies ω and $\omega + \Delta\omega$ ($\Delta\omega << \omega$) and corresponding wavelengths λ and $\lambda - \Delta\lambda$ ($\Delta\lambda << \lambda$) of same polarization, traveling along x-axis are superimposed on each other. The phase velocity and group velocity of the resultant wave are respectively given by

(*a*) $\frac{\omega\lambda}{2\pi}, \frac{\Delta\omega\lambda^2}{2\pi\Delta\lambda}$ (*b*) $\omega\lambda, \frac{\Delta\omega\lambda^2}{\Delta\lambda}$

(*c*) $\frac{\omega\Delta\lambda}{2\pi}, \frac{\Delta\omega\Delta\lambda}{2\pi}$ (*d*) $\omega\Delta\lambda, \omega\Delta\lambda$

2008

89. A parallel plate capacitor is being discharged. What is the direction of the energy flow in terms of the Poynting vector in the space between the plates?

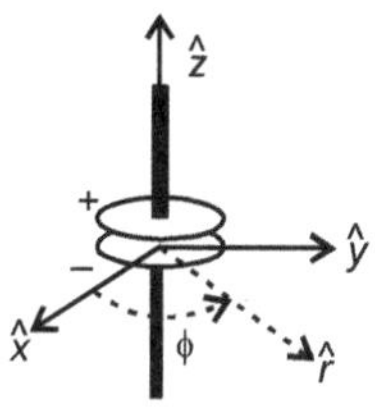

(*a*) Along the wire in the positive *z* axis

(*b*) Radially inward $(-\hat{r})$

(*c*) Radially outward $(\hat{r})$

(*d*) Circumferential (ϕ)

90. Unpolrized light falls from air to a planar air-glass interface (refractive index of glass is 1.5) and the reflected light is observed to be plane polrized. The polarization vector and the angle of incidence θ_i are

(*a*) perpendicular to the plane of incidence and $\theta_i = 42°$.

(*b*) parallel to the plane of incidence and $\theta_i = 56°$.

(*c*) perpendicular to the plane of incidence and $\theta_i = 56°$.

(*d*) parallel to the plane of incidence and $\theta_i = 42°$.

91. A plane electromagnetic wave is given by $E_0\left(\hat{x} + e^{i\delta}\hat{y}\right)\exp\{i(kz - \omega t)\}$. At a given location, the number of times $\vec{E}$ vanishes in one second is

(*a*) An integer near $\frac{\omega}{\pi}$ when $\delta = n\pi$ and zero when $\delta \neq n\pi$, *n* is integer

(*b*) An integer near $\frac{\omega}{\pi}$ and is independent of δ

(*c*) An integer near $\frac{\omega}{2\pi}$ when $\delta = n\pi$ and zero when $\delta \neq n\pi$, *n* is integer

(*d*) An integer near $\frac{\omega}{2\pi}$ and is independent of δ

92. A dielectric sphere is placed in a uniform electric field directed along the psitive *y*-axis. Which one of the following represents the correct equipotential surfaces?

(*a*)
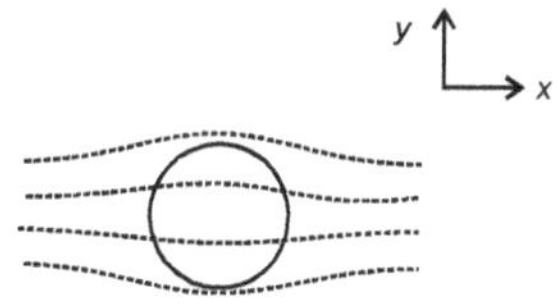

(*b*)
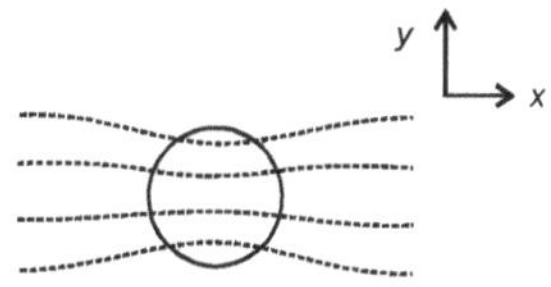

(*c*)
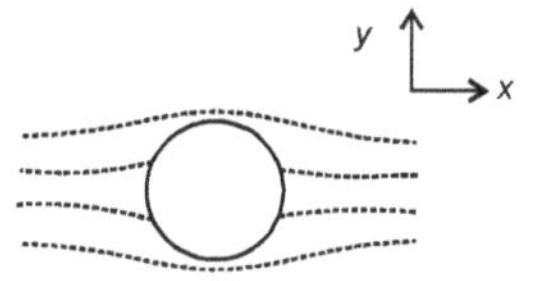

(*d*)
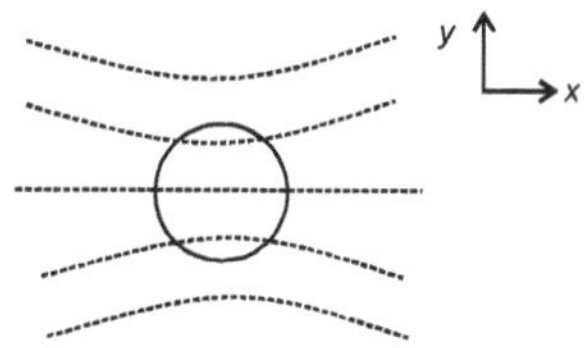

93. A rod of length L with uniform charge density λ per unit length is in the *xy*-plane and rotating about *z*-axis passing through one of its edge with an angular velocity $\vec{\omega}$ as shown in the figure below. $(\hat{r}, \hat{\phi}, \hat{z})$ refer to the unit vectors at Q, $\vec{A}$ is tghe vector potential at a distance *d* from the origin O along *z*-axis for *d* << L and $\vec{J}$ is the current density due to the motion of the rod. Which one of the following statements is correct?

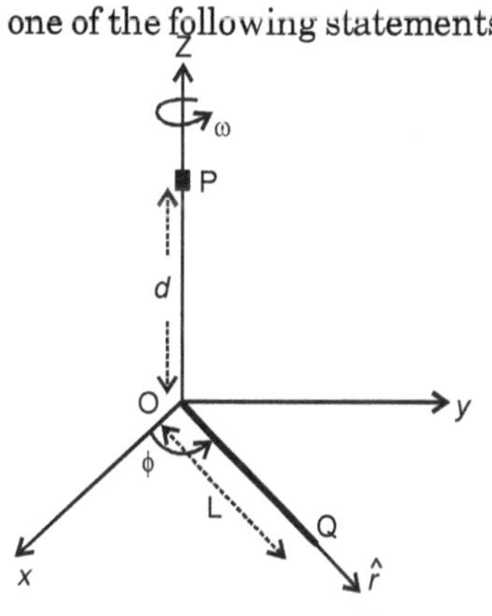

(*a*) $\vec{J}$ along $\hat{r}$; $\vec{A}$ along $\hat{z}$; $|\vec{A}| \propto \frac{1}{d}$

(*b*) $\vec{J}$ along $\hat{\phi}$; $\vec{A}$ along $\hat{\phi}$; $|\vec{A}| \propto \frac{1}{d^2}$

(*c*) $\vec{J}$ along $\hat{r}$; $\vec{A}$ along $\hat{z}$; $|\vec{A}| \propto \frac{1}{d^2}$

(*d*) $\vec{J}$ along $\hat{\phi}$; $\vec{A}$ along $\hat{\phi}$; $|\vec{A}| \propto \frac{1}{d}$

94. A circular disk of radius a one the xy plane has a surface idensity $\sigma = \frac{\sigma_0 r\cos\theta}{a}$. The electric dipole moment of the charge distribution is

(a) $\frac{\sigma_0 \pi a^4}{4}\hat{x}$

(b) $\frac{\sigma_0 \pi a^3}{4}\hat{x}$

(c) $-\frac{\sigma_0 \pi a^3}{4}\hat{x}$

(d) $-\frac{\sigma_0 \pi a^4}{4}\hat{x}$

95. An nonrelativistic charged particle moves along the positive x-axis with a constant positive acceleration $a\hat{x}$. The particle is at the origin at $t = 0$. Radiation is observed at $t = 0$ at a distant point $(0, d, 0)$ on the y-axis. Which one of the following statements is correct?

(a) The radiation is unpolarized.

(b) The radiation is plane polarized with polarization parallel to the x-axis.

(c) The radiation is plane polarized with polarization parallel to xy plane along a line inclined to the x axis.

(d) The radiation is elliptically polarized.

Common Data for Questions 96, 97:

Consider two concentric conducting spherical shells with inner and outer radii a, b and c, d as shown in the figure. Both the shells are given Q amount of positive charges.

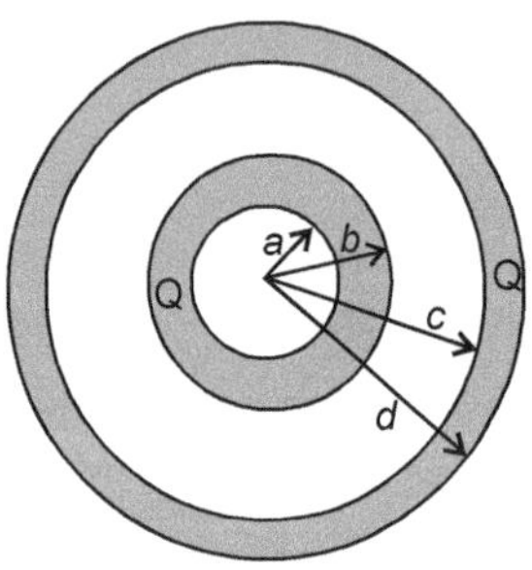

96. The electric filed in different regions are

(a) $\vec{E}=0$ for $r<a$; $\vec{E}=\frac{-Q}{4\pi\varepsilon_0 r^2}\hat{r}$ for $a<r<b$

$\vec{E} = 0$ for $b < r < c$;

$\vec{E} = \frac{Q}{4\pi\varepsilon_0 r^2}\hat{r}$ for $a < r < b$

(b) $\vec{E}=\frac{-Q}{4\pi\varepsilon_0 r^2}\hat{r}$ for $r<a$; $\overline{E}=0$ for $a<r<b$

$\vec{E} = \frac{Q}{4\pi\varepsilon_0 r^2}\hat{r}$ for $b < r < c$;

$\overline{E} = \frac{Q}{4\pi\varepsilon_0 r^2}\hat{r}$ for $r > d$

(c) $\vec{E}=\frac{-Q}{4\pi\varepsilon_0 r^2}\hat{r}$ for $r<a$; $\overline{E}=0$ for $a<r<b$

$\vec{E}=0$ for $b<r<c$; $\overline{E}=\frac{2Q}{4\pi\varepsilon_0 r^2}\hat{r}$ for $r>d$

(d) $\vec{E} = 0$ for $r < a$; $\overline{E} = 0$ for $a < r < b$

$\vec{E} = \frac{Q}{4\pi\varepsilon_0 r^2}\hat{r}$ for $b < r < c$;

$\overline{E} = \frac{2Q}{4\pi\varepsilon_0 r^2}\hat{r}$ for $r > d$

97. In order to have equal surface charge densities on the outer surface of both the shells, the following conditions should be satisfied

(a) $d = 4b$ and $c = 2a$

(b) $d = 4b$ and $c = 2a$

(c) $d = \sqrt{2b}$ and $c > a$

(d) $d > b$ and $c = \sqrt{2a}$

2007

98. Four point charges are placed in a plane at the following positions:

+ Q at (1, 0), – Q at (–1, 0) + Q at (0, 1) and –Q at (0, –1)

At large distances the electrostatic potential due to this charge distribution will be dominated by the

(a) monopole moment.

(b) dipole moment.

(c) quadrupole moment.

(d) octpole moment.

99. A charged capacitor (C) is connected in series with an inductor (L). When the displacement current reduces to zero, the nergy of the LC circuit is

(a) stored entirely in its magnetic field.

(b) stored entirely in its electric field.

(c) distribuuted equality among its electric and magnetic fields.

(d) radiated out of the circuit.

100. The elctromagnetic field due to a point charge must be described by Lienard Weichert potentials when

(*a*) the point charge is highly accelerated.

(*b*) the electric and magnetic fields are not perpendicular.

(*c*) the point charge is moving with velocity close to that of light.

(*d*) the calculation is done for the radiation zone, i.e. far away from the charge.

101. A vector field is efined everywhere $\vec{F}=\frac{y^2}{L}\hat{i}+z+\hat{k}$.

The net flux of $\vec{F}$ associated with a cube of side L, with one vertex at the origin and sides along the positive X, Y, and Z axes, is

(*a*) L^3 (*b*) $4L^3$

(*c*) $8L^3$ (*d*) $10L^3$

102. If $\vec{r}=x\hat{i}+y\hat{j}$, then

(*a*) $\vec{\nabla}\cdot\vec{r}=0$ and $\vec{\nabla}|\vec{r}|=\vec{r}$

(*b*) $\vec{\nabla}\cdot\vec{r}=2$ and $\vec{\nabla}|\vec{r}|=\vec{r}$

(*c*) $\vec{\nabla}\cdot\vec{r}=2$ and $\vec{\nabla}|\vec{r}|=\frac{\hat{r}}{r}$

(*d*) $\vec{\nabla}\cdot\vec{r}=3$ and $\vec{\nabla}|\vec{r}|=\frac{\hat{r}}{r}$

103. Consider a vector $\vec{p}=2\hat{i}+3\hat{j}+2\hat{k}$ in the coordinate system ($\hat{i},\hat{j},\hat{k}$). The axes are rotation anit-clockwise about the Y axis by an angle of 60º. The vector $\vec{p}$ in the rotate coordinate system ($\hat{i},\hat{j},\hat{k}$) is

(*a*) $(1-\sqrt{3})\hat{i}'+3\hat{j}'+(1+\sqrt{3})\hat{k}'$

(*b*) $(1-\sqrt{3})\hat{i}'+3\hat{j}'+(1-\sqrt{3})\hat{k}'$

(*c*) $(1-\sqrt{3})\hat{i}'+(3+\sqrt{3})\hat{j}'+2\hat{k}'$

(*d*) $(1-\sqrt{3})\hat{i}'+(3-\sqrt{3})\hat{j}'+2\hat{k}'$

104. A toroidal coil has N closely-wound turns. Assume the current through the coil to be I and the toroid is filled wwith a magnetic material fo relative permittivity μ_r. The magnitude of magnitic induction $\vec{B}$ inside the toroid, at a radial distance r from the axis, is given by

(*a*) $\mu_r, \mu_0 NIr$ (*b*) $\frac{\mu_r, \mu_0 NI}{r}$

(*c*) $\frac{\mu_r, \mu_0 NI}{2\pi r}$ (*d*) $2\pi\mu_r\mu_0 NIr$

105. An electromagnetic wave with $\vec{E}(z,t)=E_0\cos(\omega t-kz)\hat{i}$ is travelling in free space and crosses a disc of radius 2 m placed perpendicular to the *z*-axis. If E_0 = 60 V m^{-1}, the average power, in Watt, crossing the disc along the *z*-direction is

(*a*) 30 (*b*) 60

(*c*) 120 (*d*) 270

106. Can the following scalar and vector potentials describe an electromagnetic field.

$$\phi(\vec{x},t)=3xyz-4t$$

$$\vec{A}(\vec{x},t)=(2x-\omega t)\hat{i}+(y-2z)\hat{j}+(z-2e^{i\omega t})\hat{k}$$

where ω is a constant.

(*a*) Yes, in the Coulomb gauge.

(*b*) yes, in the Lorentz gauge.

(*c*) Yes, provided ω = 0.

(*d*) No.

Common Data For Questions 107, 108:

The Fresnel relations between the amplitudes of incident and reflected electromagnetic waves at an interface between air and a dielectric of refractive index μ, are

$$E_{\parallel}^{\text{reflected}}=\frac{\cos r-\mu\cos i}{\cos r+\mu\cos i}E_{\parallel}^{\text{incident}}$$

and $$E_{\perp}^{\text{reflected}}=\frac{\mu\cos r-\cos i}{\mu\cos r+\cos i}E_{\perp}^{\text{incident}}$$

The subscripts ∥ and ⊥ refer to polarization, parallel and normal to the plane of incidence respectively. Here, *i* and *r* are the angles of incidence and refraction respectively.

107. The condition for the reflected ray to be completely polarized is

(*a*) μ cos *i* = cos *r* (*b*) cos *i* = μ cos *r*

(*c*) μ cos *i* = – cos *r* (*d*) cos *i* = – μ cos *r*

108. For normal incident at an air-glass interface with μ = 1.5 the fraction of energy reflected is given by

(*a*) 0.40 (*b*) 0.20

(*c*) 0.16 (*d*) 0.04

Statement for Linked Answer Questions 109 and 110:

A sphhere of radius R carriers a polarization $\vec{p}=k\vec{r}$, were *k* is a constant and $\vec{r}$ is measured from the centre of the sphere.

109. The bound surface and volume charge densities are given, respectively, by

(*a*) $k|\vec{r}|$ and $3k$ (*b*) $k|\vec{r}|$ and $-3k$

(*c*) $k|\vec{r}|$ and $-4\pi kR$ (*d*) $-k|\vec{r}|$ and $4\pi kR$

110. The electric field $\vec{E}$ at a point $\vec{r}$ outside the sphere is given by

(a) $\vec{E} = 0$ (b) $\vec{E} = \frac{kR(R^2 - r^2)}{\varepsilon_0 r^3}\hat{r}$

(c) $\vec{E} = \frac{kR(R^2 - r^2)}{\varepsilon_0 r^5}\hat{r}$ (d) $\vec{E} = \frac{3k(r - R)}{4\pi\varepsilon_0 r^4}\hat{r}$

2006

111. In an electromagnetic field, which one of the following remains invariant under Lorentz transformation?

(a) $\vec{E} \times \vec{B}$ (b) $E^2 - c^2B^2$

(c) B^2 (d) E^2

112. A sphere of radius R has uniform volume charge density. The electric potential at a point r ($r < R$) is

(a) due to the charge inside a sphere of radius r only

(b) due to the entire charge of the sphere

(c) due to charge in the spherical sheli of inner and outer radii r and R, only

(d) independent of r

113. The wavefunctions of two identical particles in stated n stated n and s are given by ϕ_n, (r_1) and (r_2), respectively. The particles obey Maxwell–Boltzmann statistics. The state of the combined two particle system is expressed as

(a) $\phi_n(r_1) + \phi_s(r_2)$

(b) $\frac{1}{\sqrt{2}}[\phi_n(r_1)\phi_s(r_2) + \phi_n(r_2)\phi_s(r_1)]$

(c) $\frac{1}{\sqrt{2}}[\phi_n(r_1)\phi_s(r_2) - \phi_n(r_2)]$

(d) $\phi_n(r_1)\phi_s(r_2)$

114. A particle of mass 2 kg is moving such that at time t, its position, in metre, is given by $\vec{r}(t) = 5\hat{i} - 2t^2\hat{j}$. The angular momentum of the particle at $t = 2$ s about the origin, in kg m^2 s^{-1}, is

(a) $-40\hat{k}$ (b) $-80\hat{k}$

(c) $80\hat{k}$ (d) $40\hat{k}$

115. A system of four particles is in x-y plane. Of these, two particles each of mass m are located at (1, 1) and (–1, 1). The remaining two particles each of mass $2m$ are located at (1, 1) and (–1, – 1). The xy-component of the moment of inertia tensor of this system of particles is

(a) $10m$ (b) $-10m$

(c) $2m$ (d) $-2m$

116. A long cylindrical kept along z-axis carries a current density $\hat{J} = J_0 r\hat{k}$, where J_0 is a constant and r is the radial distance form the axis of the cylinder. Themagnetic induction $\hat{B}$ inside the conductor at a distance d from the axis of the cylinder is

(a) $\mu_0 J_0 \hat{\phi}$ (b) $-\frac{\mu_0 J_0 d}{2}\hat{\phi}$

(c) $\frac{\mu_0 J_0 d^2}{3}\hat{\phi}$ (d) $-\frac{\mu_0 J_0 d^3}{4}\hat{\phi}$

117. The vector potential in a region is given as $\vec{A}(x, y, z) = -y\hat{i} + 2x\hat{j}$. The associated magnetic induction is $\vec{B}$ is

(a) $\hat{i} + \hat{k}$ (b) $3\hat{k}$

(c) $-\hat{i} + 2\hat{j}$ (d) $-\hat{i} + \hat{j} + \hat{k}$

118. At the interface between two linear dielectrics (with dielectric constants ε_1 and ε_2), the electric field lines bend, as shown in the figures. Assume that there are no free charges at the interface. The ratio $\varepsilon_1/\varepsilon_2$ is

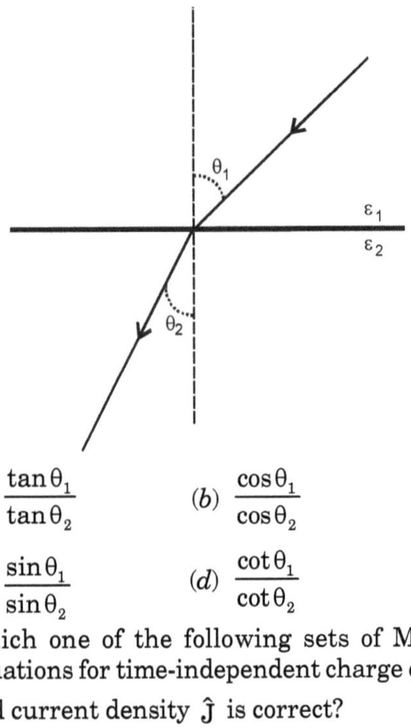

(a) $\frac{\tan\theta_1}{\tan\theta_2}$ (b) $\frac{\cos\theta_1}{\cos\theta_2}$

(c) $\frac{\sin\theta_1}{\sin\theta_2}$ (d) $\frac{\cot\theta_1}{\cot\theta_2}$

119. Which one of the following sets of Maxwell's equations for time-independent charge density ρ and current density $\hat{J}$ is correct?

(a) $\vec{\nabla}.\vec{E} = \rho/\varepsilon_0$, $\vec{\nabla}.\vec{B} = 0$, $\vec{\nabla} \times \vec{E} = -\frac{\partial \vec{B}}{\partial t}$, $\vec{\nabla} \times \vec{B} = \mu_0\varepsilon_0 \frac{\partial \vec{E}}{\partial t}$

(b) $\vec{\nabla}.\vec{E} = \rho/\varepsilon_0$, $\vec{\nabla}.\vec{B} = 0$, $\vec{\nabla} \times \vec{E} = 0$, $\vec{\nabla} \times \vec{B} = \mu_0 \hat{J}$

(c) $\vec{\nabla}.\vec{E} = 0$ (d) $\vec{\nabla}.\vec{E} = \rho/\varepsilon_0$

$\vec{\nabla}.\vec{B} = 0$ $\vec{\nabla}.\vec{B} = \mu_0\hat{J}$

$\vec{\nabla}\times\vec{E} = 0$ $\vec{\nabla}\times\vec{E} = 0$

$\vec{\nabla}\times\vec{B} = \mu_0\hat{J}$ $\vec{\nabla}\times\vec{B} = \mu_0\varepsilon_0\dfrac{\partial\vec{E}}{\partial t}$

120. A classical charged particale moving with frequency ω in a circular orbit of radius a, centred at the origin in the *x-y* plane, electromegnetic radiation. At points (*b*, 0, 0) and (0, 0, *b*), where *b* >> *a*, the electromagnetic waves are

(a) circularly polarized and elliptically polarized, respectively

(b) plane polarized and elliptically polarized, respectively

(c) plane polarized and circularly polarized, respectively

(d) circularly polarized and plane polarized, respectively

Statement for Linked Answer Questions 121 and 122:

A dielectric sphere of radius R carries polarization $\vec{P} = kr^2\hat{r}$, where *r* is the distance from the centre and *k* is a constant. In the spherical polar coordinate system, $\hat{r}, \hat{\theta}$ and $\hat{\varphi}$ are the unit vectors.

121. The bound volume charge density inside the sphere at a distance *r* from the centre is

(a) $-4kR$ (b) $-4kr$

(c) $-4kr^2$ (d) $-4kr^3$

122. The electric field inside the sphere at a distance *d* from the centre is

(a) $\dfrac{-kd^2}{\varepsilon_0}\hat{r}$ (b) $\dfrac{-kR^2}{\varepsilon_0}\hat{r}$

(c) $\dfrac{-kd^2}{\varepsilon_0}\hat{\theta}$ (d) $\dfrac{-kR^2}{\varepsilon_0}\hat{\theta}$

2005

123. Although mass-energy equivalence of special relativity allows conversion of a photon to an electron-positron pair, such a process cannot occur in free space because

(a) the mass is not conserved

(b) the energy is not conserved

(c) the momentum is not conserved

(d) the charge is not conserved

124. Three infinitely long wires are placed equally apart on the circumference of a circle of radius a, perpendicular to its plane. Two of the wires carry current I each, in the same direction, while the third carries current 2I along the direction opposite to the other two. The magnitude of the magnetic induction $\vec{B}$ at a distance r from the centre of the circle, for r > a, is

(a) 0 (b) $\dfrac{2\mu_0}{\pi}\dfrac{I}{r}$

(c) $-\dfrac{2\mu_0}{\pi}\dfrac{I}{r}$ (d) $\dfrac{2\mu_0}{\pi}\dfrac{Ia}{r^2}$

125. A solid sphere of radius R carries a uniform volume charge density ρ. The magnitude of electric field inside the sphere at a distance r from the centre is

(a) $\dfrac{r\rho}{3\varepsilon_0}$ (b) $\dfrac{R\rho}{3\varepsilon_0}$

(c) $\dfrac{R^2\rho}{r\varepsilon_0}$ (d) $\dfrac{R^3\rho}{r^2\varepsilon_0}$

126. The electric field $\vec{E}(\vec{r}, t)$ for a circularly polarized electromagnetic wave propagating along the positive z direction is

(a) $E_0(\hat{x}+\hat{y})\exp[i(kz-\omega t)]$

(b) $E_0(\hat{x}+i\hat{y})\exp[i(kz-\omega t)]$

(c) $E_0(\hat{x}+i\hat{y})\exp[i(kz+\omega t)]$

(d) $E_0(\hat{x}+\hat{y})\exp[i(kz+\omega t)]$

127. The electric (E) and magnetic (*B*) field amplitudes associated with an electromagnetic radiation from a point source behave at a distance r from the source as

(a) E = constant, B = constant

(b) $E \propto \dfrac{1}{r}$, $B \propto \dfrac{1}{r}$

(c) $E \propto \dfrac{1}{r^2}$, $B \propto \dfrac{1}{r^2}$

(d) $E \propto \dfrac{1}{r^3}$, $B \propto \dfrac{1}{r^3}$

128. Consider the following three independent cases :

(i) Particle A of charge +q moves in free space with a constant velocity $\vec{v}$ (v << speed of light)

(ii) Particle B of charge +q moves in free space in a circle of radius R with same speed v as in case (i)

(iii) Particle C having charge – q moves as in case (ii)

If the powers radiated by A, B and C are P_A, P_B, and P_C, respectively, then

(a) $P_A = 0$, $P_B > P_C$

(b) $P_A = 0$, $P_B = P_C$

(c) $P_A > P_B > P_C$

(d) $P_A = P_B = P_C$

129. If the electrostatic potential were given by $\phi = \phi_0 (x^2 + y^2 + z^2)$, where ϕ_0 is constant, then the charge density giving rise to the above potential would be

(*a*) 0 (*b*) $-6\phi_0\varepsilon_0$

(*c*) $-2\phi_0\varepsilon_0$ (*d*) $-\frac{6\phi_0}{\varepsilon_0}$

130. The work done in bringing a charge +q from infinity in free space, to a position at a distance d in front of a semi-infinite grounded metal surface is

(*a*) $-\frac{q^2}{4\pi\varepsilon_0(d)}$ (*b*) $-\frac{q^2}{4\pi\varepsilon_0(2d)}$

(*c*) $-\frac{q^2}{4\pi\varepsilon_0(4d)}$ (*d*) $-\frac{q^2}{4\pi\varepsilon_0(6d)}$

131. A plane electromagnetic wave travelling in vaccum is incident normally on a non-magnetic, non-absorbing medium of refractive index n. The incident (E_i), reflected (E_r) and transmitted (E_t) electric fields are given as,

$E_i = E \exp[i(kz - \omega t)]$,

$E_r = E_{or} \exp[i(k_r z - \omega t)]$,

$E_t = E_{0t} \exp[i(k_t z - \omega t)]$.

If E = 2 V/m and n = 1.5 then the application of appropriate boundary conditions leads to

(*a*) $E_{0r} = -\frac{3}{5}$ V/m, $E_{0t} = \frac{7}{5}$ V/m,

(*b*) $E_{0r} = -\frac{1}{5}$ V/m, $E_{0t} = \frac{8}{5}$ V/m,

(*c*) $E_{0r} = -\frac{2}{5}$ V/m, $E_{0t} = \frac{8}{5}$ V/m,

(*d*) $E_{0r} = \frac{4}{5}$ V/m, $E_{0t} = \frac{6}{5}$ V/m,

132. For a vector potential $\vec{A}$, the divergence of $\vec{A}$ is $\vec{\nabla}.\vec{A} = -\frac{\mu_0}{4\pi}\frac{Q}{r^2}$, where Q is a constant of appropriate dimension. The corresponding scalar potential $\varphi(\vec{r}, t)$ that makes $\vec{A}$ and φ Lorentz gauge invariant is

(*a*) $\frac{1}{4\pi\varepsilon_0}\frac{Q}{r}$ (*b*) $\frac{1}{4\pi\varepsilon_0}\frac{Qt}{r}$

(*c*) $\frac{1}{4\pi\varepsilon_0}\frac{Q}{r^2}$ (*d*) $\frac{1}{4\pi\varepsilon_0}\frac{Qt}{r^2}$

133. An infinitely long wire carrying a current $I(t) = I_0 \cos(\omega t)$ is placed at a distance a from a square loop of side a as shown in the figure. If the resistance of the loop is R, then the amplitude of the induced current in the loop is

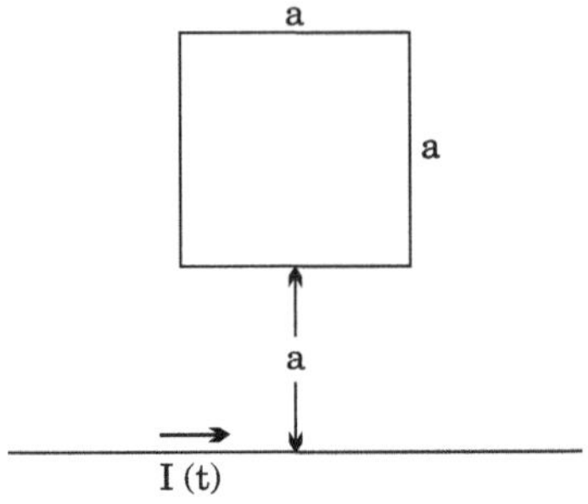

(*a*) $\frac{\mu_0}{2\pi}\frac{aI_0\omega}{R}\ln 2$ (*b*) $\frac{\mu_0}{\pi}\frac{aI_0\omega}{R}\ln 2$

(*c*) $\frac{2\mu_0}{\pi}\frac{aI_0\omega}{R}\ln 2$ (*d*) $\frac{\mu_0}{2\pi}\frac{aI_0\omega}{R}$

Statement for Linked Answer Questions 134 and 135 :

An infinitely long hollow cylinder of radius R carrying a surface charge density σ is rotated about its cylinderical axis with a constant angular speed ω

134. The magnitude of the surface current is

(*a*) $\sigma R^2 \omega$ (*b*) $2\sigma R\omega$

(*c*) $\pi\sigma R\omega$ (*d*) $2\pi\sigma R\omega$

135. The magnitude of vector potential inside the cylinder at a distance from its axis is

(*a*) $2\mu_0\sigma R\omega r$ (*b*) $\mu_0\sigma R\omega r$

(*c*) $\frac{1}{2}\mu_0\sigma R\omega r$ (*d*) $\frac{1}{4}\mu_0\sigma R\omega r$

2004

136. A charge $+q$ is kept at a distance of 2R from the center of a grounded conducting sphere of radius R. The image charge and its distance from the centre are, respectively

(*a*) $-\frac{q}{2}$ and $\frac{R}{2}$ (*b*) $-\frac{q}{2}$ and $\frac{R}{4}$

(*c*) $-q$ and $\frac{R}{2}$ (*d*) $+\frac{q}{2}$ and $\frac{R}{2}$

137. The state of polarization of light with the electric field vector

$\vec{E} = \hat{i} E_0 \cos(kz - \omega t) - \hat{j} E_0 \cos(kz - \omega t)$ is

(*a*) linearly polarized along *z*-direction

(*b*) linearly polarized at – 45° to *x*-axis

(*c*) circularly polarized

(*d*) elliptically polarized with the major axis along x-axis

138. A rigid frictionless rod rotates anticlockwise in a vertical plane with angular velocity $\vec{\omega}$. A bead of mass m moves outward along the rod with constant velocity $\vec{u_0}$. The bead will experience a coriolis force

(*a*) $2mu_0\omega\hat{\theta}$ (*b*) $-2mu_0\omega\hat{\theta}$

(*c*) $4mu_0\omega\hat{\theta}$ (*d*) $-mu_0\omega\theta$

139. A conducting sphere of radius R has charge + Q on its surface. If the charge on the sphere is doubled and its radius is halved, the energy associated with the electric field will

(*a*) increase four times

(*b*) increase eight times

(*c*) remain the same

(*d*) decrease four times

140. A conducting sphere of radius R is placed in a uniform electric field $\vec{E_0}$ directed along + *z* axis. The electric potential for outside points is given as $V_{out} = -E_0\left(1-\frac{R^3}{r^3}\right) r\cos\theta$, where *r* is the distance from the center and θ is the polar angle. The charge density on the surface of the sphere is

(*a*) $2\varepsilon_0 E_0 \cos\theta$ (*b*) $\varepsilon_0 E_0 \cos\theta$

(*c*) $3\varepsilon_0 E_0 \cos\theta$ (*d*) $\frac{\varepsilon_0}{3} E_0 \cos\theta$

141. A circular arc QTS is kept in an external magnetic field $\vec{B_0}$ as shown in figure. The arc carries a current I. The magnetic field is directed normal and into the page. The force acting on the arc is

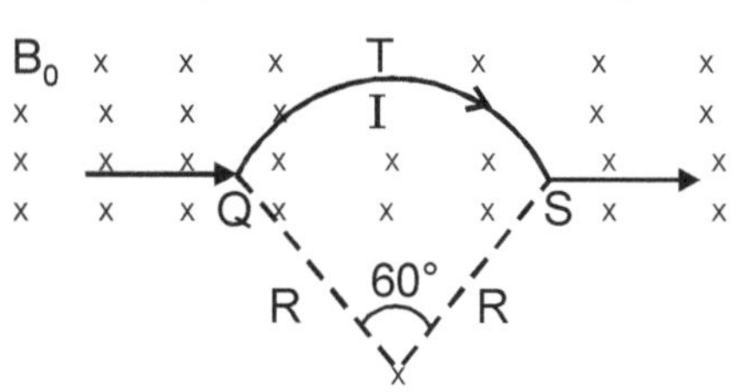

(*a*) $2IB_0R\hat{k}$ (*b*) $IB_0R\hat{k}$

(*c*) $-2IB_0R\hat{k}$ (*d*) $-IB_0R\hat{k}$

142. A plane electromagnetic wave of frequency ω is incident normally on an air-dielectric interface. he dielectric is linear, isotropic, non-magnetic and its refractive index is *n*. The reflectance (R) and transmittance (T) from theinterface are

(*a*) $R = \left(\frac{n-1}{n+1}\right)^2, T = \frac{4n}{(n+1)^2}$

(*b*) $R = -\left(\frac{n-1}{n+1}\right), T = \frac{2}{(n+1)^2}$

(*c*) $R = \left(\frac{n-1}{n+1}\right)^3, T = \frac{4n^3}{(n+1)^3}$

(*d*) $R = \left(\frac{n-1)^2}{n+1}\right), T = \frac{4n^2}{(n+1)^2}$

143. The electric field of a plane e.m. wave is $\vec{E} = \vec{E_0}$ $\exp[i(\hat{x}k\cos a + \hat{y}k\sin a - \omega t)]$. If $\hat{x}, \hat{y}$ and $\hat{z}$ are cartesian unit vectors, the wave vectors $\vec{k}$ of the e.m. wave is

(*a*) $\hat{z}k$

(*b*) $\hat{x}k\sin\alpha + \hat{y}k\cos\alpha$

(*c*) $\hat{x}k\cos\alpha + \hat{y}k\cos\alpha$

(*d*) $-\hat{z}k$

144. The dispersion relation for a low density plasma is $\omega^2 = \omega_0^2 + c^2k^2$, where w_0 is the plasma frequency and *c* is the speed of light in free space. The relationship between the group velocity (v_g) and phase velocity (v_p) is

(*a*) $V_p = V_g$ (*b*) $V_p = V_g^{1/2}$

(*c*) $V_p V_g = c^2$ (*d*) $V_g = V_p^{1/2}$

145. A Michelson interferometer is illuminated with monochromatic light. When one of the mirrors is moved through a distance of 25.3 μm, 92 fringes pass through the cross-wire. The wavelegth of the monochromatic light is

(*a*) 500 nm (*b*) 550 nm

(*c*) 600 nm (*d*) 650 nm

146. A beam of mono-energetic particles having speed *v* is described by the wave function $\psi(x) = u(x)\exp(ikx)$, where *u*(*x*) is a real function. This corresponds to a current density

(*a*) $u^2(x)v$ (*b*) *v*

(*c*) zero (*d*) $u^2(x)$

Data for Questions 147-148

Consider a comet of mass m moving in a parabolic orbit around the Sun. The closest distance between the comet and the Sun is b, the mass of the Sun is M and the universal gravitation constant is G.

147. The angular momentum of the comet is

(*a*) $M\sqrt{Gmb}$ (*b*) $b\sqrt{GmM}$

(*c*) $G\sqrt{mMb}$ (*d*) $m\sqrt{2GMb}$

148. Which one of the following is TRUE for the above system ?

(*a*) The acceleration of the comet is maximum when it is closest to the Sun

(*b*) The linear momentum of the comet is a constant

(*c*) The comet will return to the solar system after a specified period

(*d*) The kinetic energy of the comet is a constant

Data for Questions 149-150

Let $\tilde{E} = \hat{x}E_0 \exp\left[i\vec{k}.\vec{r} - \omega t\right]$, *where* $\vec{k} = \hat{z}(k\cos\phi + ik\sin\phi)$, $k = 1\ |\vec{k}|$ *and* $\hat{x}, \hat{y}$ *and* $\hat{z}$ *are cartesian unit vectors, represent an electric field of a plane electro magnetic wave of frequency* ω

149. Which one of the following statements is TRUE ?

(*a*) The magnitude of the electric field is attenuated as the wave propagates

(*b*) The energy of the e.m. wave flows along the *x*-direction

(*c*) The magnitude of the electric field of the wave is a constant

(*d*) The speed of the wave is the same as *c* (speed of light in free space)

150. The magnetic field $\tilde{B}$ of the wave is

(*a*) $\hat{y}\,\frac{k}{\omega}E_0\exp(-zk\sin\phi)\exp[i(zk\cos\phi-\omega t)]$

(*b*) $\hat{y}\,\frac{k}{\omega}E_0\exp(-zk\sin\phi)\exp(i(zk\cos\phi-\omega t+\phi)]$

(*c*) $\hat{y}\,\frac{k}{\omega}E_0\exp[i(zk\cos\phi-\omega t+\phi)]$

(*d*) $\hat{y}\,\frac{k}{\omega}E_0\exp(-zk\cos\phi)\exp[i(zk\sin\phi-\omega t)]$

2003

151. An electric charge, +Q is placed on the surface of a soild, conduction sphere of radius a. The distance measured from the centre of the sphere is denoted as r. Then

(*a*) the charge gets distributed uniformly through the volume of the sphere

(*b*) the electrostatic potential has the same value for $r < a$

(*c*) an equal and opposite charge gets induced in the bottom half of the sphere

(*d*) the electric field is given by $1/(4\pi\varepsilon_0 r^2)$ for $r < a$

152. An electric field applied along the length of a long cylinder produces a polarization P. The depolarization field produced in this configuration is

(*a*) $4\pi P/3$	(*b*) $-4\pi P/3$
(*c*) $2\pi P$	(*d*) 0

153. Which one of the following Maxwell's equations implies the absence of magnetic monopoles ?

(*a*) $\nabla\cdot E = \pi/\varepsilon_0$

(*b*) $\nabla\cdot B = 0$

(*c*) $\nabla\times E = -\partial B/\partial t$

(*d*) $\nabla\times B = (1/c^2)\,\partial B/\partial t + \mu_0 J$

154. An electromagnetic wave is propagating in free space in the z-direction. If the electric field is given by E = cos (ω t – k z) i, where ω t = c k, then the magnetic field is given by

(*a*) $B = (1/c)\cos(\omega t - kz)\,j$

(*b*) $B = (1/c)\sin(\omega t - kz)\,j$

(*c*) $B = (1/c)\cos(\omega t - kz)\,i$

(*d*) $B = (1/c)\sin(\omega t - kz)\,i$

155. The speed of a particle whose kinetic energy is equal to its rest mass energy is given by (c is the speed of light in vacuum)

(*a*) $c/3$	(*b*) $\sqrt{2}\,c/3$
(*c*) $c/2$	(*d*) $\frac{\sqrt{3}}{2}c$

156. Electromagnetic waves are propagating along a hollow, metallic waveguide whose cross-section is a square of side W. The minimum frequency of the electromagnetic waves is

(*a*) c/W	(*b*) 2c/W
(*c*) $\pi c/W$	(*d*) $\sqrt{2}\,\pi c/W$

157. Consider the given statements about E(r, t) and B(r, t), the electric and magnetic vectors respectively in a region of free space

P Both E and B are conservative vector fields

Q Both E and B are central force fields

R E and B are mutually perpendicular in the region

S Work done by B on a moving charge in the region is zero

Choose the right combination of correct statements from the following

(*a*) P and R	(*b*) R and S
(*c*) S only	(*d*) P and Q

158. An infinite conducting sheet in the x-y plane carries a surface current density K along the y-axis. The magnetic field B for z > 0 is

(*a*) $B = 0$	(*b*) $B = \mu_0 K k/z$
(*c*) $B = \mu_0 K i/2$	(*d*) $B = \mu_0 K j/(x^2 + z^2)^{0.5}$

159. A parallel beam of infrared radiation of wavelength of 1.01×10^{-6} m is incident normally on a screen with two slits 5×10^{-6} m apart and the resulting interference pattern is observed on a distant screen. What is the largest number of maxima that can be observed on the screen ?

(*a*) 4 (*b*) 9

(*c*) 13 (*d*) infinitely many

160. An external magnetic field of magnitude H is applied to a Type–I superconductor at a temperature below the transition point. Then which one of the following statements is NOT true for H less than the critical field H_C ?

(*a*) the sample is diamagnetic

(*b*) it magnetization varies linearly with H

(*c*) the lines of magnetic induction are pushed out from the sample

(*d*) the sample exhibits mixed states of magnetization near H_C

161. If the photon were to have a finite mass, then the Coulomb potential between two stationary charges separated by a distance r would

(*a*) be strictly zero beyond some distance

(*b*) fall off exponentially for large values of *r*

(*c*) fall off as $1/r^3$ for large values of *r*

(*d*) fall off as $1/r$ for large values of *r*

162. A stationary particle in free space is observed to spontaneously decay into two photons. This implies that

(*a*) the particle carries electric charge

(*b*) the spin of the particle must be greater than or equal to 2

(*c*) the particle is a boson

(*d*) the mass of the particle must be greater than or equal to the mass of the hydrogen atom

163. The masses of a hydrogen atom, neutron and $^{238}U_{92}$ are given by 1.0078, 1.0087 and 238.0508 respectively. The binding energy of $^{238}U_{92}$ is therefore approximately equal to (taking 1 a.m.u. = 931.64 MeV)

(*a*) 120 MeV

(*b*) 1500 MeV

(*c*) 1600 MeV

(*d*) 1800 MeV

Data for Questions 164 – 165 :

Consider the motion of a particle in the potential V(x) shown in the figure

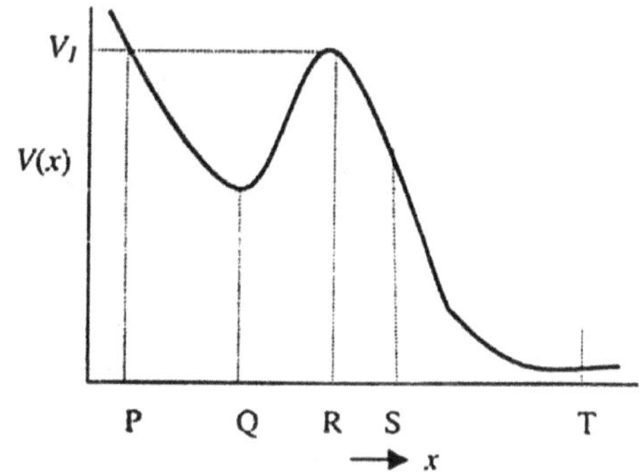

164. Suppose the particle has a total energy $E = V_1$ in the figure. Then the speed of the particle is zero when it is at

(*a*) point P (*b*) point Q

(*c*) point S (*d*) point T

165. Which one of the following statements is NOT correct about the particle ?

(*a*) It experiences no force when its position corresponds to the point Q on the curve

(*b*) It experiences no force when its position corresponds to the point R on the curve

(*c*) Its speed is the largest when it is at S

(*d*) It will be in a closed orbit between P and R if $E < V_1$

Data for Questions 166 – 167 :

Consider two conducting plates of infinite extent, one plate at z = 0 and the other at z = L, both parallel to the xy plane. The vector and scalar potential in the region between the plates is given by

$$A(r, t) = A_0\, i \cos(kz + \alpha) \cos(kct);$$

$$\phi(r, t) = 0$$

166. For this to represent a standing wave in the empty region between the plates

(*a*) $k = \pi/L$ and $\alpha = 0$

(*b*) $k = 2\pi/L$ and $\alpha = \pi/2$

(*c*) $k = \pi/(2L)$ and $\alpha = \pi/2$

(*d*) $k = \pi/2L$ and $\alpha = 0$

167. The energy density at z = 0 and t = 0 is

(*a*) 0

(*b*) $\varepsilon_0 c^2 k^2 A_0^2$

(*c*) $(1/2)\, \mu_0 A_0^2 k^2$

(*d*) $(1/2)\, \mu_0 A_0^2 k^2 + (1/2)\, \varepsilon_0 c^2 k^2 A_0^2$

2002

168. Consider an infinitely long straight cylindrical conductor of radius R with its axis along the z-direction, which carries a current of 1 A uniformly distributed over its cross section. Which of the following statements is correct ?

(*a*) $\vec{\nabla} \times \vec{B} = 0$ everywhere,

(*b*) $\vec{\nabla} \times \vec{B} = \frac{\mu_0}{\pi R^2}\hat{z}$ everwhere,

(*c*) $\vec{\nabla} \times \vec{B} = 0$ for $r > R$,

(*d*) $\vec{\nabla} \times \vec{B} = \frac{\mu_0}{\pi R^2}\hat{z}$ for $r > R$

where r is the radial distance from the axis of the cylinder.

169. Consider a set of two stationary point charges q_1 and q_2 as shown in the figure. Which of the following statements is correct ?

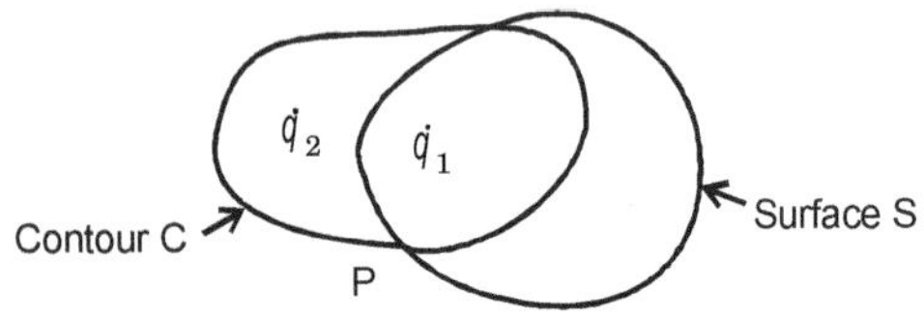

(*a*) The electric field at P is independent of q_2.

(*b*) The electric flux crossing the closed surface S is independent of q_2.

(*c*) The line integral of the electric field $\vec{E}$ over the closed contour C depends on q_1 and q_2.

(*d*) $\vec{\nabla} \cdot \vec{E} = 0$ everywhere

170. Four charges are placed at the four corners of a square of side a as shown in the figure. The electric dipole moment of this configuration is

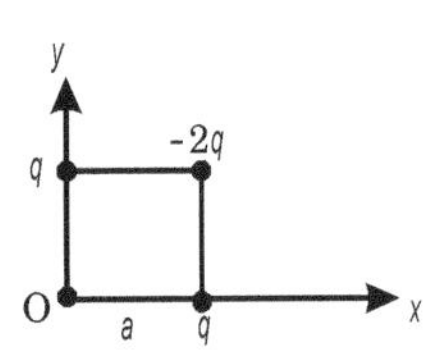

(*a*) $\vec{p} = qa\hat{i} + qa\hat{j}$

(*b*) $\vec{p} = -qa\hat{i} + qa\hat{j}$

(*c*) $\vec{p} = -qa\hat{i} - qa\hat{j}$

(*d*) $\vec{p} = qa\hat{i} - qa\hat{j}$

171. An infinitely long closely wound solenoid carries a sinusoidally varying current. The induced electric field is

(*a*) zero everywhere

(*b*) non-zero inside and zero outside the solenoid

(*c*) non-zero inside as well as outside the solenoid

(*d*) zero inside and non-zero outside the solenoid

172. A left circularly polarized light beam of wavelenfth 600 nm is incident on a crystal of thickness d and propagates perpendicular to its optic axis. The ordinary and extraordinary refractive indices of the crystal are $n_o = 1.54$ and $n_e = 1.55$ respectively. The emergent light will be right circularly polarized if d is

(*a*) 120 μm (*b*) 60 μm

(*c*) 30 μm (*d*) 15 μm

173. In a two beam interference pattern, the maximum and minimum intensity values are found to be 25 I_0 and 9 I_0 respectively, where I_0 is a constant. The intensities of the two interfering beams are

(*a*) 16 I_0 and I_0 (*b*) 5 I_0 and 3 I_0

(*c*) 17 I_0 and 8 I_0 (*d*) 8 I_0 and 2 I_0

174. A vector $\vec{A} = (5x + 2y)\hat{i} + (3y - z)\hat{j} + (2x - az)\hat{k}$ is solenoidal if the constant a has a value

(*a*) 4 (*b*) – 4

(*c*) 8 (*d*) – 8

2001

175. Two point charges Q_1 = 1 nC and Q_2 = 2 nC are kept in free space such that the distance between them is 0.1 m.

(*a*) The force on Q_2 is along the direction from Q_2 to Q_1.

(*b*) The force on Q_2 is the same in magnitude as that on Q_1.

(*c*) The force on Q_1 is attractive.

(*d*) A point charge Q_3 = – 3 nC, placed at the midpoint between Q_1 and Q_2, experiences no net force.

176. A current I flows in the anticlockwise direction through a square loop of side a lying in the xoy plane with its center at the origin. The magnetic induction at the center of the square loop is

(*a*) $\frac{2\sqrt{2}\mu_0 I}{\pi a}\hat{e}_x$ (*b*) $\frac{2\sqrt{2}\mu_0 I}{\pi a}\hat{e}_z$

(*c*) $\frac{2\sqrt{2}\mu_0 I}{\pi a^2}\hat{e}_z$ (*d*) $\frac{2\sqrt{2}\mu_0 I}{\pi a^2}\hat{e}_x$

177. Consider an electric field $\vec{E}$ existing in the interface between a conductor and free space. Then the electric field $\vec{E}$ is

(*a*) external to the conductor and normal to the conductor's surface (*b*) internal to the conductor and normal to the conductor's surface

(*c*) external to the conductor and tangential to the conductor's surface

(*d*) both external and internal to the conductor and normal to the conductor's surface

178. A coaxial cable of uniform cross-section contains an insulating material of dielectric constant 3.5. The radius of the central wire is 0.01 m and that of the sheath is 0.02 m. The capacitance per kilometer of a cable is

(*a*) 280.5 nF (*b*) 28.05 nF

(*c*) 56.10 nF (*d*) 2.805 nF

179. The xoy plane carries a uniform surface current of density $\vec{K} = 50\hat{e}_z$ A/m. The magnetic field at the point $z = -0.5$ m is

(*a*) 10×10^{-6} Wb

(*b*) 1×10^{-6} Wb

(*c*) $\pi \times 10^{-6}$ Wb

(*d*) $10\pi \times 10^{-6}$ Wb

180. Four point charges are placed at the corners of a square whose center is at the origin of a Cartesian coordinate system. A point dipole $\vec{p}$ is placed at the center of the square as shown in the figure. Then,

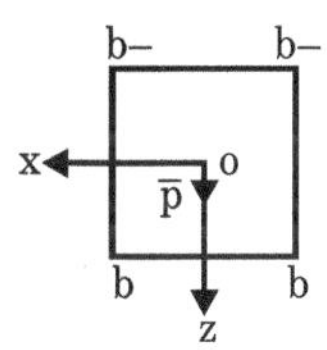

(*a*) there is no force acting on the dipole

(*b*) there is no torque about the center at O on the dipole

(*c*) the dipole has minimum energy if it is in $\hat{e}_x$ direction

(*d*) the force on the dipole is increased if the medium is replaced by another medium with larger dielectric constant

181. The electric field E(*r*, *t*) at a point *r* at time *t* in a metal due to the passage of electrons can be described by the equation

$$\nabla^2 \vec{E}\left(\vec{r}, t\right) = \frac{1}{c^2}\left[\frac{\partial^2 \vec{E}(\vec{r},t)}{\partial t^2} + \omega'^2 \vec{E}(\vec{r},t)\right]$$

where ω' is a characteristic associated with the metal and *c* is the speed of light in vacuum. The dispersion relation corresponding to the plane wave solutions

of the form $\exp\left[i\left(\vec{k}\cdot\vec{r} - \omega t\right)\right]$ is given by

(*a*) $\omega^2 = c^2k^2 - \omega'^2$

(*b*) $\omega^2 = c^2k^2 + \omega'^2$

(*c*) $\omega = ck - \omega'$

(*d*) $\omega = ck + \omega'$

182. If $u(x, y, z, t) = f(x + i\beta y - vt) + g(x - i\beta y - vt)$, where f and g are arbitrary and twice differentiable functions, is a solution of the wave equation $\frac{\partial^2 u}{\partial x^2} + \frac{\partial^2 u}{\partial y^2} = \frac{1}{c^2}\frac{\partial^2 u}{\partial t^2}$

then β is

(*a*) $\left(1 - \frac{v}{c}\right)^{1/2}$ (*b*) $\left(1 - \frac{v}{c}\right)$

(*c*) $\left(1 - \frac{v^2}{c^2}\right)^{1/2}$ (*d*) $\left(1 - \frac{v^2}{c^2}\right)$

2000

183. The space between the plates of a parallel plate capacitor is filled with two dielectric slabs of dielectric constants κ_1 and κ_2 as shown in the figure. If the capacitor is charged to a potential V, then at the interface between the two dielectrics,

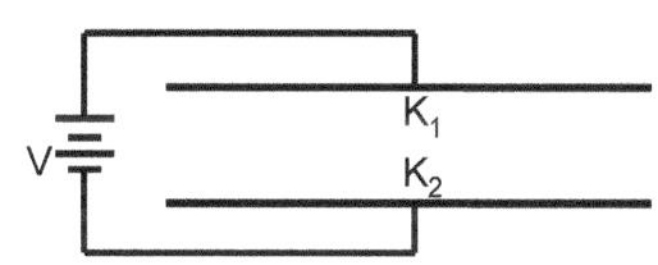

(*a*) $\vec{E}$ is discontinuous and $\vec{D}$ is continuous.

(*b*) $\vec{E}$ is discontinuous and $\vec{D}$ is discon-tinuous.

(*c*) $\vec{E}$ is continuous and $\vec{D}$ is continuous.

(*d*) $\vec{E}$ is continuous and $\vec{D}$ is discontinuous.

184. Two large parallel plates move with a constant speed v in the positive *y*-direction as shown in the figure. If both the plates have a surface charge desnity σ > 0, the magnetic field at the point P just above the top plate will have

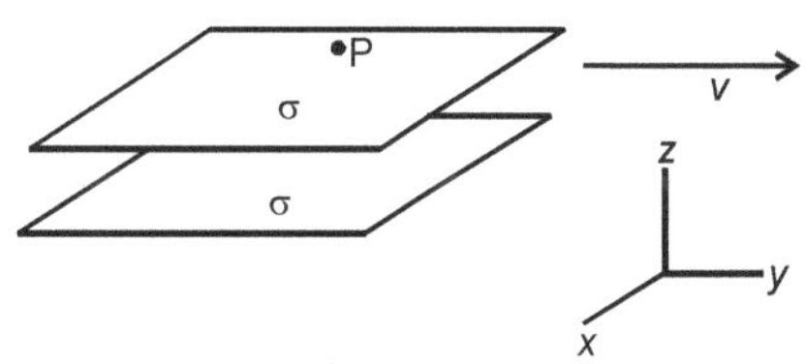

(*a*) larger magnitude than the field at the mid-point between the plates and point towards $-\hat{x}$.

(*b*) smaller magnitude than the field at the mid-point between the plates and point towards $+\hat{x}$.

(*c*) larger magnitude than the field at the mid-point between the plates and point towards $+\hat{x}$.

(*d*) smaller magnitude than the field at the mid-point between the plates and point towards $-\hat{x}$.

185. A tungsten wire of uniform cross section and high resistance is supported by two copper suports of low resistance in vacuum. The length of the wire is l. A constant current I is sent through the wire. The temperature profile (T vs. x) of the wire will look like

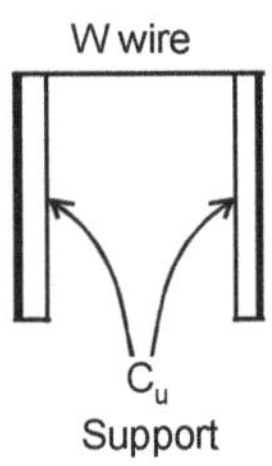

(*a*)
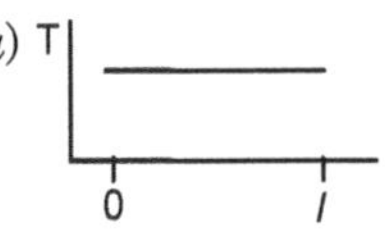

(*b*)
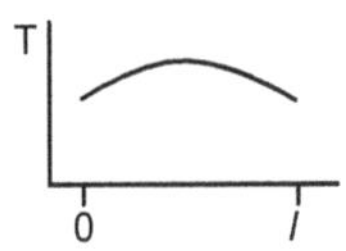

(*c*)
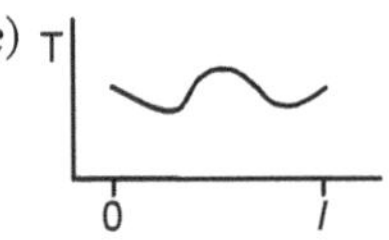

(*d*)
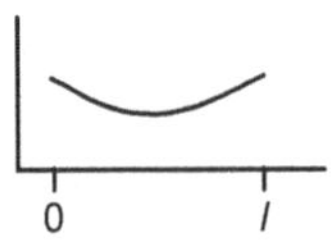

186. If $\bar{A}(t)$ is a vector of constant magnitude, which of the following is true?

(*a*) $\frac{d\bar{A}}{dt} = 0$ (*b*) $\frac{d^2\bar{A}}{dt^2} = 0$

(*c*) $\frac{d\bar{A}}{dt}\cdot\bar{A} = 0$ (*d*) $\frac{d\bar{A}}{dt}\times\bar{A} = 0$

187. A solenoid with an iron core is connected in series with a battery of emf V and it is found that a constant current I_0 passes through the solenoid. If at $t = 0$, the iron core is pulled out from the solenoid quickly in a time Δt, which one of the following could be a correct description of the current passing through the solenoid?

(*a*)

(*b*)
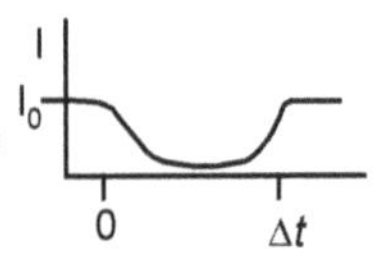

(*c*)
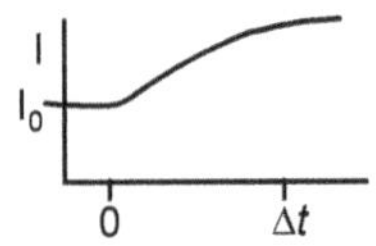

(*d*)
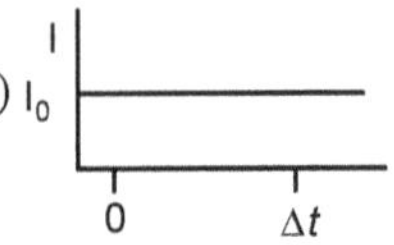

188. A point charge q is kept at the mid-point between two large parallel grounded conducting plates. Assume to gravity. The charge is displaced a little towards the right plate. The charge will now,

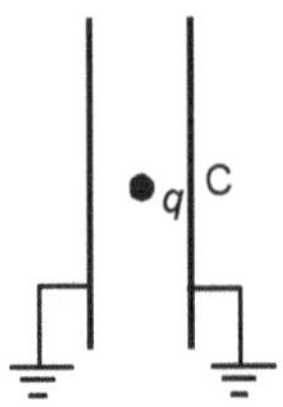

(*a*) stay where it is

(*b*) move towards the right plate

(*c*) move towards the left plate

(*d*) oscillate between the plates

189. A very long solenoid with n turns per unit length carries a current I. The magnetic field at a point, which is on its axis and its end face, is

(*a*) $\mu_0 nI$ (*b*) $(2/3)\mu_0 nI$

(*c*) $(1/3)\mu_0 nI$ (*d*) $(1/2)\mu_0 nI$

190. Three plane waves are given as

$$\bar{y}_1 = A_1\hat{y}\cos(kx-\omega t+\delta_1),$$

$$\bar{y}_2 = A_2\hat{z}\cos(kx-\omega t+\delta_2)$$

and $\bar{y}_3 = A_3\hat{y}\cos(kx-\omega t+\delta_3)$,

where δ_1, δ_2, δ_3, and A_1, A_2, A_3 are constants. If these waves are superposed pairwise, which superposition will lead to interference?

(*a*) $\bar{y}_1$ and $\bar{y}_2$

(*b*) $\bar{y}_2$ and $\bar{y}_3$

(*c*) $\bar{y}_1$ and $\bar{y}_3$

(*d*) No interference in any pair

ANSWERS

1. (c) **2.** (c) **3.** (b) **4.** (b,d) **5.** (a) **6.** (a) **7.** (8 to 8)
8. (0.24 to 0.26) **9.** (*b*) **10.** (*d*) **11.** (7 to 7) **12.** (14 to 15) **13.** (*d*) **14.** (*a*)
15. (*d*) **16.** (0 to 0) **17.** (–270 to –270 or –90 to –90 or 90 to 90 or 270 to 270) **18.** (2 to 2)
19. (*d*) **20.** (*b*) **21.** (6.26 to 6.30) **22.** (0.8 to 0.8) **23.** (*c*) **24.** (*c*)
25. (*b*) **26.** (*d*) **27.** (0.044 to 0.046) **28.** (4 to 4) **29.** (1.8 to 1.8) **30.** (6 to 6)
31. (*a*) **32.** (*c*) **33.** (*a*) **34.** (*c*) **35.** (6.5 to 6.7) **36.** (*c*)
37. (11.25 to 11.34) **38.** (41.60 to 42.00) **39.** (8.80 to 8.90) **40.** (1.72 to 1.75)
41. (*d*) **42.** 87.50 to 89.50 **43.** (*d*) **44.** (*d*) **45.** (*d*) **46.** (0.27 : 0.29)
47. (*c*) **48.** (*b*) **49.** (*c*) **50.** (11) **51.** (1.2) **52.** (*b*) **53.** (*d*) **54.** (*c*)
55. (1.49 to 1.51) **56.**(0.99 to 1.01) **57.** (2.3 to 2.4) **58.** (*b*) **59.** (*d*)
60. (*b*) **61.** (*c*) **62.** (*a*) **63.** (*c*) **64.** (0.15) **65.** (*a*) **66.** (*a*) **67.** (*d*)
68. (*a*) **69.** (*b*) **70.** (*c*) **71.** (*b*) **72.** (*a*) **73.** (*c*) **74.** (*b*) **75.** (*c*)
76. (*a*) **77.** (*a*) **78.** (*d*) **79.** (*d*) **80.** (*d*) **81.** (*c*) **82.** (*b*) **83.** (*d*)
84. (*b*) **85.** (*b*) **86.** (*c*) **87.** (*d*) **88.** (*a*) **89.** (*c*) **90.** (*c*) **91.** (*b*)
92. (*c*) **93.** (*d*) **94.** (*d*) **95.** (*c*) **96.** (*b*) **97.** (*c*) **98.** (*c*) **99.** (*a*)
100. (*a*) **101.** (*a*) **102.** (*c*) **103.** (*a*) **104.** (*c*) **105.** (*b*) **106.** (*d*) **107.** (*b*)
108. (*d*) **109.** (*b*) **110.** (*a*) **111.** (*b*) **112.** (*a*) **113.** (*d*) **114.** (*b*) **115.** (*c*)
116. (*c*) **117.** (*b*) **118.** (*a*) **119.** (*b*) **120.** (*c*) **121.** (*b*) **122.** (*a*) **123.** (*d*)
124. (*a*) **125.** (*a*) **126.** (*a*) **127.** (*c*) **128.** (*b*) **129.** (*b*) **130.** (*c*) **131.** (*c*)
132. (*d*) **133.** (*a*) **134.** (*a*) **135.** (*c*) **136.** (*a*) **137.** (*b*) **138.** (*b*) **139.** (*b*)
140. (*c*) **141.** (*b*) **142.** (*a*) **143.** (*c*) **144.** (*c*) **145.** (*b*) **146.** (*a*) **147.** (*d*)
148. (*a*) **149.** (*a*) **150.** (*b*) **151.** (*b*) **152.** (*d*) **153.** (*b*) **154.** (*a*) **155.** (*d*)
156. (*d*) **157.** (*b*) **158.** (*c*) **159.** (*d*) **160.** (*d*) **161.** (*a*) **162.** (*c*) **163.** (*d*)
164. (*a*) **165.** (*c*) **166.** (*b*) **167.** (*c*) **168.** (*c*) **169.** (*b*) **170.** (*c*) **171.** (*b*)
172. (*c*) **173.** (*a*) **174.** (*c*) **175.** (*b*) **176.** (*b*) **177.** (*a*) **178.** (*b*) **179.** (*b*)
180. (*a*) **181.** (*a*) **182.** (*c*) **183.** (*a*) **184.** (*c*) **185.** (*b*) **186.** (*c*) **187.** (*a*)
188. (*b*) **189.** (*d*) **190.** (*c*)

EXPLANATIONS

1. A/c to Poynting vector

$$\vec{S} = \int \vec{E} \times \vec{H} d\vec{S} \text{ or } < S > = \lambda B^2 K$$

$$\because \oint(\vec{E} \times \vec{H}) dS = \text{Constant}$$

$$\oint B^2 . dS = \text{Constant}$$

$$B^2 \propto \frac{1}{r^2}$$

$$B \propto \frac{1}{r}$$

So, option (c) is correct.

2. In solenoid R << L

Magnetic field is uniform inside the solenoid and zero out side.

$$\because A_\phi = (O, A_\phi, O) \text{ at } Z = \frac{L}{2} \text{ and } \vec{\nabla} \times \vec{F}(r, \phi, z)$$

$$= \hat{r}\left[\frac{1}{r} \cdot \frac{\partial F_2}{\partial \phi} - \frac{\partial F_\phi}{\partial z}\right] + \varphi\left[\frac{\partial F_r}{\partial z} - \frac{\partial F_\phi}{\partial r}\right]$$

$$+ z . \frac{1}{r}\left(\frac{\partial (rF_e)}{\partial r} - \frac{\partial F_e}{\partial r}\right)$$

A_ϕ will increase upto R. Linearly and then decrease exponentially. So option (c) is correct.

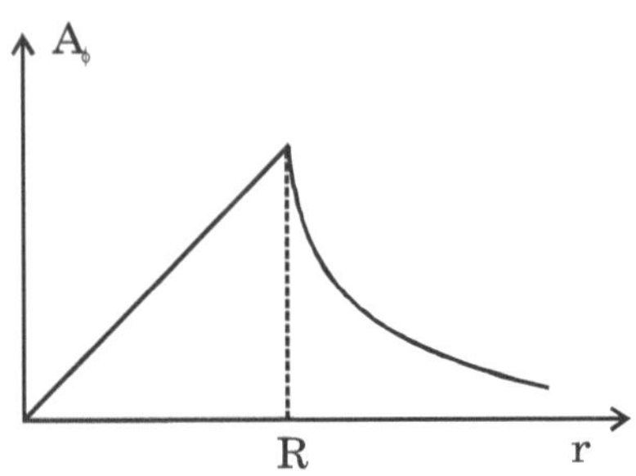

If M is the magnetization than bound current density J_b will follow the

$$\because \nabla . J_b = \nabla . (\nabla \times M)$$

$$\therefore \nabla . J_b = 0$$

$$\because \nabla . (\nabla \times M) = 0$$

So option (b) and (d) both are correct.

3. A/c to Brewester's law

$$\mu = \tan\theta$$

$$\tan\theta = \sqrt{3} \qquad \because \mu = \sqrt{3}$$

$$\theta = 60°$$

Because permeability of medium is same

So T = 1 and polarization of incident light is parallel to the plane of incidence.

So, option (b) is correct.

4. A/c to Maxwell's equation

$$\nabla . B = 0$$

If M is the magnetization than bound current density J_b will follow the

$$\because \nabla . J_b = \nabla . (\nabla \times M)$$

$$\therefore \nabla . J_b = 0 \qquad \because \nabla . (\nabla \times M) = 0$$

So, option (b) and (d) both are correct.

5. Here is the situation of T. I. R. at the boundary of medium.

Transmitted wave at

$$T_1 = \frac{4n_1 n_2}{(n_1 + n_2)^2} \Rightarrow \frac{4 \times 1.0 \times 2.2}{3.2 \times 3.2} \qquad \dots (1)$$

Electric field amplitude emerge from slab in region 2

$$E' = T_1 \times T_1 \times E_0$$

$$E' = \frac{4 \times 1.0 \times 2.2}{3.2 \times 3.2} \times \frac{4 \times 1.0 \times 2.2}{3.2 \times 3.2} \times 8$$

$$\boxed{E' = \frac{11}{1.6}} \text{ volt/cm}$$

So, option (a) is the correct answer.

6. Let a point P a distance r, electric field remain same.

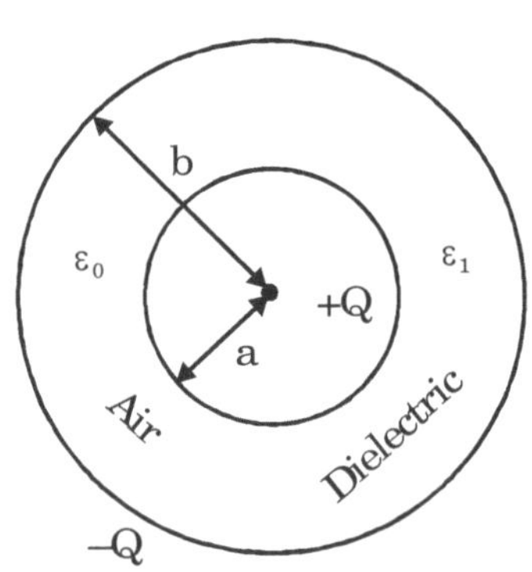

Integral form of gauss law

$$\int \vec{D} . \vec{dA} = q_f$$

$$D_1 2\pi r^2 + D_2 2\pi r^2 = Q$$

$$\begin{Bmatrix} D_1 = \text{Displacement vector for air medium} = \varepsilon_0 E \\ D_2 = \text{Displacement vector for dielectric medium} = \varepsilon E \end{Bmatrix}$$

$$\varepsilon_0 E\, 2\pi r^2 + \varepsilon E\, 2\pi r^2 = Q$$

$$E.\, 2\pi r^2 (\varepsilon + \varepsilon_0) = Q$$

$$E = \frac{Q}{2\pi r^2 (\varepsilon + \varepsilon_0)}$$

Since direction is radial

$$\boxed{E' = \frac{Q}{2\pi(\varepsilon + \varepsilon_0)}\frac{\hat{r}}{r^2}} \text{ everywhere}$$

So, option (a) is the correct answer.

7. Coulomb energy for spherical charge distribution

$$E_c = \frac{3}{5}\frac{z^2 e^2}{4\pi \epsilon_0 r}$$

$$E_c = \frac{3}{5}\frac{z^2}{r}(1.44\text{MeV} - \text{fm})$$

$$18.432\text{ MeV} = \frac{3}{5} \times \frac{z^2}{r}(1.44\text{MeV} - \text{fm})$$

$$12.8 = \frac{3}{5} \times \frac{z^2}{3}$$

$$z^2 = 64$$

$$\boxed{z = 8}$$

8. Let magnetic field between gap is B_0 and in the material is B.

Since, B = uni

$$B = \frac{2000\mu_0 \times 10^4}{5 \times 10^{-2}} \quad \ldots(1)$$

$$[\mu = \mu_r \mu_0]$$

From boundary condition

$$D_1^1 = D_2^1$$

$$\epsilon_1 E_1 = \epsilon_2 E_2$$

$$\frac{B}{\mu} = \frac{B_0}{\mu_0}$$

$$\frac{2000\mu_0 \times 10^4}{5 \times 10^{-2} \times \mu} = \frac{B}{\mu_0}$$

and $$B = \frac{2000\mu_0 \times 10^4 \times \mu_0}{5 \times 10^{-2} \times 2000\mu_0}$$

$$B = \frac{4\Pi \times 10^{-7} \times 10^6}{5}$$

$$\boxed{B = 0 - 0.25\text{T}}$$

9. We know that,

$$\frac{J_C}{J_D} = \frac{\sigma}{\omega\varepsilon}$$

$$\frac{J_C}{J_D} = \frac{\sigma}{\varepsilon_r \varepsilon_0 \omega}$$

Given $\varepsilon_r > 1$

i.e. $\frac{J_D}{J_C} > 1$ i.e. $\frac{I_D}{I_C} > 1 \Rightarrow I_D > I_C$

10.

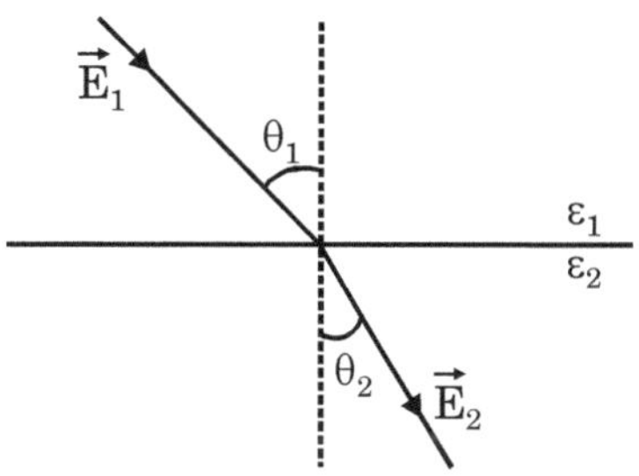

From boundary condition,

$$D_1^{\perp} = D_2^{\perp}$$

$$E_1 \varepsilon_1 \cos\theta_1 = E_2 \varepsilon_2 \cos\theta_2$$

$$\frac{\varepsilon_1}{\varepsilon_2} = \frac{E_2 \cos\theta_2}{E_1 \cos\theta_1} \quad \ldots(1)$$

$$D_1^{||} = D_2^{||}$$

$$\varepsilon E_1 \sin\theta_1 = \varepsilon E_2 \sin\theta_2$$

$$\frac{E_2}{E_1} = \frac{\sin\theta_1}{\sin\theta_2} \quad \ldots(2)$$

From (1) and (2), we get

$$\frac{\varepsilon_1}{\varepsilon_2} = \frac{\sin\theta_1 . \cos\theta_2}{\sin\theta_2 . \cos\theta_1}$$

$$\frac{\varepsilon_1}{\varepsilon_2} = \frac{\tan\theta_1}{\tan\theta_2}$$

13. Special property of a rotation matrix : Determinant of a rotation matrix is 1.

19. At $\omega t = 0$

$$\bar{\epsilon} = 4\hat{y}$$

at $\omega t = \dfrac{\pi}{2}$

$$\bar{\epsilon} = 3\sin\left(\frac{-\pi}{2}\right)\hat{x} = -3\hat{x}$$

Direction of propagation-parallel

Direction of polarization-Perpendicular

$$\Delta\phi = 90°\text{or}\frac{\pi}{2}$$

20. $\dfrac{dq}{\text{Area}} = \sigma_0 \cos\phi$

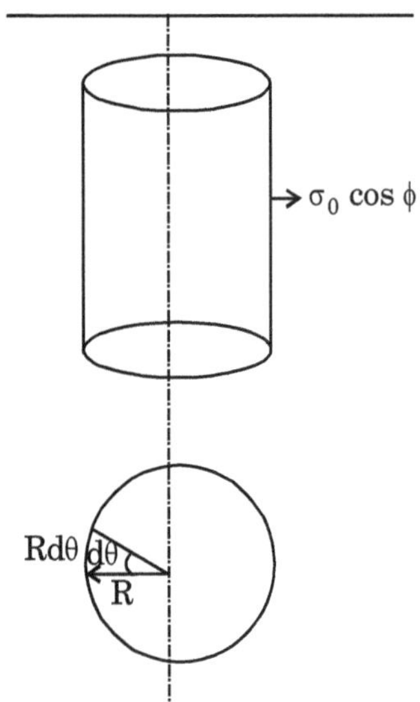

$$\frac{dq}{Rd\phi.1} = \sigma_0 \cos\phi$$

$$\frac{dq}{1} = (\sigma_0 \cos\phi)Rd\phi$$

$$d\lambda = (\sigma_0 \cos\phi)Rd\phi$$

$$d\epsilon = \frac{d\lambda}{2\pi \epsilon_0 R} = \frac{\sigma_0 \cos\gamma}{2\pi\varepsilon_0}.d\phi$$

$$\epsilon = \frac{d\lambda}{2\lambda \epsilon_0 R} = \frac{\sigma_0 \cos\phi}{2\pi \epsilon_0}.d\phi$$

$$\epsilon = \int_{0.2\pi}^{2\pi} d\epsilon.\cos\phi$$

$$\epsilon = \int_{0.2\pi}^{2\pi} \frac{\sigma_0 \cos^2\phi.d\phi}{2\pi \epsilon_0}.\frac{\sigma}{2\pi \epsilon_0}$$

$$= \int_{0.2\pi}^{2\pi} \cos^2\phi d\phi \Rightarrow \frac{\sigma_0}{2\epsilon_0}$$

21. $\phi_B = B.A$

$$e = \frac{-d\phi_B}{dt} = \frac{-d}{dt}(BA) = -A\frac{dB}{dt}$$

$$e = -A\frac{d}{dt}\left(B_0 e^{\frac{-t}{t_0}}\right) = \frac{x-1}{t_0 = 1}$$

$$e = 12.56e^{\frac{-t}{t_0}} \times 10^{-6}$$

$$I = \frac{e}{R}$$

$$I = \frac{12.51 \times 10^{-6} \times e^{\frac{-t}{t_0}}}{2}$$

$$I = 6.28 \times 10^{-6} e^{-\frac{t}{t_0}}$$

$$I = \frac{dq}{dt} \Rightarrow dv = Idt$$

$$q = \int_0^\infty Idt$$

$$= \int_0^\infty (6.28 \times 10^{-6})e^{\frac{-t}{t_0}}.dt$$

$$q = 6.28 \times 10^{-6}\left[\frac{e^{\frac{-t}{t_0}}}{\frac{-1}{t_0}}\right]_0^\infty$$

$$q = 6.28 \times 10^{-6}\,C$$

22. $\vec{E} = E_0 \cos(3y + 4z - 1.5 \times 10^9 t)\hat{x}$

$$p = \alpha \epsilon_0 E_0^2$$

$$\vec{E} = E_0 \cos(\vec{K}.\vec{r} - \omega t)x$$

$$\vec{K} = 3\hat{x} + 4\hat{z}$$

$$P = \frac{2I}{C}.\cos\theta$$

Q = Angle between z = 0 plane & wave propagation

$$\vec{K}.\hat{z} = |K| \cos\theta \Rightarrow \cos\theta = \frac{\vec{K}.\hat{z}}{|K|}$$

$$\cos\theta = \frac{(3\hat{x}+4\hat{z}).\hat{z}}{\sqrt{9+16}} = \frac{4}{5}$$

$$P = \frac{2I}{C}\times\frac{4}{5}$$

$$I = \langle s\rangle = \frac{E_0^2}{2\mu_0 C} = \frac{CE_0^2}{2\mu_0 C^2} = \frac{C \in_0 E_0^2}{2}$$

$$P = \frac{2\times 4}{5\times c}\left(\frac{C \in_0 E_0^2}{2}\right) = \frac{4}{5}\in_0 E_0^2$$

$$P = \alpha \in_0 E_0^2$$

$$P = \frac{4}{5}\in_0 E_0^2 \quad \alpha = \frac{4}{5}0.8$$

23.

$$I = \frac{1}{2}MR^2$$

$$L = I\omega = \frac{1}{2}MR^2\omega$$

$$\frac{\mu}{L} = \frac{Q}{2M} \Rightarrow \mu = \frac{Q}{2M}L = \frac{Q}{2M}\times 1MR^2\omega$$

$$\mu = \frac{QR^2\omega}{4}$$

24. $I_N = I_{N-1}\cos^2\left(\frac{\pi}{2N}\right)$

As N increases, I_N decreases.

26. $\vec{A}(s,\phi,z) = \frac{\mu_0 nI}{2}s\hat{\phi} \Rightarrow \vec{B}$

$$\vec{A}(s,\phi,z) = \frac{\mu_0 nIs}{2}(\propto \cos\phi.\hat{\phi} + \beta\sin\phi.\hat{s})$$

$$\vec{B}_1 = \vec{\nabla}\times\vec{A}\begin{vmatrix} \hat{s} & s\hat{\phi} & \hat{z} \\ \frac{\partial}{\partial s} & \frac{\partial}{\partial\phi} & \frac{\partial}{\partial z} \\ 0 & s\left(\frac{\mu_0 nIs}{2}\right) & 0 \end{vmatrix}$$

$$\vec{B}_1 = (\mu_0 nIs)\hat{z} \qquad \text{----(1)}$$

$$\vec{A}(s,\phi,z) = \frac{\mu_0 nIs}{2}\hat{\phi} + \frac{\mu_0 nIs}{2}(\propto \cos\phi.\hat{\phi} + \beta\sin\phi.\hat{s})$$

$$\vec{B}_2 = \vec{\nabla}\times\vec{A}\begin{vmatrix} \hat{s} & s\hat{\phi} & \hat{z} \\ \frac{\partial}{\partial s} & \frac{\partial}{\partial\phi} & \frac{\partial}{\partial z} \\ \frac{\mu_0 nI\beta\sin\phi}{2} & s\left(\frac{\mu_0 nIs}{2} + \frac{\mu_0 nIs}{2}\propto\cos\phi\right) & 0 \end{vmatrix}$$

$$\vec{B}_2 = \mu_0 nIs\left[1 + \alpha\cos\phi - \frac{\beta}{2}\cos\phi\right]\hat{z}$$

$$\vec{B}_1 = (\mu_0 nIs)\hat{z}$$

$$\alpha\cos\phi - \frac{\beta}{2}\cos\phi = 0$$

$$\alpha = \frac{\beta}{2}$$

27. $v = \frac{1}{4\pi\in_0}\times\frac{\vec{p}.\vec{r}}{r^2}$

$$\vec{r} = r\cos\theta\hat{z} + r\sin\theta\hat{y}$$

$$\vec{r} = r\frac{\vec{r}}{|\vec{r}|} = \cos\theta\hat{z} + \sin\theta\hat{y}$$

$$\vec{p} = \Sigma q_i \vec{r}_1$$

$$= \frac{-q}{2}d\hat{y} + \left(\frac{-q}{2}\right)d(-\hat{y}) + qd\hat{z} = qd\hat{z}$$

$$\vec{p}.\vec{r} = qd\cos\theta$$

$$v = \frac{1}{4\pi\in_0}\frac{qd\cos\theta}{r^2}$$

$$\frac{9\times10^9\times10^{-6}\times10^{-3}\times\cos 60°}{10^2} = .045v$$

28. Magnetic energy density of a straight wire

$$B \propto \frac{1}{r^2}$$

∴ Ratio of magnetic energy density

$$\frac{B_1}{B_2} = \frac{r_2^2}{r_1^2} = \frac{(2r)^2}{(r)^2} = \frac{4r^2}{r^2} = 4$$

29. Radiation pressure $(p) = \frac{XI_0}{c}$

$$= \frac{2I_0 - \frac{20}{100}I_0}{c}$$

$$= \frac{2I_0 - 0.2I_0}{c}$$

$$= \frac{I_0(2-0.2)}{c}$$

$$p = \frac{1.8I_0}{c}$$

According to question,

Radiation pressure

$$p = \frac{XI_0}{c}$$

$$\therefore \quad \frac{XI_0}{c} = \frac{1.8\,I_0}{c}$$

$$\Rightarrow \quad X = 1.8$$

30. The number of independent non-zero components of a general electromagnetic field tensor six.

31. The magnitude of electric field at the surface of the solenoid is

$$E = \frac{\mu_o nR^2}{2r}\frac{dI}{dt}$$

$$\Rightarrow \quad E = \frac{\mu_o nR^2 \alpha\omega \sin\omega t}{2R} \quad (\because \text{n} = \text{N and r} = \text{R})$$

$$\Rightarrow \quad E = \frac{1}{2}\mu_o NR\alpha\omega \sin\omega t$$

32. Suppose x a region of space that has a uniform and constant magnetic field. e.g.,

$$\vec{B} = B_0\hat{k}$$

$$\vec{B}(\vec{r}) = \vec{\nabla}\times\vec{A}(\vec{r})$$

$$= \left(\frac{\partial A\hat{i}(\vec{r})}{dx} - \frac{\partial A\hat{k}(\vec{r})}{dz}\right)$$

$$= \frac{1}{2}B_o + \frac{1}{2}B_o = B_o\hat{j}$$

$$B_o\hat{k} = B_o(r)\hat{k} = B_o x\hat{j}$$

33. Let us consider a long straight wire that has a radius a, resistance per unit length r and carries a current I.

In order to calculate the Poynting Vector, we need to find the electric and magnetic field at the surface of the wire.

Electric field, E = IR = Ir

and Magnetic field $B = \frac{\mu_0 I}{2\pi a}$

Electric field and magnetic field at the surface of the wire are perpendicular, then magnitude of the Poynting Vector

$$S = \frac{EB}{\mu_0}$$

$$= \frac{1}{\mu_0}\times Ir \times \frac{\mu_0 I}{2\pi a}$$

$$S = \frac{I^2 r}{2\pi a} \text{ (inwards)}$$

36.

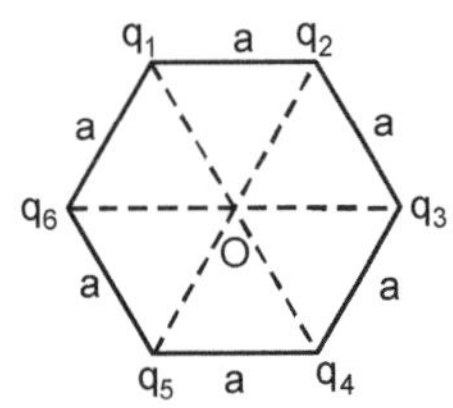

All sides of regular hexagon are equal = a

Hence charges $q_1 = q_2 = q_3 = q_4 = q_5 = q$

Electric field on the centre will be due to q_3 and field due to q_1 and q_4, q_2 and q_5 being equal and opposite cancel each other.

Hence, $E = \frac{1}{4\pi\varepsilon_0}\frac{q}{a^2}$

Electric potential at point O is $V = \frac{5q}{4\pi\varepsilon_0 a}$

37. Capacitance of parallel plate capacitor is given by

$$\begin{pmatrix} \text{dielectric constant (K)} = 10 \\ d = 10^{-6}m \\ \text{Area of square plate capacitor} = (\text{side})^2 = (1)^2 = 1 \end{pmatrix}$$

$$C = \frac{K\varepsilon_0 A}{d} = \frac{10\times 8.854\times 10^{-12}\times(1)^1}{1\times 10^{-6}}$$

$$= 88.54 \times 10^{-6}$$

As we know that $V = \frac{|Q|}{C} = \frac{1}{88.54\times 10^{-6}}$

$= 0.01129\times 10^6 = 11.29 \times 10^3 V = 11.29$ kV

38. As we know that

$$n_i \sin\theta_c = n_t \sin 90°$$

($n_i = 1.5, n_t = 1$ as refractive index of vacuum is 1)

$$\therefore \; 1.5 \sin\theta_c = 1$$

$$\sin\theta_c = \frac{1}{4.5} = 0.6666$$

$$\theta_c = \sin^{-1}(0.6666)$$

$$= 41.80.$$

39. Given

Electric field (E_0) in x free space = |V|m

$\varepsilon_0 = 8.854 \times 10^{-12}$ F/m.

Pressure exerted on the mirror (P) $= \varepsilon_0 E_0^2$

$$\therefore \; P = \varepsilon_0 E_0^2$$

$$= 8.854 \times 10^{-12} \times (1)^2$$

$$= 8.854 \times 10^{-12} \text{ Pa.}$$

40.

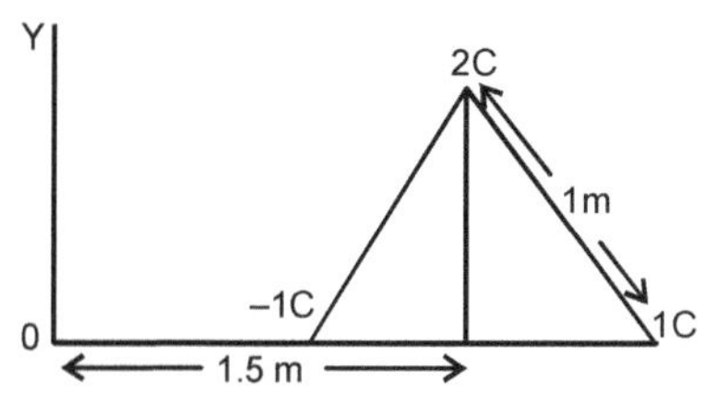

There are two triangles. The hypotenuse of both these triangles = 1 dm,

$$\text{Bottom} = \frac{1dm}{2}$$

From pythoyoms theorem are calculate the third side

$$H^2 = P^2 + B^2$$

$$id^2 = x^2 + \left(\frac{1d}{2}\right)^2$$

$$x = \sqrt{(1d)^2 + \left(\frac{1d}{2}\right)^2}$$

$$= \sqrt{1d + \frac{Id}{4}} = \frac{\sqrt{3}d}{2}$$

Electric dipole moment is given by

$$P = (2q) \times x = 2q \times \frac{\sqrt{3}d}{2}$$

$$= \sqrt{3}qd = \sqrt{3}$$

$$= 1.732 \text{ cm.}$$

41. The axis of the solenoid is along z direction

$\hat{r}$ and $\hat{\theta}$ are the usual radial and polar directions in cylinderical polar cooardinates

$\vec{B} = B_r\hat{r} + B_\theta\hat{\theta} + B_z\hat{z}$ is the magnetic field outside the solenoid.

$$B_r = 0, B_\theta = 0, B_z \neq 0.$$

42. A uniform volume charge density is placed inside a conductor

$$\rho = 10^{-2}\ \Omega\text{m}$$

Given : $\varepsilon_0 = 8.854 \times 10^{-12}$ F/m

We know that

$$T_m = \frac{\varepsilon_m}{\sigma_m}$$

$$\sigma = \frac{1}{\rho} = \frac{1}{10^{-2}}\ 10^2 \text{ sm}$$

$$\varepsilon_m = \varepsilon_{rm} e_0$$

ε_{rm} is dielectric constants = 1

$$\therefore \quad T_m = \frac{8.854 \times 10^{-12} \times 10^{-2}}{8.854 \times 10^{-14}}$$

$$= 88.54 \times 10^{-15} \text{ second}$$

$$= 88.54 \text{ femto second.}$$

43. Electric field

$$E = \frac{\sigma}{\varepsilon}\hat{z} = \frac{\sigma}{\varepsilon_0 K}\hat{z}$$

Polarization $\vec{P} = \varepsilon_0 \times e\vec{E}$

$$= \varepsilon_0(K-1)\frac{\sigma}{\varepsilon_0 K}\hat{z}$$

$$= \frac{\sigma(K-1)}{K}\hat{z}$$

$$\sigma_1 = \vec{P}.\hat{z} = \frac{\sigma(K-1)}{K}$$

44. $\vec{E}(z, t) = E_0 \cos(kz - \omega t)i$

$$\vec{B} = \frac{1}{c}\hat{z} \times \vec{E}(z,t)$$

$$= \frac{E_0}{c}\cos(kz - \omega t)\hat{j}$$

The poynting vector of the wave will be,

$$\vec{S} = \frac{1}{\mu_0}(\vec{E} \times \vec{B})$$

$$= \frac{1}{\mu_0}[E_0^2 \cos^2(kz - \omega t)\hat{k}]$$

$$= c\varepsilon_0 E_0^2 \cos^2(kz - \omega t)\hat{k}$$

45. x-y plane is the boundary free space and a magnetic material with relative permeability μ_r

In free space $B = B_x\hat{i} + B_z\hat{k}$

In material only field will change along x direction $\hat{z}$ will not change

So in medium

$$\bar{B} = \frac{\mu}{\mu_0}B_x\hat{i} + B_z\hat{k}$$

$$\bar{B} = \frac{\mu_r\mu_0}{\mu_0}B_x\hat{i} + B_z\hat{k}$$

$$\bar{B} = \mu_r B_x\hat{i} + B_z\hat{k}$$

46. For $|\phi_2>$,

$$\text{Probability} = \frac{2^2}{1^2+2^2+3^2}$$

$$= \frac{4}{1+4+9} = \frac{4}{14} = 0.28$$

47. $$e^{-1} = e^{-2\alpha x_0}$$

$$\Rightarrow \quad x_0 = \frac{1}{2\alpha}$$

48.

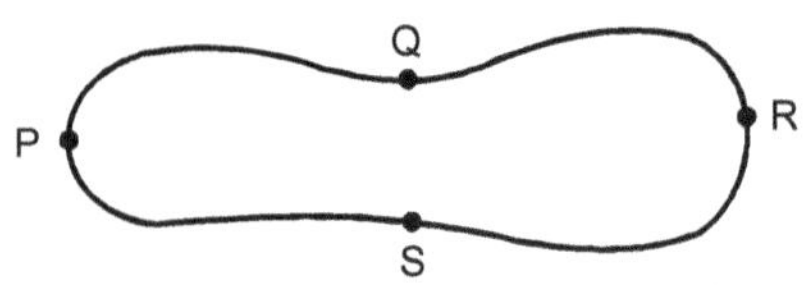

given magnetic flux through the loop PQRS P is ϕ.

if $\int_P^R A\cdot dl = \phi_1$ (along PQR),

$$\int_R^P A\cdot dl = \phi_2 = ?$$

$$\oint A\cdot dl = \phi \text{ (around close loop)}$$

And

then $\phi = \phi_1 + \phi_2$

$$\oint A\cdot dl = \int_P^R A\cdot dl + \int_R^P A\cdot dl$$

therefore, $\phi_2 = \phi - \phi_1$

(option (*b*) is correct)

50. $n = \frac{360}{\theta} - 1$

$$= \frac{360}{30} - 1 = 11$$

51. $d\sin\theta = n\lambda$

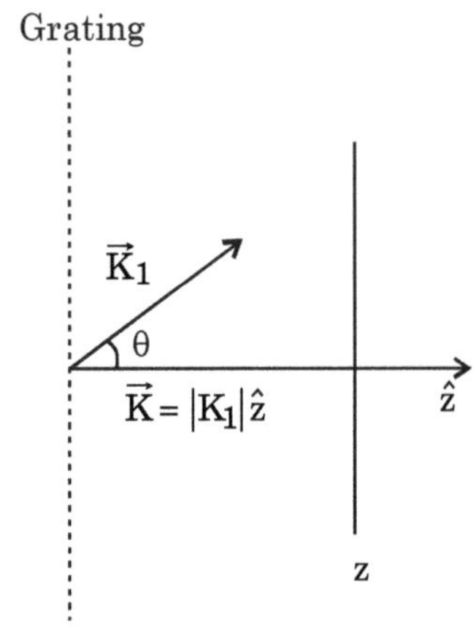

$$\Rightarrow d = \frac{\lambda}{\sin\theta}, \qquad \because n = 1$$

and $\vec{K}_1 = |\vec{K}_1|\left[\frac{1}{2}\hat{x} + \frac{\sqrt{3}}{2}\hat{z}\right]$

$$\Rightarrow \sin\theta = \frac{\vec{K}\times\vec{K}_1}{|\vec{K}_1||\vec{K}|}$$

$$= \frac{\hat{z}\times\left(\frac{1}{2}\hat{x} + \frac{\sqrt{3}}{2}\hat{z}\right)}{\sqrt{\frac{1}{4}+\frac{3}{4}}\times\sqrt{\frac{1}{4}+\frac{3}{4}}} = \frac{1}{2}$$

$$\Rightarrow \theta = 30°$$

So, $d = \frac{600}{\sin 30}\text{nm} = 1200 \text{ nm} = 1.2 \text{ µm}$

52. Incident wave: $(\hat{x} + i\hat{y})E_0e^{i\theta}$

$= [E_0\cos\theta\hat{x} - E_0\sin\theta\hat{y}]$

Left circular polarisation with phase angle

$\phi_1 = -\theta = \theta i\pi$

Emergent wave: $(\hat{x} - i\hat{y})E_0e^{i\theta}$

$= [E_0\cos\theta\hat{x} + E_0\sin\theta\hat{y}]$

Right circular polarisation with phase angle,

$\phi_1 = +\theta = \theta e^{i\theta}$.

Thus there is phase change of π and hence path difference is $\frac{\lambda}{n}$.

53. $\oint \vec{E}.d\vec{l} = -\int \frac{\partial \vec{B}}{\partial t}.d\vec{a};$

For $r < R$, $|\vec{E}|2\pi r$

$$= -\mu_0 n \frac{dI}{dt}\int_{r^1=0}^{r} 2\pi r^1 dr^1$$

$$= -\mu_0 n \frac{dI}{dt} \times \frac{2\pi r^2}{2}$$

$$\Rightarrow |\vec{E}| = \frac{-1}{2}\mu_0 n \frac{dI}{dt} r$$

For $r > R$, $|\vec{E}|2\pi r = -\mu_0 n \frac{dI}{dt}\int_{r^1=0}^{R} 2\pi r^1 dr^1$

$$= -\mu_0 n \frac{dI}{dt}\frac{2\pi R^2}{2}$$

$$\Rightarrow |\vec{E}| = -\frac{1}{2r}\mu_0 n \frac{dI}{dt} R^2$$

54. The angle between the reflected and refracted wave is 90°

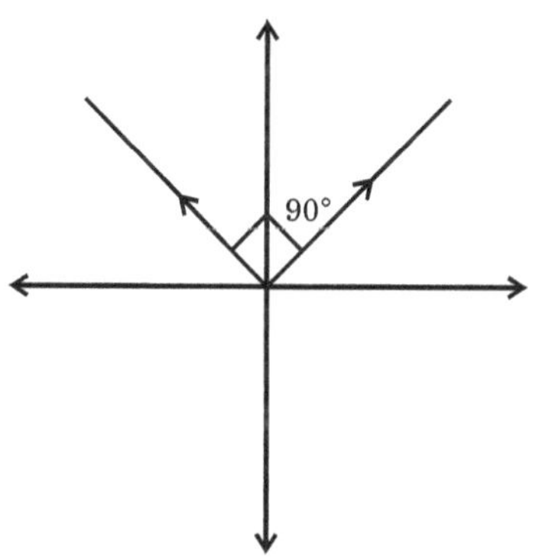

55. $E = \hat{x}\ 10\cos\left(6\pi \times 10^7 t - 0.4\pi z\right) V/m$

Now, we have $\omega = kv_p$ where v_p is phase velocity

$$\therefore\ v_p = \frac{\omega}{k} = \frac{6\pi \times 10^7}{0.4\pi}$$

$$= \frac{6}{4} \times 10^8$$

$$= 1.5 \times 10^8$$

$$v_p = 1.5 \times 10^8 \text{ m/sec}$$

56. $\vec{A} = \alpha x\hat{x} + 2y\hat{y} - 3z\hat{z}$

According to Coulomb gauge $\overline{\nabla}.\overline{A} = 0$

$\therefore\ \alpha + 2 - 3 = 0$

$\therefore\ \alpha = 1$

57.

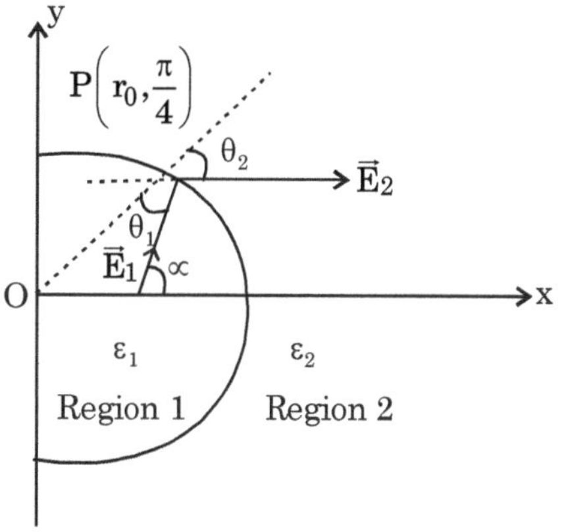

$\because \vec{E}_1 = 7\hat{e}_r - 3\hat{e}_\phi$

$$\Rightarrow E_x = \left(7\hat{e}_r - 3\hat{e}_\phi\right)\cdot\hat{x}$$

$$= 7\cos 45° + 3\sin 45° = \frac{10}{\sqrt{2}}$$

$$\Rightarrow E_y = \left(7\hat{e}_r - 3\hat{e}_\phi\right)\cdot\hat{y}$$

$$= 7\sin 45° - 3\sin 45° = \frac{4}{\sqrt{2}}$$

Thus $\vec{E}_1$ makes an angle $\propto = \tan^{-1}\left(\frac{E_y}{E_x}\right)$

$$= \tan^{-1}\left(\frac{4}{10}\right) = 21.8°$$

$$\therefore\ \frac{\tan\theta_2}{\tan\theta_1} = \frac{\varepsilon_2}{\varepsilon_1} = \frac{\tan 4.5}{\tan 23.2}$$

$$= \boxed{2.32}$$

where $\theta_1 = \propto - 45°$

$\theta_2 = 45°$

58. We know that

$$\sqrt{\frac{I_1}{I_2}} = \frac{\sqrt{\frac{I_{max}}{I_{min}}} + 1}{\sqrt{\frac{I_{max}}{I_{min}}} - 1}$$

Given $I_{max} = 9I_0$

$I_{min} = I_0$

$$\therefore \quad \sqrt{\frac{I_1}{I_2}} = \frac{\sqrt{\frac{9\,I_0}{I_0}}+1}{\sqrt{\frac{9\,I_0}{I_0}}-1} \Rightarrow \frac{\sqrt{9}+1}{\sqrt{9}-1}$$

$$\therefore \quad \sqrt{\frac{I_1}{I_2}} = \frac{4}{2} = 2$$

$$\therefore \quad I_1 = 4\,I_2$$

from option, if $I_2 = I_0$ then $I_1 = 4\,I_0$

Hence intensities of individual waves are

$4\,I_0$ and I_0

62. Potential

$$v(r) = \frac{1}{4\pi\varepsilon_0}\left[\int_{-a}^{a}\frac{\rho(x^1)}{x}dx^1 + \int_{-a}^{+a}\frac{\rho(x^1)}{x^2}x^1dx^1 + \int_{-a}^{+a}\frac{\rho(x^1)}{x^3}x^{12}dx^1 + \ldots\right]$$

First term total charge,

$$Q_T = \int \rho(x^1)dx^1$$

$$Q\int_{-x_0}^{x_0}\delta(x^1 - x_0)dx^1 - Q\int_{-x_0}^{x_0}\delta(x^1 + x_0)dx$$

$$= Q - Q = 0$$

Second term Dipole Moment,

$$P = \int x^1\rho(x^1)dx^1$$

$$= Q\int_{-x_0}^{x_0}x^1\delta(x^1 - x_0)dx^1 - Q\int_{-x_0}^{x_0}x^1\delta(x^1 + x_0)dx^1$$

$$= Qx_0 - Qx(-x_0) = 2Qx_0$$

$$v = \frac{2Qx_0}{4\pi\varepsilon_0 x^2}$$

$$\Rightarrow \vec{E} = \frac{-\partial v}{\partial x}\hat{x} = \frac{4Qx_0}{4\pi\varepsilon_0 x^3}\hat{x}$$

$$= \frac{4Qx_0}{4\pi\varepsilon_0(2x_0)^3}\hat{x} = \frac{Q}{8\pi\varepsilon_0 x_0^2}\hat{x}$$

64. $\oint \vec{A}.\vec{dl} = \int_S(\vec{\nabla}\times\vec{A})\vec{da} = \int_S \vec{B}.\vec{da} = BA\cos 60°$

$$= 0.6\times\left(\frac{1}{\sqrt{2}}\right)^2\times\frac{1}{2}$$

$$= 0.15 \text{ T.m}^2$$

65. Electric field is given by

$$\vec{E} = \dot{x}E_0\cos(\omega t - kz)$$

so, average energy density

$$uE = \frac{1}{2}\varepsilon_0\, zE^2 >$$

$$= \frac{1}{2}E_0\,E_0^2 < \cos^2(\omega t - kz) >$$

$$= \frac{1}{2}E_0\,E_0^2 \times \frac{1}{2} \qquad \left(\because \overline{\cos^{-2}}\theta = \frac{1}{2}\right)$$

$$= \frac{1}{4}\varepsilon_0\,E_0^2$$

66. $\mu_1 = \frac{3}{2}$ and $\mu_2 = 1$

The reflection coefficient

$$R = \left(\frac{\mu_1 - \mu_2}{\mu_1 + \mu_2}\right)^2 = \left(\frac{\frac{3}{2}-1}{\frac{3}{2}+1}\right)^2 = \left(\frac{\frac{1}{2}}{\frac{5}{2}}\right)^2 = \frac{1}{25}$$

so reflectivity = 0.04 or 4%

67. Given $\phi(z, t) = 0$ and $\vec{A}(z,t) = \hat{i}\,t_2$

Now $\vec{E} = -\vec{\nabla}\phi - \frac{\partial\vec{A}}{\partial t}$

$$\Rightarrow 0\ \frac{-\partial}{\partial t}(\hat{i} + z) = -\hat{i}\,z$$

and $\vec{B} = \nabla\times\vec{A} = \left(\hat{i}\frac{\partial}{\partial x} + \hat{j}\frac{\partial}{\partial y} + \hat{k}\frac{\partial}{\partial z}\right)\times(\hat{i}\,tz)$

$$= 0 - \hat{k}\frac{\partial}{\partial y}(tz) + \hat{j}\,\frac{\partial}{\partial z}(tz) = 0 - 0 + \hat{j}t$$

$$= \hat{j}t$$

68. From Maxwell's equation

$$\nabla \times E = \frac{-\delta B}{\delta t}$$

$$\vec{\nabla} \times \vec{E} = \begin{vmatrix} \hat{i} & \hat{j} & \hat{k} \\ \frac{\partial}{\partial x} & \frac{\partial}{\partial y} & \frac{\partial}{\partial 2} \\ 0 & Ey & 0 \end{vmatrix}$$

$\because$ $E(x, z) = 10j \exp[i(6x + 8z)]$

$$\vec{\nabla} \times \vec{E} = \hat{i}\left(\frac{-\partial Ey}{\partial 2}\right) + \hat{j}(0) + \hat{k}\left(\frac{\partial Ey}{\partial x}\right)$$

$$= -10i \exp[i(6x+8z)] \times 8\hat{i} + 10\hat{k} \exp [i(6x+8z)] \times 6i$$

$$= 10[6\hat{k} - 8\hat{i})i \exp[i(6x+8z)] \quad ...(i)$$

Taking option (*a*),

$$\frac{\partial B}{\partial t} = \frac{1}{C}(6\hat{k} - 8z)\exp[i(6x+8z)] \times e^{-i10ct} \times (-i)10\,C$$

$$\frac{-\partial B}{\partial t} = 10(6\hat{k} - 8\hat{i})i\exp[i(6x+8z)]e^{-10ict}$$

$$\frac{-\partial B}{\partial t} = 10(6\hat{k} - 8\hat{i})i\exp[i(6x+8z)] \quad ...(ii)$$

From equation (*i*) and (*ii*) $\vec{\nabla} \times \vec{E} = \frac{-\partial B}{\partial t}$

69. Since, there is no free charge at the interface so, normal component of $\vec{D}$ is continuous at the boundary.

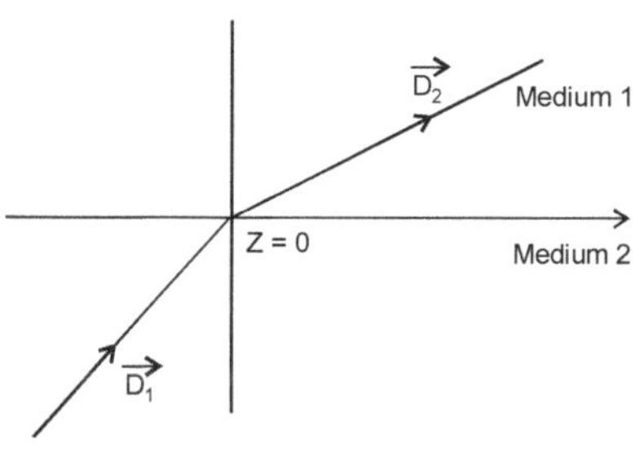

$$D_1^{\perp} = D_2^{\perp}$$

$$\varepsilon_1 E_1^{\perp} = \varepsilon_2 E_2^{\perp}$$

$$E_1^{\perp} = \frac{2 \times 5}{5} = 2$$

$$\therefore \quad \vec{E}_2 = 2\hat{i} + 3\hat{j} + 2\hat{k}$$

$$\vec{D}_2 = \varepsilon_2 \vec{E}_2 = 5\varepsilon_0(2\hat{i} - 3\hat{j} + 2\hat{k})$$

$$= \varepsilon_0(10\hat{i} - 15\hat{j} + 10\hat{k})$$

70. $E_{dipule} = \frac{1}{4\pi\varepsilon_0}, \frac{qd}{r^3}$

For $r = a, E = \frac{1}{4\pi\varepsilon_0}, \frac{qd}{a^3}$

$$\Rightarrow \quad d = 4\pi e_0 a^3 \frac{E}{q} = \frac{1}{9\times10^9} \frac{x(0.5)^2 \times 10^{-30} \times 30 \times 10^5}{1.6\times10^{-19}}$$

$$= 2.60 \times 10^{-16} m$$

71. $E = \frac{1}{4\pi\varepsilon_0} \frac{qd}{a^3}$

$$p = qd = (4\pi\varepsilon_0 a^3) \times E = \alpha.E$$

Polarizability

$$\alpha = 4\pi\varepsilon_0 a^3$$

$$\alpha = \frac{qd}{E} = \frac{1.6\times10^{-19} \times 2.60\times10^{-16}}{30\times10^5}$$

$$\cong 1.4 \times 10^{-41}\ C^2\text{–}m/N$$

72. The grounded infinite conducting plane behaves as plane mirror and form thc imagcs of charges *g* and 2*g* as

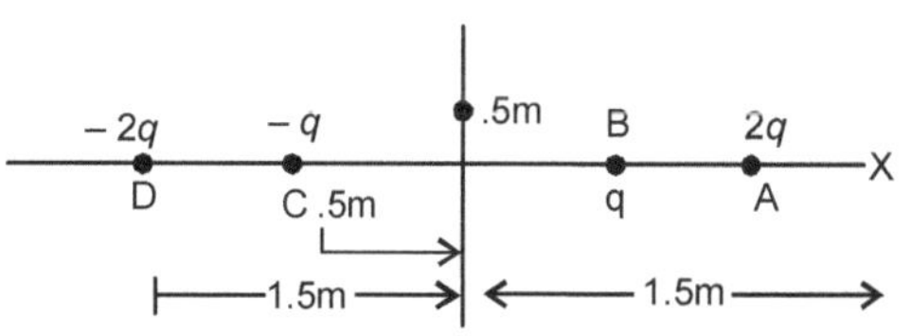

$$\vec{F} = \vec{F}_{AB} + \vec{F}_{CB} + \vec{F}_{DB}$$

$\vec{F}_{AB}$ and $\vec{F}_{DB}$ are in same direction

hence $\theta = 0°$

$$F = F_{AB} + F_{CB} + F_{DB}$$

$$F = \frac{1}{4\pi\varepsilon_0}\left[\frac{2q^2}{1^2} + \frac{q^2}{1^2} + \frac{2q^2}{2^2}\right]$$

$$F = \frac{1}{4\pi\varepsilon_0}\left[3q^2 + \frac{q^2}{2}\right] = \frac{1}{4\pi\varepsilon_0} \frac{7q^2}{2}$$

73. $\vec{k} = K\hat{T}$

Hence $By = 0$

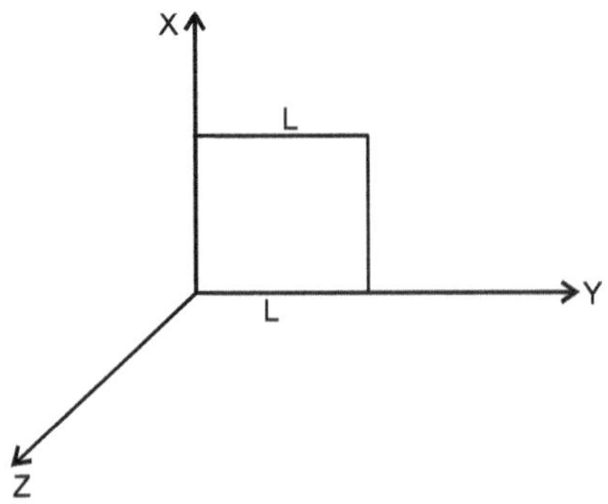

Since vertical component along (Z – axis) + × is cancelled by the corresponding filament at $-x$, hence $B_2 = 0$

Consider a rectangular Amperial loop

$z = 0$

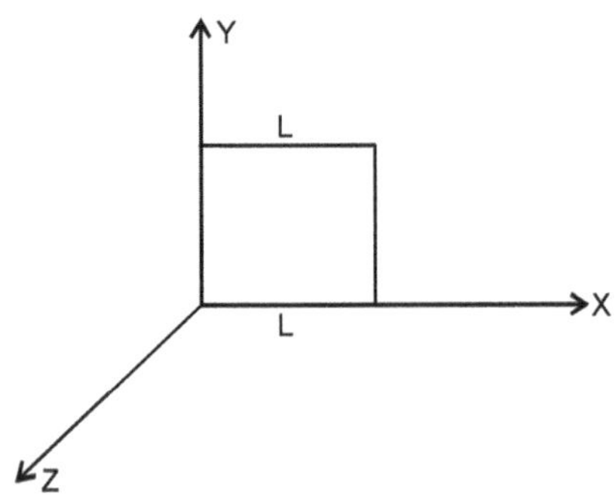

Applying Ampere's law, we get

$$\oint \vec{B}\cdot\vec{dl} = 2B^2 = \mu_0 I_{enc} = \mu_0 kL$$

$$\Rightarrow \quad B = \frac{\mu_0 k}{2}$$

In vector form,

$$B = \begin{cases} \frac{\mu_0 k}{2}\hat{i} & \text{for } z < 0 \\ -\frac{\mu_0 k}{2}\hat{i} & \text{for } z > 0 \end{cases}$$

74. Let $\vec{F}$ be the force on the dipole then torque at a position $\vec{r}$ is

$$\vec{\tau} = \vec{r}\times\vec{F}$$

But $\quad \vec{F} = \nabla(\vec{m}\cdot\vec{B})$

$$\Rightarrow \quad \vec{\tau} = \vec{r}\times\vec{\nabla}(\vec{m}\cdot\vec{B})$$

75. $\vec{B} = \vec{\nabla}\times\vec{A}$

$$\vec{B} = \begin{vmatrix} \hat{i} & \hat{j} & \hat{k} \\ \frac{\partial}{\partial x} & \frac{\partial}{\partial y} & \frac{\partial}{\partial z} \\ Ax & Ay & Az \end{vmatrix} \quad \hat{i}\left(\frac{2Ax}{\partial x} - \frac{\partial Ax}{\partial y}\right)$$

$$\Rightarrow Bx\hat{i} + By\hat{j} + Bz\hat{k}$$

$$= +\hat{j}\left(\frac{\partial Ax}{\partial z} - \frac{\partial Az}{\partial Ax}\right) + \hat{k}\left(\frac{\partial Ay}{\partial x} - \frac{\partial Ax}{\partial y}\right)$$

$\because$ B_0 is not along z-direction

$$\Rightarrow \quad B_2 = 0$$

$$\Rightarrow \frac{\partial Ay}{\partial x} - \frac{\partial Ax}{\partial y} = 0$$

Only $\vec{A} = \left(\frac{B_0 x}{2}, \frac{B_0 y}{0}, 0\right)$

Satisfies the above condition.

76. Poisson's equation in spherical coordinate is

$$\nabla^2 l = \frac{1}{r^2}\frac{\partial}{\partial r}\left(r^2\frac{\partial l}{\partial r}\right) + \frac{1}{\partial^2 \sin\theta}\frac{\partial}{\partial\theta}$$

$$\left(\sin\theta\frac{\partial l}{\partial\theta}\right) + \frac{1}{r^2\sin^2\theta}\frac{\partial^2 l}{\partial\phi^2} = \frac{-\rho}{\varepsilon_0}$$

$\because$ ρ is independent of ϕ

$$\Rightarrow \quad \frac{\partial^2 l}{\partial\phi^2} = 0$$

In this case $l(r, \theta)$ can be written as

$$l(r, \theta) = -E_0\left(r - \frac{a^3}{r^2}\right)\cos\theta \qquad \ldots (i)$$

The contribution of induced charge is

$$= \frac{E_0 a^3}{r^2}\cos\theta$$

Induced charge density using equation (i)

$$\sigma(\theta) = -\varepsilon_0\left.\frac{\partial l}{\partial r}\right|_{r=0} = \cos\varepsilon_0 E_0\left.\left(1 + \frac{2q^3}{r^3}\right)\right|_{r=a}$$

$$\sigma(\theta) \quad = 3\varepsilon_0 E_0 x\cos\theta$$

$$\sigma(\theta)|_{\theta = 30°} = 3\varepsilon_0 E_0 \times \cos 30° = \frac{3\sqrt{3}}{2}\varepsilon_0 E_0$$

77. Here $\vec{B}(x,g,z,t) = B_0 \sin\left[\frac{(x+4)k}{\sqrt{z}} + wt\right]\hat{k}$

Here $\vec{B}$ is along positive z-axis i.e., $\hat{k}$

$\hat{k}$ is along $\frac{-(i+j)}{\sqrt{z}}$ *i.e.,* $\frac{-\hat{i}}{\sqrt{z}} - \frac{\hat{j}}{\sqrt{z}}$

$\because$ $\vec{E}$ is perpendicular to $\hat{k}$

Hence, must be along $\left(\frac{\hat{i}-\hat{j}}{\sqrt{z}}\right)$

Now, $\because B_0 = \frac{E_0}{C}$

$\Rightarrow$ $E_0 = B_0 C$

Hence

$$\vec{E} = CB_0 \sin\left[(x+y)\frac{k}{\sqrt{z}} + wt\right]\left[\frac{\hat{i}-\hat{j}}{\sqrt{z}}\right]$$

78. $\because$ $\vec{S} = \frac{\vec{F}\times\vec{B}}{\mu_0}$

Hence average $(\vec{S})$ is given by

$$(\vec{S}) = -\frac{CB_0^2}{2\mu_0}\times\frac{(\hat{i}+\hat{j})}{\sqrt{2}} \quad (\because \vec{S} \text{ is along } \hat{k})$$

$$\Rightarrow \quad (\vec{S}) = -\frac{lB_0^2}{2\mu_0}\frac{(\hat{i}+\hat{j})}{\sqrt{2}}$$

79. $V(r) = \left[\frac{1}{r}\int_V \rho d\tau + \frac{1}{r^2}\int \rho\cos\theta d\tau + \ldots\right]$

I[st] term,

$$\int \rho d\tau = \int_0^a\int_0^\pi\int_0^{2\pi} \rho_0\left(a^2 - r^2\right)\cos\theta \times r^2 \sin\theta dr d\theta d\phi$$

$= 0$

IInd term,

$$\int \rho\cos\theta dr = \int_0^a\int_0^\pi\int_0^{2\pi} \rho_0\left(a^2 - r^2\right)\cos^2\theta \times r^2 \sin\theta dr d\theta d\phi \neq 0$$

$$\Rightarrow V \propto \frac{1}{r^2}$$

$$\Rightarrow E \propto \frac{1}{r^3}$$

80. The interaction energy of two magnetic dipoles separated by a displacement r is given by

$$\dot{v} = \frac{\mu_0}{4\pi}\frac{1}{r^3}\left[\vec{m}_1,\vec{m}_2 - 3(m,r)(\vec{m}_2\cdot r)\right]$$

Here $m_1 = m_2 = m$ and $r = d$

Since angle between m_1 and m_2 is 90°

$$U = \frac{\mu_0}{4\pi}\frac{1}{d^3}\left[0 - 3\times\frac{m}{\sqrt{2}}\times\frac{m}{\sqrt{2}}\right]$$

$$\Rightarrow \quad U = -\frac{3m^2\mu_0}{8\pi d^3}$$

82. $\vec{S} = \frac{1}{\mu_0}[\vec{E}\times\vec{B}]$

Here μ_0 is negative and $\vec{E}\times\vec{B}$ is the direction of $\hat{k}$. Hence $\vec{S}$ and $\hat{k}$ are antiparallel to each other.

83. Since flux is changing by two types, one due to varying magnetic field (B) (+) and other due to velocity of conductor let V_{emf} is the induced emf, then

$$V_{emf} = V_1 + V_2$$

$$V_1 = \int_S \frac{-\partial\vec{B}}{\partial t}\cdot\vec{ds}$$

and $\quad V_2 = +\oint_L \vec{V}\times\vec{B}\cdot\vec{dL}$

$$V_{emf} = -\int_S \frac{\partial\vec{B}}{dt}\cdot\vec{ds} + \oint_L \vec{V}\times\vec{B}\cdot\vec{dL}$$

84. Applying ampere's law, we get

$$\oint \vec{H}\cdot d\vec{l} = len\, C$$

Here $\vec{H} = \frac{10^4}{r}\left[\frac{1}{\alpha^2}\sin(\alpha r) - \frac{r}{\alpha}\cos(\alpha r)\right]\hat{\phi}$

$$d\vec{l} = rd\hat{\phi}$$

and ϕ goes from $0 \to 2\pi$

$$\Rightarrow 10^4\left(\frac{1}{\alpha^2}\sin(\alpha r) - \frac{r}{\alpha}\cos(\alpha r)\right)_{r=a}\int_0^{2\pi} d\phi = len\, C$$

$$\Rightarrow 10^4\left(\frac{1}{\alpha^2}\sin(\alpha a) - \frac{a}{\alpha}\cos\alpha a\right)\times 2\pi = len\, C$$

Now, $\alpha = \frac{\pi}{2a}$

$$10^4\left[\frac{1}{\left(\frac{\pi}{2a}\right)^2}\sin\frac{\pi}{2} - \frac{2a^2}{\pi}\cos\frac{\pi}{2}\right]\times 2\pi = len C$$

$$\frac{4a^2}{\pi^2}\times 2\pi\times 10^4 = len C$$

$$len C = \frac{800}{\pi}$$

85. We konw that

$$\vec{B} = \nabla \times \vec{A} = \begin{vmatrix} \hat{i} & \hat{j} & \hat{k} \\ \frac{\partial}{\partial x} & \frac{\partial}{\partial y} & \frac{\partial}{\partial z} \\ y^2 & x^2 & 0 \end{vmatrix}$$

$$= \left| \frac{\partial}{\partial x} x^2 - \frac{\partial}{dy} y^2 \right| \hat{k}$$

$$= (2x - 2y)\hat{k}$$

Now $\vec{\nabla} \times \vec{B} = \mu_0 J$ (Ampere's law)

$$\vec{J} = \frac{1}{\mu_0} \left| \vec{\nabla} \times \vec{B} \right|$$

$$\Rightarrow \frac{1}{\mu_0} \begin{vmatrix} \hat{i} & \hat{j} & \hat{k} \\ \frac{\partial}{\partial x} & \frac{\partial}{\partial y} & \frac{\partial}{\partial z} \\ 0 & 0 & 2x - 2y \end{vmatrix}$$

$$\Rightarrow \frac{1}{\mu_0}\left[-2\hat{i} - 2\hat{j}\right] = \frac{-2}{\mu_0}\left[\hat{i} + \hat{j}\right]$$

86. $\vec{E} = 20\sin\left[10^8 t + kz\right]\hat{j}$

Comparing with standerd equation

$$\vec{E} = E_0 \sin(wt - kz)i$$

$$E_0 = 20,\ w = 10^8$$

Now $\vec{B} = \frac{\hat{k} \times \vec{E}}{w}$

$$\Rightarrow \vec{B} = \mu_0 \vec{H}$$

$$\Rightarrow \vec{H} = \frac{1}{\mu_0 w}(\hat{k} + \vec{E})$$

$$\vec{H} = \frac{20}{10^8 \mu_0} \sin(10^8 t - kz)(\hat{k} \times \hat{j})$$

$$= -\frac{20}{10^8 \mu_0} \sin(10^8 t - kz)\hat{i}$$

87. $\vec{P} = (5z^2 + 7)\hat{k}$

$$\rho_b = \vec{\nabla} \cdot \vec{P} = -10z$$

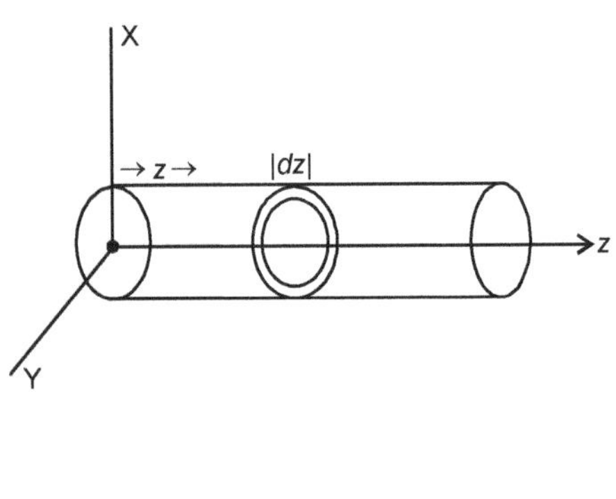

$$dv = \pi r^2 dz$$

$$dQ = \rho_b dv = -10\pi r^2 z\, dz$$

Let Q is the total bound charge

$$Q = \int_V dQ = -10\pi r^2 \int_0^L z\, dz$$

$$\Rightarrow -10\pi r^2 \times \frac{L^2}{2} = -5\pi r^2 L^2$$

89. $\vec{E} = E(t)(-\hat{k})$

$$\vec{B} = B(t)(-\phi)$$

Hence $C = \frac{1}{2\mu_0}(\vec{E} \times \vec{B}) = \frac{E(t)B(t)}{2\mu_0}(\hat{k} \times \hat{\phi})$

$$= \frac{E(t)B(H)}{2\mu_0}(\hat{r})$$

90. $\theta_i = \tan^{-1}(n)$

$$\theta_i = \tan^{-1}(1.5) = 56°$$

93. $\because$ $<\psi^*|E|\psi> = 20\ eV$

$$<\psi^*|E_0|\psi_0> = 10\ eV$$

and $<\psi_1^*|E_1|\psi> = 30\ eV$

Only option (*d*) satisfied this condition

$$\frac{1}{2}(\psi_0^*|E_0|\psi_0) + \frac{1}{2}(\psi_1^*|E_1|\psi_1)$$

$$= \frac{1}{2}10 + \frac{1}{2}30 = 20\ eV$$

98.

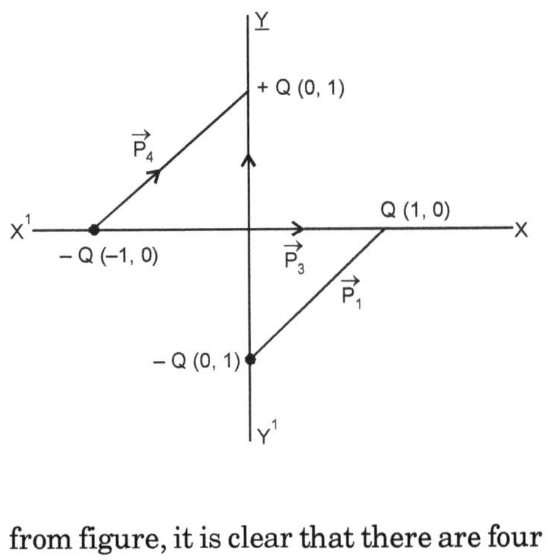

from figure, it is clear that there are four dipoles. Hence, at large distances the electrostatic potential is due to quadrupole moment.

99. The displacement current density, $J = \mu_0 \in_0 \frac{\partial E}{\partial t}$.
Let A is the plate area of capacitor. Displacement current

$$I_D = JA = \mu_0 \in_0 \frac{A\partial \vec{E}}{\partial t}$$

Now, $I_D = 0 \Rightarrow \mu_0 \in_0 \frac{A\partial \vec{E}}{\partial t} = 0$

$$\frac{\partial \vec{E}}{\partial t} = 0$$

But $E = \frac{\sigma}{2\in_0} = \frac{Q}{2\in_0 A}$

$$\frac{1}{2\in_0 A}\frac{\partial Q}{\partial t} = 0$$

$$\frac{1}{2\in_0 A} \times I = 0$$

I = 0, hence capacitor is fully discharged. Thus total Energy appears across the inductor and entirely stored in Magnetic field.

100. The situation is shown in figure below:

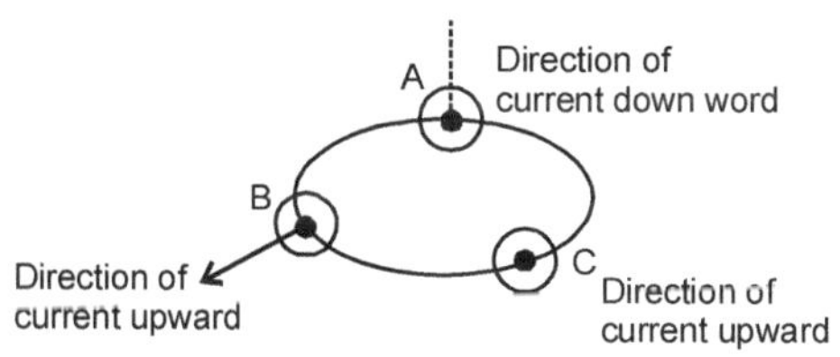

Here A, B, C are three infinite wires. The current in A is 21 (downward) and in B and C are l and l upward. Using Ampere's law on thc circumference of this circle (ABC).

$$\oint \vec{B} \cdot \vec{dl} = \mu_0 I_{enc}.$$

$$B \cdot 2\pi a = \mu_0 [-2I + I + I]$$

$$B = 0$$

101. $\vec{F} = \frac{y^2}{L}\hat{i} + Z\hat{k}$

$$\vec{\nabla} \cdot \vec{F} = \frac{\partial}{\partial x}F_x + \frac{\partial}{\partial y}F_y + \frac{\partial}{\partial Z}F_Z$$

$$\frac{\partial}{\partial X}\frac{y^2}{L} + \frac{\partial}{\partial y}0 + \frac{\partial}{\partial z}z = 1$$

Using Gauss's divergence theorem

$$\oint_A \vec{F} \cdot d\vec{a} = \oint_V \vec{\nabla} \cdot \vec{F} dV$$

Now, flux $Q = \oint_V 1 \cdot dV = L^3$

Hence, none option is correct.

102. $\vec{r} = x\hat{i} + y\hat{j}$

(1) $\vec{\nabla} \cdot \vec{\partial} = \left(\frac{\partial}{\partial x}\hat{i} + \frac{\partial}{\partial y}\hat{j} + \frac{\partial}{\partial Z}\hat{k}\right) \cdot (x\hat{i} + y\hat{j})$

$$= \frac{\partial}{\partial x}x + \frac{\partial}{\partial y}y + 0 = 2$$

(2) $|\vec{r}| = \sqrt{x^2 + y^2}$

Hence

$$\vec{\nabla}|\vec{r}| = \left(\frac{\partial}{\partial x}\hat{i} + \frac{\partial}{\partial y}\hat{j} + \frac{\partial}{\partial Z}\hat{k}\right)\left(\sqrt{x^2 + y^2}\right)$$

$$= \left[\hat{i}\frac{\partial}{\partial x}\left(\sqrt{x^2 + y^2}\right) + \frac{\partial}{\partial y}\left(\sqrt{x^2 + y^2}\right)j\right]$$

$$= \hat{i}\frac{1}{2\sqrt{x^2 + y^2}} \times 2x + \frac{1}{\sqrt[2]{x^2 + y^2}} \times 2yj$$

$$= \frac{xi + yj}{\sqrt{x^2 + y^2}} = \frac{\vec{r}}{|\vec{r}|}$$

103. $\vec{p} = 2\hat{i} + 3\hat{j} + 2\hat{k}$

Since axes are roted about y-axis hence y coordinate remains in variant.

$$\vec{p'} = x\hat{i}' + y\hat{j}' + Z\hat{k}'$$

$$x = 2\cos 60° - 2\sin 60° = 1 - \sqrt{3}$$

$$Z = 2\cos 60° + 2\sin 60° = 1 + \sqrt{3}$$

Hence $\vec{p'} = \left(1 - \sqrt{3}\right)\hat{i}' + 3\hat{j}' + \left(1 + \sqrt{3}\right)\hat{k}'$

104. For a tordial coil of N turns with relative permittivity μ_r, $\vec{B}$ inside is given by

$$\vec{B}(r) = \frac{\mu_0 \mu_r NI}{2\pi r}(\hat{\phi})$$

105. $\vec{E}(z, t) = E_0 \cos(\omega t - kz)\,\hat{i} = E_x$

$$\vec{H}(z, t) = H_0 \cos(\omega t - kz)\,\hat{j} = H_y$$

Magnitude of poynting vector $|\vec{S}|$

$$S = E_0 H_0 \cos^2(\omega t - kz)$$

Average $< s > = E_0 H_0 < \cos^2(\omega t - kz) >$

$$<s> = \frac{E_0 H_0}{2} \quad (\because E_0 = B_0 C)$$

$$E_0 = \mu_0 H_0 C$$

$$<s> = \frac{E_0^2}{2\mu_0 C} = \frac{E_0^2}{2\times 4\pi\times 10^{-7}\times 3\times 10^8}$$

$$<s> = \frac{E_0^2}{2\mu_0 \pi}$$

for $E_0 = 60$

$$<s> = \frac{(60)^2}{2\mu_0\pi} = \frac{3600}{2\mu_0\pi} = \frac{15}{\pi}$$

$$\text{Average power} = <s> \times \text{Area} = \frac{15}{\pi}\times\pi\times(2)^2$$

106. $\phi(x, t) = 3xyz - 4t$

$$\vec{A}(x,t) = (2x - wt)\hat{i} + (y - 2z)\hat{j} + (z - 2e^{iwt})\hat{k}$$

$$\text{Div } \vec{A} = \vec{\nabla}\cdot\vec{A} = 2 + 1 + 1 = 4$$

$$\frac{\partial\phi}{\partial t} = -4$$

$$\text{Now, Div } \vec{A} + \mu_0 \in_0 \frac{\partial\phi}{\partial t} = 4 - 4\mu_0 \in_0 \neq 0$$

Hence, it is not an electromagnetic field.

109. Volume charge density

$$\rho_v = \vec{\nabla}\vec{P} = -K(\vec{\nabla}\cdot\vec{r})$$

$$\rho_v = -K\times 3 \qquad (\because \vec{\nabla}\cdot\vec{r} = 3)$$

$$\rho_v = -3K$$

Surface charge density

$$= \frac{\left((-3K)\times\frac{4}{3}\pi R^3\right)}{4\pi R^2} = -KR$$

$$= |KR| = K|\vec{r}|$$

111. $E^2 - c^2B^2$ remain invariant under L.T.

112. Due to the change inside a sphere of radius r only.

114. Angular momentum

$$\bar{L} = m(\bar{r}\times\bar{p}) = m(\bar{r}\times\bar{r})$$

$$\bar{r}(t) = \left.\begin{matrix} 5\hat{i} - 2t^2\hat{j} \\ \bar{r}(t) = -4t\hat{j}\end{matrix}\right\} \quad m = 2 \text{ kg.}$$

$$\bar{L} = -80\hat{k}$$

115.

$$I_{XY} = \sum_I m_i n_i y_i$$

$$= 2m.\ 1.1 + m.(-1).\ 1 + 2m.(-1).(-1) + m.(1).(-1)$$

$$= 4m - 2m$$

$$= 2m$$

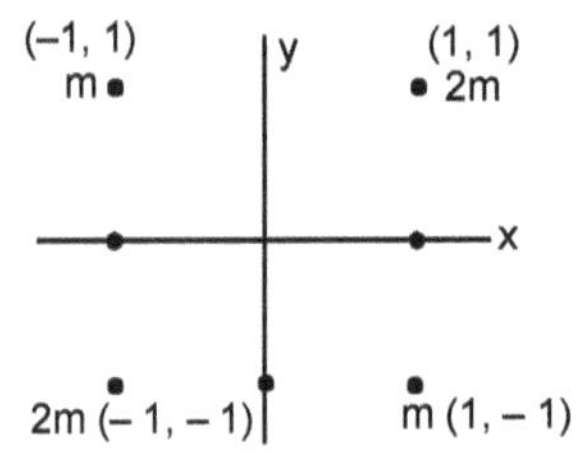

116.

$$\bar{B} = \frac{\mu_0 J_0 d^2}{3}\hat{\phi}$$

Current density,

$$\bar{J} = J_0 r\hat{k}$$

Current within a distance d,

$$I = \int_{r=0}^{d}\int_{\phi=0}^{2\pi} \bar{J}.\overline{ds}$$

$$= \int_0^d\int_0^{2\pi} (J_0\, r.r\, d\phi\, dr)\left(\widehat{k.k.}\right)$$

$$[\because \overline{ds} = \hat{k}\, r d\phi\, dr]$$

$$= 2\pi J_0 \frac{d^3}{3}$$

From Ampere's Law, $\oint_c \bar{B}.d\bar{l} = \mu_0 I$

Here loop c is a circle of radius r, B is the magnetic field at r.

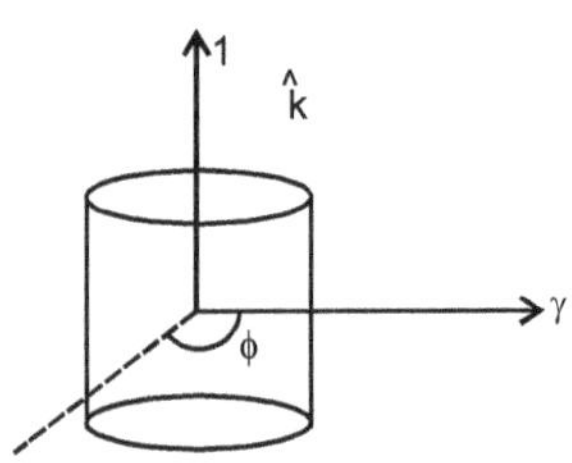

$$\therefore \quad 2\pi r.\ B = \mu_0 2\pi J_0 \frac{d^3}{3}$$

$$\Rightarrow \quad B_{r=d} = \frac{\mu_0 J_0 d^2}{3}$$

117. $\bar{A}(x,y,z) = -\ y\hat{i} + 2x\hat{j}$

$\therefore$ Magnetic field,

$$\bar{B} = \bar{\nabla}\times\bar{A} = \begin{vmatrix} \hat{i} & \hat{j} & \hat{k} \\ \partial_x & \partial_y & \partial_z \\ -y & 2x & 0 \end{vmatrix} = 3\hat{k}$$

118. From boundary conditions,

$D_{1n} = D_{2n}$ (normal component of disp. vector)

$E_{1t} = E_{2t}$ (tangential component of electric field)

$$\left.\begin{aligned} D_1\cos\theta_1 &= D_2\cos\theta_2 \\ \therefore\ E_1\sin\theta_1 &= E_2\sin\theta_2 \end{aligned}\right\}|D| = \varepsilon|E|$$

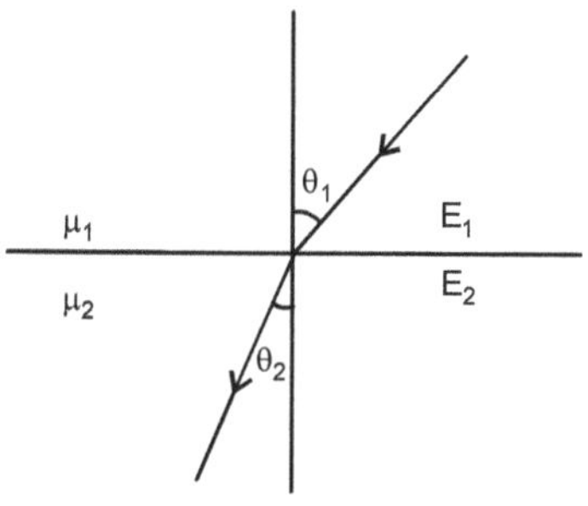

$\Rightarrow \quad \varepsilon_1 \cot\theta_1 = e_2 \cot\theta_2,$

$$\frac{\varepsilon_1\cot\theta_1}{\varepsilon_2\cot\theta_2} = 1$$

$$\Rightarrow \quad \frac{\varepsilon_1}{\varepsilon_2} = \frac{\tan\theta_1}{\tan 0_2}$$

119. For time independent charge density

$$\frac{\partial\bar{B}}{\partial t} = 0,\ \frac{\partial\bar{D}}{\partial t} = 0.$$

Thus Maxwell's equations are

$$\bar{\nabla}.\bar{E} = \frac{\rho}{\varepsilon_0}$$

$$\bar{\nabla}.\bar{B} = 0$$

$$\bar{\nabla}\times\bar{E} = 0$$

$$\bar{\nabla}\times\bar{B} = \mu_0\bar{J}$$

120. Plane polarised and circularly polarized, respectively

121. $\vec{\nabla}.\vec{P} = \vec{\nabla}\cdot(kr^2\hat{r}) = \vec{\nabla}\cdot(kr\vec{r}) = k\left\{3r + \vec{r}\cdot\frac{\vec{r}}{r}\right\} = 4kr$

$\rho = -\vec{\nabla}\cdot\vec{P} = -4kr.$

123. For conservation,

$$h\upsilon = 2ymc^2$$

But here $h\upsilon < 2ym^2$

So, energy is not conserved.

124. From ampere circuital law

$$\phi B.dl = \mu_o i$$

where, i = Net current passing through the loop

Now $i = I + I - 2I = 0$

$\therefore \quad B.2\pi r = 0$

$\Rightarrow \quad B = 0$

125.
$$E = \frac{1}{4\pi\varepsilon_0}\frac{qr}{R^3}$$

$$= \frac{q}{\frac{4}{3}\pi R^3} \times \frac{r}{3\varepsilon_0}$$

$$= \frac{\rho\times r}{3\varepsilon_0}$$

$$= \frac{r\rho}{3\varepsilon_0}$$

where, ρ = charge density = $\frac{\text{charge}}{\text{volume}}$

128. Since charge moving with constant velocity radiates zero power.

129.
$$E = -\nabla\phi = -\phi_o 2\left[x\hat{i} + y\hat{i} + z\hat{x}\right]$$

$\therefore$
$$= \epsilon_0\nabla.E$$

$$= -2\ \epsilon_0\ \phi_0\nabla.(x\hat{i} + y\hat{i} + z\hat{x})$$

$$= -6\phi_0\ \epsilon_0$$

131.
$$n = \frac{V_1}{V_2} = 1.5$$

$$\therefore \quad E_{oi} = \left(\frac{V_1 - V_2}{V_2 + V_1}\right)E_{oi}$$

$$= -\frac{2}{5}\text{V/m}$$

$$E_{ot} = \left(\frac{2V_2}{V_1 + V_2}\right)E = \frac{8}{5}\text{V/m}.$$

132. We know, $\nabla . A = \text{div } A = \dfrac{-\mu_0 \theta}{4\pi r^2}$

Field equation in Lorentz gauge is

$$\text{div } A + \mu \varepsilon \frac{\partial \phi}{\partial t} = 0$$

$$\Rightarrow \quad \frac{\partial \phi}{\partial t} = -\frac{1}{\mu \varepsilon_0} \text{div} A$$

$$= -\frac{1}{\mu \varepsilon_0} . \frac{-\mu_0 \rho \varphi}{4\pi r^2}$$

$$= \frac{Q}{4\pi \varepsilon_0 r^2}$$

$$\therefore \quad \phi = \frac{Qt}{4\pi \varepsilon_0 r^2}$$

134. $\oint B . dt = \mu_0 I$

$$B \times 2\pi R = \mu_0 \frac{\sigma dA}{dt}$$

$$\left[\because \quad I = \frac{d\phi}{dt} = \frac{d(\sigma A)}{dt} = \frac{\sigma dA}{dt} \right]$$

$\therefore$ Surface current, $J_s = M = \dfrac{B}{\mu_0} = \dfrac{\sigma}{2\pi R} \dfrac{dA}{dt}$

Now, $A = \pi R^2$

$$\therefore \quad \frac{dA}{dt} = 2\pi R \frac{dR}{dt} = 2\pi R 2\omega$$

$$\therefore \quad J_s = M = \frac{\sigma}{2\pi R} \times 2\pi R^2 \omega = \sigma R^2 \omega$$

135. We know, $A = \dfrac{\mu_0}{4\pi} \displaystyle\int \dfrac{J_0 . dA}{r}$

where, $A = \pi r^2$,

$\Rightarrow \quad dA = 2\pi r dr$

$$\therefore \quad A = \frac{\mu_0}{4\pi} \sigma R \omega \int \frac{2\pi r dr}{r}$$

$$= \frac{\mu_0}{2} \sigma R \omega \int dr$$

$$= \frac{\mu_0}{2} \sigma R \omega r$$

136. The potential inside and on it is zeero

By Poisson's equation, $\phi = \left(\dfrac{q}{r} - \dfrac{q'}{r'} \right) \dfrac{1}{4\pi\varepsilon}$

Here $-\dfrac{q}{4\pi\varepsilon_0 r}$ is potential due to image charge

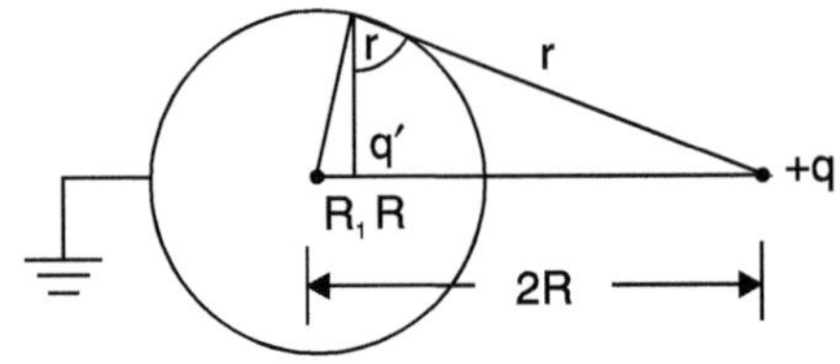

$$\Rightarrow \quad \frac{q}{4\pi\varepsilon_0 r} = \frac{-q'}{4\pi\varepsilon_0 r'}$$

$$\Rightarrow \quad \frac{q}{r} = \frac{-q'}{r'}$$

$$= \frac{R^2 + (2R)^2 - 4R^2 \cos\theta}{R^2 + R_1^2 - 2R_1 R \cos\theta}$$

So condition

$$R^2 = 2RR_1$$

$$\Rightarrow \quad R_1 = \frac{R}{2}$$

and $\quad q' = -q \sqrt{\dfrac{R/2}{2R}} = -q \sqrt{\dfrac{1}{4}} = -\dfrac{q}{2}.$

137. Here, $\vec{E} = \hat{i} E_0 \cos(kz - \omega t) - \hat{j} E_0 \cos(kz - \omega t)$

The magnitude of amplitude and phase both are same but direction of amplitude is different, so electric field vectors $\vec{E}$ is linearly polarized at $-45°$ to x-axis.

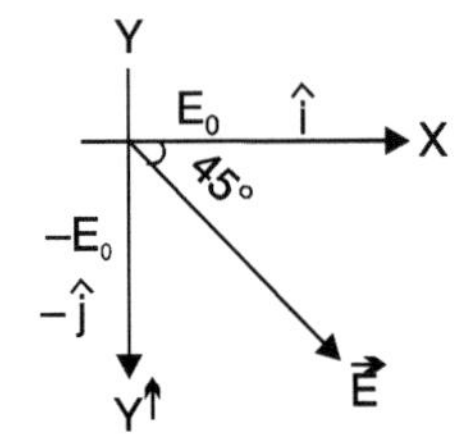

138. Coriolis force $F = -2m(\vec{v} \times \vec{\omega})$

$$F = -2m u \omega \hat{\theta} \qquad [\because \quad v = u.]$$

139. $q_1 \to Q, \quad q_2 = 2Q, \quad r_1 \to r$ and $r_2 \to \dfrac{r}{2}$

$$\because \quad \frac{E_2}{E_1} = \frac{Q_2}{Q_1} . \frac{r_1^2}{r_2^2}$$

$$\Rightarrow \quad E_2 = E_1 \left(\frac{2Q}{Q} \times \frac{r^2}{r^2/4} \right) = \mathbf{8E_1}.$$

140. $$V_0 = -E_0\left(1-\frac{R^3}{r^3}\right) r\cos\theta$$

Since $E = \frac{\sigma}{\varepsilon_0}$, therefore surface charge density,

$\sigma = \varepsilon_0 E$

Since $E = -\nabla V$

$$\Rightarrow \quad \sigma = -\varepsilon_0 \left.\frac{\partial V}{\partial r}\right|_{r=R}$$

$$= -\varepsilon_0\left(-E_0\cos\theta - \frac{2R^3}{r^3}E_0\cos\theta\right)_{r=R}$$

$$= \varepsilon_0(E_0\cos\theta + 2E_0\cos\theta)$$

$$= \mathbf{3\varepsilon_0 E_0 \cos\theta.}$$

141. $$F = \int_{60^\circ}^{120^\circ} dF\sin\theta$$

We know, $\vec{F} = I(\vec{l}\times\vec{B})$

or $\quad d\vec{F} = IB\,dl\sin\theta$

$$\therefore \quad F = \int_{60^\circ}^{120^\circ} IB\sin\theta\, dl$$

where, $dl = Rd\theta$

$$\therefore \quad F = \int_{60^\circ}^{120^\circ} IBR\sin\theta\, d\theta$$

$$= IBR[-\cos\theta]_{60^\circ}^{120^\circ}$$

$$= IBR\left[-\left(-\frac{1}{2}\right)+\left(\frac{1}{2}\right)\right] = IBR$$

Hence force acting on the arc is IBR $\hat{k}$.

142. $$R + T = 1$$

By option (*a*) $R = \left(\frac{n-1}{n+1}\right)^2$

and $\quad T = \frac{4n}{(n+1)^2}$

$$R + T = \left(\frac{n-1}{n+1}\right)^2 + \frac{4n}{(n+1)^2}$$

$$= \frac{n^2-2n+1+4n}{(n+1)^2}$$

$$= \frac{(n^2+2n+1)}{(n+1)^2} = \frac{(n+1)^2}{(n+1)^2} = \mathbf{1.}$$

143. $E = E_\circ \exp[i(xk\cos\alpha + yk\sin\alpha - \omega t)]$

$\because \vec{k} = xk\cos\alpha + yk\cos\alpha$

wave vectors of the electromagnetic wave

$$\therefore \vec{r}\cdot\vec{k} = (x\hat{x} + y\hat{y})\cdot(xk\cos\alpha + yk\cos\alpha)$$

$$= \hat{x}\,k\cos\alpha + \hat{y}k\sin\alpha.$$

144. $$\omega^2 = \omega_0^2 + c^2k^2$$

$$v_p = \frac{\omega}{k}$$

$$\Rightarrow \quad v_p^2 = \frac{\omega^2}{k^2} = \left(\frac{\omega_0^2 + c^2k^2}{k^2}\right)$$

$$\Rightarrow \quad v_p = \frac{\sqrt{\omega_0^2 + c^2k^2}}{k}$$

$$v_g = \frac{d\omega}{dk} = \frac{d}{dk}\left(\sqrt{\omega_0^2 + c^2k^2}\right)$$

$$= \frac{1}{2\sqrt{\omega_0^2 + c^2k^2}} \times 2c^2k$$

$$= \frac{c^2k}{\sqrt{\omega_0^2 + c^2k^2}}$$

$$\Rightarrow \quad v_g = \frac{c^2}{v_p}$$

$$\Rightarrow \quad v_pv_g = c^2.$$

145. $$2d = n\lambda \quad ...(i)$$

If shift by x, then n change to $(n + N)$

$$2(d+x) = (n+N)\lambda \quad ...(ii)$$

Substracting equation (*i*) from equation (*ii*), we get

$$2x = N\lambda$$

$$\Rightarrow \quad \lambda = \frac{2x}{N} = \frac{2\times 25.3\ \mu m}{92}$$

$$= 0.55\ \mu m. = 550\ nm.$$

146. Since $\psi(x) = u(x)\exp(ikx)$

$$\Rightarrow \quad \psi^*(x) = u(x)\exp(-ikx)$$

$$J = \frac{i\hbar}{2m}[\psi\nabla\psi^* - \psi^*\nabla\psi]$$

$$= -\frac{i\hbar}{2m}[-iku^2 - iku^2]$$

$$= -\frac{i^2\hbar k u^2}{2m}(2) = u^2(x)\frac{\hbar k}{m}$$

$$J = u^2(x)\,V.$$

147. $$\in = \sqrt{\left\{1+\frac{2EJ^2}{mk^2}\right\}}$$

The closest distance between the comet and sun is b

$$r_{min} = b = \frac{J^2}{mk(1+\in)}$$

For parabola path, $\in = 1$

$$\therefore \quad b = \frac{J^2}{2mk}$$

$$\Rightarrow \quad J^2 = 2bmk \quad \Rightarrow J = \sqrt{2bmk}$$

But $$F(r) = +\frac{GmM}{r^2} \quad [\because K = GmM]$$

$$\therefore \quad J = \sqrt{2bm(GmM)}$$

$$= m\sqrt{2GMb}\,.$$

148. F_{max} where r_{min}, so than ma_{max} when r_{min} and acceleration is max when r_{min}.

149. Magnitude of the electric field is attenuated as the wave propagates.

150. $$\vec{\nabla}\times\vec{E} = -\frac{\partial\vec{B}}{\partial t}$$

$$[\because E = \hat{x}E_0 \exp[i(\vec{k}\cdot\vec{r} - \omega t)]$$

Since $k = \hat{z}(k\cos\phi + ik\sin\phi)$

$$\Rightarrow E = \hat{x}E_0\exp[iz(k\cos\phi + ik\sin\phi) - i\omega t]$$

$$\nabla\times E = \begin{vmatrix} \hat{x} & \hat{y} & \hat{z} \\ \frac{\partial}{\partial x} & \frac{\partial}{\partial y} & \frac{\partial}{\partial z} \\ E_0\exp[iz(k\cos\phi + ik\sin\phi) - i\omega t] & 0 & 0 \end{vmatrix}$$

$$= \hat{y}\left[E_0\frac{\partial}{\partial z}\{\exp[iz(k\cos\phi + ik\sin\phi) - i\omega t]\}\right]$$

$$= \hat{y}\ [iE_0(k\cos\phi + ik\sin\phi)$$
$$\{\exp[iz(k\cos\phi + ik\sin\phi) - i\omega t]\}]$$

$$= \hat{y}\ [iE_0(k\cos\phi + ik\sin\phi)$$
$$\{\exp[izk\cos\phi - zk\sin\phi - i\omega t]\}]$$

$$\Rightarrow B = -\int(\nabla\times E)\,dt$$

$$= -\int\hat{y}[iE_0k(\cos\phi + i\sin\phi)$$
$$\exp(iz(izk\cos\phi - z\,k\sin\phi)$$
$$\exp^{-i\omega t}\ dt$$

$$= \hat{y}\frac{E_0}{\omega}k(\cos\phi + i\sin\phi)$$
$$\exp(izk\cos\phi - zk\sin\phi - i\omega t)$$

$$= \hat{y}\frac{E_0}{\omega}k\{\exp^{i\phi}\exp^{(izk\cos\phi - i\omega t)}\}\exp^{(-zk\sin\phi)}$$

$$B = \hat{y}\frac{k}{\omega}E_0\exp^{(-zk\sin\phi)}\exp^{\{i(zk\cos\phi - \omega t + \phi)\}}.$$

151. Charge (+Q) gets distributed uniformly over the surface of the sphere. We know that electrostatic potential inside a solid, conducting sphere is constant.

155. $$K.E. = \text{rest mass energy}$$

$$\therefore \quad mc^2 = \text{Total energy}$$
$$= KE + \text{rest mass energy}$$

$$\Rightarrow \quad mc^2 = 2\,m_0c^2$$

$$\Rightarrow \quad \frac{m_0}{\sqrt{1-\frac{v^2}{c^2}}} = 2m_0$$

$$\frac{v^2}{c^2} = 1 - \frac{1}{4}$$

$$\therefore \quad v^2 = \frac{3}{4}c^2$$

160. There is no mixed state in Type 1 superconductor.

162. Photons are bosons (integral spin)

163. $BE = [92\times1.0078 + (238-92)\times1.0087 - 238.0508)\times931.64$

165. Speed is largest at T.

166. A = 0 at Z = 0 and Z = L

$$\therefore \quad k = \frac{2\pi}{L},\ \alpha = \frac{\pi}{2}$$

167 $B = \text{curl}\,A = A_0\cos(kz)$

168. Current density, $\tau = 0$ for $r > R$

172. The emergent light will be right circularly polarised if the path difference of $\frac{\lambda}{2}$ be introduced.

Then $(n_e - n_0)\times d = \frac{\lambda}{2}$

$$\Rightarrow \quad d = \frac{600\times10^{-9}}{2\times0.01} = \mathbf{30\ \mu m}$$

173. If the amplitues of interferring beam be a_1 and a_2.

Then $I_{max} = (a_1 + a_2)^2 = 25\, I_0$

and $I_{min} = (a_1 - a_2)^2 = 9\, I_0$

$\Rightarrow \quad a_1 + a_2 = 5\sqrt{I_0}$

$a_1 + a_2 = 3\sqrt{I_0}$

$\therefore \quad a_1 = 4\sqrt{I_0}$ and $a_2 = \sqrt{I_0}$

$I_1 = 16\, I_0$ and $I_2 = I_0$

174. A vector $\bar{A}$ is called solenoidal if

$$\nabla \cdot \bar{A} = 0 \qquad ...(i)$$

Now $\nabla \cdot \bar{A} = 5 + 3 - a$

Putting value of $\nabla \cdot \bar{A}$ in equation (i), we get

$$a = 8$$

176. Field due to one side of loop at

$$O = \frac{\mu_0 I}{4\pi\left(\frac{a}{2}\right)}(2\sin 45°)$$

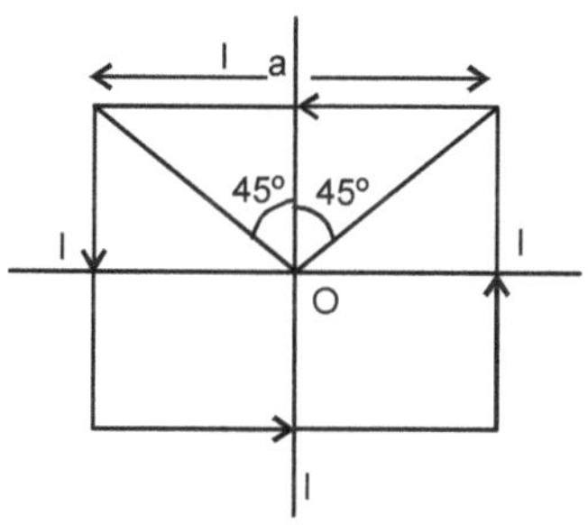

Field at O due to all four sides is along unit vector $\hat{e}_z$

$$\therefore \quad \text{Total field} = 4.\frac{\mu_0 I}{4\pi\left(\frac{a}{2}\right)}(2\sin 45°)$$

$$= \frac{2\sqrt{2}\mu_0 I}{\pi a}$$

178. $C = \dfrac{2\pi\varepsilon_0\varepsilon_r}{\ln\frac{b}{a}}$

$$= \frac{2\pi\varepsilon_0 \times 3.5}{ln\, 2} = \frac{2\pi \times 8.85 \times 10^{-12} \times 3.5}{\ln 2}$$

$= 28.05$ nF

180. Force on positive and negative charges of dipole balance each other.

181. Putting $\bar{E}(r, t) = E_0 \exp\{i(\bar{k}\,.\,\bar{r}) - \omega t\}$

into the given expression, we get

$$c^2 k^2 = \omega^2 + \omega'^2$$

$$\Rightarrow \quad \omega^2 = c^2 k^2 - \omega'^2$$

182. Put value of u in the given equation.

183. At an dielectric interface normal component of $\bar{D}$ vector is continuous and tangential component of $\bar{E}$ vector is continous

184. At mid points the field due to plates are oppositely directed.

186. We have, $\bar{A}.\bar{A} = |\bar{A}|^2$

$$\Rightarrow \quad 2\bar{A}.\frac{d\bar{A}}{dt} = 0$$

$$\Rightarrow \quad \frac{d\bar{A}}{dt}.\bar{A} = 0$$

187. Due to decrease in L flux enclosed by coil decreases. By Lenz's law, current must increase. Since the source is D.C., steady state current must be I_0.

188. Using method of images, we get following equivalent system of charges

$$-q + q - q$$

190. Waves must have same plane of vibration.

4 CHAPTER

Quantum Physics and Applications

2021

1. A matter wave is represented by the wave function
$\Psi(x, y, z, t) = Ae^{i(4x+3y+5z-10\pi t)}$
where A is a constant. The unit vector representing the direction of propagation of this matter wave is

(a) $\frac{4}{5\sqrt{2}}\hat{x} + \frac{3}{5\sqrt{2}}\hat{y} + \frac{1}{\sqrt{2}}\hat{z}$

(b) $\frac{3}{5\sqrt{2}}\hat{x} + \frac{4}{5\sqrt{2}}\hat{y} + \frac{1}{5\sqrt{2}}\hat{z}$

(c) $\frac{1}{5\sqrt{2}}\hat{x} + \frac{3}{5\sqrt{2}}\hat{y} + \frac{1}{\sqrt{2}}\hat{z}$

(d) $\frac{1}{5\sqrt{2}}\hat{x} + \frac{4}{5\sqrt{2}}\hat{y} + \frac{3}{5\sqrt{2}}\hat{z}$

2. Three non-interacting bosonic particles of mass m each, are in a onedimensional infinite potential well of width a. The energy of the third excited state of the system is $x \times \frac{h^2\pi^2}{ma^2}$. The value of x (in integer) is ______.

3. If $\vec{a}$ and $\vec{b}$ are constant vectors, $\vec{r}$ and $\vec{p}$ are generalized positions and conjugate momenta, respectively, then for the transformation $Q = \vec{a}\cdot\vec{p}$ and $P = \vec{b}\cdot\vec{r}$ to be canonical, the value of $\vec{a}\cdot\vec{b}$ (in integer) is ________.

4. Consider a spin $S = h/2$ particle in the state $|\phi\rangle = \frac{1}{3}\begin{bmatrix} 2+i \\ 2 \end{bmatrix}$. The probability that a measurement finds the state with $S_x = +h/2$ is

(a) 5/18 (b) 11/18
(c) 15/18 (d) 17/18

5. Consider a particle in a one-dimensional infinite potential well with its walls at $x = 0$ and $x = L$. The system is perturbed as shown in the figure.

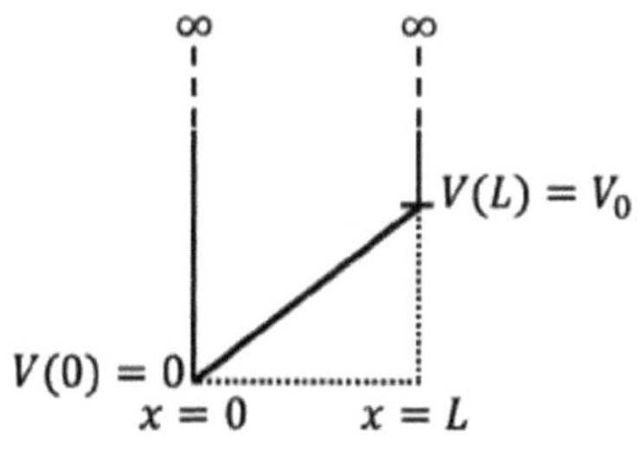

The first order correction to the energy eigenvalue is

(a) $\frac{V_0}{4}$ (b) $\frac{V_0}{3}$

(c) $\frac{V_0}{2}$ (d) $\frac{V_0}{5}$

6. Consider a state described by $\psi(x, t) = \psi_2(x, t) + \psi_4(x, t)$, where $\psi_2(x, t)$ and $\psi_4(x, t)$ are respectively the second and fourth normalized harmonic oscillator wave functions and ω is the angular frequency of the harmonic oscillator. The wave function $\psi(x, t = 0)$ will be orthogonal to $\psi(x, t)$ at time t equal to

(a) $\frac{\pi}{2\omega}$ (b) $\frac{\pi}{\omega}$

(c) $\frac{\pi}{4\omega}$ (d) $\frac{\pi}{6\omega}$

7. Consider the potential $U(r)$ defined as

$$U(r) = -U_0\frac{e^{-\alpha r}}{r}$$

where α and U_0 are real constants of appropriate dimensions. According to the first Born approximation, the elastic scattering amplitude calculated with $U(r)$ for a (wave-vector) momentum transfer q and $\alpha \to 0$, is proportional to

(*Useful integral:* $\int_0^\infty \sin(qr)e^{-\alpha r}dr = \frac{q}{\alpha^2 + q^2}$)

(a) q^{-2} (b) q^{-1}
(c) q (d) q^2

8. P and Q are two Hermitian matrices and there exists a matrix R, which diagonalizes both of them, such that $RPR^{-1} = S_1$ and $RQR^{-1} = S_2$, where S_1 and S_2 are diagonal matrices. The correct statement(s) is(are)

(a) All the elements of both matrices S_1 and S_2 are real.
(b) The matrix PQ can have complex eigenvalues.
(c) The matrix QP can have complex eigenvalues.
(d) The matrices P and Q commute.

9. For a two-nucleon system in spin singlet state, the spin is represented through the Pauli matrices σ_1, σ_2 for particles 1 and 2, respectively.
The value of $(\sigma_1 . \sigma_2)$ (in integer) is ______.

10. The normalized radial wave function of the second excited state of hydrogen atom is

$$R(r) = \frac{1}{\sqrt{24}} a^{-3/2} \frac{r}{a} e^{-r/2a}$$

where a is the Bohr radius and r is the distance from the center of the atom. The distance at which the electron is most likely to be found is $y \times a$. The value of y (in integer) is _____.

2020

11. A real, invertible 3×3 matrix M has eigenvalues λ_i, (i = 1, 2, 3) and the corresponding eigenvectors are $|e_i\rangle, (i = 1, 2, 3)$ respectively. Which one of the following is correct ?

(*a*) $M|e_i\rangle = \frac{1}{\lambda_i}|e_i\rangle$, for i = 1, 2, 3

(*b*) $M^{-1}\langle e_i| = \frac{1}{\lambda_i}\langle e_i|$, for i = 1, 2, 3

(*c*) $M^{-1}|e_i\rangle = \lambda_i |e_i\rangle$, for i = 1, 2, 3

(*d*) The eigenvalues of M and M^{-1} are not related.

12. A quantum particle is subjected to the potential

$$V(x) = \begin{cases} \infty, & x \le -\frac{a}{2} \\ 0, & -\frac{a}{2} < x < \frac{a}{2} \\ \infty, & x \ge \frac{a}{2}. \end{cases}$$

The ground state wave function of the particle is proportional to

(*a*) $\sin\left(\frac{\pi x}{2a}\right)$ (*b*) $\sin\left(\frac{\pi x}{a}\right)$

(*c*) $\cos\left(\frac{\pi x}{2a}\right)$ (*d*) $\cos\left(\frac{\pi x}{a}\right)$

13. Let $\hat{a}$ and $\hat{a}^\dagger$, respectively denote the lowering and raising operators of a one-dimensional simple harmonic oscillator.

Let $|n\rangle$ be the energy eigenstate of the simple harmonic oscillator. Given that $|n\rangle$ is also an eigenstate of $\hat{a}^\dagger\hat{a}^\dagger\hat{a}\hat{a}$, the corresponding eigenvalue is

(*a*) n (n – 1) (*b*) n (n + 1)

(*c*) $(n + 1)^2$ (*d*) n^2

14. Particle A with angular momentum $j = \frac{3}{2}$ decays into two particles B and C with angular momenta j_1 and j_2, respectively. If $\left|\frac{3}{2},\frac{3}{2}\right\rangle_A = \alpha|1,1\rangle_B \otimes \left|\frac{1}{2},\frac{1}{2}\right\rangle_C$, the value of α is ______.

15. $\hat{S}_x$ denotes the spin operator defined as $\hat{S}_x = \frac{\hbar}{2}\begin{pmatrix} 0 & 1 \\ 1 & 0 \end{pmatrix}$. Which one of the following is correct?

(a) The eigenstates of spin operator $\hat{S}_x$ are

$|\uparrow\rangle_x \equiv \begin{pmatrix} 1 \\ 0 \end{pmatrix}$ and $|\downarrow\rangle_x \equiv \begin{pmatrix} 0 \\ 1 \end{pmatrix}$

(b) The eigenstates of spin operator $\hat{S}_x$ are

$|\uparrow\rangle_x \equiv \frac{1}{\sqrt{2}}\begin{pmatrix} 1 \\ -1 \end{pmatrix}$ and $|\uparrow\rangle_x \equiv \frac{1}{\sqrt{2}}\begin{pmatrix} 1 \\ 1 \end{pmatrix}$

(c) In the spin state $\frac{1}{2}\begin{pmatrix} 1 \\ \sqrt{3} \end{pmatrix}$, upon the measurement of $\hat{S}_x$, the probability for obtaining $|\uparrow\rangle_x$ is $\frac{1}{4}$

(d) In the spin state $\frac{1}{2}\begin{pmatrix} 1 \\ \sqrt{3} \end{pmatrix}$, upon the measurement of $\hat{S}_x$, the probability for obtaining $|\uparrow\rangle_x$ is $\frac{2+\sqrt{3}}{4}$.

16. The radial wave function of a particle in a central potential is given by $R(r) = A\frac{r}{a}\exp\left(-\frac{r}{2a}\right)$, where A is the normalization constant and a is positive constant of suitable dimensions. If γa is the most probable distance of the particle from the force center, the value of γ is _____.

17. A free particle of mass M is located in a three-dimensional cubic potential well with impenetrable walls. The degeneracy of the fifth excited state of the particle is _____.

18. Consider the Hamiltonian

$\hat{H} = \hat{H}_0 + \hat{H}'$where

$$\hat{H}_0 = \begin{pmatrix} E & 0 & 0 \\ 0 & E & 0 \\ 0 & 0 & E \end{pmatrix}$$

and $\hat{H}'$ is the time independent perturbation given by

$\hat{H} = \begin{pmatrix} 0 & k & 0 \\ k & 0 & k \\ 0 & k & 0 \end{pmatrix}$, where k > 0. If, the maximum energy eigenvalue of $\hat{H}$ is 3 eV corresponding to E = 2eV, the value of k (rounded off to three decimal places) in eV is ______.

19. A hydrogen atom is in an orbital angular momentum state $|l, m = l\rangle$. If $\vec{L}$ lies on a cone which makes a half angle 30° with respect to the z-axis, the value of l is ______.

20. The Planck's energy density distribution is given by $u(\omega) = \dfrac{\hbar\omega^3}{\pi^2 c^3 \left(e^{\hbar\omega/k_B T} - 1\right)}$. At long wavelengths, the energy density of photons in thermal equilibrium with cavity at temperature T varies as T^α, where α is ______.

2019

21. Consider a one–dimensional gas of N Non-interacting particles of mass m with the Hamiltonian for a single particle given by,

$$H = \frac{p^2}{2m} + \frac{1}{2}m\omega^2\left(x^2 + 2x\right)$$

The high temperature specific heat in units of R = Nk_B (k_B is the Boltzmann constant) is

(*a*) 1 (*b*) 1.5

(*c*) 2 (*d*) 2.5

22. An electric field $\vec{E} = E_0\hat{z}$ is applied to a Hydrogen atom in n= 2 excited state. Ignoring spin the n = 2 state is fourfold degenerate, which in the $|l, m\rangle$ basis are given by $|0,0\rangle, |1,1\rangle, |1,0\rangle$ and $|1,-1\rangle$. If H' is the interaction Hamiltonian corresponding to the applied electric field, which of the following matrix elements is nonzero?

(*a*) $\langle 0,0|H'|0,0\rangle$ (*b*) $\langle 0,0|H'|1,1\rangle$

(*c*) $\langle 0,0|H'|1,0\rangle$ (*d*) $\langle 0,0|H'|1-1\rangle$

23. For a spin $\frac{1}{2}$ particle, let $|\uparrow\rangle$ and $|\downarrow\rangle$ denote its spin up and down states, respectively. if

$$|a\rangle = \frac{1}{\sqrt{2}}\left(|\uparrow\rangle|\downarrow\rangle + |\downarrow\rangle|\uparrow\rangle\right) \text{ and}$$

$$|b\rangle = \frac{1}{\sqrt{2}}\left(|\uparrow\rangle|\downarrow\rangle - |\downarrow\rangle|\uparrow\rangle\right)$$

are composite states of two such particles, which of the following statements is ture for their toatl spin S?

(*a*) S = 1 for $|a\rangle$ and $|b\rangle$ is not an eigenstate of the operator $\hat{S}^2$

(*b*) $|a\rangle$ is not an eigenstate of the operator $\hat{S}^2$, and S = for $|b\rangle$

(*c*) S = 0 for $|a\rangle$, and S = 1 for $|b\rangle$

(*d*) S = 1 for $|a\rangle$ and S = 0 for $|b\rangle$

24. The Hamiltonian for a quantum harmonic oscillator of mass m in three dimensions is

$$H = \frac{p^2}{2m} + \frac{1}{2}m\omega^2 r^2$$

where ω is the angular frequency. The expectation value of r^2 in the first excited state of the oscillator in units of $\frac{\hbar}{m\omega}$ (rounded off to one decimal place) is____________

25. Let $|\psi_1\rangle = \begin{pmatrix} 1 \\ 0 \end{pmatrix}$ $|\psi_2\rangle = \begin{pmatrix} 0 \\ 1 \end{pmatrix}$ represent two possible states of a two–level quantum system. The state obtained by the incurrent superposition of $|\psi_1\rangle$ and $|\psi_2\rangle$ is given by a density matrix that is defined as $\rho \equiv |\psi_1\rangle\langle\psi_1| + c_2|\psi_2\rangle\langle\psi_2|$. If $C_1 = 0.4$ and $C_2 = 0.6$, the matrix element ρ_{22} (rounded off to one decimal place) is____________

26. The wave function ψ (x) of a particle is as shown below

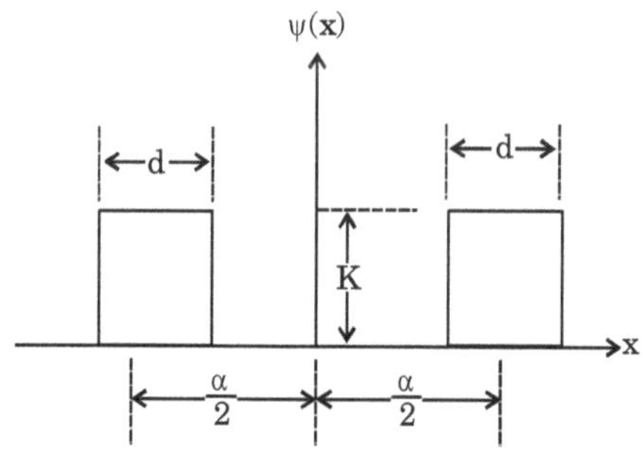

Here K is a constant, and $\alpha > d$. The position uncertainty $\langle \Delta x \rangle$ of the particle is

(*a*)

(*a*) $\sqrt{\dfrac{\alpha^2 + 3d^2}{12}}$ (*b*) $\sqrt{\dfrac{3\alpha^2 + d^2}{12}}$

(*c*) $\sqrt{\dfrac{d^2}{6}}$ (*d*) $\sqrt{\dfrac{d^2}{24}}$

27. Consider the motion of a particle along the x–axis in a potential $V(x) = F\,|x|$. Its ground state energy E_0 is estimated using the uncertainty principle. Then E_0 is proportional to

(*a*) $F^{\frac{1}{3}}$ (*b*) $F^{\frac{1}{2}}$

(*c*) $F^{\frac{2}{5}}$ (*d*) $F^{\frac{2}{3}}$

28. The Hamiltonian operatore for a tow level quantum system is $H = \begin{pmatrix} E_1 & 0 \\ 0 & E_2 \end{pmatrix}$. If the state of the system at t = 0 is given by $|\psi(0)\rangle = \frac{1}{\sqrt{2}}\begin{pmatrix} 1 \\ 1 \end{pmatrix}$ then $|\langle \psi(0)|\psi(t)\rangle|^2$ at a later time t is

(*a*) $\frac{1}{2}\left(1 + e^{-(E_1 - E_2)t/\hbar}\right)$

(*b*) $\frac{1}{2}\left(1 - e^{-(E_1 - E_2)t/\hbar}\right)$

(*c*) $\frac{1}{2}\left(1 + \cos\left[(E_1 - E_2)t/\hbar\right]\right)$

(*d*) $\frac{1}{2}\left(1 - \cos\left[(E_1 - E_2)t/\hbar\right]\right)$

29. Consider a potential barrier V (x) of the form:

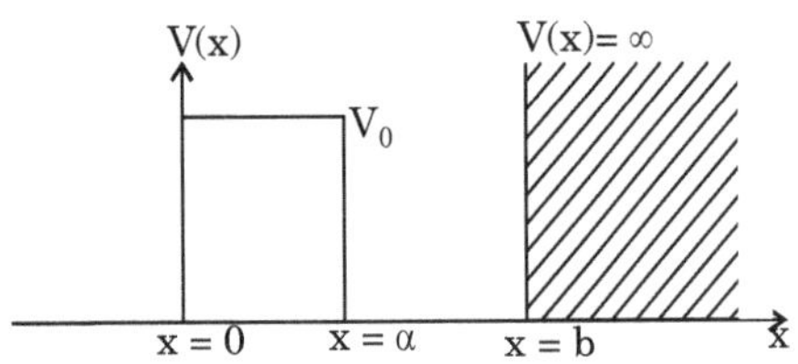

where V0 is a constant . For paricles of energy E < V0 incident on this barrier form the left, which of the following schematic diagrams best represents the probability density $|\psi(x)|^2$ as a function of x.

(*a*)

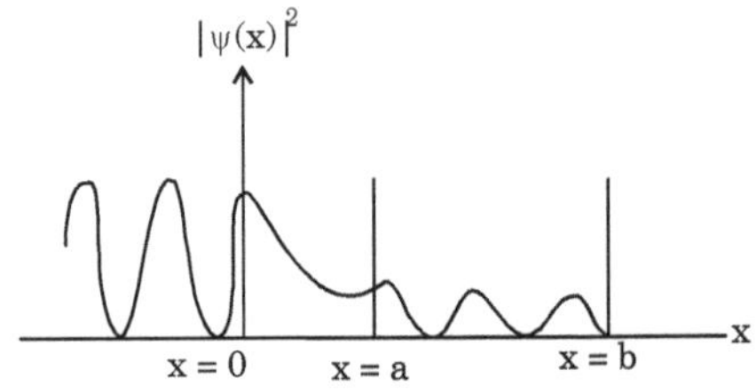

(*b*)

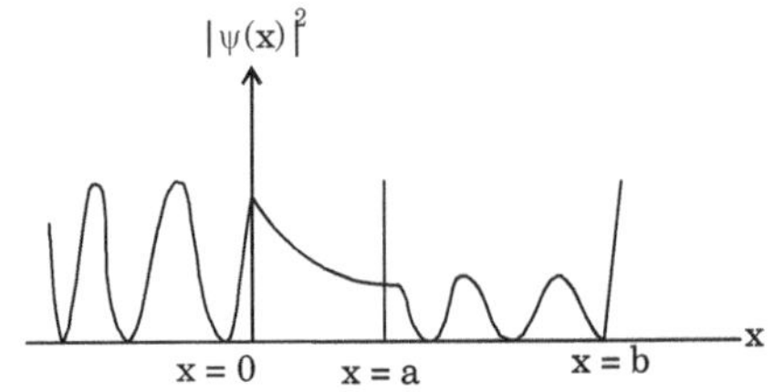

(*c*)

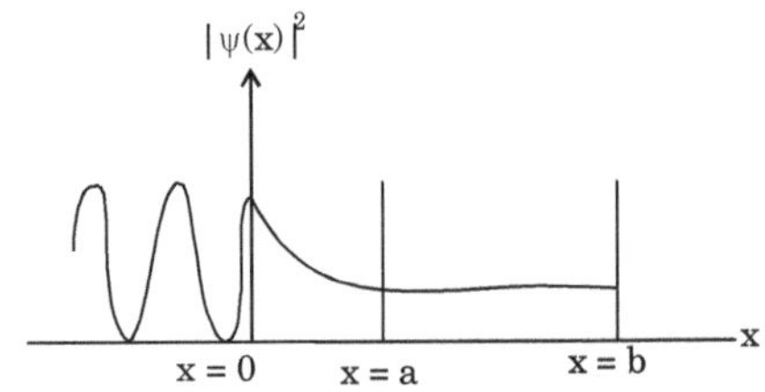

(*d*)

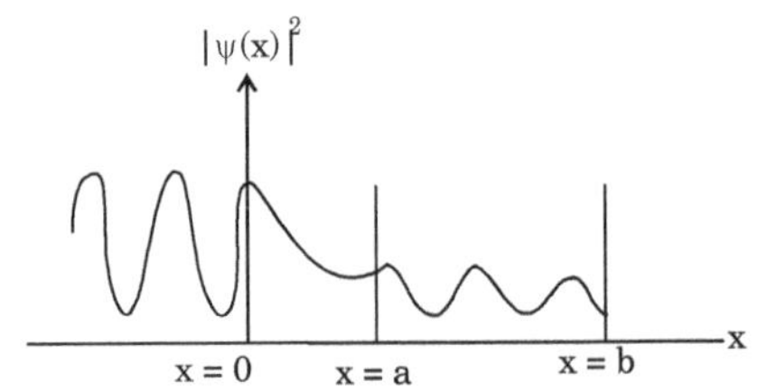

30. Electrons with spin in the z -direction ($\hat{z}$) are passed through a Stern-Gerlach (SG) set up with the magnetic field at θ=60° from $\hat{z}$ .The fraction of electrons that will emerge with their spin parallel to the magnetic field in the SG set up (rounded off to two decimal places) is ______________

$$\left[\sigma_x \begin{pmatrix} 0 & 1 \\ 1 & 0 \end{pmatrix}, \sigma_y = \begin{pmatrix} 0 & -i \\ i & 0 \end{pmatrix}, \sigma_z = \begin{pmatrix} 1 & 0 \\ 0 & -1 \end{pmatrix}\right]$$

31. The Hamiltonian of a system is $H = \begin{pmatrix} 1 & \varepsilon \\ \varepsilon & -1 \end{pmatrix}$ with $\varepsilon << 1$. The fourth order contribution to the ground state energy of H is $\gamma\varepsilon^4$. The value of γ (rounded off to three decimal places) is ________

2018

32. The eigenvalues of a Hermitian matrix are all

(*a*) real (*b*) imaginary

(*c*) of modulus one (*d*) real and positive

33. Which one of the following represents the $3p$ radial wave function of hydrogen atom? (a_0 is the Bohr radius)

(a)
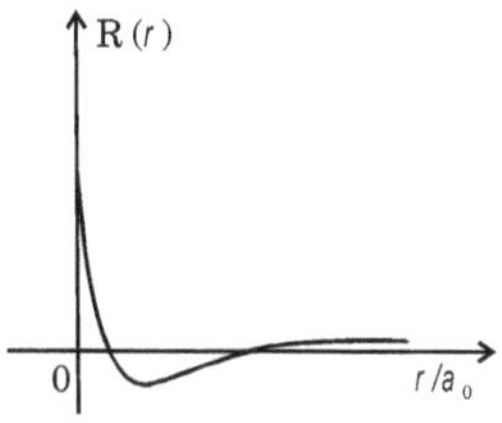

(b)
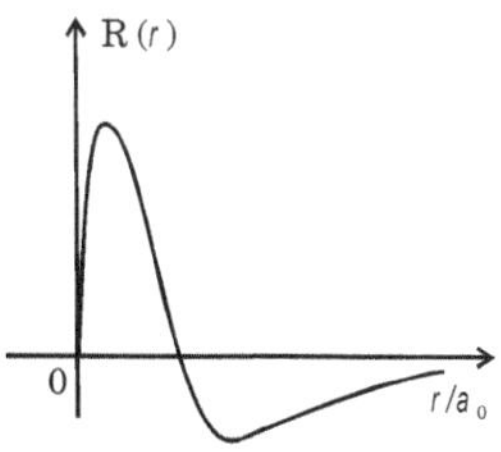

(c)
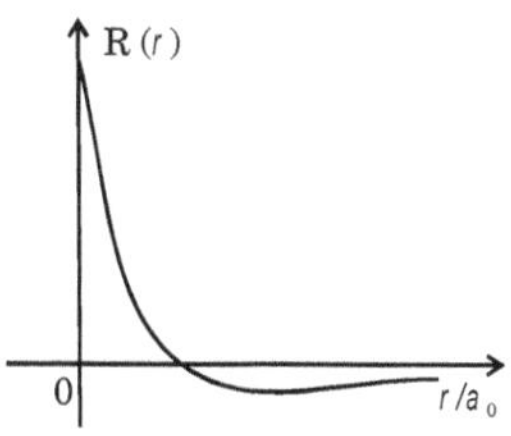

(d)
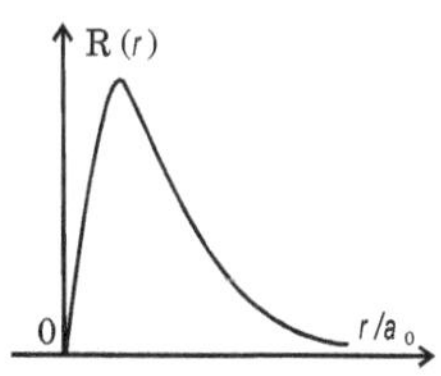

34. In spherical polar coordinates (r, θ, ϕ), the unit vector $\hat{\theta}$ at $(10, \pi/4, \pi/2)$ is

(a) $\hat{k}$ (b) $\frac{1}{\sqrt{2}}(\hat{j}+\hat{k})$

(c) $\frac{1}{\sqrt{2}}(-\hat{j}+\hat{k})$ (d) $\frac{1}{\sqrt{2}}(\hat{j}-\hat{k})$

35. If H is the Hamiltonian for a free particle with mass m, the commutator $[x,[x,H]]$ is

(a) $\hbar^2/m$ (b) $-\hbar^2/m$

(c) $-\hbar^2/(2m)$ (d) $\hbar^2/(2m)$

36. Three particles are to be distributed in four non-degenerate energy levels. The possible number of ways of distribution :

(i) for distinguishable particles, and

(ii) for identical Bosons, respectively, is

(a) (i) 24, (ii) 4

(b) (i) 24, (ii) 20

(c) (i) 64, (ii) 20

(d) (i) 64, (ii) 16

37. For the transformation

$$Q = \sqrt{2q}\, e^{-1+2\alpha} \cos p, P = \sqrt{2q}\, e^{-\alpha-1} \sin p$$

(where α is a constant) to be canonical, the value of α is________.

38. The ground state energy of a particle of mass m in an infinite potential well is E_0. It changes to $E_0(1+\alpha\times10^{-3})$, when there is a small potential bump of height $V_0 = \frac{\pi^2\hbar^2}{50mL^2}$ and width $a = L/100$, as shown in the figure. The value of α is________ (up to two decimal places).

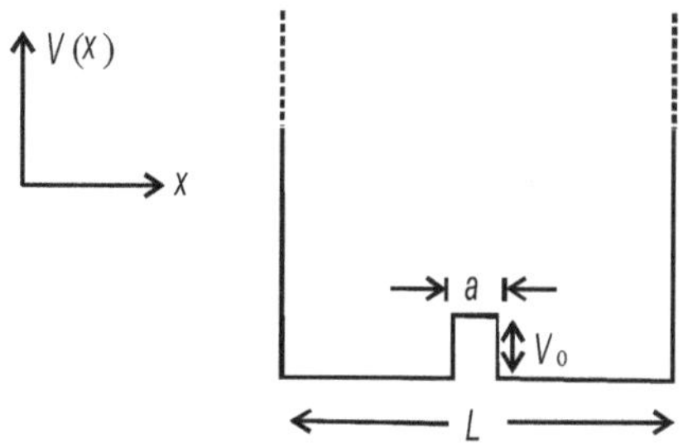

39. A two-state quantum system has energy eigenvalues $\pm\varepsilon$ corresponding to the normalized states $|\psi_\pm\rangle$. At time $t = 0$, the system is in quantum state $\frac{1}{\sqrt{2}}\left[|\psi_+\rangle + |\psi_-\rangle\right]$. The probability that the system will be in the same state at $t = h/(6\varepsilon)$ is________ (up to two decimal places).

2017

40. The wavefunction of which orbital is spherically symmetric :

(a) p_x (b) p_y

(c) s (d) d_{xy}

41. The Compton wavelength of a proton is ______ fm. (up to two decimal places). ($m_p = 1.67\times10^{-27}$ kg, $h = 6.626\times10^{-34}$ Js, $e = 1.602\times10^{-19}$ C, $c = 3\times10^8$ ms^{-1})

42. A one dimensional simple harmonic oscillator with Hamiltonian $H_0 = \frac{p^2}{2m} + \frac{1}{2}k\,x^2$ is subjected to a small perturbation, $H_1 = \alpha x + \beta x^3 + \gamma x^4$. The first order correction to the ground state energy is dependent on

(a) only β (b) α and γ

(c) α and β (d) only γ

43. For the Hamiltonian $H = a_0 I + \vec{b}.\vec{\sigma}$ where $a_0 \in R, \vec{b}$ is a real vector, I is the 2 × 2 identity matrix, and $\vec{\sigma}$ are the Pauli matrices, the ground state energy is

(a) $|b|$ (b) $2a_0 - |b|$

(c) $a_0 - |b|$ (d) a_0

44. The degeneracy of the third energy level of a 3-dimensional isotropic quantum harmonic oscillator is

(a) 6 (b) 12

(c) 8 (d) 10

45. The electronic ground state energy of the Hydrogen atom is – 13.6 eV. The highest possible electronic energy eigenstate has an energy equal to

(a) 0 (b) 1 eV

(c) + 13.6 eV (d) ∞

46. A free electron of energy 1 eV is incident upon a one-dimensional finite potential step of height 0.75 eV. The probability of its reflection from the barrier is _______ (up to two decimal places).

47. Consider a one-dimensional potential well of width 3 nm. Using the uncertainty principle $\left(\Delta x.\Delta p \geq \hbar/2\right)$, an estimate of the minimum depth of the well such that it has at least one bound state for an electron is (m_e = 9.31 × 10^{-31} kg, h = 6.626 × 10^{-34} J s, e = 1.602 × 10^{-19} C) :

(a) 1 μeV (b) 1 meV

(c) 1 eV (d) 1 MeV

48. Let X be a column vector of dimension $n > 1$ with at least one non-zero entry. The number of non-zero eigenvalues of the matrix $M = XX^T$ is

(a) 0 (b) n

(c) 1 (d) $n - 1$

49. Using Hund's rule, the total angular momentum quantum number J for the electronic ground state of the nitrogen atom is

(a) $\frac{1}{2}$ (b) $\frac{3}{2}$

(c) 0 (d) 1

50. Which one of the following operators is Hermitian?

(a) $i\frac{\left(p_x x^2 - x^2 p_x\right)}{2}$ (b) $i\frac{\left(p_x x^2 + x^2 p_x\right)}{2}$

(c) $e^{ip_x a}$ (d) $e^{-ip_x a}$

51. Consider two particles and two non-degenerate quantum levels 1 and 2. Level 1 always contains a particle. Hence, what is the probability that level 2 also contains a particle for each of the two cases:
(i) when the two particles are distinguishable and
(ii) when the two particles are bosons?

(a) (i) $\frac{1}{2}$ and (ii) $\frac{1}{3}$ (b) (i) $\frac{1}{2}$ and (ii) $\frac{1}{2}$

(c) (i) $\frac{2}{3}$ and (ii) $\frac{1}{2}$ (d) (i) 1 and (ii) 0

52. The geometric cross-section of two colliding protons at large energies is very well estimated by the product of the effective sizes of each particle. This is closest to

(a) 10 b (b) 10 mb

(c) 10 μb (d) 10 pb

2016

53. Which of the following magnetic vector potentials gives rise to a uniform magnetic field $B_0\hat{k}$?

(a) $B_0 z\hat{k}$ (b) $-B_0 x\hat{j}$

(c) $\frac{B_0}{2}\left(-y\hat{i} + x\hat{j}\right)$ (d) $\frac{B_0}{2}\left(y\hat{i} + x\hat{j}\right)$

54. A one-dimensional linear chain of atoms contains two types of atoms of masses m_1 and m_2 (where $m_2 > m_1$), arranged alternately. The distance between successive atoms is the same. Assume that the harmonic approximation is valid. At the first Brillouin zone boundary, which of the following statements is correct?

(a) The atoms of mass m_2 are at rest in the optical mode, while they vibrate in the acoustical mode.
(b) The atoms of mass m_1 are at rest in the optical mode, while they vibrate in the acoustical mode.
(c) Both types of atoms vibrate with equal amplitudes in the optical as well as in the acoustical modes.
(d) Both types of atoms vibrate, but with unequal, non-zero amplitudes in the optical as well as in the acoustical modes.

55. A particle moving under the influence of a central force $\vec{F}(\vec{r}) = -k\vec{r}$ ((where $\vec{r}$ is the position vector of the particle and k is a positive constant) has non-zero angular momentum. Which of the following curves is a possible orbit for this particle?

(a) A straight line segment passing through the origin.
(b) An ellipse with its center at the origin.
(c) An ellipse with one of the foci at the origin.
(d) A parabola with its vertex at the origin.

56. If $\vec{s}_1$ and $\vec{s}_2$ are the spin operators of the two electrons of a He atom, the value of $\langle \vec{s}_1 \cdot \vec{s}_2 \rangle$ for the ground state is

(a) $-\frac{3}{2}\hbar^2$ (b) $-\frac{3}{4}\hbar^2$

(c) 0 (d) $\frac{1}{4}\hbar^2$

57. Let V_i be the ith component of a vector field $\vec{V}$, which has zero divergence. If $\partial_j \equiv \partial/\partial x_j$, the expression for $\in_{ijk}\in_{lmk} \partial_j \partial_l V_m$ is equal to

(a) $-\partial_j\partial_k V_i$ (b) $\partial_j\partial_k V_i$

(c) $\partial^2_j V_i$ (d) $-\partial^2_j V_i$

58. A particle of rest mass M is moving along the positive x-direction. It decays into two photons γ_1 and γ_2 as shown in the figure. The energy of γ_1 is 1 GeV and the energy of γ_2 is 0.82 GeV. The value of M (in units of GeV/c²) is ________. (Give your answer upto two decimal places)

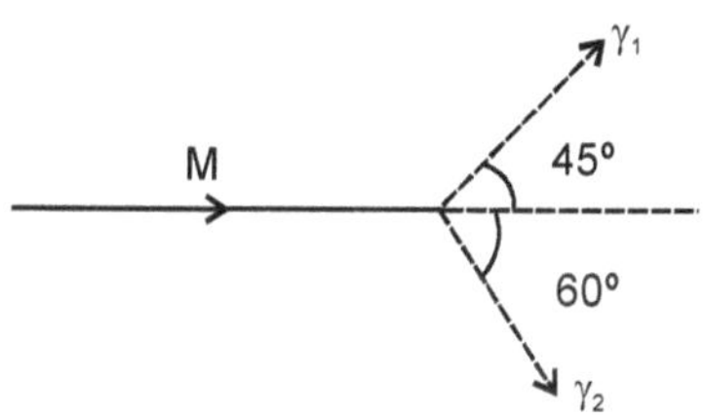

59. If x and p are the x components of the position and the momentum operators of a particle respectively, the commutator $[x^2, p^2]$ is

(a) $i\hbar(xp - px)$ (b) $2i\hbar(xp - px)$

(c) $i\hbar(xp + px)$ (d) $2i\hbar(xp + px)$

60. Let $|l, m\rangle$ be the simultaneous eigenstates of L^2 and L_z. Here $\vec{L}$ is the angular momentum operator with Cartesian components (L_x, L_y, L_z), l is the angular momentum quantum number and m is the azimuthal quantum number. The value of $\langle 1,0|(L_x + iL_y)|1,-1\rangle$ is

(a) 0 (b) $\hbar$

(c) $\sqrt{2}\hbar$ (d) $\sqrt{3}\hbar$

61. For the parity operator P, which of the following statements is NOT true?

(a) $P^\dagger = P$ (b) $P^2 = -P$

(c) $P^2 = I$ (d) $P^\dagger = P^{-1}$

2015

62. The Pauli matrices for three spin-1/2 particles are $\vec{\sigma}_1, \vec{\sigma}_2$ and $\vec{\sigma}_3$, respectively. The dimension of the Hilbert space required to define an operator $\hat{O} = \vec{\sigma}_1 . \vec{\sigma}_2 \times \vec{\sigma}_3$ is ________

63. Let $\vec{L}$ and $\vec{p}$ be the angular and linear momentum operators, respectively, for a particle. The commutator $[L_x, p_y]$ gives

(a) $-i\hbar p_z$ (b) 0

(c) $i\hbar p_x$ (d) $i\hbar p_z$

64. A particle is confined in a box of length L as shown below.

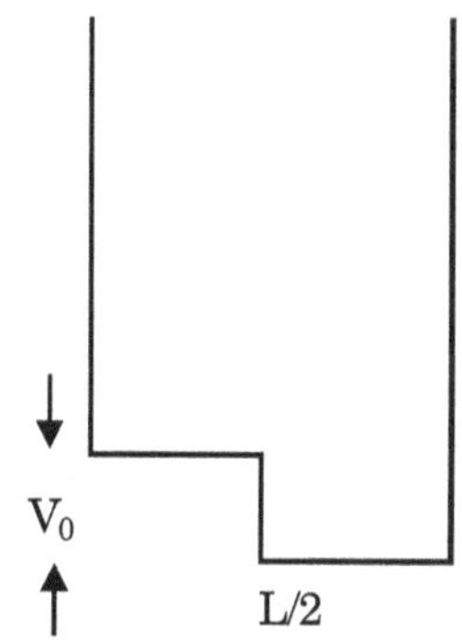

If the potential V_0 is treated as a perturbation, including the first order correction, the ground state energy is

(a) $E = \frac{\hbar^2\pi^2}{2mL^2} + V_0$ (b) $E = \frac{\hbar^2\pi^2}{2mL^2} - \frac{V_0}{2}$

(c) $E = \frac{\hbar^2\pi^2}{2mL^2} + \frac{V_0}{4}$ (d) $E = \frac{\hbar^2\pi^2}{2mL^2} + \frac{V_0}{2}$

65. In a rigid-rotator of mass M, if the energy of the first excited state is 1 meV, then the fourth excited state energy (in meV) is ________

66. The average energy U of a one dimensional quantum oscillator of frequency ω and in contact with a heat bath at temperature T is given by

(a) $U = \frac{1}{2}\hbar\omega \cot h\left(\frac{1}{2}\beta\hbar\omega\right)$

(b) $U = \frac{1}{2}\hbar\omega \sin h\left(\frac{1}{2}\beta\hbar\omega\right)$

(c) $U = \frac{1}{2}\hbar\omega \tan h\left(\frac{1}{2}\beta\hbar\omega\right)$

(d) $U = \frac{1}{2}\hbar\omega \cos h\left(\frac{1}{2}\beta\hbar\omega\right)$

67. Consider a system of eight non-interacting, identical quantum particles of spin-3/2 in a one dimensional box of length L. The minimum excitation energy of the system, in units of $\frac{\pi^2\hbar^2}{2mL^2}$ is ________

68. Let the Hamiltonian for two spin $-\frac{1}{2}$ particles of equal masses m, momenta $\vec{p}_1$ and $\vec{p}_2$ and positions $\vec{r}_1$ and $\vec{r}_2$ be

$$H = \frac{1}{2m}p_1^2 + \frac{1}{2m}p_2^2 + \frac{1}{2}m\omega^2\left(r_1^2 + r_2^2\right) + k\,\vec{\sigma}_1\cdot\vec{\sigma}_2$$

where $\vec{\sigma}_1$ and $\vec{\sigma}_2$ denote the corresponding Pauli matrices, $\hbar\omega = 0.1$ eV and $k = 0.2$ eV. If the ground state has net spin zero, then the energy (in eV) is ________

69. Suppose a linear harmonic oscillator of frequency ω and mass m is in the state

$|\psi\rangle = \frac{1}{\sqrt{2}}\left[|\psi_0\rangle + e^{i\frac{\pi}{2}}|\psi_1\rangle\right]$ at $t = 0$ where $|\psi_0\rangle$ and $|\psi_1\rangle$ are the ground and the first excited states, respectively. The value of $\langle\psi|x|\psi\rangle$ in the units of $\sqrt{\frac{\hbar}{m\omega}}$ at t = 0 is ________

2014

70. The recoil momentum of an atom is p_A when it emits an infrared photon of wavelength 1500 nm, and it is p_B when it emits a photon of visible wavelength 500 nm. The ratio $\frac{p_A}{p_B}$ is

(*a*) 1 : 1 (*b*) 1 : $\sqrt{3}$

(*c*) 1 : 3 (*d*) 3 : 2

71. The ground state and the first excited state wave functions of a one dimensional infinite potential well are Ψ_1 and Ψ_2, respectively. When two spin-up electrons are placed in this potential, which one of the following, with x_1 and x_2 denoting the position of the two electrons, correctly represents the space part of the ground state wave function of the system?

(*a*) $\frac{1}{\sqrt{2}}[\psi_1(x_1)\psi_2(x_1) - \psi_1(x_2)\psi_2(x_2)]$

(*b*) $\frac{1}{\sqrt{2}}[\psi_1(x_1)\psi_2(x_2) + \psi_1(x_2)\psi_2(x_1)]$

(*c*) $\frac{1}{\sqrt{2}}[\psi_1(x_1)\psi_2(x_1) + \psi_1(x_2)\psi_2(x_2)]$

(*d*) $\frac{1}{\sqrt{2}}[\psi_1(x_1)\psi_2(x_2) - \psi_1(x_2)\psi_2(x_1)]$

72. If $\vec{L}$ is the orbital angular momentum and $\vec{S}$ is the spin angular momentum, then, $\vec{L}\cdot\vec{S}$ does NOT commute with

(*a*) S_z (*b*) L^2

(*c*) S^2 (*d*) $(\vec{L}+\vec{S})^2$

73. The energy, ε_k for band electrons as a function of the wave vector, k in the first Brillouin zone $(-\pi/a \le k \le \pi/a)$ of a one dimensional monatomic lattice is shown as (a is lattice constant)

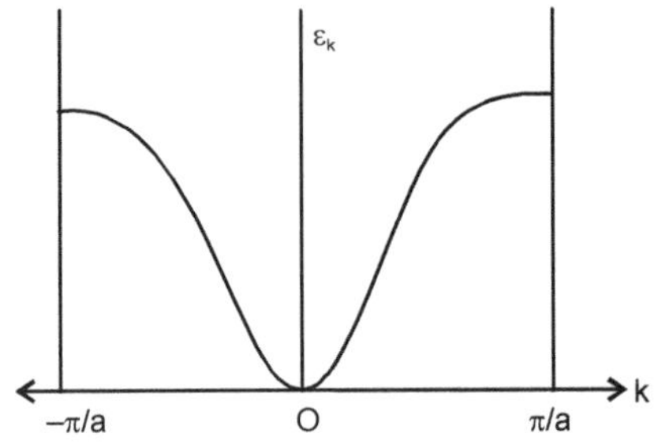

The variation of the group velocity v_k is most appropriately represented by

(*a*)
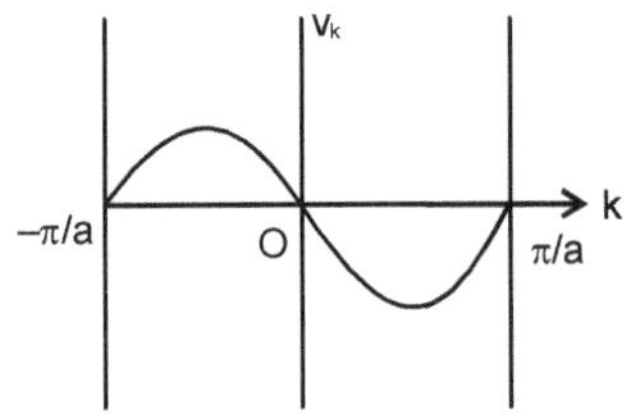

(*b*)
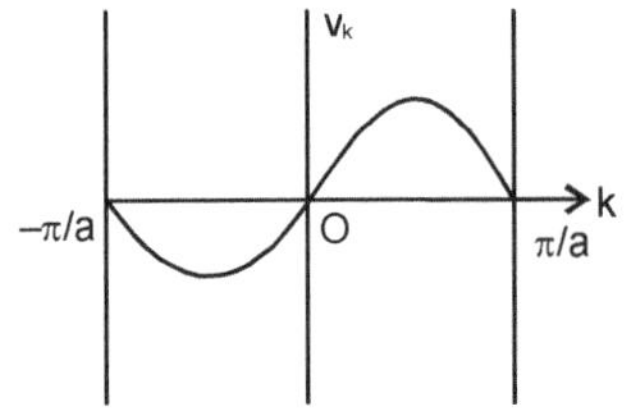

(*c*)
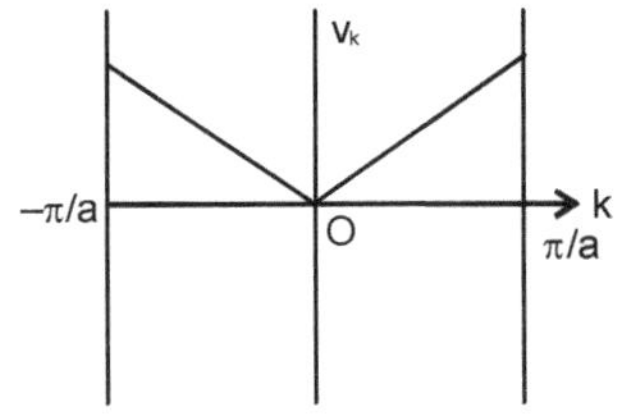

(*d*)
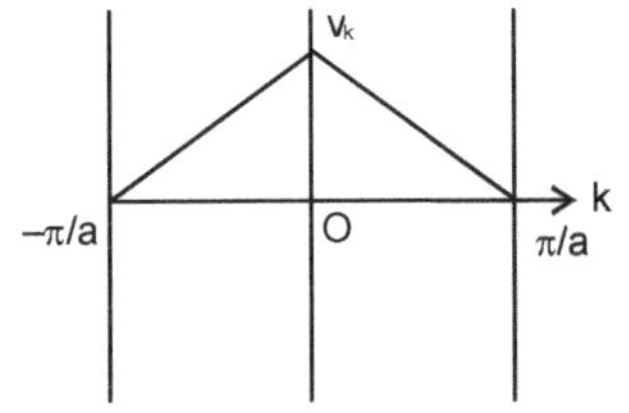

74. An electron in the ground state of the hydrogen atom has the wave function

$$\psi(\vec{r}) = \frac{1}{\sqrt{\pi a_0^3}} e^{-(r/a_0)}$$

where a_0 is constant. The expectation value of the operator $\hat{Q} = z^2 - r^2$, where $z = r\cos\theta$ is

(Hint: $\int_0^{\infty} e^{-ar} r^n dr = \frac{r(n)}{a^{n+1}} = \frac{(n-1)!}{a^{n+1}}$)

(*a*) $-a_0^2/2$ (*b*) $-a_0^2$

(*c*) $-3a_0^2/2$ (*d*) $-2a_0^2$

75. A particle of mass m is subjected to a potential,

$$V(x, y) = \frac{1}{2} m\omega^2 (x^2 + y^2),$$

$$-\infty \le x \le \infty, \ -\infty \le y \le \infty$$

The state with energy $4\hbar\omega$ is g-fold degenerate. The value of g is _______.

76. A hydrogen atom is in the state

$$\psi = \sqrt{\frac{8}{21}}\psi_{200} - \sqrt{\frac{3}{7}}\psi_{310} + \sqrt{\frac{4}{21}}\psi_{321},$$

where n, l, m in ψ_{nlm} denote the principal, orbital and magnetic quantum numbers, respectively. If $\vec{L}$ is the angular momentum operator, the average value of L^2 is _______ $\hbar^2$.

77. ψ_1 and ψ_2 are two orthogonal states of a spin $\frac{1}{2}$ system. It is given that

$$\psi_1 = \frac{1}{\sqrt{3}}\begin{pmatrix}1\\0\end{pmatrix} + \sqrt{\frac{2}{3}}\begin{pmatrix}0\\1\end{pmatrix},$$

where $\begin{pmatrix}1\\0\end{pmatrix}$ and $\begin{pmatrix}0\\1\end{pmatrix}$ represent the spin-up and spin-down states, respectively. When the system is in the state ψ_2, its probability to be in the spin-up state is _________.

78. A particle is confined to a one dimensional potential box with the potential

$V(x) = 0, \ 0 < x < a$

$= \infty$, otherwise

If the particle is subjected to a perturbation, within the box, $W = \beta x$, where β is a small constant, the first order correction to the ground state energy is

(*a*) 0 (*b*) $a\beta/4$

(*c*) $a\beta/2$ (*d*) $a\beta$

79. A one dimensional harmonic oscillator is in the superposition of number states, $|n\rangle$, given by

$$|\psi\rangle = \frac{1}{2}|2\rangle + \frac{\sqrt{3}}{2}|3\rangle.$$

The average energy of the oscillator in the given state is _____________ $\hbar\omega$.

80. If L_+ and L_- are the angular momentum ladder operators, then, the expectation value of $(L_+L_- + L_-L_+)$, in the state $|l = 1, m = 1\rangle$ of an atom is ____ $\hbar^2$.

2013

81. Which one of the following commutation relations is NOT CORRECT ? Here, symbols have their usual meanings.

(*a*) $[L^2, L_z] = 0$

(*b*) $[L_x, L_y] = i\hbar L_z$

(*c*) $[L_z, L_+] = \hbar L_+$

(*d*) $[L_z, L_-] = \hbar L_-$

82. The degenerate eigenvalue of the matrix

$$\begin{bmatrix} 4 & -1 & -1 \\ -1 & 4 & -1 \\ -1 & -1 & 4 \end{bmatrix}$$ is (your answer should be an integer) _______

83. A spin-half particle is in a linear superposition $0.8|\uparrow\rangle + 0.6|\downarrow\rangle$ of its spin-up and spin-down states. If $|\uparrow\rangle$ and $|\downarrow\rangle$ are the eigenstates of σ_z then what is the expectation value, up to one decimal place, of the operator $10\sigma_z + 5\sigma_x$? Here, symbols have their usual meanings.__________

84. Consider the wave function $Ae^{ikr}(r_0/r)$, where A is the normalization constant. For $r = 2r_0$, the magnitude of probability current density up to two decimal places, in units of $(A^2\hbar k/m)$, is ______.

Common Data for Questions 85 and 86:

To the given unperturbed Hamiltonian

$$\begin{bmatrix} 5 & 2 & 0 \\ 2 & 5 & 0 \\ 0 & 0 & 2 \end{bmatrix},$$

we add a small perturbation given by

$$\varepsilon\begin{bmatrix} 1 & 1 & 1 \\ 1 & 1 & -1 \\ 1 & -1 & 1 \end{bmatrix},$$

where ε is a small quantity.

85. The ground state eigenvector of the unperturbed Hamiltonian is

(a) $(1/\sqrt{2}, 1/\sqrt{2}, 0)$ (b) $(1/\sqrt{2}, -1/\sqrt{2}, 0)$

(c) $(0, 0, 1)$ (d) $(1, 0, 0)$

86. A pair of eigenvalues of the perturbed Hamiltonian, using first order perturbation theory, is

(a) $3 + 2\varepsilon, 7 + 2\varepsilon$ (b) $3 + 2\varepsilon, 2 + \varepsilon$

(c) $3, 7 + 2\varepsilon$ (d) $3, 2 + 2\varepsilon$

2012

87. A particle of mass m is confined in a two dimensional square well potential of dimension a. This potential $V(x, y)$ is given by

$$V(x, y) = 0 \text{ for } -a < x < a \text{ and } -a < y < a$$
$$= \infty \text{ elsewhere}$$

The energy of the first excited state for this particle is given by,

(a) $\frac{\pi^2 h^2}{ma^2}$ (b) $\frac{2\pi^2 h^2}{ma^2}$

(c) $\frac{5\pi^2 h^2}{2ma^2}$ (d) $\frac{4\pi^2 h^2}{ma^2}$

88. Consider the wavefunction $\Psi = \Psi(\vec{r}_1, \vec{r}_2)\chi_s$ for a fermionic system consisting of two spin-half particles. The spatial part of the wavefunction is given by,

$$\psi(\vec{r}_1, \vec{r}_2)\chi = \frac{1}{\sqrt{2}}[\phi_1(\vec{r}_1)\phi_2(\vec{r}_2) + \phi_2(\vec{r}_1)\phi_1(\vec{r}_2)]$$

where ϕ_1 and ϕ_2 are single particle states. The spin part χ_s of the wavefunction with spin states $\alpha(+½)$ and $\beta(-½)$ should be

(a) $\frac{1}{\sqrt{2}}(\alpha\beta + \beta\alpha)$ (b) $\frac{1}{\sqrt{2}}(\alpha\beta - \beta\alpha)$

(c) $\alpha\alpha$ (d) $\beta\beta$

89. A particle is constrained to move in a truncated harmonic potential well ($x > 0$) as shown in the figure. Which one of the following statements is CORRECT?

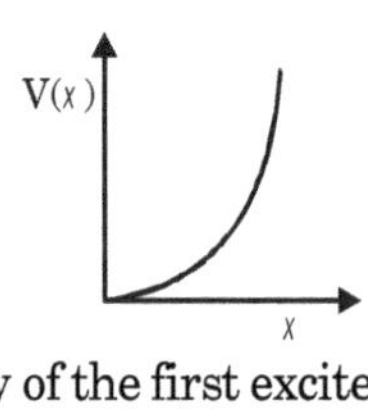

(a) The parity of the first excited state is even

(b) The parity of the ground state is even

(c) The ground state energy is $\frac{1}{2}\hbar\omega$

(d) The first excited state energy is $\frac{7}{2}\hbar\omega$

90. Consider a system in the unperturbed state described by the Hamiltonian, $H_0 = \begin{pmatrix} 1 & 0 \\ 0 & 1 \end{pmatrix}$. The system is subjected to a perturbation of the form $H' = \begin{pmatrix} \delta & \delta \\ \delta & \delta \end{pmatrix}$, where $\delta \ll 1$. The energy eigenvalues of the perturbed system using the first order perturbation approximation are

(a) 1 and $(1 + 2\delta)$ (b) $(1 + \delta)$ and $(1 - \delta)$

(c) $(1 + 2\delta)$ and $(1 - 2\delta)$ (d) $(1 + \delta)$ and $(1 - 2\delta)$

91. The ground state wavefunction for the hydrogen atom is given by

$$\psi_{100} = \frac{1}{\sqrt{4\pi}}\left(\frac{1}{a_0}\right)^{3/2} e^{-r/a_0},$$

where a_0 is the Bohr radius.

The plot of the radial probability density, $P(r)$ for the hydrogen atom in the ground state is

(a) 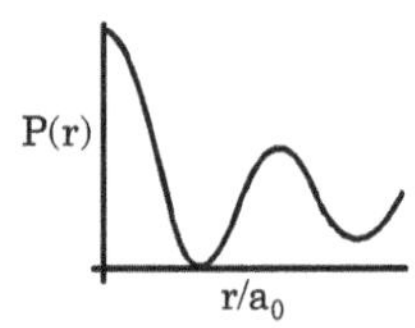

(b) 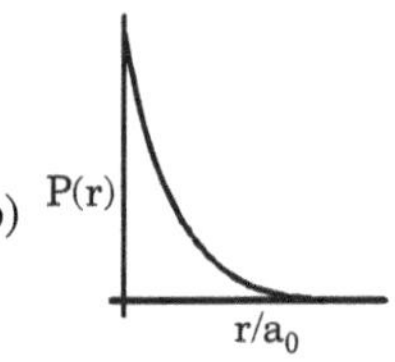

(c) 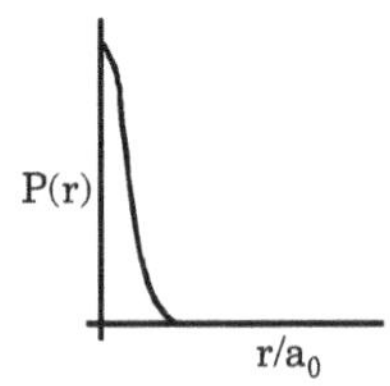

(d) 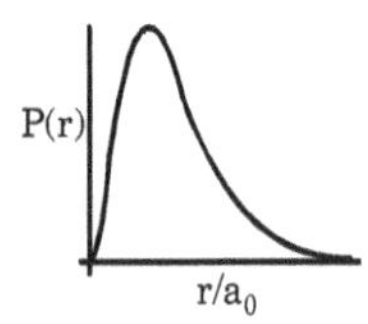

COMMON DATA QUESTIONS

Common Data Q. 92 and 93 :

The wavefunction of a particle moving in free space is given by, $\psi = e^{ikx} + 2e^{-ikx}$

92. The energy of the particle is

(a) $\frac{5\hbar^2 k^2}{2m}$ (b) $\frac{3\hbar^2 k^2}{4m}$

(c) $\frac{\hbar^2 k^2}{2m}$ (d) $\frac{\hbar^2 k^2}{m}$

93. The probability current density for the real part of the wavefunction is

(a) 1 (b) $\frac{\hbar k}{m}$

(c) $\frac{\hbar k}{2m}$ (d) 0

2011

94. Two matrices A and B are said to be similar if $B = P^{-1} AP$ for some invertible matrix P. Which of the following statements is NOT TRUE?

(*a*) Det A = Det B

(*b*) Trace of A = Trace of B

(*c*) A and B have the same eigenvectors

(*d*) A and B have the same eigenvalues

95. A Carnot cycle operates on a working substance between two reservoirs at temperatures T_1 and T_2, with $T_1 > T_2$. During each cycle an amount of heat Q_1 is extracted from the reservoir at T_1 and an amount Q_2 is delivered to the reservoir at T_2. Which of the following statements is INCORRECT?

(*a*) work done in one cycle is $Q_1 - Q_2$

(*b*) $\frac{Q_1}{T_1} = \frac{Q_2}{T_2}$

(*c*) entropy of the hotter reservoir decreases

(*d*) entropy of the universe (consisting of the working substance and the two reservoirs) increases

96. A 3×3 matrix has elements such that its trace is 11 and its determinant is 36. The eigenvalues of the matrix are all known to be positive integers. The largest eigenvalue of the matrix is

(*a*) 18 (*b*) 12

(*c*) 9 (*d*) 6

97. An electron with energy E is incident from left on a potential barrier, given by

$$V(x) = 0 \text{ for } x < 0$$
$$= V_0 \text{ for } x > 0$$

as shown in the figure.

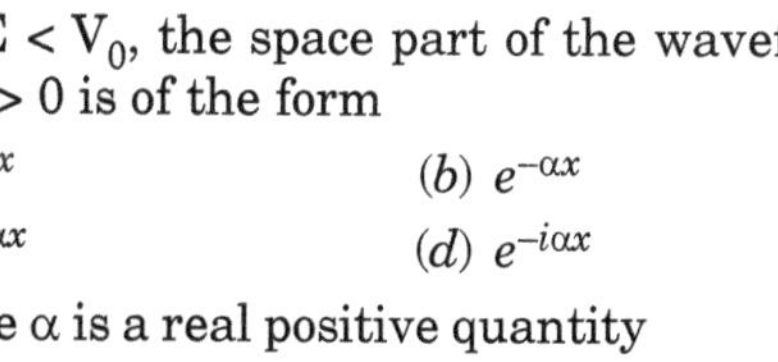

For $E < V_0$, the space part of the wavefunction for $x > 0$ is of the form

(*a*) $e^{\alpha x}$ (*b*) $e^{-\alpha x}$

(*c*) $e^{i\alpha x}$ (*d*) $e^{-i\alpha x}$

where α is a real positive quantity

98. If L_x, L_y and L_z are respectively the *x*, *y* and *z* components of angular momentum operator L, the commutator $[L_xL_y, L_z]$ is equal to

(*a*) $i\hbar(L_x^2 + L_y^2)$ (*b*) $2i\hbar L_z$

(*c*) $i\hbar(L_x^2 - L_y^2)$ (*d*) 0

99. The normalized ground state wavefunction of a hydrogen atom is given by $\psi(r) = \frac{1}{\sqrt{4\pi}} \frac{2}{a^{3/2}} e^{-r/a}$, where *a* is the Bohr radius and *r* is the distance of the electron from the nucleus, located at the origin. The expectation value $\left\langle \frac{1}{r^2} \right\rangle$ is

(*a*) $\frac{8\pi}{a^2}$ (*b*) $\frac{4\pi}{a^2}$

(*c*) $\frac{4}{a^2}$ (*d*) $\frac{2}{a^2}$

100. The normalized eigenstates of a particle in a one-dimensional potential well

$$V(x) = \begin{cases} 0 & \text{if } 0 \le x \le a \\ \infty & \text{otherwise} \end{cases}$$

are given by

$$\psi_n(x) = \sqrt{\frac{2}{a}} \sin\left(\frac{n\pi x}{a}\right),$$

where $n = 1,2,3.......$

The particle is subjected to a perturbation

$$V'(x) = V_0 \cos\left(\frac{\pi x}{a}\right) \text{ for } 0 \le x \le \frac{a}{2}$$
$$= 0 \qquad \text{otherwise}$$

The shift in the ground state energy due to the perturbation, in the first order perturbation theory, is

(*a*) $\frac{2V_0}{3\pi}$ (*b*) $\frac{V_0}{3\pi}$

(*c*) $-\frac{V_0}{3\pi}$ (*d*) $-\frac{2V_0}{3\pi}$

LINKED ANSWER QUESTIONS

Statement for Linked Answer Questions 101 and 102:

In a one-dimensional harmonic oscillator, φ_0, φ_1 and φ_2 are respectively the ground, first and the second excited states. These three states are normalized and are orthogonal to one another.

ψ_1 and ψ_2 are two states defined by

$$\psi_1 = \varphi_0 - 2\varphi_1 + 3\varphi_2$$
$$\psi_2 = \varphi_0 - \varphi_1 + \alpha\varphi_2$$

where α is a constant.

101. The value of α for which ψ_2 is orthogonal to ψ_1 is

(*a*) 2 (*b*) 1

(*c*) -1 (*d*) -2

102. For the value of α determined in Q.52, the expectation value of energy of the oscillator in the state ψ_2 is

(*a*) $\hbar\omega$ (*b*) $3\hbar\omega/2$

(*c*) $3\hbar\omega$ (*d*) $9\hbar\omega/2$

2010

103. The eigen values of the matrix $\begin{pmatrix} 2 & 3 & 0 \\ 3 & 2 & 0 \\ 0 & 0 & 1 \end{pmatrix}$ are

(*a*) 5, 2, -2 (*b*) -5, -1, 1

(*c*) 5, 1, -1 (*d*) -5, 1, 1

104. Which of the following is an allowed wavefunction for a particle in a bound state? N is a constant and α, $\beta > 0$.

(*a*) $\psi = N\dfrac{e^{-\alpha r}}{r^3}$

(*b*) $\psi = N(1 - e^{-\alpha r})$

(*c*) $\psi = Ne^{-\alpha x}e^{-\beta(x^2+y^2+z^2)}$

(*d*) $\psi = \begin{cases} \text{non-zero constant} & \text{if } r < R \\ 0 & \text{if } r > R \end{cases}$

105. A particle is confined within a spherical region of radius one femtometer (10^{-15}m). Its momentum can be expected to be about

(*a*) $20\dfrac{keV}{c}$ (*b*) $200\,\dfrac{keV}{c}$

(*c*) $200\dfrac{MeV}{c}$ (*d*) $\dfrac{GeV}{c}$

106. The ground state wavefunction of deuteron is in a superposition of *s* and *d* states. Which of the following is NOT true as a consequence?

(*a*) It has a non-zero quadruple moment

(*b*) The neutron-proton potential is non-central

(*c*) The orbital wavefunction is not spherically symmetric

(*d*) The Hamiltonian does not conserve that total angular momentum

107. A particle of mass *m* is confined in the potential

$$V(x) = \begin{cases} \dfrac{1}{2}m\omega^2x^2 & \text{for } x < 0, \\ \infty & \text{for } x \le 0 \end{cases}$$

Let the wavefunction of the particle be given by

$$\psi(x) = -\frac{1}{\sqrt{5}}\psi_0 + \frac{2}{\sqrt{5}}\psi_1,$$

where ψ_0 and ψ_1 are the eigenfunctions of the ground state and the first excited state respectively. The expectation value of the energy is

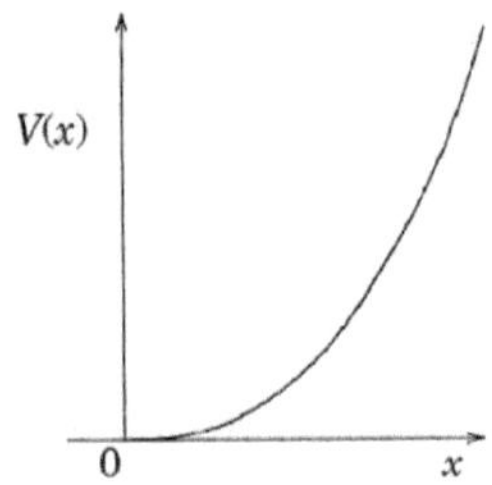

(*a*) $\dfrac{31}{10}\hbar\omega$

(*b*) $\dfrac{25}{10}\hbar\omega$

(*c*) $\dfrac{13}{10}\hbar\omega$

(*d*) $\dfrac{11}{10}\hbar\omega$

108. For a spin-s particle, in the eigen basis of $\vec{S}^2$, S_z the expectation value $\langle sm|\vec{S}_x^2|sm\rangle$ is

(*a*) $\dfrac{\hbar\{s(s+1)-m^2\}}{2}$

(*b*) $\hbar^2\{s(s+1) - 2m^2\}$

(*c*) $\hbar^2\{s(s+1) - m^2\}$

(*d*) $\hbar^2 m^2$

109. A particle of mass *m* is confined in an infinite potential well:

$$V(x) = \begin{cases} 0 & \text{if } 0 < x < L, \\ \infty & \text{otherwise.} \end{cases}$$

It is subjected to a perturbing potential

$$V_P(x) = V_0 \sin\left(\frac{2\pi x}{L}\right)$$

within the well. Let $E^{(1)}$ and $E^{(2)}$ be the corrections to the ground state energy in the first and second order in V_0,

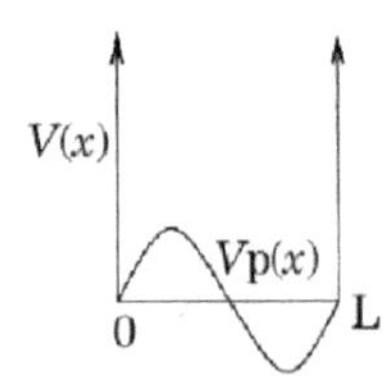

(*a*) $E^{(1)} = 0$; $E^{(2)} < 0$

(*b*) $E^{(1)} > 0$; $E^{(2)} = 0$

(*c*) $E^{(1)} = 0$; $E^{(2)}$ depends on the sign of V_0

(*d*) $E^{(1)} < 0$; $E^{(2)} < 0$

2009

110. The eigenvalues of the matrix $A = \begin{pmatrix} 0 & i \\ i & 0 \end{pmatrix}$ are

(*a*) real and distinct
(*b*) complex and distinct
(*c*) complex and coinciding
(*d*) real and coinciding

111. σ_i (i = 1, 2, 3) represent the Pauli spin matrices. Which one of the following is NOT true?

(*a*) $\sigma_i\sigma_j + \sigma_j\sigma_i = 2\delta_{ij}$
(*b*) Tr $(\sigma_i) = 0$
(*c*) The eigenvalues of σ_i are ± 1
(*d*) det $(\sigma_i) = 1$

112. Which one of the functions given below represents the bound state eigenfunction of the operator $-\frac{d^2}{dx^2}$ in the region, $0 \le x < \infty$, with the eigenvalue –4?

(*a*) $A_0 e^{2x}$ (*b*) $A_0 \cosh 2x$
(*c*) $A_0 e^{-2x}$ (*d*) $A_0 \sinh 2x$

113. The de Broglie wavelength of particles of mass m with average momentum p at a temperature T in three dimensions is given by

(*a*) $\lambda = \frac{h}{\sqrt{2mk_BT}}$ (*b*) $\lambda = \frac{h}{\sqrt{3mk_BT}}$

(*c*) $\lambda = \frac{h}{\sqrt{2k_BT}}$ (*d*) $\lambda = \frac{h}{\sqrt{2m}}$

114. Let $|n\rangle$ and $|p\rangle$ denote the isospin state with $I = \frac{1}{2}, I_3 = \frac{1}{2}$ and $I = \frac{1}{2}, I_3 = -\frac{1}{2}$ of a nucleon respectively. Which one of the following two-nucleon states has $I = 0, I_3 = 0$?

(*a*) $\frac{1}{\sqrt{2}}(|nn\rangle - |pp\rangle)$ (*b*) $\frac{1}{\sqrt{2}}(|nn\rangle + |pp\rangle)$

(*c*) $\frac{1}{\sqrt{2}}(|np\rangle - |pn\rangle)$ (*d*) $\frac{1}{\sqrt{2}}(|np\rangle + |pn\rangle)$

115. The spin function of a free particle, in the basis in which S_z is diagonal, can be written as $\begin{pmatrix} 1 \\ 0 \end{pmatrix}$ and $\begin{pmatrix} 0 \\ 1 \end{pmatrix}$ with eigenvalues $+\frac{\hbar}{2}$ and $-\frac{\hbar}{2}$, respectively. In the given basis, the normalized eigenfunction of S_y with eigenvalue $-\frac{\hbar}{2}$

(*a*) $\frac{1}{\sqrt{2}}\begin{pmatrix} 1 \\ i \end{pmatrix}$ (*b*) $\frac{1}{\sqrt{2}}\begin{pmatrix} 0 \\ i \end{pmatrix}$

(*c*) $\frac{1}{\sqrt{2}}\begin{pmatrix} i \\ 0 \end{pmatrix}$ (*d*) $\frac{1}{\sqrt{2}}\begin{pmatrix} i \\ 1 \end{pmatrix}$

116. $\hat{A}$ and $\hat{B}$ represent two physical characteristics of a quantum system. If $\hat{A}$ is Hermitian , then for the product $\hat{A}\hat{B}$ to be Hermitian, it is sufficient that

(*a*) $\hat{B}$ is Hermitian
(*b*) $\hat{B}$ is anti-Hermitian
(*c*) $\hat{B}$ is Hermitian and $\hat{A}$ and $\hat{B}$ commute
(*d*) $\hat{B}$ is Hermitian and $\hat{A}$ and $\hat{B}$ anti-commute

117. A particle is in the normalized state $|\psi\rangle$ which is a superposition of the energy eigenstates $|E_0 = 10eV\rangle$ and $|E_1 = 30eV\rangle$. The average value of energy of the particle in the state $|\psi\rangle$ is 20 eV. The state $|\psi\rangle$ is given by

(*a*) $\frac{1}{2}|E_0 = 10eV\rangle + \frac{\sqrt{3}}{4}|E_1 = 30eV\rangle$

(*b*) $\frac{1}{\sqrt{3}}|E_0 = 10eV\rangle + \sqrt{\frac{2}{3}}|E_1 = 30eV\rangle$

(*c*) $\frac{1}{2}|E_0 = 10eV\rangle - \frac{\sqrt{3}}{4}|E_1 = 30eV\rangle$

(*d*) $\frac{1}{\sqrt{2}}|E_0 = 10eV\rangle - \frac{1}{\sqrt{2}}|E_1 = 30eV\rangle$

Common data Q. 118 and 119

A free particle of mass m moves along the x direction. At $t = 0$, the normalized wave function of the particle is given by $\psi(x, 0) = \frac{1}{(2\pi\alpha)^{1/4}}\exp\left[-\frac{x^2}{4\alpha^2} + ix\right]$, where α is a real constant.

118. The expectation value of the momentum, in this state is

(*a*) $\hbar\alpha$ (*b*) $\hbar\sqrt{\alpha}$

(*c*) α (*d*) $\hbar/\sqrt{\alpha}$

119. The expectation value of the particle energy is

(*a*) $\frac{\hbar^2}{2m}\frac{1}{2\alpha^{3/2}}$ (*b*) $\frac{\hbar^2}{2m}\alpha^2$

(*c*) $\frac{\hbar^2}{2m}\frac{4\alpha^2+1}{4\alpha^{3/2}}$ (*d*) $\frac{\hbar^2}{8m\alpha^{3/2}}$

2008

120. For arbitrary matrices E, F, G and H, if EF – FE = 0 then Trace (EFGH) is equal to

(*a*) Trace (HGFE)

(*b*) Trace(E).Trace(F).Trace(G).Trace(H)

(*c*) Trace (GFEH)

(*d*) Trace (EGHF)

121. Let $|\psi_0\rangle$ denote the ground state of the hydrogen atom. Choose the correct statement from those given below

(*a*) $[L_X, L_Y]|\psi_0\rangle = 0$ (*b*) $J^2|\psi_0\rangle = 0$

(*c*) $\vec{L}.\vec{S}|\psi_0\rangle \neq 0$ (*d*) $[S_x, S_y,]|\psi_0\rangle = 0$

122. For a physcial system , two observables O_1 and O_2 are known to be compatible. Choose the correct implication from amongst those given below:

(*a*) Every eigenstate of O_1 must necessarily be an eigenstate of O_2.

(*b*) Every non-degenerate eigenstate of O_1 must necessarily be an eigenstate of O_2.

(*c*) When observation of O_1 is carried out on an arbitrary state $|\psi\rangle$ of the physical system, a subsequent observation of O_2 leads to an umabiguos result.

(*d*) Observation of O_1 andO_2, carried out on an arbitrary state $|\psi\rangle$ of the physical system, lead to the identical results irrespective of the order in which the observations are made.

123. A particle is placed in a one dimensional box of size L along the x-axis $(0 < x < L)$. Which of the following is true?

(*a*) In the ground state, the probability of finding the particle in the interval $\left(\frac{L}{4}, \frac{3L}{4}\right)$ is half.

(*b*) In the first excited state, the probability of finding the particle in the interval $\left(\frac{L}{4}, \frac{3L}{4}\right)$ is half. This also holds for states with $n = 4, 6, 8$....

(*c*) For an arbitrary $|\psi\rangle$, the probability of finding the particle in the left half of the well is half.

(*d*) In the ground state, the particle has a definit momentum.

COMMON DATA QUESTIONS

Common Data for Questions 124, 125, 126:

A beam of identical particles of mass m and energy E is incident from left on a potential barrier of width L (between $0 < x < L$) and height V_0 as shown in the figure $(E < V_0)$.

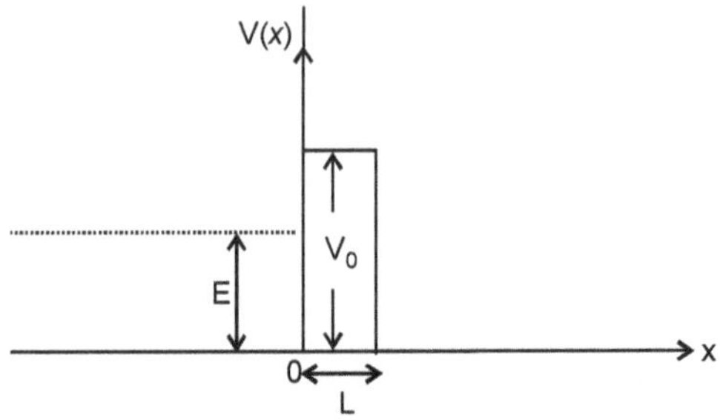

For $x > L$, there is tunneling with a transmission coefficient $T > 0$. Let A_0, A_R and A_T denote the amplitudes for the incident, reflected and the transmitted waves, respectively.

124. Throughout $0 < x < L$, the wave-function

(*a*) can be chosen to be real.

(*b*) is exponentially decaying.

(*c*) is generally complex.

(*d*) is zero.

125. Let the probability current associated with the incident wave be S_0. Let R be the reflection coefficient. Then

(*a*) the probability current vanishes in the classically forbidden region.

(*b*) the probability current is TS_0 for $x > L$.

(*c*) for, $x < 0$, the probability current is $S_0(I + R)$

(*d*) for $x > L$, the probability current is complex.

126. The ratio of the reflected to the incident amplitude A_R/A_0 is

(*a*) $1 - A_T/A_0$

(*b*) $\sqrt{(1-T)}$ in magnitude

(*c*) a real negative number

(*d*) $\sqrt{\left(1-\left|\frac{A_T}{A_0}\right|^2\right)\frac{E}{V_0 - E}}$

2007

127. Match the following

P :Franck-Hertz experiment

Q : Hartree-Fock method

R :Stern-Gerlach experiment

S :Franck-Condon principle

1. electronic excitation of molecules
2. wave function of atoms
3. spin angular momentum of atoms
4. energy levels in atoms

(*a*) P-4, Q-2, R-3, S-1 (*b*) P-1, Q-4, R-3, S-2

(*c*) P-3, Q-2, R-4, S-1 (*d*) P-4, Q-1, R-3, S-2

128. The wave function of a particle,moving in a one-dimensional time-independent potential V(x), is given by $\psi(x)$, is given $\psi(X) = e^{-iax+b}$, where a and b are constants. This means that the potential V(x) is of the form

(*a*) $V(x) \propto x$ (*b*) $V(x) \propto x^2$

(*c*) $V(x) = 0$ (*d*) $V(x) \propto e^{-ax}$

129. The energy levels of a particle of mass m in a potential of the from

$$V(x) = \infty, \quad x \le 0$$
$$= \frac{1}{2}m\omega^2x^2, \quad x > 0$$

are given, in terms of quantum number n = 0, 1, 2, 3......, by

(*a*) $\left(n+\frac{1}{2}\right)\hbar\omega$ (*b*) $\left(2n+\frac{1}{2}\right)\hbar\omega$

(*c*) $\left(2n+\frac{3}{2}\right)\hbar\omega$ (*d*) $\left(n+\frac{3}{2}\right)\hbar\omega$

130. For a particle of mass m in a one-dimensional harmonic oscillatoor potential of the form $V(x) = \frac{1}{2}m\omega^2x^2$, the first excited energy eigenstate is $\psi(x) = xe^{-ax^2}$. The value of a is

(*a*) $m\omega / 4\hbar$ (*b*) $m\omega / 3\hbar$

(*c*) $m\omega / 2\hbar$ (*d*) $2m\omega / 3\hbar$

131. If $[x, p] = i\hbar$, the value of $[x^3, p]$ is

(*a*) $2i\hbar x^2$ (*b*) $-2i\hbar x^2$

(*c*) $3i\hbar x^2$ (*d*) $-3i\hbar x^2$

132. There are only three bound states for a particle of mass m in a one-dimensional potential well of the form shown in the figure. The depth V_0 of the potential statisfies

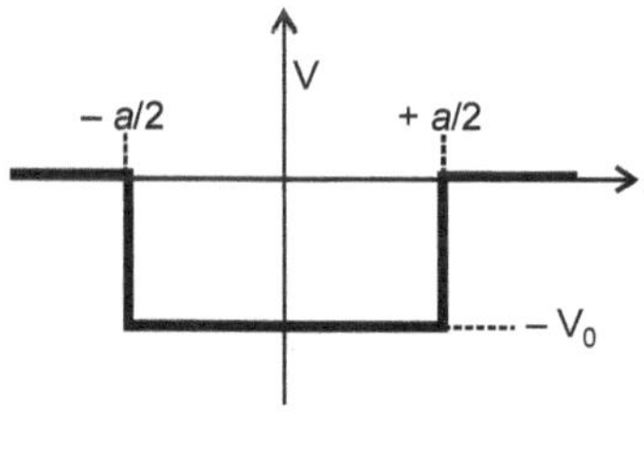

(*a*) $\frac{2\pi^2\hbar^2}{ma^2} < V_0 < \frac{9\pi^2\hbar^2}{2ma^2}$ (*b*) $\frac{\pi^2\hbar^2}{ma^2} < V_0 < \frac{2\pi^2\hbar^2}{ma^2}$

(*c*) $\frac{2\pi^2\hbar^2}{ma^2} < V_0 < \frac{8\pi^2\hbar^2}{ma^2}$ (*d*) $\frac{2\pi^2\hbar^2}{ma^2} < V_0 < \frac{50\pi^2\hbar^2}{ma^2}$

133. An atomic state of hydrogen is represented by the following wave function:

$$\Psi(r,\theta,\varphi) = \frac{1}{\sqrt{2}}\left(\frac{1}{a_0}\right)^{3/2}\left(1-\frac{r}{2a_0}\right)e^{-r/2a_0}\cos\theta.$$

where a_0 is a constant. The quantum number of the state are

(*a*) $l = 0, m = 0, n = 1$

(*b*) $l = 1, m = 1, n = 2$

(*c*) $l = 1, m = 0, n = 2$

(*d*) $l = 2, m = 0, n = 3$

134. Three operators X, Y and Z satisfy the commutation relations

$[X, Y] = i\hbar Z$, $[Y, Z] = i\hbar X$ and $[Z, X] = i\hbar Y$.

The set of all possibel eigenvalues of the operator Z, in units of $\hbar$, is

(*a*) $\{0, \pm1, \pm2, \pm3, ...\}$

(*b*) $\left\{\frac{1}{2}, 1, \frac{3}{2}, 2, \frac{5}{2}, ...\right\}$

(*c*) $\left\{0, \pm\frac{1}{2}, \pm1, \pm\frac{3}{2}, \pm2, \pm\frac{5}{2}, ...\right\}$

(*d*) $\left\{-\frac{1}{2}, +\frac{1}{2}\right\}$

COMMON DATA QUESTIONS

Common Data For Questions 135, 136, 137:

A partiicle of mass m is confined in the ground state of a one-dimensional box, extending from $x = -2L$ to $x = \pm 2L$. The wavefunction of the particle in the state is $\psi(x) = \psi_0 \cos\frac{\pi x}{4L}$ **where ψ_0 is a constant.**

135. The normalization factor ψ_0 of this wave function is

(*a*) $\sqrt{\frac{2}{L}}$ (*b*) $\sqrt{\frac{1}{4L}}$

(*c*) $\sqrt{\frac{1}{2L}}$ (*d*) $\sqrt{\frac{1}{L}}$

136. The energy eigenvalue corresponding to this state is

(*a*) $\frac{h^2\pi^2}{2mL^2}$ (*b*) $\frac{h^2\pi^2}{4mL^2}$

(*c*) $\frac{h^2\pi^2}{16mL^2}$ (*d*) $\frac{h^2\pi^2}{8L^2}$

137. The expection value

(*a*) $\sqrt{\frac{2}{L}}$ (*b*) $\sqrt{\frac{1}{4L}}$

(*c*) $\sqrt{\frac{1}{2L}}$ (*d*) $\sqrt{\frac{1}{L}}$

2006

138. A particle is moving in a spherically symmetric potential $V(r) = \alpha r^2$, where α is a positive constant. In a stationary state, the expectation value of the kinetic energy <T> of the particle is

(a) $\langle T\rangle = \langle V\rangle$ (b) $\langle T\rangle = 2\langle V\rangle$

(c) $\langle T\rangle = 3\langle V\rangle$ (d) $\langle T\rangle = 4\langle V\rangle$

139. A particle of mass m is represented by the wavefunction $\Psi(x) = Ae^{ikx}$, where k is the wavevector and A is a constant. The magnitude of the probability current density of the particle is

(a) $|A|^2 \frac{\hbar k}{m}$ (b) $|A|^2 \frac{\hbar k}{2m}$

(c) $|A|^2 \frac{(\hbar k)^2}{m}$ (d) $|A|^2 \frac{(\hbar k)^2}{2m}$

140. A one-dimensional harmonic oscillator is in the state

$$\Psi(x) = \frac{1}{\sqrt{14}}[3\psi_0(x) - 2\psi_1(x) + \psi_2(x)],$$

where $\Psi_0(x)$, $\Psi_1(x)$ and $\Psi_2(x)$ are the ground, first excited and second excited states, respectively. The probability of finding the oscillator in the ground state is

(a) 0 (b) $\frac{3}{\sqrt{14}}$

(c) $\frac{9}{14}$ (d) 1

141. The wavefunction of a particle in a one-dimensinal potential at time $t = 0$ is

$$\psi(x, t = 0) = \frac{1}{\sqrt{15}}[2\psi_0(x) - \psi_1(x)],$$

where $\Psi_0(x)$ and $\Psi_1(x)$ are the ground and the first excited states of the particle with corresponding energies E_0 and E_1. The wavefunction of the particle at a time t is

(a) $\frac{1}{\sqrt{5}} e^{-i(E_0E_1)t/2\hbar}[2\psi_0(x) - \psi_1(x)]$

(b) $\frac{1}{\sqrt{5}} e^{-E_0 t/\hbar}[2\psi_0(x) - \psi_1(x)]$

(c) $\frac{1}{\sqrt{5}} e^{-iE_1 t/\hbar}[2\psi_0(x) - \psi_1(x)]$

(d) $\frac{1}{\sqrt{15}}\left[2\psi_0(x)e^{-E_0t/\hbar} - \psi_1(x)e^{-iE1t/4}\right]$

142. In hydro genic states, the probability of finding an eelctron at $r = 0$ is

(a) zero in state $\phi_{1s}(r)$

(b) non-zero in state $\phi_{1s}(r)$

(c) zero in state $\phi_{2s}(r)$

(d) zero in state $\phi_{2p}(r)$

2005

143. The parities of the wave functions (*i*) cos (kx), and (*ii*) tan h(kx) are

(a) (*i*) odd, (*ii*) odd (b) (*i*) even, (*ii*) even

(c) (*i*) odd, (*ii*) even (d) (*i*) even, (*ii*) odd

144. The commutator, $[L_z, Y_{lm}(\theta, \phi)]$ where L_2 is the z component of the orbital angular momentum and $Y_{lm}(\theta, \phi)$ is a spherical harmonic, is

(a) $l(l+1)\hbar$ (b) $-m\hbar$

(c) $m\hbar$ (d) $+l\hbar$

145. A system in a normalized state $|\psi\rangle = c_1|\alpha_1\rangle + c_2|\alpha_2\rangle$, with $|\alpha_1\rangle$ and $|\alpha_2\rangle$ representing two different eigenstates of the system, requires that the constants c_1 and c_2 must satisfy the condition

(a) $|c_1|.|c_2| = 1$ (b) $|c_1| + |c_2| = 1$

(c) $(|c_1| + |c_2|)^2 = 1$ (d) $|c_1|^2 + |c_2|^2 = 1$

146. A one dimensional harmonic oscillator carrying a charge $-q$ is placed in a uniform electric field $\vec{E}$ along the positive x-axis. The corresponding Hamiltonian operator is

(a) $\frac{\hbar^2}{2m}\frac{d^2}{dx^2} + \frac{1}{2}kx^2 + qEx$

(b) $\frac{\hbar^2}{2m}\frac{d^2}{dx^2} + \frac{1}{2}kx^2 - qEx$

(c) $-\frac{\hbar^2}{2m}\frac{d^2}{dx^2} + \frac{1}{2}kx^2 + qEx$

(d) $-\frac{\hbar^2}{2m}\frac{d^2}{dx^2} + \frac{1}{2}kx^2 - qEx$

147. The L_β line of X-rays emitted from an atom with principal quantum numbers $n = 1, 2, 3, ...$, arises from the transition

(a) $n = 4 \rightarrow n = 2$

(b) $n = 3 \rightarrow n = 2$

(c) $n = 5 \rightarrow n = 2$

(d) $n = 3 \rightarrow n = 1$

148. For an electron in hydrogen atom, the states are characterized by the usual quantum numbers n, l, m_l. The electric dipole transition between any two states requires that

(a) $\Delta l = 0, \Delta m_l = 0, \pm 1$

(b) $\Delta l = \pm 1, \Delta m_l = \pm 1, \pm 2$

(c) $\Delta l = \pm 1, \Delta m_l = 0, \pm 1$

(d) $\Delta l = \pm 1, \Delta m_l = 0, \pm 2$

149. The de Broglie wavelength λ for an electron of energy 150 eV is

(a) 10^{-8} m (b) 10^{-10} m

(c) 10^{-12}m (d) 10^{-14}m

150. A particle is incident with a constant energy E on a one-dimensional potential barrier as shown in the figure. The wavefunctions in regions I and II are respectively

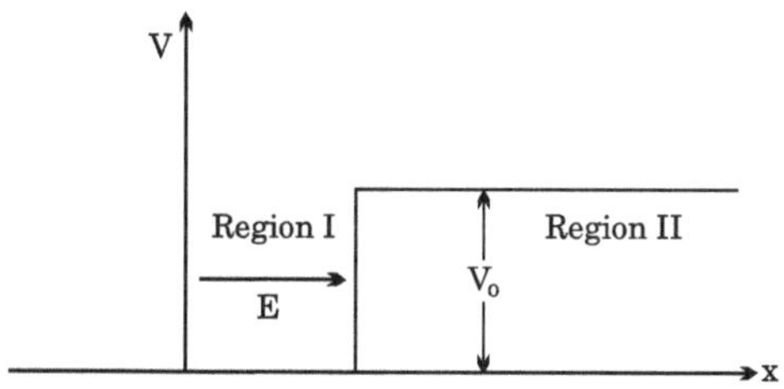

(a) decaying, oscillatory

(b) oscillatory, oscillatory

(c) oscillatory, decaying

(d) decaying, decaying

151. The expectation value of the z coordinate, $\langle z \rangle$, in the ground state of the hydrogen atom (wavefunction : $\psi_{100}(r) = A\, e^{-r/a_0}$, where A is the normalization constant and a_0 is the Bohr radius), is

(a) a_0 (b) $\frac{a_0}{2}$

(c) $\frac{a_0}{4}$ (d) 0

152. The degeneracy of the $n = 2$ level for a three dimensional isotropic oscillator is

(a) 4 (b) 6

(c) 8 (d) 10

153. For a spin 1/2 particle, the expectation value of $s_x s_y s_z$, where s_x, s_y and s_z are spin operators, is

(a) $\frac{i\hbar^3}{8}$ (b) $-\frac{i\hbar^3}{8}$

(c) $\frac{i\hbar^3}{16}$ (d) $-\frac{i\hbar^3}{16}$

154. If the probability that x lies between x and x + dx is $p(x)\,dx = ae^{-ax}\,dx$, where $0 < x < \infty$, $a > 0$, then the probability that x lies between x_1 and x_2 $(x_2 > x_1)$ is

(a) $\left(e^{-ax_1} - e^{-ax_2}\right)$ (b) $a\left(e^{-ax_1} - e^{-ax_2}\right)$

(c) $e^{-ax_2}\left(e^{-ax_1} - e^{-ax_2}\right)$ (d) $e^{-ax_1}\left(e^{-ax_1} - e^{-ax_2}\right)$

Statement for Linked Answer Questions 155 and 156 :

A particle is scattered by a spherically symmetric potential. In the centre of mass (CM) frame the wavefunction of the incoming particle is $\psi = A\, e^{ikz}$ where k is the wavevector and A is a constant.

155. If f (θ) is an angular function then in the asymptotic region the scattered wavefunction has the form

(a) $\frac{A f(\theta)\, e^{ikr}}{r}$ (b) $\frac{A f(\theta)\, e^{-ikr}}{r}$

(c) $\frac{A f(\theta)\, e^{ikr}}{r^2}$ (d) $\frac{A f(\theta)\, e^{-ikr}}{r^2}$

156. The differential scattering cross section $\sigma(\theta)$ in CM frame is

(a) $\sigma(\theta) = |A|^2 \frac{|f(\theta)|^2}{r^2}$

(b) $\sigma(\theta) = |A|^2 |f(\theta)|^2$

(c) $\sigma(\theta) = |f(\theta)|^2$

(d) $\sigma(\theta) = |A|\,|f(\theta)|$

2004

157. The time-independent Schrodinger equation of a system represents the conservation of the

(a) total binding energy of the system

(b) total potential energy of the system

(c) total kinetic energy of the system

(d) total energy of the system

158. In a hydrogen atom, the accidental or Coulomb degeneracy for the $n = 4$ state is

(a) 4 (b) 16

(c) 18 (d) 32

159. The wave function of a spin-less particle of mass m in a one-dimensional potential V(x) is $\psi(x) = A \exp(-\alpha^2 x^2)$ corresponding to an eigenvalue $E_0 = \hbar^2 \alpha^2/m$. The potential V(x) is

(a) $2E_0(1 - \alpha^2 x^2)$ (b) $2E_0(1 + \alpha^2 x^2)$

(c) $2E_0 \alpha^2 x^2$ (d) $2E_0(1 + 2\alpha^2 x^2)$

160. Two spin-1/2 fermions having spins $\vec{S_1}$ and $\vec{S_2}$ interact via a potential $V(r) = \vec{S_1} \cdot \vec{S_2}\, V_0(r)$. The contributions of this potential in the singlet and triplet states, respectively, are

(a) $-\frac{3}{2}V_0(r)$ and $\frac{1}{2}V_0(r)$

(b) $\frac{1}{2}V_0(r)$ and $-\frac{3}{2}V_0(r)$

(c) $\frac{1}{4}V_0(r)$ and $-\frac{3}{4}V_0(r)$

(d) $-\frac{3}{4}V_0(r)$ and $\frac{1}{4}V_0(r)$

161. The wave function of a one-dimensional harmonic oscillator is $\psi_0 = A\exp(-\alpha^2x^2/2)$ for the ground state $E_0 = \hbar\omega/2$, where $\alpha^2 = \mu\omega/\hbar$. In the presence of a perturbing potential of $E_0(\alpha x/10)^4$, the first order change in the ground state energy is

[Given : $\Gamma(x+1) = \int_0^\infty t^x \exp(-t)\, dt$]

(a) $(\frac{1}{2}E_0)\, 10^{-4}$ (b) $(3E_0)\, 10^{-4}$

(c) $(\frac{3}{4}E_0)\, 10^{-4}$ (d) $(E_0)\, 10^{-4}$

Data for Questions 162-163

A particle is confined to the region $0 < x < L$ in one dimension.

162. If the particle is in the first excited state, then the probability of finding the particle is maximum at

(a) $x = \frac{L}{6}$ (b) $x = \frac{L}{2}$

(c) $x = \frac{L}{3}$ (d) $x = \frac{L}{4}$ and $\frac{3L}{4}$

163. If the particle is in the lowest energy state, then the probability of finding the particle in the region $0 < x < L/4$ is

(a) $\frac{1}{4} - \frac{1}{(2\pi)}$ (b) $\frac{1}{4}$

(c) $\frac{1}{4} + \frac{1}{(2\pi)}$ (d) $\frac{1}{2}$

Data for Questions 164-165

The one-electron states for non-interacting electrons confined in a cubic box of side a are $\varepsilon_0 < \varepsilon_1 < \varepsilon_2 < \varepsilon_3 < \varepsilon_4$ etc.

164. The energy of the lowest state is

(a) Zero (b) $\frac{\hbar^2\pi^2}{2ma^2}$

(c) $\frac{\hbar^2\pi^2}{ma^2}$ (d) $\frac{3\hbar^2\pi^2}{2ma^2}$

165. The degeneracy (including spin) of the level ε_3 is

(a) 2 (b) 4

(c) 6 (d) 8

2003

166. Given a wave with the dispersion relation $\omega = ck + m$ for $k > 0$ and $m > 0$, which one of the following is true ?

(a) The group velocity is greater than the phase velocity

(b) The group velocity is less than the phase velocity

(c) The group velocity and the phase velocity are equal

(d) There is no definite relation between the group velocity and the phase velocity

167. Which of the following is a valid normalized wave function for a particle in a one dimensional infinite potential well of width L centered at $x = 0$?

(a) $(2/L)\,[\cos(2\pi x/L) + \sin(2\pi x/L)]$

(b) $(2/L)^{1/2} \sin[n\pi x/L]$ for odd n

(c) $(2/L)^{1/2} \cos[n\pi x/L]$ for odd n

(d) $(2/L)\cos(\pi x/L)$

168. The commutator $[x, P^2]$, where x and P are position and momentum operators respectively, is

(a) $2i\hbar P$ (b) $-i\hbar P$

(c) $2i\hbar xP$ (d) $-2i\hbar xP$

169. A spin half particle is in the state $S_z = \hbar/2$. The expectation values of S_x, S^2_x, S_y, S_y^2 are given by

(a) $0, 0, \hbar^2/4, \hbar^2/4$ (b) $0, \hbar^2/4, \hbar^2/4, 0$

(c) $0, \hbar^2/4, 0, \hbar^2/4$ (d) $\hbar^2/4, \hbar^2/4, 0, 0$

170. The hyperfine splitting of the spectral lines of an atom is due to

(a) the coupling between the spins of two or more electrons

(b) the coupling between the spins and the orbital angular momenta of the electrons

(c) the coupling between the electron spins and the nuclear spin

(d) the effect of external electromagnetic fields

171. Consider the energy E in the first Brillouin zone as a function of the magnitude of the wave vector k for a crystal of lattice constant a. Then

(*a*) the slope of E versus k is proportional to the group velocity

(*b*) the slope of E versus k has its maximum value at $|k| = \pi/a$

(*c*) the plot of E versus k will be parabolic in the interval $(-\pi/a) < k < (\pi/a)$

(*d*) the slope of E versus k is non-zero for all k the interval $(-\pi/a) < k < (\pi/a)$

172. The energy density of states of an electron in a one dimensional potential well of infinitely high walls is (the symbols have their usual meaning)

(*a*) $L\sqrt{m}/[\pi \hbar \sqrt{(2E)}]$ (*b*) $Lm/(\pi\hbar \sqrt{E})$

(*c*) $Lm/[\pi \hbar \sqrt{(2E)}]$ (*d*) $L\sqrt{m}/(2\pi\hbar E)$

Data for Questions 173 – 174 :

A particle is located in a three diamensional cubic well of width L with impenetrable walls

173. The sum of the energies of the third and the fourth levels is

(*a*) $10\pi^2 \hbar^2/mL^2$ (*b*) $10\pi^2 \hbar^2/3mL^2$

(*c*) $11\pi^2 \hbar^2/2mL^2$ (*d*) $15\pi^2 \hbar^2/2mL^2$

174. The degeneracy of the fourth level is given by

(*a*) 1 (*b*) 2

(*c*) 3 (*d*) 4

Data for Questions 175 – 176 :

The normalized wave functions Ψ_1 and Ψ_2 correspond to the ground state and the first excited state of a particle in a potential. You are given the information that the operator $\hat{A}$ acts on the wave functions as $\hat{A}\Psi_1 = \Psi_2$ and $\hat{A}\Psi_2 = \Psi_1$

175. The expectation value of A for the state $\Psi = (3\Psi_1 + 4\Psi_2)/5$ is

(*a*) –0.32 (*b*) 0.0

(*c*) 0.75 (*d*) 0.96

176. Which of the following are eigen functions of $\hat{A}^2$?

(*a*) Ψ_1 and Ψ_2 (*b*) Ψ_2 and not Ψ_1

(*c*) Ψ_1 and not Ψ_2 (*d*) neither Ψ_1 nor Ψ_2

177. An electron in a time independent potential is in a state which is the superposition of the ground state (E_0 = 11 eV) and the first excited state (E_1 = 1 eV). The wave function of the electron will repeat itself with a period of

(*a*) 3.1×10^{-18} s (*b*) 2.1×10^{-15} s

(*c*) 1.2×10^{-12} s (*d*) 1.0×10^{-9} s

178. A particle has the wave function $\Psi(x, t) = A(\exp(i\omega t)\cos(kx)$. Which one of the following is correct?

(*a*) This is an eigenstate of both energy and momentum

(*b*) This is an eigenstate of momentum and not energy

(*c*) This is an eigenstate of energy and not momentum

(*d*) This is not an eigenstate of energy or momentum

179. A free particle with energy E whose wave-function is a plane wave with wavelength λ enters a region of constant potential V > 0 where the wavelength of the particle is 2λ. The ratio (V/E) is

(*a*) 1/2 (*b*) 2/3

(*c*) 3/4 (*d*) 4/5

2002

180. If R_1 is the value of the Rydberg constant assuming mass of the nucleus to be infinitely large compared to that of an electron, and if R_2 is the Rydberg constant taking nuclear mass to be 7500 times the mass of the electron, then the ratio R_2/R_1 is

(*a*) a little less than unity

(*b*) a little more than unity

(*c*) infinitely small

(*d*) infinitely large

181. If the wave function of a particle trapped in space between $x = 0$ and $x = L$ is given by

$\psi(x) = A\sin\left(\frac{2\pi x}{L}\right)$, where A is a constant,

for which value(s) of x will the probability of finding the particle be the maximum ?

(*a*) $\frac{L}{4}$ (*b*) $\frac{L}{2}$

(*c*) $\frac{L}{6}$ and $\frac{L}{3}$ (*d*) $\frac{L}{4}$ and $\frac{3L}{4}$

182. If σ is the total cross-section and $f(\theta)$, θ being the angle of scattering, is the scattering amplitude for a quantum mechanical elastic scattering by a spherically symmetric potential, then which of the following is true ? Note that k is the magnitude of the wave vector along the $\hat{z}$ direction.

(*a*) $\sigma = |f(\theta)|^2$

(*b*) $\sigma = \frac{4\pi}{k}|f(\theta = 0)|^2$

(*c*) $\sigma = \frac{4\pi}{k} \times$ Imaginary part of $[f(\theta = 0)]$

(*d*) $\sigma = \frac{4\pi}{k}|f(\theta)|^2$

183. An electron propagating along the x-axis passes through a slit of width $\Delta y = 1$ nm. The uncertainty in the y-component of its velocity after passing through the slit is

(a) 7.322×10^5 m/s (b) 1.166×10^5 m/s
(c) 3.436×10^5 m/s (d) 2.326×10^4 m/s

184. $\hat{A}$ and $\hat{B}$ are two quantum mechanical operators. If $[\hat{A}, \hat{B}]$ stands for the commu-tator of $\hat{A}$ and $\hat{B}$, then $[[\hat{A},\hat{B}],[\hat{B},\hat{A}]]$ is equal to

(a) $\hat{A}\hat{B}\hat{A}\hat{B} - \hat{B}\hat{A}\hat{B}\hat{A}$
(b) $\hat{A}(\hat{A}\hat{B} - \hat{B}\hat{A}) - \hat{B}(\hat{B}\hat{A} - \hat{A}\hat{B})$
(c) zero
(d) $([\hat{A},\hat{B}])^2$

185. An electron is in a state with spin wave function

$$\phi_s = \begin{pmatrix} \sqrt{3}/2 \\ 1/2 \end{pmatrix}$$

in the S_z representation. What is the probability of finding the z-component of its spin along the $-\hat{z}$ direction?

(a) 0.75 (b) 0.50
(c) 0.35 (d) 0.25

186. If the wavelength of the first line of the Balmer series in the hydrogen spectrum is λ, then the wavelength of the first line of the Lyman series is

(a) $(27/5)\lambda$ (b) $(5/27)\lambda$
(c) $(32/27)\lambda$ (d) $(27/32)\lambda$

2001

187. A spinless particle moves in a central potential $V(r)$.

(a) The kinetic energy and the potential energy of the particle cannot simultaneously have sharp values.
(b) The total energy and the potential energy of the particle can simultane-ously have sharp values.
(c) The total energy and the square of the orbital angular momentum about the origin cannot simultaneously have sharp values.
(d) The total energy of the particle can have only discrete eigenvalues.

188. Which of the following functions represents acceptable wave function of a particle in the range $-\infty \le x \le \infty$.

(a) $\varphi(x) = A \tan x$, $A > 0$
(b) $\varphi(x) = B \cos x$, B real
(c) $\varphi(x) = C \exp(-D/x^2)$, $C > 0$, $D < 0$
(d) $\varphi(x) = E\, x \exp(-Fx^2)$, $E, F > 0$

189. A quantum harmonic oscillator is in the energy eigenstate $|n\rangle$. A time independent perturbation $1\,\lambda(a^t a)^2$ acts on the particle, where λ is a constant of suitable dimensions and a and a^t are lowering and raising operators respectively. Then the first order energy shift is given by

(a) λn (b) $\lambda^2 n$
(c) λn^2 (d) $(\lambda n)^2$

190. x and p are two operators which satisfy $[x, p] = i$. The operators X and P are defined as
$X = x \cos\phi + p \sin\phi$ and
$Y = -x \sin\phi + p \cos\phi$,
for ϕ real. Then $[X, Y]$ equals

(a) 1 (b) -1
(c) i (d) $-i$

191. A quantum particle of mass m is confined to a square region in xoy-plane whose vertices are given by (0, 0), (L, 0), (L, L) and (0, L). Which of the following represents an admissible wave function of the particle (for l, m, n positive integers)?

(a) $\frac{2}{L} \sin\left(\frac{n\pi x}{L}\right) \cos\left(\frac{m\pi y}{L}\right)$

(b) $\frac{2}{L} \cos\left(\frac{l\pi x}{L}\right) \cos\left(\frac{n\pi y}{L}\right)$

(c) $\frac{2}{L} \sin\left(\frac{m\pi x}{L}\right) \sin\left(\frac{n\pi y}{L}\right)$

(d) $\frac{2}{L} \cos\left(\frac{n\pi x}{L}\right) \sin\left(\frac{l\pi y}{L}\right)$

192. Let $\vec{L} = (L_x, L_y, L_z)$ denote the orbital angular momentum operators of a particle and let $L_+ = L_x + iL_y$ and $L_- = L_x - iL_y$. The particle is in an eigenstate of L^2 and L_z eigenvalues $\hbar^2(l+1)$ and $\hbar l$ respectively. The expectation value of L_+L_- in this state is

(a) $l\hbar^2$ (b) $2l\hbar^2$
(c) 0 (d) $l\hbar$

193. A normalized state of a particle moving in a potential $V(x)$ is given by

$$\psi(x, t) = \sum_{n=0}^{\infty} C_n \phi_n(x) e^{-E_n t/\hbar}$$

where $\phi_n(x)$'s are the normalized eigenfunctions of the particle corres-ponding to the energies E_n's. Then

(a) $\sum_{n=0} |C_n|^2 = 1$

(b) The average energy of the particle in the state $\psi(x, t)$ is $\sum_{n=\infty}^{\infty} |C_n| E_n$

(*c*) $\psi(x, t)$ is an eigenfunction of the Hamiltonian of the particle.

(*d*) $\psi(x, t)$ is an eigenfunction of the momentum operator.

2000

194. Let E_1, E_2, E_3 be the respective ground state energies of the following potentials:
Which one of the following is correct?

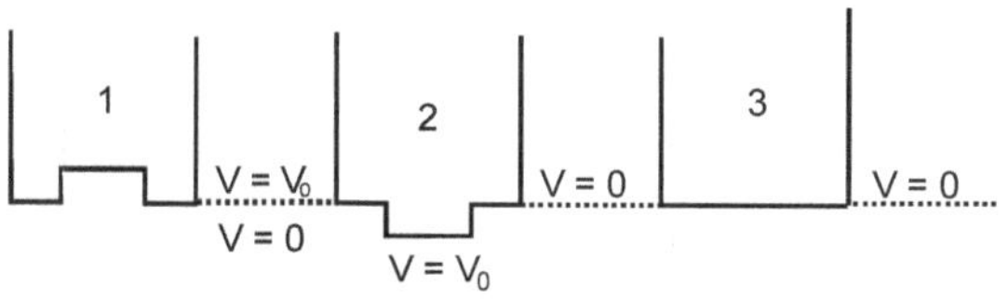

(*a*) $E_1 < E_2 < E_3$ (*b*) $E_3 < E_1 < E_2$
(*c*) $E_2 < E_3 < E_1$ (*d*) $E_2 < E_1 < E_3$

195. A particle is scattered by a central potential. If the dominant contribution to the scattering is from the *p*-wave, the differential cross section is
(*a*) isotropic
(*b*) proportional to $\cos^2\theta$
(*c*) proportional to $\cos\theta \cos^2\phi$
(*d*) proportional to $\sin^2\theta \sin^2\phi$

196. Let a particle move ina potential field given by

$$V(x) = \begin{cases} \left(\frac{1}{2}\right) m\omega^2 x^2 & \text{for } x > 0 \\ \infty & \text{for } x \le 0 \end{cases}$$

The allowed energies of this particle are
(*a*) $(n + 1/2)\hbar\omega$
(*b*) $(2n + 3/2)\hbar\omega$
(*c*) $(2n + 1/2)\hbar\omega$
(*d*) $(n + 5/2)\hbar\omega$

197. Bound state eigenfunctions of an attractive finite range smooth potential behave for larger *r* as:
(*a*) $\exp(-r/r_0)$ where r_0 is a positive constant
(*b*) (I/r^n) where $n > 0$
(*c*) constant
(*d*) $\exp(ikr)/r$

198. An admissible potential between the proton and the neutron in a deutron is
(*a*) Coulomb
(*b*) Harmonic oscillator
(*c*) Finite square well
(*d*) Infinite square well

ANSWERS

1. (a) **2.** (6 to 6) **3.** (–1 to –1) **4.** (d) **5.** (c) **6.** (a) **7.** (a) **8.** (a,d)
9. (–3 to –3) **10.** (4 to 4) **11.** (*b*) **12.** (*d*) **13.** (*a*) **14.** (1 to 1) **15.** (*d*) **16.** (4 to 4)
17. (6 to 6) **18.** (0.706 to 0.708) **19.** (3 to 3) **20.** (1 to 1) **21.** (*a*) **22.** (*c*) **23.** (*d*)
24. (2.5 to 2.5) **25.** (06 to 0.6) **26.** (*b*) **27.** (*d*) **28.** (*c*) **29.** (*a*) **30.** (0.25 to 0.25)
31. (0.125 to 0.125) **32.** (*a*) **33.** (*b*) **34.** (*d*) **35.** (*b*) **36.** (*c*) **37.** (2 to 2) **38.** (0.78 to 0.82)
39. (0.25 to 0.25) **40.** (*c*) **41.** 1.30 to 1.34 **42.** (*d*) **43.** (*c*) **44.** (*a*) **45.** (*d*) **46.** 0.10 to 0.12
47. (*b*) **48.** (*c*) **49.** (*b*) **50.** (*a*) **51.** (*c*) **52.** (*b*) **53.** (*c*) **54.** (*a*) **55.** (*b*) **56.** (*b*) **57.** (*d*)
58. (1.40 : 1.45) **59.** (*d*) **60.** (*c*) **61.** (*b*) **62.** (8) **63.** (*d*) **64.** (*d*) **65.** (10) **66.** (*a*) **67.** (5)
68. (–0.3) **69.** (0) **70.** (*c*) **71.** (*d*) **72.** (*a*) **73.** (*b*) **74.** (*d*) **75.** (3.99 to 4.01) **76.** (1.99 to 2.01)
77. (0.66 to 0.68) **78.** (*c*) **79.** (3.2 to 3.3) **80.** (1.99 to 2.01) **81.** (*d*) **82.** (5) **83.** (7.6) **84.** (0.25)
85. (*c*) **86.** (*c*) **87.** (*c*) **88.** (*b*) **89.** (*d*) **90.** (*a*) **91.** (*d*) **92.** (*c*) **93.** (*d*) **94.** (*c*) **95.** (*b*)
96. (*c*) **97.** (*b*) **98.** (*c*) **99.** (*d*) **100.** (*a*) **101.** (*c*) **102.** (*d*) **103.** (*c*) **104.** (*d*) **105.** (*d*) **106.** (*d*)
107. (*c*) **108.** (*a*) **109.** (*c*) **110.** (*b*) **111.** (*d*) **112.** (*c*) **113.** (*b*) **114.** (*c*) **115.** (*c*) **116.** (*c*) **117.** (*d*)
118. (*b*) **119.** (*c*) **120.** (*a*) **121.** (*a*) **122.** (*a*) **123.** (*c*) **124.** (*a*) **125.** (*b*) **126.** (*a*) **127.** (*b*) **128.** (*c*)
129. (*a*) **130.** (*c*) **131.** (*c*) **132.** (*c*) **133.** (*b*) **134.** (*d*) **135.** (*c*) **136.** (*d*) **137.** (*d*) **138.** (*a*) **139.** (*a*)
140. (*c*) **141.** (*d*) **142.** (*b*) **143.** (*d*) **144.** (*c*) **145.** (*d*) **146.** (*d*) **147.** (*b*) **148.** (*c*) **149.** (*a*) **150.** (*c*)
151. (*d*) **152.** (*b*) **153.** (*b*) **154.** (*a*) **155.** (*a*) **156.** (*c*) **157.** (*d*) **158.** (*b*) **159.** (*c*) **160.** (*a*) **161.** (*c*)
162. (*d*) **163.** (*a*) **164.** (*d*) **165.** (*c*) **166.** (*b*) **167.** (*c*) **168.** (*c*) **169.** (*c*) **170.** (*c*) **171.** (*c*) **172.** (*a*)
173. (*a*) **174.** (*c*) **175.** (*d*) **176.** (*a*) **177.** (*b*) **178.** (*a*) **179.** (*c*) **180.** (*a*) **181.** (*d*) **182.** (*c*) **183.** (*b*)
184. (*c*) **185.** (*d*) **186.** (*b*) **187.** (*a*) **188.** (*d*) **189.** (*c*) **190.** (*c*) **191.** (*c*) **192.** (*a*) **193.** (*a*) **194.** (*c*)
195. (*b*) **196.** (*a*) **197.** (*a*) **198.** (*d*)

EXPLANATIONS

1. Unit vector representing the direction of propagation is

$$= \frac{4\hat{x} + 3\hat{y} + 5\hat{z}}{\sqrt{4^2 + 3^2 + 5^2}}$$

$$= \frac{4\hat{x} + 3\hat{y} + 5\hat{z}}{\sqrt{16 + 9 + 25}}$$

$$= \frac{4}{\sqrt{50}}\hat{x} + \frac{3}{\sqrt{50}}\hat{y} + \frac{5}{\sqrt{50}}\hat{z}$$

$$= \frac{4}{5\sqrt{2}}\hat{x} + \frac{3}{5\sqrt{2}}\hat{y} + \frac{1}{\sqrt{2}}\hat{z},$$

So, option (a) is correct.

2. For third Exited stat

$E = 2E_0 + 9E_0$

$$\left[E_0 = \frac{\pi^2\hbar^2}{2mL^2}\right]$$

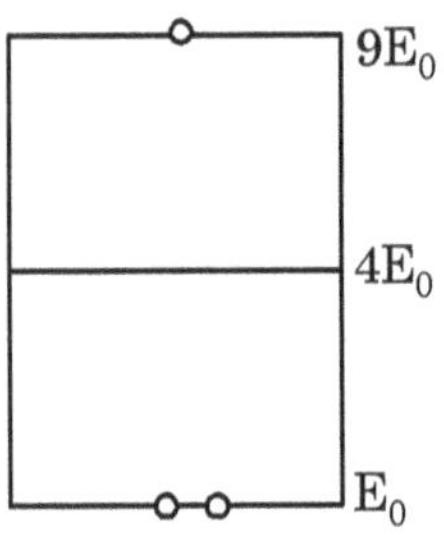

So,

$E = 11\, E_0$

$$E = \frac{11\pi^2\hbar^2}{2mL}$$

So, $\boxed{x = 5.5}$

3. Given

$Q = \vec{a}.\vec{p}$

$P = \vec{b}.\vec{r}$

For Canonical transformation

$[Q, P] = 1$ ...(1)

We know that

$[Q, P] = \left[\vec{a}.\vec{p}, \vec{b}.\vec{r}\right]$

$[Q, P] = [a_x\, p_x + a_y\, p_y + a_z\, p_z,\ b_x\, x + b_y y + b_z Z]$

$[Q, P] = a_x\, b_x\, [p_x.x] + a_y b_y\, [p_y.y] + a_z\, b_z\, [p_z, z]$

$[Q, P] = -[a_x\, b_x + a_y\, b_y + a_z\, b_z]$

$[Q, P] = -\vec{a}\,.\,\vec{b}$

From (1)

$-\vec{a}\,.\,\vec{b} = 1$

$\boxed{\vec{a}\,.\,\vec{b} = -1}$

4. $|\phi\rangle = \frac{1}{3}\begin{bmatrix} 2+i \\ 2 \end{bmatrix}$

Since, given state is normalised

$|\phi\rangle = C_1|\uparrow\rangle + C_2|\downarrow\rangle$

$$|\phi\rangle = C_1\frac{1}{2}\begin{pmatrix}1\\1\end{pmatrix} + C_2\frac{1}{\sqrt{2}}\begin{pmatrix}1\\-1\end{pmatrix}$$

$$C_1 = \frac{1}{3\sqrt{2}}(1\quad 1)\begin{pmatrix}2+i\\2\end{pmatrix}$$

$$= \frac{1}{3\sqrt{2}}(2 + i + 2)$$

$$C_1 = \frac{1}{3\sqrt{2}}(4 + i)$$

$$P\left(+\frac{h}{2}\right) = |C_1|^2$$

$$= \frac{1}{9 \times 2}|4 + i|^2 = \frac{17}{18}$$

Hence, option (d) is the correct answer.

6. $\psi(x, t = 0) = \psi_2(x, t = 0) + \psi_4(x, t = 0)$

$|\psi> = |2> + |4>$

Normalised state

$$|\psi> = \frac{1}{\sqrt{2}}[|2> + |4>]$$

after time t

$$|\psi\,(x, t\ |> = \frac{1}{\sqrt{2}}\left[e^{-iE_2\frac{t}{h}}|2> + e^{-iE_4\frac{t}{h}}|4>\right]$$

According to the question

$<\psi\,(x, t = 0)\ |\ \psi\,(x, t)> = 0$

therefore

$$=\frac{1}{2}\left\langle\left(<2|+<4|\right)\left(e^{-iE_2\frac{t}{h}}|2>+e^{-iE_4\frac{t}{h}}|4>\right)\right\rangle$$

$$=\frac{1}{2}\left(e^{-iE_2\frac{t}{h}}+e^{-iE_4\frac{t}{h}}\right)=0$$

$$e^{-iE_2\frac{t}{h}}=-e^{-iE_4\frac{t}{h}}$$

$$e^{-i(E_4-E_2)\frac{t}{h}}=-1$$

$$e^{-i(E_4-E_2)\frac{t}{h}}=e^{i\pi}$$

and $e^{\frac{-i(2h\omega t)}{h}}=e^{i\pi}$

By comparing exponents

$$\boxed{t=\frac{\pi}{2\omega}}$$

7. Given

$$U(r)=-U_0\frac{e^{-dr}}{r}=V(r)$$

Using Born Approximation

$$f(\theta,\phi)=\frac{-2m}{h^2q}\int_0^\infty rv(r)\sin(qr)dr$$

$$=\frac{-2m}{h^2q}\int_0^\infty\left(-U_0\frac{e^{-\infty}}{r}\right)r\sin qr\,dr$$

$$=\frac{-2mU_0}{h^2q}\int_0^\infty e^{-dr}\sin(qr)\,dr$$

$$=\frac{2mU_0}{h^2q}\left(\frac{q}{\propto^2+q^2}\right)$$

$$F(\theta,\phi)=\frac{2mU_0}{h^2\left(\propto^2+q^2\right)}$$

$F(\theta,\phi)\propto q^{-2}$

8. Given that,

$P^+=P,\ Q^+=Q$

$RPR^{-1}=S_1,\ RQR^{-1}=S_2$

P & Q both have real eigen values then both matrices S_1 & S_2 have all real elements,

Now, $S_1S_2=RPR^{-1}RQR^{-1}$

$=RP(R^{-1}R)QR^{-1}$

$=RPQR^{-1}$ ...(1)

$S_2S_1=RQR^{-1}RPR^{-1}$

$=RQ(R^{-1}R)PR^{-1}$

$=RQPR^{-1}$ ...(2)

We know that diagonal matrices follows the cumulative Law of multiplication, then

$S_1S_2=S_2S_1$

From equation (1) & (2)

$RPQR^{-1}=RQPR^{-1}$

$PQ=QP$

$[P,Q]=0$

i.e, P & Q commute

Since P & Q are hermitian matrices and commute therefore product of P & Q is also hermitian then product matrix have real eigen values.

9. Total spin of system

$$\vec{\sigma}=\vec{\sigma_1}+\vec{\sigma_2}$$

$$(\vec{\sigma},\vec{\sigma})=(\vec{\sigma_1}+\vec{\sigma_2}).(\vec{\sigma_1}+\vec{\sigma_2})$$

$$\sigma^2=\sigma_1^2+\sigma_2^2+2\vec{\sigma_1},\vec{\sigma_2}$$

$$\vec{\sigma_1},\vec{\sigma_2}=\frac{\sigma^2-\sigma_1^2+\sigma_2^2}{2}$$

We know that tor sight singlet state, total spin $\sigma=0$

$$\vec{\sigma_1}\cdot\vec{\sigma_2}=\frac{-1}{2}\left[\sigma_{1x}^2+\sigma_{1y}^2+\sigma_{1z}^2+\sigma_{2x}^2+\sigma_{2y}^2+\sigma_{2z}^2\right]$$

$$=\frac{-1}{2}[I+I+I+I+I+I]$$

[We know that, $\sigma_x^2=\sigma_y^2=\sigma_z^2=I$]

$$\vec{\sigma_1}.\vec{\sigma_2}=-3I$$

$$\boxed{\vec{\sigma_1}.\vec{\sigma_2}=-3}$$

10. $P(r)=\frac{1}{\sqrt{24}}\left(a^{\frac{-3}{2}}\right)\frac{r}{a}e^{\frac{-r}{2a}}$

For most probable Value

$$\frac{\partial}{\partial r}\left(r^2|R(r)|^2\right)=0$$

$$\frac{\partial}{\partial r}\left(\frac{1}{24}a^{-3}\frac{r^4}{a^2}e^{-\frac{r}{a}}\right)=0$$

$$\frac{1}{24a^5}\left(4r_e^{3-\frac{r}{a}} - \frac{1}{a}r_e^{4-\frac{r}{a}}\right) = 0$$

$$r^3 e^{-\frac{r}{a}}\left(4 - \frac{1\times r}{a}\right) = 0$$

or, $4 - \frac{r}{a} = 0$

$\boxed{r = 4a}$

By comparing

$\boxed{y = 4}$

11. $$AX = \lambda X$$

$$\hat{A}\,|\,e_i\rangle = \lambda_i|e_i\rangle$$

$$\hat{A}\,|\,e_1\rangle = \lambda_1|e_1\rangle$$

$$\hat{A}\,|\,e_2\rangle = \lambda_2|e_2\rangle$$

$$\hat{A}\,|\,e_3\rangle = \lambda_3|e_3\rangle$$

$$\hat{A}^{-1} \Rightarrow \frac{1}{\lambda_1}, \frac{1}{\lambda_2}, \frac{1}{\lambda_3}$$

$$\hat{A}^{-1}\,|\,e_i\rangle = \lambda_i^{-1}\,|\,e_i\rangle$$

$$= \frac{1}{\lambda_i}\,|\,e_i\rangle$$

12. $$V = \begin{cases} \infty & x < -\frac{a}{2} \\ 0 & -\frac{a}{2} < x < \frac{a}{2} \\ \infty & x > \frac{a}{2} \end{cases}$$

Ground state wave is proportional to

$$\sqrt{\frac{2}{a}}\cos\left(\frac{\pi x}{a}\right)$$

13. $a^+a^+\hat{a}\hat{a}\,|\,n\rangle$

$$a|n\rangle = \sqrt{n}\,|\,n-1\rangle$$

$$a^+|n\rangle = \sqrt{n+1}\,|\,n+1\rangle$$

$$a^+a^+a\sqrt{n}\,|\,n-1\rangle$$

$$a^+a^+\sqrt{n}\sqrt{n-1}\,|\,n-2\rangle$$

$$\sqrt{n}\sqrt{n-1}\sqrt{n-1}\,|\,n+1\rangle$$

$$\sqrt{n}\sqrt{n-1}\sqrt{n-1}\sqrt{n}\,|\,n\rangle$$

$$\langle n\,|\,a^+a^+\hat{a}\hat{a}\langle n\,| = ??$$

$$(n-1)\,n|n\rangle$$

$$\langle n|n(n-1)|n\rangle$$

$$n(n-1)\langle n|n\rangle$$

$$\Rightarrow n(n-1)$$

14. $J = \frac{3}{2}$; decay in to a particle B and C have angular momentum J_1 and J_2

$$\left|\frac{3}{2},\frac{3}{2}\right\rangle = \alpha|1,1\rangle_B \otimes \left|\frac{1}{2},\frac{1}{2}\right\rangle$$

$$\frac{3}{2} \to \left(1+\frac{1}{2}\right)\left(1-\frac{1}{2}\right) \to \frac{3}{2}, \frac{1}{2}$$

$$\alpha = 1$$

$$\frac{3}{2} = 1 + \frac{1}{2} = \frac{3}{2}$$

16. $$R(r) = A\frac{r}{a}\exp\left(\frac{-r}{2a}\right)$$

$$P = A^2\left(\frac{r^2}{a^2}\right)\exp\left(\frac{-r}{a}\right)r^2$$

For most probable $\frac{dP}{dr} = 0$

$$0 = \frac{A^2}{r^2}\left[r^4\left(-\frac{1}{a}\right)e^{-\frac{r}{a}} + 4r^3e^{-\frac{r}{a}}\right]$$

By solving, we get

$$r = 4a$$

$$\therefore \quad \gamma = 4$$

19. $$\langle l, m_l = l\rangle$$

$$L_z = |L|\cos\theta \qquad \text{...(1)}$$

$$L_z = m_l\hbar$$

$$\cos\theta = \frac{m_l\hbar}{\hbar\sqrt{l(l+1)}}$$

$$\theta = 30°$$

$$\cos 30° = \frac{l\hbar}{\hbar\sqrt{l^2+l}}$$

$$\frac{\sqrt{3}}{2} = \frac{1}{\sqrt{l^2+1}}$$

By solving we get, $l = 3$

20. $$u(\upsilon) = \frac{\hbar^3 \upsilon^3}{c^3 \exp\left[\left(\frac{h\upsilon}{kT}\right) - 1\right]}$$

$$T^4 \propto 4$$

$$\alpha = 4$$

21. $$E = N\left(\frac{1}{2}K_BT + \frac{1}{2}K_BT\right)$$

$$= NK_BT$$

$$C_v = \frac{dE}{dT} = (1)NK_B$$

22. Stark effect in Hydrogen Atom

n = 2

10, 0 >, | 1, –1 >, 11, 1>

$$\langle lm'|H'|lm\rangle = \begin{cases} \text{zero}, & m' \neq m \\ \text{non-zero}, & m' = m \end{cases}$$

$$\langle l'\, m'|H'|l, m\rangle = \begin{cases} \text{zero, if} & l' = l \\ \text{non-zero}, & l' \neq l \end{cases}$$

25. $$\rho = C_1\begin{pmatrix}1\\0\end{pmatrix}^{\begin{bmatrix}1 & 0\\1 & 0\end{bmatrix}} + C_2\begin{pmatrix}0\\1\end{pmatrix}^{\begin{bmatrix}0 & 1\\0 & 1\end{bmatrix}}$$

$$= 0.4\begin{bmatrix}1 & 0\\0 & 0\end{bmatrix} + 0.6\begin{bmatrix}0 & 0\\0 & 1\end{bmatrix}$$

$$= \begin{bmatrix}0.4 & 0\\0 & 0\end{bmatrix} + \begin{bmatrix}0 & 0\\0 & 0.6\end{bmatrix} = \begin{bmatrix}0.4 & 0\\0 & 0.6\end{bmatrix}$$

$$P_{22} = 0.6$$

27. $$E_0 = \alpha\left[\left(n - \frac{1}{2}\right)\hbar\sqrt{\frac{\pi}{2m \propto}}\frac{\sqrt{\frac{1}{\nu} + \frac{3}{2}}}{\frac{1}{\nu} + 1}\right]^{\left(\frac{2\nu}{\nu+2}\right)}$$

$$\alpha = F, \nu = 1$$

$$E_0 = F\left[\left(n - \frac{1}{2}\right)\hbar\sqrt{\frac{\pi}{2mF}}\frac{\sqrt{1 + \frac{3}{2}}}{1 + 1}\right]^{\left(\frac{2}{1+2}\right)}$$

$$E_0 = \alpha F\left[\frac{1}{\sqrt{F}}\right]^{\frac{2}{3}}$$

$$E_0 = \alpha F^{\frac{2}{3}}$$

28. $$H = \begin{pmatrix}\epsilon_1 & 0\\0 & \epsilon_2\end{pmatrix}, |\psi(o)\rangle = \frac{1}{\sqrt{2}}\begin{pmatrix}1\\1\end{pmatrix}$$

$$|\langle\psi(0)1\psi(t)\rangle|^2 = ?$$

$$|\psi(0)\rangle = \frac{1}{\sqrt{2}}\begin{pmatrix}0\\1\end{pmatrix} + \frac{1}{\sqrt{2}}\begin{pmatrix}0\\1\end{pmatrix}$$

$$|\psi(t)\rangle = e^{-\frac{i\hat{H}}{\hbar}(t-t_0)}|\psi(0)\rangle$$

$$|\psi(t)\rangle = \frac{1}{\sqrt{2}}e^{-\frac{i\epsilon_1 t}{\hbar}}\begin{pmatrix}1\\0\end{pmatrix} + \frac{1}{\sqrt{2}}e^{-\frac{i\epsilon_2 t}{\hbar}}\begin{pmatrix}0\\1\end{pmatrix}$$

$$\langle\psi(t)|\psi\rangle(t)\rangle = \left\{\frac{1}{\sqrt{2}}(1 \quad 0) + \frac{1}{\sqrt{2}}(0 \quad 1)\right\}$$

$$\frac{1}{\sqrt{2}}e^{-\frac{i\epsilon_1 t}{\hbar}}\begin{pmatrix}1\\0\end{pmatrix} + \frac{1}{\sqrt{2}}e^{-\frac{i\epsilon_1 t}{\hbar}}\begin{pmatrix}1\\0\end{pmatrix} + \frac{1}{\sqrt{2}}e^{-\frac{i\epsilon_2 t}{\hbar}}\begin{pmatrix}0\\1\end{pmatrix}$$

$$\langle\psi(0)|\psi(t)\rangle = \frac{1}{2}e^{-\frac{\epsilon_1 t}{\hbar}} + 0 + 0 + \frac{1}{2}e^{-\frac{\epsilon_2 t}{\hbar}}$$

$$\langle\psi(0)|\psi(t)\rangle = \frac{1}{2}\left(e^{-\frac{\epsilon_1 t}{\hbar}} + e^{-\frac{\epsilon_2 t}{\hbar}}\right)$$

$$|\langle\psi(0)|\psi(t)\rangle|^2$$

$$= \frac{1}{4}\left(e^{-\frac{\epsilon_1 t}{\hbar}} + e^{-\frac{\epsilon_2 t}{\hbar}}\right)\left(e^{-\frac{\epsilon_1 l}{\hbar}} + e^{-\frac{\epsilon_2 t}{\hbar}}\right)$$

$$= \frac{1}{4}\left[1 + e^{-\frac{(\epsilon_1-\epsilon_2)t}{\hbar}} + e^{-\frac{-i(\epsilon_2-\epsilon_1)t}{\hbar}} + 1\right]$$

$$= \frac{1}{4}\left[2 + 2x\left\{\frac{e^{i\frac{(\epsilon_1-\epsilon_2)t}{\hbar}} + e^{-\frac{-i(\epsilon_2-\epsilon_1)t}{\hbar}}}{2}\right\}\right]$$

$$= \frac{1}{4}\left[1 + e^{i\frac{(\epsilon_1-\epsilon_2)t}{\hbar}} + e^{-\frac{-i(\epsilon_2-\epsilon_1)t}{\hbar}} + 1\right]$$

$$= \frac{1}{4}\left[2 + 2x\left\{\frac{e^{\frac{i(\epsilon_2-\epsilon_1)t}{\hbar}} + e^{-\frac{-i(\epsilon_2-\epsilon_1)t}{\hbar}}}{2}\right\}\right]$$

$$= \frac{1}{4}\left[2 + 2\cos(\epsilon_2 - \epsilon_1)\frac{t}{\hbar}\right]$$

$$= \frac{1}{2}\left[1 + \cos(\epsilon_2 - \epsilon_1)\frac{t}{\hbar}\right]$$

29.

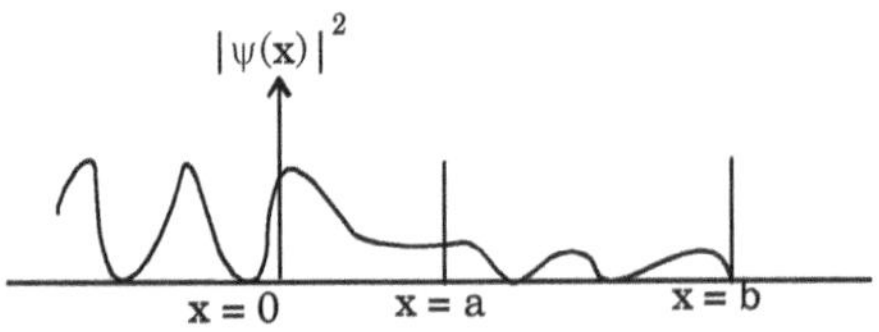

31. $|H - I\lambda| = 0$

$$\left|\begin{pmatrix} 1 & \epsilon \\ \epsilon & -1 \end{pmatrix} - \begin{pmatrix} \lambda & 0 \\ 0 & \lambda \end{pmatrix}\right| = 0$$

$$\begin{vmatrix} 1-\lambda & \epsilon \\ \epsilon & -1-\lambda \end{vmatrix} = 0$$

$$\lambda = (1+\epsilon^2)^{\frac{1}{2}}$$

$$(1+\epsilon^2)^{\frac{1}{2}} = 1 + \frac{1}{2}\epsilon^2 + \frac{1}{2}\left(\frac{1}{2}-1\right)\epsilon^4 + ...$$

$$= 1 + \frac{\epsilon^2}{2} - \frac{1}{8}\epsilon^4 + ...$$

$$\boxed{\frac{-1}{8} = -0.128}$$

32. The eigenvalues of a Hermitian matrix are all real numbers and the average value of any physical variable for any state at any time must be real number.

33. An orbital is a mathematical function called a wave function that describes an electron in an atom. The wave functions of the atomic orbitals can be expressed as the product of a radial wave function R and an angular wave function.

The no. of maxima in the radial distribution function is equal to n–1

34. We first plot the point Q on xy – plane with polar coordinate (10, π/4). We then rotate $\overrightarrow{OQ}$ in the vertical direction (i.e, the rotation remains in the plane spanned by z-axis and line OQ) till its angle with z-axis reaches π/2.

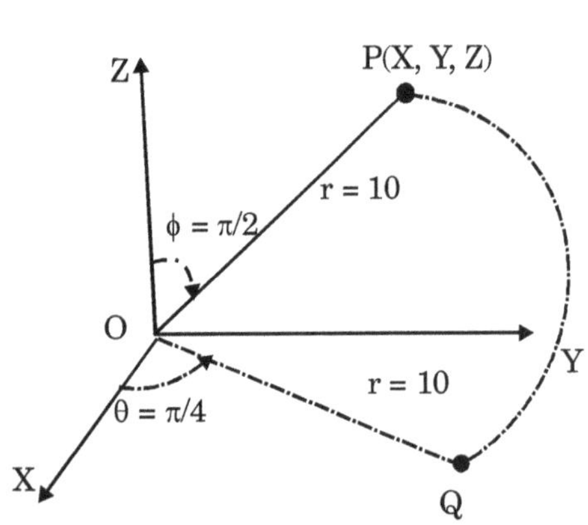

$$x = r\sin\phi\cos\theta = 10\sin\left(\frac{\pi}{4}\right)\cos\left(\frac{\pi}{2}\right) = 0\left(\because \cos\frac{\pi}{2} = 0\right)$$

$$y = r\sin\phi\sin\theta = 10\sin\left(\frac{\pi}{4}\right)\sin\left(\frac{\pi}{2}\right) = \frac{10}{\sqrt{2}}$$

$$\text{and } z = r\sin\phi = 10\cos\left(\frac{\pi}{4}\right) = \frac{10}{\sqrt{2}}$$

$$\text{Unit vector } \hat{\theta} = \frac{1}{10}\left(i \times 0 + j \times \frac{10}{\sqrt{2}} - \hat{k} \times \frac{10}{\sqrt{2}}\right)$$

$$= \frac{1}{\sqrt{2}}(\hat{j} - \hat{k})$$

35. For free particle - The particle is not bound by any potential energy. So the potential is zero and this Hamiltion is the simplest.

For one dimension; $H = -\frac{\hbar^2}{2m}\cdot\frac{\partial^2}{\partial x^2}$ and

For three dimensions; $H = \frac{\hbar^2}{2m}\nabla^2$

Here, the commutator $[x\,(x, H)] = \frac{\hbar^2}{m}$

36. (*i*) For distinguishable particle–

No. of particles = 3

The distributed in four non-degenerate energy levels

$$= (4)^3 = 64$$

(*ii*) For indentical Bosons

$$\Omega = \frac{(n_i + g_i - 1)!}{n_i!(g_i - 1)!}$$

$$= \frac{(3+4-1)!}{3!(4-1)!}$$

$$= \frac{6!}{3!\ 3!} = 20$$

37. Given that

$$Q = \sqrt{2q}e^{-1+2\alpha}\cos p$$

$$P = \sqrt{2q}e^{-\alpha-1}\sin p$$

we know that transformation P (p, q) and Q(p, q) is canonical if

$$\frac{\partial Q}{\partial q}\frac{\partial P}{\partial p} - \frac{\partial Q}{\partial p}\frac{\partial P}{\partial q} = 1$$

$$\Rightarrow e^{-2+\alpha}\left(\cos^2 p+\sin^2 p\right)=1$$

$$\Rightarrow e^{-2+\alpha}=1$$

On taking logarithm to the base e both sides, we get

$$(-2+\alpha)\ln_e e=\ln_e 1$$

$$\Rightarrow \alpha = 2$$

which is the required value of α

40. The wave function of S orbital is spherically symmetric. As the wave function of S orbital will depend upon the distance from nucleus and does not depend upon direction. Whenever a subshell is filled or half filled the total wave function is spherically symmetric.

41. The compton wavelength of a proton is

$$\lambda = \frac{h}{m_p C}$$

$$= \frac{6.624\times10^{-34}}{(1.67\times10^{-27})(3\times10^{8})}$$

$= 1.32 \times 10^{-15}$ m

$m_p = 1.67 \times 10^{-27}$ kg

$C = 3 \times 10^8$ m/s

42. Hamiltonian of one dimensional Harmonic oscillator is

$$H_0 = \frac{p^2}{2m}+\frac{1}{2}kx^2$$

The first order correction to ground state is dependent only on γ.

43. Hamiltonian $H = a_0 I + \vec{b}.\vec{\sigma}$

where $a_0 \in R, \vec{b}$ is a real vector. The ground state energy is $a_0 - |b|$

Because the eigen value of the Hamiltonian

$\lambda\pm = E_0 \pm \Delta$ (Real constant)

Ground state energy is given by

$|G> = E = E_0 - \Delta = a_0 - |b|$.

44. The degeneracy of third energy level of a 3-dimensional is isotropic quantum harmonic oscillator is given by :

$$g(n) = \frac{1}{2}(n+1)(n+2)$$

for third energy level $n = 2$

$$g(2) = \frac{1}{2}(2+1)(2+2)$$

$$= \frac{1}{2}\times 12 = 6$$

45. The electronic ground plate energy of hydrogen atom is – 13.6 eV. The highest possible electronic energy eigen state has energy equal to ∞.

$n = \infty$ .... E = 0

$n = 4$ .... E = – 0.85 eV

$n = 3$ .... E = – 1.51 eV

$n = 2$ .... E = – 3.40 eV

....

$n = 1$ .... E = – 13.6 eV

46. Given:- E = 1 eV

$V_0 = 0.75$ eV.

Probability of reflection from the potential barrier is given by:-

$$R = \left|\frac{K_1 - K_2}{K_1 + K_2}\right|^2$$

where $K_1 = \sqrt{\frac{2mE}{h}}$

$$K_1 = \sqrt{\frac{2\times9.1\times10^{-31}\times1.6\times10^{-19}}{1.055\times10^{-34}}}$$

and $K_2 = \sqrt{\frac{2m(E-V_0)}{h}} = 5.114971$

$$\sqrt{\frac{2\times9.1\times10^{-31}(0.25)\times1.6\times10^{-19}}{1.055\times10^{-34}}} = 2.557485$$

$$\therefore \quad R = \left(\frac{5.114971-2.557485}{5.114971+2.557485}\right)^2$$

$$\frac{2.5574}{7.6724} = (0.3333)^2 = 0.108 = 0.10.$$

47. Given :

$m_e = 9.31 \times 10^{-31}$ kg

$h = 6.624 \times 10^{-34}$ Js, $e = 1.6 \times 10^{-19}$ C.

We know that

Energy of one dimensional potential well is given by

$$E = \frac{h^2}{8mL^2}$$

$$= \frac{(6.624 \times 10^{-34})^2}{8 \times 9.1 \times 10^{-31} \times (3 \times 10^{-9})^2}$$

$$= \frac{43.903876 \times 10^{-68}}{655.2 \times 10^{-49}}$$

$$= 0.067008357 \times 10^{-19} \times 1.6 \times 10^{-19}$$

$$= 0.1072 \times 10^{-3}$$

= 1.0 meV

48. $X = \begin{bmatrix} 1 \\ 0 \\ 0 \end{bmatrix}$, $X^T = [1, 0\ 0]$

$$M = XX^T = \begin{bmatrix} 1 & 0 & 0 \\ 0 & 0 & 0 \\ 0 & 0 & 0 \end{bmatrix}_{3\times 3}$$

The number of non-zero eigen value of matrix M $= XX^T$ is 1.

49. Hund's Rule

Ground state of Nitrogen ($z = 7$)

According to Hund's Second Rule maximum J doe's not play a role because only $l = 0$ state is totally antisymeleric since the shell is half filled we complete to the lonest $j = |l - s| = \left|0 - \frac{3}{2}\right| = \frac{3}{2}$.

According to Hund's first rule maximum total states us to complete the three electrons spins to $s = \frac{3}{2}$

This is symmetric spin.

50. Hermitian operator is a deep adjoint operator

So $\frac{i(P_x x^2 - x^2 P_x)}{2}$ is a self conjugate operator.

$$< \psi | H | \psi > = \int_{-\infty}^{\infty} \psi^*(x) H \psi(x) dx$$

$$= < \psi | H | \psi > = < H^* \psi | \psi >$$

$$\int \left(\frac{i(P_x x^2 - x^2 Px)}{2} \right)^* \psi dx$$

$$\frac{i}{2} \int \left(\frac{P_x x^2 - x^2 Px}{2} \right)^* \psi dx$$

and after integrating by parts

$$-\frac{i}{2} \int \psi^* \left(\frac{P_x x^2 - x^2 Px}{2} \right) dx$$

$$\int \psi^* i \left(\frac{P_x x^2 - x^2 P_x}{2} \right) \psi\, dx$$

$$= \frac{i(P_x x^2 - x^2 P_x)}{2}$$

51. Two particles and two non-degenerati quantum levels 1 and 2.

Probability that level 2 contains a particle

(i) When the two particles are distinguishable.

Total no. of states far 2 distinguishable particle = 3

Probability of level 2 contains a particle

$$= \frac{3}{2} \left(\frac{n_i}{g_i} \right)$$

(ii) When two particles are basens.

Total no. of states for two basen particles = 4

1	2
A	B
B	A

Probability of level 2 contains a basen particle

$$= \frac{2}{4} = \frac{1}{2}.$$

Solution is $\left(\frac{2}{3} \text{ and } \frac{1}{2} \right)$

52. The geometric cross-section of two colliding protons at large energies is very well estimates by the product of the effective sizes of each particle. Hence it is closest to 10 mb.

53. $B = \nabla \times A$

$$B = \nabla \times \frac{B_0}{2}(-y\hat{i} + x\hat{j})$$

$$B = \left(\hat{i}\frac{\partial}{\partial x} + \hat{j}\frac{\partial}{\partial y} + \hat{k}\frac{\partial}{\partial z} \right) \times \frac{B_0}{2}(-y\hat{i} + x\hat{j})$$

$$= \frac{B_0}{2}\hat{k} + \frac{B_0}{2}\hat{k}$$

$$B = B_0 \hat{k}$$

56. $$s_1 = \frac{1}{2},\ s_2 = \frac{1}{2}$$

$$s = \sqrt{s(s+1)}\,\hbar$$

$$s = \sqrt{\frac{1}{2}\left(\frac{1}{2}+1\right)}\hbar = \sqrt{\frac{3}{4}}\hbar$$

$$\langle s_1\, s_2 \rangle = -\frac{3}{4}\hbar^2$$

57. $\epsilon_{ijk}\epsilon_{lmk}\,\partial_j\partial_l V_m = \delta^i_m\partial_l\partial_l = \delta^{ij}_{lm}\partial_l\partial_l V_m = \delta$

$$= \left(\delta^i_l\delta^k_m - \delta^l_m\delta^j_l\right)$$

$$= \partial_m\partial^i V_m - \delta^i_m\partial_l\partial_l V_m$$

$$= \partial_m\partial^i V_m - \partial_l\partial_l V_m$$

$$\epsilon^{ijk}\epsilon_{lmk}\,\partial_l\partial_l V_m = \left(\delta^i_l\delta^j_m - \delta^i_m\delta^j_l\right)$$

$$= \delta^i_l\delta^j_m\partial_j\partial_l$$

$$= \partial_m\partial^i V_m$$

when $i = m$

$$\epsilon_{ijk}\epsilon_{lmk}\,\partial_l\partial_l V_m = -\partial_j^2 V_i$$

58. $$\sqrt{p^2c^2 + M^2c^4} = E \qquad ...(1)$$

Also, $E = E_1 + E_2$ ...(2)

$\therefore$ $E = 1 + 0.82 = 1.82$ GeV

Also, $$p = \frac{E_1\cos\theta_1}{c} + \frac{E_2\cos\theta_2}{c}$$

$$= \frac{1}{c}\frac{1}{\sqrt{2}} + \frac{1}{c}\times 0.820\times\frac{1}{\sqrt{2}}$$

$$= \frac{1.11}{c}\text{GeV}$$

$\Rightarrow$ $pc = 1.11$

$\Rightarrow$ $p^2c^2 = 1.21$

Putting in (1)

$1.21 + M^2c^4 = 3.312$

$\Rightarrow$ $M^2c^4 = 2.077$

$\Rightarrow$ $M = \sqrt{2.077} = 1.40$

59. $$[x^2 p^2] = 2i\hbar\,(xp_x + p_x x)$$

We know $[\hat{x}, \hat{p}_x] = i\hbar x$

$$[\hat{x}^2, \hat{p}_x] = [\hat{x}\,\hat{x}\ \ p_x]$$

$$= [\hat{x}\ p_x]\hat{x} + \hat{x}[\hat{x}\ \hat{p_x}]$$

$$= 2i\hbar\,\hat{x}$$

$$[x, p_x^2] = [\hat{x}\ \hat{p}_x\ \hat{p}_x]$$

$$= [\hat{x}\ \hat{p}_x]\hat{p}_x + \hat{p}_x[\hat{x}\ \hat{p}_x]$$

$$= 2i\hbar\ p_x + i\hbar\,\hat{p}_x$$

$$= 2i\hbar\ p_x$$

$$[x^2 p_x^2] = [x.x\ \hat{p}_x^2\ ]$$

$$= [x\ \hat{p}_x^2]x + \hat{x}[\hat{x}\ \hat{p}_x^2]$$

$$= 2i\hbar\ p_x\ x + x\,2i\hbar\,p_x$$

$$[x^2\ \hat{p}_x^2] = 2i\hbar\,(p_x\ x + x\,p_x)$$

60. $L_x + iL_y = L_+$

$\langle l^1, m' | L_+ | l, m\rangle$

$$= \langle 1, 0\ | L_+ |\ 1, -1\rangle$$

$$= \hbar\sqrt{l(l+1) - m(m+1)}$$

$$= \hbar\sqrt{2-0} = \sqrt{2}\,\hbar$$

61. According to parity rule,

$$P^2 \neq -P$$

62. $\sigma_2 \times \sigma_3$ has dimension of 4 and $\sigma_1 \cdot \sigma_2 \times \sigma_3$ has dimension of $2 \times 4 = 8$

63. $\vec{L}$ and $\vec{p}$ be the angular and linear momentum operators, respectively, for a particle

$$\left[L_x\ p_y\right] = ?$$

$$L = r \times p$$

$$= \begin{vmatrix} i & j & k \\ x & y & z \\ p_x & p_y & p_z \end{vmatrix}$$

$L = (yp_z - zp_y)\hat{i} + (zp_x - xp_z)j + (xp_y - yp_x)k$ i.e. x component will be

$$L_x = (yp_z - zp_y)$$

then

$$\left[L_x, p_y\right] = \left[yp_z - zp_y, p_y\right]$$

$$= \left[yp_z, p_y\right] - \left[zp_y, p_y\right]$$

$$= y[p_z, p_y] + [y, p_y]p_z - [z, p_y]p_y - z[p_y, p_y]$$

$$\quad\ \ \| \qquad\qquad \| \qquad\qquad \| \qquad\qquad \|$$

$$\quad\ \ 0 \qquad\quad i\hbar p_z \qquad\ \ 0 \qquad\qquad 0$$

$$\left[L_x, p_y\right] = i\hbar\,p_z$$

64. Ist order perturbation energy,

$$E_n : E_n^{(0)} + E_n^{(1)}$$

$$E_n = E_n^{(0)} + <\phi_n|H_P'|\phi_n>$$

where, $$E_n^{(1)} = <\phi_n^0|H_P'|\phi_n^0>$$

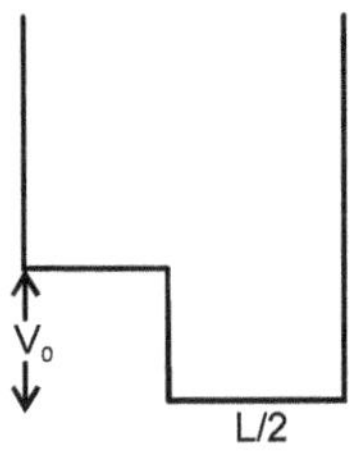

Here, $$H' = \begin{cases} V_0 & 0 < x < \dfrac{L}{2} \\ 0 & x > \dfrac{L}{2} \end{cases}$$

$$E_n^{(1)} = \frac{2}{L}V_0\int_0^{L/2}\sin^2\left(\frac{\pi x}{L}\right)dx$$

$$= \frac{2}{2L}V_0\int_0^{L/2}\left(1-\cos\left(\frac{2\pi x}{L}\right)\right)dx$$

$$= \frac{2V_0}{2L}\left[x - \frac{\sin\frac{2\pi x}{L}}{\left(\frac{2\pi}{L}\right)}\right]_0^{L/2}$$

$$= \frac{2V_0}{2L}\left[\frac{L}{2}-0\right] = \frac{V_0}{2}$$

then ground state energy,

$$E = \frac{\pi^2\hbar^2}{2mL^2} + \frac{V_0}{2}$$

65. Energy of a rigid rotator is given by (Quantum Mechanically)

$$E_{rot} = J(J+1)\frac{h^2}{8\pi^2 I}$$

where J – rotational quantum number

and I – moment of inertia

Given, for first excited state

$$E_{rot}' = 2\frac{h^2}{8\pi^2 I}$$

$$1\text{ meV} = 2\frac{h^2}{8\pi^2 I}$$

$$\frac{h^2}{8\pi^2 I} = \frac{1}{2}\text{mev}$$

For fourth excited state

$$E_{rot}^4 = 4(5)\frac{h^2}{8\pi^2 I}$$

$$= 20\times\frac{1}{2}\text{mev}$$

$$E_{rot}^4 = 10\text{ mev}$$

66. $$1\psi\rangle = \frac{1}{2}\left[1\psi_0\rangle + e^{\frac{i\pi}{2}}|\psi_1\rangle\right]$$

$$\langle\psi|x|\psi\rangle = \langle\psi|\sqrt{\frac{\hbar}{2m\omega}}\left(a+a^+\right)|\psi\rangle$$

$$\Rightarrow \langle\psi|x|\psi\rangle = \sqrt{\frac{\hbar}{2m\omega}}\left(\langle\psi|a|\psi\rangle + \langle\psi|a^+|\psi\rangle\right)$$

$$\because a|\psi\rangle = \frac{1}{\sqrt{2}}e^{\frac{i\pi}{2}}|\psi_0\rangle$$

and $$a^+|\psi\rangle = \frac{1}{\sqrt{2}}\left[|\psi_1\rangle + \sqrt{2}e^{\frac{i\pi}{2}}|\psi_2\rangle\right]$$

$$\Rightarrow \langle\psi|x|\psi\rangle$$

$$= \sqrt{\frac{\hbar}{2m\omega}}\left(\frac{1}{2}\frac{1}{2}e^{\frac{i\pi}{2}}\langle\psi_0|\psi_0\rangle + \frac{1}{\sqrt{2}}\frac{1}{\sqrt{2}}e^{\frac{-i\pi}{2}}\langle\psi_1|\psi_1\rangle\right)$$

$$\Rightarrow \langle\psi|x|\psi\rangle = \sqrt{\frac{\hbar}{2m\omega}}(0) = 0$$

67. Spin $\frac{3}{2}$

$\Rightarrow$ degeneracy = (2s + 1)

$$= 2\times\frac{3}{2}+1 = 4$$

$$E_{ground} = 4\times\frac{\pi^2\hbar^2}{2mL^2} + 4\times\frac{4\pi^2\hbar^2}{2mL^2}$$

$$= 20\frac{\pi^2\hbar^2}{2mL^2}$$

$$E_{excited}^{Ist} = 4\times\frac{\pi^2\hbar^2}{2mL^2}+3\times\frac{4\pi^2\hbar^2}{2mL^2}+1\times\frac{9\pi^2\hbar^2}{2mL^2}$$

$$= 25\frac{\pi^2\hbar^2}{2mL^2}$$

Now Minimum excitation energy,

$$\Delta E = E_{excited}^{Ist} - E_{ground}$$

$$= 25\frac{\pi^2\hbar^2}{2mL^2} - 20\frac{\pi^2\hbar^2}{2mL^2}$$

$$= 5\cdot\frac{\pi^2\hbar^2}{2mL^2}$$

$$= 5$$

68. $H = \frac{1}{2m}P_1^2 + \frac{1}{2m}P_2^2 + \frac{1}{2}m\omega^2\left(r_1^2 + r_2^2\right) + K\vec{\sigma}_1\cdot\vec{\sigma}_2$

$$\vec{\sigma} = \vec{\sigma}_1 + \vec{\sigma}_2$$

$$\Rightarrow \vec{\sigma}^2 = \vec{\sigma}_1{}^2 + \vec{\sigma}_2{}^2 + 2\vec{\sigma}_1\cdot\vec{\sigma}_2$$

$$\Rightarrow 2\vec{\sigma}_1\cdot\vec{\sigma}_2 = \vec{\sigma}^2 - 2\vec{\sigma}_1{}^2 - \vec{\sigma}_2{}^2$$

$$\Rightarrow 2\vec{\sigma}_1\cdot\vec{\sigma}_2 = 0 - 3I - 3I = -6I$$

$$\Rightarrow 2\vec{\sigma}_1\cdot\vec{\sigma}_2 = -6I$$

$$\Rightarrow \vec{\sigma}_1\cdot\vec{\sigma}_2 = -3$$

Now Energy, $E = 2\times\frac{3}{2}\hbar\omega + K(-3)$

$= 3 \times (0.1) + (0.2)(-3)$

$= -0.3ev$

70. $P_A = \frac{h}{\lambda_A}$ and $P_B = \frac{h}{\lambda_B}$

$$\therefore \quad \frac{P_A}{P_B} = \frac{\frac{h}{\lambda_A}}{\frac{h}{\lambda_B}} = \frac{\lambda_B}{\lambda_A}$$

$$= \frac{500nm}{1500nm} = \frac{1}{3}$$

$$\frac{P_A}{P_B} = \frac{1}{3}$$

71.

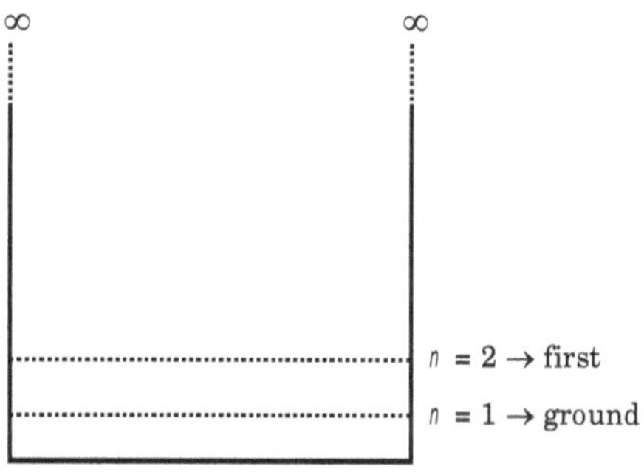

The space part of ground state wave function is given by

$$= \frac{1}{\sqrt{2}}[\psi_1(x_1)\psi_2(x_2) - \psi_1(x_2)\psi_2(x_1)]$$

72. $J = L + S$

$J^2 = (L + S).(L + S) = L^2 + S^2 + 2\,L.S$

$\vec{L}.\vec{S}$ commute with L^2, S^2, J^2, $(L + S)^2$ but not S_z

$$[\vec{L}.\vec{S},S_z] \Rightarrow [L_x S_x + L_y S_y + L_z S_z, S_z]$$
$$= [L_x S_x, S_z] + [L_y S_y, S_z] + [L_z S_z, S_z]$$
$$\Rightarrow L_x [S_x, S_z] + [L_x, S_z] S_x + L_y [S_y, S_z] + [L_y, S_z] S_y + L_z [S_z, S_z] + [L_z, S_z] S_z$$

Then $[L_x S_z] = [L_y S_z] = [L_z S_z]$

$= [S_z S_z] = 0$

$= L_x [S_x, S_z] + L_y [S_y, S_z]$

$$[\vec{L}.\vec{S},S_z] = L_x(-i\hbar S_y) + i\hbar S_y S_x$$
$$= i\hbar [L_y S_x - L_x S_y]$$

$[\vec{L},\vec{S}.S_z] = i\hbar (S \times L)_z$ It is now zero

73.

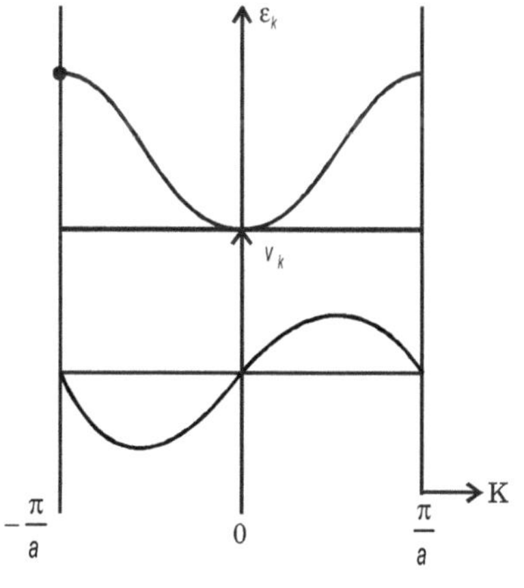

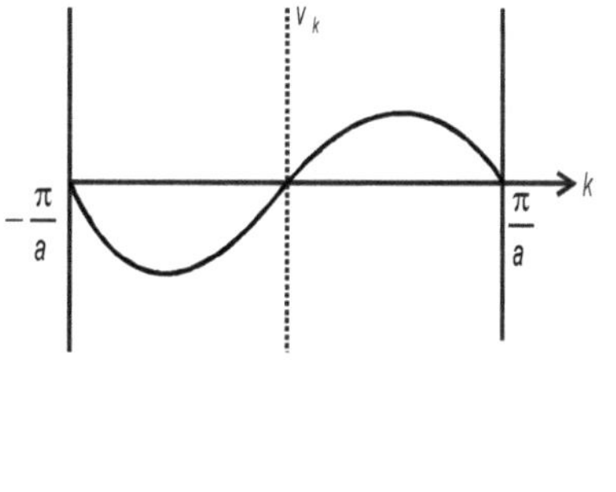

74. Wave function,

$$\psi(\vec{r}) = \frac{1}{\sqrt{\pi a_0^3}} e^{-(r/a_0)}$$

Now $\hat{Q} = z^2 - r^2$ and $z = r\cos\theta$ (given)

$$\therefore <z^2 - r^2> = <r^2\cos^2\theta - r^2>$$
$$= <r^2\cos^2\theta> - <r^2>$$
$$= <r^2><\cos^2\theta> - <r^2> \quad ...(i)$$

Now, $<r^2> = \frac{n^2 a_0^2}{2}[5n^2 + 1 - 3l(l+1)]$

For $n = 1$ and $l = 0$,

$$<r^2> = \frac{a_0^2}{2}[5+1]$$
$$= \frac{6a_0^2}{2}$$
$$= 3a_0^2 \quad ...(ii)$$

Putting the value of $<r^2>$ from equation (*ii*) in equation (*i*) we get

$$<z^2 - r^2> = 3a_0^2 \times \frac{1}{2} - 3a_0^2$$
$$= \frac{-3a_0^2}{2}$$

Hence expectation value of the operator

$$\hat{Q} = z^2 - r^2 = \frac{-3a_0^2}{2}$$

75. $V(x, y) = \frac{1}{2}mw^2(x^2 + y^2)$

$$E_{n_{xxy}} = (n_x + n_y + 1)\hbar w$$

$$E_{n_{xxy}} = 4\hbar w$$

$$\left.\begin{array}{l} n_x = 0 \,;\, n_y = 3 \\ n_x = 3 \,;\, n_y = 0 \\ n_x = 1 \,;\, n_y = 2 \\ n_x = 2 \,;\, n_y = 1 \end{array}\right] g = 4$$

The value of g is 4

76. $\psi = -\sqrt{\frac{8}{21}}\psi_{200} - \sqrt{\frac{3}{7}}\psi_{310} + \sqrt{\frac{4}{21}}\psi_{321}$

$<\psi | L^2 | \psi>$

$$= \frac{8}{21} \times 0 + \frac{3}{7} \times 1(2)\hbar^2 + \frac{4}{21} \times 2(2+1)\hbar^2$$

$$= \frac{6\hbar^2}{7} + \frac{24\hbar^2}{21} = \left(\frac{18+24}{21}\right)\hbar^2 = \left(\frac{42}{21}\right)\hbar^2$$

Answer $<\psi | L^2 | \psi> = 2\hbar^2$

77. $$\varphi_1 = \frac{1}{\sqrt{3}}\begin{pmatrix}1\\0\end{pmatrix} + \sqrt{\frac{2}{3}}\begin{pmatrix}0\\1\end{pmatrix}$$

$$\chi_{1/2} = \begin{pmatrix}1\\0\end{pmatrix} \quad \chi_{-1/2} = \begin{pmatrix}0\\1\end{pmatrix}$$

$$\psi = a\chi_{1/2} + b\chi_{-1/2}$$

According to orthogenality condition

$$\int \psi_2 \psi_1^* \, dc = 0$$

$$\frac{a}{\sqrt{3}} + \sqrt{\frac{2}{3}}\, b = 0 \Rightarrow a = -\sqrt{2}\, b$$

Now, $(a)^2 + (b)^2 = 1$

$\therefore (2 + 1)(b)^2 = 1$

$$\therefore \quad (b)^2 = \frac{1}{3}$$

$$\therefore \quad |b| = \frac{1}{\sqrt{3}}$$

Value of $a = -\sqrt{\frac{2}{3}}$

When the system is in the state ψ_2, its probability to be in the spin-up sate is $\frac{2}{3}$.

78. $W = \beta x \qquad \phi_0 = \sqrt{\frac{2}{a}}\sin\frac{\pi x}{a}$

$$<\phi_0 | w | \phi_0> = \frac{2\beta}{a}\int_0^a x\sin^2\frac{\pi x}{a}dx$$

$$= \frac{\beta}{a}\int_0^a x\left(2\sin^2\frac{\pi x}{a}\right)dx$$

$$= \frac{\beta}{a}\int_0^a x\left[-\cos^2\frac{\pi x}{a} + 1\right]dx$$

$$= \frac{\beta}{a}\left[\int_0^a x\,dx - \int_0^a x\cos\frac{2\pi x}{a}dx\right]$$

$$< W > = \frac{\beta a}{2}$$

79. $|\psi\rangle = \frac{1}{2}|2\rangle + \frac{\sqrt{3}}{2}|3\rangle$

$$< H > = \Sigma P_i \varepsilon_i = P_2 \varepsilon_2 + P_3 \varepsilon_3$$

Now

$$\varepsilon_n = \left(n + \frac{1}{2}\right)\hbar\omega$$

$$\varepsilon_2 = \frac{5}{2}\hbar\omega$$

$$\varepsilon_3 = \frac{7}{2}\hbar\omega$$

$$< H > = \frac{1}{4} \times \frac{5}{2}\hbar\omega + \frac{3}{4} \times \frac{7}{2}\hbar\omega = \frac{26}{8}\hbar\omega$$

$$< H > = \frac{13}{4}\hbar\omega = 3.25\ \hbar\omega$$

$\therefore < H > = 3.25\ \hbar\omega$

80. $[L_+ L_- + L_- L_+) = 2\left(L^2 - L_z^{\ 2}\right)$

$$= 2\left[\ell(\ell+1) - m_\ell^{\ 2}\right]\hbar^2$$

if $\ell = 1$, m = 1,

$= 2\hbar^2$

82. $\begin{bmatrix} 4 & -1 & -1 \\ -1 & 4 & -1 \\ -1 & -1 & 4 \end{bmatrix}$,

Fir eigen value $|A - \lambda I| = 0$

$$\begin{bmatrix} 4-\lambda & -1 & -1 \\ -1 & 4-\lambda & -1 \\ -1 & -1 & 4-\lambda \end{bmatrix}$$

$$\Rightarrow (4-\lambda)\left((4-\lambda)^2 - 1^2\right) + 1\left((-1)(4-\lambda) - 1\right] - 1$$
$$\left(1 + (4-\lambda)\right] = 0$$

$$\Rightarrow (4-\lambda)\left[(4-\lambda)^2 - 1\right] + (-4+\lambda-1) - 5 - \lambda = 0$$

$$\Rightarrow (4-\lambda)\left(16 + \lambda^2 - 8\lambda - 1\right] + (-5+\lambda) - 5 + \lambda = 0$$

$$\Rightarrow 64 + 4\lambda^2 - 32\lambda - 4 - 16\lambda - \lambda^3 + 8\lambda^2 + \lambda - 10 + 2\lambda = 0$$

$$\Rightarrow -\lambda^3 + 12\lambda^2 - 45\lambda + 50 = 0$$

$$\Rightarrow \lambda^3 - 12\lambda^2 + 45\lambda - 50 = 0$$

$$\Rightarrow (\lambda-2)(\lambda^2 - 10\lambda + 25) = 0$$

$$\Rightarrow (\lambda-2)(\lambda-5)(\lambda-5) = 0$$

$\lambda = 2,\ \lambda = 5,\ \lambda = 5$

For degenerate λ should be 5

83. $|\psi\rangle = 0.8|\uparrow\rangle + 0.6|\downarrow\rangle$

$$= 0.8\begin{pmatrix} 1 \\ 0 \end{pmatrix} + 0.6\begin{pmatrix} 0 \\ 1 \end{pmatrix} = \begin{pmatrix} 0.8 \\ 0.6 \end{pmatrix}$$

Operator $A = 10\sigma_3 + 5\sigma_x$

$$= 10\begin{pmatrix} 1 & 0 \\ 0 & -1 \end{pmatrix} + 5\begin{pmatrix} 0 & 1 \\ 1 & 0 \end{pmatrix}$$

$$\Rightarrow\ A = \begin{pmatrix} 10 & 5 \\ 5 & -10 \end{pmatrix}$$

$$\langle A \rangle = \langle \psi | A | \psi \rangle$$

$$= (0.8\ \ 0.6)\begin{pmatrix} 10 & 5 \\ 5 & -10 \end{pmatrix}\begin{pmatrix} 0.8 \\ 0.6 \end{pmatrix}$$

$$= (8.8 - 1.2) = 7.6$$

84. $\vec{J} = |\psi|^2 \frac{\hbar k}{m}$

$$= |A|^2 \left|\frac{r_0}{r}\right|^2 \frac{\hbar k}{m}$$

$$\Rightarrow J = |A|^2 \left|\frac{r_0}{2r_0}\right|^2 \frac{\hbar k}{m}$$

$$\Rightarrow J = |A|^2 \frac{\hbar k}{4m}$$

$$= (0.25)|A|^2 \frac{\hbar k}{m}$$

85. $H_0 = \begin{bmatrix} 5 & 2 & 0 \\ 2 & 5 & 0 \\ 0 & 0 & 2 \end{bmatrix}$,

$$H_P = \varepsilon\begin{bmatrix} 1 & 1 & 1 \\ 1 & 1 & -1 \\ 1 & -1 & 1 \end{bmatrix}$$

Eigen value of H_0 is $E_1 = 2$, $E_2 = 3$, $E_3 = 7$ and the eigen vectors corresponds to

$$|\phi_1\rangle = \begin{pmatrix} 0 \\ 0 \\ 1 \end{pmatrix}, |\phi_2\rangle = \frac{1}{2}\begin{pmatrix} 1 \\ -1 \\ 0 \end{pmatrix}, |\phi_3\rangle = \frac{1}{\sqrt{2}}\begin{pmatrix} 1 \\ 1 \\ 0 \end{pmatrix}$$

86. $E_1' = \langle\phi_1|H_P|\phi_1\rangle = 1\varepsilon \Rightarrow E_1 = 2 + 1\varepsilon$

$E_2' = \langle\phi_2|H_P|\phi_2\rangle$

$$= \frac{1}{\sqrt{2}}(1 \ \ -1 \ \ 0)\cdot\varepsilon\begin{pmatrix}1 & 1 & 1\\ 1 & 1 & -1\\ 1 & -1 & 1\end{pmatrix}\cdot\frac{1}{\sqrt{2}}\begin{pmatrix}1\\ -1\\ 0\end{pmatrix}$$

$$= \varepsilon(0 \ \ 0 \ \ 1)\begin{pmatrix}1\\ -1\\ 0\end{pmatrix} = 0$$

$E_3' = \langle\phi_3|H_P|\phi_3\rangle$

$$= \frac{1}{\sqrt{2}}(1 \ \ 1 \ \ 0)\cdot\varepsilon\begin{pmatrix}1 & 1 & 1\\ 1 & 1 & -1\\ 1 & -1 & 1\end{pmatrix}\cdot\frac{1}{\sqrt{2}}\begin{pmatrix}1\\ 1\\ 0\end{pmatrix}$$

$$= \varepsilon\cdot\frac{1}{2}(2 \ \ 2 \ \ 0)\begin{pmatrix}1\\ 1\\ 0\end{pmatrix}$$

$$\Rightarrow E_3' = \frac{1}{2}(4\varepsilon) = 2\varepsilon$$

$E_1 = 2 + 1\varepsilon$, $E_2 = 3 + 0\varepsilon$, $E_3 = 7 + 2\varepsilon$.

87. Two dimensional potential function is given by

$$V(x, y) = 0 \quad -\frac{a}{2} < x < \frac{a}{2}$$
$$= \infty \text{ else where}$$

The energy of the particle is

$$E\,(nx, ny) = \left(\frac{\pi^2 h^2}{2m}\right)\left(\frac{n_x^2}{L_x^2} + \frac{n_y^2}{L_y^2}\right)$$

For 1st excited state $n_x = 1$, $n_y = 2$ and

$$L_x = L_y = a$$

$$E(1, 2) = \frac{\pi^2 h^2}{2m}\left(\frac{1}{a^2} + \frac{4}{a^2}\right)$$

$$= \frac{5\pi^2 h^2}{2ma^2}$$

88. The fermionie have Anti symmetric wavefunctions and for anti symmetric wavefunction spatial part

$$= \frac{1}{\sqrt{2}}(\alpha\beta - \beta\alpha)$$

89. For half harmonic oscillator well

$$E_n = (2_n + \frac{3}{2})h\omega$$

For first excited state $n = 1$

$$E = \frac{7}{2}\ h\omega$$

90. Unperturbed Hamiltonian $H_0 = \begin{vmatrix}1 & 0\\ 0 & 1\end{vmatrix}$

Perturbed Hamiltonian,

$$H_1 = \begin{vmatrix}\delta & \delta\\ \delta & \delta\end{vmatrix}$$

The eigenvalues of total Hamiltonian

$$H = H_0 + H^1 = \begin{vmatrix}\delta+1 & \delta\\ \delta & \delta+1\end{vmatrix}$$

Diagonalising

$$H \Rightarrow |H - \lambda| = 0$$

$$\begin{vmatrix}\delta+1 & \delta\\ \delta & \delta+1\end{vmatrix} - \lambda\begin{vmatrix}1 & 0\\ 0 & 1\end{vmatrix} = 0$$

$$\begin{vmatrix}\delta+1-\lambda & \delta\\ \delta & \delta+1-\lambda\end{vmatrix} = 0$$

$$(\delta + 1 - \lambda)^2 - \delta^2 = 0$$

$\Rightarrow \quad \delta + 1 - \lambda = \pm\delta$

Now, $\quad \delta + 1 - \lambda = +\delta$

$\Rightarrow \quad \lambda = 1$

$\quad \delta + 1 - \lambda = -\delta$

$\Rightarrow \quad \lambda = 1 + 2\delta$

91. The radial probability density

$$P(r) = r^2|\psi 100|^2$$

$$= r^2 e^{-2r/a_0}$$

initially the pilot increases attains maximum value due to r^2 and then it starts decreasing exponentially

92. Given wave function

$$\psi = e^{ikx} + 2e^{-ikx}$$

Schrodinger equation.

$$-\frac{h^2}{2m}\frac{d^2}{dx^2}\psi = E\psi$$

$$\Rightarrow \quad -\frac{h^2}{2m}\frac{d^2}{dx^2}[e^{ikx} + 2e^{-ikx}] = E\psi$$

$$\Rightarrow \quad -(ik) = \frac{h^2}{2m}\frac{d}{dx}[e^{ikx} - 2e^{-ikx}] = E\psi$$

$$= Nk_B \log 2$$

93. The probability current density

$$J(x, t) = \frac{h}{m} en[\psi * \nabla\psi]$$

So real part ofψ $J(n, t) = 0$

94. If A and P be sequare matrices of the same type and if P be inversible then the matrices A and B $= P^{-1}$ AP have the same characteristics roots

Let $B = P^{-1} AP$

$$B - \lambda e = P^{-1} AP - \lambda l$$
$$= P^{-1} AP - P^{-1} \lambda lp$$
$$= P^{-1}(a - \lambda l)P$$

Where, l is identity matrix, $|B - \lambda l| = 0$ is called the characterstics equation and λ' 0 are the characterstic roots or eigen values

$$\therefore \quad |B - \lambda l| = |P^{-1}(A - \lambda l) P|$$
$$= |P^{-1}|\ |A - \lambda l|\ |P|$$
$$= |A - \lambda l|\ |P^{-1}|\ |P|$$
$$= |A\lambda l|\ |PP^{-1}|$$
$$= |A - \lambda l|\ |l|$$
$$= |A - \lambda l| \qquad (\because |l| = 1)$$

Then the matrices A and B ($= P^{-1}$ AP) have the same characterstics equations and hence charactertics roots or eigen values. Since the sum of the eigen values of a matrix is equal to the trace of the maxtrix is equal to determinant of matrix, hence third alternative is incorrect.

95. The efficiency of an engine is defined as the ratio of the network done to the head absorbed during one complete cycle.

$$\text{efficiency } x = \frac{\text{useful work done}}{\text{Heat absorbed}} \qquad \dots (i)$$

After one complete cycle, the engine returns to its original state. Therefore, there will be no change in its internal energy,

i.e., $\Delta V = 0$ From the first law, we have

$$\Delta V = Q_1 - Q_2 - w = 0$$

or $\quad w = Q - Q_2 \qquad \dots (ii)$

Let T_1 and T_2 be the temperature of the head reservoirs such that

$$T_1 > T_2$$

the working substance, say a gas, is contained with in a cylinder fitted with a frictionless piston.

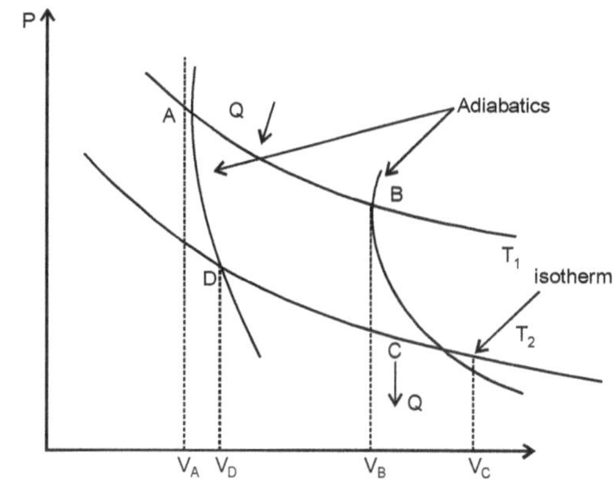

carnot cycle on p-v diagram

(*i*) We places the cylinder in thermal contact with the hot reservoir and let the gas udergoes reversible isothermal expansion, let Q_1 head flow from the reservoir in to the gas in the process we indicates this changes as A to B on the P – V diagram

$$\text{Work done } w_1 = \int_A^B P\, dv$$

$$w_1 = Q_1 = nRT_1 \int_{V_A}^{V_B} \frac{dv}{V}$$

$$= nRT_1 4n\left(\frac{V_B}{V_A}\right) \qquad \dots(iv)$$

$V_B, V_A \rightarrow$ volumes

A, B $\rightarrow$ states

(*ii*) Next the temperature falls from T_1 to T_2

$$\text{Work done } w_2 = \int_B^C P\, dv$$

Using the adiabatic relation $P^{V^\gamma} = k$, we get

$$w_2 = k\int_{V_B}^{V_C} \frac{dv}{V^\gamma} = \frac{k}{1-\gamma}\left(\frac{1}{V_C^{\gamma-1}} - \frac{1}{V_B^{\gamma-1}}\right)$$

$$= \frac{1}{1-\gamma}(P_C V_C - P_B V_B)$$

$$= nR\frac{(T_1 - T_2)}{(\gamma - 1)} \qquad \dots(v)$$

using $P_C V_C^\gamma = k = P_B V_B^\gamma$

(*iii*) We place the cylinder in thermal contact with the cold reservoir at T_2 and compress the gas.

$$\text{Workdone } w_3 = Q_2 = \int_C^P P dv$$

$$= n\, RT_2 \ln\left(\frac{V_P}{V_C}\right) \quad ...(vi)$$

$$(iv)\ w_4 = \int_D^A P dv = \frac{nR}{\gamma - 1}(T_2 - T_1) = -w_2 \quad ...(vii)$$

Since $T_2 < T_1$, we $\rightarrow$ –ve

$$w = \oint P\,dv = w_1 + w_2 - w_3 - w_4$$

Since $w_2 = w_4$ we have

$$w = w_1 - w_3$$

$$= nRT_1 \ln\left(\frac{V_B}{V_A}\right) - nRT_2 \ln\left(\frac{V_D}{V_C}\right) \quad ...(viii)$$

$$T_1 V_B^{\gamma-1} = T_2 V_C^{\gamma-1}$$

or $$\frac{T_1}{T_2} = \left(\frac{V_C}{V_B}\right)^{\gamma-1} \quad ...(ix)$$

Similarly for D and A we can write

$$T_2 V_D^{\gamma-1} = T_1 V_A^{\gamma-1}$$

or $$\frac{T_1}{T_2} = \left(\frac{V_D}{V_A}\right)^{\gamma-1} \quad ...(x)$$

we comparing equation (*ix*) and (*x*), we get

$$\left(\frac{V_C}{V_B}\right)^{\gamma-1} = \left(\frac{V_D}{V_A}\right)^{\gamma-1} \quad ...(xi)$$

or $$\frac{V_B}{V_A} = \frac{V_C}{V_D}$$

Using this result in equation (*viii*), we get

$$w = nR(T_1 - T_2) \ln\left(\frac{V_B}{V_A}\right) \quad ...(xii)$$

$$n = \frac{w}{Q_1} = \frac{nR}{nRT_1} \frac{(T_1 - T_2)\ln\left(\frac{V_B}{V_A}\right)}{\ln\left(\frac{V_B}{V_A}\right)}$$

$$= \frac{(T_1 - T_2)}{T_1} = 1 - \frac{T_2}{T_1} \quad ...(xiii)$$

On comparing (*iii*) and (*xiii*)

$$\frac{T_1}{T_2} = \frac{Q_1}{Q_2} \quad ...(xiv)$$

i.e., for an ideal gas the ratio $\frac{Q_1}{Q_2}$ also depends on temperatures T_1 and T_2

96. We know that for any matrix

1. The product of eigen values is equal to determinant of the matrix
2. The sum of eigen values is equal to trace of that matrix.

If λ_1, λ_2 and λ_3 are 3 eigen vlaue of 3×3 matrix, then

$$\lambda_1 + \lambda_2 + \lambda_3 = 11$$

$$\lambda_1 \lambda_2 \lambda_3 = 36$$

largest eigen vlaue of the matrix is 9.

97. Here $v(x) = v_0$ for $x > 0$ and $E < v_0$

Applying schrodinger equation in region II

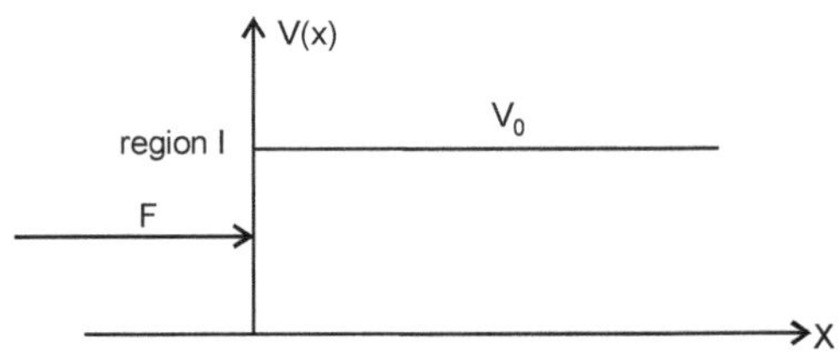

$$\frac{d^2\psi}{dx^2} + \frac{2m}{\hbar^2}(E - v_0)\psi = 0$$

$\because$ $E < v_0$

$$\Rightarrow \frac{d^2\psi}{dx^2} - \frac{2m}{\hbar^2}[v_0 - E]\psi = 0$$

Let $\alpha^2 = \frac{2m}{\hbar^2}(v_0 - E)$ $\quad$ ($\because \alpha$ is real)

$$\frac{d^2\psi}{dx^2} - \alpha^2\psi = 0$$

$\Rightarrow$ $\alpha^2 > 0$

$$\psi = Ae^{ax} + Be^{-ax}$$

$\because$ ψ must becomes ω when $x = \infty$. Hence its not possible wave function thus ψ must be Be^{-ax}.

98. $(L_x, L_y, L_z) = (L_x L_z)L_y + L_x[L_y, L_z]...$ (*i*)

$\because$ $[A, B, C] = (A, C)B + A(B, C)$

Equation (*i*) can be written as

$$(L_x, L_y, L_z) = -(L_z, L_x)L_y + L_x[L_y, L_z]$$

Now, $(L_z, L_x) = i\hbar\, L_y$ and $[L_y, L_z] = i\hbar L_x$

$$(L_x, L_y, L_z) = i\hbar\, Ly^2 + i\hbar\, Lx^2$$

$$= i\hbar\, [Lx^2 - Ly^2]$$

99. Expectation value of $\frac{1}{r^2}$ is given by

$$\left(\frac{1}{r^2}\right) = \int \psi^*(r)\frac{1}{r^2}\psi(r)dv$$

Here $\psi(r) =$

$$\psi^*(r) = \frac{1}{\sqrt{4}\pi}\frac{2}{a^{3/2}}e^{-r/a} \qquad (\because \psi \text{ is real})$$

and volumed $v = 4\pi r^2\, dr$

$$\left(\frac{1}{r^2}\right) = \int_0^{\infty}\frac{1}{\sqrt{4\pi}}\frac{2}{\frac{a^3}{2}}e^{-r/a}\times\frac{1}{r^2}\times\frac{1}{\sqrt{4\pi}}$$

$$\times\frac{2}{\frac{a^3}{2}}\times e^{-r/a}\times 4\pi r^2 dr$$

$$\left(\frac{1}{r^2}\right) = \frac{4}{a^3}\int_0^{\infty}e^{-2r/a}dr$$

$$\because \int_0^{\infty}e^{-ax}x^n dx = \frac{n!}{\alpha^{n+1}}$$

$$\Rightarrow \left(\frac{1}{r^2}\right) = \frac{4}{a^3}\pi\frac{0!}{\left(\frac{2}{a}\right)^{0+1}} = \frac{4}{a^3}\times\frac{a}{2} \qquad (\because 0! = 1)$$

$$\left(\frac{1}{r^2}\right) = \frac{2}{a^2}$$

100. $E_1' = \int_0^{\frac{a}{2}}\psi_1' V^1(x)\psi_1 dx$

$$= \frac{2}{a}\int_0^{a/2}\sin^2\left(\frac{\pi x}{a}\right)V_0\cdot\cos\left(\frac{\pi x}{a}\right)dx$$

$$= \frac{2}{a}V_0\left.\frac{\sin^3\frac{\pi x}{a}}{3\frac{\pi}{a}}\right|_0^{a/2}$$

$$= \frac{2V_0}{3\pi}$$

101. Here $\psi_1 = \phi_0 - 2\phi_1 + 3\phi_2$

$\psi_2 = \phi_0 - \phi_1 + \alpha\phi_2$

ψ_2 is orthogonal to ψ_1

Hence, $\int \psi_1, \psi_2{}^* d\tau = 0$

$$\Rightarrow \int\phi_0^2 d\tau + 2\int\phi_1^2 d\tau + 3\alpha\int\phi_2^2 d\tau = 0$$

$\because$ ϕ_1, ϕ_2 and ϕ_3 are orthogonal to each other.

Hence $\int\phi_0\phi_1 d\tau = \int\phi_1\phi_2 d\tau$

$$= \int\phi_0\phi_3 d\tau = 0$$

Now since ϕ_0, ϕ_1 and ϕ_2 are normalized.

Hence, $\int\phi_0^2 d\tau = \int\phi^2 d\tau = \int\phi_2^2 d\tau = 1$

$\Rightarrow$ $1 + 2 + 3\alpha = 0$

$\Rightarrow$ $\boxed{\alpha = -1}$

102. For $\alpha = -1$

$\psi_2 = \phi_0 - \phi_1 - \phi_2$

Expectation vlaue in ψ_2 state is

$(E) = (\psi_2{}^* \mid E \mid \psi_2)$

$$(E) = (\phi_0 | E_0 | \phi_0 >< \phi_1 | E_1 | \phi_1 > + < \phi_2 | E_2 | \phi_2 >$$

Using orthogonal property,

Now, $\quad E_n = \left(\frac{n+1}{2}\right)h\omega$

$$(E_0) = \frac{1}{2}h\omega,\ (E_1) = \frac{3}{2}h\omega$$

and $\quad (E_2) = \frac{5}{2}h\omega$

$$\Rightarrow \quad E = (E_0) + (E_1) + (E_2)$$

$$= \frac{1}{2}h\omega + \frac{3}{2}h\omega + \frac{5}{2}h\omega$$

$$= \frac{9}{2}h\omega$$

103. The characterstics equation of the matrix A is

$|A - \lambda I| = 0$

$$A = \begin{bmatrix} 2 & 3 & 0 \\ 3 & 2 & 0 \\ 0 & 0 & 1 \end{bmatrix}$$

$$\text{Hence, } |A - \lambda I| = \begin{vmatrix} 2-\lambda & 3 & 0 \\ 3 & 2-\lambda & 0 \\ 0 & 0 & 1-\lambda \end{vmatrix} = 0$$

or $(2 - \lambda)(z - 3\lambda + \lambda^2) - 3[3 - 3\lambda] + 0 = 0$

or $(4 - 6\lambda + 2\lambda^2.2\lambda + 3\lambda^2 - \lambda^3) - 9 + 9\lambda = 0$

or $\lambda^3 - 5\lambda^2 - \lambda + 5 = 0$

After solving the above cubic equation, we get

$\lambda = 5, 1, -1$

107. Since given potential is for SHO

Hence $E_n = \left(n+\frac{1}{2}\right)h\omega$

$$\psi(x) = -\frac{1}{\sqrt{5}}\psi_0 + \frac{2}{5}\psi_1$$

Expectation values of energy is

$$(\psi^* | E | \psi) = \frac{1}{5}\left(\psi_0 | E_0 | \psi_0^* | + 4 | 5(\psi_1 | E_1 |)\psi_1^*\right)$$

$$= \frac{1}{5}(E_0) + \frac{4}{5}(E_1)$$

$$(E) = \frac{1}{5}\times\frac{h\omega}{2}+\frac{4}{5}+\frac{3h\omega}{2}$$

$$= \frac{13h\omega}{10}$$

108. $\left(sm\left|\vec{S}_x^2\right|sm\right)$

$\because \quad \vec{S}^2 = \vec{S}_x^2 + \vec{S}_y^2 + \vec{S}_z^2$

$$\left(sm\left|\vec{S}^2 - \vec{S}_y^2 - \vec{S}_z^2\right|sm\right)$$

$$\left(sm\left|\vec{S}^2\right|sm\right) - \left(sm\left|S_y^2\right|sm > - < sm\left|S_z^2\right|sm\right)$$

$$\Rightarrow \frac{hs(s+1)}{2} - 0 - \frac{m^2h}{2}$$

$$\Rightarrow h(s(t+1) - m^2)\frac{h}{2}$$

109. We have

$$(E^1) = \int_0^L \frac{2V_0}{L}\sin^2\frac{\pi x}{L}\sin\frac{2\pi x}{L}dx$$

On solving, we get

$$(E_1) = 0$$

Now, $(E_2) = V_0 + \frac{(E^1)^2}{E_2 - E_1} = V_0 \quad (\because E^1 = 0)$

110. Hence $A = \begin{vmatrix} 0 & i \\ i & 0 \end{vmatrix}$

For eigen values, $|A - \lambda I| = 0$

$$\begin{vmatrix} -\lambda & i \\ i & -\lambda \end{vmatrix} = 0$$

$\Rightarrow \quad \lambda^2 - i^2 = 0$

$\Rightarrow \quad \lambda = \pm i$

Hence, eigen values are complex and distinct.

112. $\because$ $2x$, $\cos h\ 2x$ and $\sin h\ 2x$ do not remain finite at ∞. Hence there are not possible.

$$\frac{-d^2}{dx^2}\left(A_0e^{-2x}\right) = \frac{-d}{dx}\left[\frac{d}{dx}\left(A_0e^{-2x}\right)\right]$$

$$= 2\frac{d}{dx}\left(A_0e^{-2x}\right)$$

$$= -4\left(A_0e^{-2x}\right)$$

113. $KE = \frac{3}{2}K_BT$

$\because \quad P^2 = 2m\ KE$

$$P = \sqrt{2m(KE)}$$

$$= \sqrt{2m\times\frac{3}{2}K_BT}$$

$$= \sqrt{3mK_BT}$$

de Broglie wavelength $\lambda = \frac{h}{P}$

$$= \frac{h}{\sqrt{3mK_BT}}$$

115. In the given basis wave function of S_y can be written as $\begin{bmatrix} 0 \\ i \end{bmatrix}$ and $\begin{bmatrix} i \\ 0 \end{bmatrix}$ hence normalized eigen function will be $\frac{1}{\sqrt{2}}\begin{bmatrix} i \\ 0 \end{bmatrix}$.

116. $\hat{A}$ and $\hat{B}$ represent two physical characteristics of a quantum system if $\hat{A}$ is Hermitian then product $\hat{A}\,\hat{B}$ Hermitian if $\hat{B}$ is Hermitian than $\hat{A}$ and $\hat{B}$ commute.

117. $\because$

$$<\psi^* |E| \psi> = 2 \text{ eV}$$

$$<\psi_0^* |E_0| \psi_0> = 10 \text{ eV}$$

$$<\psi_1^* |E_1| \psi_1> = 30 \text{ eV}$$

Only option (d) satisfies this condition as

$$\frac{1}{2}\left(\psi_0^*|E_0|\psi_0\right) + \frac{1}{2}\left(\psi_1^*|E_1|\psi_1\right)$$

$$= \frac{1}{2}10 + \frac{1}{2}30 = 20 \text{ eV}$$

120. $\because$ $EF - FE = 0$

$\Rightarrow$ $EF = FE$

Trace EF = Trace FE

Also, Trace (EFGH) = Trace (FEGH)

$\because$ Trace i.e., sum of eigen values remains invariant on transpose

Hence, Trace (FEGH) = Trace (HGFE)

121. $$(\psi_0) = \frac{1}{\sqrt{\pi a_0^3}} e^{-r/a_0}$$

Now $(L_x, L_y) = ih, lz$ and $L_z = -ih\frac{\partial}{\partial\phi}$

$$\Rightarrow (L_x, L_y)(\psi_0) = ihL_z \frac{1}{\sqrt{\pi a_0^3}} e^{-r/a_0}$$

$$= -ih^2 \frac{\partial}{\partial\phi}\left(\frac{1}{\sqrt{\pi a_0^3}} e^{-r/a_0}\right) = 0$$

$[(Lx, Ly) \mid \psi_0] = 0$

123. $$\psi_n = \sqrt{\frac{2}{L}} \sin\frac{n\pi x}{L}$$

$$P = \int_0^{1/2} \psi_n^* \psi_n x\, dx$$

$$P = \frac{2}{L}\int_0^{1/2} \sin 2\frac{n\pi x}{L} dx$$

$$= \frac{2}{L}\int_0^{L/2} \frac{\left(1-\cos\frac{m\pi x}{L}\right)}{2} dx$$

$$\Rightarrow \quad = \frac{1}{2}(x)^{L/0} - \frac{1}{2n\pi}\left[\sin\frac{2n\pi x}{L}\right]_0^{L/2}$$

$$\Rightarrow \quad = \frac{1}{L}\times\frac{L}{2} - \frac{L}{2\pi n}(\sin n\pi - \sin 0)$$

$$= \frac{1}{2} - 0 \quad (\because \sin n\pi = \sin 0° = 0)$$

$$= \frac{1}{2}$$

124. $E < V_0$

Applying Schrodinger equation from

$0 < X < L$

$$\frac{d^2\psi}{dx^2} + \frac{2m}{h^2}(E - V_0)\psi = 0$$

$$\frac{d^2\psi}{dx^2} - \frac{2m}{h^2}(V_0 - E)\psi = 0$$

$$\psi = A e^{kx} + B e^{-kx}$$

Where $k = \sqrt{\frac{2m}{h^2}(V_0 - E)}$

Wave function $0 < X < L$ is real.

126. $\because$ $R + T = 1$

$$\frac{A_R}{A_0} + \frac{A_T}{A_0} = 1$$

$$\frac{A_R}{A_0} = 1 - \frac{A_T}{A_0}$$

128. The given wave function is

$$\psi(x) = e^{-iax+b} = e^b e^{-iax} = Be^{-iax}$$

This wave function is possible when

$$V(x) = 0$$

129. Since, the given potential is for the harmonic oscillator. Hence, energy levels are described by

$$E_n = \left(n + \frac{1}{2}\right)\hbar\omega$$

130. $\psi(x) = xe^{-ax^2}$

$$\frac{d\psi}{dx} = e^{-ax^2} - 2ax^2e^{-ax^2}$$

$$\frac{d^2\psi}{dx^2} = -2axe^{-ax^2} - 4axe^{-ax^2} + 4a^2x^3e^{-ax^2}$$

$$\frac{d^2\psi}{dx^2} = \left(4a^2x^3 - 6ax\right)e^{-ax^2}$$

The Eigen values for harmonic oscillation is

$E_n = \left(n + \frac{1}{2}\right)\hbar w$ for first excited state

$E = E_1 = \frac{3}{2}\hbar w$ and given potential is $V = \frac{mw^2x^2}{4}$

Applying Schrodinger equation

$$\frac{d^2\Psi}{dx^2} + \frac{2m}{\hbar^2}[E - V]\psi = 0$$

$$\left(4a^2x^3 - 6ax\right)e^{-ax^2} + \frac{2m}{\hbar}\left[\frac{3}{2}\hbar w - \frac{mw^2x^2}{2}\right]xe^{-ax^2} = 0$$

$$4a^2x^3 - 6a + \frac{3mw}{\hbar} - \frac{m^2w^2}{\hbar^2}x^3 = 0$$

$$4a^2x^2 - 6a + \frac{3mw}{\hbar} - \frac{m^2w^2}{\hbar^2}x^2 = 0$$

$$\left[4a^2 - \frac{m^2w^2}{\hbar^2}\right]x^2 + \left(\frac{2mw}{\hbar} - 6a\right) = 0$$

$$x \neq 0 \Rightarrow 4a^2 - \frac{m^2w^2}{\hbar^2} = 0 \quad ...(i)$$

and $$\frac{3mw}{\hbar} - 6a = 0 \quad ...(ii)$$

$$\pm\frac{mw}{2\hbar} = a$$

and $a = \dfrac{mw}{2\hbar}$ by Eq. (*ii*)

Hence, only $a = \dfrac{mw}{2\hbar}$ Satisfies both Eq. (*i*) and (*ii*)

131. It is given that $[X, P] = i\hbar$

we know that

$$[x^n, P] = i\hbar\, n\, x^{n-1}$$

$$[x^3, P] = i\hbar \times 3 \times x^{3-1} = 3\, i\hbar\, x^2$$

132. For bound state in this case

$$V_0 - |E| = \frac{n^2\pi^2\hbar^2}{2ma^2}$$

$$V_0 = \frac{n^2\pi^2\hbar^2}{2ma^2} + |E|$$

∵ There are only three bound states

Hence, $n = 2$ to $n = 4$

$$\frac{2\pi^2\hbar^2}{ma^2} < V_0 < \frac{8\pi^2\hbar^2}{ma^2}$$

133. The given function

$$\psi(r,\theta,\phi) = \frac{1}{\sqrt{2}}\left(\frac{1}{a_0}\right)^{3/2}\left(1 - \frac{r}{2a_0}\right)e^{-\frac{r}{2a_0\cos\theta}}$$

is the state function of H-atom is $2p^2$ state.

Hence $n = 2$, $l = 1$ and $m = 1$

134.

$$[x, y] = i\hbar Z$$

$$[y, z] = i\hbar x$$

$$[z, x] = i\hbar y$$

Hence x, y and z are $\hat{s}_x, \hat{s}_y$ and $\hat{s}_z$ and in term of $\hbar$, S_z can take $+\frac{1}{2}$ and $-\frac{1}{2}$ eigen values.

136. $$\psi = \frac{1}{\sqrt{2L}}\cos\frac{\pi x}{4L}$$

$$\frac{d^2\psi}{dx^2} + \frac{2mE}{\hbar^2}\psi = 0$$

$$-\left(\frac{\pi}{4L}\right)^2\psi + \frac{2mE}{\hbar_2}\psi = 0$$

$$\frac{\pi^2}{16L^2} = \frac{2mE}{\hbar}$$

$$E = \frac{\pi^2\hbar^2}{32mL^2}$$

137. $$< P^2 > = \int_{-2L}^{2L}\frac{1}{\sqrt{2L}}\cos\frac{\pi x}{4L}P^2\frac{1}{\sqrt{2L}}\cos\frac{\pi x}{4L}dx$$

$$= \frac{-\hbar^2}{2L}\times 2\times\int_0^{2L}\cos\frac{\pi x}{4L}\frac{d^2}{dx^2}\left(\cos\frac{\pi x}{4L}\right)dx$$

$$= \frac{\hbar^2}{L}\times\frac{\pi^2}{16L^2}\int_0^{2L}\cos^2\frac{\pi x}{4L}dx$$

$$= \frac{\hbar^2}{16L^3}\times\pi^2\times 2L$$

$$= \frac{\pi^2\hbar^2}{8L^2}$$

138. $$\langle T\rangle = \langle V\rangle$$

$$V(r) = \alpha r^2, \qquad \langle V\rangle \simeq \frac{k_B T}{2}$$

$$T = \frac{1}{2}mv^2, \qquad \langle T\rangle = \frac{k_B T}{2}$$

$$\langle T\rangle = \langle V\rangle$$

139. $$\Psi = Ae^{ikx}$$

Probability current density

$$\bar{J} = \frac{i\hbar}{2m}\left[\Psi\nabla\Psi^* - \Psi^*\nabla\Psi\right]$$

$$= \frac{i\hbar}{2m}\left[A^{ikx}\left(-Aike^{-ikx}\right) - Ae^{-ikx}(Aike^{ikx})\right]$$

$$= \frac{\hbar k}{m}|A|^2$$

140. $$\Psi(x) = \frac{1}{\sqrt{14}}[3\Psi_0(x) - 2\Psi_1(x) + \Psi_2(x)]$$

The ground state wave function is Ψ_0.

∴ The probability of finding the oscillator in ground state in

$$P_{g.s} = |\langle\Psi | \Psi_0\rangle|^2$$

$$= \left(\frac{3}{\sqrt{14}}\right)^2 = \frac{9}{14}$$

141. $\Psi(n, t = 0) = \frac{1}{\sqrt{5}}[2\psi_0 - \Psi_1]$

$$\Psi(n, t = t) = \frac{1}{\sqrt{5}}\left[2\psi_0 e^{-iE_0 t/\hbar} - \Psi_1 e^{-iE1t/\hbar}\right]$$

142. $\phi_{1_s}^{(r)} = \phi_{10}(r) = 2\left(\frac{z}{a_0}\right)^{3/2} e - \frac{zr}{a_0}$

$$\phi_{2s}^{(r)} = \phi_{20}(r) = \frac{1}{2\sqrt{2}}\left(\frac{z}{a_0}\right)^{3/2}\left(1 - \frac{zr}{a_0}\right)e^{-\frac{zr}{2a_0}}$$

143. Parities of wave function $f(x)$ is

even if $f(x) = f(-x)$

and, odd if $f(x) = -f(-x)$

$$f(x) = \cos kx$$

$$f(-x) = \cos(-kx)$$

$$= \cos kx = f(x)$$

$\therefore$ $\cos kx$ – Even parity

$$\tan h\,(kx) = \frac{e^{2x} + e^{-2x}}{e^{2x} - e^{-2x}}$$

$$\tan h\,(-kx) = \frac{e^{-2x} + e^{2x}}{e^{-2x} - e^{2x}}$$

$$= -\tan h(kx)$$

$\therefore$ $f(x) = -f(-x)$, hence odd parity

144. $\psi = Y_{lm}(\theta, \phi) = \theta\,\phi$

$$\hat{L}_z \psi = -i\hbar \frac{\partial\psi}{\partial\phi}$$

$$= -i\hbar \frac{\partial\gamma(\theta_1\phi)}{\partial\phi}$$

$$= -i\hbar\,\theta \frac{\partial\Phi}{\partial\phi}$$

Now $\Phi = Ae^{im\phi}$

$$\therefore \quad \frac{d\Phi}{d\phi} = mAie^{im\phi} = im\,\Phi$$

$$\therefore \quad \hat{L}_z \psi = -i\hbar\,(im)\,\theta\,\Phi$$

$$= m\hbar\, Y_{lm}(\theta, \Phi) = m\hbar\psi$$

$$\therefore \quad \hat{L}_z = m\hbar$$

146. Here, $H = \Sigma q_j p_j - L$

Now, $L = T - V$

where, $T = \frac{1}{2} m\dot{x}^2$

and $V = \frac{1}{2} kx^2 - qE\,x$

$$\therefore \quad L = \frac{1}{2} m\dot{x}^2 - \frac{1}{2} kx^2 + qE\,x$$

$$\therefore \quad p_j = \frac{\partial L}{\partial \dot{q}_j} = m\dot{x}$$

$$\Rightarrow \quad \dot{x} = \frac{p_j}{m}$$

or $p_j = m\dot{x}$

or $\dot{x} = \frac{1}{m} - i\hbar \frac{\partial}{2x}$

$$\therefore \quad H = \dot{x}p_j - L$$

$$= m\dot{x}^2 - \left(\frac{1}{2}m\dot{x}^2 - \frac{1}{2}kx^2 + qEx\right)$$

$$= \frac{1}{2} m\dot{x}^2 + \frac{1}{2} kx^2 - qE\,x$$

$$= \frac{1}{2} m \left(-\frac{i\hbar}{m}\frac{\partial}{\partial x}\right)^2 + \frac{1}{2}kx^2 - qE\,x$$

$$= -\frac{\hbar^2}{2m}\frac{d^2}{dx^2} + \frac{1}{2}kx^2 - qE\,x$$

148. Transition rule allows that for dipole transition to take place, the parity of final state eigen function must be different from initial state eigen function, i.e. parity must change

$\therefore$ $\Delta l = \pm 1$, and $\Delta m_l = 0, \pm 1$

149. Use $\lambda = \frac{h}{\sqrt{2mE}}$

154. $P(x_1, x_2) = \int_{x_1}^{x_2} p(x)dx = \int_{x_1}^{x_2} ae^{-ax}\,dx$

$$= \frac{a\left[-e^{-ax}\right]_{x_1}^{x_2}}{a} = e^{-ax_1} - e^{-ax_2}$$

157. The time-independent Schrodinger equation

$\because \quad H\psi = E\psi \quad ...(i)$

$\because \quad H = T + V = E \text{ (energy)}$

$(T + V)\psi = E\psi$

or $\quad T\psi = (E - V)\psi$

or $\quad -\dfrac{\hbar^2}{2m}\dfrac{d^2\psi}{dx^2} = (E - V)\psi$

or $\quad \dfrac{d^2\psi}{dx^2} + \dfrac{2m}{\hbar^2}(E - V)\psi = 0$

So total energy of system is conserved.

158. Coulomb degeneracy is n^2

$\therefore$ $n = 4$, so degeneracy is 16.

159. $H\psi = E\psi$

$$\Rightarrow \left[-\frac{\hbar^2}{2m}\nabla^2 + v\right]\psi = E\psi$$

$\because \psi = A\exp(-\alpha^2 x^2)$

and $\dfrac{\partial\psi}{\partial x} = -2\alpha^2 Ax\exp(-\alpha^2 x^2)$

$$\frac{\partial^2\psi}{\partial x^2} = -2\alpha^2 A[1 - 2\alpha^2 x^2]\exp(-\alpha^2 x^2)$$

$$\Rightarrow -\frac{\hbar^2}{2m}[(-2)\alpha^2 A(1 - 2\alpha^2 x^2)$$

$$\exp(-\alpha^2 x^2] + V\psi = E_0\psi$$

$$\Rightarrow \frac{\alpha^2\hbar^2}{m}(1 - 2\alpha^2 x^2)\psi + V\psi = E_0\psi$$

$$\Rightarrow \frac{\alpha^2\hbar^2}{m}(1 - 2\alpha^2 x^2) + V = E_0 \quad [\because \quad E_0 = \frac{\hbar^2\alpha^2}{m}]$$

$\Rightarrow E_0(1 - 2\alpha^2 x^2) + V = E_0$

$\Rightarrow V = E_0 - E_0(1 - 2\alpha^2 x^2) = \mathbf{2E_0\alpha^2 x^2}.$

161. $\psi_0 = A\exp\left(-\dfrac{\alpha^2 x^2}{2}\right)$

Since $\displaystyle\int_{-\infty}^{\infty}\psi^*\psi\, dx = 1$

$$\Rightarrow A^2\int_{-\infty}^{\infty}\exp(-\alpha^2 x^2)\, dx = 1$$

$$A^2 = \frac{1}{\int_{-\infty}^{\infty}\exp(-\alpha^2 x^2)\, dx}$$

Put $\alpha x = y$, $\Rightarrow \alpha dx = dy$

$$\therefore \quad A^2 = \frac{1}{\int_{-\infty}^{\infty}\frac{1}{\alpha}\exp(-y^2)\, dy}$$

But $\displaystyle\int_{-\infty}^{\infty} e^{-y^2}\, dy = \sqrt{\pi}$

$$\Rightarrow \quad A^2 = \frac{1}{\frac{1}{\alpha}\sqrt{\pi}} = \left(\frac{\alpha}{\sqrt{\pi}}\right)$$

$$\because \quad H = H_0 + E_0\left(\frac{\alpha x}{10}\right)^4 = H_0 + H'$$

First order ground state energy

$$E = \langle \psi^* | H' | \psi \rangle$$

$$= \int_{-\infty}^{\infty}\psi^* H'\psi\, dx$$

$$= 2\int_0^{\infty}\psi^* H\psi\, dx$$

$$= \frac{2E_0\alpha^2}{10^4}\left(\frac{\alpha}{\sqrt{\pi}}\right)\int_0^{\infty} x^4\exp(-\alpha^2 x^2)\, dx$$

Let $\alpha^2 x^2 = t$, $\Rightarrow x = \dfrac{\sqrt{t}}{\alpha}$

$\Rightarrow 2\alpha^2 x\, dx = dt,$

$\Rightarrow dx = \dfrac{dt}{2\alpha^2 x}$

$$E = \frac{2E_0\alpha^4}{10^4\sqrt{\pi}}\int_0^{\infty}\frac{t^2}{\alpha^4}\exp(-t)\times\frac{dt}{2\alpha^2\left(\frac{\sqrt{t}}{\alpha}\right)}$$

$$= \frac{E_0}{10^4\sqrt{\pi}}\int_0^{\infty} t^{3/2}\exp(-t)\, dt$$

Now by using formula $\Gamma(x+1) = \displaystyle\int_0^{\infty} t^x e^{-t}\, dt$

$$E = \frac{E_0}{10^4\sqrt{\pi}}\times\left|\frac{5}{2}\right.$$

$$= \frac{E_0}{10^4\sqrt{\pi}}\times\frac{3}{2}\times\frac{1}{2}\times\sqrt{\pi}$$

$$= \left(\frac{3}{4}E_0\right)10^{-4}$$

162. $\psi_n = \sqrt{\frac{2}{L}} \sin\left(\frac{n\pi x}{L}\right)$

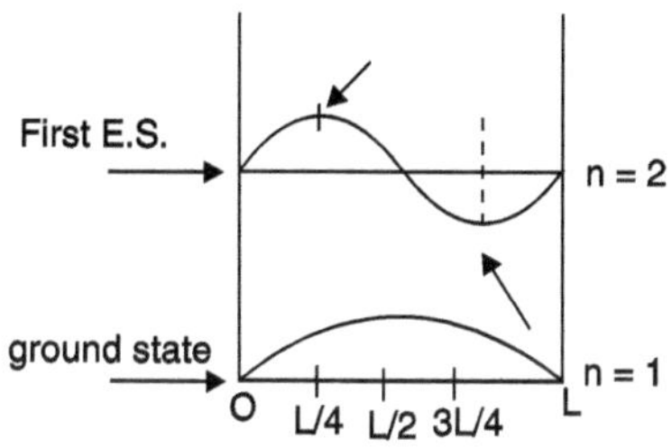

Since $n = 2,\ \psi_2 = \sqrt{\frac{2}{L}} \sin\left(\frac{2\pi x}{L}\right)$

So by figure, probability of finding the particle is maximum by wave function at $x = \frac{L}{4}$ and $\frac{3L}{4}$.

163. In lowest energy state, $n = 1$

$$\psi = \sqrt{\frac{2}{L}} \sin\left(\frac{\pi x}{L}\right)$$

$$P = \int_{x_1}^{x_2} |\psi|^2\, dx$$

$$= \int_0^{L/4} \psi^* \psi\, dx$$

$$= \frac{2}{L} \int_0^{L/4} \sin^2\left(\frac{\pi x}{L}\right) dx$$

$$= \frac{2}{L} \int_0^{L/4} \frac{1}{2}\left(1 - \frac{\cos 2\pi x}{L}\right) dx$$

$$= \frac{1}{L} \int_0^{L/4}\left(1 - \frac{\cos 2\pi x}{L}\right) dx$$

$$= \frac{1}{L} \left\{\frac{L}{4} - \left(\frac{L}{2\pi}\sin\frac{2\pi x}{L}\right)\right\}_0^{L/4}$$

$$= \frac{1}{4} - \frac{1}{2\pi}\left\{\sin\frac{\pi}{2} - \sin 0\right\}$$

$$= \frac{1}{4} - \left(\frac{1}{2\pi}\right).$$

164. For cubic box,

$$\varepsilon_0 = \frac{h^2}{8\,ma^2}({n_1}^2 + {n_2}^2 + {n_3}^2)$$

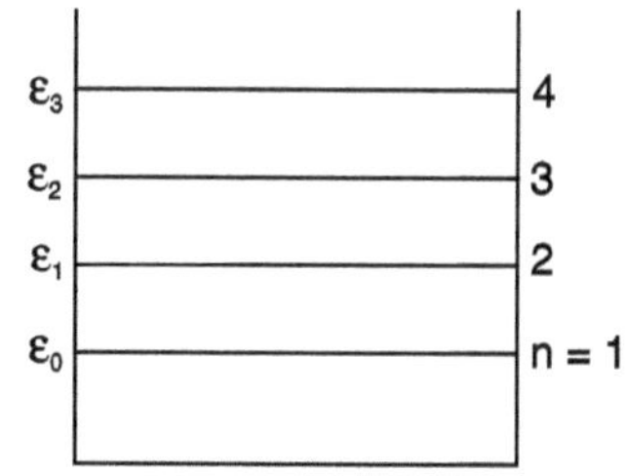

At lowest state, $n_1 = n_2 = n_3 = 1$

Since $\hbar = \left(\frac{h}{2\pi}\right)$

$$\Rightarrow \quad \varepsilon_0 = \frac{3h^2}{2ma^2}.$$

165. $E = \frac{\hbar\pi^2}{2ma^2}(n_1^2 + n_2^2 + n_3^2)$

$$\because \quad E_3 = \frac{11\hbar^2\pi^2}{2ma^2},$$

$n = 3$, so combination are $\underbrace{(3\,1\,1), (1\,3\,1), (1\,1\,3)}_{3\text{ orbitals}}$

and 3 including spin total degeneracy = 3 + 3 = 6.

166. $\omega = ck + m$

$$V_{\text{phase}} = \frac{\omega}{k} = c + \frac{m}{k}$$

$$V_{\text{group}} = \frac{d\omega}{dk} = c$$

167. $\psi = A \sin kz + B \cos kz \ldots k^2 = \frac{2m}{\hbar^2} E$

$$\psi = \sqrt{\frac{2}{L}} \sin\left(\frac{n\pi x}{L}\right) \text{ for even } n$$

$$\psi = \sqrt{\frac{2}{L}} \cos\left(\frac{n\pi x}{L}\right) \text{ for odd } n$$

168. $[x, P^2] = [x. P_x]\, P_x - P_x\, [P_x, x]$

$$= [x. P_x]\, P_x + P_x\, [x, P_x]$$

$$= i\hbar\ P_x + P_x\ \hat{i}\hbar$$

$$\left[\hat{i}, \hat{p}_x\right]^2 = 2i\hbar\, P_x$$

172. $E_n = \frac{n^2h^2}{8ml^2}$

$$\Rightarrow \quad n = \left(\frac{8ml^2E_n}{h^2}\right)^{1/2}$$

$$g(E) = \frac{dn}{dE}$$

175. $\int \Psi^* A\Psi d\alpha = \frac{3\times4+4\times3}{25} = 0.96$

176. $A^2\Psi_1 = A\Psi_2 = \Psi_1$

$A^2\Psi_2 = A\Psi_1 = \Psi_2$

177. $\Delta T \sim \frac{\hbar}{\Delta E} = 2.1 \times 10^{-15}$ s

179. $E = \frac{p^2}{2m} = \frac{h^2}{\lambda^2 .2m}$

$E - V = \frac{h^2}{2^2\lambda^2 \cdot 2m}$

$V/E = 3/4$

180. $R \propto \mu$ (Reduced mass)

181. $\rho \propto |\psi|^2$

$|\psi|^2$ is maximum when

$$\frac{2\pi x}{L} = \frac{n\pi}{2};\ n = 1, 3, 5,...$$

183. By Heisenberg's uncertainty principle

$$\Delta p_y \approx \frac{\hbar}{\Delta y}$$

$$\Delta v_y = \frac{\hbar}{m\Delta y}$$

$$= \frac{6.6\times10^{-34}}{9.1\times10^{-31}\times10^{-9}\times2\pi}$$

$$= 1.16 \times 10^{-5} \text{ m/s}$$

184. $[(\hat{A}\hat{B}-\hat{B}\hat{A}),(\hat{B}\hat{A}-\hat{A}\hat{B})]$

$= [\hat{A}\hat{B}, (\hat{B}\hat{A}-\hat{A}\hat{B})] - [\hat{B}\hat{A},(\hat{B}\hat{A}-\hat{A}\hat{B})]$

$= [\hat{A}\hat{B},\hat{B}\hat{A}]+[\hat{B}\hat{A},\hat{A}\hat{B}]$

$= [\hat{A}\hat{B},\hat{B}\hat{A}]-[\hat{A}\hat{B}, \hat{B}\hat{A}]$

$= 0$

185. $\phi_s = \begin{bmatrix}\frac{\sqrt{3}}{2}\\ \frac{1}{2}\end{bmatrix} = \frac{\sqrt{3}}{2}[S_z \uparrow]+\frac{1}{2}[S_z \downarrow]$

$\therefore$ Required probability $= \left(\frac{1}{2}\right)^2 = \frac{1}{4} = 0.25$

186. For first line of Balmer series

$$\frac{1}{\lambda} = R\left(\frac{1}{4}-\frac{1}{9}\right)$$

$$\Rightarrow \quad R = \frac{36}{5\lambda}$$

$\therefore$ Wavelenth of the first line, λ_L of the Lyman series is given by

$$\frac{1}{\lambda_L} = R\left(1-\frac{1}{4}\right) = \frac{36}{5\lambda}\times\frac{3}{4} = \frac{27}{5\lambda}$$

$$\Rightarrow \quad \lambda_L = \frac{5\lambda}{27}$$

187. K.E. is function of momentum and P.E. is function of position in this case. Heisenberg's uncertainty principle momentum and position can't be determined accurately simultaneously.

189. We have $a^t|n\rangle = \sqrt{n+1}|n+1\rangle$

and $a|n\rangle = \sqrt{n}|n-1\rangle$

$\therefore\ \lambda(a^+ a\, a^+ a)|\, n\rangle = \lambda\sqrt{n}\left(a^+ a\, a^+\right)|n-1\rangle$

$= \lambda\sqrt{n}\sqrt{n}\left(a^+ a\right)|n\rangle$

$= \lambda n\sqrt{n}|n-1\rangle$

$= \lambda n^2|n\rangle$

$\Rightarrow \quad \lambda(a^+a)^2|n\rangle = \lambda n^2|n\rangle$

190. $[X, Y] = [x\cos\phi + p\sin\phi), (-x\sin\phi + p\cos\phi)]$

$= [x\cos\phi, -x\sin\phi]$

$+ [x\cos\phi, p\cos\phi] + [p\sin\phi, -x\sin\phi] + [p\sin\phi, p\cos\phi]$

$= \cos^2\phi\,[x, p] + \sin^2\phi\,[x, p]$

$= i$

191. Wave function must vanish at boundaries as particle is confined within given boundaries.

192. $L + L = L_x^2 + L_y^2$

$= L^2 - L_2$

$\therefore$ Expectation value of $L_+ L_-$

$<\Psi\,|L_+ L_-|\,\Psi> = Y\,|L^2 - L_z^2|\psi>$

$= \hbar^2 l(l+1) - \hbar^2 l^2$

$= \hbar^2 l$

195. $\sigma(\theta) = |f(\theta)|^2$

$= \frac{1}{k^2}\left|\sum_{l=0}^{\infty}\left(2l + 1e^{i\delta l}\sin\delta_t p_l(\cos\theta)\right)\right|^2$

$\propto |p_l(\cos\theta)|^2$

$\propto \cos^2\theta$

5 CHAPTER

Thermodynamics and Statistical Physics

2021

1. For a finite system of Fermions where the density of states increases with energy, the chemical potential

(a) decreases with temperature

(b) increases with temperature

(c) does not vary with temperature

(d) corresponds to the energy where the occupation probability is 0.5

2. Consider a single one-dimensional harmonic oscillator of angular frequency ω, in equilibrium at temperature $T = (k_B\beta)^{-1}$. The states of the harmonic oscillator are all non-degenerate having energy $E_n\left(n+\frac{1}{2}\right)\hbar\omega$ with equal probability, where n is the quantum number. The Helmholtz free energy of the oscillator is

(a) $\frac{\hbar\omega}{2}+\beta^{-1}\ln[1-\exp(\beta\hbar\omega)]$

(b) $\frac{\hbar\omega}{2}+\beta^{-1}\ln[1-\exp(-\beta\hbar\omega)]$

(c) $\frac{h\omega}{2}+\beta^{-1}\ln[1+\exp(-\beta\hbar\omega)]$

(d) $\beta^{-1}\ln[1-\exp(-\beta\hbar\omega)]$

3. A system of two atoms can be in three quantum states having energies 0, ε and 2ε. The system is in equilibrium at temperature $T = (k_B\beta)^{-1}$. Match the following **Statistics** with the **Partition function.**

Statistics	**Partition function**
CD: Classical (distinguishable particles)	**Z1:** $e^{-\beta\varepsilon} + e^{-2\beta\varepsilon} + e^{-3\beta\varepsilon}$
CI: Classical (indistinguishable particles)	**Z2:** $1 + e^{-\beta\varepsilon} + 2e^{-2\beta\varepsilon} + e^{-3\beta\varepsilon} + e^{-4\beta\varepsilon}$
FD: Fermi-Dirac (indistinguishable	**Z3:** $1 + 2e^{-\beta\varepsilon} + 3e^{-2\beta\varepsilon} + 2e^{-3\beta\varepsilon} + e^{-4\beta\varepsilon}$
BE: Bose-Einstein	**ZA:** $\frac{1}{2} + e^{-\beta\varepsilon} + \frac{3}{2}e^{-2\beta\varepsilon} + 2e^{-3\beta\varepsilon} + \frac{1}{2}e^{-4\beta\varepsilon}$

(a) CD:Z1, CI:Z2, FD:Z3, BE:Z4

(b) CD:Z2, CI:Z3, FD:Z4, BE:Z1

(c) CD:Z3, CI:Z4, FD:Z1, BE:Z2

(d) CD:Z4, CI:Z1, FD:Z2, BE:Z3

4. The free energy of a ferromagnet is given by $F = F_0 + a_0(T - T_C)M^2 + bM^4$, where F_0, a_0, and b are positive constants, M is the magnetization, T is the temperature, and T_C is the Curie temperature. The relation between M^2 and T is best depicted by

(a)
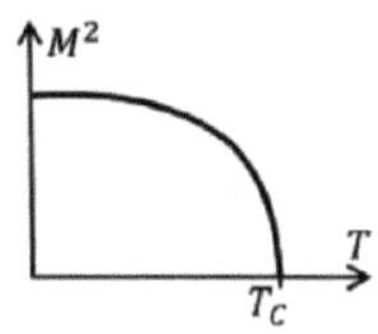

(b)
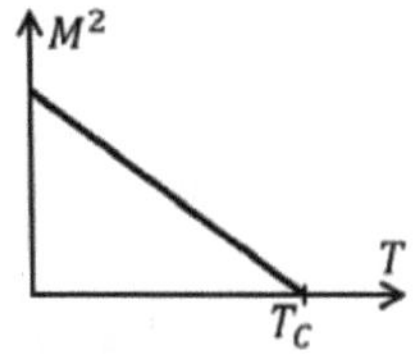

(c)
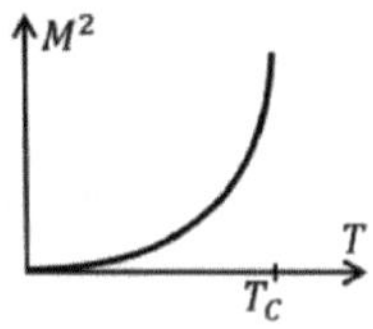

(d)
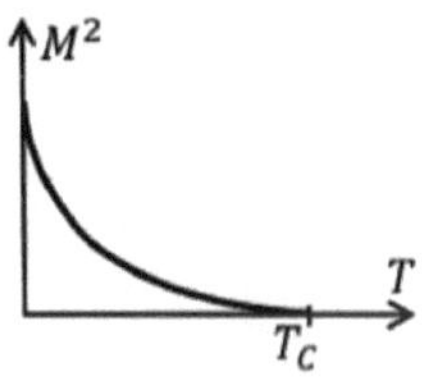

2020

5. Choose the correct statement related to the Fermi energy (E_F) and the chemical potential (μ) of a metal.

(*a*) $\mu = E_F$ only at 0 K

(*b*) $\mu = E_F$ at finite temperature

(*c*) $\mu < E_F$ at 0 K

(*d*) $\mu > E_F$ at finite temperature

6. As shown in the figure, an ideal gas is confined to chamber A of an insulated container, with vacuum in chamber B. When the plug in the wall separating the chambers A and B is removed, the gas fills both the chambers. Which one of the following statements is true?

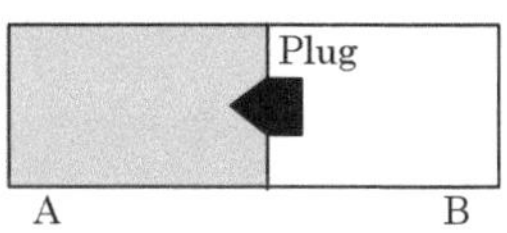

(a) The temperature of the gas remains unchanged

(b) Internal energy of the gas decreases

(c) Temperature of the gas decreases as it expands to fill the space in chamber B

(d) Internal energy of the gas increases as its atoms have more space to move around

7. The internal energy U of a system is given by $U(S, V)\ \lambda V^{\frac{-2}{3}} S^2$, where λ is a constant of appropriate dimensions; V and S denote the volume and entropy, respectively. Which one of the following gives the correct equation of state of the system?

(a) $\frac{PV^{\frac{1}{3}}}{T^2} = \text{constant}$ (b) $\frac{PV}{T^{\frac{1}{3}}} = \text{constant}$

(c) $\frac{P}{V^{\frac{1}{3}}T} = \text{constant}$ (d) $\frac{PV^{\frac{2}{3}}}{T} = \text{constant}$

8. An electron in a hydrogen atom is in the state n = 3, l = 2, m = –2. Let $\hat{L}_y$ denote the y-component of the orbital angular momentum operator. If $\left(\Delta\hat{L}_y\right)^2 = \alpha\hbar^2$, the value of α is ______.

9. Consider a gas of hydrogen atoms in the atmosphere of the Sun where the temperature is 5800 K. If a sample from this atmosphere contains 6.023×10^{23} of hydrogen atoms in the ground state, the number of hydrogen atoms in the first excited state is approximately 8×10^n, where n is an integer. The value of n is ______.

(Boltzmann constant : 8.617×10^{-5} eV/K)

10. For a gas of non-interacting particle, the probability that a particle has a speed v in the interval v to v + dv is given by

$$f(v)\,dv = 4\pi v^2 dv\left(\frac{m}{2\pi k_B T}\right)^{\frac{3}{2}} e^{-mv^2} / 2k_B T$$

If E is the energy of a particle, then the maximum in the corresponding energy distribution in units of E/k_BT occurs at ______ (rounded off to one decimal place).

2019

11. A learge number N of ideal bosons, each of mass m, are trapped in a three–dimensional potential $V(r) = \frac{m\omega^2 r^2}{2}$. The bosonic system is kept at temperature T which is much lower than the Bose–Einstein condensation temperature T_C. The chemical potential (μ) satisfies

(a) $\mu \le \frac{3}{2}\hbar\omega$ (b) $2\hbar\omega > \mu > \frac{3}{2}\hbar\omega$

(c) $3\hbar\omega > \mu > 2\hbar\omega$ (d) $\mu = 3\hbar\omega$

12. At tempereture T Kelvin (K), the value of the Fermi function at an energy 0.5 eV above the Fermi energy is 0.01. Then T, to the nearest integer, is________

($k_B = 8.62 \times 10^{-5}$ eV/K)

13. In a thermally insulated container, 0.01 kg of ice at 273 K is mixed with 0.1 kg of water at 300 K. Neglecting the specific heat of the container, the change in the entropy of the system in J/K on attaining thermal equilibrium (rounded off to two decimal places) is______________

(Specific heat of water is 4.2 kJ/kg-K and the latent heat of ice is 335 kJ/kg).

14. Consider two systems A and B each having two distinguishable particles. In both the systems, each particle can exist in states with energies 0, 1, 2 and 3 units with equal probability. The total energy of the combined system is 5 units. Assuming that the system A has energy 3 units and the system B has energy 2 units, the entropy of the combined system is $k_B \ln \lambda$. The value of λ is__________

2018

15. A microcanonical ensemble consists of 12 atoms with each taking either energy 0 state, or energy ε state. Both states are non-degenerate.

If the total energy of this ensemble is 4ε, its entropy will be ________ k_B (up to one decimal place), where k_B is the Boltzmann constant.

16. An air-conditioner maintains the room temperature at 27°C while the outside temperature is 47°C. The heat conducted through the walls of the room from outside to inside due to temperature difference is 7000 W. The minimum work done by the compressor of the air-conditioner per unit time is ________ W.

17. Two solid spheres A and B have same emissivity. The radius of A is four times the radius of B, and temperature of A is twice the temperature of B. The ratio of the rate of heat radiated from A to that from B is ________.

18. The partition function of an ensemble at a temperature T is $Z = \left(2\cosh\frac{\varepsilon}{k_BT}\right)^N$, where k_B is the Boltzmann constant. The heat capacity of this ensemble at $T = \frac{\varepsilon}{k_B}$ is $X\,Nk_B$, where the value of X is ________ (up to two decimal places).

19. The quantum effects in an ideal gas become important below a certain temperature T_Q when de Broglie wavelength corresponding to the *root mean square* thermal speed becomes equal to the inter-atomic separation. For such a gas of atoms of mass 2×10^{-26} kg and number density $6.4\times10^{25}\ \text{m}^{-3}$, T_Q = ________ $\times10^{-3}$ K (up to one decimal place).

($k_B = 1.38\times10^{-23}$ J/K, $h = 6.6\times10^{-34}$ J-s)

2017

20. A reversible Carnot engine is operated between temperatures T_1 and T_2 ($T_2 > T_1$) with a photon gas as the working substance. The efficiency of the engine is

(*a*) $1-\frac{3T_1}{4T_2}$ (*b*) $1-\frac{T_1}{T_2}$

(*c*) $1-\left(\frac{T_1}{T_2}\right)^{3/4}$ (*d*) $1-\left(\frac{T_1}{T_2}\right)^{4/3}$

21. Water freezes at 0°C at atmospheric pressure (1.01×10^5 Pa). The densities of water and ice at this temperature and pressure are 1000 kg/m^3 and 934 kg/m^3 respectively. The latent heat of fusion is 3.34×10^5 J/kg. The pressure required for depressing the melting temperature of ice by 10°C is ______ GPa. (up to two decimal places)

22. Consider N non-interacting, distinguishable particles in a two-level system at temperature T. The energies of the levels are 0 and ε, where ε > 0. In the high temperature limit ($k_B T >> \varepsilon$), what is the population of particles in the level with energy ε ?

(*a*) $\frac{N}{2}$ (*b*) N

(*c*) $\frac{N}{4}$ (*d*) $\frac{3N}{4}$

2016

23. The total power emitted by a spherical black body of radius R at a temperature T is P_1. Let P_2 be the total power emitted by another spherical black body of radius R/2 kept at temperature 2T. The ratio, P_1/P_2 is ______. (Give your answer upto two decimal places)

24. The entropy S of a system of N spins, which may align either in the upward or in the downward direction, is given by $S = k_B N[p \ln p + (1 - p) \ln (1 - p)]$. Here k_B is the Boltzmann constant. The probability of alignment in the upward direction is p. The value of p, at which the entropy is maximum, is ________. (Give your answer upto one decimal place)

25. For a system at constant temperature and volume, which of the following statements is correct at equilibrium?

(*a*) The Helmoholtz free energy attains a local minimum.

(*b*) The Helmoholtz free energy attains a local maximum.

(*c*) The Gibbs free energy attains a local minimum.

(*d*) The Gibbs free energy attains a local maximum.

26. N atoms of an ideal gas are enclosed in a container of volume V. The volume of the container is changed to 4V, while keeping the total energy constant. The change in the entropy of the gas, in units of $NK_B \ln 2$, is ______, where k_B is the Boltzmann constant.

27. Consider a metal which obeys the Sommerfeld model exactly. If E_F is the Fermi energy of the metal at T = 0 K and R_H is its Hall coefficient, which of the following statements is correct?

(*a*) $R_H \propto E_r^{3/2}$

(*b*) $R_H \propto E_F^{2/3}$

(*c*) $R_H \propto E_F^{-3/2}$

(*d*) R_H is independent of E_F

28. A two-dimensional square rigid box of side L contains six non-interacting electrons at T = 0 K. The mass of the electron is m. The ground state energy of the system of electrons, in units of $\frac{\pi^2\hbar^2}{2mL^2}$ is ________.

29. The Fermi energies of two metals X and Y are 5 eV and 7 eV and their Debye temperatures are 170 K and 340 K, respectively. The molar specific heats of these metals at constant volume at low temperatures can be written as $(C_V)_X = \gamma_X T + A_X T^3$ and $(C_V)_Y = \gamma_Y T + A_Y T^3$, where γ and A are constants. Assuming that the thermal effective mass of the electrons in the two metals are same, which of the following is correct?

(*a*) $\frac{\gamma_X}{\gamma_Y}=\frac{7}{5}, \frac{A_X}{A_Y}=8$ (*b*) $\frac{\gamma_X}{\gamma_Y}=\frac{7}{5}, \frac{A_X}{A_Y}=\frac{1}{8}$

(*c*) $\frac{\gamma_X}{\gamma_Y}=\frac{5}{7}, \frac{A_X}{A_Y}=\frac{1}{8}$ (*d*) $\frac{\gamma_X}{\gamma_Y}=\frac{5}{7}, \frac{A_X}{A_Y}=8$

2015

30. In Bose-Einstein condensates, the particles

(*a*) have strong interparticle attraction

(*b*) condense in real space

(*c*) have overlapping wave functions

(*d*) have large and positive chemical potential

31. The energy dependence of the density of states for a two dimensional non-relativistic electron gas given by, $g(E) = CE^n$, where C is constant. The value of n is ________

32. Consider a system of N non-interacting spin-½ particles, each having a magnetic moment μ, is a magnetic field $\vec{B} = B\hat{z}$. If E is the total energy of the system, the number of accessible microstates Ω is given by

(*a*) $\Omega = \dfrac{N!}{\frac{1}{2}\left(N - \frac{E}{\mu B}\right)!\,\frac{1}{2}\left(N + \frac{E}{\mu B}\right)!}$

(*b*) $\Omega = \dfrac{\left(N - \frac{E}{\mu B}\right)!}{\left(N + \frac{E}{\mu B}\right)!}$

(*c*) $\Omega = \frac{1}{2}\left(N - \frac{E}{\mu B}\right)!\,\frac{1}{2}\left(N + \frac{E}{\mu B}\right)!$

(*d*) $\Omega = \dfrac{N!}{\left(N + \frac{E}{\mu B}\right)!}$

33. For a black body radiation in a cavity, photons are created and annihilated freely as a result of emission and absorption by the walls of the cavity. This is because

(*a*) the chemical potential of the photons is zero

(*b*) photons obey Pauli exclusion principle

(*c*) photons are spin –1 particles

(*d*) the entropy of the photons is very large

34. The entropy of a gas containing N particles enclosed in a volume V is given by $S = Nk_B \ln\left(\frac{aVE^{3/2}}{N^{5/2}}\right)$, where E is the total energy, a is a constant and k_B is the Boltzmann constant. The chemical potential μ of the system at a temperature T is given by

(*a*) $\mu = -k_B T\left[\ln\left(\frac{aVE^{3/2}}{N^{5/2}}\right) - \frac{5}{2}\right]$

(*b*) $\mu = -k_B T\left[\ln\left(\frac{aVE^{3/2}}{N^{5/2}}\right) - \frac{3}{2}\right]$

(*c*) $\mu = -k_B T\left[\ln\left(\frac{aVE^{3/2}}{N^{3/2}}\right) - \frac{5}{2}\right]$

(*d*) $\mu = -k_B T\left[\ln\left(\frac{aVE^{3/2}}{N^{3/2}}\right) - \frac{3}{2}\right]$

35. Given that the Fermi energy of gold is 5.54 eV, the number density of electrons is _____ $\times 10^{28}$ m^{-3} (upto one decimal place)

Mass of electron = 9.11×10^{-31} kg; $h = 6.626 \times 10^{-34}$ J.s; 1 eV = 1.6×10^{-19}J)

2014

36. For a gas under isothermal conditions, its pressure P varies with volume V as $P \propto V^{-5/3}$. The bulk modulus B is proportional to

(*a*) $V^{-1/2}$ (*b*) $V^{-2/3}$

(*c*) $V^{-3/5}$ (*d*) $V^{-5/3}$

37. At a given temperature, T, the average energy per particle of a non-interacting gas of two-dimensional classical harmonic oscillators is ________$k_B T$ (k_B is the Boltzman constant).

38. The plot of specific heat versus temperature across the super-conducting transition temperature (T_C) is most appropriately represented by

(*a*) 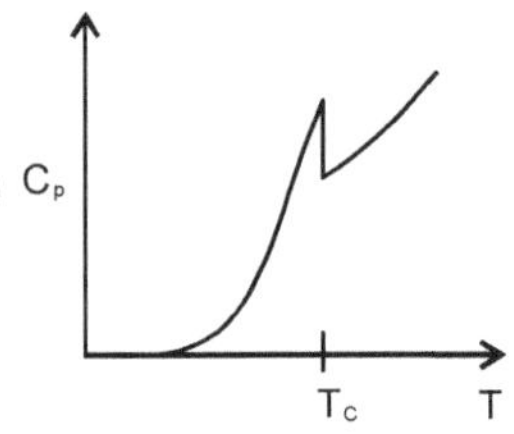

(*b*) 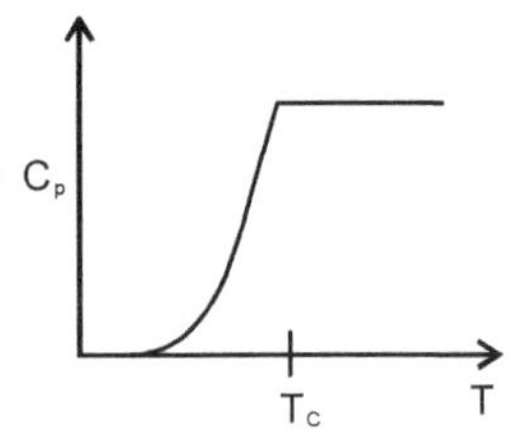

(*c*) 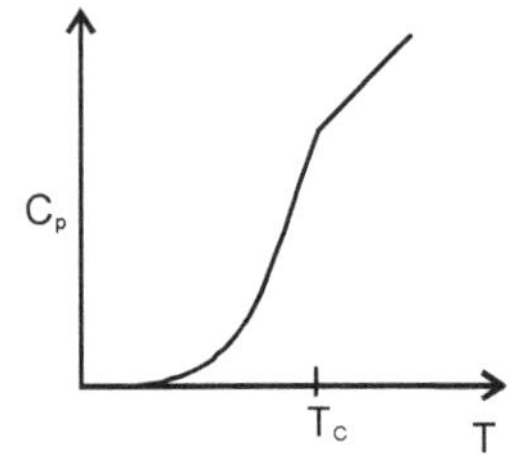

(*d*) 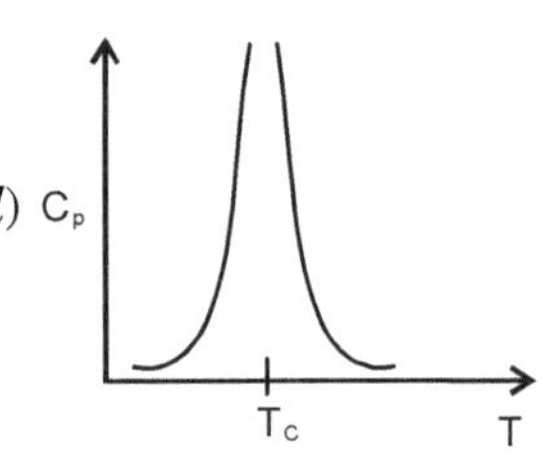

39. For a free electron gas in two dimensions, the variation of the density of states, N(E) as a function of energy E, is best represented by

(*a*) N(E) 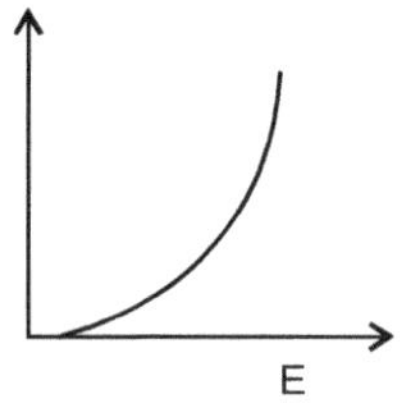

(*b*) N(E) 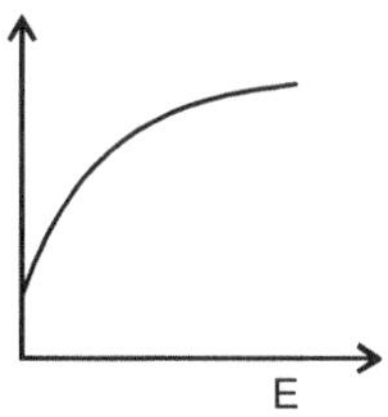

(*c*) N(E) 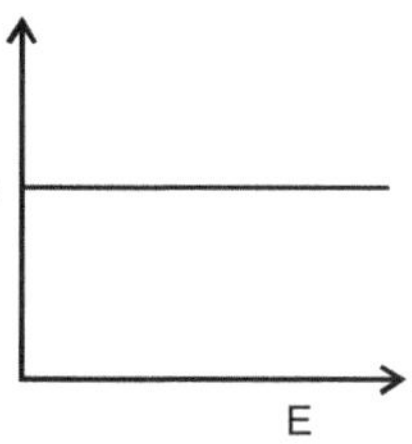

(*d*) N(E) 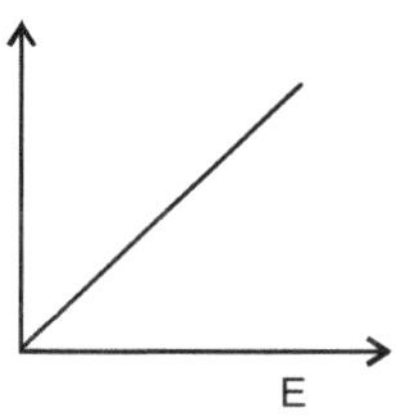

40. For a system of two bosons, each of which can occupy any of the two energy levels 0 and and ε, the mean energy of the system at a temperature T with $\beta = \dfrac{1}{k_B T}$ is given by

(*a*) $\dfrac{\varepsilon e^{-\beta\varepsilon} + 2\varepsilon e^{-2\beta\varepsilon}}{1 + 2e^{-\beta\varepsilon} + e^{-2\beta\varepsilon}}$ (*b*) $\dfrac{1 + \varepsilon e^{-\beta\varepsilon}}{2e^{-\beta\varepsilon} + e^{-2\beta\varepsilon}}$

(*c*) $\dfrac{2\varepsilon e^{-\beta\varepsilon} + \varepsilon e^{-2\beta\varepsilon}}{2 + e^{-\beta\varepsilon} + e^{-2\beta\varepsilon}}$ (*d*) $\dfrac{\varepsilon e^{-\beta\varepsilon} + 2\varepsilon e^{-2\beta\varepsilon}}{2 + e^{-\beta\varepsilon} + e^{-2\beta\varepsilon}}$

41. The Hamiltonian of a particle of mass m is given by $H = \dfrac{p^2}{2m} - \dfrac{\alpha q^2}{2}$. Which one of the following figures describes the motion of the particle in phase space?

(*a*) 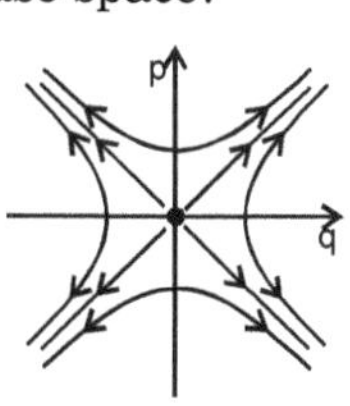 (*b*) 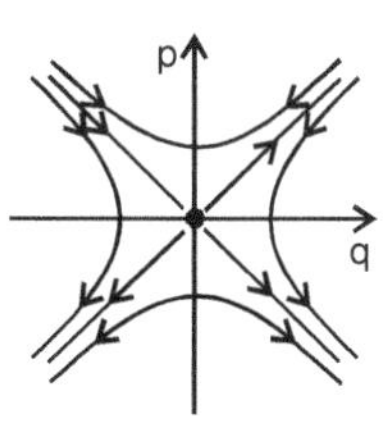

(*c*) 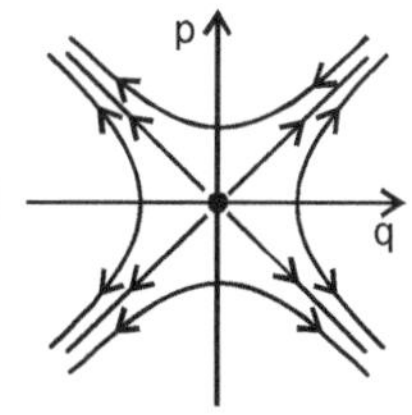 (*d*) 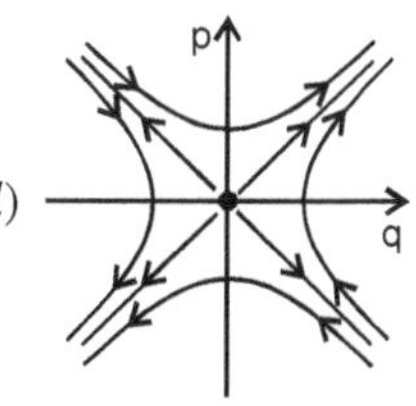

42. Consider a system of 3 fermions, which can occupy any of the 4 available energy states with equal probability. The entropy of the system is

(*a*) $k_B \ln 2$ (*b*) $2k_B \ln 2$

(*c*) $2k_B \ln 4$ (*d*) $3k_B \ln 4$

2013

43. If Planck's constant were zero, then the total energy contained in a box filled with radiation of all frequencies at temperature T would be (k is the Boltzmann constant and T is nonzero)

(*a*) Zero (*b*) Infinite

(*c*) $\dfrac{3}{2}kT$ (*d*) kT

44. Across a first order phase transition, the free energy is

(*a*) proportional to the temperature

(*b*) a discontinuous function of the temperature

(*c*) a continuous function of the temperature but its first derivative is discontinuous

(*d*) such that the first derivative with respect to temperature is continuous

45. Two gases separated by an impermeable but movable partition are allowed to freely exchange energy. At equilibrium, the two sides will have the same

(*a*) pressure and temperature

(*b*) volume and temperature

(*c*) pressure and volume

(*d*) volume and energy

46. The entropy function of a system is given by $S(E) = aE(E_0 - E)$ where a and E_0 are positive constants. The temperature of the system is

(*a*) negative for some energies

(*b*) increases monotonically with energy

(*c*) decreases monotonically with energy

(*d*) Zero

47. Consider a linear collection of N independent spin 1/2 particles, each at a fixed location. The entropy of this system is (k is the Boltzmann constant)

(*a*) Zero (*b*) Nk

(*c*) $\dfrac{1}{2}Nk$ (*d*) $Nk \ln(2)$

48. The number of distinct ways of placing four indistinguishable balls into five distinguishable boxes is ____________.

49. Consider a gas of atoms obeying Maxwell-Boltzmann statistics. The average value of $e^{i\vec{a}.\vec{p}}$ over all the momenta $\vec{p}$ of each of the particles (where $\vec{a}$ is a constant vector and a is its magnitude, m is the mass of each atom, T is temperature and k is Boltzmann's constant) is,

(a) One (b) Zero

(c) $e^{-\frac{1}{2}a^2mkT}$ (d) $e^{-\frac{1}{3}a^2mkT}$

2012

50. The isothermal compressibility, k of an ideal gas at temperature T_0 and volume V_0, is given by

(a) $-\frac{1}{V_0}\frac{\partial V}{\partial P}\Big|_{T_0}$ (b) $\frac{1}{V_0}\frac{\partial V}{\partial P}\Big|_{T_0}$

(c) $-V_0\frac{\partial P}{\partial V}\Big|_{T_0}$ (d) $V_0\frac{\partial P}{\partial V}\Big|_{T_0}$

51. For an ideal Fermi gas in three dimensions, the electron velocity V_F at the Fermi surface is related to electron concentration n as,

(a) $V_F \propto n^{2/3}$

(b) $V_F \propto n$

(c) $V_F \propto n^{1/2}$

(d) $V_F \propto n^{1/3}$

52. A classical gas of molecules, each of mass m, is in thermal equilibrium at the absolute temperature, T. The velocity components of the molecules along the Cartesian axes are v_x, v_y and v_z. The mean value of $(v_x + v_y)^2$ is

(a) $\frac{k_B T}{m}$ (b) $\frac{3k_B T}{2m}$

(c) $\frac{1k_B T}{2m}$ (d) $\frac{2k_B T}{m}$

53. The total energy, E of an ideal non-relativistic Fermi gas in three dimensions is given by $E \propto \frac{N^{5/3}}{V^{2/3}}$ where N is the number of particles and V is the volume of the gas.

Identify the CORRECT equation of state (P being the pressure),

(a) $PV = \frac{1}{3}E$ (b) $PV = \frac{2}{3}E$

(c) $PV = E$ (d) $PV = \frac{5}{3}E$

54. Consider a system whose three energy levels are given by 0, ε and 2ε. The energy level ε is two-fold degenerate and the other two are non-degenerate. The partition function of the system with $\beta = \frac{1}{k_B T}$ is given by

(a) $1 + 2e^{-\beta\varepsilon}$ (b) $2e^{-\beta\varepsilon} + e^{-2\beta\varepsilon}$

(c) $(1 + e^{-\beta\varepsilon})^2$ (d) $1 + e^{-\beta\varepsilon} + e^{-2\beta\varepsilon}$

55. A paramagnetic system consisting of N spin-half particles, is placed in an external magnetic field.It is found that N/2 spins are aligned parallel and the remaining N/2 spins are aligned antiparallel to the magnetic field. The statistical entropy of the system is,

(a) $2Nk_B$ In 2 (b) $\frac{N}{2}k_B$ In 2

(c) $\frac{3N}{2}k_B$ In 2 (d) Nk_B In 2

56. At a certain temperature T, the average speed of nitrogen molecules in air is found to be 400 m/s. The most probable and the root mean square speeds of the molecules are, respectively,

(a) 355 m/s, 434 m/s (b) 820 m/s, 917 m/s

(c) 152 m/s, 301 m/s (d) 422 m/s, 600 m/s

2011

57. A Carnot cycle operates on a working substance between two reservoirs at temperatures T_1 and T_2, with $T_1 > T_2$. During each cycle an amount of heat Q_1 is extracted from the reservoir at T_1 and an amount Q_2 is delivered to the reservoir at T_2. Which of the following statements is INCORRECT?

(a) work done in one cycle is $Q_1 - Q_2$

(b) $\frac{Q_1}{T_1} = \frac{Q_2}{T_2}$

(c) entropy of the hotter reservoir decreases

(d) entropy of the universe (consisting of the working substance and the two reservoirs) increases

58. In a first order phase transition, at the transition temperature, specific heat of the system

(a) diverges and its entropy remains the same

(b) diverges and its entropy has finite discontinuity

(c) remains unchanged and its entropy has finite discontinuity

(d) has finite discontinuity and its entropy diverges

59. If the isothermal compressibility of a solid is $\kappa_T = 10^{-10}\ (\text{Pa})^{-1}$, the pressure required to increase its density by 1% is approximately

(*a*) 10^4 Pa (*b*) 10^5 Pa

(*c*) 10^8 Pa (*d*) 10^{10} Pa

60. A system of N non-interacting and distinguishable particles of spin 1 is in thermodynamic equilibrium. The entropy of the system is

(*a*) $2k_B \ln N$ (*b*) $3k_B \ln N$

(*c*) $Nk_B \ln 2$ (*d*) $Nk_B \ln 3$

61. A system has two energy levels with energies ε and 2ε. The lower level is 4-fold degenerate while the upper level is doubly degenerate. If there are N non-interacting classical particles in the system, which is in thermodynamic equilibrium at temperature T, the fraction of particles in the upper level is

(*a*) $\dfrac{1}{1+e^{-\varepsilon/k_BT}}$ (*b*) $\dfrac{1}{1+2e^{\varepsilon/k_BT}}$

(*c*) $\dfrac{1}{2e^{\varepsilon/k_BT}+4e^{2\varepsilon/k_BT}}$ (*d*) $\dfrac{1}{2e^{\varepsilon/k_BT}-4e^{2\varepsilon/k_BT}}$

2010

62. Which of the following atoms cannot exhibit Bose-Einstein condensation, even in principle?

(*a*) $^1H_1$

(*b*) $^4He_2$

(*c*) $^{23}Na_{11}$

(*d*) $^{40}K_{19}$

63. For a two-dimensional free electron gas, the electronic density n, and the Fermi energy E_F, are related by

(*a*) $n = \dfrac{(2mE_F)^{3/2}}{3\pi^2\hbar^3}$ (*b*) $n = \dfrac{mE_F}{\pi\hbar^2}$

(*c*) $n = \dfrac{mE_F}{2\pi\hbar^2}$ (*d*) $n = \dfrac{2^{3/2}(mE_F)^{1/2}}{\pi\hbar}$

64. CO_2 molecule has the first few energy levels uniformly separated by approxima-tely 2.5 meV. At a temperature of 300K, the ratio of the number of molecules in the 4th excited state to the number in the 2nd excited state is about

(*a*) 0.5 (*b*) 0.6

(*c*) 0.8 (*d*) 0.9

65. Which among the following sets of Maxwell relations is correct? (U - internal energy, H-enthalpy, A - Helmholtz free energy and G - Gibbs free energy)

(*a*) $T = \left(\dfrac{\partial U}{\partial V}\right)_S$ and $P = \left(\dfrac{\partial U}{\partial S}\right)_V$

(*b*) $V = \left(\dfrac{\partial H}{\partial P}\right)_S$ and $T = \left(\dfrac{\partial H}{\partial S}\right)_P$

(*c*) $P = -\left(\dfrac{\partial G}{\partial V}\right)_T$ and $V = \left(\dfrac{\partial G}{\partial P}\right)_S$

(*d*) $P = -\left(\dfrac{\partial A}{\partial S}\right)_T$ and $S = -\left(\dfrac{\partial A}{\partial P}\right)_V$

Common Data for Questions 66 and 67:

The partition function for a gas of photons is given by

$$\ln Z = \frac{\pi^2}{45}\frac{V(k_BT)^3}{\hbar^3C^3}$$

66. The specific heat of the photon gas varies with temperature as

(*a*)

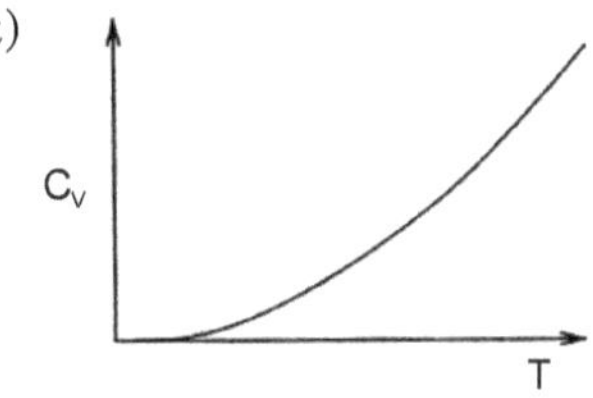

(*b*)

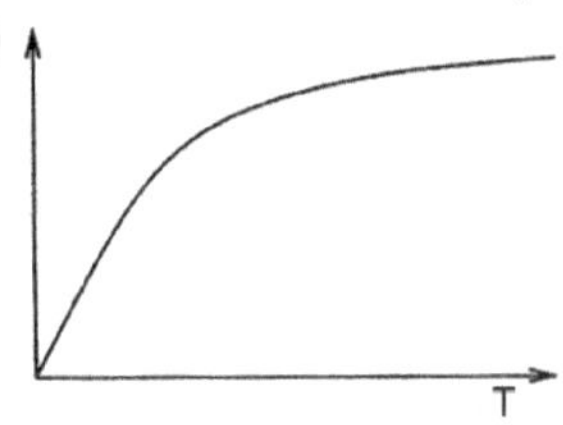

(*c*)

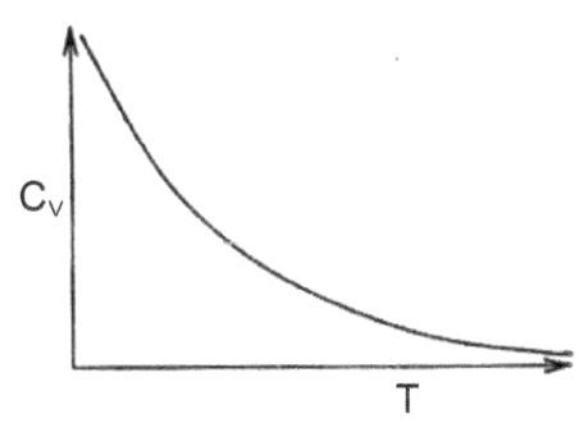

(*d*)

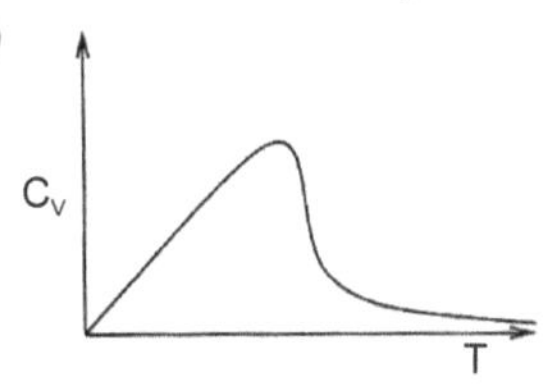

67. The pressure of the photon gas is

(*a*) $\dfrac{\pi^2}{15}\dfrac{V(k_BT)^3}{\hbar^3c^3}$ (*b*) $\dfrac{\pi^2}{8}\dfrac{V(k_BT)^4}{\hbar^3c^3}$

(*c*) $\dfrac{\pi^2}{45}\dfrac{V(k_BT)^4}{\hbar^3c^3}$ (*d*) $\dfrac{\pi}{45}\dfrac{V(k_BT)^{3/2}}{\hbar^3c^3}$

2009

68. Identify which one is a first order phase transition?

(*a*) A liquid to gas transition at its critical temperature.

(*b*) A liquid to gas transition close to its triple point.

(*c*) A paramagnetic to ferromagnetic transition in the absence of a magnetic field.

(*d*) A metal to superconductor transition in the absence of a magnetic field.

69. The probability that an energy level ε at a temperature T is unoccupied by a fermion of chemical potential μ is given by

(*a*) $\frac{1}{e^{(\varepsilon-\mu)/k_B T}+1}$ (*b*) $\frac{1}{e^{(\varepsilon-\mu)/k_B T}-1}$

(*c*) $\frac{1}{e^{(\mu-\varepsilon)/k_B T}+1}$ (*d*) $\frac{1}{e^{(\mu-\varepsilon)/k_B T}-1}$

70. For a Fermi gas of N particles in three dimensions at T = 0 K, the Fermi energy, E_F is proportional to

(*a*) $N^{2/3}$ (*b*) $N^{3/2}$

(*c*) N^3 (*d*) N^2

COMMON DATA QUESTIONS

Common data Q. 71 and 72 :

Consider a two level quantum system with energies $\varepsilon_1 = 0$ and $\varepsilon_2 = \varepsilon$.

71. The Helmholtz free energy of the system is given by

(*a*) $-k_B T \ln\left(1+e^{-\varepsilon/k_B T}\right)$ (*b*) $k_B T \ln\left(1+e^{-\varepsilon/k_B T}\right)$

(*c*) $\frac{3}{2}k_B T$ (*d*) $\varepsilon - k_B T$

72. The specific heat of the system is given by

(*a*) $\frac{\varepsilon}{k_B T}\frac{e^{-\varepsilon/k_B T}}{\left(1+e^{-\varepsilon/k_B T}\right)^2}$ (*b*) $\frac{\varepsilon^2}{k_B T^2}\frac{e^{-\varepsilon/k_B T}}{\left(1+e^{-\varepsilon/k_B T}\right)}$

(*c*) $-\frac{\varepsilon^2 e^{-\varepsilon/k_B T}}{\left(1+e^{-\varepsilon/k_B T}\right)^2}$ (*d*) $\frac{\varepsilon^2}{k_B T^2}\frac{e^{-\varepsilon/k_B T}}{\left(1+e^{-\varepsilon/k_B T}\right)^2}$

2008

73. Thermodynamic variables of a system can be volume V, pressure P, temperature T, number of particles N, internal energy E and chemical potential μ, etc. For a system to be specified by Microanonical (MC), Canoncial (CE) and Grand Canonical (GC) ensembles, the parameters required for the respective ensembles are

(*a*) MC: (N, V, T); CE: (E, V, N); GC: (V, T, μ)

(*b*) MC: (E, V, N); CE: (N, V, T); GC: (V, T, μ)

(*c*) MC: (V, T, μ); CE: (N, V, T); GC: (E, V, N)

(*d*) MC: (E, V, N); CE: (V, T, μ); GC: (N, V, T)

74. The pressure versus temperature diagram of a given system at certain low temperature range is found to be parallel to the temperature axis in the liquid-to-solid transition region. The change in the specific volume remains constant in the region. The conclusion one can get from the above is

(*a*) the entropy of solid is zero in this temperature region.

(*b*) the entropy increases when the system goes from liquid to solid phase in this temperature region.

(*c*) the entropy decreases when the system transforms from liquid to solid phase in this region of temperature.

(*d*) the change in entropy is zero in the liquid-to-solid transition region.

75. The radial wave function of the electrons in the state of $n = 1$ and $l = 0$ in a hydrogen atom is $R_{10} = \frac{2}{a_0^{3/2}}\exp\left(\frac{r}{a_0}\right)$, a_0 is the Bohr radius. The most probable value r for and electron is

(*a*) a_0 (*b*) $2a_0$

(*c*) $4a_0$ (*d*) $8a_0$

76. An exact measurement of the position of a simple harmonic oscillator (SHO) is made with the result $x = x_0$. [The SHO has energy levels $E_n (n = 0, 1, 2,.....)$ and associated normalized wave-fuunctions ψ_n]. subsequently, an exact measurement of energy E is made using the general notation $Pr(E = E')$ dentoing the prbability that a result E′ is obtained for this measurement, the follwoiing statements are written.

Which one of the following statements is correct?

(*a*) $Pr(E = E_0) = 0$

(*b*) $Pr(E = E_n) = 1$ for some value of n.

(*c*) $Pr(E = E_n) \propto \psi_n (X)$

(*d*) $Pr(E > E'') > 0$ for any E″.

77. Two identical particles have to be distributed among three energy levels. Let r_B, r_F and r_C represent the ratios of probability of finding two particles to that of finding one particle in a given energy state. The subscripts B, F and C correspond to whether the particles are bosons, fermions and classical particles, respectively. The, $r_B : r_F : r_C$ is equal to

(*a*) $\frac{1}{2}:0:1$ (*b*) $1:\frac{1}{2}:1$

(*c*) $1:\frac{1}{2}:\frac{1}{2}$ (*d*) $1:0:\frac{1}{2}$

78. A photon gas is at thermal equilibrium at temperature T. The mean number of photons in an energy state $\varepsilon = \hbar\omega$ is

(*a*) $\exp\left(\frac{\hbar\omega}{k_B T}\right)+1$ (*b*) $\exp\left(\frac{\hbar\omega}{k_B T}\right)-1$

(*c*) $\left(\exp\left(\frac{\hbar\omega}{k_B T}\right)+1\right)^{-1}$ (*d*) $\left(\exp\left(\frac{\hbar\omega}{k_B T}\right)-1\right)^{-1}$

79. Consider a system of N atoms of an ideal gas of type A at temperature T and volume V. It is kept in diffusive contact with another system of N atoms of another ideal gas of type B at the same temperature T and volume V. Once the combined system reaches equilibrium,

(*a*) the total ntropy of the final system is the same as the sum of the entropy of the individual system always.

(*b*) the entropy of mixing $2Nk_B \ln 2$.

(*c*) the entropy of the final system iis less than that of sum of the initial entropies of the two gases.

(*d*) the entropy of mixing is non-zero when the atoms A and B are of the same type.

80. Consider a syystem of two non-interacting classical particles which can occupy any of the three energy levels with energy value E = 0, ε and 2ε having degeneracies *g*(E) = 1, 2 and 4 respectively. The eman energy of the system is

(*a*) $\varepsilon\left(\frac{4\exp\left(-\varepsilon/k_B T\right)+8\exp\left(2\varepsilon/k_B T\right)}{1+2\exp\left(-\varepsilon/k_B T\right)+4\exp\left(-2\varepsilon/k_B T\right)}\right)$

(*b*) $\varepsilon\left(\frac{2\exp\left(-\varepsilon/k_B T\right)+8\exp\left(-2\varepsilon/k_B T\right)}{1+2\exp\left(-\varepsilon/k_B T\right)+4\exp\left(-2\varepsilon/k_B T\right)}\right)$

(*c*) $\varepsilon\left(\frac{2\exp\left(-\varepsilon/k_B T\right)+4\exp\left(-2\varepsilon/k_B T\right)}{1+2\exp\left(-\varepsilon/k_B T\right)+4\exp\left(-2\varepsilon/k_B T\right)}\right)^2$

(*d*) $\varepsilon\left(\frac{\exp\left(-\varepsilon/k_B T\right)+2\exp\left(-2\varepsilon/k_B T\right)}{1+\exp\left(-\varepsilon/k_B T\right)+\exp\left(-2\varepsilon/k_B T\right)}\right)$

Statement for Linked Answer Questions 81 and 82 :

Consider a two dimensional electron gas of N electrons of mass m each in a system of size L × L.

81. The density of states between energy ε and ε + *d*ε is

(*a*) $\frac{4\pi L^2 m}{h^2}d\varepsilon$ (*b*) $\frac{4\pi L^2 m}{h^2}\frac{1}{\sqrt{\varepsilon}}d\varepsilon$

(*c*) $\frac{4\pi L^2 m}{h^2}\sqrt{\varepsilon d\varepsilon}$ (*d*) $\frac{4\pi L^2 m}{h^2}\varepsilon d\varepsilon$

82. The ground state energy E_0 of the system in terms of the Fermi energy E_F and the number of electrons N is given by

(*a*) $\frac{1}{3}NE_F$ (*b*) $\frac{1}{2}NE_F$

(*c*) $\frac{2}{3}NE_F$ (*d*) $\frac{3}{5}NE_F$

2007

83. A heat pump working on the Carnot cycle maintains the inside temperature of a house at 22ºC by supplying 450 kJs^{-1}. If the outside temperature is 0º C, the heat taken, in kJs^{-1}, from the outside air is approximately

(*a*) 487 (*b*) 470

(*c*) 467 (*d*) 417

84. The vapour pressure *p* (in mm of Hg) of a solid, at temperature T, is expressed by ln *p* = 23 – 3863/T and that of its liquid phase by ln *p* = 19 – 3063/T. The triple poing (in Kelving) of the material is

(*a*) 185 (*b*) 190

(*c*) 195 (*d*) 200

85. The free energy for a photon gas is given by F = – $\left(\frac{a}{2}\right)$ VT^{-4}, where *a* is a constant.

The entropy S and the pressure P of the photon gas are

(*a*) $S=\frac{4}{3}aVT^3, P=\frac{a}{3}T^4$

(*b*) $S=\frac{1}{3}aVT^4, P=\frac{4a}{3}T^3$

(*c*) $S=\frac{4}{3}aVT^4, P=\frac{a}{3}T^3$

(*d*) $S=\frac{1}{3}aVT^3, P=\frac{4a}{3}T^4$

Statement for Linked Answer Questions 86 & 87:

An ensemble of quantum harmonic oscillators is kept at a finite temperature $T = 1/k_B\beta$.

86. The partition function of a single oscillator with energy levels $\left(n+\frac{1}{2}\right)\hbar\omega$ is given by

(*a*) $Z = \frac{e^{-\beta\hbar\omega/2}}{1-e^{-\beta\hbar\omega}}$ (*b*) $Z = \frac{e^{-\beta\hbar\omega/2}}{1+e^{-\beta\hbar\omega}}$

(*c*) $Z = \frac{1}{1-e^{-\beta\hbar\omega}}$ (*d*) $Z = \frac{1}{1+e^{-\beta\hbar\omega}}$

87. The average number of energy quanta of the oscillators is given

(*a*) $<n>= \frac{1}{e^{-\beta\hbar\omega}-1}$ (*b*) $<n>= \frac{e^{-\beta\hbar\omega}}{e^{-\beta\hbar\omega}-1}$

(*c*) $<n>= \frac{1}{e^{-\beta\hbar\omega}+1}$ (*d*) $<n>= \frac{e^{-\beta\hbar\omega}}{e^{-\beta\hbar\omega}+1}$

2006

88. The free energy of a photon gas enclosed in a volume V is given by $F = -\frac{1}{13}aVT^{-4}$ where a is a constant and T is the temperature of the gas. The chemical potential of the photon gas is

(*a*) 0 (*b*) $\frac{4}{3}aVT^3$

(*c*) $\frac{1}{3}aT^{-4}$ (*d*) aVT^{-4}

89. Each of the two isolated vessels, A and B of fixed volumes, contains N molecules of a perfect monatomic gas at a pressure P. The temperasture of A and B are T_1 and T_2, respectively. The two vessels are brought into termal contact. At equilibrium, the change in entropy is

(*a*) $\frac{3}{2}Nk_B \ln\left[\frac{T_1^2+T_2^2}{4T_1T_2}\right]$

(*b*) $Nk_B \ln\left(\frac{T_2}{T_1}\right)$

(*c*) $\frac{3}{2}Nk_B \ln\left[\frac{(T_1+T_2)^2}{4T_1T_2}\right]$

(*d*) $2Nk_B$

90. The internal energy of n moles of a gas is given $E = E = \frac{3}{2}nRT - \frac{a}{V}$, where V is the volume of the gas at temperature T and a is a positive constant. One mole of the gas in state (T_1, V_1) is allowed to expand adiabatically into vacuum to a final state (T_2, V_2). The temperature T_2 is

(*a*) $T_1 + Ra\left(\frac{1}{V_2}+\frac{1}{V_1}\right)$ (*b*) $T_1 - \frac{2}{3}Ra\left(\frac{1}{V_2}-\frac{1}{V_1}\right)$

(*c*) $T_1 + \frac{2}{3}Ra\left(\frac{1}{V_2}-\frac{1}{V_1}\right)$ (*d*) $T_1 - \frac{1}{3}Ra\left(\frac{1}{V_2}-\frac{1}{V_1}\right)$

91. The mean internal of a one-dimensional classical harmonic oscillator in equilibrium with a heat bath of temperature T is

(*a*) $\frac{1}{2}k_BT$ (*b*) k_BT

(*c*) $\frac{3}{2}k_BT$ (*d*) $3k_BT$

92. A monatomic crystalline solid comprises of N atoms, out of which n atoms are in interstitial positions. If the available interstitial sites are N′, then number of possible microstates is

(*a*) $\frac{(N'+n)!}{n!N!}$

(*b*) $\frac{N!}{n!(N'+n)!}\frac{N!}{n!(N'+n)!}$

(*c*) $\frac{N!}{n!(N'-n)!}$

(*d*) $\frac{N!}{n!(N'-n)!}\frac{N!}{n!(N'-n)!}$

2005

93. If the equation of state for a gas with internal energy U is $pV = \frac{1}{3}U$, then the equation for an adiabatic process is

(*a*) $pV^{1/3}$ = constant (*b*) $pV^{2/3}$ = constant

(*c*) $pV^{4/3}$ = constant (*d*) $pV^{3/5}$ = constant

94. The total number of accessible states of N noninteracting particles of spin 1/2 is

(*a*) 2^N (*b*) N^2

(*c*) $2^{N/2}$ (*d*) N

95. The pressure for a non-interacting Fermi gas with internal energy U at temperature T is

(*a*) $p = \frac{3}{2}\frac{U}{V}$ (*b*) $p = \frac{2}{3}\frac{U}{V}$

(*c*) $p = \frac{3}{5}\frac{U}{V}$ (*d*) $p = \frac{1}{2}\frac{U}{V}$

96. A system of non-interacting Fermi particles with Fermi energy E_F has the density of states proportional to $\sqrt{E}$, where E is the energy of a particle. The average energy per particle at temperature T = 0 is

(a) $\frac{1}{6}E_F$ (b) $\frac{1}{5}E_F$

(c) $\frac{2}{5}E_F$ (d) $\frac{3}{5}E_F$

97. If the partition function of a harmonic oscillator with frequency ω at a temperature T is $\frac{kT}{\hbar\omega}$, then the free energy of N such independent oscillators is

(a) $\frac{3}{2}NkT$ (b) $kT\ln\frac{\hbar\omega}{kT}$

(c) $NkT\ln\frac{\hbar\omega}{kT}$ (d) $NkT\ln\frac{\hbar\omega}{2kT}$

98. The partition function of two Bose particles each of which can occupy any of the two energy levels 0 and ε is

(a) $1+e^{-2\varepsilon/kT}+2e^{-\varepsilon/kT}$ (b) $1+e^{-2\varepsilon/kT}+e^{-\varepsilon/kT}$

(c) $2+e^{-2\varepsilon/kT}+e^{-\varepsilon/kT}$ (d) $e^{-2\varepsilon/kT}+e^{-\varepsilon/kT}$

99. A one dimensional random walker takes steps to left or right with equal probability. The probability that the random walker starting from origin is back to origin after N even number of steps is

(a) $\frac{N!}{\left(\frac{N}{2}\right)!\left(\frac{N}{2}\right)!}\left(\frac{1}{2}\right)^N$ (b) $\frac{N!}{\left(\frac{N}{2}\right)!\left(\frac{N}{2}\right)!}$

(c) $2N!\left(\frac{1}{2}\right)^{2N}$ (d) $N!\left(\frac{1}{2}\right)^N$

100. The number of states for a system of N identical free particles in a three dimensional space having total energy between E and E + δE ($\delta E << E$), is proportional to

(a) $\left(E^{\frac{3N}{2}-1}\right)\delta E$ (b) $E^{\frac{N}{2}}\delta E$

(c) $NE^{\frac{1}{2}}\delta E$ (d) $E^N\delta E$

2004

101. For any process, the second law of thermodynamics requires that the change of entropy of the universe be

(a) positive only (b) positive or zero

(c) zero only (d) negative or zero

102. The dimension of phase space of ten rigid diatomic molecules is

(a) 5 (b) 10

(c) 50 (d) 100

103. The specific heat of an ideal Fermi gas in 3-dimension at very low temperatures (T) varies as

(a) T (b) $T^{3/2}$

(c) T^2 (d) T^3

104. Which one of the following is a first order phase transition ?

(a) Vaporization of a liquid at its boiling point

(b) Ferromagnetic to paramagnetic

(c) Normal liquid He to superfluid He

(d) Superconducting to normal state

105. In the region of co-existence of a liquid and vapor phases of a material

(a) C_P and C_V are both infinite

(b) C_V and β $\left[=\frac{1}{V}\left(\frac{\partial V}{\partial T}\right)_P\right]$ are both finite

(c) C_V and K $\left[=-\frac{1}{V}\left(\frac{\partial V}{\partial P}\right)_T\right]$ are both finite

(d) C_P, β and K are all infinite

106. Which one of the following statements is TRUE ?

(a) Magnetic tapes are made of Iron

(b) Permanent magnets are made from ferrites

(c) Ultrasonic transducers are made from quartz crystals

(d) Optoelectronic devices are made from soft ferrites

107. Which one of the following statements is NOT TRUE ?

(a) Entropy decreases markedly on cooling a superconductor below the critical temperature, T_C

(b) The electronic contribution to the heat capacity in the superconducting state has an exponential form with an argument proportional to T^{-1}, suggestive of an energy gap

(c) A type I superconductor is a perfect diamagnet

(d) Critical temperature of superconductors does not vary with the isotopic mass

108. The form factor $F(\vec{q}) = \int \exp(i\vec{q}\cdot\vec{r}/\hbar)\rho(\vec{r})\,d^3r$ of Rutherford scattering is obtained by choosing a delta function for the charge density $\rho(\vec{r})$. The value of the form factor is

(a) unity (b) infinity

(c) zero (d) undefined

Data for Q. 109-110 :

An ensemble of N three-level systems with energies $\varepsilon = -\varepsilon_0, 0, +\varepsilon_0$ is in thermal equilibrium at temperature T. Let $\beta = (k_BT)^{-1}$.

109. If $\beta\varepsilon_0 = 2$, the probability of finding the system in the level $\boldsymbol{\varepsilon} = 0$ is

(*a*) (cosh 2)/2 (*b*) $(\cosh 2)^{-1}$

(*c*) $(2 \cosh 2)^{-1}$ (*d*) $(1 + 2 \cosh 2)^{-1}$

110. The free energy of the system at high temperatures (i.e., $x \equiv \beta\varepsilon_0 << 1$) is approximately

(*a*) $-NK_B Tx^2$

(*b*) $-NK_BT[\ln 2 + x^2/2]$

(*c*) $-NK_BT [\ln 3 + x^2/3]$

(*d*) $-Nk_BT \ln 3$

2003

111. A piston containing an ideal gas is originally in the state X (see figure). The gas is taken through a thermal cycle X → Y → X as shown

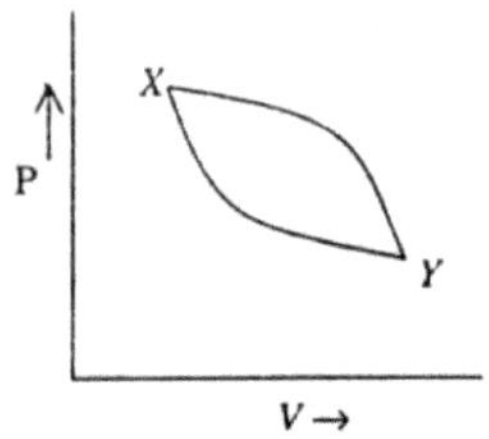

The work done by the gas is positive if the direction of the thermal cycle is

(*a*) clockwise

(*b*) counter-clockwise

(*c*) neither clockwise nor counter-clockwise

(*d*) clockwise from X → Y and counter-clockwise from Y → X

112. A second order phase transition is one in which

(*a*) the plot of entropy as a function of temperature shows a discontinuity

(*b*) the plot of specific heat as a function of temperature shows a discontinuity

(*c*) the plot of volume as a function of pressure shows a discontinuity

(*d*) the plot of comprehensibility as a function of temperature is continuous

113. The temperature of a cavity of fixed volume is doubled. Which of the following is true for the black-body radiation inside the cavity ?

(*a*) its energy and the number of photons both increase 8 times

(*b*) its energy increases 8 times and the number of photons increases 16 times

(*c*) its energy increases 16 times and the number of photons increases 8 times

(*d*) its energy and the number of photons both increase 16 times

114. A sample of ideal gas with initial pressure P and volume V is taken through an isothermal expansion proceed during which the change in entropy is found to be ΔS. The Universal gas constant is R. Then the work done by the gas is given by

(*a*) $(PV\Delta S)/(nR)$ (*b*) $nR\Delta S$

(*c*) PV (*d*) $(P\Delta S)/(nRV)$

115. Hydrogen molecules (mass m) are in thermal equilibrium at a temperature T. Assuming classical distribution of velocity, the most probable speed at room temperature is

(*a*) $(k_BT)/m$ (*b*) $2k_BT/m$

(*c*) $(\sqrt{2k_BT/m})$ (*d*) $m/(\sqrt{2k_BT})$

116. A ferromagnetic material has a Curie temperature 100 K. Then

(*a*) its susceptibility is doubled when it is cooled from 300 K to 200 K

(*b*) all the atomic magnets in it get oriented in the same direction above 100 K

(*c*) the plot of inverse susceptibility versus temperature is linear with a slope T_C

(*d*) the plot of its susceptibility versus temperature is linear with an intercept T_C

Data for Q. 117 – 119 :

A system consists of three spin-half particles, the z components of whose spins $S_z(1)$, $S_z(2)$ and $S_z(3)$ can take value +1/2 and – 1/2. The total spin of the system is $S_z = S_z(1) + S_z(2) + S_z(3)$

117. The total number of possible micro-states of this system is

(*a*) 3 (*b*) 6

(*c*) 7 (*d*) 8

118. The total number of micro-states with $S_z = 1/2$ is

(*a*) 3 (*b*) 5

(*c*) 6 (*d*) 7

119. Consider an ensemble of systems where each microstate has equal probability. The ensemble average of S_z is

(*a*) –1/2 (*b*) 0

(*c*) 1/2 (*d*) 3/2

Data for Q. 120 – 121 :

A gas of N particles is enclosed in a volume V at a temperature T. The logarithm of the partition function is given by $\ln Z = N \ln \{(V - bN)(k_B T)^{3/2}\}$ *where b is a constant with appropriate dimensions*

120. If P is the pressure of the gas, the equation of state is given by

(*a*) $P(V - bN) = Nk_B T$ (*b*) $P(V - bN) = k_B T$

(*c*) $P(V - b) = Nk_B T$ (*d*) $P(V - b) = Nk_B T$

121. The internal energy of the gas is given by

(*a*) $U = (1/2)\, k_B T$ (*b*) $U = Nk_B T$

(*c*) $U = (3/2)\, Nk_B T$ (*d*) $U = 2N\, k_B T$

2002

122. In a classical micro-canonical ensemble for a system of N non-interacting particles, the fundamental volume in phase space which is regarded as "equivalent to one micro-state" is

(*a*) h^{3N} (*b*) h^{2N}

(*c*) h^N (*d*) h

where h is the Planck's constant.

123. Which of the following conditions should be satisfied by the temperature T of a system of N non-interacting particles occupying a volume V, for Bose-Einstein condensation to take place ?

(*a*) $T < \dfrac{h^2}{2\pi m k_B}\left\{\dfrac{N}{V\varsigma\left(\frac{3}{2}\right)}\right\}^{3/2}$

(*b*) $T < \dfrac{h^2}{2\pi m k_B}\left\{\dfrac{V}{N\varsigma\left(\frac{3}{2}\right)}\right\}^{3/2}$

(*c*) $T < \dfrac{h^2}{2\pi m k_B}\left\{\dfrac{N}{V\varsigma\left(\frac{3}{2}\right)}\right\}^{1/2}$

(*d*) $T < \dfrac{h^2}{2\pi m k_B}\left\{\dfrac{V}{N\varsigma\left(\frac{3}{2}\right)}\right\}^{1/2}$

where m is the mass of each particle of the system, k_B is the Boltzmann constant, h is the Planck's constant and ς is the well known Zeta function.

124. Which of the following statements is true ?

(*a*) In a micro-canonical ensemble the total number of particles N and the energy E are constants while in a canonical ensemble N and temperature T are constants.

(*b*) In a micro-canonical ensemble the total number of particles N is a constant but the energy E is variable while in a canonical ensemble N and T are constants.

(*c*) In a micro-canonical ensemble N and E are constants while in a canonical ensemble N and T both vary.

(*d*) In a micro-canonical ensemble N and E are constants while in a canonical ensemble N is a constant but T varies.

125. The mean free path of the particles of a gas at a temperature T_0 and pressure p_0 has a value λ_0. If the pressure is increased to 1.5 p_0 and the temperature is reduced to 0.75 T_0, the mean free path

(*a*) remains unchanged

(*b*) is reduced to half

(*c*) is doubled

(*d*) is equal to 1.125 λ_0

2001

126. Two particles are said to be distinguishable when

(*a*) the average distance between them is large compared to their de Broglie wavelengths.

(*b*) the average distance between them is small compared to their de Broglie wavelengths.

(*c*) they have overlapping wavepackets.

(*d*) their total wave function is symmetric under particle exchange.

127. The rotational partition function for a diatomic molecule of moment of inertia I at a temperature T is given by

(*a*) $\dfrac{Ik_B T}{\hbar^2}$ (*b*) $\dfrac{2Ik_B T}{\hbar^2}$

(*c*) $\dfrac{3Ik_B T}{\hbar^2}$ (*d*) $\dfrac{Ik_B T}{2\hbar^2}$

2000

128. A closed tall jar containing air and a fly is placed on a sensitive weighting machine. When the flyh is stationary, the reading of the weighing machine is W. IF the fly starts flying with some upwards acceleration, the reading of the machine will be

(*a*) W

(*b*) > W

(*c*) < W

(*d*) directly proportional to the acceleration

129. The order of magnitude of the number of air molecules in a room of volume 50 m^3 at STP is

(*a*) 10^{27} (*b*) 10^{24}

(*c*) 10^{30} (*d*) 10^{20}

130. An amount of heat Q is transferred from a heat reservoir at temperature T_A to another heat reservoir at temperature T_B. What is the change in the entropy ΔS of the combined system?

(*a*) $Q\left(\frac{1}{T_B}-\frac{1}{T_A}\right)$

(*b*) $Q\left(\frac{1}{T_A}+\frac{1}{T_B}\right)$

(*c*) $\frac{Q}{\sqrt{T_B T_A}} \ln \frac{T_B}{T_A}$

(*d*) $Q\left(\frac{1}{T_A}-\frac{1}{T_B}\right)$

131. In a canonical ensemble

(*a*) the energy and the temperature are constants

(*b*) the entropy and the energy are constants

(*c*) the temperature and the density are constants

(*d*) the density and the entropy are constants

ANSWERS

1. (a,d) **2.** (b) **3.** (c) **4.** (b) **5.** (*a*) **6.** (*a*) **7.** (*a*) **8.** (1 to 1) **9.** (14 to 15)
10. (0.5 to 0.5) **11.** (*a*) **12.** (1260 to 1266) **13.** (1.02 to 1.04) **14.** (12 to 12)
15. (6.1 to 6.3) **16.** (466 to 467) **17.** (256 to 256) **18.** (0.41 to 0.43) **19.** (78 to 90)
20. (*b*) **21.** 0.15 to 0.19 **22.** (*a*) **23.** (0.25 : 0.25) **24.** (0.5 : 0.5) **25.** (*a*) **26.** (2 : 2)
27. (*c*) **28.** (24 : 24) **29.** (*a*) **30.** (*c*) **31.** (0) **32.** (*a*) **33.** (*a*) **34.** (*a*) **35.** (5.9 to 6.0)
36. (*d*) **37.** (1.99 to 2.01) **38.** (*a*) **39.** (*c*) **40.** (*a*) **41.** (*d*) **42.** (*b*) **43.** (*b*) **44.** (*b*)
45. (*a*) **46.** (*a*) **47.** (*d*) **48.** (70) **49.** (*c*) **50.** (*a*) **51.** (*d*) **52.** (*d*) **53.** (*d*) **54.** (*c*)
55. (*d*) **56.** (*c*) **57.** (*b*) **58.** (*b*) **59.** (*a*) **60.** (*c*) **61.** (*a*) **62.** (*a*) **63.** (*c*) **64.** (*c*)
65. (*b*) **66.** (*b*) **67.** (*c*) **68.** (*b*) **69.** (*c*) **70.** (*a*) **71.** (*a*) **72.** (*b*) **73.** (*b*) **74.** (*c*)
75. (*a*) **76.** (*a*) **77.** (*d*) **78.** (*d*) **79.** (*d*) **80.** (*d*) **81.** (*a*) **82.** (*d*) **83.** (*d*) **84.** (*d*)
85. (*a*) **86.** (*a*) **87.** (*c*) **88.** (*a*) **89.** (*c*) **90.** (*c*) **91.** (*b*) **92.** (*d*) **93.** (*c*) **94.** (*d*)
95. (*c*) **96.** (*d*) **97.** (*b*) **98.** (*c*) **99.** (*a*) **100.** (*c*) **101.** (*b*) **102.** (*d*) **103.** (*a*) **104.** (*a*)
105. (*a*) **106.** (*c*) **107.** (*a*) **108.** (*a*) **109.** (*d*) **110.** (*d*) **111.** (*d*) **112.** (*b*) **113.** (*c*) **114.** (*a*)
115. (*c*) **116.** (*a*) **117.** (*d*) **118.** (*a*) **119.** (*b*) **120.** (*a*) **121.** (*c*) **122.** (*a*) **123.** (*a*) **124.** (*a*)
125. (*b*) **126.** (*a*) **127.** (*b*) **128.** (*a*) **129.** (*a*) **130.** (*a*) **131.** (*c*)

EXPLANATIONS

1. We know that,

Chemical potential relates with fermi energy.

$$\mu(T) = E_F(0)\left(1 - \frac{(kT)^2}{E_F(0)} + \ldots\ldots\ldots\right)$$

It is clear, as we increase T, $\mu(T)$ decrease, so option (a) is correct.

Occupation probability,

$$f(E) = \frac{1}{e^{\frac{E-E_f}{kT}} + 1}$$

at fermi energy = E = E_f = μ

$$\text{So, } f(E) = \frac{1}{1+1} \Rightarrow \frac{1}{2} \Rightarrow 0.5$$

So, option (d) is also correct.

2. We know that the partition function for 1-D harmonic oscillator is

$$Z = \frac{1}{e^{\frac{\hbar\omega}{2}\beta} - e^{\frac{\hbar\omega}{2}\beta}}$$

$$Z = \frac{e^{\frac{-\hbar\omega}{2}\beta}}{1 - e^{-h\omega\beta}}$$

Helmholtz free energy is

$$F = \frac{-1}{\beta}\ln Z$$

$$F = \frac{-1}{\beta}\ln\left[\frac{e^{-\left(\frac{\hbar\omega}{2}\beta\right)}}{1-e^{-\hbar\omega\beta}}\right]$$

$$F = \frac{1}{\beta}\frac{\hbar\omega\beta}{2}+\frac{1}{\beta}\ln\left[1-\exp(-\hbar\omega\beta)\right]$$

$$F = \frac{\hbar\omega}{2}+\beta^{-1}\ln\left[1-\exp(-\hbar\omega\beta)\right]$$

Hence, option (b) is correct answer.

3. Given Energies are O, E and 2E

CD: Classical (distinguishable particles)

Partition function

0	∈	2∈	Total
•	–	–	0
–	•	–	∈
–	–	•	2∈

$$Z^{MB}_{dist} = \left(e^{0}+e^{-\beta E}+e^{-2\beta E}\right)^2$$

$$Z^{MB}_{dist} = \left(1+e^{-\beta\epsilon}+e^{-2\beta\epsilon}\right)^2$$

$$= 1+e^{-2\beta\epsilon}+e^{-4\beta\epsilon}+2e^{-\beta\epsilon}+2e^{-3\beta\epsilon}+2e^{-2\beta\epsilon}$$

$$= 1+2e^{-\beta\epsilon}+3e^{-2\beta\epsilon}+2e^{-3\beta\epsilon}+e^{-4\beta\epsilon}$$

CI: Classical (Intinguishable particle)

Partition function

$$Z^{MB}_{Indist} = \frac{Z^{MB}_{dist.}}{2!}$$

$$= \frac{\left(1+2e^{-\beta\epsilon}+3e^{-2\beta\epsilon}+2e^{-3\beta\epsilon}+e^{-4\beta\epsilon}\right)}{2}$$

$$= \frac{1}{2}+e^{-\beta\epsilon}+\frac{3}{2}e^{-2\beta\epsilon}+e^{-3\beta\epsilon}+\frac{1}{2}e^{-4\beta\epsilon}$$

FD : fermidral

Partition function

0	∈	2∈	Total Energy
•	•	–	∈
•	–	•	2∈
–	•	•	3∈

$$Z^{FD} = e^{-\beta\epsilon}+e^{-2\beta\epsilon}+e^{-3\beta\epsilon}$$

BE: Bose – Einstein

Partition function

$$Z^{\beta E} = e^{-\beta\epsilon}+e^{-2\beta\epsilon}+e^{-3\beta\epsilon}+e^{0}+e^{-2\beta\epsilon}+e^{-4\beta\epsilon}$$

0	∈	2∈	Total Energy
•	•	–	∈
•	–	•	2∈
–	•	•	3∈
••	–	–	0
–	••	–	2∈
–	–	••	4∈

$$Z^{\beta E} = 1+e^{-\beta E}+2e^{-2\beta E}+e^{-3\beta E}+e^{-4\beta E}$$

So, option (c) is correct answer.

4. Free energy is given by

$f = f_0 + a_0(T - T_c)M^2 + bM^4$

Differentiate with respect to M

$$\frac{dF}{dM} = 0 + 2Ma_0(T - T_c) + 4bM^3$$

Put $\frac{dF}{dM} = 0$

$2Ma_0(T - T_c) + 4bM^3 = 0$

$2M[a_0T - a_0T_c + 2bM^2] = 0$

$$M^2 + \frac{a_0}{2b}T - \frac{a_0T_c}{2b} = 0$$

This is the linear equation in M^2 and T

So, graph should be straight line.

Hence, option (b) is correct answer.

5.
$$u(\upsilon) = \frac{\hbar^3\upsilon^3}{c^3\exp\left(\frac{h\upsilon}{kT}-1\right)}$$

At T = OK; $\mu = E_F$

6. It is an example of free expansion process.

In free expansion process,

$$dU = 0$$
$$\delta W = 0$$
$$\delta Q = 0$$
$$dT = 0$$

7.
$$U = \lambda V^{-\frac{2}{3}}S^2$$

$$dU = -PdV + TdS$$

$$P = \left(\frac{\partial U}{\partial V}\right)_S \text{ and } T = \left(\frac{\partial U}{\partial S}\right)_V$$

$$P = -\frac{5}{3}\lambda.V^{-\frac{5}{3}}S^2 \quad \text{... (1)}$$

$$T = 2\lambda V^{-\frac{2}{3}}.S \quad \text{... (2)}$$

$$\frac{PV^{\frac{1}{3}}}{T^2} = \text{constant}$$

8.

$$(\Delta L_y) = \langle L_y^2\rangle - \langle L_y\rangle^2$$

$$\langle L_y\rangle = 0$$

$$\langle L_y^2\rangle = \frac{1}{2}\left[L^2 - L_z^2\right]$$

$$= \frac{\hbar^2}{2}\left[l(l+1) - m^2\right]$$

$$= \frac{\hbar^2}{2}[2\times3-4]$$

$$= 1\hbar^2$$

9. For n-atom, $N_A = 6.023 \times 10^{23}$

$M_2 = ??$

$T = 5800$ K; $M_2 = 3 \times 10^{21}$

$$\frac{N_1}{N_2} = e^{\frac{E_2-E_1}{kT}}$$

$E_1 = 13.6$ eV

$E_2 = -3.4$ eV

$N_2 = 8.4 \times 10^{14}$

11. $v(r) = \dfrac{m\omega^2 r^2}{2}$

$N_0 \to \infty$, K = 0 state in B.E.C.

$$N_0 = \frac{1}{e^{-1\beta\mu}-1}$$

When $N_0 \to \infty$

$e^{-1\beta\mu} - 1 \Rightarrow \mu = 0$

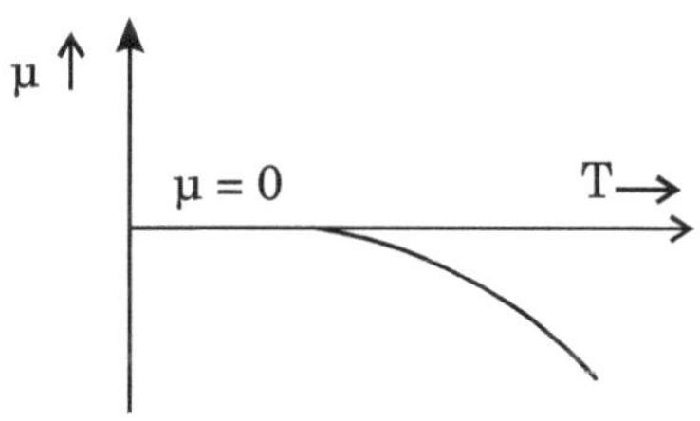

$$\epsilon_n = \left(n+\frac{3}{2}\right)\hbar\omega$$

$$\epsilon_0 = \frac{3}{2}\hbar\omega$$

$$\mu \le \frac{3}{2}\hbar\omega$$

12. $F = (E) = \dfrac{n(E)}{g(E)} = 0.01$

$E - E_f = 0.5$ eV

T = ?

$$f(E) = \frac{1}{e^{\alpha}e^{\beta E}+1}\quad \alpha = \frac{-E_F}{KT}, \beta = \frac{1}{KT}$$

$$f(E) = \frac{1}{e^{\left(\frac{E-E_f}{KT}\right)}+1} = 0.01$$

$$\frac{1}{e^{\left(\frac{0.5}{8.62\times10^{-5}T}\right)}+1} = 0.01$$

By solving we get, T = 1262 K

13. $Q_{\text{gained by the ice}}$

$= Q_{\text{lost by the water}}$

$mL + mC(T - 273) = mC\,(300 - T)$

$.01 \times 335 + .01 \times 4.2\,(T - 273) = 0.1 \times 4.2\,(300 - T)$

By solving, we get

T = 290.3 K

$$(\Delta s)_{ice} = \frac{dQ}{T} = \frac{mL}{T} + mc\int_{T_i}^{T_f}\frac{dT}{T}$$

$$\frac{mL}{T} + mC\ln\left(\frac{T_f}{T_i}\right)$$

$$(\Delta s)_{ice} = 12.27 + 2.58\ \text{J/K}$$

$$(\Delta s)_{water} = mC\int_{T_i}^{T_f}\frac{dT}{T} = mC\ln\left(\frac{T_f}{T_i}\right)$$

$$= 0.1\times4.2\times1000\ln\left(\frac{290.3}{300}\right) = -13.8\text{J/K}$$

$$(\Delta s)_{total} = (\Delta s)_{ice} + (\Delta s)_{water} = -1.05\text{J/K}$$

14. For system having $E_A = 3$, $\Omega_1 = 4$

For system having $E_B = 2$, $\Omega_2 = 3$

$E_A = 3$, $E_B = 2$

$E = E_A + E_B = 5$

$S = S_1 + S_2 = K\ln\Omega_1 + K\ln\Omega_2$

$= K\ln\Omega_1\Omega_2$

$= K\ln(4 \times 3)$

$= K\ln(12)$

20. The efficiency of carnot engine is given by

$$\eta = 1 - \frac{T_1}{T_2}\ \text{As}\ (T_2 > T_1)$$

21. Latent heat of fusion = 3.34×10^5 J/kg

From the Relation

$$\frac{dP}{dT} = \frac{lfn}{TV_{SL}}$$

$$\int_{P_1}^{P_2} dP = \frac{lfn}{V_{SP}}\ln\frac{T_2}{T_1}$$

$$P_2 - P_1 = \frac{3.34\times10^5}{1000^{-1} - 934^{-1}}\ln\frac{263}{273}$$

$$P_2 - P_1 = 176386364.8$$

$P_2 - 101325 = 176386364.8$

$$P_2 = 176487689.8 \text{ Pa}$$
$$= 0.176 \text{ GPa}$$

22. Given :

N non-interacting, distinguishable particles in a two level system at temperature T. The population of the particle in the level with the energy E is $\frac{N}{2}$.

We know that $\bar{n}p = \frac{ni}{gi} = e^{\frac{1}{\epsilon i - \mu 0/KT}+1} = \frac{N}{2}$.

'Ni' is the no. of non interacting particles 'gi' is the no. of sublevels.

23.
$$P \propto A^2 T^4$$
$$P_1 = R_1 T_1^4$$
$$P_2 = R_2^2 T^4$$
$$\frac{P_1}{P_2} = \frac{R_1^2 T^4}{R_2^2 T_2^4} = \frac{R^2 T^4}{\left(\frac{R}{2}\right)^2 (2T)^4}$$

$(\because T_1 = T2, T_2 = 2T \text{ and } R_1 = R, R_2 = \frac{R}{2}$

$$= \frac{1}{4} = 0.25.$$

24. $S = -k_B N\,[p \ln p + (1-p)\ln(1-p)]$

For maxima/minima

$$\frac{dS}{dp} = 0$$

or, $\ln p + \frac{p\times 1}{p} - \ln(1-p) + (1-p)\frac{1}{(1-p)} \times (-1) = 0$

or, $\ln p + 1 - \ln(1-p) - 1 = 0$

or, $\ln\left(\frac{p}{1-p}\right) = 0$

or $p = 1 - p$

or $2p = 1$

or $p = 0.5$

25. $dF = -SdT - PdV$

Hence, The Helmholtz free energy attains a local minimum.

26. $S_1 = -Nk_B \ln 1$

$S_2 = -Nk_B \ln 4$

$\Delta S = S_2 - S_1 = Nk_B \ln 4 = 2\,Nk_B \ln 2$

27.
$$R_H \propto \frac{1}{n_e}$$

and
$$E_F = \frac{\hbar^2}{2m}(3\pi^2 n)^{2/3}$$

$$\therefore \quad n = \left(\frac{2m}{\hbar^2}\right)^{3/2}\left(\frac{E_F}{3\pi^2}\right)^{3/2}$$

$$\therefore \quad R_H \propto E_F^{-3/2}$$

28. $E = 2\times\frac{(1^2+1^2)\pi^2\hbar^2}{2mL^2} + 4\times\left(\frac{2^2+1^2}{2mL^2}\right)\pi^2\hbar^2$

$$\therefore \quad E = \frac{2(2)\pi^2\hbar^2}{2mL^2} + \frac{4\times 5}{2mL^2}\pi^2\hbar^2 = \frac{24\,\pi^2\hbar^2}{2mL^2}$$

29. Heat capacity is defined as

$$C_V = rT + AT^3$$
$$r = \frac{3}{2}Nk_B^2\frac{1}{E_F}$$
$$\frac{r_x}{r_y} = \frac{\frac{3}{2}Nk_B^2\,1/E_{F_x}}{\frac{3}{2}Nk_B^2\,1/E_{F_y}}$$

or
$$\frac{r_x}{r_y} = \frac{E_{F_y}}{E_{F_x}} = \frac{7}{5}$$

$$A = 234\,Nk_B.\frac{1}{\theta_D^3}$$

$$\frac{A_x}{A_y} = \frac{234\,Nk_B\frac{1}{\theta_{D_x}^3}}{234\,Nk_B\frac{1}{\theta_{D_y}^3}} = \left(\frac{\theta_{D_y}}{\theta_{D_x}}\right)^3 = \left(\frac{340}{170}\right)^3$$

$$= 2^3 = 8$$

Thus $\frac{r_x}{r_y} = \frac{7}{5}$,

$$\frac{A_x}{A_y} = 8$$

30. In Bose – Einstein condensates. The particles have overlapping wavefunctions because they don't obey Pauli's principle.

31. $g(E) = CE^n$

For two dimensional,

$$g(E) = \frac{A\pi}{n^2}\left(\sqrt{2m}\right)^2 = \frac{A\pi 2m}{n^2}$$

i.e, $g(E) \propto E^0$

So $\boxed{n = 0}$

32. Number of Microstate is $N_{C_{n_1}}$, where n_1 is number of particle in $\frac{+1}{2}$ state and.

$n_2 = (N - n_1)$ is number of state of in $\frac{-1}{2}$ state.

where $n_1 = \frac{1}{2}\left(N - \frac{E}{\mu B}\right), n_2 = \frac{1}{2}\left(N + \frac{E}{\mu B}\right)$

So, Number of Microstate

i.e. $r = \dfrac{N!}{\frac{1}{2}\left(N - \frac{E}{\mu_B}\right)!\frac{1}{2}\left(N + \frac{E}{\mu_B}\right)!}$

34. $\left(\frac{\partial G}{\partial T}\right)_P = -S = -NK_B \ln\left(\frac{aVE^{3/2}}{N^{5/2}}\right)$

$\because\ S = NK_B \ln\left(\frac{aVE^{3/2}}{N^{5/2}}\right)$

$\Rightarrow\ G = -NK_B T \ln\left(\frac{aVE^{3/2}}{N^{5/2}}\right) + \ln A$

$$\Rightarrow\ \mu = \left(\frac{\partial G}{\partial N}\right) = -\left[K_B T \ln\left(\frac{aVE^{3/2}}{N^{5/2}}\right) + NK_B T \frac{N^{5/2}}{aVE^{3/2}} \cdot \frac{\left(\frac{-5}{2}\right)}{(N^{7/2})} \times aVE^{3/2}\right]$$

$$\mu = -K_B T\left[\ln\left(\frac{aVE^{\frac{3}{2}}}{N^{\frac{5}{2}}}\right) - \frac{5}{2}\right]$$

36. $P \propto V^{-5/3}$

Bulk modulus $B = -V\dfrac{dP}{dV}$

$\therefore\ B = -V\left(\frac{-5}{3}\right)V^{-8/3}\quad\left[\because \frac{dP}{dV} = \frac{-5}{3}V^{-8/3}\right]$

$\therefore\ B = \frac{5}{3}V^{-5/3}$

Hence $B \propto V^{-5/3}$

37. According to equipartition theorem

$$E = \frac{p_x^2}{2m_1} + \frac{p_y^2}{2m_2} + \frac{1}{2}kk_p^2 + \frac{1}{2}ky^2$$

$$\downarrow\quad\downarrow\quad\downarrow\quad\downarrow$$

$$\frac{1}{2}k_B T\quad \frac{1}{2}k_B T\quad \frac{1}{2}k_B T\quad \frac{1}{2}k_B T$$

$\langle E \rangle = 2k_B T$

Every term give the $\frac{1}{2}k_B T$

Hence $\langle E \rangle = 2k_B T$

38.

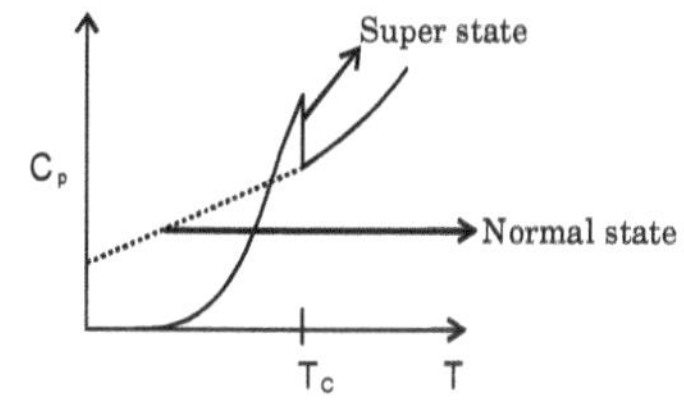

At the low temperature the specific heat of a normal metal has the form $AT + BT^3$, where the Linear term does the electronic excitations and the cubic term is due to lattice vibration. Below the super conducting critical temperature this behaviour is substantially alter. As the temperature drops below T_c (in magnetic field) the specific heat jumps to a higher value and slowly decreases. Eventually falling well below the value one world expect for a normal metal.

39.

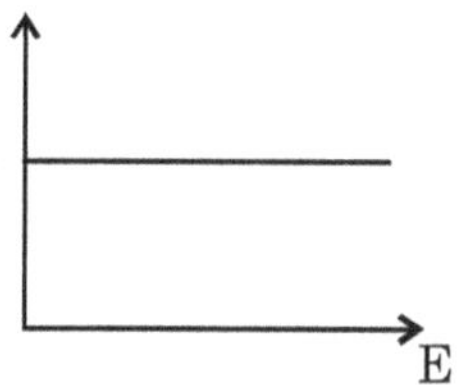

$N(E) \propto E^{\circ}$

$\downarrow$

Density of state in 2 – 0

40. There are three possibility of Boson particles

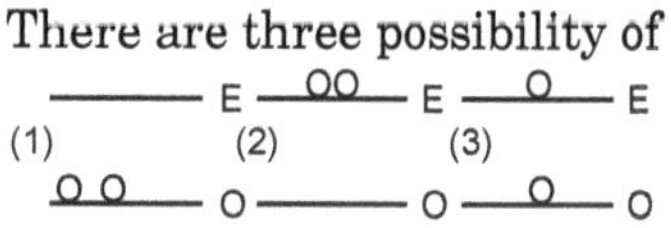

Partition function for Bosons particle,

$$Z = \sum_i e^{-N\beta\epsilon_i}$$

Where N is the number of particle,

$$Z = e^{-2\beta \times 0} + e^{-2\beta\epsilon} + e^{-1 \times \beta \times 0} + e^{-\beta\epsilon}$$

$$= 1 + e^{-2\beta\epsilon} + 1 + e^{-\beta\epsilon}$$

$$Z = 2 + e^{-\beta\epsilon} + e^{-2\beta\epsilon}$$

Mean energy,

$$E = -\frac{\partial}{\partial\beta}\ln Z$$

$$E = -\frac{\partial}{\partial\beta}\ln[2 + e^{-\beta\epsilon} + e^{-2\beta\epsilon}]$$

$$\therefore\ E = \frac{\epsilon e^{-\beta\epsilon} + 2\epsilon e^{-2\beta\epsilon}}{2 + e^{-\beta\epsilon} + e^{-2\beta\epsilon}}$$

41. $H = \frac{p^2}{2m} - \frac{\alpha q^2}{2}$

The motion of the particle in phase space

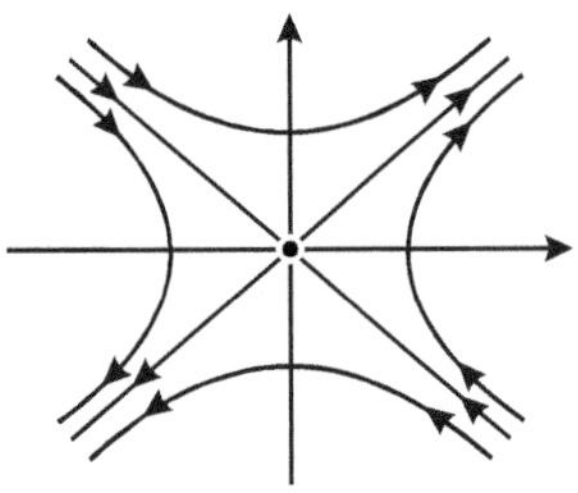

$$V = -\frac{\alpha q^2}{2}$$

$$F = -\frac{dV}{dq} = \alpha q$$

The equation of motion

$$m\frac{d^2q}{dt^2} = \alpha q$$

$$\frac{d^2q}{dt^2} = \frac{\alpha}{m}q$$

$$\frac{d^2q}{dt^2} = \beta^2 q \qquad \because \beta = \sqrt{\frac{\alpha}{m}}$$

put $y = \frac{dq}{dt}$ then $\frac{dy}{dt} = \beta^2 q$

$$\frac{\frac{dy}{dt}}{\frac{dq}{dt}} = \frac{\beta^2 q}{y}$$

$$\frac{dy}{dq} = \frac{\beta^2 q}{y}$$

Integrating the equation

we get

$$y^2 - \beta^2 q^2 = C$$

This equation represent the hyperbola.

If at time $t = 0$, $x = x_0$ and $y = y_0$ the solution of differentiating equation is given as

$$x = x_0 \cosh \beta t + \frac{y_0}{\beta}\sinh \beta t$$

$$y = \beta x_0 \sinh \beta t + y_0 \cosh \beta t$$

There are two asymptotes $y = \pm \beta x$ to the family of hyperbola.

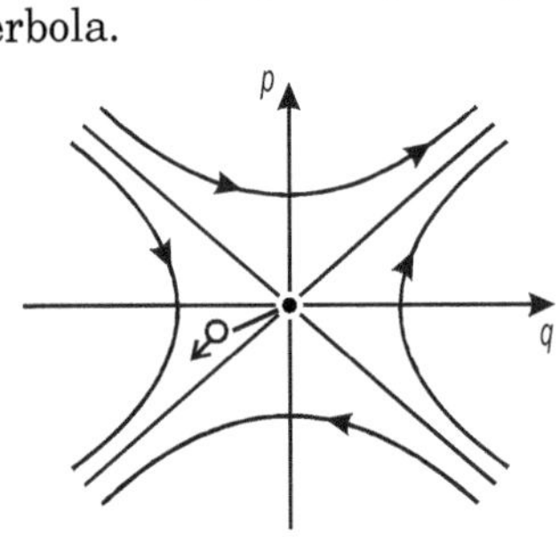

42. The number of ways that 3 termions will adjust in 4 available energy is $^4C_3 = 4$

So, entropy is k_B In $4 = 2k_B \ln 2$

43. If Plank's constant are zero, then the system behaved as a classical system and the Energy is KT

46. From first and second law of Thermodynamics Td.SdV – PdV

$$\Rightarrow dS = \frac{1}{T}(dV - PdV)$$

$$\Rightarrow \left(\frac{\partial S}{\partial E}\right)_V = \frac{1}{T} \qquad \because E = U$$

$S(E) = aE(E_0 - E) - aE$

$= a(E_0 - 2E)$

$$\Rightarrow T = \frac{1}{a(E_0 - 2E)}$$

47. There are two micro-state possible for $\delta\int \sin\frac{1}{2}$ particle, So entropy is given by NK ln (2).

49. $$\left\langle e^{\vec{P}.\vec{a}}\right\rangle = \int_{-\infty}^{+\infty}\int_{-\infty}^{+\infty}\int_{-\infty}^{+\infty} f(P_x.P_y.P_z)e^{\vec{P}.\vec{a}}dp_x dp_y dp_z$$

where $f(P_x, P_y, P_z)$ is maxwell probability distribution at Temperature T.

$$\left\langle e^{\vec{P}.\vec{a}}\right\rangle = \int_{-\infty}^{+\infty} A_x e^{\frac{-Px^2}{2mkT}} e^{P_x a_x} dP_x \int_{-\infty}^{+\infty} A_y e^{\frac{-P_y^2}{2mkT}}$$

$$e^{P_y a_y}dP_y \int_{-\infty}^{+\infty} Az \times e^{\frac{-P_z^2}{2mkT}} e^{P_z P_z} dP_z$$

$$\Rightarrow \left\langle e^{\vec{P}.\vec{a}}\right\rangle = e^{\frac{-(ax^2+ay^2+az^2)mkT}{2}} \int_{-\infty}^{+\infty} Axe^{\frac{-(P_x - mKTax)^2}{2mKT}}$$

$$dPx \times \int_{-\infty}^{+\infty} Aye^{\frac{-(P_y - mKTa_y)^2}{2mKT}-\infty} \times \int_{-\infty}^{+\infty} Aze^{\frac{-(P_z - mKTa_z)^2}{2mKT}dPz}$$

$$\left\langle e^{\vec{P}.\vec{a}}\right\rangle = e^{\frac{-(ax^2+ay^2+az^2)mKT}{2}} .|.|.|$$

$$= e^{-\frac{1}{2}a^2 mKT}$$

50. Bulk modulus or isothermal incompressi-bility of an ideal

$$= \frac{-\partial P}{\partial V / V_0}\Bigg|_{To}$$

Hence, compressibility $= \frac{-1}{V_0}\frac{\partial V}{\partial P}\Bigg|_{To}$

51. Fermi energy

$$E_F = \frac{1}{2}mv F^2 = \frac{h^2}{2m}(2\pi^2 n)^{2/3}$$

or $V^2F \mathcal{L} n\frac{2}{3}$ or $VF \mathcal{L} n\frac{1}{3}$

52. $<(V_x + V_y)^2> = <Vx^2> + <Vy^2> + 2<VxVy>$

Putting $<V_x^2> = <Vy^2> = <V_2^2> = \frac{RT}{m}$

and $<Vx\,Vy> = 0$

$$<(V_x + V_y)^2> = \frac{RT}{m} + \frac{RT}{m} = 0 = \frac{2KT}{m}$$

53. Let $E = K/V^{2/3}$ ...(*i*)

Now, $P = \frac{-\partial E}{\partial V} = \frac{-\partial}{\partial V}\left(\frac{K}{V^{2/3}}\right)$

$$= \frac{2}{3}KV\frac{-5}{3} = \frac{K}{V\frac{5}{3}}$$

or $P = \frac{2K}{3V \times V^{2/3}}$, $PV = \frac{2}{3}\frac{K}{V^{2/3}} = \frac{2}{3}E$

using equation (*i*)

54. For 2 fold degeneracy of energy level

The partition function

$$z = e^{-\beta.0} + 2e^{-\beta t} + e^{-\beta.2t}$$

$$= 1 + 2e^{-\beta t} + (e^{-\beta t})^2 = (1 + e^{-\beta t})^2$$

55. For spin -half system, number of microstates

$$\Omega = (25 + 1)^N = 2^N$$

$\therefore$ Entropy $S = k_B \log \Omega = k_B \log_2 N$

$$= Nk_B \log 2$$

59. Since $P = \frac{1}{kT} = 10^{10} Pa$

For isothermal process

P^v = constent $\left[\because V\alpha\frac{1}{P}\right]$

$$\frac{P}{P} = \text{constant}$$

$$\frac{P_2}{P_1} = \frac{P_2}{P_1}$$

$$\Rightarrow P_2 = \frac{P_2}{P_2} \times P_1$$

$$P_2 = \frac{\left(\rho_1 + \frac{P_1}{100}\right)}{\rho_1} = 10^{10} \approx 10^{10} P_a$$

60. $S = k_B \sum_i \text{In}\,\Omega, \Omega = 3$ is number of microstate,

$S = 1: S_Z = -1, 0, 1$

The entropy of the system is $NK_B \ln 3$

61. Partition function,

$$Z = 4e^{\frac{-E}{kT}} + 2e^{\frac{-E}{kT}}$$

$$\Rightarrow P(2\varepsilon) = \frac{2e^{\frac{-2\varepsilon}{kT}}}{4e^{\frac{-\varepsilon}{kT}} + 2e^{\frac{-2\varepsilon}{kT}}}$$

$$= \frac{1}{1 + 2^{\frac{\varepsilon}{kT}}}$$

62. For Bose – Einstein Condensation :

No. of electron + No. of proton + no. of Neutron = Even

For $^{30}K_{19}$

No. of proton = 19, No. of electron = 19,

No. of neutron = 11.

19 + 19 + 11 = 49, this is odd. So it will not exhibit Bose – Einstein Condensation.

63. $n = \int_0^{E_f} g(E) + (E)dE,\ g(E)dE = \frac{2m}{h^2}dE$

At T = 0,

$$f(E) = \begin{cases} 1, \text{ if } E > E_F \\ 0, \text{if } E > E_F \end{cases}$$

$$\Rightarrow n = \frac{2mE_f}{n^2} = \frac{mE_f}{2\pi^2\hbar^2}$$

65. dH = TdS + VdP

$$\Rightarrow \left(\frac{\partial H}{\partial S}\right)_P = T,$$

$$\left(\frac{\partial H}{\partial P}\right)_S = V.$$

66. Here $\log z = \frac{\pi^2}{45}V\frac{(k_B T)^3}{h^3C^3}$

Putting $K_B T = \frac{1}{\beta}$, we get

$$\log z = \frac{\pi^2}{45}\frac{V}{h^3C^3}\frac{1}{\beta^3}$$

$$U = -\frac{\partial}{\partial\beta}\log z = -\frac{\pi^2}{45}\frac{V}{h^3C^3}\frac{(-3)}{\beta^4}$$

$$= \frac{\pi^2}{15}\frac{V}{h^3C^3}(k_BT)^4 \qquad \because \beta = \frac{1}{k_BT}$$

$$C_V = \left(\frac{\partial V}{\partial T}\right)_V = \frac{4\pi^2}{15}\frac{Vk_B^4}{h^3C^3}T^3$$

$$\Rightarrow \quad C_V \propto T^3$$

67. $$P = \frac{\log z}{\beta} = k_BT \times \frac{\pi^2}{45}\frac{V(k_BT)^3}{h^3C^3}$$

$$= \frac{\pi^2}{45}\frac{V(k_BT)^4}{h^3C^3}$$

68. The determinate equation can be written as

$$\begin{vmatrix} 2C - M_1w^2 & -C(1+\exp(-ika)) \\ -C[1+exp(ika)] & 2C - M_2w^2 \end{vmatrix} = 0$$

For $k = 0$ this becomes

$$\begin{vmatrix} 2C - M_1w^2 & -2C \\ -2C & 2C - M_2w^2 \end{vmatrix} = 0$$

$$\Rightarrow 2C(M_2 + M_1) = M_1M_2w^2$$

$$w = \sqrt{2C\left[\frac{1}{M_1} + \frac{1}{M_2}\right]}$$

70. The fermi energy is given by

$$E_F = \frac{h^2}{2m}(3\pi^2N)^{2/3}$$

$$E_F \propto N^{2/3}$$

71. $$F = -k_BT\log\left[e^{-E_1/K_BT} + e^{-E_2/K_BT}\right]$$

$$\because \quad E_1 = 0, E_2 = E$$

$$F = -k_BT\log\left[1 + e^{-E/k_BT}\right]$$

72. $$\frac{\partial F}{\partial T} = \frac{e^{-E/k_BT}}{(1+e^{-E/k_BT})} \times \left(\frac{-E}{k_BT^2}\right) \times (-k_BT)$$

$$= \frac{E}{T}\left(\frac{e^{-E/k_BT}}{(1+e^{-E/k_BT})}\right)$$

$$C_V = \frac{E}{k_BT} \times \frac{\partial F}{\partial T} = \frac{E^2}{k_BT^2}\frac{e^{-E/k_BT}}{(1+e^{-E/k_BT})}$$

75. Here $R_{10} = \frac{2}{a_0^{3/2}}\exp\left(\frac{-r}{a_0}\right)$

$$P(r)\,dr = R_{10} \quad R_{10}^* \quad r_2dr$$

$$P(r)\,dr = \frac{4}{a_0^3}\exp\left(\frac{-2r}{a_0}\right)r^2dr \quad \left(\because R_{10} = R_{10}^*\right)$$

$$P(r) = \frac{4}{a_0^3}\exp\left(\frac{-2r}{a_0}\right)^{r^2}$$

For most probable value of r

$$\frac{dP(r)}{dr} = 0$$

$$\frac{4}{a_0^3}\left[r^2\exp\left(\frac{-2r}{a_0}\right)\left(\frac{-2}{a_0}\right)\right] + 2r\exp\left(\frac{-2r}{a_0}\right) = 0$$

$$4/a_0^3\exp\left(\frac{-2r}{a_0}\right)\left(\frac{2r-2r^2}{a_0}\right) = 0$$

$$\Rightarrow 2r - \frac{2r^2}{a_0} = 0$$

$$\Rightarrow r = a_0$$

80. $$0 * e^{-0/k_BT} + 2e^{-E/k_BT}$$

$$= 0 * 2e^{-E/k_BT} + 4e^{-2E/k_BT}$$

$$= 2\left(e^{-E/k_BT} + 2e^{-2E/k_BT}\right)$$

$$\text{Mean energy} = \frac{E\left(\exp\left(\frac{E}{k_BT}\right) + 2\exp\left(\frac{-2E}{k_BT}\right)\right)}{\left(1+\exp\left(\frac{-E}{k_BT}\right) + \exp\left(\frac{-2E}{k_BT}\right)\right)}$$

83. The efficiency of carnot heat engine may be defined as

$$\eta = \frac{Q_1 - Q_2}{Q_1} = \frac{T_1 - T_2}{T_1}$$

Where, Q_1 is the heat taken from the source by working substance and Q_2 is the heat given by working substance to sink respectively. T_1 and T_2 are the temperatures of source and sink respectively.

Hence, in question

Q_1 = 450 KJ/S, Q_2 = ?

T_1 = 22 + 273 = 295K

T_2 = 273K

$$\text{Hence, } \frac{450 - Q_2}{450} = \frac{295 - 273}{295}$$

$$Q_2 = \frac{273 \times 450}{295} = 416.44\,\text{KJ/S}$$

$$\cong 417\ \text{KJ/S}$$

84. At triple point the vapour pressures of solid and liquid phases are equal. Hence

$$23 - \frac{3863}{T} = 19 - \frac{3063}{T}$$

or
$$4T = 800$$

$$T = 200\ \text{K}$$

85. $$F = -\frac{9}{3}VT^{-4}$$

$$P = -\left(\frac{\partial F}{\partial V}\right)_T = \frac{9}{3}T^4$$

$$S = \left(\frac{\partial F}{\partial T}\right)_V = \frac{4}{3}aVT^3$$

88. Chemical potential,

$$\mu = -\frac{1}{\beta}\frac{\partial}{\partial N}(1nz) = \frac{\partial}{\partial N}(F) = 0$$

as F is not an explicit function of N.

89. If T_f be the final temperature, $T_1 > T_f > T_2$

At equilibrium, $mc(T_1 - T_f) = mc(T_f - T_2)$

c = specific heat, m = mass.

$$\Rightarrow \quad T_f = \frac{T_1 + T_2}{2}$$

∴ Entropy change

$$\Delta S = \frac{3}{2}Nk_B \ln\left[\frac{T_f}{T_i}\right]$$

$$= \frac{3}{2}Nk_B\left[\ln\left(\frac{T_f}{T_1}\right) + \ln\left(\frac{T_f}{T_2}\right)\right]$$

$$= \frac{3}{2}Nk_B\left[\ln\left(\frac{T_f^2}{T_1\, T_2}\right)\right]$$

$$= \frac{3}{2}Nk_B \ln\left[\frac{(T_1 + T_2)^2}{4T_1T_2}\right]$$

90. Adiabatic expansion, $V_2 > V_1$

$$\therefore \quad \frac{1}{V^2} < \frac{1}{V_1}$$

$$\therefore \quad \left(\frac{1}{V_2} - \frac{1}{V_1}\right) < 0$$

In adiabatic expansion,

$T_2 < T_1$ as the internal energy decreases for doing work (by the gas)

$$\therefore \quad T_2 - T_1 < 0$$

For answers, (a), (b) and (d) are correct

$$T_2 > T_1$$

only, $\frac{2}{3}Ra\left(\frac{1}{V_2} - \frac{1}{V_1}\right) = -ve.$

So, $\quad T_2 - T_1 < 0$

or $\quad T_2 < T_1$

91. $H = \frac{1}{2}mv^2 + \frac{1}{2}m\omega^2x^2$

$$\langle H\rangle = \frac{1}{2}k_BT + \frac{1}{2}k_BT = k_BT$$

92. Number of ways for choosing n atoms. Out of N atoms, is = ${}^{N}C_n = \frac{N!}{n!(N-n)!}$

Number of ways for putting n atoms in N' sits, is

$$= {}^{N'}C_n = \frac{N'}{n!(N-n)!}$$

∴ Total number of microstate,

$$\Omega = \frac{N!}{n!(N-n)!}\,\frac{N'!}{n!(N'-n)!}$$

93. We know, $pV = \frac{1}{3}U$

or, $\quad U = 3pV = 3RT$

$$\therefore \quad C_V = \frac{\partial U}{\partial T} = 3R$$

and $\quad C_P = C_V + R = 4R$

Now, $\quad V = \frac{C_P}{C_V} = \frac{4}{3}$

Hence, equation for adiabetic processis, $pV^{4/3}$ = constant.

94. Since the particle of spin $\frac{1}{2}$ are fermious & obey Pauli exclusion principle so, not more that 1 particle can be placed in any one state. Hence, number of accessible states = N

96. We know $\bar{E}_o = \frac{U_0}{N}$

where, $\quad U_0 = \frac{\pi}{5}\left[\frac{8ma^2}{n^2}\right]^{3/2} [E_F]^{5/2}$

and $\quad N = \frac{\pi}{3}\left[\frac{8\,m}{n^2}\right] V\, E_F^{3/2}$

$$\therefore \quad \bar{E}_o = \frac{3}{5}E_F$$

97.

$$F = -k\,T\,ln\,z$$

$$= k\,T \ln\frac{xT}{h\omega}$$

$$= k\,T \ln\frac{hw}{kT}$$

98. Partition function, $z = \Sigma\, e^{-\frac{\varepsilon_i n_i}{kT}}$

where, ε_i = Energy of ith quantam electron occupied by n_i particles

$$\therefore \quad z = e^{-\frac{0}{kT}\times 2} + e^{-\frac{\varepsilon}{kT}\times 2} + e^{-\frac{\varepsilon}{kT}} + e^{-\frac{0}{kT}}$$

$$= 2 + e^{\frac{-2\varepsilon}{kT}} + e^{\frac{-\varepsilon}{kT}}$$

99. Since, $\quad P = \frac{N!}{\frac{1}{2}(N+m)!\ \frac{1}{2}(N-m)!}\left(\frac{1}{2}\right)N$

Here, $\quad m = 0$

$$\therefore \quad P = \frac{N!}{\left(\frac{N}{2}\right)!\left(\frac{N}{2}\right)!}\cdot\left(\frac{1}{2}\right)^N$$

101. In reversible process, change in entropy is zero but in irreversible process, change in entropy is positive. So that for any process, the change of entropy of the universe will be positive and zero.

102. One diatomic molecule have 10 phase space. So 10 diatomic have = 10 × 10 = 100 phase space.

103. $C_V = AT + BT^3$

Here BT^3 term due to the thermal vibration in the lattice of the solid which take place at high temperature but at low temperature the magnitude of AT is more than BT^3 term. So the AT term is more dominant at low temperature.

$$C_V \propto T \text{ at low temperature.}$$

104. In vaporization of liquid at its boiling point in which only phase is changed. It is first order phase transition but in other three choices, the structure of material is changed, these are second order phase transition.

105. At the region of co-existance of a liquid and vapour phase C_v is finite but C_P, B and K are all infinite.

107. We know, $T_C M^{-1/2}$ = constant

$$\Rightarrow \quad T_C \propto \frac{1}{M^{-1/2}}$$

$$T_C \propto M^{1/2}$$

108. $$F(q) = \int \exp\left(\frac{i\vec{q}.\vec{r}}{\hbar}\right)\rho(\vec{r})d^3r \quad \ldots (i)$$

Since, delta function $\int f(r)\,\delta(r - r_0)\,d^3r = f(r_0)$

If $r_0 = 0$, then $\int f(r)\,\delta(r - 0)\,d^3r = f(0)$

so that in equation (i), $r_0 = 0$

then here $f(r) = \exp\left(\frac{i\vec{q}.\vec{r}}{\hbar}\right)$

$$\int \exp\left(i.q.\frac{\vec{r}}{h}\right)\rho(\vec{r})\,d^3r = f(0)$$

$$= \exp\left(\frac{i\,q.0}{\hbar}\right) = e^0 = 1 \text{ (unity)}$$

109. $$p(\varepsilon_i) = \frac{e^{-\beta\varepsilon_i}}{\sum_i e^{-\beta\varepsilon_i}}$$

But $\varepsilon = 0, -\varepsilon_0, +\varepsilon_0$

$$\Rightarrow \quad p(0) = \frac{e^{\circ}}{(e^{\beta\varepsilon_0} + e^0 + e^{-\beta\varepsilon_0})}$$

$$\Rightarrow \quad p(0) = \frac{1}{(e^{\beta\varepsilon_0} + 1 + e^{-\beta\varepsilon_0})}$$

Put $\beta\varepsilon_0 = 2$

so $$p(0) = \frac{1}{(1 + e^2 + e^{-2})}$$

From formula,

$$\cosh 2 = \left(\frac{e^2 + e^{-2}}{2}\right)$$

or $$p(0) = \frac{1}{(1 + 2\cosh 2)} = (1 + 2\cosh 2)^{-1}.$$

110. $F = -Nk_BT \ln z$

Since $z = \Sigma e^{-\beta\varepsilon_i} = 1 + e^{-\beta\varepsilon_0} + e^{+\beta\varepsilon_0}$

Put $\beta\varepsilon_0 = x$

Since $z = 1 + e^{-x} + e^x$

Put high temperature $(\beta\varepsilon_0 << 1) \Rightarrow x << 1$

$$e^{-x} \approx (1 - x) \quad \text{and} \quad e^x \approx (1 + x)$$

$$z = 1 + (1 - x) + (1 + x) = 3$$

$$\Rightarrow \quad F = -Nk_BT \ln 3.$$

113. $E \propto T^4$

114. $PV = nRT$

$$\Delta S = \frac{\Delta Q}{T}$$

$$= \frac{PV\Delta S}{nR}$$

116. $$\chi = \frac{C}{T - T_C}$$

where, T_C = Curie T

120. $P(V - b) = RT$

$= Nk_BT$

121. $U = \frac{3}{2} N_B T$

122. Volume = h^f

where f = degree of freedom

System of N non-particles can be considered as a particle with 3N degree of freedom in phase space.

123. Mean free path λ is given by

$$\lambda = \frac{kT}{\sqrt{2}\pi d^2 p}$$

127. Energy per degree of freedom $= \frac{1}{2}kT$

∴ Rotation energy of a diametic molecule

$$= 2 \times \frac{1}{2}kT$$

Also rational energy of the molecule in a given state $J = \frac{\hbar^2}{2I} J(J+1)$

$$\therefore \quad 2 \times \frac{1}{2} kT = \frac{\hbar}{2I} J(J+1)$$

$$\Rightarrow \quad J(J+1) = \frac{2kTI}{\hbar^2}$$

which is the required partition function.

129. At N.T.P., 22.4 litre contains 6.023×10^{23} molecules

i.e.i 22.4 litre = 6.023×10^{23}

$$22.4 \text{ m}^3 = 6.023 \times 10^{26}$$

$$50 \text{ m}^3 = \frac{6.023}{22.4} \times 50 \times 10^{26}$$

131. In a connonical ensemble particles can exchange energy but have constant N,V and T.

■■

6 CHAPTER Atomic and Molecular Physics

2021

1. Assume that ^{13}N $(Z = 7)$ undergoes first forbidden β^+ decay from its ground state with spin-parity J_i^π, to a final state J_f^π. The possible values for J_i^π and J_f^π, respectively, are

(a) $\frac{1^-}{2}, \frac{5^+}{2}$ (b) $\frac{1^+}{2}, \frac{5^+}{2}$

(c) $\frac{1^-}{2}, \frac{1^-}{2}$ (d) $\frac{1^+}{2}, \frac{1^-}{2}$

2. Among the term symbols

$^4S_1$, $^2D_{7/2}$, $^3S_1$ and $2D_{5/2}$

choose the option(s) possible in the *LS* coupling notation.

(a) $^4S_1$ (b) $^2D_{7/2}$

(c) $^3S_1$ (d) $^2D_{5/2}$

3. To sustain lasing action in a three-level laser as shown in the figure, necessary condition(s) is(are)

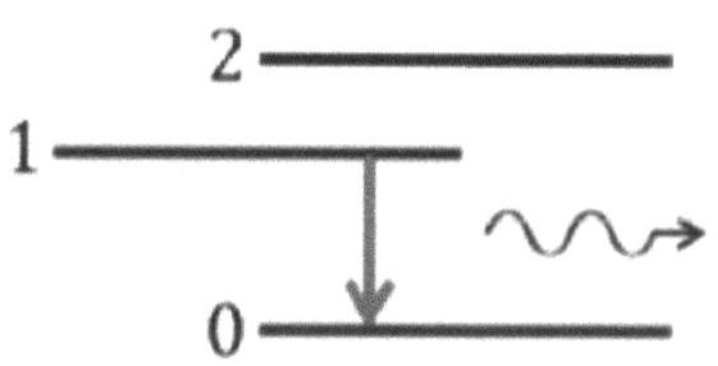

(a) lifetime of the energy level 1 should be greater than that of energy level 2

(b) population of the particles in level 1 should be greater than that of level 0

(c) lifetime of the energy level 2 should be greater than that of energy level 0

(d) population of the particles in level 2 should be greater than that of level 1

4. A light source having its intensity peak at the wavelength 289.8 nm is calibrated as 10,000 K which is the temperature of an equivalent black body radiation. Considering the same calibration, the temperature of light source (in K) having its intensity peak at the wavelength 579.6 nm (rounded off to the nearest integer) is _____.

5. The spacing between two consecutive S-branch lines of the rotational Raman spectra of hydrogen gas is 243.2 cm^{-1}. After excitation with a laser of wavelength 514. 5 nm, the Stoke's line appeared at 17611.4 cm^{-1} for a particular energy level. The wavenumber (rounded off to the nearest integer), in cm^{-1}, at which Stoke's line will appear for the next higher energy level is _______.

6. The transition line, as shown in the figure, arises between $^2D_{3/2}$ and $^2P_{1/2}$ states without any external magnetic field. The number of lines that will appear in the presence of a weak magnetic field (in integer) is _____.

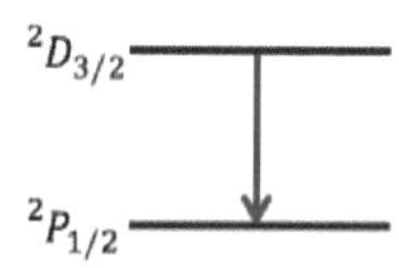

7. Consider the atomic system as shown in the figure, where the Einstein *A* coefficients for spontaneous emission for the levels are $A_{2\to 1} = 2 \times 10^7$ s^{-1} and $A_{1\to 0} = 10^8$ s^{-1}. If 10^{14} atoms/cm^3 are excited from level 0 to level 2 and a steady state population in level 2 is achieved, then the steady state population at level 1 will be $x \times 10^{13}$ cm^{-3}. The value of x (in integer) is _____.

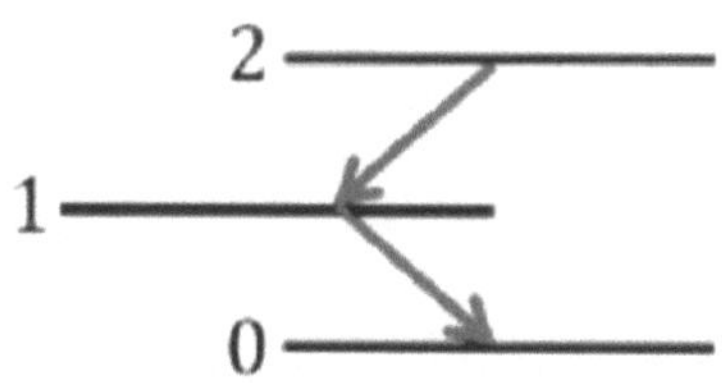

8. The spin $\vec{S}$ and orbital angular momentum $\vec{L}$ of an atom precess about $\vec{J}$, the total angular momentum. $\vec{J}$ precesses about an axis fixed by a magnetic field $\vec{B_1} = 2B_0\hat{z}$, where B_0 is a constant. Now the magnetic field is changed to $\vec{B_2} = B_0\left(\hat{x} + \sqrt{2}\hat{y} + \hat{z}\right)$. Given the orbital angular momentum quantum number l = 2 and spin quantum number s = 1/2, θ is the angle between $\vec{B_1}$ and $\vec{J}$ for the largest possible values of total angular quantum number j and its z-component j_z. The value of θ (in degree, rounded off to the nearest integer) is _____.

9. The spin-orbit effect splits the $^2P \rightarrow ^2S$ transition (wavelength, λ = 6521 Å) in Lithium into two lines with separation of Δλ = 0.14 A. The corresponding positive value of energy difference between the above two lines, in eV, is $m \times 10^{-5}$. The value of m (rounded off to the nearest integer) is _____.

(Given: Planck's constant, h = 4.125×10^{-15}eV s
Speed of light, $c = 3 \times 10^8$ m s^{-1})

2020

10. The total angular momentum j of the ground state of the $^{17}_{8}O$ nucleus is

(*a*) $\frac{1}{2}$ (*b*) 1

(c) $\frac{3}{2}$ (d) $\frac{5}{2}$

11. Consider a diatomic molecule formed by identical atoms. If E_V and E_e represent the energy of the vibrational nuclear motion and electronic motion respectively, then in terms of the electronic mass m and nuclear mass M, $\frac{E_V}{E_e}$ is proportional to

(*a*) $\left(\frac{m}{M}\right)^{\frac{1}{2}}$ (*b*) $\frac{m}{M}$

(*c*) $\left(\frac{m}{M}\right)^{\frac{3}{2}}$ (*d*) $\left(\frac{m}{M}\right)^{2}$

12. A hydrogenic atom is subjected to a strong magnetic field. In the absence of spinorbit coupling, the number of doubly degenerate states created out of the d-level is ______.

2019

13. The spin –orbit interaction term of an electron moving in a central field is written as $f(r)\vec{l}\cdot\vec{s}$, where r is the radial distance of the electron form the origin. IF an electron moves inside a uniformly charged sphere, then

(*a*) f (r) = contant (*b*) $f(r) \propto r^{-1}$

(*c*) $f(r) \propto r^{-2}$ (*d*) $f(r) \propto r^{-3}$

14. The ground state electronic configuration of the rare–earth ion (Nd^{3+}) is [Pd] 4f3 5s25p6. Assuming LS coupling, the Lande g-factor of this ion is $\frac{8}{11}$. The effective magnetic moment in units of Bohr magneton mB (rounded off to two decimal places) is__________

2018

15. Given the following table, which one of the following correctly matches the experiments from Group I to their inferences in Group II?

Group I	Group II
P : Stern-Gerlach experiment	**1 :** Wave nature of particles
Q : Zeeman effect	**2 :** Quantization of energy of electrons in the atoms
R : Frank-Hertz experiment	**3 :** Existence of electron spin
S : Davisson-Germer experiment	**4 :** Space quantization of angular momentum

(*a*) P-2, Q-3, R-4, S-1 (*b*) P-1, Q-3, R-2, S-4
(*c*) P-3, Q-4, R-2, S-1 (*d*) P-2, Q-1, R-4, S-3

16. The expression for the second overtone frequency in the vibrational absorption spectra of a diatomic molecule in terms of the harmonic frequency ω_e and anharmonicity constant x_e is

(*a*) $2\omega_e(1-x_e)$ (*b*) $2\omega_e(1-3x_e)$
(*c*) $3\omega_e(1-2x_e)$ (*d*) $3\omega_e(1-4x_e)$

17. Match the physical effects and order of magnitude of their energy scales given below, where $\alpha = \frac{e^2}{4\pi\varepsilon_0\hbar c}$ is fine structure constant; m_e and m_p are electron and proton mass, respectively.

Group I	Group II
P : Lamb shift	**1 :** $\sim O(\alpha^2 m_e C^2)$
Q : Fine structure	**2 :** $\sim O(\alpha^4 m_e C^2)$
R : Bohr energy	**3 :** $\sim O(\alpha^4 m_e^2 C^2/m_p)$
S : Hyper finestructure	**4 :** $\sim O(\alpha^5 m_e C^2)$

(*a*) P-3, Q-1, R-2, S-4
(*b*) P-2, Q-3, R-1, S-4
(*c*) P-4, Q-2, R-1, S-3
(*d*) P-2, Q-4, R-1, S-3

18. The intrinsic/permanent electric dipole moment in the ground state of hydrogen atom is (a_0 is the Bohr radius)

(*a*) $-3ea_0$ (*b*) zero
(*c*) ea_0 (*d*) $3ea_0$

19. Which one of the following is an allowed electric dipole transition?

(*a*) $^1S_0 \rightarrow {}^3S_1$ (*b*) $^2P_{3/2} \rightarrow {}^2D_{5/2}$

(*c*) $^2D_{5/2} \rightarrow {}^2P_{1/2}$ (*d*) $^3P_0 \rightarrow {}^5D_0$

20. The term symbol for the electronic ground state of oxygen atom is

(*a*) $^1S_0$ (*b*) $^1D_2$

(*c*) $^3P_0$ (*d*) $^3P_2$

21. An atom in its singlet state is subjected to a magnetic field. The Zeeman splitting of its 650 nm spectral line is 0.03 nm. The magnitude of the field is ______ Tesla (up to two decimal places).

$$(e = 1.60 \times 10^{-19}\,\text{C}, m_e = 9.11 \times 10^{-31}\,\text{kg}, c = 3.0 \times 10^8\,\text{ms}^{-1})$$

2017

22. J^P for the ground state of the $^{13}C_6$ nucleus is

(*a*) 1^+ (*b*) $\frac{3^-}{2}$

(*c*) $\frac{3^+}{2}$ (*d*) $\frac{1^-}{2}$

23. Which one of the following gases of diatomic molecules is Raman, infrared, and NMR active?

(*a*) $^1H - {}^1H$ (*b*) $^{12}C - {}^{16}O$

(*c*) $^1H - {}^{35}Cl$ (*d*) $^{16}O - {}^{16}O$

24. Positronium is an atom made of an electron and a positron. Given the Bohr radius for the ground state of the Hydrogen atom to be 0.53 Angstroms, the Bohr radius for the ground state of positronium is ______ Angstroms. (up to two decimal places).

2016

25. The molecule $^{17}O_2$ is

(*a*) Raman active but not NMR (nuclear magnetic resonance) active.

(*b*) Infrared active and Raman active but not NMR active.

(*c*) Raman active and NMR active.

(*d*) Only NMR active.

26. Which of the following transitions is NOT allowed in the case of an atom, according to the electric dipole radiation selection rule?

(*a*) 2s-1s (*b*) 2p-1s

(*c*) 2p-2s (*d*) 3d-2p

27. In a Young's double slit experiment using light, the apparatus has two slits of unequal widths. When only slit-1 is open, the maximum observed intensity on the screen is $4I_0$. When only slit-2 is open, the maximum observed intensity is I_0. When both the slits are open, an interference pattern appears on the screen. The ratio of the intensity of the principal maximum to that of the nearest minimum is ________.

28. A hydrogen atom is in its ground state. In the presence of a uniform electric field $\vec{E} = E_0\hat{z}$, the leading order change in its energy is proportional to $(E_0)^n$. The value of the exponent n is ______.

2015

29. A coin is tossed thrice. Let X be the event that head occurs in each of the first two tosses. Let Y be the event that a tail occurs on the third toss. Let Z be the event that two tails occur in three tosses. Based on the above information, which one of the following is TRUE?

(*a*) X and Y are not independent

(*b*) Y and Z are dependent

(*c*) Y and Z are independent

(*d*) X and Z are independent

30. Match the phrases in Group I and Group II and identify the correct option.

Group I

(P) Electron spin resonance (ESR)

(Q) Nuclear magnetic resonance (NMR)

(R) Transition between vibrational states of a molecule

(S) Electronic transition

Group II

(*i*) radio frequency

(*ii*) visible range frequency

(*ii*) microwave frequency

(*iv*) far-infrared range

(*a*) (P-*i*), (Q-*ii*), (R-*iii*), (S-*iv*)

(*b*) (P-*ii*), (Q-*i*), (R-*iv*), (S-*iii*)

(*c*) (P-*iii*), (Q-*iv*), (R-*i*), (S-*ii*)

(*d*) (P-*iii*), (Q-*i*), (R-*iv*), (S-*ii*)

31. The excitation wavelength of laser in a Raman effect experiment is 546 *nm*. If the stokes' line is observed at 552 *nm*, then the wavenumber of the anti-Stokes' line (in cm^{-1}) is ______

32. The number of permitted transitions from $^2P_{3/2} \rightarrow {}^2S_{1/2}$ in the presence of a weak magnetic field is ________

2014

33. The number of normal Zeeman splitting components of $^1P \rightarrow {}^1D$ transition is

(*a*) 3 (*b*) 4

(*c*) 8 (*d*) 9

34. For Nickel, the number density is 8×10^{23} atoms/cm^3 and electronic configuration is $1s^2\, 2s^2\, 2p^6\, 3s^2\, 3p^6\, 3d^8\, 4s^2$. The value of the saturation magnetization of Nickel in its ferromagnetic state is_____ $\times 10^9$ A/m.

(Given the value of Bohr magneton $\mu_B = 9.21 \times 10^{-21}$ Am^2)

35. The moment of inertia of a rigid diatomic molecule A is 6 times that of another rigid diatomic molecule B. If the rotational energies of the two molecules are equal, then the corresponding values of the rotational quantum numbers J_A and J_B are

(*a*) $J_A = 2, J_B = 1$

(*b*) $J_A = 3, J_B = 1$

(*c*) $J_A = 5, J_B = 0$

(*d*) $J_A = 6, J_B = 1$

36. The intensity of a laser in free space is 150 *m*W/*m*^{2}. The corresponding amplitude of the electric field of the laser is _______ V/*m*.

($\varepsilon_0 = 8.854 \times 10^{-12}$ C^2 / N.m^2)

37. The emission wavelength for the transition $^1D_2 \rightarrow {}^1F_3$ is 3122 Å. The ratio of populations of the final to the initial states at a temperature 5000 K is ($h = 6.626 \times 10^{-34}$ J.s, $c = 3 \times 10^8$ m/s, $k_B = 1.380 \times 10^{-23}$ J/K)

(*a*) 2.03×10^{-5}

(*b*) 4.02×10^{-5}

(*c*) 7.02×10^{-5}

(*d*) 9.83×10^{-5}

2013

38. The number of spectral lines allowed in the spectrum for the $3^2D \rightarrow 3^2p$ transition in sodium is ____________.

39. In a normal Zeeman effect experiment, spectral splitting of the line at the wavelength 643.8 nm corresponding to the transition 5 $^1D_2 \rightarrow$ 5 $^1P_1$ of cadmium atoms is to be observed. The spectrometer has a resolution of 0.01 nm. The minimum magnetic field needed to observe this is ($m_e = 9.1 \times 10^{-31}\, kg, e = 1.6 \times 10^{-19}\, C, c = 3 \times 10^8\, m/s$)

(*a*) 0.26 *T*

(*b*) 0.52 *T*

(*c*) 2.6 *T*

(*d*) 5.2 *T*

40. The spacing between vibrational energy levels in CO molecule is found to be 8.44×10^{-2} *e*V. Given that the reduced mass of CO is 1.14×10^{-26} kg, Planck's constant is 6.626×10^{-34} Js and 1 eV = 1.6×10^{-19} J. The force constant of the bond in CO molecule is

(*a*) 1.87 N/m (*b*) 18.7 N/m

(*c*) 187 N/m (*d*) 1870 N/m

LINKED ANSWER QUESTIONS

Statement for Linked Answer Q. 41 and 42:

In the Schmidt model of nuclear magnetic moments, we have,

$$\vec{\mu} = \frac{eh}{2Mc}\left(g_l \vec{l} + g_s \vec{S}\right)$$

where the symbols have their usual meaning

41. For the case J = l +1/2, where J is the total angular momentum, the expectation value of $\vec{S} \cdot \vec{J}$ in the nuclear ground state is equal to,

(*a*) (J – 1)/2 (*b*) (J + 1)/2

(*c*) J/2 (*d*) – J/2

42. For the O^{17} nucleus (A = 17, Z = 8), the effective magnetic moment is given by,

$$\vec{\mu}_{eff} = \frac{e\hbar}{2Mc} g\vec{J},$$

where g is equal to, (g_s = 5.59 for proton and – 3.83 for neutron)

(*a*) 1.12 (*b*) – 0.77

(*c*) – 1.28 (*d*) 1.28

2012

43. The ground state of sodium atom (^{11}Na) is a $S_{1/2}$ state. The difference is energy levels arising in the presence of a weak external magnetic field B, given in terms of Bohr magneton, μ_B, is

(*a*) $\mu_B B$ (*b*) $2\mu_B B$

(*c*) $4\mu_B B$ (*d*) $6\mu_B B$

44. The first Stokes line of a rotational Raman spectrum is observed at 12.96 cm^{-1}. Considering the rigid rotor approximation, the rotational constant is given by

(*a*) 6.48 cm^{-1} (*b*) 3.24 cm^{-1}

(*c*) 2.16 cm^{-1} (*d*) 1.62 cm^{-1}

45. Match the typical spectroscopic regions specified in **Group I** with the correspon-ding type of transitions in **Group II**.

Group I

(P) Infra-red region

(Q) Ultraviolet-visible region

(R) X-ray region

(S) γ-ray region

Group II

(*i*) electronic transitions involving valence electrons

(*ii*) nuclear transitions

(*iii*) vibrations transitions of molecules

(*iv*) transitions involving inner shell electrons

(*a*) (P, *i*); (Q, *iii*); (R, *ii*); (S, *iv*)

(*b*) (P, *ii*); (Q, *iv*); (R, *i*); (S, *iii*)

(*c*) (P, *iii*); (Q, *i*); (R, *iv*); (S, *ii*)

(*d*) (P, *iv*); (Q, *i*); (R, *ii*); (S, *iii*)

46. The terms $\{j_1, j_2\}_J$ arising from $2s^13d^1$ electronic configuration in j-j coupling scheme are

(*a*) $\left\{\frac{1}{2},\frac{3}{2}\right\}_{2,1}$ and $\left\{\frac{1}{2},\frac{5}{2}\right\}_{3,2}$

(*b*) $\left\{\frac{1}{2},\frac{1}{2}\right\}_{1,0}$ and $\left\{\frac{1}{2},\frac{3}{2}\right\}_{2,1}$

(*c*) $\left\{\frac{1}{2},\frac{1}{2}\right\}_{1,0}$ and $\left\{\frac{1}{2},\frac{5}{2}\right\}_{3,2}$

(*d*) $\left\{\frac{3}{2},\frac{1}{2}\right\}_{2,1}$ and $\left\{\frac{1}{2},\frac{5}{2}\right\}_{3,2}$

47. The equilibrium vibration frequency for an oscillator is observed at 2990 cm^{-1}. The ratio of the frequencies corresponding to the first and the fundamental spectral lines is 1.96. Considering the oscillator to be anharmonic, the anharmonicity constant is

(*a*) 0.005 (*b*) 0.02

(*c*) 0.05 (*d*) 0.1

2011

48. The population inversion in a two level laser material **CANNOT** be achieved by optical pumping because

(*a*) the rate of upward transitions is equal to the rate of downward transitions

(*b*) the upward transitions are forbidden but downward transitions are allowed

(*c*) the upward transitions are allowed but downward transitions are forbidden

(*d*) the spontaneous decay rate of the higher level is very low

49. An atom with one outer electron having orbital angular momentum l is placed in a weak magnetic field. The number of energy levels into which the higher total angular momentum state splits, is

(*a*) $2l + 2$ (*b*) $2l + 1$

(*c*) $2l$ (*d*) $2l - 1$

50. For a multi-electron atom, l, L and S specify the one-electron orbital angular momentum, total orbital angular momentum and total spin angular momentum, respectively. The selection rules for electric dipole transition between the two electronic energy levels, specified by l, L and S are

(*a*) $\Delta L = 0, \pm 1$; $\Delta S = 0$; $\Delta l = 0, \pm 1$

(*b*) $\Delta L = 0, \pm 1$; $\Delta S = 0$; $\Delta l = \pm 1$

(*c*) $\Delta L = 0, \pm 1$; $\Delta S = \pm 1$; $\Delta l = 0, \pm 1$

(*d*) $\Delta L = 0, \pm 1$; $\Delta S = \pm 1$; $\Delta l = \pm 1$

51. The lifetime of an atomic state is 1 nanosecond. The natural line width of the spectral line in the emission spectrum of this state is of the order of

(*a*) 10^{-10} eV (*b*) 10^{-9} eV

(*c*) 10^{-6} eV (*d*) 10^{-4} eV

52. The degeneracy of an excited state of nitrogen atom having electronic configuration $1s^22s^22p^23d^1$ is

(*a*) 6 (*b*) 10

(*c*) 15 (*d*) 150

53. The far infrared rotational absorption spectrum of a diatomic molecule shows equidistant lines with spacing 20 cm^{-1}. The position of the first Stokes line in the rotational Raman spectrum of this molecule is

(*a*) 20 cm^{-1} (*b*) 40 cm^{-1}

(*c*) 60 cm^{-1} (*d*) 120 cm^{-1}

2010

54. To detect trace amounts of a gaseous species in a mixture of gases, the preferred probing tools is

(*a*) Ionization spectroscopy withX-rays

(*b*) NMR spectroscopy

(*c*) ESR spectroscopy

(*d*) Laser spectroscopy

55. A collection of N atoms is exposed to a strong resonant electromagnetic radiation with N_g atoms in the ground state and N_e atoms in the excited state, such that $N_g + N_e = N$. This collection of two-level atoms will have the following population distribution:

(*a*) $N_g << N_e$ (*b*) $N_g >> N_e$

(*c*) $N_g \approx N_e \approx N/2$ (*d*) $N_g - N_e \approx N/2$

56. Two states of an atom have definite parities. An electric dipole transition between these states is

(*a*) Allowed if both the states have even parity

(*b*) Allowed if both the states have odd parity

(*c*) Allowed if the two sates have opposite parities

(*d*) Not allowed unless a static electric field is applied

57. The spectrum of radiation emitted by a black body at a temperature 1000 K peaks in the

(*a*) Visible range of frequencies

(*b*) Infrared range of frequencies

(*c*) Ultraviolet range of frequencies

(*d*) Microwave range of frequencies

58. The three principal moments of inertia of a methanol (CH_3OH) molecule have the property $I_x = I_y = I$ and $I_z \neq I$. The rotational energy eigenvalues are

(*a*) $\frac{\hbar^2}{2I}l(l+1)+\frac{\hbar^2 m_l^2}{2}\left(\frac{1}{I_z}-\frac{1}{I}\right)$

(*b*) $\frac{\hbar^2}{2I}l(l+1)$

(*c*) $\frac{\hbar^2 m_l^2}{2}\left(\frac{1}{I_z}-\frac{1}{I}\right)$

(*d*) $\frac{\hbar^2}{2I}l(l+1)+\frac{\hbar^2 m_l^2}{2}\left(\frac{1}{I_z}+\frac{1}{I}\right)$

59. Match the typical spectra of stable molecules with the corresponding wave-number range

1. Electronic spectra (*i*) 10^6 cm^{-1} and above

2. Rotational spectra (*ii*) $10^5 - 10^6$ cm^{-1}

3. Molecular dissociation (*iii*) $10^0 - 10^2$ cm^{-1}

(*a*) 1 – *ii*, 2 – *i*, 3 – *iii*, (*b*) 1 – *ii*, 2 – *iii*, 3 – *i*,

(*c*) 1 – *iii*, 2 – *ii*, 3 – *i*, (*d*) 1 – *i*, 2 – *ii*, 3 – *iii*,

2009

60. Group I lists some physical phenomena while Group II gives some physical parameters. Match the phenomena with the corresponding parameter.

Group I	Group II
P. Doppler Broadening	1. Moment of inertia
Q. Natural Broadening	2. Refractive index
R. Rotational spectrum	3. Lifetime of the energy level
S. Total internal reflection	4. Pressure

(*a*) P– 4, Q – 3, R–1, S – 2

(*b*) P – 3, Q – 2, R – 1, S – 4

(*c*) P –2, Q – 3, R – 4, S – 1

(*d*) P –1, Q – 4, R – 2, S – 3

61. The separation between the first Stokes and corresponding anti-Stokes lines of the rotational Raman spectrum in terms of the rotational constant, B is

(*a*) 2B (*b*) 4B

(*c*) 6B (*d*) 12B

62. In a diatomic molecule, the internuclear separation of the ground and first excited electronic state are the same as shown in the figure. If the molecule is initially in the lowest vibrational state of the ground state, then the absorption spectrum will appear as

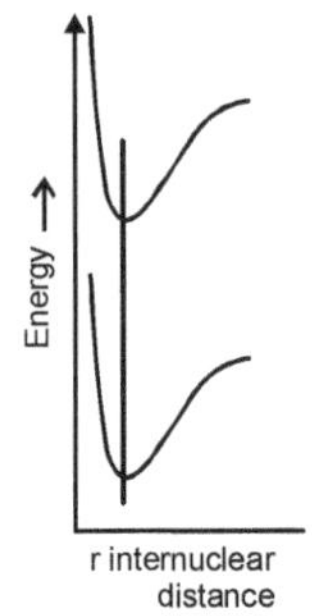

(*a*)
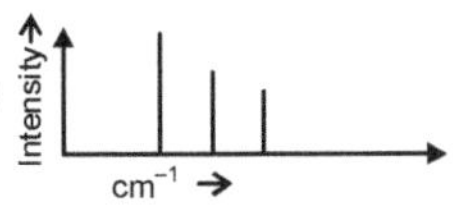

(*b*)
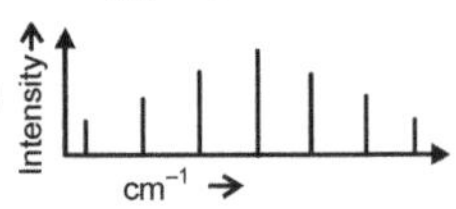

(*c*)
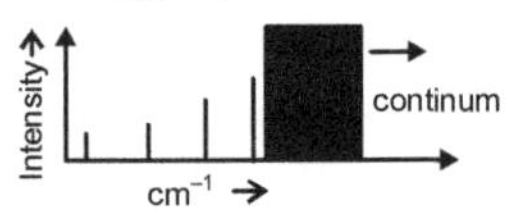

(*d*)
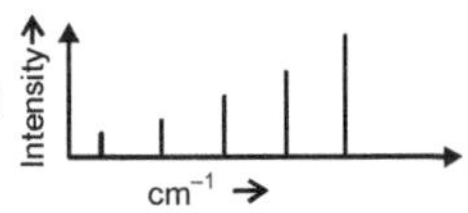

63. Five energy levels of a system including the ground state are shown below. Their lifetimes and the allowed electric dipole transitions are also marked.

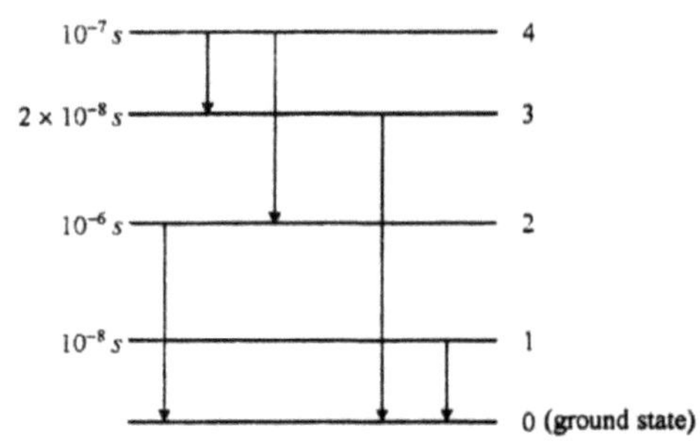

Which one of the following transitions is the most suitable for a continuous wave (CW) laser?

(*a*) $1 \to 0$ (*b*) $2 \to 0$

(*c*) $4 \to 2$ (*d*) $4 \to 3$

64. Cesium has a nuclear spin of 7/2. The hyperfine spectrum of the D lines of the cesium atom will consist of

(*a*) 10 lines (*b*) 4 lines

(*c*) 6 lines (*d*) 14 lines

Common data Q. 65 and 66 :

Consider the Zeeman splitting of a single electron system for the $3d \rightarrow 3p$ electric dipole transition.

65. The Zeeman spectrum is

(*a*) randomly polarized

(*b*) only π polarized

(*c*) only σ polarized

(*d*) both π and σ polarized

66. The fine structure line having the longest wavelength will split into

(*a*) 17 components (*b*) 10 components

(*c*) 8 components (*d*) 4 components

2008

67. The last two terms of the electronic configuration of manganese (Mn) atom is $3d^5 4s^2$. The term factor of Mn^{4+} ion is

(*a*) $^4D_{1/2}$ (*b*) $^4F_{3/2}$

(*c*) $^3F_{9/2}$ (*d*) $^3D_{7/2}$

68. The coherence length of laser light is

(*a*) directly proportional to the length of the active lasting medium.

(*b*) directly proportional to the width of the spectral line.

(*c*) inversely proportionla to the width of the spectral line.

(*d*) inverssely proportional to the length of the active lasing medium.

69. Consider the combined system of proton and electron in the hydrogen atom in its (electronic) ground state. Let I denote the quantum number associated with the total angular momentum and let $<9\pi>$ denote the magnitude of the expectation value of the net magnetic moment in the state. Which of the following pairs represents a possible state of the system (μ_B is Bohr magneton)?

(*a*) $I = 0$ $<9\pi> = 0$

(*b*) $I = \frac{1}{2}$, $<9\pi> = 1\mu_B$

(*c*) $I = 1$, $<9\pi> \approx 1\mu_B$

(*d*) $I = 0$, $<9\pi> = 2\mu_B$

70. Three consecutive absorption lines at 64.275 cm^{-1}, 77.130 cm^{-1} and 89.958 cm^{-1} have been observed in a microwave spectrum for a linear rigid diatomic molecule. The moments of inertia I_A and I_B are (I_A is with respect to the bond axis passing through the centre of mass and I_B is with respect to an axis passing through the centre of mass and perpendicular to bond axis)

(*a*) both equal to $\frac{\hbar^2}{12.855hc}$ gm cm^2

(*b*) zero and $\frac{\hbar^2}{12.855hc}$ gm cm^2

(*c*) both equal to $\frac{\hbar^2}{6.427hc}$ gm cm^2

(*d*) zero and $\frac{\hbar^2}{6.427hc}$ gm cm^2

71. A pure rotational Raman spectrum of a linear diatomic molecule is recorded using electromagnetic radiation of frequency v_e. The frequency of two consecutive Stockes lines are

(*a*) $v_e - 10B$, $v_e - 14B$

(*b*) $v_e - 2B$, $v_e - 4B$

(*c*) $v_e + 10B$, $v_e + 14B$

(*d*) $v_e + 2B$, $v_e + 4B$

72. Which one of the following statement is **INCORRECT** in vibrational spectroscopy with anharmonicity?

(*a*) The selection rule for vibrational spectroscopy is $\Delta v = \pm 1, \pm 2, ...$

(*b*) Anharmonicity leads to multiple absorption lines.

(*c*) The intesities of hot band lines are stronger than the fundamental absorption.

(*d*) The frequency of hot band lines are smaller than the fundamental absorption.

73. The molecular spectra of two linear molecules O-C-O and O-C-S are recorded in the microwave region. Which one of the following statement is correct?

(*a*) Both the molecules would show absorption lines.

(*b*) Both the molecules would not show absorption lines.

(*c*) O-C-O would show absorption lines, but not O-C-S.

(*d*) O-C-S would show absorption lines, but not O-C-O.

74. When the refractive indes μ of the active medium changes by $\Delta\mu$ in a laser resonator of length L, the change in the spectral spacing between the longitudinal modes of the laser is (c is the speed of light in free space)

(*a*) $\frac{c}{2(\mu + \Delta\mu)L}$. (*b*) $\frac{c}{2\Delta\mu L}$.

(*c*) $\frac{c}{2L}\frac{\Delta\mu}{\mu(\mu + \Delta\mu)}$. (*d*) zero.

2007

75. The D_1 and D_2 lines of Na ($3^2P_{1/2} \rightarrow 3^2S_{1/2}$, $3^2P_{3/2} \rightarrow 3^2S_{1/2}$) will split on the application of a weak magnetic field into

(a) 4 and 6 lines respectively.

(b) 3 lines each.

(c) 6 and 4 lines respectively.

(d) 6 lines each.

76. In a He-Ne laser, the laser transition takes place in

(a) He only.

(b) Ne only.

(c) Ne first, then in He.

(d) He first, then in Ne.

77. The allowed states for He ($2p^2$) configuration are

(a) $^1S_0, ^3S_1, ^1P_1, ^3P_{0,1,2}, ^1D_2$ and $^3D_{1,2,3}$

(b) $^1S_0, ^3P_{0,1,2}$ and $^1D_2$

(c) $^1P_1$ and $^3P_{0,1,2}$

(d) $^1S_0$ and $^1P_1$

78. A system has energy level E_0, $2E_0$, $3E_0$,....., where the excited states are triply degenerate. Four non-interacting bosons are placed in this system. If the total energy of these bosons is 5 E_0, the number of microstates is

(a) 2 (b) 3

(c) 4 (d) 5

79. In accordance with the selection rules of electric dipole transitions, the 4^3P_1 state of helium can decay by photon emission to the states

(a) $2^1S_0, 2^1P_1$ and 2^1D_2

(b) $3^1P_1, 3^1D_2$ and 3^1S_0

(c) $3^3P_2, 3^3D_3$ and 3^3P_0

(d) $2^3S_1, 3^3D_2$ and 3^3D_1

80. If an atom is in the $^3D_3$ state, the angle between its orbital and spine angular momentum vectors ($\vec{L}$ and $\vec{S}$) is

(a) $\cos^{-1}\frac{1}{\sqrt{3}}$ (b) $\cos^{-1}\frac{2}{\sqrt{3}}$

(c) $\cos^{-1}\frac{1}{2}$ (d) $\cos^{-1}\frac{\sqrt{3}}{2}$

81. The hyperfine structure of Na($3^2P_{3/2}$) with nuclear spin I = 3/2 has

(a) 1 state. (b) 2 states.

(c) 3 states. (d) 4 states.

82. The allowed rotational energy levels of a rigid hetero-nuclear diatomic molecule are expressed as ε_J = BJ(J + 1), whhhere B is the rotational constant and J is a rotational quantum number.

In a system of such diatomic molecules of reduced mass μ, some ofo the atoms of one element are replaced by a heavier isotope, such that the reduced mass is changed to 1.05μ. In the rotational spectrum of the system, the shift in the spectral line, corresponding to a transition J = 4 → J = 5, is

(a) 0.475 B (b) 0.50 B

(c) 0.95 B (d) 1.0 B

83. The number of fundamental vibrational modes of CO_2 molecule is

(a) four : 2 are Raman active and 2 are infrared active.

(b) four : 1 is Raman active and 3 are infrared active.

(c) three : 1 is Raman active and 2 are infrared active.

(d) three : 2 are Raman active and 1 is infrared active.

84. A peice of paraffin is placed in a uniform magnetic field H_0. The sample contains hydrogen nuclei of mass m_p. wbhich interact only with external magnetic filed. An additional oscillating magnetic field is appllied to observe resonance absorption. If g_1 is the g-factor of the hydrogen nucleus, the frequency, at which resonance absorption takes place, is given by

(a) $\frac{3g_1eH_0}{2\pi m_p}$ (b) $\frac{3g_1eH_0}{4\pi m_p}$

(c) $\frac{g_1eH_0}{2\pi m_p}$ (d) $\frac{g_1eH_0}{4\pi m_p}$

2006

85. Which one of the following relations in true for Pauli matrices σ_x, σ_y and σ_z?

(a) $\sigma_x\sigma_y = \sigma_y\sigma_x$ (b) $\sigma_x\sigma_y = \sigma_z$

(c) $\sigma_x\sigma_y = i\sigma_z$ (d) $\sigma_x\sigma_y = -\sigma_y\sigma_x$

86. The target of an X-ray tube is subjected to an excitation voltage V. The wavelength of the emitted X-rays is proportional to

(a) $\frac{1}{\sqrt{V}}$ (b) $\sqrt{V}$

(c) $\frac{1}{V}$ (d) V

87. The principal series o spectral lines of lithium is obtained by transitions between

(*a*) nS and 2P, $n > 2$ (*b*) nD and 2P, $n > 2$

(*c*) nP and 2S, $n > 2$ (*d*) nF and 3D, $n > 3$

88. In the microwave specturm of identical rigid diatomic molecules, the separation between the spectral lines is recorded to be 0.7143 cm^{-1}. The moment of inertia of the molecule, in kg m^2, is

(*a*) 2.3×10^{-36} (*b*) 2.3×10^{-40}

(*c*) 7.8×10^{-42} (*b*) 7.8×10^{-46}

89. Which one of the following electronic transitions in Neon is NOT responsible for LASER action in a helium-neon laser?

(*a*) $6s \to 5p$ (*b*) $5s \to 4p$

(*c*) $5s \to 3p$ (*b*) $4s \to 3p$

90. In the linear Stark effect, the application of an electric field

(*a*) completely lifts the degeneracy of $n = 2$ level on hydrogen atom and splits $n = 2$ level into four levels

(*b*) partially lifts the degeneracy of $n = 2$ level on hydrogen atom and splits $n = 2$ level into three levels

(*c*) partially lifts the degeneracy of $n = 2$ level on hydrogen atom and splits $n = 2$ level into two level

(*d*) does not affect the $n = 2$ levels

91. In hyperfine interaction, there is coupling between the electron angular momentum $\vec{J}$ and nuclear angular momentum $\vec{I}$, forming resultant angular momentum $\vec{F}$. The selectiion rules for the corresponding quantum number F in hyperfine transitions are

(*a*) $\Delta F = \pm 2$ only (*b*) $\Delta F = \pm 1$ only

(*c*) $\Delta F = 0, \pm 1$ (*d*) $\Delta F = \pm 1, \pm 2$

92. A vibrational-electronic spectrum of homonuclear binary molecules, involving electronic ground state ε'' and excited ε', exhibits a continuum at $\bar{\upsilon}$ cm^{-1}. If the total energy of the dissociated atoms in the exicted state exceeds the total energy of the dissociated atoms in the ground state by E_{ex} cm^{-1}, then dissociation energy of the molecule in the ground state is

(*a*) $(\bar{\upsilon} + E_{ex})/2$ (*b*) $(\bar{\upsilon} + E_{ex})/2$

(*c*) $(\bar{\upsilon} - E_{ex})$ (*d*) $\sqrt{(\bar{\upsilon}^2 - E_{ex}^2)}$

93. The NMR spectrum of ethanol (CH_3 CH_2 CH) comprises of three bunches of spectral lines. The number of spectral lines in the bunch corresponding to CH_2 group is

(*a*) 1 (*b*) 2

(*a*) 3 (*b*) 4

94. The energy $E(\vec{k})$ of electrons of wavevector $\vec{k}$ in a solid is given by $E(\vec{k}) = Ak^2 + Bk^4$, where A and B are constants. The effective mass of the electron at $|\vec{k}| = k_0$ is

(*a*) Ak_0^2 (*b*) $\frac{\hbar^2}{2A}$

(*c*) $\frac{\hbar^2}{2A + 12Bk_0^2}$ (*d*) $\frac{\hbar^2}{Bk_0^2}$

Statement for Linked Answer Questions 95 and 96:

The powder diffraction pattern of a body centred cubic crystal is recorded by using Cu K_a X-rays of wavelength 1.54 A

95. If the (002) planes diffract at 60º, then lattice parameter is

(*a*) 2.67 Å (*b*) 3.08 Å

(*c*) 3.56 Å (*d*) 5.34 Å

96. Assuming the atomic mass of the constituent atoms to be 50.94 amu, then density of the crystal in units of kg m^{-3} is

(*a*) 3.75×10^3 (*b*) 4.45×10^3

(*c*) 5.79×10^3 (*d*) 8.89×10^3

2005

97. Match the following and choose the correct combination

	Group 1 *Characteristics*	Group 2 *Element*
P.	Atomic configuration $1s^2 2s^2 2p^6 3s^2 3p^6$	1. Na
Q.	Strongly electropositive	2. Si
R.	Strongly electronegative	3. Ar
S.	Covalent bonding	4. Cl

(*a*) P-1, Q-2, R-3, S-4 (*b*) P-3, Q-2, R-4, S-1

(*c*) P-3, Q-1, R-4, S-2 (*d*) P-3, Q-4, R-1, S-2

98. An atom emits a photon of wavelength $\lambda = 600$ nm by transition from an excited state of lifetime 8×10^{-9} s. If Δv represents the minimum uncertainty in the frequency of the photon, the fractional width $\frac{\Delta \upsilon}{\upsilon}$ of the spectral line is of the order of

(*a*) 10^{-4} (*b*) 10^{-6}

(*c*) 10^{-8} (*d*) 10^{-10}

99. The sodium doublet lines are due to transitions from $^2P_{3/2}$ and $^2P_{1/2}$ levels to $^2S_{1/2}$ level. On application of a weak magnetic field, the total number of allowed transitions becomes

(*a*) 4 (*b*) 6
(*c*) 8 (*d*) 10

100. A three level system of atoms has N_1 atoms in level E_1, N_2 in level E_2, and N_3 in level E_3 ($N_2 > N_1 > N_3$, and $E_1 < E_2 < E_3$). Laser emission is possible between the levels

(*a*) $E_3 \to E_1$ (*b*) $E_2 \to E_1$
(*c*) $E_3 \to E_2$ (*d*) $E_2 \to E_3$

101. In a Raman scattering experiment, light of frequency v from a laser is scattered by diatomic molecules having the moment of inertia I. the typical Raman shifted frequency depends on

(*a*) v and I (*b*) only v
(*c*) only I (*d*) neither v nor I

102. For a diatomic molecule with the vibrational quantum number n and rotational quantum number J, the vibrational level spacing $\Delta E_n = E_n - E_{n-1}$ and the rotational level spacing $\Delta E_J = E_J - E_{J-1}$ are approximately

(*a*) ΔE_n = constant, ΔE_J = constant
(*b*) ΔE_n = constant, $\Delta E_J \propto J$
(*c*) $\Delta E_n \propto n$, $\Delta E_J \propto J$
(*d*) $\Delta E_n \propto n$, $\Delta E_J \propto J^2$

103. The typical wavelengths emitted by diatomic molecules in purely vibrational and purely rotational transitions are respectively in the region of

(*a*) infrared and visible
(*b*) visible and infrared
(*c*) infrared and microwave
(*d*) microwave and infrared

104. In a two electron atomic system having orbital and spin angular momenta l_1, l_2 and s_1, s_2 respectively, the coupling strengths are defined as $\Gamma_{l_1l_2}, \Gamma_{s_1s_2}$, $\Gamma_{l_1s_1}, \Gamma_{l_2s_2}$, $\Gamma_{l_1s_2}$ and $\Gamma_{l_2s_1}$. For the jj coupling scheme to be applicable, the coupling strengths MUST satisfy the condition

(*a*) $\Gamma_{l_1l_2}, \Gamma_{s_1s_2} > \Gamma_{l_1s_1}, \Gamma_{l_2s_2}$
(*b*) $\Gamma_{l_1s_1}, \Gamma_{l_2s_2} > \Gamma_{l_1l_2}\ \Gamma_{s_1s_2}$
(*c*) $\Gamma_{l_1s_2}, \Gamma_{l_2s_1} > \Gamma_{l_1l_2}\ \Gamma_{s_1s_2}$
(*d*) $\Gamma_{l_1s_2}, \Gamma_{l_2s_1} > \Gamma_{l_1s_1}\ \Gamma_{l_2s_2}$

Statement for Linked Answer Questions 105a & 105 b :

Lead has atomic weight of 207.2 amu and density of 11.35 gm cm^{-3}

105a. Number of atoms per cm^3 for lead is

(*a*) 1.1×10^{25} (*b*) 3.3×10^{22}
(*c*) 1.1×10^{22} (*d*) 3.3×10^{25}

105b. If the energy of vacancy formation in lead is 0.55 eV/atom, the number of vacancies/cm^3 at 500K is

(*a*) 3.2×10^{16} (*b*) 3.2×10^{19}
(*c*) 9.5×10^{19} (*d*) 9.5×10^{16}

2004

106. The spectral terms for a certain electronic configuration are given by ^3D, ^1D, ^3P, ^1P, ^5S, ^3S. The term with the lowest energy is

(*a*) ^5S (*b*) ^3P
(*c*) ^3D (*d*) ^3S

107. The degeneracy of the spectral term ^3F is

(*a*) 7 (*b*) 9
(*c*) 15 (*d*) 21

108. The Lande g factor for the level $^3D_3$ is

(*a*) $\frac{2}{3}$ (*b*) $\frac{3}{2}$
(*c*) $\frac{3}{4}$ (*d*) $\frac{4}{3}$

109. All vibrations producing a change in the electric dipole moment of a molecule yield

(*a*) Raman spectra (*b*) Infrared spectra
(*c*) Ultra-violet spectra (*d*) X-ray spectra

110. The L, S and J quantum numbers corresponding to the ground state electronic configuration of Boron (Z = 5) are

(*a*) L = 1, S = 1/2, J = 3/2
(*b*) L = 1, S = 1/2, J = 1/2
(*c*) L = 1, S = 3/2, J = 1/2
(*d*) L = 0, S = 3/2, J = 3/2

111. The degeneracies of the J-states arising from the ^3P term with spin-orbit interaction are

(*a*) 1, 3, 5 (*b*) 1, 2, 3
(*c*) 3, 5, 7 (*d*) 2, 6, 10

112. Assuming that the L-S coupling scheme is valid, the number of permited transitions from $^2P_{3/2}$ to $^2S_{1/2}$ due to a weak magnetic field is

(*a*) 2 (*b*) 4
(*c*) 6 (*d*) 10

113. Consider the pure rotational spectrum of a diatomic rigid rotor. The separation between two consecutive lines $(\Delta\overline{V})$ in the spectrum

(*a*) is directly proportional to the moment of inerital of the rotor

(*b*) in inversely proportional to the moment of inertia of the rotor

(*c*) dependns on the angular momentum

(*d*) is directly proportional to the square of the interatomic separation

114. Light of wavelegth 1.5 μm incident on a material with a characteristic Raman frequency of 20×10^{12} Hz results in a Stokes-shifted line of wavelength [Given : $c = 3 \times 10^8$ m.s^{-1}]

(*a*) 1.47 μm (*b*) 1.57 μm

(*c*) 1.67 μm (*d*) 1.77 μm

115. Consider black body radiation in a cavity maintained at 2000 K. If the volume of the cavity is reversibly and adiabatically increased from 10 cm^3 to 640 cm^3, the temperature of the cavity changes to

(*a*) 800 K (*b*) 700 K

(*c*) 600 K (*d*) 500 K

2003

116. The spectral term for the atom with 70% filled subshell and only S = 3/2 is

(*a*) $^3P_0$

(*b*) $^4F_{9/2}$

(*c*) $^3F_{1/2}$

(*d*) $^4P_{1/2}$

117. The dielectric constant of water is 80. However its refractive index is 1.75 invalidating the expression $n = \varepsilon^{1/2}$. This is because

(*a*) the water molecule has a permanent dipole moment

(*b*) the boiling point of water is 100° C

(*c*) the two quantities are measured in different experiments

(*d*) water is transparent to visible light

118. A parallel beam of electrons of a given momentum pass through a screen S_1 containing a slit and then produces a diffraction pattern on a screen S_2 placed behind it. The width of the central maximum observed on the screen S_2 can be increased by

(*a*) decreasing the distance between the screens S_1 and S_2

(*b*) increasing the width of the slit in screen S_1

(*c*) decreasing the momentum of the electrons

(*d*) increasing the momentum of the electrons

119. The vibrational spectrum of a molecule exhibits a strong line with P and R branches at a frequency ν_1 and a weaker line at a frequency ν_2. The frequency ν_3 is not shown up. Its vibrational Raman spectrum shows a strongly polarized line at frequency ν_3 and no feature at ν_1 and ν_2

(*a*) the molecule could be linear

(*b*) the molecule lacks a center of inversion

(*c*) ν_1 arises from a symmetric stretching mode

(*d*) ν_3 arises from a bending mode

120. Three values of rotational energies of molecules are given below in different units

P 10 cm^{-1}

Q 10^{-23} J

R 10^4 MHz

Choose the correct arrangement in the increasing order of energy

(*a*) P, Q, R (*b*) R, Q, P

(*c*) R, P, Q (*d*) Q, R, P

121. The short wavelength cut off of the continuous X-ray spectrum from a nickel target is 0.0825 nm. The voltage required to be applied to an X-ray tube is

(*a*) 0.15 KV (*b*) 1.5 KV

(*c*) 15 KV (*d*) 150 KV

122. The spin-orbit coupling constant for the upper state of sodium atom which emits D lines of wave numbers 16956.2 and 16973.4 cm^{-1} is

(*a*) 15 cm^{-1} (*b*) 11.4 cm^{-1}

(*c*) 12.5 cm^{-1} (*d*) 15.1 cm^{-1}

123. Consider the following statements about molecular spectra

P CH_4 does not give pure rotational Raman lines

Q SF_6 could be studied by rotational Raman spectroscopy

R N_2 shows infrared absorption spectrum

S CH_3CH_3 shows vibrational Raman and infrared absorption lines

T H_2O_2 shows pure rotational spectrum

Choose the right combination of correct statements

(*a*) P and Q (*b*) P, R and T

(*c*) P, S and T (*d*) Q and R

124. The point group symmetrics of the three molecules shown in Figs. 1–3 are respectively

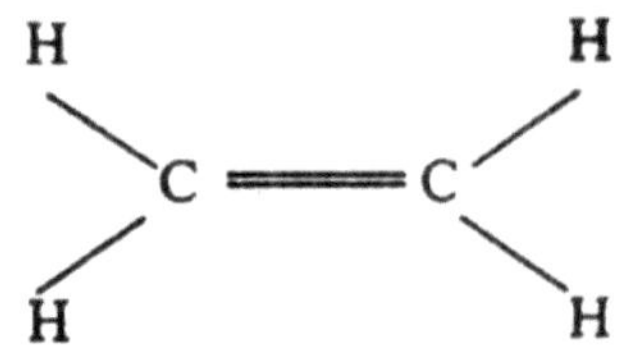

Fig.1

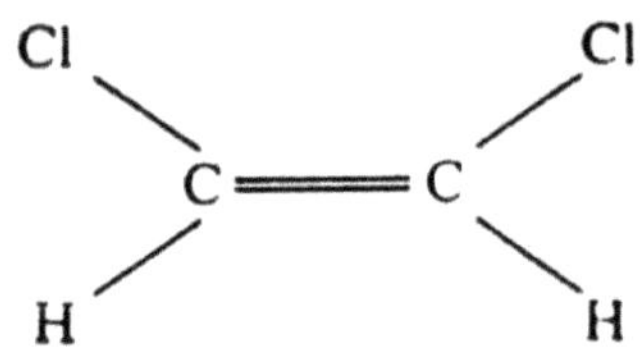

Fig.2

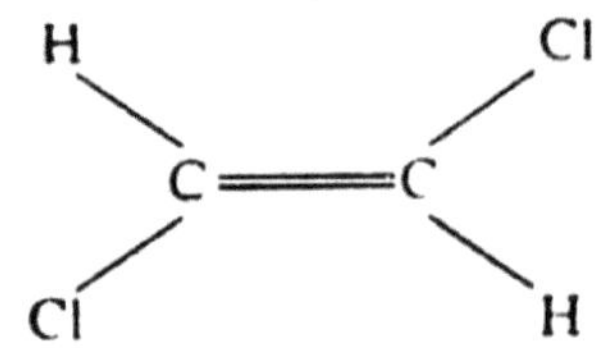

Fig.3

[notation : C_{2v} = 2 mm; C_{2h} = 2/m; D_{2h} = mmm]

(*a*) C_{2h}, C_{2v}, C_{2h} (*b*) C_{2v}, C_{2h}, C_{2h}

(*c*) D_{2h}, C_{2v}, C_{2h} (*d*) C_{2v}, D_{2h}, C_{2h}

125. Which one of the following statements concerning the Compton effect is NOT correct ?

(*a*) The wavelength of the scattered photon is greater than or equal to the wavelength of the incident photon

(*b*) The electron can acquire a kinetic energy equal to the energy of the incident photon

(*c*) The energy of the incident photon equals to the kinetic energy of the electron plus the energy of the scattered photon

(*d*) The kinetic energy acquired by the electron is the largest when the incident and scattered photons move in opposite directions

Data for Q. 126 – 127 :

In the presence of an inhomogeneous weak magnetic field, spectral lines due to transitions between two sets of states were observed

(1) $^5l_5 \rightarrow {}^5H_4$ *and (2)* $^2D_{5/2} \rightarrow {}^2P_{3/2}$

126. The types of Zeeman effect observed in (1) and (2) respectively are

(*a*) normal, normal

(*b*) anomalous, anomalous

(*c*) anomalous, normal

(*d*) normal, anomalous

127. The number of levels into which each of the above four terms split into respectively is

(*a*) 6, 4, 10, 8 (*b*) 4, 6, 10, 12

(*c*) 11, 9, 6, 4 (*d*) 9, 5, 12, 10

2002

128. In a Stern-Gerlach experiment, the magentic field is in +*z* direction. A particle comes out of this experiment in $\left|+\hat{z}\uparrow\right\rangle$ state. Which of the following statements is true ?

(*a*) The particle has a definite value of the *y*-component of the spin angular momentum.

(*b*) The particle has a definite value of the square of the spin angular momentum.

(*c*) The particle has a definite value of the *x*-component of spin angular momentum.

(*d*) The particle has definite values of *x*-and *y*-components of spin angular momentum.

129. A laser beam of wavelength 600 nm with a circular cross section having a radius of 10 mm falls normally on a lens of radius 20 mm and focal length 10 cm. The radius of the focussed spot is approximately

(*a*) 0.3 μm (*b*) 0.6 μm

(*c*) 3.0 μm (*d*) 6.0 μm

2001

130. X-rays were produced using Cobalt (Z = 27) as target. It was observed that the X-ray spectrum contained a strong K_α line of wavelength 0.1785 nm and a weak K_α line of wavelength 0.1930 nm. Then, the weak K_α line is due to an impurity whose atomic number is

(*a*) 25 (*b*) 26

(*c*) 28 (*d*) 30

131. A sample of Silicon of thickness 200 μm is doped with 10^{23} Phosphorous atoms per m^3. If the sample is kept in a magnetic field of 0.2 Wb/m^2 and a current of 1 mA is passed through the sample, the Hall voltage produced is

(*a*) 62.5 μV (*b*) –6.25 μV

(*c*) +6.25 μV (*d*) – 62.5 μV

132. The distance between the adjacent atomic planes in $CaCO_3$ is 0.3 nm. The smallest angle of Bragg scattering for 0.03 nm X-ray is

(*a*) 2.9° (*b*) 1.5°

(*c*) 0.29° (*d*) 5.8°

133. The number of hyperfine components obser-ved in the electronic transition $^2p_{1/2} \rightarrow {}^2S_{1/2}$ of an atom with nuclear spin $\frac{1}{2}$ is

(*a*) 3 (*b*) 4

(*c*) 6 (*d*) 5

134. Which of the following reactions violates lepton number conservation ?

(*a*) $e^+ + e^- \rightarrow \upsilon + \overline{\upsilon}$ (*b*) $e^- + p \rightarrow \upsilon + n$

(*c*) $e^+ + n \rightarrow p + \upsilon$ (*d*) $\mu^- \rightarrow e^- + \upsilon + \overline{\upsilon}$

135. The Lande *g*-factor for the $^3P_1$ level of an atom is

(*a*) 1/2 (*b*) 3/2

(*c*) 5/2 (*d*) 7/2

136. The pure rotational levels of a molecule in the far-infrared region follows the formula F(J) = BJ (J + 1), where F(J) is the energy of the rotational level with quantum number J and B is the rotational constant. The lowest rotational energy gap in rotational Raman spectrum is

(*a*) 2B (*b*) 4B

(*c*) 6B (*d*) 8B

137. The total number of Zeeman components observed in an electronic transition $^2D_{5/2} \rightarrow {}^2P_{3/2}$ of an atom in a weak field is

(*a*) 4 (*b*) 6

(*c*) 12 (*d*) 10

2000

138. The energy E of K_α X-rays emitted from targets of different atomic number Z varies as

(*a*) Z^2 (*b*) $Z^{2/3}$

(*c*) Z (*d*) $Z^{1/2}$

139. In a Stem-Gerlach experiment the atomic beam whose angular momentum state is to be determined, must travel through

(*a*) homogeneous radio frequency magnetic field

(*b*) homogeneous static magnetic field

(*c*) inhomogeneous static magnetic field

(*d*) inhomogeneous radio frequency magnetic field.

140. If a spin I particle is in the state $|m = 0\rangle$ with respect to a quantization axis $\hat{n}$, which of the following is correct?

(*a*) $\langle \overline{S} \rangle = 0$ (*b*) $\langle \overline{S} \rangle = \hat{n}$

(*c*) $\langle \overline{S} \rangle = \sqrt{2}\ \hat{n}$ (*d*) $\langle \overline{S} \rangle = -\hat{n}$

141. A spin ½ particle with $g > 0$, is subjected to a magnetic field $\overline{H} = H_0\hat{x}$. If the quantization axis is along + $\hat{z}$, then the minimum energy eigenstate is given by,

(*a*) $|\downarrow\rangle$ (*b*) $\frac{1}{\sqrt{2}}(|\uparrow\rangle - |\downarrow\rangle)$

(*c*) $|\uparrow\rangle$ (*d*) $\frac{1}{\sqrt{2}}(|\uparrow\rangle + |\downarrow\rangle)$

142. Magnetic long range order is typically exhibited by

(*a*) noble metals (*b*) alkali metals

(*c*) inert gas solids (*d*) transition metals

143. X-rays with a wave vector $\vec{K}$ are scattered from a simple cubic lattice with lattice spacing $a = 2\pi$. The scattered X-rays have wave vector $\vec{K}$. The possible values of $\Delta K_1 = K_x - K_x'$ for which there are peaks in the scattered intensity are

(*a*) $0 \leq \Delta K_1 < \pi/4$

(*b*) ΔK_x = integer

(*c*) ΔK_x = interger multiple of 2π

(*d*) ΔK_x = integer/2*x*

144. Under the LS coupling scheme, the possible spectral terms $^{2s+1}L_f$ for the electronic configuration $2s^1 3s^1$ are

(*a*) $^2S_{1/2}$, $^2P_{3/2}$, $^2P_{1/2}$ (*b*) $^1S_0$, $^3S_1$

(*c*) $^1S_0$, $^1S_1$, $^3S_0$, $^3S_1$, (*d*) $^3S_0$, $^3S_1$

145. Which of the following is the spectroscopic ground state $^{2z+1}L_f$ for Mn^{3+} ions of electronic configuration $1s^2 2s^2 2p 3s^2 3p^6 3\ d^4$ predicted by Hund's rules?

(*a*) $^5D_0$ (*b*) $^5D_4$

(*c*) $^5D_3$ (*d*) $^5D_2$

146. The Bohr model gives the value for the ionisation potential of Li^{2+} ion as

(*a*) 13.6 eV (*b*) 27.2 eV

(*c*) 40.8 eV (*d*) 122.4 eV

ANSWERS

1. (a)	**2.** (c,d)	**3.** (a, b)	**4.** (5000 to 5000)	**5.** (17367 to 17371)			**6.** (6 to 6)		**7.** (2 to 2)	
8. (92 to 93)		**9.** (3 to 5)		**10.** (*d*)	**11.** (*a*)	**12.** (3 to 3)		**13.** (*a*)	**14.** (3.61 to 3.63)	
15. (*c*)	**16.** (*d*)	**17.** (*c*)	**18.** (*b*)	**19.** (*b*)	**20.** (*d*)	**21.** (1.51 to 1.53)		**22.** (*d*)	**23.** (*c*)	
24. 0.99 to 1.10		**25.** (*c*)	**26.** (*a*)	**27.** (9 : 9)	**28.** (2 : 2)	**29.** (1844.3 to 1844.5)			**30.** (*d*)	
31. (18513 to 18519)			**32.** (6)	**33.** (*a*)	**34.** (40 to 43)		**35.** (*b*)	**36.** (10.58 to 10.70)		
37. (*c*)	**38.** (3)	**39.** (*b*)	**40.** (*c*)	**41.** (*b*)	**42.** (*b*)	**43.** (*b*)	**44.** (*c*)	**45.** (*d*)	**46.** (*a*)	**47.** (*b*)
48. (*a*)	**49.** (*b*)	**50.** (*c*)	**51.** (*c*)	**52.** (*b*)	**53.** (*c*)	**54.** (*c*)	**55.** (*)	**56.** (*c*)	**57.** (*b*)	**58.** (*a*)
59. (*a*)	**60.** (*a*)	**61.** (*d*)	**62.** (*a*)	**63.** (*b*)	**64.** (*a*)	**65.** (*d*)	**66.** (*d*)	**67.** (*b*)	**68.** (*a*)	**69.** (*a*)
70. (*b*)	**71.** (*a*)	**72.** (*c*)	**73.** (*d*)	**74.** (*c*)	**75.** (*a*)	**76.** (*b*)	**77.** (*b*)	**78.** (*b*)	**79.** (*d*)	**80.** (*a*)
81. (*d*)	**82.** (*a*)	**83.** (*c*)	**84.** (*d*)	**85.** (*d*)	**86.** (*c*)	**87.** (*c*)	**88.** (*d*)	**89.** (*a*)	**90.** (*b*)	**91.** (*c*)
92. (*c*)	**93.** (*d*)	**94.** (*c*)	**95.** (*c*)	**96.** (*a*)	**97.** (*c*)	**98.** (*b*)	**99.** (*d*)	**100.** (*b*)	**101.** (*c*)	**102.** (*b*)
103. (*c*)	**104.** (*b*)	**105a.** (*b*)	**105b.** (*a*)	**106.** (*a*)	**107.** (*d*)	**108.** (*d*)	**109.** (*b*)	**110.** (*b*)	**111.** (*a*)	**112.** (*c*)
113. (*b*)	**114.** (*c*)	**115.** (*d*)	**116.** (*b*)	**117.** (*a*)	**118.** (*c*)	**119.** (*a*)	**120.** (*b*)	**121.** (*c*)	**122.** (*a*)	**123.** (*c*)
124. (*b*)	**125.** (*d*)	**126.** (*b*)	**127.** (*c*)	**128.** (*b*)	**129.** (*c*)	**130.** (*b*)	**131.** (*d*)	**132.** (*a*)	**133.** (*b*)	**134.** (*c*)
135. (*b*)	**136.** (*b*)	**137.** (*c*)	**138.** (*a*)	**139.** (*c*)	**140.** (*c*)	**141.** (*d*)	**142.** (*d*)	**143.** (*b*)	**144.** (*b*)	**145.** (*a*)
146. (*d*)										

EXPLANATIONS

1. For $^{13}_{7}N$

N = 6 and Z = 7

it gives

$$\left({}^1S_{\frac{1}{2}}\right)^2 \left({}^1P_{\frac{3}{2}}\right)^4 \left({}^1P_{\frac{1}{2}}\right)^2$$

$$j = \frac{1}{2}, l = 1$$

$$j^P = \frac{1^-}{2}$$

For first forbidden $\Delta J = \pm 2$ (Parity Change)

Hence, option (a) is the correct answer.

2. For 4_{S_1}

$2s + 1 = 4, l = 0, J = 1$

$$s = \frac{3}{2}$$

but, $j = |l + s|$ to $|l - s|$

$$j = \frac{3}{2} \quad \text{(Not possible)}$$

For $2_{D_{\frac{7}{2}}}$

$$(2s + 1) = 2, l = 2, j - \frac{7}{2}$$

$$s = \frac{1}{2}$$

but $j = |l + s|$ to $|l - s|$

$$j = \frac{5}{2}, \frac{3}{2} \quad \text{(Not possible)}$$

For 3_{S_1}

$(2s + 1) = 3, l = 0, j = 1$

$s = 1$

but $j = |l + s|$ to $|l - s|$

$j = 1$ (Possible)

For $2_{D_{\frac{5}{2}}}$,

$$(2s + 1) = 2, l = 2, j = \frac{5}{2}$$

$$s = \frac{1}{2}$$

but $j = |l + s|$ to $|l - s|$

$$j = \frac{5}{2}, \frac{3}{2} \quad \text{(Possible)}$$

3. Population Inversion is the necessary condition for lasor action therefore

Population of the particle in level 1 should be greater than that of level 0.

4. We know that from wein's Relation

$\lambda_{Peak} T = b$

and we can write

$\lambda_1 T_1 = \lambda_2 T_2$

$$T_2 = \frac{\lambda T_1}{\lambda_2}$$

$$T_2 = \frac{289.8 \times 10000}{579.6}$$

$$\boxed{T_2 = 5000 \text{ K}}$$

5. Given that

$\nabla \bar{v} = 4B = 243.2 \text{ cm}^{-1}$

$\lambda = 514.5$ nm

$\bar{v}_S = 17611.4 \text{ cm}^{-1}$

Now the Next higher energy level

$$\bar{v}_S{}' = v_s + \Delta\bar{v}$$

$$\bar{v}_S{}' = (17611.4 + 243.2)$$

$$= 17854.6 \text{ cm}^{-1}$$

$$\boxed{\bar{v}_S{}' \approx 17855 \text{ cm}^{-1}}$$

6.

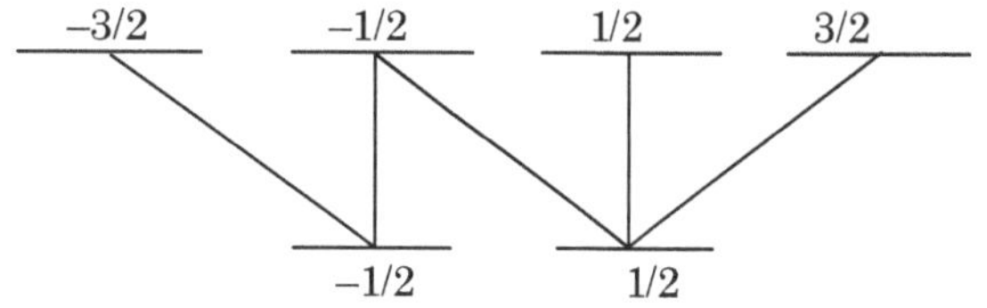

Selection rules are $\Delta mj = +1, 0$

So, No. of lines are = 6

7. Given that,

$A_{21} = 2 \times 107 \text{ s}^{-1}$

$A_{21} = 10^8 \text{ s}^{-1}$

$N_2 = 10^{14} \text{ cm}^{-3}$

We know that

$$\frac{dN_1}{dt} = N_2A_{21} - N_1A_{12}$$

at equilibrium state

$$\frac{dN_1}{dt} = 0$$

$$N_2\, A_{21} = N_1\, A_{12}$$

$$N_1 = N_2 \frac{A_{21}}{A_{12}}$$

$$N_1 = 10^{14} \times \frac{2\times 10^7}{10^8}$$

$$\boxed{N_1 = 2\times 10^{13}\,\text{cm}^3}$$

So, $\boxed{x = 2}$

8. $l = 2,\ s = \frac{1}{2}$

$$j = |l+s| \text{ to } |l-s|$$

$$j = \frac{5}{2}, \frac{3}{2}$$

longest $j = \frac{5}{2}$

$$\cos\theta = \frac{J_z}{J}$$

$$\cos\theta = \frac{m_j h}{\sqrt{j(j+1)}h}$$

$$\cos\theta = \frac{m_j}{\sqrt{j(j+1)}}$$

$$\cos\theta = \frac{\frac{5}{2}}{\sqrt{\frac{5}{2}\left(\frac{5}{2}+1\right)}}$$

$$\cos\theta = \sqrt{\frac{5}{7}}$$

$$\theta = 32.32^\circ$$

$$\boxed{\theta \simeq 32}$$

9. We know that

$$E = \frac{hc}{\lambda}$$

$$\Delta E = -\frac{hc}{\lambda^2}\Delta\lambda$$

taking positive value

$$\Delta E = \frac{hc}{\lambda^2}\Delta\lambda$$

$$\Delta E = \frac{4.125\times 10^{-15}\,\text{eV}\times 3\times 10^8\,\text{m/s}\times 0.14\times 10^{-10}\,\text{m}}{6521\times 6521\times 10^{-20}\,\text{m}^2}$$

$$\Delta E = 4.07 \times 10^{-5}\ \text{eV}$$

$$\boxed{\Delta E \approx 4\times 10^{-5}\,\text{eV}}$$

13. $v(r) = \frac{KQ}{2R}\left(3 - \frac{r^2}{R^2}\right)$ for $r < R$

$$\frac{\partial v}{\partial r} = \frac{-KQr}{R^3}$$

$$f(r) = \frac{1}{r}.\frac{\partial v}{\partial r} = \frac{-KQ}{R^3}$$

$f(r)$ = constant.

14. $S = \frac{1}{2} + \frac{1}{2} + \frac{1}{2} = \frac{3}{2}$

$$l = 3$$

$$j = l + s = 3 + \frac{3}{2} = \frac{9}{2}$$

$$j = l \pm s$$

$$j = (l - s) \text{ to } (l + s)$$

$$j = \left(3 - \frac{3}{2}\right) \text{to} \left(3 + \frac{3}{2}\right)$$

$$j = \frac{3}{2}, \frac{5}{2}, \frac{7}{2}, \frac{9}{2}$$

$$\frac{\mu_J}{J} = g\frac{e}{2m}$$

$$\mu J = \frac{ge}{2m} = \frac{ge}{2m}\sqrt{j(j+1)}\hbar$$

$$\mu_J = g\sqrt{j(j+1)}\mu_B$$

$$\mu_J = \frac{8}{11}\sqrt{\frac{9}{2}\left(\frac{9}{2}+1\right)};\mu_B = 3.62\mu_B$$

15. (*i*) Stern-Gerlach experiment - Existence of electron spin

(*ii*) Zeeman effect - Space quantization of angular momentum

(*iii*) Frank-Heartz experiment - Quantization of energy of electrons in the atoms.

(*iv*) Davisson-Germer experiment-wave nature of particles.

16. For second overton frequency is n = 0 → n = 3

$$w_{0\to3} = w_3 - w_0 = \left[w_e\left(\frac{7}{2}\right) - w_e x_e\left(\frac{7}{2}\right)^2\right] - \left[w_e\left(\frac{1}{2}\right) - w_e x_e\left(\frac{1}{2}\right)^2\right]$$

$$w_{0-3} = [3w_e - 12w_e x_e]$$

$$w_{0-3} = 3w_e[1-4x_e]$$

17. The physical effects and order of magnitude of their energy scales matching as given below.

Group - I	Group - II
Lamb shift	$\sim O(\alpha^5 m_e c^2)$
Fine structure	$\sim O(\alpha^4 m_e c^2)$
Bohr energy	$\sim O(\alpha^2 m_e c^2)$
Hyperfine structure	$\sim O(\alpha^4 m_e c^2/m_p)$

18. The permanent electric dipole moment in the ground state of hydrogen atom is zero.

19. $$^2P_{3/2} = \left[L = 1, S = \frac{1}{2}, J = \frac{3}{2}\right]$$

and $$^2D_{5/2} = \left[L = 2, S = \frac{1}{2}, J = \frac{5}{2}\right]$$

$$^2P_{3/2} \longrightarrow {}^2D_{5/2}$$

This transition allowed electric dipole transition

20. Oxygen atoms have 8 electrons and the shell structure is 2, 6.

The ground state electronic configuration of ground state gaseous neutral oxygen is $2s^2$, $2p^4$ and the term symbol is $^3p_2$

22. JP for the ground state of $^{13}C_6$ nuclues is-

$$\left(\frac{1S_1}{2}\right)^2\left(\frac{1P_3}{2}\right)^4\left(\frac{1P_1}{2}\right)^1$$

Ground state have $J^\pi = \frac{1}{2}$

Parity = $(-1)^1 = (-1)^1 = -1$

Hence, J^P for the ground state of $^{13}C_6 = \frac{1}{2}$

23. ^1H - ^{35}Cl is the diatomic molecule which is Raman, infrared and NMR active.

AS it has xx, xy, yy symmetry. It changes its dipole moment as a result of the vibration that occurs when IR radiation is absorbed.

24. Bohr radius for the ground state of hydrogen atom = 0.53Å

Bohr radius for the ground state of positronium

= 2 × Bohr radius for hydrogen atom.

= 2 × 0.53Å = 1.06Å

25. Molecule $^{17}O_2$ can not absorb infrared as there is no change in dipole moment during vibration. Thus $^{17}O_2$ is infrared inactive.

26. The allowed electric dipole transitions are those involving a change is parity

For 1S-2S $l = 0$ both state have same parity so it is not allowed.

27. $$\frac{I_{max}}{I_{min}} = \frac{(\sqrt{I_1}+\sqrt{I_2})^2}{(\sqrt{I_1}-\sqrt{I_2})^2} = \left(\frac{\sqrt{4I_0}+\sqrt{I_0}}{\sqrt{4I_0}-\sqrt{I_0}}\right)^2 = 9$$

28. For ground state $n = 1$

For next leading order i.e. 1st excited state

$$n = 2$$

So $$(E_0)^n = (E_0)^2$$

29. When an X-ray beam incident on a metal sheet or slab, the intensity of transmitted beam is given by

$$I = I_0 e^{-\mu x}$$

where,

μ = linear absorption coefficient

and x is the thickness of metal sheet or slab

I_0 (intensity of x-rays)

Then $$\frac{I}{I_0} = e^{-\mu x}$$

$$e^{\mu x} = \frac{I_0}{I}$$

$$\mu x = \ln\left(\frac{I_0}{I}\right)$$

$$\mu = \frac{1}{x}\ln\left(\frac{I_0}{I}\right)$$

$$= \frac{1}{2\times10^{-3}}\ln\left(\frac{1}{.025}\right)$$

$$= \frac{1}{2\times10^{-3}}\ln(40)$$

$$\mu = 1844.439$$

30. Electron spin resonance (ESR) falls on microwave frequency region of Electromagnetic spectrum. Nuclear magnetic resonance (NMR) falls on radi frequency region of E.M. spectrum.

Transition between vibrational state of a molecule lie in far-infrared and IR region.

And Electronic transition occur in ultraviolet and visible frequency range.

So option (*d*) is correct.

31. In Raman spectroscopy, lines having wave length greater then that of incident wavelength are called stokes line and those having shorter wavelength, antistokes lines.

i.e.

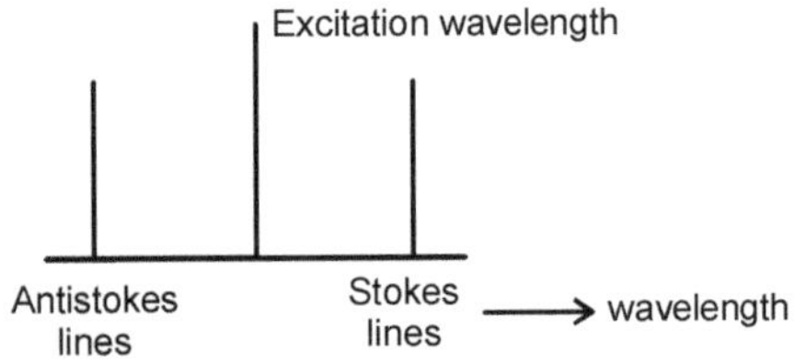

these Antistokes and stokes lines are symmetrically displaced about parent lines.

i.e., stokes line observed at = 552 nm

then Antistokes line would be at = 540 nm

and wavenumber equivalent to this

$$\bar{v} = \frac{1}{\lambda} = \frac{1}{540\text{nm}}$$

$$= .0018518 \times 10^{+9} \text{ m}$$

$$= 18518.519 \ (\text{cm}^{-1})$$

32. In the presence of weak magnetic field the permitted transition from

$2p_{3/2} \to 2s_{1/2}$ is as

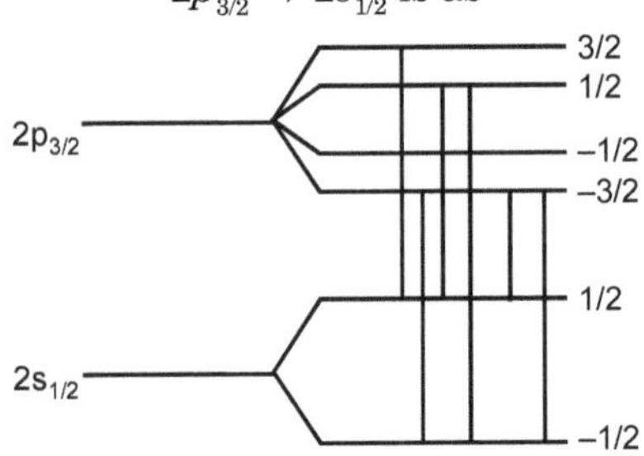

as $\Delta M_J = 0, \ \pm 1$ (selection rule)

but $M_J = 0 \not\leftrightarrow M_J = 0$ if $\Delta J = 0$

therefore permitted transition is 6.

33. This is singlet transition.

$(2s + 1)$

$= 2 \times 1 + 1$

$= 3$

34. Saturation magnetization,

$M_S = 0.60 \ \mu_B N$

$= 0.60 \times 9.21 \times 10^{-21} \times 8 \times 10^{23} \times 10^6$

$= 44.208 \times 10^8$ A/m

$M_S = 4.4208 \times 10^9$ A/m

35.
$$E_A = \frac{\hbar^2}{2I_A} J_A(J_A + 1)$$

$$E_B = \frac{\hbar^2}{2I_B} J_B(J_B + 1)$$

Given $\quad I_A = 6I_B$ and $E_A = E_B$

$$\frac{\hbar^2}{2}\frac{J_A(J_A+1)}{I_A} = \frac{\hbar^2}{2}\frac{J_B(J_B+1)}{I_B}$$

$$J_A(J_A + 1) = 6 \ J_B(J_B + 1)$$

Now from options,

If we put $\quad J_A = 3$ and $J_B = 1$ then

$3 \times 4 = 6 \times 2$

$\therefore \quad$ L.H.S. = R.H.S.

Hence $\quad J_A = 3$ and $J_B = 1$

36. $I = \frac{1}{2} C \varepsilon_0 E_0^2$

$$E_0^2 = \frac{2I_0}{C\varepsilon_0} = \frac{2 \times 150 \times 10^{-3}}{3 \times 10^8 \times 8.854 \times 10^{-12}}$$

$$= \frac{2 \times 150 \times 10^{-3} \times 10^{12-8}}{3 \times 8.854} = \frac{300 \times 10}{3 \times 8.854} = \frac{1000}{8.854}$$

$E_0^2 = 112.943 = (10.6)^2$

$E_0 = 10.6$ V/m

37.
$$\frac{P_2}{P_1} = \exp\left(-\frac{(E_2 - E_1)}{K_B T}\right) = \exp\left(-\frac{hc}{\lambda K_B T}\right)$$

$$\therefore \frac{P_2}{P_1} = \exp\left[-\frac{\{6.626 \times 10^{-34} \times 3 \times 10^8\}}{3122 \times 10^{-10} \times 1.38 \times 10^{-23} \times 5000}\right]$$

$$\left[\because 1A^0 = 10^{-20} \text{m}\right]$$

$$\therefore \frac{P_2}{P_1} = \exp\left[-\frac{6.626 \times 3}{3.122 \times 1.38 \times 5} \times 10^{-34+8+10+23-6}\right]$$

$$\therefore \frac{P_2}{P_1} = \exp\left[-\frac{6.626 \times 3 \times 10}{3.122 \times 1.38 \times 5}\right]$$

$$\therefore \frac{P_2}{P_1} = \exp\left[-\frac{39.756}{3.122 \times 1.38}\right]$$

$$\therefore \frac{P_2}{P_1} = \exp(-9.23)$$

$$\therefore \frac{P_2}{P_1} = 9.83 \times 10^{-5}$$

38. The numbers of Zeeman Component for

${}^2D_{\frac{5}{2}} \to {}^2P_{\frac{3}{2}}$ transition = 12

The numbers of Zeeman components for

${}^2D_{\frac{3}{2}} \to {}^2P_{\frac{3}{2}}$ transition = 10.

The numbers of Zeeman components for

$^2D_{\frac{3}{2}} \to {}^2P_{\frac{1}{2}}$ transition = 6.

39. Separation of Zeeman components,

$$\Delta\nu = \frac{eB}{4\pi m}$$

$$\Rightarrow \Delta\lambda = \frac{\lambda^2}{C}\Delta\nu = \frac{\lambda^2}{C}\frac{eB}{4\pi m}$$

$$B = \frac{4\pi m C}{e}\frac{\Delta\lambda}{\lambda^2}$$

$$= \frac{4\times3.14\times9.1\times10^{-31}\times3\times10^{8}}{1.6\times10^{-19}}\times\frac{0.01\times10^{-9}}{\left(643.8\times10^{-9}\right)^2}$$

$= 0.514T$

40. The energy of the quantum H . O is

$$E = h\nu\left(n+\frac{1}{2}\right), n = 0,1,2,...$$

The frequency of oscillation is $\nu = \frac{1}{2\pi}\sqrt{\frac{K}{\mu}}$

where K = spring constant and
μ = reduced mass
The energy levels are equally spaced with energy

separation of $\Delta E = h\nu = \frac{h}{2\pi}\sqrt{\frac{K}{\mu}}$

$$K = \left(\frac{2\pi}{h}\Delta E\right)^2 \mu$$

$$= \left[\frac{2\times3.14}{6.626\times10^{-34}}\times8.44\times10^{-2}\times1.6\times10^{-19}\right]^2 \times 1.14\times10^{-26}$$

= 186.7 N/m.

43. For 25V $^1/_2$ state

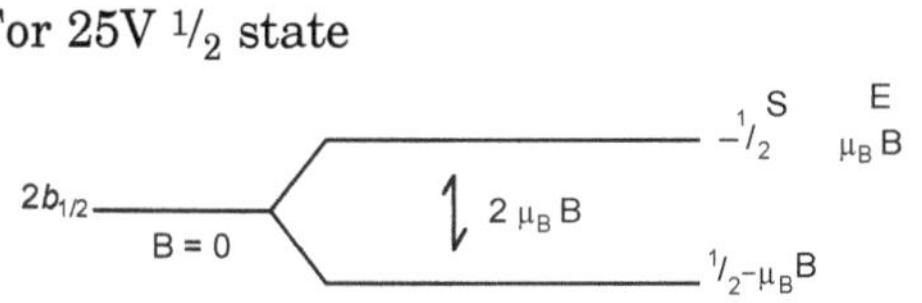

$$E = -g\mu B\,\vec{j}\cdot\vec{B}$$

$$= -2\,\mu g\left(-\frac{1}{2}\right)B$$

$$= \mu_B\,B \text{ for } \left(j = -\frac{1}{2}\right)$$

Again $E^1 = -g\mu_B\,\vec{j},\vec{B} = -2\mu_B\left(\frac{1}{2}\right)^B$

$$= \mu_B\,B\left(\text{for } \hat{j} = \frac{1}{2}\right)$$

Hence difference in energy levels.

$$= \mu_B\,B - (-\mu_B\,B) = -2\mu_B\,B$$

44. For the first Stokes line of rotational Raman spectrum 6B = 12.96

or $B = 2.16\ cm^{-1}$

46. Given electronic configuration $2S^1\,3d^1$

$$|L - S| \leq j \leq |L + S|$$

For $2S^1$ $L_1 = 0,\ S_1 = \frac{1}{2},\ j_1 = \frac{1}{2}$

for $3d^1,\ L_2 = 2,\ S_2 = \frac{1}{2}$

$\Rightarrow$ $j_2 = \frac{3}{2}, \frac{5}{2}$

Now, $\left|\hat{j}_1 - \hat{j}_2\right| \leq \hat{j} \leq \left|\hat{j}_1 + \hat{j}_2\right|$

For $\hat{j}_1 = \frac{1}{2}$ $\hat{j}_2 = \frac{3}{2}$ $\hat{j} = 1,2$

For $\hat{j}_1 = \frac{1}{2}$ $\hat{j}_2 = \frac{5}{2}$ $\Rightarrow \hat{j} = 2,3$

$\therefore$ $\left[\hat{j}_1, \hat{j}_2\right] = \left[\frac{1}{2}, \frac{3}{2}\right], \left[\frac{1}{2}, \frac{5}{2}\right] 2,3$

47. $W_e\,(1 - 2X_e) = 2990\ cm^{-1}$

and $\frac{2W_e(1-3X_e)}{W_e(1-2X_e)} = 1.96$

$$\Rightarrow \frac{1-3X_e}{1-2X_e} = 0.98$$

$$\Rightarrow \boxed{X_e = 0.02}$$

50. The selection rules for the electic-dipole transitions in multi electron atoms are closely similar to the selection rule for the an electron atom.

(*i*) Most of the transitions occur in which one electron jumps at a time and such that its *l* value change by one unit i.e.,

$$\Delta l = \pm 1$$

According to the laporte rule the parity of the configuration.

Must change in an electric-diapole transition.

(*ii*) There is no restriction on the total quantum numbers n of either electron.

(*iii*) For the atom as a whole, the quantum numbers L.S., and J must change as follows :

$$\Delta L = 0, \pm 1 \text{ (In one electron atoms)}$$

$\Delta L = 0$ is not allowed

$\Delta S = 0$

$\Delta J = 0, \pm 1$ but $J = 0 \not\leftrightarrow J = 0$

51. The product of the uncertainty Δx in the position of a body at some instant and uncertainty ΔP in its momentum on same instant is equal to or greater then

$$\Delta x \, \Delta P > h$$

The uncertainity relation $\Delta x \, \Delta P \geq h$ can also be written in terms of the congugate pair of quantities energy E and time t. We know that the kinetic energy of a particle is

$$E = \frac{1}{2} mV^2 = \frac{P^2}{2m}$$

where P is the linear momentum (P = mV)

$$\text{Thus,} \quad \Delta F = \frac{P\Delta P}{m}$$

$$\text{But,} \quad P = mV = m\frac{\Delta x}{\Delta t}$$

$$\therefore \quad \Delta E = \frac{\Delta x}{\Delta t}\Delta P$$

$$\text{or,} \quad \Delta E \, \Delta t = \Delta x \Delta P$$

$$\Delta x \, \Delta P > h$$

We can write $\Delta E \, \Delta t \Rightarrow \dfrac{h}{2\pi}$

Since $\Delta x \, \Delta P \Rightarrow h$

We can write $\Delta E \, \Delta t = \dfrac{h}{2\pi}$

In question $\Delta t = 10^{-9}$ S

$$\text{Hence} \quad \Delta E = \frac{6.6 \times 10^{-34}}{2 \times 3.14 \times 10^{-19}} J$$

$$= \frac{6.6 \times 10^{-34}}{6.28 \times 10^{-9} \times 1.6 \times 10^{-19}}$$

$$\approx 0.65 \times 10^{-6} eV \approx 10^{-6} eV$$

52. The electronic configuration of N-atom is $1s^2, 2s^2, 2p^2, 3d^1$,

For d-shell $l = 2$, $s = \pm\dfrac{1}{2}$, $j = l \pm \Delta$

$$= 2 \pm \frac{1}{2} = \frac{5}{2}, \pm\frac{3}{2}$$

Hence degeneracy of $\dfrac{5}{2}$ state is

$$(2j + 1) = 2 \times \frac{5}{2} + 1 = 6$$

Degeneracy of $\dfrac{3}{2}$ state is

$$\left(2 \times \frac{3}{2} + 1\right) = 4$$

Thus total degeneracy = 6 + 4 = 10

53. To determine the possible energies of rotation of the molecule. We have to solve the schrodingers equation of a rigid rotator which is

$$\nabla^2\psi + \frac{8\pi^2 \mu m}{h^2} E\psi = 0$$

The potential energy term V has been taken zero because r is fixed, μ is the reduced mass of diatomic molecule, after getting solution, we find that the eigen functions ψ are single valued finite and continous only for certain values of E, given by

$$E = \frac{h^2}{8\pi^2 l} J(J+1)$$

Where J is called rotational quantum number which can take the integral value and $l (= \mu r^2)$ is the moment of inertia of equivalent single point mass.

From classical mechanics, we have

$$E = \frac{1}{2} I\omega^2$$

$$\text{and} \quad L = I\omega, \; L = \sqrt{2EI}$$

From quantum mechanics the angular momentum L of the system in the quantum state J is given by

$$L = \hbar\sqrt{J(J+1)}$$

$$\text{Hence} \quad \omega = \frac{L}{l} = \frac{\hbar}{l}\sqrt{J(J+1)}$$

$$\text{or} \quad V_{rot} = \frac{\hbar}{4\pi^2 l}\sqrt{J(J+1)}$$

In terms of wave no. the energy equation can be written as

$$F(J) = \frac{E}{hC} = \frac{h}{8\pi^2 I_C} J(J+1) = BJ(J+1)$$

Where $B = \dfrac{h}{8\pi^2 l_C}$ is known as rotational constant.

Substituting J = 0, 1, 2, ... we get

$$F(J) = 0, 2B, 6B, 12B$$

$$V = F(J') - F(J'')$$

$$= BJ'(J'+1) - BJ''(J''+1)$$

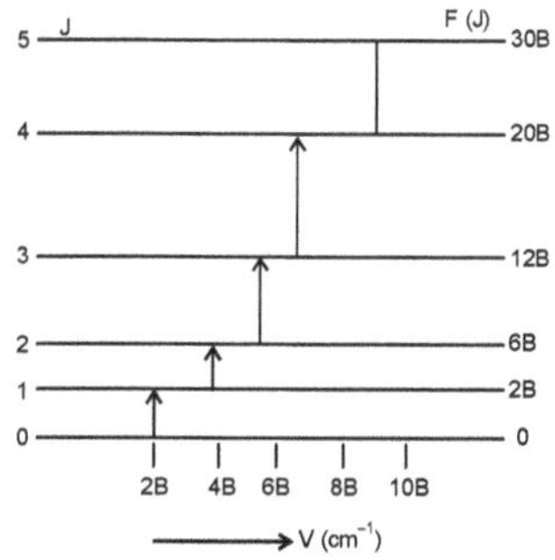

For absorption

The rotation transitions follow the following selection rule.

$\Delta J = \pm 1$

$\therefore \quad J' > J''$

$\Rightarrow \quad J' = J'' + 1$

So, $\quad V = 2B(J'' + 1)$

For simplicity we put J' = J and we have substituting

$V = 2B(J + 1)$

$J = 0, 1, 2$

$V = 2B, 4B, 6B, 8B$

Thus the absorption spectrum of a rigid rotator is expected to consist of a series of equidistant lines with constent separation 2B, provided transitions start from various energy levels J = 0, 1, 2, 3, ... as shown

According to question

$2B = 20\ cm^{-1}$

$\Rightarrow \quad B = 10\ cm^{-1}$

From the solution of we know that the distence of first stoke lines from Rayleigh line is 6B.

$6 \times 10 = 60\ cm^{-1}$

55. In the two lair population inversion is possible to achieve at any power level. The maximum possible situation can be.

$$Ng \approx Ne \approx \frac{N}{2}$$

58. CH_3OH is an example of symmetric rotator, where $I_x = I_y \neq I_z$, $(I_x + I_y = I)$ and $(I_z \neq I)$

The classical expression for energy is

$$E = \frac{1}{2I}\left(J_x^2 + J_y^2\right) + \frac{1}{2I_z}J_z^2.$$

This can be expressed in terms of,

$J^2 = J_x^2 + J_y^2 + J_z^2$ by adding and subtracting J_z^2

$$E = \frac{1}{2I}J^2 + \left(\frac{1}{2I_z} - \frac{1}{2I}\right)J_z^2.$$

Quantum Mechanically,

$$E = \frac{\hbar^2}{2I}J(J+1) + \frac{\hbar^2 m_J^2}{2}\left(\frac{1}{I_2} - \frac{1}{I}\right)$$

67. Last two terms $mn = 3ds, 4s^2$

$Mn^{4+} = 3d^1$ for d subshell

$l = 2, s = \pm \frac{1}{2}$

$j = 2 \pm \frac{1}{2} = \frac{5}{2}, \frac{3}{2}$

$2j + 1 = 2 \times \frac{3}{2} + 1 = 4$

$3d^1$ electron lie in $\frac{3}{2}$ angular momentum

69. $\mu_B = \mu_e - \mu_P \qquad (\because \mu_e = \mu_P)$

$\mu_B = 0$

Now angular momentum and spin momentum of electron and proton are equal and opposite.

$l = [l_P + s_P + (l_e + s_e)]$

$l = l_P + s_P + l_e + s_e$

$l_P = -l_P$ and $s_P = -s_e$

$l = 0$

70. $\bar{v} = F(\sigma' -) - F(J'')$

$= B(J')(J'+1) - BJ''(J''+1)$

$J' > J''$

$v = 2B(\sigma + 1)$

Substituting J = 0, 1, 2

$\overline{V} = 2B, 4B, 6B, 8B$

$$B = \frac{h}{8\pi^2 \theta C}$$

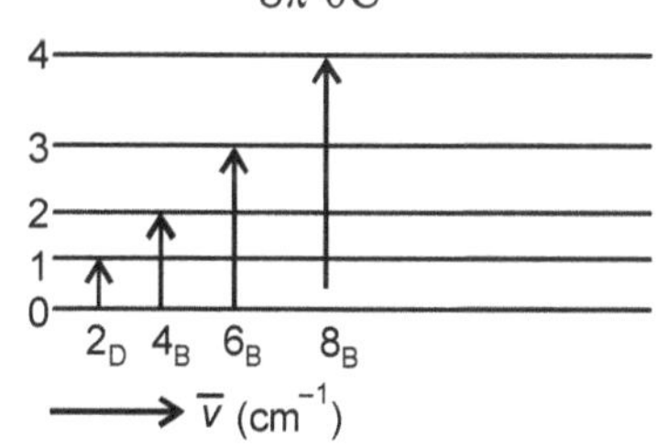

$2B = (77.130 - 64.275)\ cm^{-1}$

$= (89.985 - 77.130)\ cm^{-1}$

$$12.855\ cm^{-1} = \frac{2h}{8\pi^2 l_B C}$$

$$l_B = \frac{2h}{8\pi^2 12.855\, hC}\ g\ cm^2$$

and $\quad I_A = 0$

73. O–C–O have no permanent dipole moment. Hence, it does not show absorption lines but O–C–S has permanent dipole moment. Hence it will show absorption lines.

$$\vec{A} = \frac{2\pi(\vec{b}\times\vec{c})}{\vec{a},\vec{b}\times\vec{c}}$$

$$\vec{A} = \frac{2\pi}{\frac{a^3}{2}}\begin{vmatrix} \hat{x} & \hat{y} & \hat{z} \\ \frac{-9}{2} & \frac{a}{2} & \frac{a}{2} \\ \frac{a}{2} & \frac{a}{2} & \frac{a}{2} \end{vmatrix}$$

$$= \frac{4\pi}{a^3}\left[\hat{x}\left(\frac{a^2}{4}+\frac{a^2}{4}\right)+\hat{y}\left(\frac{a^2}{4}+\frac{a^2}{4}\right)\right]$$

$$= \frac{4\pi}{a^3}\times\frac{a^2}{2}(\hat{x}+\hat{y})$$

$$\vec{A} = \frac{2\pi}{a}(\hat{x}+\hat{y})$$

$$\vec{B} = 2\pi\frac{\vec{c},\vec{a}}{\vec{a},\vec{b}\times\vec{c}}$$

$$\Rightarrow \frac{4\pi}{a^3}\begin{vmatrix} \hat{x} & \hat{y} & \hat{z} \\ \frac{a}{2} & \frac{-a}{2} & \frac{a}{2} \\ \frac{a}{2} & \frac{a}{2} & \frac{a}{2} \end{vmatrix}$$

$$= \frac{4\pi}{a^3}\left(\hat{x}\left(\frac{a^2}{4}-\frac{a^2}{4}\right)-\hat{y}\left(\frac{-a^2}{4}-\frac{a^2}{4}\right)+\hat{z}\left(\frac{a^2}{4}+\frac{a^2}{4}\right)\right)$$

$$\Rightarrow \frac{4\pi}{a^3}\begin{bmatrix} \hat{x} & \hat{y} & \hat{z} \\ \frac{a}{2} & \frac{a}{2} & \frac{-a}{2} \\ \frac{-a}{2} & \frac{a}{2} & \frac{a}{2} \end{bmatrix}$$

$$= \left(\frac{a^2}{4}+\frac{a^2}{4}\right)\hat{x}+\hat{y}\left(\frac{a^2}{4}-\frac{a^2}{4}\right)+\hat{z}\left(\frac{a^2}{4}+\frac{a^2}{4}\right)$$

$$= \frac{4\pi}{a^3}\left(\frac{a^2}{2}+\frac{a^2}{2}\hat{z}\right) = \frac{2\pi}{a}(\hat{x}+\hat{z})$$

74. $$\mu = \frac{c}{v},\ v = \frac{c}{\mu}$$

$$t = \frac{L}{v}, \frac{L\mu}{c} \text{ and frequency is}$$

$$\upsilon = \frac{1}{t} = \frac{c}{L\mu}$$

When $\mu \to \mu + \Delta\mu$, $\Delta \to \upsilon^1$

$$\upsilon^1 = \frac{c}{\mu+\Delta\mu}$$

$$\therefore\quad t^1 = \frac{L}{v^1} = \frac{L(\mu\pm\Delta\mu)}{c}$$

$$\upsilon^1 = \frac{c}{L(\mu\pm\Delta\mu)}$$

$$\Delta\upsilon = v - v' = \frac{c}{L}\left[\frac{1}{\mu}-\frac{1}{\mu+\Delta\mu}\right] = \frac{c}{L}\left[\frac{\Delta\mu}{\mu(\mu+\Delta\mu)}\right]$$

75. $3^2\,P_{1/2} \to 3^2\,S_{1/2}$

The levels for $P_{1/2} = 2(^1/_2) + 1 = 2$

The levels for $S_{1/2} = 2(^1/_2 + 1) = 2$

Hence this transition breaks up in 4-states.

The transition $3^2\,P_{1/2} \to 3^2\,S_{1/2}$

$$\text{is break up} = \left\{2\left(\frac{3}{2}\right)+1\right\}+\left\{2\left(\frac{1}{2}\right)+1\right\}$$

$$= 4 + 2 = 6$$

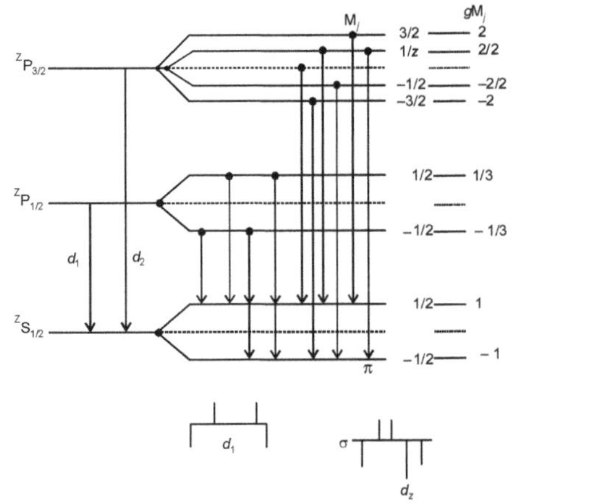

Applying the selection rules ΔJM_j = 0, ±1. We obtain four Zeeman Components for the d_1 line and six components for d_2 line. The components correspondings to ΔM_j = 0 are polarised with electric vector perpendicular to the field (σ-components)

76. In He-Ne laser, the laser to transition takes place in Ne-only.

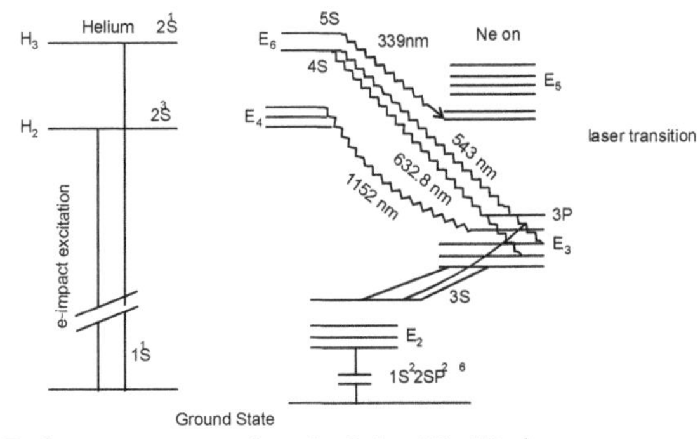

Relevant energy level of the He-Ne laser.

77. The spectral terms for equivalent electrons like He ($2p^2$) can be obtained by briet scheme. In this scheme we write all the possible values of M_L which can be formed by the combination of Ml_1 and Ml_2. For this we write the values Ml_1 and Ml_2 in a row and in a column respectively. The sums M_L are written below Ml_1 and to the left of Ml_2. These nine values of M_L from three sets divided by the L-shaped (dotted) lines.

m_1	1	0	–1	
M_L =	2	1	0	1
M_L =	1	0	– 1	0
M_L =	0	– 1	– 2	– 1
	S	P	d	m_2

(Two equivalent Electrons)
$l_1 = 1; l_2 = 1$

These sets are

2	1	0	– 1	–2	(I set)
	1	0	–1		(II set)
		0			(III set)

These sets of M_L-values correspond to L = 2, 1 and 0 respectively *i.e.* to one d, one p and one s term. The spin of two electrons can be combined to form either s = 0 (singlets) or S = 1 (triplets). For S = 1 both electrons have the same spin quantum numbers M_s. and hence they must differ in this values of Ml, therefore. We cannot combine any table with S = 1 (because diagonal corresponds to equal values of Ml_1 and Ml_2) thus with S =1 (triplet) we are limited to the following M_L values

1 0 –1 (II set)

Which are the components of a term with L = 1. This corresponds to a 3P term or a $3P_{0,1,2}$ multiplet. when S = 0, the electron differ in their spin quantum numbers and there is no restriction on the value of M_L which may be combined with this value at S. We have only the remaining I and III sets to combine with S = 0 these sets are components at terms with L = 2 and L = 0 respectively. Hence, they corresponds to d and S terms.

Thus two equivalent p-electrons give ${}^1s_1\,{}^1d_1$ terms or ${}^1s_0$, ${}^1d_2$ and ${}^3p_{0,1,2}$ multiplets.

79. Using the obtuse angled triangle formed by $\vec{L},\vec{S}$ and $\vec{J}$ we have by cosine Law.

$$\cos(\vec{L}\cdot\vec{S}) = \frac{L(L+1)+S(S+1)-J(J+1)}{2\sqrt{L(L+1)\times S(S+1)}}$$

Hence,

$$\cos(\vec{L}\cdot\vec{S}) = \frac{2(2+1)+1(1+1)-3(3+1)}{2\sqrt{2(2+1)\times1(1+1)}}$$

$$= \frac{6+2-12}{2\sqrt{12}} = \frac{-4}{2\times2\sqrt{3}} = -\frac{1}{\sqrt{3}}$$

$$(\vec{L}\cdot\vec{S}) = \cos^{-1}\left(\frac{-1}{\sqrt{3}}\right)$$

81. For Na ($3^2\,P_{3/2}$) with nuclear spin $l = 3/2$

$J = 3/2, l = 3/2$

$F = 3, 2, 1, 0$

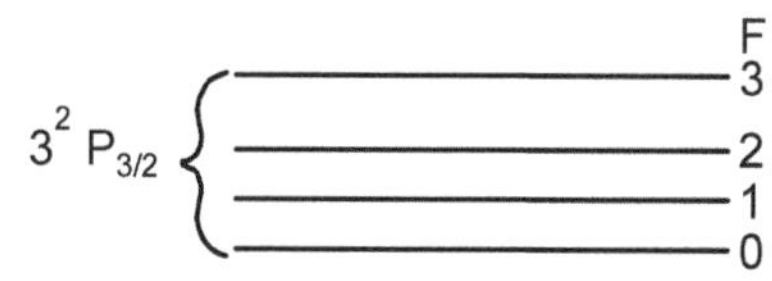

This term is splitted into four hyperfine structure components as shown in adjacent figure.

82. Wave number of emitted/absorbed radiation is given as

$$\vec{V} = 2B(J+1)$$

where J represent the rotational quantum number of lower level (in case at absorption) and

$B = \dfrac{h}{8\pi^2 lc}$ is known as rotational constant

For the reduced mass μ and transition $J = 4 \to 5$, the wave number v is

$$\vec{v} = 2\left(\frac{h}{8\pi^2\mu r^2 c}\right)(4+1) = 10B(l = \mu r^2)$$

$$\vec{v} = 2\left(\frac{h}{8\pi^2 1.05\mu r^2 c}\right)(4+1)$$

$$= \frac{10B}{1.05} = 9.524B.$$

Hence, the shift in spectral line, corresponding to a transition due to replacing of heavier isotopes is

$$\overline{V} - \overline{V}' = 10B - 9.524B$$

$$= 0.476B.$$

84. The interaction energy is given by

$$\Delta E = \frac{eh}{4\pi m} gB(\Delta M_j)$$

here $B = H_0$ and $m = m_p$ and $g = g_1$

$$\Delta E = \frac{eh}{4\pi m_p} g_1 H_0 \times 1 \quad \because \Delta M_j = 1$$

Hence, resonance frequency

$$v = \frac{\Delta E}{h} = \frac{eg_1 H_0}{4\pi m_p}$$

85. For Pauli matrices, σ_x, σ_y, σ_z

$$\sigma_i, \sigma_j = -\sigma_j \cdot \sigma_i$$

$$\begin{bmatrix}1 & 0\\0 & -1\end{bmatrix}\begin{bmatrix}0 & 1\\1 & 0\end{bmatrix} = \begin{bmatrix}0 & 1\\-1 & 0\end{bmatrix}$$

$$= -\begin{bmatrix}0 & 1\\1 & 0\end{bmatrix}\begin{bmatrix}1 & 0\\0 & -1\end{bmatrix}$$

86. For X-ray tube. $eV = \frac{hc}{\lambda}$

Thus, $\lambda = \frac{hc}{eV} \propto \frac{1}{V}$

87. The principle series of lines in the atomic lithum is for $nP \longrightarrow 2S$, $n > 2$ transitions

88. Rotational energy,

$$E = \frac{\hbar^2}{2I}(\gamma+1)\gamma$$

$$\Delta E = \frac{\hbar^2}{2I}[(\gamma+1)\gamma - \gamma(\gamma-1)] = \frac{\hbar^2}{I}\gamma$$

$$\Delta E = h_c\Delta\bar{v},$$

Here, $\Delta\bar{v} = 0.7143.\ cm^{-1} = 71.43$ wave number drift.

$$I \simeq \frac{\hbar^2}{hc\Delta v} \simeq 7.8\times10^{-46}$$

89. $6s \to 5p$ transition has nothing to do with.

He-Ne LASER action.

90. Linear stark effect,

$n = 2,\quad l = 0, \pm 1$

Level is splitted into 3 levels

nth level is splitted into $(2n-1)$ levels in Linear Stark effect.

91. $F = J + I$

I – nuclearspin angular momentum

J – electronic spin angular momentum

Dipole selection rules,

$\Delta J = 0, \pm 1$; $\Delta I = \pm 1$; $\Delta F = 0, \pm 1$

92. $(\bar{v} - E_{ex})$

94. Effective mass, $m^* = h^2\left(\frac{d^2E}{dR^2}\right)^{-1}_{R_0}$

Since, $E(k) = Ak^2 + Bk^4$

$\therefore m^* = \frac{h^2}{2A + 12Bk_0^2}$

95. $2d_{hkl}\sin\theta = n\lambda$; $d_0 = d_{khl}\sqrt{h^2+k^2+l^2}$

$(002) \to (001)$ plane $= \frac{\lambda}{2\sin\theta}\sqrt{h^2+k^2+l^2}$

$$d_0 = \frac{1.54}{2\sin 60^\circ}\sqrt{0^2+0^2+1^2}$$

98. $\Delta E.\Delta t \sim h$

$$E = \frac{hc}{\lambda}$$

$$\Delta E = \frac{hc}{\lambda}\frac{\Delta\lambda}{\lambda}$$

$$\therefore \frac{hc}{\lambda}\frac{\Delta\lambda}{\lambda}.\Delta t \sim h$$

Now, $c = \nu\lambda$

$\therefore \nu\Delta\lambda + \lambda\Delta\nu = 0$

$\Rightarrow \nu\Delta\lambda = -\lambda\Delta\nu$

$\therefore \frac{\Delta\lambda}{\lambda} = -\frac{\Delta\nu}{\nu}$

$\therefore \frac{ch}{\lambda}\frac{\Delta\nu}{\nu}\Delta t \sim h$

$$\frac{\Delta\nu}{\nu} \sim \frac{h}{\Delta t}\frac{\lambda}{hc} \sim \frac{\lambda}{\Delta tc}$$

$$= \frac{600\times10^{-9}}{8\times10^{-9}\times3\times10^{8}}$$

$\therefore \frac{\Delta\nu}{\nu} \sim 10^{-6}$

99.

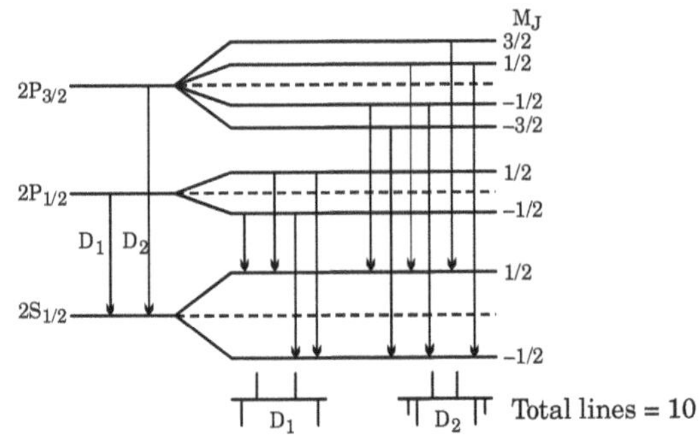

Selection Rule $DM_J = 0, \pm 1$

100. Laser emission involves population inversion.

102. $\Delta E_n = h\nu_0 =$ constant

$\Delta E_j = \frac{h}{2\pi I}(J+1)$ or, $\Delta E_j \propto J$

105a. We know, $V = \frac{M}{\rho}$

where M is mass and ρ is density.

$\therefore V = \frac{207.2}{11.35}$

$\therefore$ Number of atoms/cm^3

$= \frac{N_o}{V}$ (Avagadro number)

$= \frac{6.023\times10^{23}}{207.2}\times11.35$

$= 3.3\times10^{22}$

105b. We know, $n = N\exp\left(\frac{-E_V}{K_BT}\right)$

when, N = Total number of atoms/ cm^3 in crystal.

n = Number of vacancies

E_v = Energy of vacancy formation

T = Temprature (in Kelvin)

K_B = Boltzmann constant

$$\therefore\ n = 3.3 \times 10^{22} \exp\left(\frac{-0.55}{8.02 \times 10^{-5} \times 500}\right)$$
$$= 3.3 \times 10^{22} \exp(-13.7)$$
$$\cong 10^{-6} \times 3.3 \times 10^{22} \cong 3.3 \times 10^{16}$$

106. ^5S because the term which have largest S value is lowest energy term

In $^5S \Rightarrow (2J + 1) = 5 \Rightarrow J = 2$

and $3_S \Rightarrow (2J + 1) = 3 \Rightarrow J = 1$

so that $5_{S(E)} < 3_{S(E)}$.

107. $\because$ 3_F multiplicity $(2S + 1) = 3 \Rightarrow S = 1$

For F, the value L = 3

and $J = |L - S|, \ldots(L + S)$

$\Rightarrow J = 2, 3, 4$

$\Rightarrow$ $^3F_2$, $^3F_3$ and $^3F_4$

Each J level is still (2J + 1) fold degenerate

$J = 2, \quad (2J + 1) = 5$

$J = 3, \quad (2J + 1) = 7$

$J = 4, \quad (2J + 1) = 9.$

Total degeneracy of ^3F term=21

108. Since $^3D_3 = {}^MD_J$

Here, $(2S + 1) = 3$, $S = 1$ and $J = 3$

For D term, L = 2

$$g = 1 + \frac{J(J+1) + S(S+1) - L(L+1)}{2J(J+1)}$$
$$= 1 + \frac{3(3+1) + 1(1+1) - 2(2+1)}{2 \times 3(3+1)} = \frac{4}{3}.$$

109. Infrared spectra.

110. $B(Z = 5) = 1s^2\, 2s^2\, 2p^1$

Term $^2P_{1/2}$ (ground) and $^2P_{3/2}$

So one e^- in the $^2P_{1/2}$ state means

$(2S + 1) = 2$

$\Rightarrow S = \frac{1}{2}$, for P orbit, L = 1

$J = \frac{1}{2}$.

111. ^3P term multiplicity $(2S + 1) = 3$, $\Rightarrow S = 1$

For P orbit $L = 1$, $J = |L - S|, \ldots\ldots(L + S)$

$J = 0, 1, 2$, so $^3P_0$, $^3P_1$, $^3P_2$

Degeneracy (2J + 1) for $\left.\begin{array}{l} J = 0 \Rightarrow 1 \\ J = 1 \Rightarrow 3 \\ J = 2 \Rightarrow 5 \end{array}\right\}$

112.

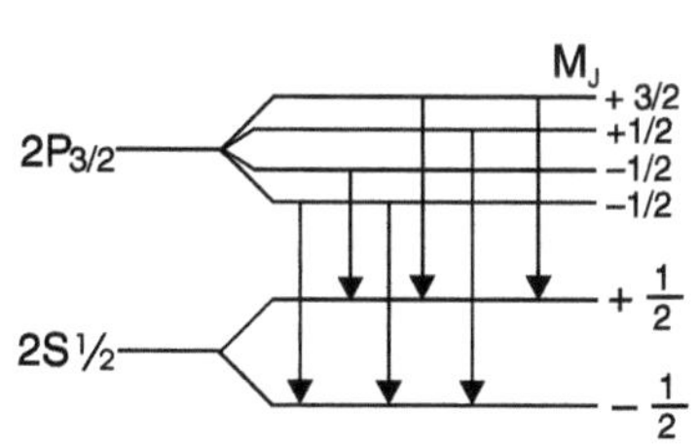

Selection rule.

$$\Delta M_J = 0, \pm 1$$

Total 6 transitions.

113. $$\Delta\bar{v} = \frac{\hbar}{2\pi Ic} \qquad [\because \quad I = mR^2]$$
$$\Delta\bar{v} \propto \frac{1}{I} \propto \frac{1}{R^2}.$$

114. The wave number corresponding to incident line

$$\bar{\nu} = \frac{1}{1.5\,\mu m} = 666666.66 \text{ metre}^{-1}$$

finding Raman shift

$$\lambda = \frac{c}{\nu} = \frac{3 \times 10^8}{20 \times 10^{12}}$$
$$= 1.5 \times 10^{-5} = 15\ \mu m$$

so, the wave number of Raman shift

$$\Delta\bar{v} = \frac{1}{15\,\mu m} = 66666.66 \text{ metre}^{-1}$$

Stoke linem

$$\bar{v}_{stoke} = \bar{v} - \Delta\bar{v}$$
$$= (666666.66 - 66666.66)\ m^{-1}$$
$$= 600000\ m^{-1}$$
$$\Rightarrow \quad \lambda_{st} = \frac{1}{\bar{\nu}_{st}} = \frac{1}{600000}$$
$$\lambda_{stoke} = 1.666 \times 10^{-6} = \mathbf{1.67\ \mu m.}$$

115. $TV^{r-1} = \text{constant}$

For $r = \frac{4}{3} = 1.66$,

$TV^{1/3} = \text{constant}$

$\Rightarrow T_1V_1^{1/3} = T_2V_2^{1/3}$

$$\Rightarrow \quad T_2 = T_1\left(\frac{V_1}{V_2}\right)^{1/3}$$
$$= 2000 \times \left(\frac{10}{640}\right)^{1/3}$$
$$= 2000 \times \left(\frac{1}{64}\right)^{1/3}$$
$$= 2000 \times \frac{1}{4} = \mathbf{500\ K.}$$

116. For $^4F_{9/2}$,

$$S = \frac{3}{2}, L = 3, J = \frac{9}{2}$$

$$\text{Multiplicy} = \frac{3}{2} \times 2 + 1$$
$$= 4\ (L > S)$$

117. Water shows infrared absorption.

120. $\text{Energy}_R < \text{Energy}_Q < \text{Energy}_P$

121. $V = \dfrac{hc}{e\lambda} = 15 \text{ kV}$

125. $\Delta\lambda \propto (1 - \cos\phi)$

126. For $S \neq 0$, effect is anomalous

127. $J = 5$, $\pm 5, \pm 4, \pm 3, \pm 2, \pm 1, 0$

Number of levels = 11

$J = 4$, $\pm 4, \pm 3, \pm 2, \pm 1, 0$

Number of levels = 9

$J = 5/2$, $\pm 5/2, \pm 3/2, \pm 1/2$

Number of levels = 6

$J = 3/2$, $\pm 3/2, \pm 1/2$

Number of levels = 4

128. $\theta \simeq \dfrac{\lambda}{D}$

$R = \theta \times d$

$= \dfrac{600 \times 10^{-9} \times 10 \times 10^{-2}}{20 \times 10^{-3}}$

$= 3 \times 10^{-6}\text{m}$

$= \mathbf{3\ \mu m}$

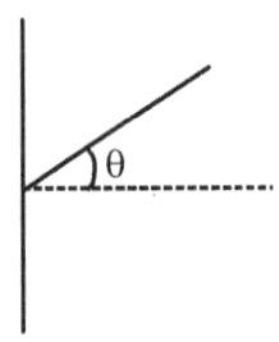

132. For K_α line, $\dfrac{1}{\lambda} \propto (Z-1)^2$

133. Hall Voltage,

$$V = \frac{BJd}{ne}$$

$$\Rightarrow \quad V = \frac{B\left(\dfrac{I}{A}\right)d}{ne} = \frac{BI}{t\,n\,e} \quad (\therefore \dot{A} = d \times t)$$

134. $2a \sin\theta = \lambda$

$$\Rightarrow \quad \sin\theta = \frac{\lambda}{2a} = \frac{0.03 \times 10^{-9}}{2 \times 0.3 \times 10^{-9}}$$

$\theta = 29°$

135. Clearly, $T = \dfrac{1}{2}$ and $I = \dfrac{1}{2}$

For $2_{p_{1/2}} \rightarrow$ The allowed value of hyper fine structure quantum number are

$F = (J + 1), |J + I - 1|, ..., |J - I|$

$= 1, 0$

136. For e^+, $L = -1$

For n and p, $L = 0$

For v, $L = 1$

137. $$g = 1 + \frac{J(J+1) + S(S+1) - L(L+1)}{2J(J+1)}$$

For $^3p_1$ $J = 1, S = 1, L = 1$

$\therefore \quad g = 1 + \dfrac{2}{4} = \mathbf{\dfrac{3}{2}}$

138. In case of Raman spectra selection rule is

$\Delta J = 0, \pm 2$

140. $\hat{S}\Psi = C_i \Psi;$

where $C_i = 1, 0, -1$

$\langle S \rangle = \sqrt{1^2 + 1^2 + 0^2}$

$= \sqrt{2}$

144. $r_1 = \dfrac{1}{2}$

$r_2 = \dfrac{1}{2}$

$\therefore \quad S = 0, 1$

$l_1 = 0, l_2 = 0$

$\therefore \quad L = 0$

For $S = 0, L = 0$; $J = 1$ and therefore term in $3S_1$

146. For hydrogen ionisation potential is 13.6 eV

Since $E_n \propto Z^2$

$\therefore$ I.P. for $Li^{++} = (3)^2 \times 13.6$

$= \mathbf{122.4\ eV}$

■■

7 CHAPTER Solid State Physics

2021

1. Choose the graph that best describes the variation of dielectric constant (ε_r) with temperature (T) in a ferroelectric material.

(T_c is the curie temperature)

(a)

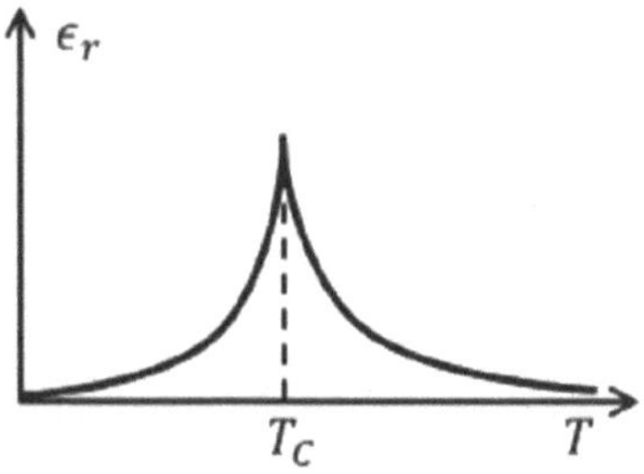

(b)

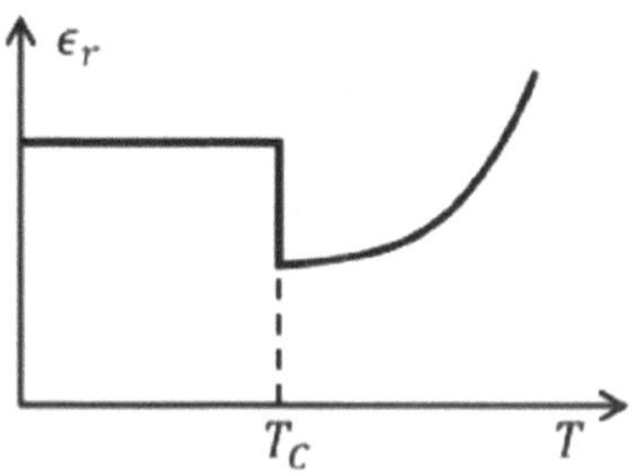

(c)

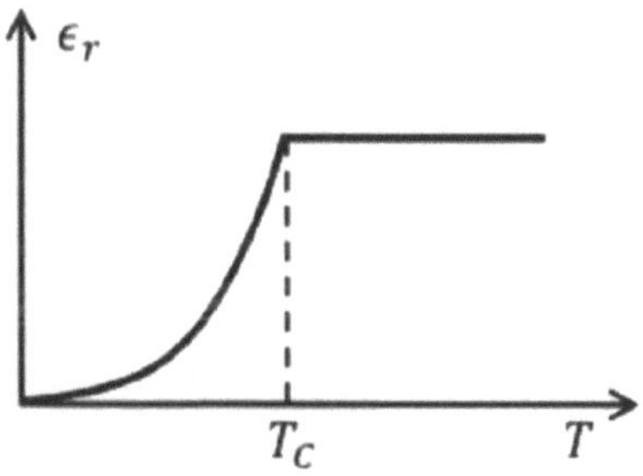

(d)

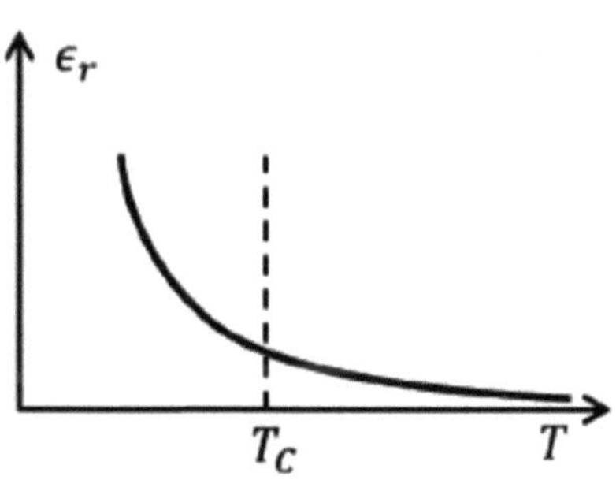

2. As shown in the figure, two metal-semiconductor junctions are formed between an n-type semiconductor S and metal M. The work functions of S and M are φ_S and φ_M, respectively with φ_M and φ_S.

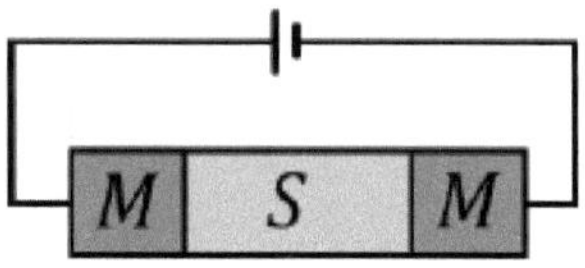

The I–V characteristics (on linear scale) of the junctions is best represented by

(a)

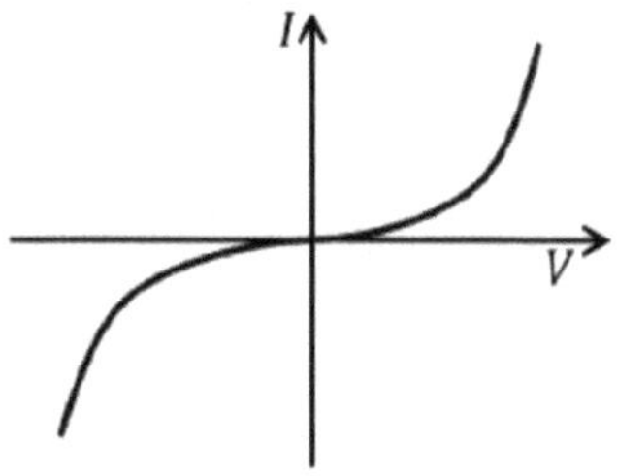

(b)

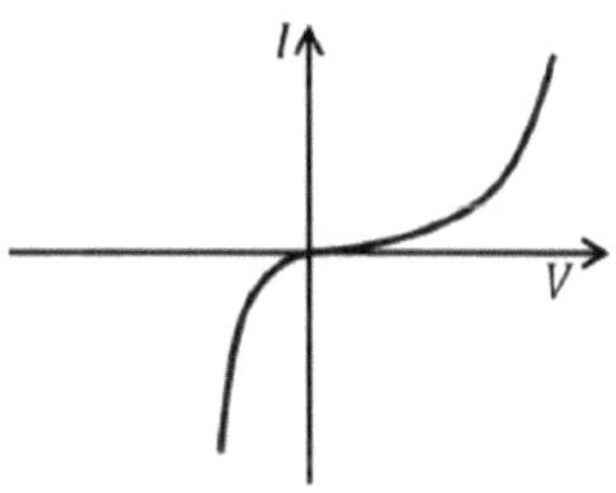

(c)

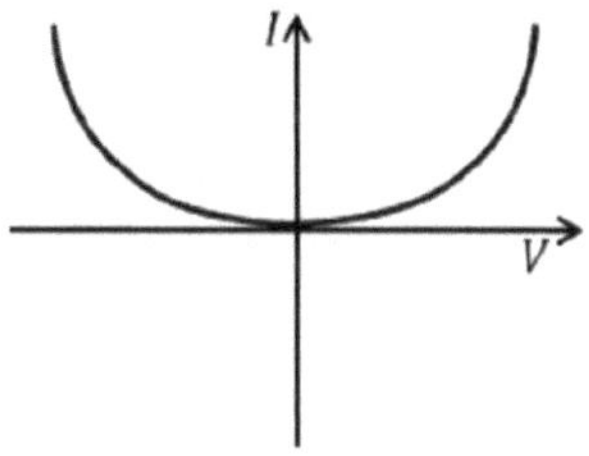

(d)

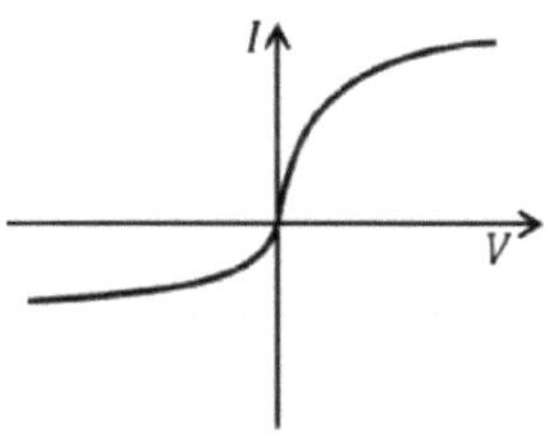

3. Choose the correct statement from the following.
(a) Silicon is a direct band gap semiconductor. '
(b) Conductivity of metals decreases with increase in temperature.
(c) Conductivity of semiconductors decreases with increase in temperature.
(d) Gallium Arsenide is an indirect band gap semiconductor.

4. A two-dimensional square lattice has lattice constant a. k represents the wavevector in reciprocal space. The coordinates (k_x, k_y) of reciprocal space where band gap(s) can occur, are

(a) (0, 0) (b) $\left(\pm\frac{\pi}{a}, \pm\frac{\pi}{a}\right)$
(c) $\left(\pm\frac{\pi}{a}, \pm\frac{\pi}{1.3a}\right)$ (d) $\left(\pm\frac{\pi}{3a}, \pm\frac{\pi}{a}\right)$

5. The donor concentration in a sample of n-type silicon is increased by a factor of 100. Assuming the sample to be non-degenerate, the shift in the Fermi level (in me V) at 300 K (rounded off to the nearest integer) is –
(Given: k_BT = 25 meV at 300 K)

6.
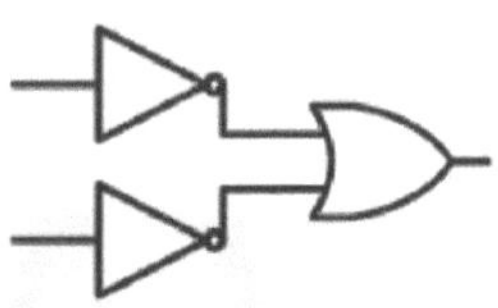

The above combination of logic gates represents the operation
(a) OR (b) NAND
(c) AND (d) NOR

7. In a semiconductor, the ratio of the effective mass of hole to electron is 2 : 11 and the ratio of average relaxation time for hole to electron is 1 : 2. The ratio of the mobility of the hole to electron is
(a) 4 : 9 (b) 4 : 11
(c) 9 : 4 (d) 11 : 4

8. As shown in the figure, inverse magnetic susceptibility ($1/\chi$) is plotted as a function of temperate (T) for three different materials in paramagnetic states.

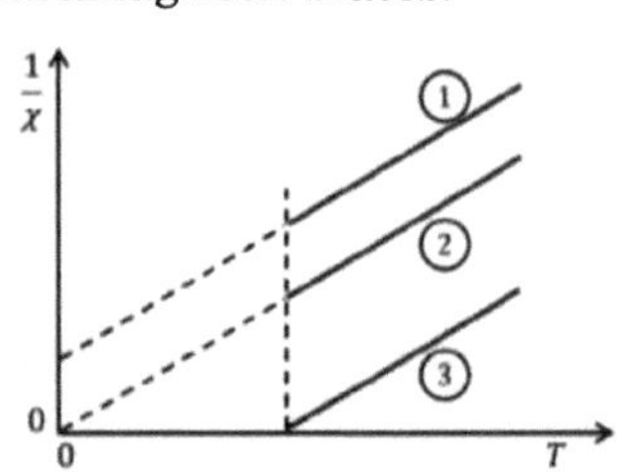

(Curie temperature of ferromagnetic material = T_C
Nė el temperature of antiferromagnetic material = T_N)
Choose the correct statement from the following
(a) Material 1 is antiferromagnetic ($T < T_N$), 2 is paramagnetic, and 3 is ferromagnetic ($T < T_C$).
(b) Material 1 is paramagnetic, 2 is antiferromagnetic ($T < T_N$), and 3 is ferromagnetic ($T < T_C$).
(c) Material 1 ferromagnetic ($T < T_C$), 2 is antiferromagnetic ($T < T_N$), and 3 is paramagnetic.
(d) Material 1 is ferromagnetic ($T < T_C$), 2 is paramagnetic, and 3 is antiferromagnetic ($T < T_N$).

9. A linear charged particle accelerator is driven by an alternating voltage source operating at 10 MHz. Assume that it is used to accelerate electrons. After a few drift-tubes, the electrons attain a velocity 2.9×10^8 m s^{-1}. The minimum length of each drift-tube, in m, to accelerate the electrons further (rounded off to one decimal place) is ______.

10. Consider an atomic gas with number density $n = 10^{20}$ m^{-3}, in the ground state at 300 K. The valence electronic configuration of atoms is f^7. The paramagnetic susceptibility of the gas $\chi = m \times 10^{-11}$. The value of m (rounded off to two decimal places) is ____.
(Given: Magnetic permeability of free space $\mu_0 = 4\pi \times 10^{-7}$ H m^{-1}
Bohr magneton $\mu_B = 9.274 \times 10^{-24}$Am^2
Boltzmann constant $k_B = 1.3807 \times 10^{-23}$ J K^{-1})

2020

11. Which one of the following is a universal logic gate?
(*a*) AND
(*d*) NOT
(*c*) OR
(*d*) NAND

12. Which one of the following is the correct binary equivalent of the hexadecimal F6C?
(a) 0110 1111 1100
(*b*) 1111 0110 1100
(*c*) 1100 0110 1111
(*d*) 0110 1100 0111

13. The number of distinct ways the primitive unit cell can be constructed for the two dimensional lattice as shown in the figure is ______.

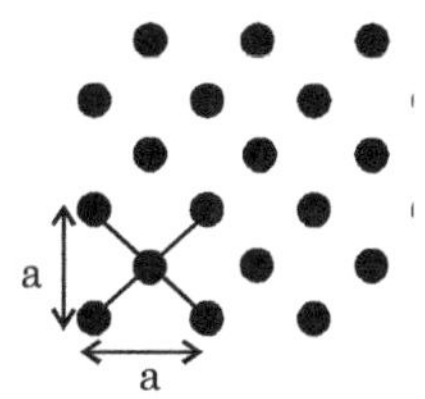

14. The input voltage (V_{in}) to the circuit shown in the figure is 2 cos (100t) V. The output voltage (V_{out}) is $2\cos\left(100t - \frac{\pi}{2}\right)$ V. If R = 1 kΩ, the value of C (in μF) is

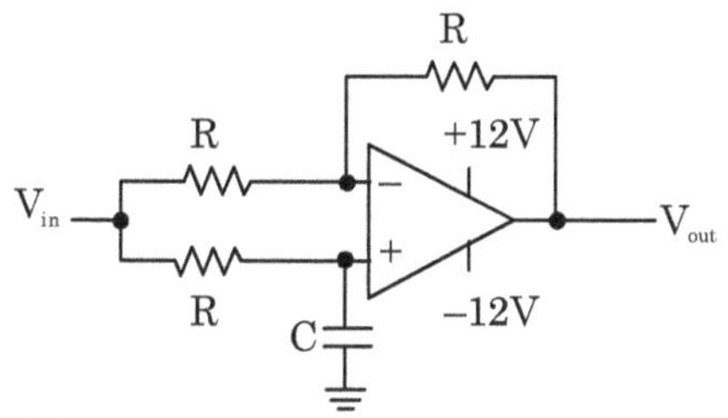

(a) 0.1
(b) 1
(c) 10
(d) 100

15. Consider a 4-bit counter constructed out of four flip-flops. It is formed by connecting the J and K inputs to logic high and feeding the Q output to the clock input of the following flip-flop (see the figure). The input signal to the counter is a series of square pulses and the change of state is triggered by the falling edge. At time t = t_0 the outputs are in logic low state ($Q_0 = Q_1 = Q_2 = Q_3 = 0$). Then at t = t_1, the logic state of the outputs is

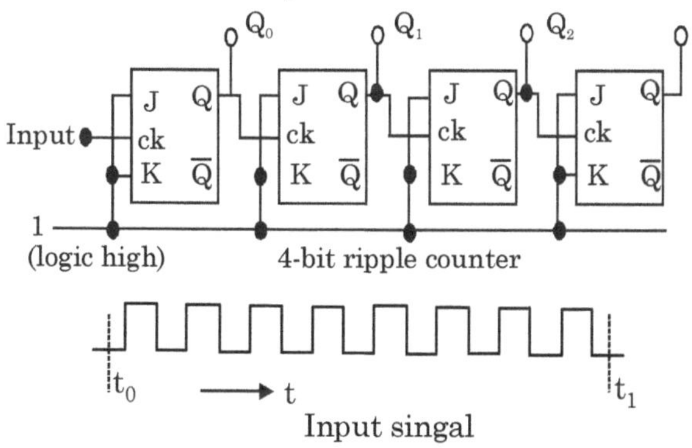

(*a*) $Q_0 = 1, Q_1 = 0, Q_2 = 0$ and $Q_3 = 0$
(*b*) $Q_0 = 0, Q_1 = 0, Q_2 = 0$ and $Q_3 = 1$
(*c*) $Q_0 = 1, Q_1 = 0, Q_2 = 1$ and $Q_3 = 0$
(*d*) $Q_0 = 0, Q_1 = 1, Q_2 = 1$ and $Q_3 = 1$

16. Consider a one -dimensional non-magnetic crystal with one atom per unit cell. Assume that the valence electrons (i) do not interact with each other and (ii) interact weakly with ions. If n is the number of valence electrons per unit cell, then at 0 K,

(a) the crystal is metallic for any value of n
(b) the crystal is non-metallic for any value of n
(c) the crystal is metallic for even values of n
(d) the crystal is metallic for odd values of n

17. Consider a two dimensional crystal with 3 atoms in the basis. The number of allowed optical branches (n) and acoustic branches (m) due to the lattice vibrations are

(a) (n, m) = (2, 4)
(b) (n, m) = (3, 3)
(c) (n, m) = (4, 2)
(d) (n, m) = (1, 5)

18. Consider the circuit given in the figure. Let the forward voltage drop across each diode be 0.7 V. The current I (in mA) through the resistor is _____.

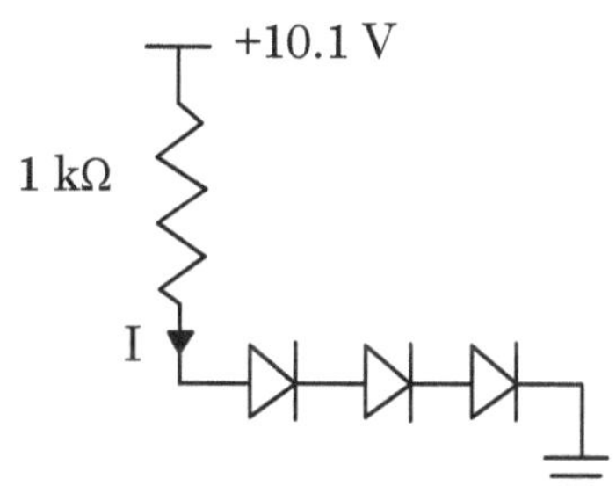

19. Consider a simple cubic monoatomic Bravais lattice which has a basis with vectors $\vec{r_1} = 0, \vec{r_2} = \frac{a}{4}(\hat{x} + \hat{y} + \hat{z})$, a is the lattice parameter. The Bragg reflection is observed due to the change in the wave vector between the incident and the scattered beam as given by $\vec{K} = n_1\vec{G}_1 + n_2\vec{G}_2 + n_3\vec{G}_3$, where $\vec{G}_1$, $\vec{G}_2$, and $\vec{G}_3$ are primitive reciprocal lattice vectors. For n_1 = 3, and n_2 = 3 and n_3 = 2, the geometrical structure factor is ______.

2019

20. The relative magnetic permeability of a type–I superconductor is

(*a*) 0 (*b*) –1
(*c*) 2π (*d*) $\frac{1}{4\pi}$

21. For the following circuit, what is the magnitude of V_{out} if V_{in} = 1.5 V?

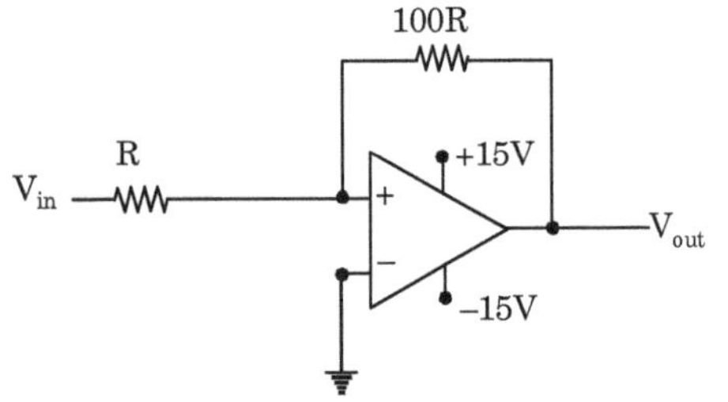

(*a*) 0.015 V (*b*) 0.15 V
(*c*) 15 V (*d*) 150 V

22. In order to estimate the specific heat of phonons, the appropriate method to apply would be

(*a*) Einstein model for acoustic phorious and Debye model for optical phonons
(*b*) Einstein model for optical phorious and Debye model for acoustic phonons
(*c*) Einstein model for both optical and acoustic phonons
(*d*) Debye model for both optical and acoustic phonons

23. Consider a three–dimensional crystal of N inert gas atoms. The total energy is given by U (R) = 2 $N_{\in}\left[p\left(\frac{\sigma}{R}\right)^{12}-q\left(\frac{\sigma}{R}\right)^{6}\right]$ where p= 12.13, q = 14.45, and R is the nearest neighbour distance between two atoms. The Two constants, ∈ and R, have the dimensions of energy fand length, respectively. The equilibrium separation between two nearest neithbour atoms in units of σ (rounded off to tow decimal places) is_________

24. The energy –wavevector (E – k) dispersion relation for a particle in two dimensions is E = Ck, where C is a constant. If its density of states D (E) is proportional to E^p then the value of p is______

25. A conventional type–I superconductor has a critical temperature of 4.7 K at zero magnetic field and a critical magnetic field of 0.3 Tesla at 0K. The critical field in Tesla at 2 K (rounded off to three decimal places) is____________

26. Consider the following Boolean expression:

$$\left(\bar{A}+\bar{B}\right)\left[\overline{A\left(B+C\right)}\right]+A\left(\bar{B}+\bar{C}\right)$$

It can be represented by a single three-input logic gate. Identify the gate.

(*a*) AND (*b*) OR
(*c*) XOR (*d*) NAND

27. A 3-bit analog-to-digital converter is designed to digitize analog signals ranging from 0 V to 10 V. For this converter, the binary output corresponding to an input of 6 V is

(*a*) 011 (*b*) 101
(*c*) 100 (*d*) 010

28. A practical of mass m moves in a lattice along the x–axis in a periodic potential V(x) = V (x+d) with periodicity d. The corresponding Brillouin zone extends from – k_0 to k_0 with these two k–points being equivalent. If a weak force F in the x–direction is applied to the particle, it starts a periodic motion with time period T. Using the equation of motion $F=\frac{dp_{crystal}}{dt}$ for a particle moving in band, where $p_{crystal}$ is the crystal momentum of the particle, the period T is found to be (h is Planck constant)

(*a*) $\sqrt{\frac{2md}{F}}$ (*b*) $2\sqrt{\frac{2md}{F}}$

(*c*) $\frac{2h}{Fd}$ (*d*) $\frac{h}{Fd}$

29. For the following circuit, the correct logic values for the entries X_2 and Y_2 in the truth table are

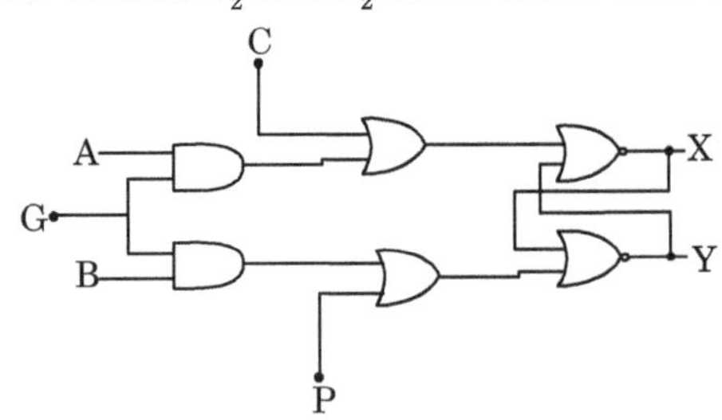

G	A	B	P	C	X	Y
1	0	1	0	0	0	1
0	0	0	1	0	X2	Y2
1	0	0	0	1	0	1

(*a*) 1 and 0 (*b*) 0 and 0
(*c*) 0 and 1 (*d*) 1 and 1

30. For a given load resistance R_L = 4.7 ohm, the power transfer efficiencies $\left(\eta=\frac{P_{load}}{P_{total}}\right)$ of a dc voltage source and a dc current source with intrenal resistance R_1 and R_2, respectively, are equal. The product R_1R_2 in units of ohm² (rounded off to one decimal place) is__________

31. In a certain two–dimensional lattice, the energy dispersion of the electrons is

$$\varepsilon(\vec{k})=-2t\left[\cos k_x\alpha+2\cos\frac{1}{2}k_x\alpha\cos\frac{\sqrt{3}}{2}k_y\alpha\right]$$

where $\vec{k} = (kx, ky)$ denotes the wave vector, a is the lattice constant and t is a constant in units of eV. In this lattice the effective mass tensor m_{ij} of electrons calculated at the center of the Brillouim zone has the form

$m_{ij} = \frac{\hbar^2}{t\alpha^2}\begin{pmatrix} \alpha & 0 \\ 0 & \alpha \end{pmatrix}$ the value of a (rounded off to three decimal palces is)__________

2018

32. For the given unit cells of a two dimensional square lattice, which option lists all the primitive cells?

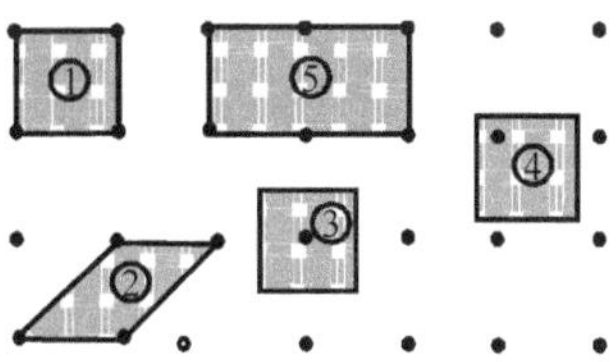

(a) ①and② (b) ①,②and③

(c) ①,②,③and④ (d) ①,②,③,④and⑤

33. The logic expression $\bar{A}BC + \bar{A}\bar{B}C + AB\bar{C} + A\bar{B}\bar{C}$ can be simplified to

(a) A XOR C (b) A AND $\bar{C}$

(c) 0 (d) 1

34. At low temperatures (T), the specific heat of common metals is described by (with α and β as constants)

(a) $\alpha T + \beta T^3$ (b) βT^3

(c) $\exp(-\alpha / T)$ (d) $\alpha T + \beta T^5$

35. In a 2-to-1 multiplexer as shown below, the output X = A_0 if C = 0, and X = A_1 if C = 1.

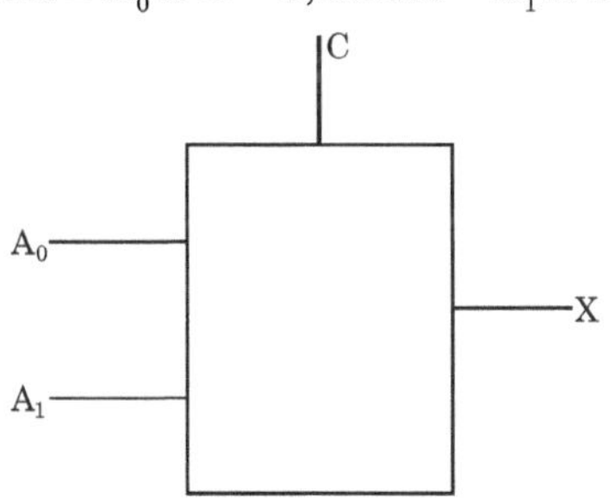

Which one of the following is the correct implementation of this multiplexer?

(a)
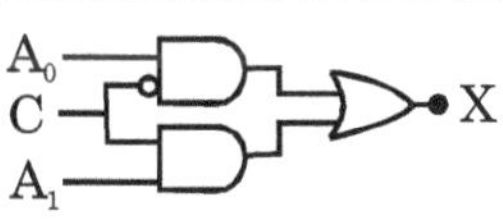

(b)
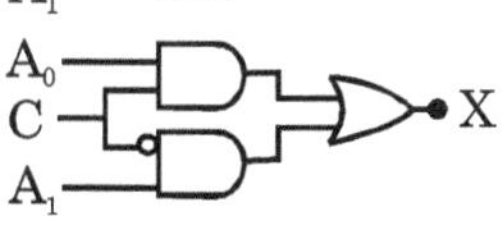

(c)
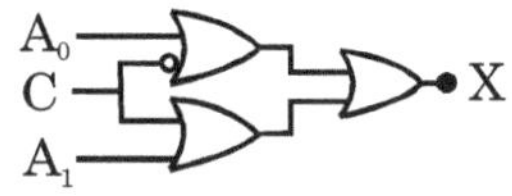

(d)
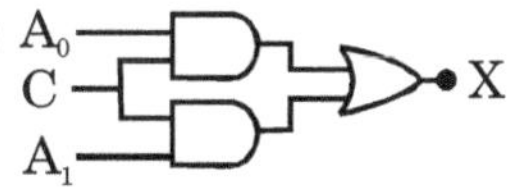

36. The high temperature magnetic susceptibility of solids having ions with magnetic moments can be described by $\chi \propto \frac{1}{T+\theta}$ with T as absolute temperature and θ as constant. The three behaviors i.e. paramagnetic, ferromagnetic and anti-ferromagnetic are described, respectively, by

(a) $\theta < 0, \theta > 0, \theta = 0$ (b) $\theta > 0, \theta < 0, \theta = 0$

(c) $\theta = 0, \theta < 0, \theta > 0$ (d) $\theta = 0, \theta > 0, \theta < 0$

37. For an operational amplifier (ideal) circuit shown below,

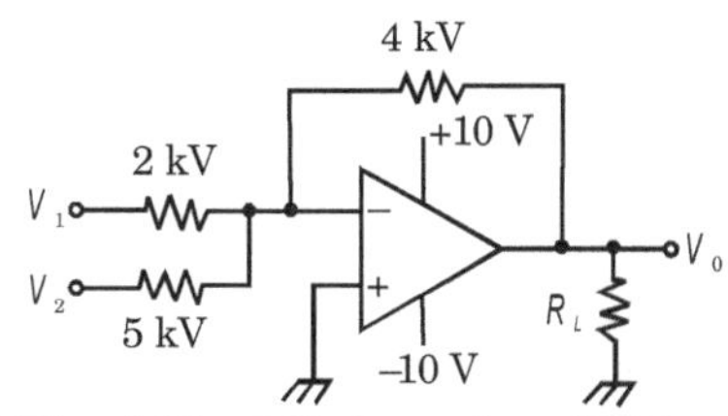

if V_1 = 1 V and V_2 = 2 V, the value of V_0 is ______ V (up to one decimal place).

38. If X is the dimensionality of a free electron gas, the energy (E) dependence of density of states is given by $E^{\frac{1}{2}X - Y}$, where Y is ________.

39. The energy dispersion for electrons in one dimensional lattice with lattice parameter a is given by $E(k) = E_0 - \frac{1}{2}W \cos ka$, where W and E_0 are constants. The effective mass of the electron near the bottom of the band is

(a) $\frac{2\hbar^2}{Wa^2}$ (b) $\frac{\hbar^2}{Wa^2}$

(c) $\frac{\hbar^2}{2Wa^2}$ (d) $\frac{\hbar^2}{4Wa^2}$

40. Amongst electrical resistivity (ρ), thermal conductivity (κ), specific heat (C), Young's modulus (Y), and magnetic susceptibility (χ), which quantities show a sharp change at the superconducting transition temperature?

(a) ρ, κ, C, Y (b) ρ, C, χ

(c) ρ, κ, C, χ (d) κ, Y, χ

41. A p-doped semiconductor slab carries a current I = 100 mA in a magnetic field B = 0.2 T as shown. One measures V_y = 0.25 mV and V_x = 2 mV. The mobility of holes in the semiconductor is ________ $m^2V^{-1}s^{-1}$ (up to two decimal places).

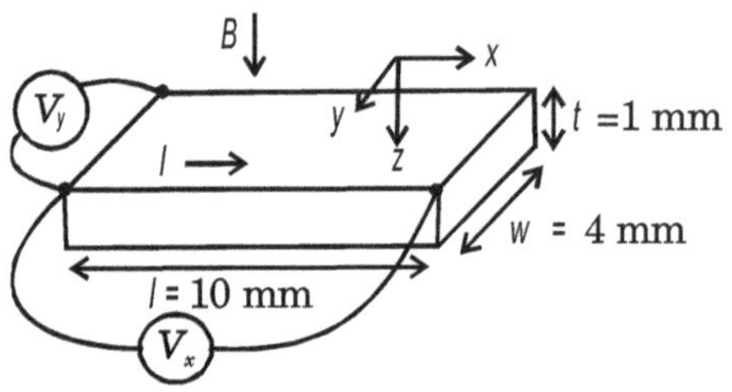

42. An n-channel FET having Gate-Source switch-off voltage $V_{GS(OFF)} = -2$ V is used to invert a 0 – 5 V square-wave signal as shown. The maximum allowed value of R would be ________ kΩ (up to two decimal places).

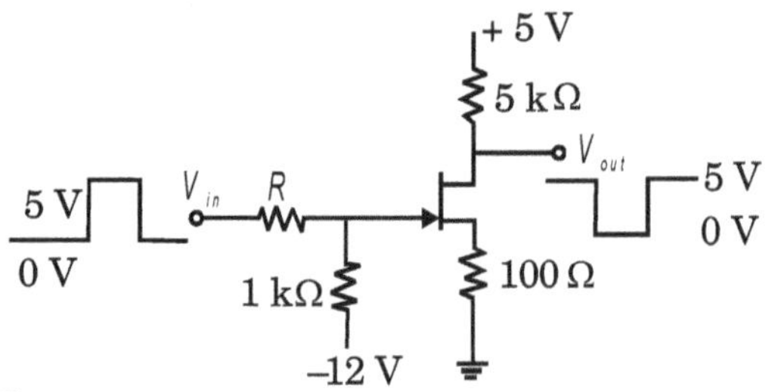

2017

43. The best resolution that a 7 bit A/D convertor with 5 V full scale can achieve is _______ mV. (up to two decimal places).

44. In the figure given below, the input to the primary of the transformer is a voltage varying sinusoidally with time. The resistor R is connected to the centre tap of the secondary. Which one of the following plots represents the voltage across the resistor R as a function of time?

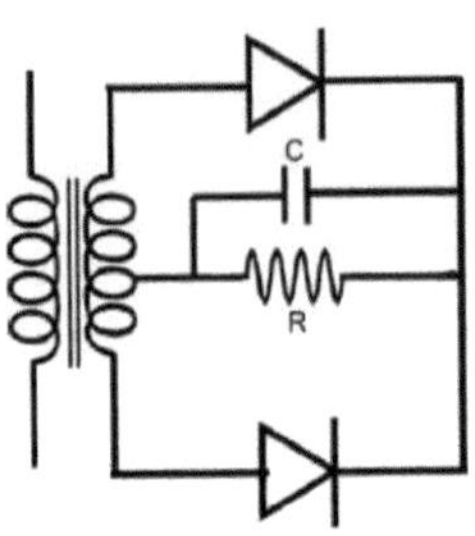

(*a*) 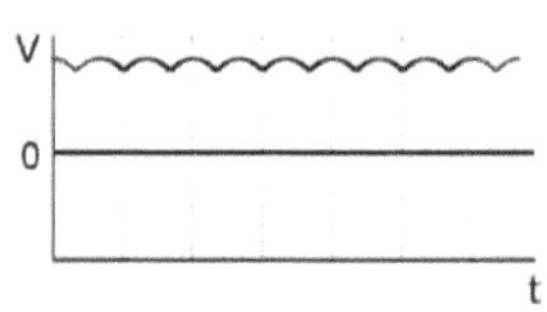

(*b*) 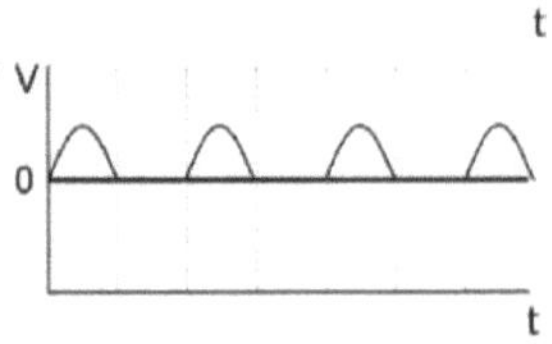

(*c*) 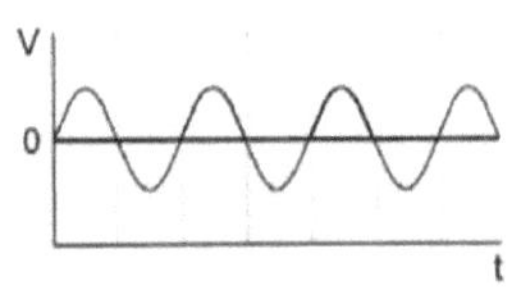

(*d*) 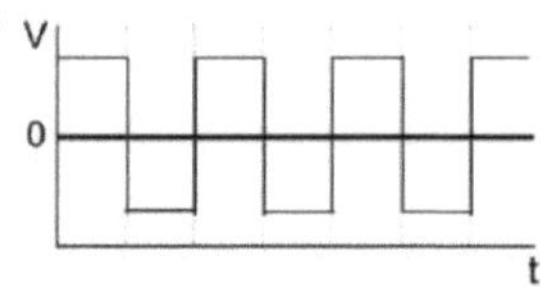

45. The atomic mass and mass density of Sodium are 23 and 0.968 g cm^{-3}, respectively. The number density of valence electrons is ______ $\times 10^{22}$ cm^{-3}. (Up to two decimal places.) (Avogadro number, $N_A = 6.022 \times 10^{23}$).

46. Consider a one-dimensional lattice with a weak periodic potential $U(x) = U_0 \cos\left(\frac{2\pi x}{a}\right)$. The gap at the edge of the Brillouin zone $\left(k = \frac{\pi}{a}\right)$ is :

(*a*) U_0 (*b*) $\frac{U_0}{2}$

(*c*) $2U_0$ (*d*) $\frac{U_0}{4}$

47. Consider a triatomic molecule of the shape shown in the figure below in three dimensions. The heat capacity of this molecule at high temperature (temperature much higher than the vibrational and rotational energy scales of the molecule but lower than its bond dissociation energies) is :

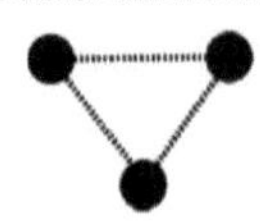

(*a*) $\frac{3}{2}k_B$ (*b*) $3\,k_B$

(*c*) $\frac{9}{2}k_B$ (*d*) $6\,k_B$

48. The minimum number of NAND gates required to construct an OR gate is :

(*a*) 2 (*b*) 4

(*c*) 5 (*d*) 3

49. Consider a 2-dimensional electron gas with a density of 10^{19} m^{-2}. The Fermi energy of the system is ________ eV (up to two decimal places). ($m_e = 9.31 \times 10^{-31}$ kg, $h = 6.626 \times 10^{-34}$ Js, $e = 1.602 \times 10^{-19}$ C)

50. The total energy of an inert-gas crystal is given by $E(R) = \frac{0.5}{R^{12}} - \frac{1}{R^6}$ (in eV), where R is the inter-atomic spacing in Angstroms. The equilibrium separation between the atoms is ______ Angstroms. (up to two decimal places).

51. Consider a metal with free electron density of 6×10^{22} cm^{-3}. The lowest frequency electromagnetic radiation to which this metal is transparent is 1.38×10^{16} Hz. If this metal had a free electron density of 1.8×10^{23} cm^{-3} instead, the lowest frequency electromagnetic radiation to which it would be transparent is _______ $\times 10^{16}$ Hz. (up to two decimal places).

52. The energy density and pressure of a photon gas are given by $u = aT^4$ and $P = u/3$, where T is the temperature and a is the radiation constant. The entropy per unit volume is given by $\alpha a T^3$. The value of α is _______. (up to two decimal places).

53. The real space primitive lattice vectors are $\vec{a}_1 = a\hat{x}$ and $\vec{a}_2 = \frac{a}{2}\left(\hat{x} + \sqrt{3}\hat{y}\right)$. The reciprocal space unit vectors $\vec{b}_1$ and $\vec{b}_2$ for this lattice are, respectively

(*a*) $\frac{2\pi}{a}\left(\hat{x} - \frac{\hat{y}}{\sqrt{3}}\right)$ and $\frac{4\pi}{a\sqrt{3}}\hat{y}$

(*b*) $\frac{2\pi}{a}\left(\hat{x} + \frac{\hat{y}}{\sqrt{3}}\right)$ and $\frac{4\pi}{a\sqrt{3}}\hat{y}$

(*c*) $\frac{2\pi}{a\sqrt{3}}\hat{x}$ and $\frac{4\pi}{a}\left(\frac{\hat{x}}{\sqrt{3}} + \hat{y}\right)$

(*d*) $\frac{2\pi}{a\sqrt{3}}\hat{x}$ and $\frac{4\pi}{a}\left(\frac{\hat{x}}{\sqrt{3}} - \hat{y}\right)$

54. For the transistor amplifier circuit shown below with $R_1 = 10$ kΩ, $R_2 = 10$ kΩ, $R_3 = 1$ kΩ, and $\beta = 99$. Neglecting the emitter diode resistance, the input impedance of the amplifier looking into the base for small ac signal is _______ kΩ. (up to two decimal places.)

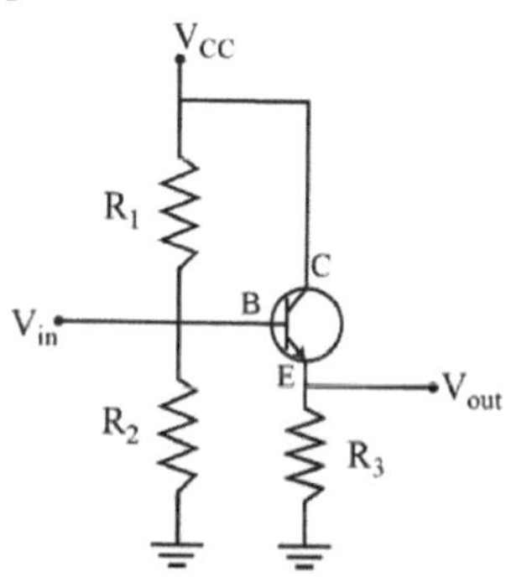

55. Consider an ideal operational amplifier as shown in the figure below with $R_1 = 5$ kΩ, $R_2 = 1$ kΩ, $R_L = 100$ kΩ. For an applied input voltage V = 10 mV, the current passing through R_2 is _______ μA. (up to two decimal places).

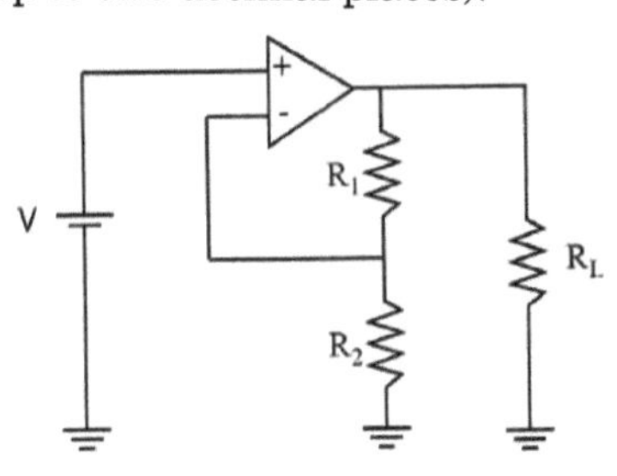

2016

56. The number density of electrons in the conduction band of a semiconductor at a given temperature is 2×10^{19} m^{-3}. Upon lightly doping this semiconductor with donor impurities, the number density of conduction electrons at the same temperature becomes 4×10^{20} m^{-3}. The ratio of majority to minority charge carrier concentration is _______.

57. A solid material is found to have a temperature independent magnetic susceptibility, $\chi = C$. Which of the following statements is correct?

(*a*) If C is positive, the material is a diamagnet.

(*b*) If C is positive, the material is a ferromagnet.

(*c*) If C is negative, the material could be a type I superconductor.

(*d*) If C is positive, the material could be a type I superconductor.

58. Atoms, which can be assumed to be hard spheres of radius R, are arranged in an fcc lattice with lattice constant a, such that each atom touches its nearest neighbours. Take the center of one of the atoms as the origin. Another atom of radius r (assumed to be hard sphere) is to be accommodated at a position (0, a/2, 0) without distorting the lattice. The maximum value of r/R is _______. (Give your answer upto two decimal places)

59. The energy vs. wave vector (E – k) relationship near the bottom of a band for a solid can be approximated as $E = A(ka)^2 + B(ka)^4$, where the lattice constant a = 2.1 Å. The values of A and B are 6.3×1^{-19} and 3.2×10^{-20} respectively. At the bottom of the conduction band, the ratio of the effective mass of the electron to the mass of free electron is _______. (Give your answer upto two decimal places)

(Take $\hbar = 1.05 \times 10^{-34}$ J-s, mass of free electron $= 9.1 \times 10^{-31}$ kg)

2015

60. The lattice parameters a, b, c of an orthorhombic crystal are related by $a = 2b = 3c$. In units of a the interplanar separation between the (110) planes is _______ (upto three decimal places)

61. Which one of the following DOES NOT represent an exclusive OR operation for inputs A and B

(*a*) $(A+B)\overline{AB}$

(*b*) $A\overline{B} + B\overline{A}$

(*c*) $(A+B)(\overline{A} + \overline{B})$

(*d*) A +B) AB

62. The dispersion relation for phonons in a one dimensional monatomic Bravais lattice with lattice spacing a and consisting of ions of masses M is given by $\omega(k) = \sqrt{\frac{2C}{M}[1-\cos(ka)]}$, where ω the frequency of oscillation, k is the wavevector and C is the spring constant. For the long wavelength modes ($\lambda >> a$), the ratio of the phase velocity to the group velocity is ______

63. In a Hall effect experiment, the Hall voltage for an intrinsic semiconductor is negative. This is because (symbols carry usual meaning)

(*a*) $n \approx p$ (*b*) $n > p$

(*c*) $\mu_e > \mu_h$ (*d*) $m_e^* > m_h^*$

64. The band gap of an intrinsic semiconductor is $E_g = 0.72$ eV and $m_h^* = 6m_e^*$. At 300 K, the Ferm level with respect to the edge of the valence band (in eV) is at ____ (upto three decimal place) $K_B = 1.38 \times 10^{-23}JK^{-1}$

65. Which one of the following represents the electron occupancy for a superconductor in its normal and superconducting states ?

(*a*)

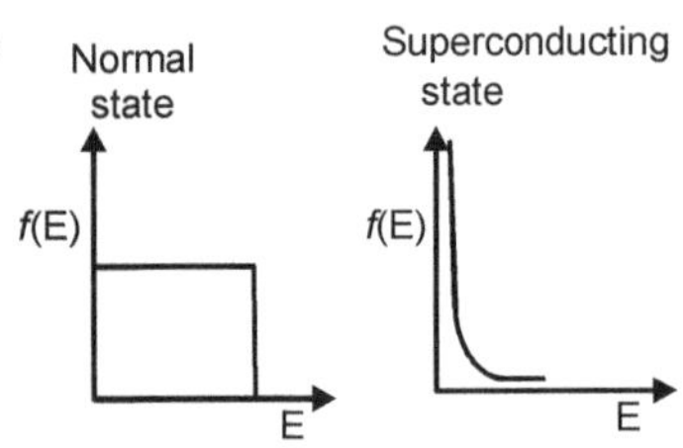

(*b*)

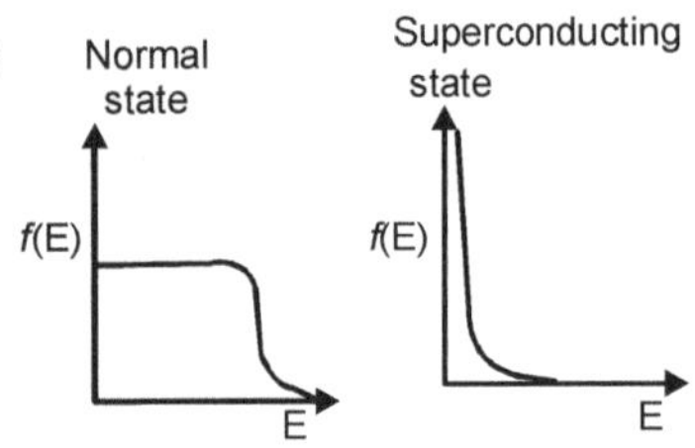

(*c*)

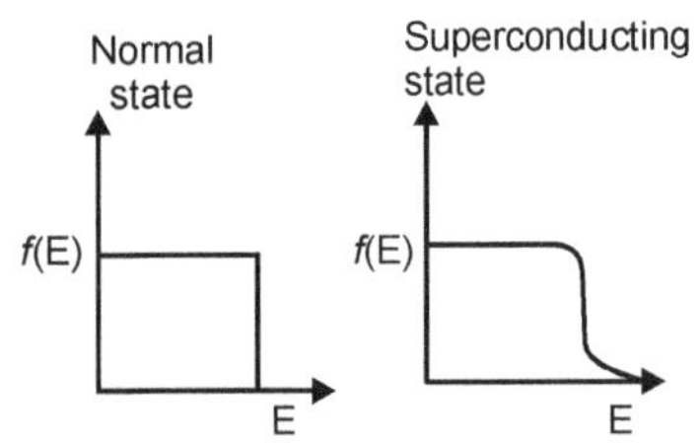

(*d*)

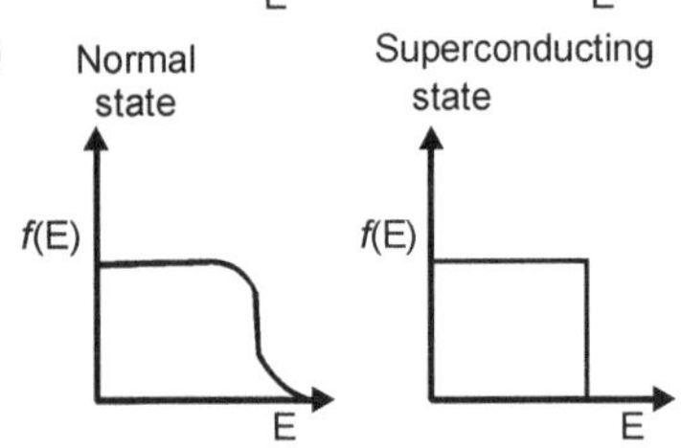

2014

66. The Miller indices of a plane passing through the three points having coordinates (0,0,1) (1,00), $\left(\frac{1}{2},\frac{1}{2},\frac{1}{4}\right)$ are

(*a*) (212) (*b*) (111)

(*c*) (121) (*d*) (211)

67. The donor concentration in a sample of n-type silicon is increased by a factor of 100. The shift in the position of the Fermi level at 300K, assuming the sample to be non degenerate is ______ meV.

($k_BT = 25\ meV$ at 300 K)

68. Neutrons moving with speed 10^3 m/s are used for the determination of crystal structure. If the Bragg angle for the first order diffraction is 30^0, the interplanar spacing of the crystal is _________Å. (Given : $m_n = 1.675 \times 10^{-27}$ kg, $h = 6.626 \times 10^{-34}$ J.s)

69. In order to measure a maximum of 1V with a resolution of 1mV using a n-bit A/D converter, working under the principle of ladder network, the minimum value of n is ____________ .

2013

70. A phosphorous doped silicon semicond-uctor (doping density: $10^{17}/cm^3$) is heated from 100°C to 200°C. Which one of the following statements is CORRECT ?

(*a*) Position of Fermi level moves towards conduction band

(*b*) Position of dopant level moves towards conduction band

(*c*) Position of Fermi level moves towards middle of energy gap

(*d*) Position of dopant level moves towards middle of energy gap

71. Considering the BCS theory of superconductors, which one of the following statements is NOT CORRECT?

(h is the Planck's constant and e is the electronic charge)

(*a*) Presence of energy gap at temperatures below the critical temperature

(*b*) Different critical temperatures for isotopes

(*c*) Quantization of magnetic flux in superc-onducting ring in the unit of $\left(\frac{h}{e}\right)$

(*d*) Presence of Meissner effect

72. A lattice has the following primitive vectors (in Å) : $\vec{a} = 2(\hat{j} + \hat{k})$, $\vec{b} = 2(\hat{k} + \hat{i})$, $\vec{c} = 2(\hat{i} + \hat{j})$.
The reciprocal lattice corresponding to the above lattice is

(*a*) BCC lattice with cube edge of $\left(\frac{\pi}{2}\right)$ Å^{-1}
(*b*) BCC lattice with cube edge of (2π) Å^{-1}
(*c*) FCC lattice with cube edge of $\left(\frac{\pi}{2}\right)$ Å^{-1}
(*d*) FCC lattice with cube edge of (2π) Å^{-1}

73. The total energy of an ionic solid is given by an expression $E = -\frac{\alpha e^2}{4\pi\varepsilon_0 r} + \frac{B}{r^9}$ where α is Madelung constant, r is the distance between the nearest neighbours in the crystal and B is a constant. If r_0 is the equilibrium separation between the nearest neighbours then the value of B is

(*a*) $\frac{\alpha e^2 r_0^8}{36\pi\varepsilon_0}$ (*b*) $\frac{\alpha e^2 r_0^8}{4\pi\varepsilon_0}$
(*c*) $\frac{2\alpha e^2 r_0^{10}}{9\pi\varepsilon_0}$ (*d*) $\frac{\alpha e^2 r_0^{10}}{36\pi\varepsilon_0}$

74. A proton is confined to a cubic box, whose sides have length 10^{-12} *m*. What is the minimum kinetic energy of the proton? The mass of proton is 1.67×10^{-27} *kg* and Planck's constant is 6.63×10^{-34} *Js*.

(*a*) 1.1×10^{-17} *J* (*b*) 3.3×10^{-17} *J*
(*c*) 9.9×10^{-17} *J* (*d*) 6.6×10^{-17} *J*

2012

75. Identify the CORRECT energy band diagram for Silicon doped with Arsenic. Here CB, VB, E_D and E_F are conduction band, valence band, impurity level and Fermi level, respectively.

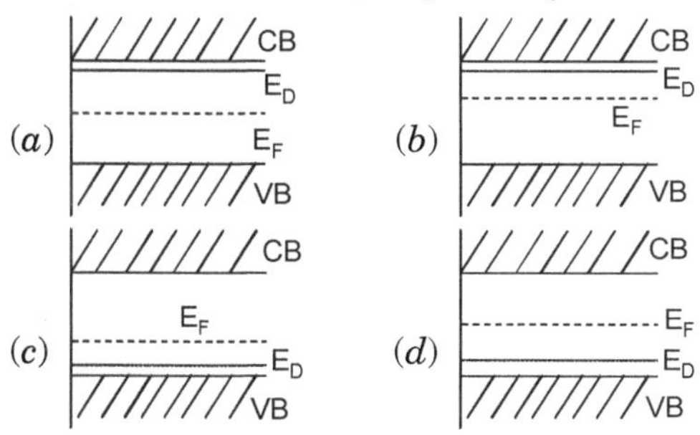

76. Which one of the following CANNOT be explained by considering a harmonic approximation for the lattice vibrations in solids?

(*a*) Debye's T^3 law
(*b*) Dulong Petit's law
(*c*) Optical branches in lattices
(*d*) Thermal expansion

77. Inverse susceptibility $(1/\chi)$ as a function of temperature, T for a material undergoing paramagnetic to ferromagnetic transition is given in the figure, where O is the origin. The values of the Curie constant, C, and the Weiss molecular field constant, λ in CGS units, are

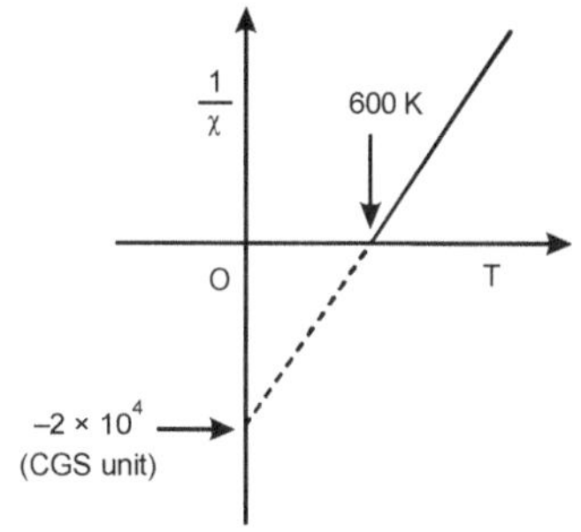

(*a*) $C = 5 \times 10^{-5}$, $\lambda = 3 \times 10^{-2}$
(*b*) $C = 3 \times 10^{-2}$, $\lambda = 5 \times 10^{-5}$
(*c*) $C = 3 \times 10^{-2}$, $\lambda = 2 \times 10^{4}$
(*d*) $C = 2 \times 10^{4}$, $\lambda = 3 \times 10^{-2}$

78. A simple cubic crystal with lattice parameter a_c undergoes transition into a tetragonal structure with lattice parameters

$a_t = b_t = \sqrt{2}\, a_c$ and $c_t = 2a_c$, below a certain temperature. The ratio of the interplanar spacings of (1 0 1) planes for the cubic and the tetragonal structures is

(*a*) $\sqrt{\frac{1}{6}}$ (*b*) $\frac{1}{6}$
(*c*) $\sqrt{\frac{3}{8}}$ (*d*) $\frac{3}{8}$

79. Consider the following circuit in which the current gain β_{dc} of the transistor is 100.

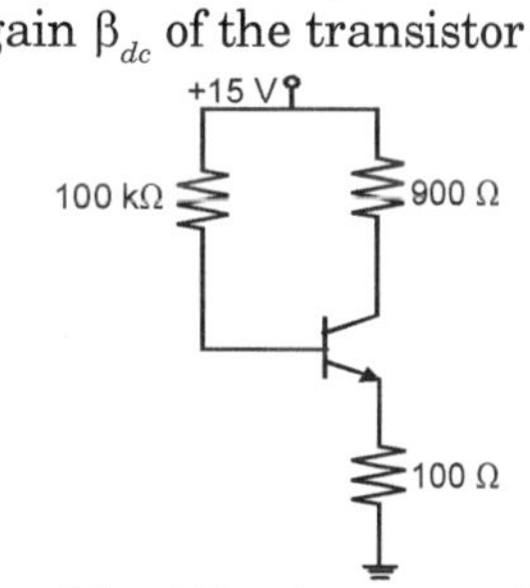

Which one of the following correctly represents the load line (collector current I_C with respect to collector-emitter voltage V_{CE}) and Q-point of this circuit?

(*a*)
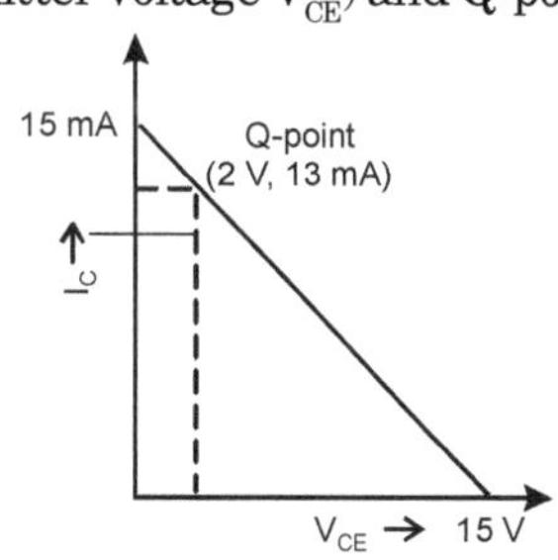

(b)

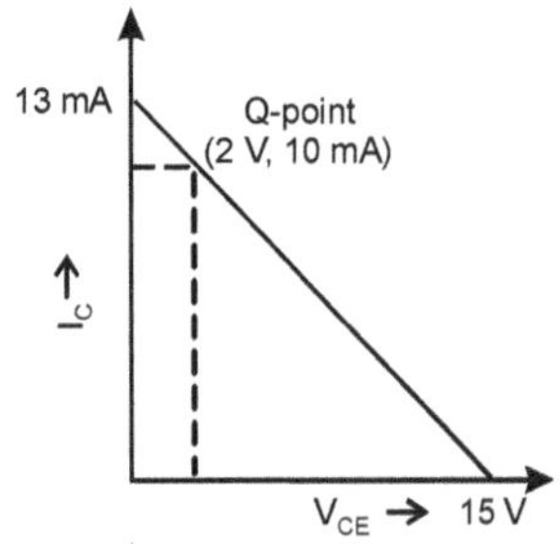

(c)

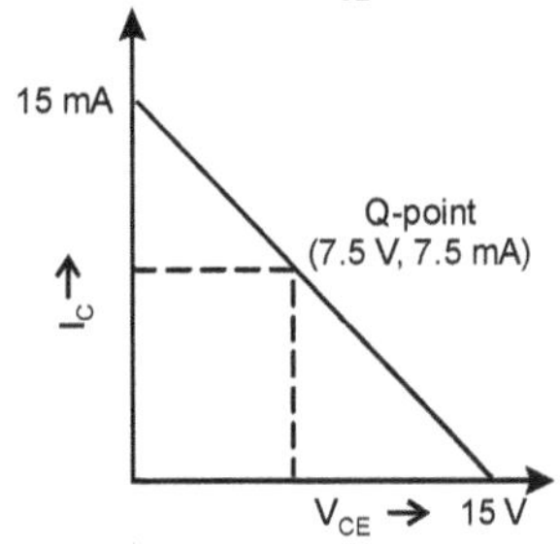

(d)

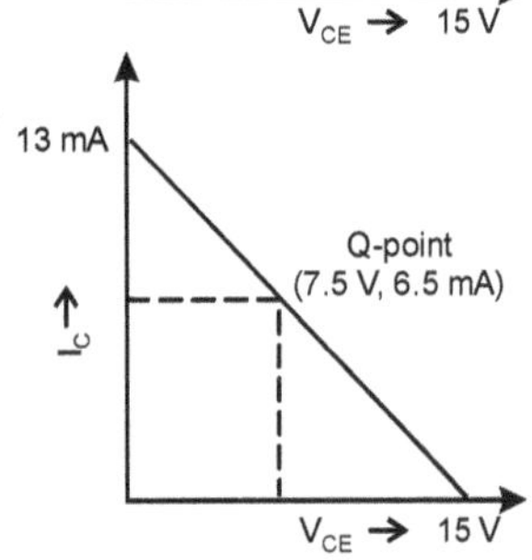

Common Data Q. 80 and 81 :

The dispersion relation for a one dimensional monatomic crystal with lattice spacing a, which interacts via nearest neighbour harmonic potential is given by

$$\omega = A\left|\sin\frac{Ka}{2}\right|,$$

where A is a constant of appropriate unit.

80. The group velocity at the boundary of the first Brillouin zone is

(a) 0 (b) 1

(c) $\sqrt{\frac{Aa^2}{2}}$ (d) $\frac{1}{2}\sqrt{\frac{Aa^2}{2}}$

81. The force constant between the nearest neighbour of the lattice is (M is the mass of the atom)

(a) $\frac{MA^2}{4}$ (b) $\frac{MA^2}{2}$

(c) MA^2 (d) $2\ MA^2$

2011

82. The temperature (T) dependence of magnetic susceptibility (χ) of a ferro-magnetic substance with a Curie temperature (T_c) is given by

(a) $\frac{C}{T-T_c}$, for $T < T_c$

(b) $\frac{C}{T-T_c}$, for $T > T_c$

(c) $\frac{C}{T+T_c}$, for $T > T_c$

(d) $\frac{C}{T+T_c}$, for all temperatures

where C is constant.

83. The order of magnitude of the energy gap of a typical superconductor is

(a) 1 MeV (b) 1 KeV

(c) 1 eV (d) 1 meV

84. For a three-dimensional crystal having N primitive unit cells with a basis of p atoms, the number of optical branches is

(a) 3 (b) $3p$

(c) $3p - 3$ (d) $3N - 3p$

85. For an intrinsic semiconductor, m^*_e and m^*_h are respectively the effective masses of electrons and holes near the corresponding band edges. At a finite temperature, the position of the Fermi level

(a) depends on m^*_e but not on m^*_h

(b) depends on m^*_h but not on m^*_e

(c) depends on both m^*_e and m^*_h

(d) depends neither on m^*_e nor on m^*_h

86. A metal with body centered cubic (bcc) structure shows the first (i.e. smallest angle) diffraction peak at a Bragg angle of $\theta = 30°$. The wavelength of X-ray used is 2.1 Å. The volume of the PRIMITIVE unit cell of the metal is

(a) 26.2 (Å)^3 (b) 13.1 (Å)^3

(c) 9.3 (Å)^3 (d) 4.6 (Å)^3

2010

87. The valence electrons do not directly determine the following property of a metal.

(a) Electrical conductivity

(b) Thermal conductivity

(c) Shear modulus

(d) Metallic lustre

88. Consider X-ray diffraction from a crystal with a face-centered-cubic (fcc) lattice. The lattice plane for which there is NO diffraction peak is

(a) (2, 1, 2) (b) (1, 1, 1)

(c) (2, 0, 0) (d) (3, 1, 1)

89. The Hall coefficient, R_H, of sodium depends on

(*a*) the effective charge carrier mass and carrier density

(*b*) the charge carrier density and relaxation time

(*c*) the charge carrier density only

(*d*) the effective charge carrier mass

90. The Bloch theorem states that within a crystal, the wavefunction, $\psi(\vec{r})$, of an electron has the form

(*a*) $\psi(\vec{r}) = u(\vec{r})e^{i\vec{k}.\vec{r}}$ where $u(\vec{r})$ is an arbitrary function and $\vec{k}$ is an arbitrary vector

(*b*) $\psi(\vec{r}) = u(\vec{r})e^{i\vec{G}.\vec{r}}$ where $u(\vec{r})$ is an arbitrary function and $\vec{G}$ is a reciprocal lattice vector

(*c*) $\psi(\vec{r}) = u(\vec{r})e^{i\vec{G}.\vec{r}}$ where $u(\vec{r}) = u(\vec{r} + \vec{\Lambda})$, $\vec{\Lambda}$ is a lattice and $\vec{G}$ is a reciprocal lattice vector

(*d*) $\psi(\vec{r}) = u(\vec{r})e^{i.\vec{k}.\vec{r}}$ where $u(\vec{r}) = u(\vec{r} + \vec{\Lambda})$, $\vec{\Lambda}$ is a lattice vector and $\vec{k}$ is an arbitrary vector

91. In an experiment involving a ferromagnetic medium, the following observations were made. Which one of the plots does NOT correctly represent the property of the medium? (T_C is the Curie temperature)

(*a*)
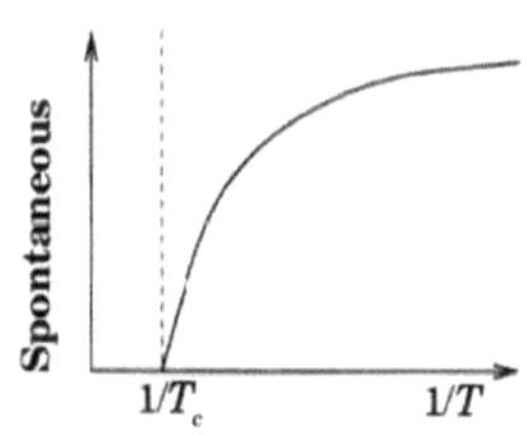

(*b*)
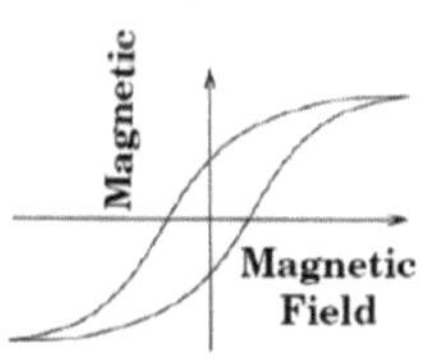

(*c*)
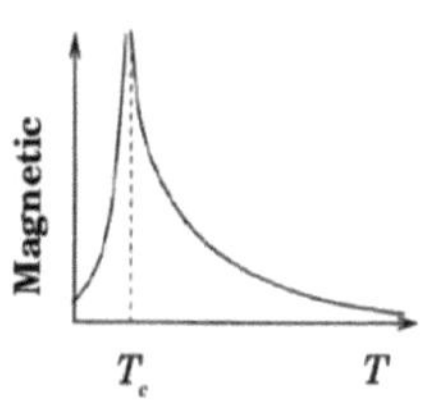

(*d*)
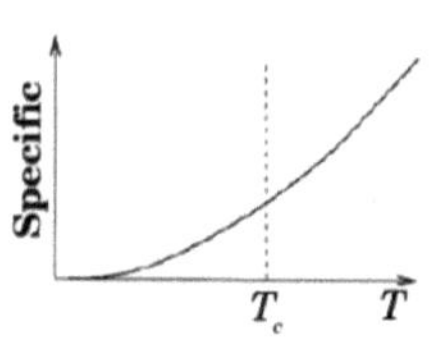

92. The thermal conductivity of a given material reduces when it undergoes a transition from its normal state to the superconducing state. The reason is:

(*a*) The Cooper pairs cannot transfer energy to the lattice

(*b*) Upon the formation of Cooper pairs, the lattice becomes less efficient in heat transfer

(*c*) The electrons in the normal state lose their ability to transfer heat because of their coupling to the Cooper pairs

(*d*) The heat capacity increases on transition to the superconducting state leading to a reduction in thermal conductivity

93. Far away from any of the resonance frequencies of a medium, the real part of the dielectric permittivity is

(*a*) Always independent of frequency

(*b*) Monotonically decreasing with frequency

(*c*) Monotonically increasing with frequency

(*d*) A non-monotonic function of frequency

2009

94. An electrostatic field $\vec{E}$ exists in a given region R. Choose the WRONG statement.

(*a*) Circulation of $\vec{E}$ is zero

(*b*) $\vec{E}$ can always be expressed as the gradient of a scalar field

(*c*) The potential difference between any two arbitrary points in the region R is zero

(*d*) The work done in a closed path lying entirely in R is zero

95. The Common Mode Rejection Ratio (CMRR) of a differential amplifier using an operational amplifier is 100 dB. The output voltage for a differential input of 200 μV is 2 V. The common mode gain is

(*a*) 10 (*b*) 0.1

(*c*) 30 dB (*d*) 10 dB

96. In an insulating solid which one of the following physical phenomena is a consequence of Pauli's exclusion principle?

(*a*) Ionic conductivity

(*b*) Ferromagnetism

(*c*) Paramagnetism

(*d*) Ferroelectricity

97. A superconducting ring is cooled in the presence of a magnetic field below its critical temperature (T_C). The total magnetic flux that passes through the ring is

(*a*) zero (*b*) $n\frac{h}{2e}$

(*c*) $\frac{nh}{4\pi e}$ (*d*) $\frac{ne^2}{hc}$

98. In a cubic crystal, atoms of mass M_1 lie on one set of planes and atoms of mass M_2 lie on planes interleaved between those of the first set. If C is the force constant between nearest neighbour planes, the frequency of lattice vibrations for the optical phonon branch with wavevector $k = 0$ is

(*a*) $\sqrt{2C\left(\frac{1}{M_1}+\frac{1}{M_2}\right)}$

(*b*) $\sqrt{C\left(\frac{1}{2M_1}+\frac{1}{M_2}\right)}$

(*c*) $\sqrt{C\left(\frac{1}{M_1}+\frac{1}{2M_2}\right)}$

(*d*) 0

99. The dependence of the magnetic susceptibility (χ) of a material with temperature (T) can be represented by $\chi \propto \frac{1}{T-\theta}$, where θ is the Curie-Weiss temperature. The plot of magnetic susceptibility versus temperature is sketched in the figure, as curves P, Q and R with curve Q having $\theta = 0$. Which one of the following statements is correct?

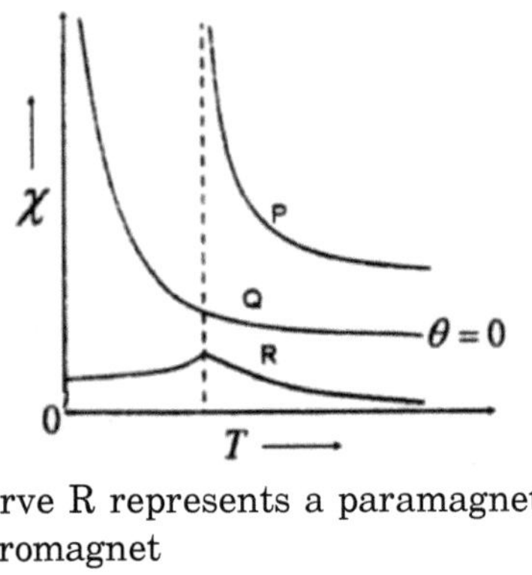

(*a*) Curve R represents a paramagnet and Q a ferromagnet

(*b*) Curve Q represents a ferromagnet and P an antiferromagnet

(*c*) Curve R represents an anti-ferromagnet and Q a paramagnet

(*d*) Curve R represents an anti-ferromagnet and Q a ferromagnet

100. The dielectric constant of a material at optical frequencies is mainly due to

(*a*) ionic polarizability

(*b*) electronic polarizability

(*c*) dipolar polarizability

(*d*) ionic and dipolar polarizability

101. An electron of wavevector $\vec{k}_e$, velocity $\vec{v}_e$ and effective mass m_e is removed from a filled energy band. The resulting hole has wavevector $\vec{k}_h$, velocity $\vec{v}_h$, and effective mass m_h. Which one of the following statements is correct?

(*a*) $\vec{k}_h = -\vec{k}_e;\ \vec{v}_h = -\vec{v}_e;\ m_h = -m_e$

(*b*) $\vec{k}_h = \vec{k}_e;\ \vec{v}_h = \vec{v}_e;\ m_h = m_e$

(*c*) $\vec{k}_h = \vec{k}_e;\ \vec{v}_h = -\vec{v}_e;\ m_h = -m_e$

(*d*) $\vec{k}_h = -\vec{k}_e;\ \vec{v}_h = \vec{v}_e;\ m_h = -m_e$

Statement for Linked Answer Questions 102 and 103 :

The primitive translation vectors of the face centered cubic (*fcc*) lattice are

$$\hat{a}_1 = \frac{a}{2}(\hat{j}+\hat{k});\ \hat{a}_2 = \frac{a}{2}(\hat{i}+\hat{k});\ \hat{a}_3 = \frac{a}{2}(\hat{i}+\hat{j}).$$

102. The primitive translation vectors of the *fcc* reciprocal lattice are

(*a*) $\hat{b}_1 = \left(\frac{2\pi}{a}\right)(-\hat{i}+\hat{j}+\hat{k});$

$\hat{b}_2 = \left(\frac{2\pi}{a}\right)(\hat{i}-\hat{j}+\hat{k});\ \hat{b}_3 = \left(\frac{2\pi}{a}\right)(\hat{i}+\hat{j}-\hat{k})$

(*b*) $\hat{b}_1 = \left(\frac{\pi}{a}\right)(-\hat{i}+\hat{j}+\hat{k});$

$\hat{b}_2 = \left(\frac{\pi}{a}\right)(\hat{i}-\hat{j}+\hat{k});\ \hat{b}_3 = \left(\frac{\pi}{a}\right)(\hat{i}+\hat{j}-\hat{k})$

(*c*) $\hat{b}_1 = \left(\frac{\pi}{2a}\right)(-\hat{i}+\hat{j}+\hat{k});$

$\hat{b}_2 = \left(\frac{\pi}{2a}\right)(\hat{i}-\hat{j}+\hat{k});\ \hat{b}_3 = \left(\frac{\pi}{2a}\right)(\hat{i}+\hat{j}-\hat{k})$

(*d*) $\hat{b}_1 = \left(\frac{3\pi}{a}\right)(-\hat{i}+\hat{j}+\hat{k});$

$\hat{b}_2 = \left(\frac{3\pi}{a}\right)(\hat{i}-\hat{j}+\hat{k});\ \hat{b}_3 = \left(\frac{3\pi}{a}\right)(\hat{i}+\hat{j}-\hat{k})$

103. The volume of the primitive cell of the *fcc* reciprocal lattice is

(*a*) $4\left(\frac{2\pi}{a}\right)^3$ (*b*) $4\left(\frac{\pi}{a}\right)^3$

(*c*) $4\left(\frac{\pi}{2a}\right)^3$ (*d*) $4\left(\frac{3\pi}{a}\right)^3$

2008

Statement for Linked Answer Questions 104 and 105:

The following circuit contains three operational amplifiers and resistors.

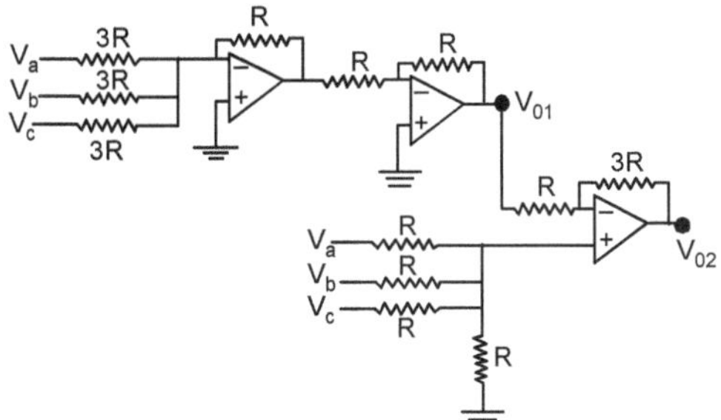

104. The output voltage at the end of second operational amplifier V_{01} is

(*a*) $V_{01} = 3(V_a + V_b + V_c)$

(*b*) $V_{01} = -\frac{1}{3}(V_a + V_b + V_c)$

(*c*) $V_{01} = \frac{1}{3}(V_a + V_b + V_c)$

(*d*) $V_{01} = \frac{4}{3}(V_a + V_b + V_c)$

105. The output V_{02} (at the end of third op amp) of the above circuit is

(*a*) $V_{02} = 2(V_a + V_b + V_c)$

(*b*) $V_{02} = 3(V_a + V_b + V_c)$

(*c*) $V_{02} = -\frac{1}{2}(V_a + V_b + V_c)$

(*d*) Zero

106. Metallic monvalent sodium crystallizes in body centeredcubic structure. If the length of the unit cell is 4×10^{-8} cm, the concentration of conduction electrons in metallic sodium is

(*a*) 6.022×10^{23} cm^{-3}

(*b*) 3.125×10^{22} cm^{-3}

(*c*) 2.562×10^{21} cm^{-3}

(*d*) 1.250×10^{20} cm^{-3}

107. The plot of inverse magnetic susceptibility $\frac{1}{\chi}$ versus temperature T of an antiferro-magnetic sample corresponds to

(*a*)

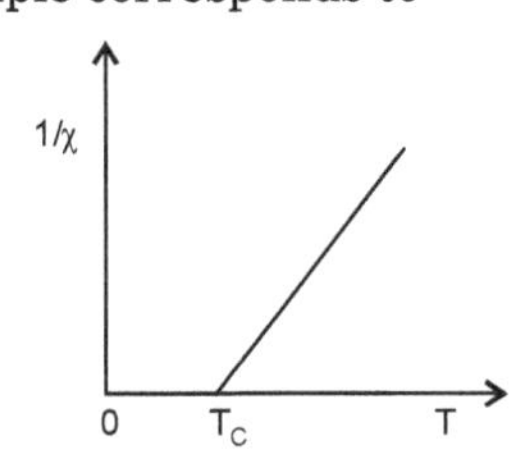

(*b*)

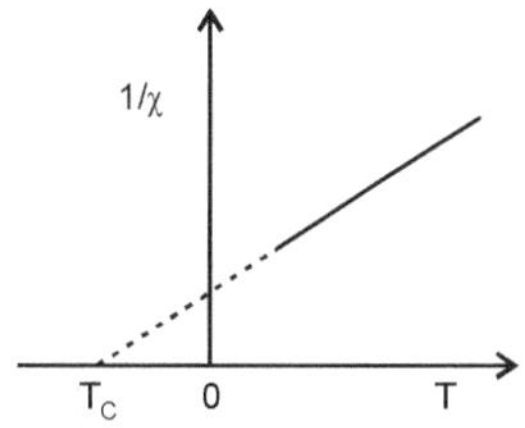

(*c*)

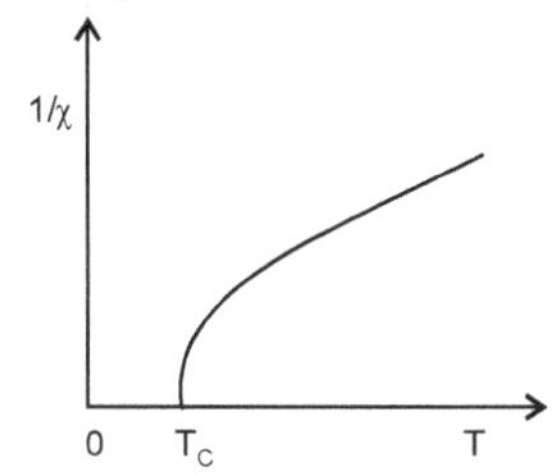

(*d*)

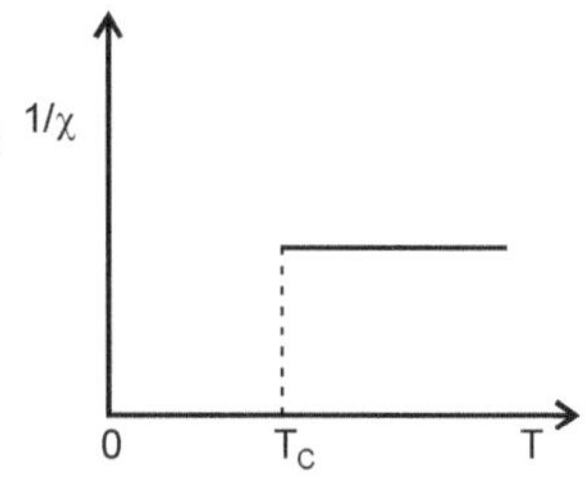

108. A common emitter transistor amplifier circuit is operated under a fixed bias. In this circuit, the operating point

(*a*) remains fixed with an increase in temperature.

(*b*) moves towards cut-off region with an increase in temperature.

(*c*) moves towards the saturation region with a decrease in temperature.

(*d*) moves towards the saturation region with an increase in temperature.

109. Under normal operating conditions, the gate terminal of an n-channel juction field effect transistor (JFET) and an n-channel metal oxide semiconductor field effectg transistor (MOSFET) are

(*a*) both biased with positive potentials.

(*b*) both biased with negative potentials.

(*c*) biased with positive and negative potentials, respectively.

(*d*) biased with negative and positive potentials, respectively.

110. At time $t = 0$, a charge distribution $\rho(\vec{r},0)$ exists within an ideal homogeneous conductor of permittivity ε and conductivity σ. At a later time $\rho(\vec{r},t)$ is given by

(*a*) $\rho(\vec{r},t) = \rho(\vec{r},0)\exp\left(-\frac{\sigma t}{\varepsilon}\right)$

(*b*) $\rho(\vec{r},t)=\dfrac{\rho(\vec{r},0)}{1+(\theta t/\varepsilon)^2}$

(*c*) $\rho(\vec{r},t)=\rho(\vec{r},0)\exp\left[-\left(\dfrac{\sigma t}{\varepsilon}\right)^2\right]$

(*d*) $\rho(\vec{r},t)=\rho(\vec{r},0)\exp\dfrac{\varepsilon}{\sigma t}\sin\left(\dfrac{\sigma t}{\varepsilon}\right)$

111. The primitive translation vectors of the body centered cubic lattice are $\vec{a}=\frac{a}{2}(\hat{x}+\hat{y}-\hat{z})$, $\vec{b}=\frac{a}{2}(-\hat{x}+\hat{y}+\hat{z})$ and $\vec{c}=\frac{a}{2}(\hat{x}-\hat{y}+\hat{z})$. The primitive translation vectors $\vec{A},\vec{B}$ and $\vec{C}$ of the reciprocal lattice are

(*a*) $\vec{A}=\frac{2\pi}{a}(\hat{x}-\hat{y});\vec{B}=\frac{2\pi}{a}(\hat{y}+\hat{z});\vec{C}=\frac{2\pi}{a}(\hat{x}+\hat{z})$

(*b*) $\vec{A}=\frac{2\pi}{a}(\hat{x}+\hat{y});\vec{B}=\frac{2\pi}{a}(\hat{y}-\hat{z});\vec{C}=\frac{2\pi}{a}(\hat{x}+\hat{z})$

(*c*) $\vec{A}=\frac{2\pi}{a}(\hat{x}+\hat{y});\vec{B}=\frac{2\pi}{a}(\hat{y}+\hat{z});\vec{C}=\frac{2\pi}{a}(\hat{x}-\hat{z})$

(*d*) $\vec{A}=\frac{2\pi}{a}(\hat{x}+\hat{y});\vec{B}=\frac{2\pi}{a}(\hat{y}+\hat{z});\vec{C}=\frac{2\pi}{a}(\hat{x}+\hat{z})$

112. The structure factor of a single cell of identical atoms of form factor f is given by

$S_{hkl}=f\sum_j \exp\left(-i2\pi\left(x_jh+y_jk+z_jl\right)\right)$

where (x_j, y_j, z_j) is the coordinate of an atom, and hkl are the Miller indices. Which one of the following statement is correct ofor the diffraction peaks of body centered cubuic (BCC) and face centered cubic (FCC) lattices?

(*a*) BCC : (200); (110); (222)
FCC : (111); (311); (400)

(*b*) BCC : (210); (110); (222)
FCC : (111); (311); (400)

(*c*) BCC : (200); (110); (222)
FCC : (111); (211); (400)

(*d*) BCC : (200); (210); (222)
FCC : (111); (2 11); (400)

113. The lattice specific heat C of a crystalline solid can be obtained using the Dulong Petit model, Einstein model and Debye model. At low temperature $\hbar\omega >> k_BT$, which one of the following statements is true (a and A are constants)

(*a*) Dulong Petit : $C \propto \exp\left(\frac{-a}{T}\right)$; Einstein : C = .constant; Debye: $C \propto \left(\frac{T}{A}\right)^3$

(*b*) Dulong Petit : C = .constant; Einstein : $C \propto \left(\frac{T}{A}\right)^3$; Debye: $C \propto \exp\left(\frac{-a}{T}\right)$

(*c*) Dulong Petit : C = .constant; Einstein : $C \propto \frac{e^{-a/T}}{T^2}$; Debye: $C \propto \left(\frac{T}{A}\right)^3$

(*d*) Dulong Petit : $C \propto \left(\frac{T}{A}\right)^3$; Einstein : $C \propto \frac{e^{-a/T}}{T^2}$; Debye: C = .constant

114. A linear diatomic lattice of lattice constant a with masses M and m (M > m) are coupled by a force constant C. The dispersion relation is given by

$$\omega_\pm^2=C\left(\frac{M+m}{Mm}\right)\pm\left[C^2\left(\frac{M+m}{Mm}\right)^2-\frac{4C^2}{Mm}\sin^2\frac{ka}{2}\right]^{1/2}$$

Which one of the following statements is **incorrect?**

(*a*) The atoms vibrating in transverse mode correspond to the optical branch.

(*b*) The maximum frequency of the acoustic branch depends on the mass of the lighter atom m.

(*c*) The dispersion of frequency in the optical branch is smaller than that in the acoustic branch.

(*d*) No normal modes exist in the acoustic branch for any frequency greater than the maximum frequency at $k = \frac{\pi}{a}$.

115. The kinetic energy of a free elctron at a corner of the first Brillouin zone of a two dimensional square lattice is larger than that of an electron at the mid-point of a side of the zone by a factor b. The value of b is

(*a*) $b = \sqrt{2}$ (*b*) $b = 2$
(*c*) $b = 4$ (*d*) $b = 8$

116. An intrinsic semiconductor with mass of a hole m_h and mass of an electron m_e is at a finite temperature T. If the top of the valence band energy is E_v and the bottome of the conduction band energy is E_c, the Fermi energy of the semiconductor is

(*a*) $E_F=\left(\frac{E_v+E_c}{2}\right)-\frac{3}{4}k_BT\ln\left(\frac{m_h}{m_e}\right)$

(*b*) $E_F=\left(\frac{k_BT}{2}\right)+\frac{3}{4}(E_v+E_c)\ln\left(\frac{m_h}{m_e}\right)$

(*c*) $E_F=\left(\frac{E_v+E_c}{2}\right)+\frac{3}{4}k_BT\ln\left(\frac{m_h}{m_e}\right)$

(*d*) $E_F=\left(\frac{k_BT}{2}\right)-\frac{3}{4}(E_v+E_c)\ln\left(\frac{m_h}{m_e}\right)$

117. An a.c. voltage of 200 V_{rms} is applied to the primary of a 10 : 1 step-down transformer. The secondary of the transformer is centre tapped and connected to a full wave rectifier with a load resistance. The d.c. votlage appearing across the load is

(*a*) $\frac{22}{\pi}$ (*b*) $\frac{31}{\pi}$

(*c*) $\frac{62}{\pi}$ (*d*) $\frac{44}{\pi}$

118. Let I_1 and I_2 represent mesh currents in the loop *abcda* and *befcb* respectively. The correct expression describing Kirchoff's voltage loop law in one of the following loops is,

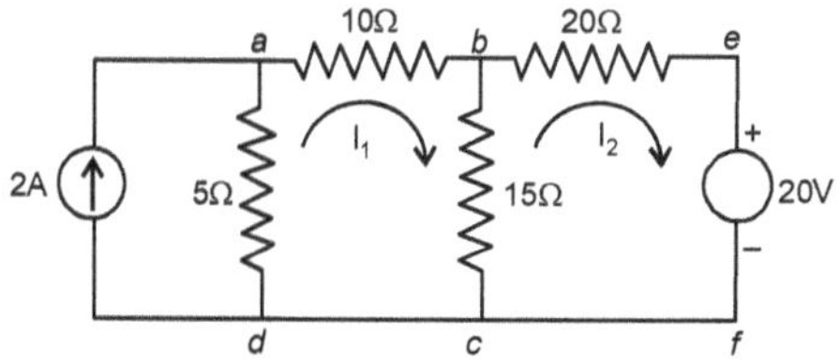

(*a*) $30I_1 - 15I_2 = 10$ (*b*) $-15I_1 + 20I_2 = -20$

(*c*) $30I_1 - 15I_2 = -10$ (*d*) $-15I_1 + 20I_2 = 20$

119. The simplest logic gate circuit corresponding to the Boolean expression, Y = P + PQ is

(*a*) 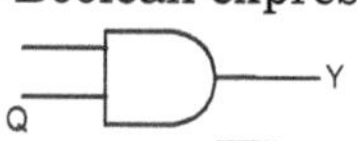

(*b*) 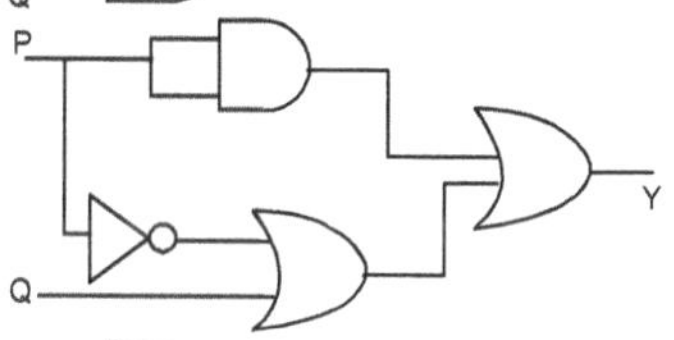

(*c*) 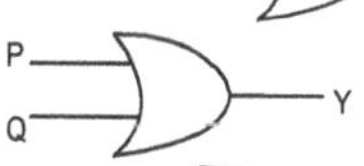

(*d*) 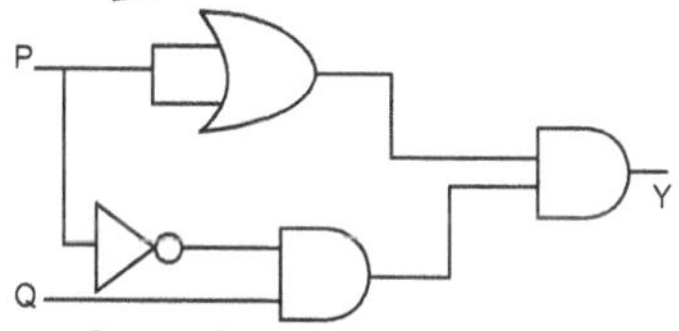

120. An analog voltage V is converted into 2-bit binary number. The minimum number of comparators required and their reference voltages are

(*a*) $3, \left(\frac{V}{4}, \frac{V}{2}, \frac{3V}{4}\right)$ (*b*) $3, \left(\frac{V}{3}, \frac{2V}{3}, V\right)$

(*c*) $4, \left(\frac{V}{5}, \frac{2V}{5}, \frac{3V}{5}, \frac{4V}{5}\right)$ (*d*) $4, \left(\frac{V}{4}, \frac{V}{2}, \frac{3V}{5}, V\right)$

121. The following circuit (whe re $R_L >> R$) performs the operation of

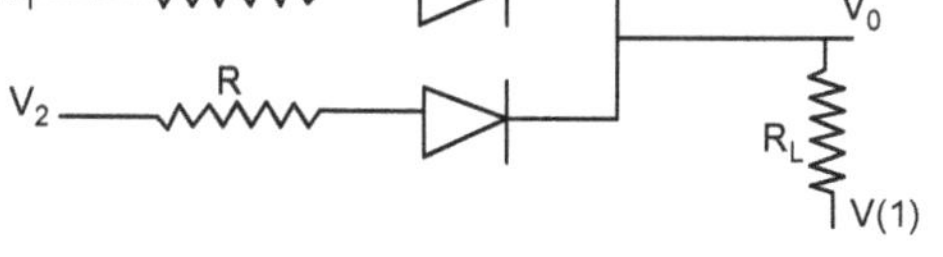

(*a*) OR gate for a negative logic system

(*b*) NAND gate for a negative logic system

(*c*) AND gate for a positive logic system

(*d*) AND gate for a negative logic system

122. In the T type master-slave JK flip flop is shown along with the clock and input waveforms. The Q_n output of flip flop was zero initially. Identify the correct output waveform.

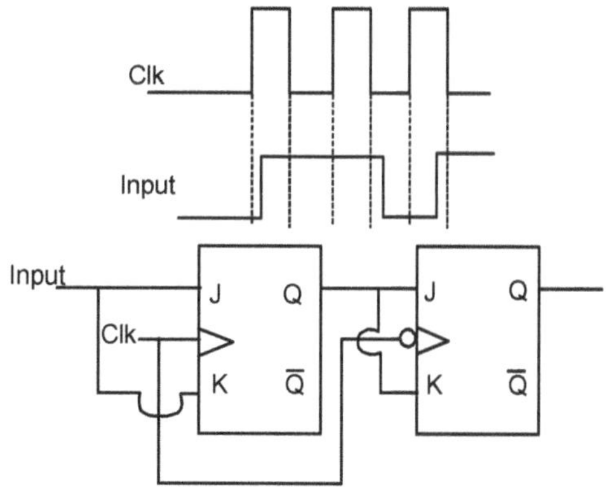

(*a*) 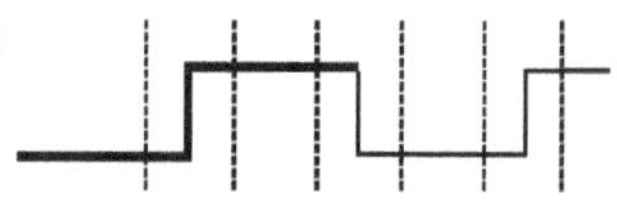

(*b*) 

(*c*) 

(*d*) 

2007

Statement for Linked Answer Questions 123 & 124:

The atomic density of a solid is 5.85 × 10^{28} m^{-3}. Its electrical resistivity is 1.6 × 10^{-8}Ω m.

Assume that electrical conduction is described by the Drude model (classical theory), and that each atom contributes one conduction electron.

123. The drift mobility (in $m^2V^{-1}s^{-1}$) of the conduction electrons is

(*a*) 6.67 × 10^{-3} (*b*) 6.67 × 10^{-6}

(*c*) 7.63 × 10^{-3} (*d*) 7.63 × 10^{-6}

124. The relaxation time (mean free time), in seconds, of the conduction electrons is

(*a*) 3.98 × 10^{-15} (*b*) 3.79× 10^{-14}

(*c*) 2.84 × 10^{-12} (*d*) 2.64 × 10^{-11}

125. The partition function of a single gas molecule is Z_a. The partition function of N such non-interacting gas molecules is then given by

(a) $\frac{(Z_a)^n}{N!}$ (b) $(Z_a)^N$

(c) $N(Z_a)$ (d) $\frac{(Z_a)^n}{N}$

126. A solid superconductor is placed in an external magnetic field and then cooled below its critical temperature. The superconductor

(a) retains its magnetic flux because the surface current supports it.

(b) expels out its magnetic flux because it behaves like a paramagnetic material.

(c) expels out its magnetic flux because it behaves like an anti-ferromagnetic material.

(d) expels out its magnetic flux because the surface current induces a field in the direction opposite to the applied magnetic field.

127. A particle with energy E is in a time-indepndent double well potential as shown in the figure. Which of the following statements about the particle is NOT correct?

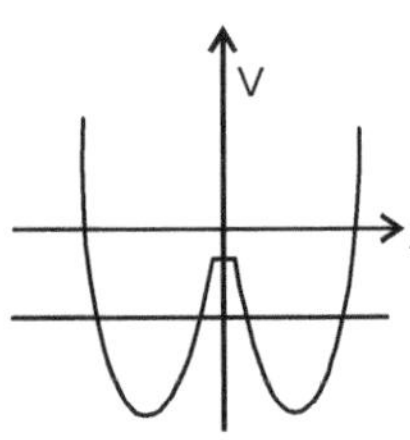

(a) Thge particle will always be in a bound state.

(b) The probability of finding the particle in one well will be time-dependent.

(c) The particle will be confined to any one of the wells.

(d) The particle can tunnel from one well too the other, and back.

128. It is necessary to apply quantum statistics to a system of particles if

(a) there is substantial overlap between the wavefunctions of the particles.

(b) the mean free path of the particles is comparable to the inter-particle separation.

(c) the particles have identical mass and charge.

(d) the particles are interacting.

129. When liquid oxygen is poured down close to a strong bar magnet, the oxygen steam is

(a) repelled towards the lower field because it is diamagnetic.

(b) attracted towards the higher field because it is diamagnetic.

(c) repelled towards the lower field because it is paramagnetic.

(d) attracted towards the higher field because it is paramagnetic.

130. The solid phase of an element follows van der Waals bonding with inter-atomic potential $V(r) = -\frac{P}{r^6} + \frac{Q}{r^{12}}$, where P and Q are constants. The bond length can be expressed as

(a) $\left(\frac{2Q}{P}\right)^{-6}$ (b) $\left(\frac{Q}{P}\right)^{-6}$

(c) $\left(\frac{P}{2Q}\right)^{-6}$ (d) $\left(\frac{P}{Q}\right)^{-6}$

131. Consider the atomic packing factor (APF) of the following crystal structures:

P. Simple Cubic

Q. Body Centred Cubic

R. Face Centred Cubic

S. Diamond

T. Hexagonal Close Packed

Which two of the above structures have equal APF?

(a) P and Q (b) S and T

(c) R and S (d) R and T

132. In a powder differation pattern recorded from a face-centred cubic sample using x-rays, the first peak appears at 30°. The second peak will appear at

(a) 32.8° (b) 33.7°

(c) 34.8° (d) 35.63°

133. Variation of electrical resistivity ρ with temperature T of three solids is sketched (on different scales) in the figure as curves P, Q and R.

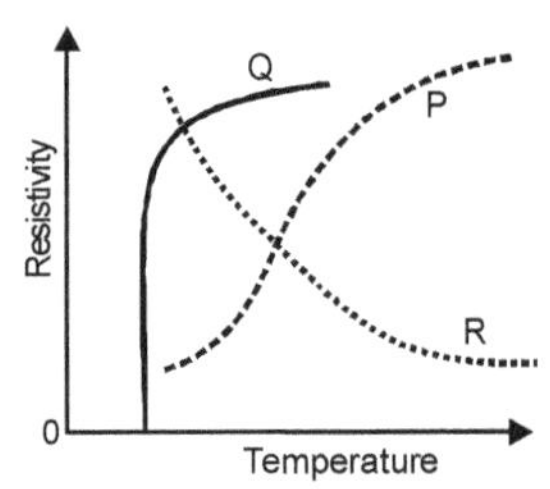

Which one of the following statements describes the variations most appropriately?

(a) P is for a superconductor, and R for a semiconductor.

(b) Q is for a superconductor, and P for a conductor.

(*c*) Q is for a superconductor, and R for a conductor.

(*d*) R is for a superconductor, and P for a conductor.

134. An extrinsic semiconductor sample of cross-section A and length L is doped in such a way thaht the doping concetration varies as $N_D(x) = N_0 \exp\left(-\frac{x}{L}\right)$, where N_0 is a constant. Assume that the mobility μ of the majority carriers remains constant. The resistance R of the sample is given by

(*a*) $R = \frac{L}{A\mu e N_0}[\exp(1.0) - 1]$

(*b*) $R = \frac{L}{\mu e N_0}[\exp(1.0) - 1]$

(*c*) $R = \frac{L}{A\mu e N_0}[\exp(-1.0) - 1]$

(*d*) $R = \frac{1}{A\mu e N_0}$

2006

135. Which one of the following is NOT a correct statement about semiconductors?

(*a*) The electrons and holes have different mobilities in a semiconductor

(*b*) In an *n*-type semiconductor, the Fermi level lies closer to the conduction band edge

(*c*) Silicon is a direct band gap semiconductor

(*d*) Silicon has diamond structure

136. Which one the following axes of rotational symmetry is NOT permissible in single crystals?

(*a*) two–fold axis (*b*) three–fold axis

(*c*) four–fold axis (*d*) five–fold axis

137. A system of N localized, non-interacting spin $\frac{1}{2}$ ions of magnetic moment μ each is kept in an external magnetic field H. If the system is in equilibrium at temperature T, then Helmholtz free energy of the system is

(*a*) $Nk_BT\ \ln\left(\cosh\frac{\mu H}{k_BT}\right)$

(*b*) $-Nk_BT\ \ln\left(2\cosh\frac{\mu H}{k_BT}\right)$

(*c*) $Nk_BT\left(2\cosh\frac{\mu H}{k_BT}\right)$

(*d*) $-Nk_BT\ \ln\left(2\sinh\frac{\mu H}{k_BT}\right)$

138. The phase diagram of a free particle of mass *m* and kinetic energy E, moving in one-dimensional box with perfectly elastic walls at $x = 0$ and $x = L$, is given by

(*a*)

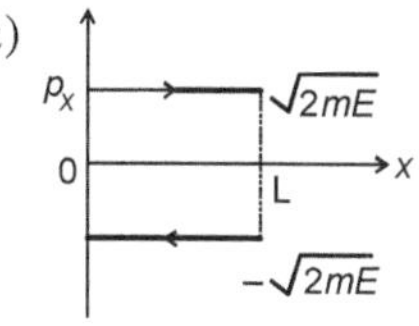

(*b*)

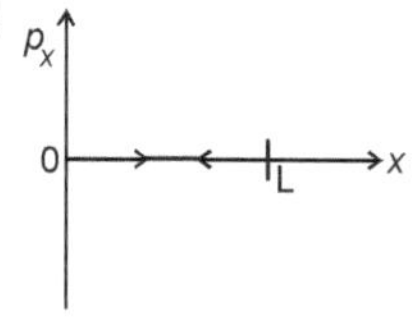

(*c*)

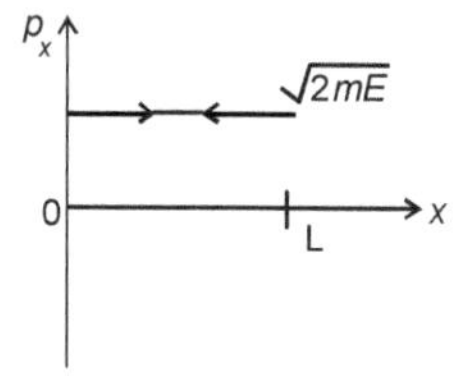

(*d*)

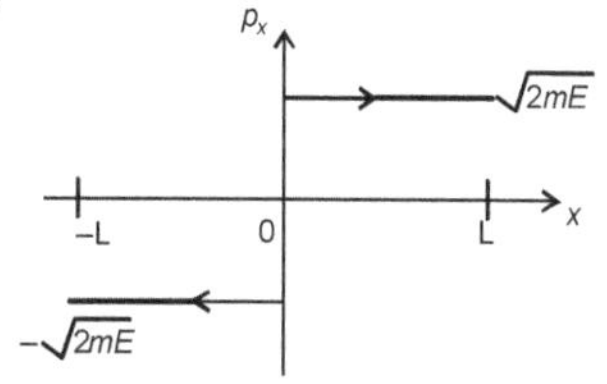

139. Which one of the following statements is NOT correct about the Brillouin zones (BZ) of a square lattices with constant *a*?

(*a*) The first BZ is a square of sides $2\pi/a$ in $k_x - k_y$ plane

(*b*) The areas of the first BZ and third BZ are the same

(*c*) The *k*-points are equidistant in k_x as well as in k_y directions

(*d*) The area of the second BZ is twice that of the first BZ

140. In a crystal of N primitive cells, each cell contains two monovalent atoms. The highest occupied energy band of the crystal is

(*a*) one-fourth filled

(*b*) one-third filled

(*c*) half filled

(*d*) completely filled

141. If the number density of a free electron gas changes from 10^{28} to 10^{26} electrons/m^3, the value of plasma frequency (in Hz) changes from 5.7×10^{15} to

(*a*) 5.7×10^{13} (*b*) 5.7×10^{14}

(*c*) 5.7×10^{16} (*d*) 5.7×10^{17}

142. Which one of the following statements about superconductors in NOT true?

(*a*) A type I superconductor is completely diamagnetic

(*b*) A type II superconductor exhibits Meissner effect upto the second critical magnetic field (H_{C_2})

(*c*) A type II superconductor exhibits zero resistance upto the second critical magnetic field

(*d*) Both type I and type II superconductors exhibit sharp fall in resistance at the superconducting transition temperature.

143. Two dielectric materials A and B exhibit both ionic and orientational polarizabilities. The variation of their susceptibilities χ (= $\varepsilon_r - 1$) with temperature T is shown in the figure, where ε_r is the relative dielectric constant. It can be inferred from the figure that

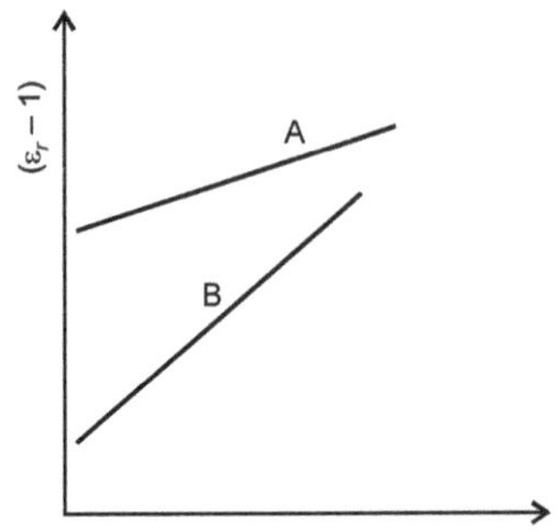

(*a*) A is more polar and it has a smaller value of ionic polarizability than that of B

(*b*) A is more polar and it has a higher value of ionic polarizability than that of B

(*c*) B is more polar and it has a higher value of ionic polarizability than that of A

(*d*) B is more polar and it has a smaller value of ionic polarizability than that of A.

2005

144. In crystallographic notations the vector $\overline{OP}$ in the cubic cell shown in the figure is

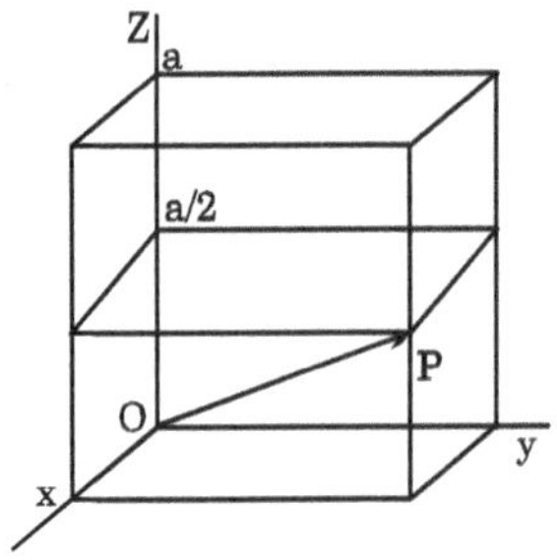

(*a*) [221] (*b*) [122]

(*c*) [121] (*d*) [112]

145. A junction field effect transistor behaves as a

(*a*) voltage controlled current source

(*b*) voltage controlled voltage source

(*c*) current controlled voltage source

(*d*) current controlled current source

146. The circuit shown can be used as

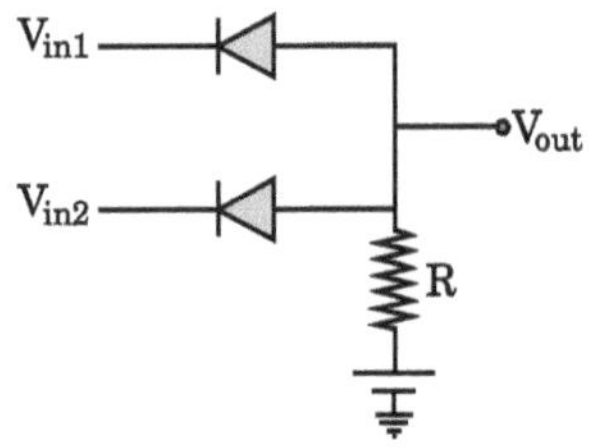

(*a*) NOR gate (*b*) OR gate

(*c*) NAND gate (*d*) AND gate

147. The energy of a ferromagnet as a function of magnetization M is given by

$$F(M) = F_0 + 2(T - T_c) M^2 + M^4, \quad F_0 > 0.$$

The number of minima in the function F(M) for $T > T_c$ is

(*a*) 0 (*b*) 1

(*c*) 3 (*d*) 4

148. For a closed packed BCC structure of hard spheres, the lattice constant a is related to the sphere radius R as

(*a*) $a = 4R/\sqrt{3}$ (*b*) $a = 4R\sqrt{3}$

(*c*) $a = 4R\sqrt{2}$ (*d*) $a = 2R\sqrt{2}$

149. An n-type semiconductor has an electron concentration of 3×10^{20} m^{-3}. If the electron drift velocity is 100 ms^{-1} in an electric field of 200 Vm^{-1}, the conductivity (in Ω^{-1}m^{-1}) of this material is

(*a*) 24 (*b*) 36

(*c*) 48 (*d*) 96

150. Density of states of free electrons in a solid moving with an energy 0.1 eV is given by 2.15×10^{21} eV^{-1} cm^{-3}. The density of states (in eV^{-1} cm^{-3}) for electrons moving with an energy of 0.4 eV will be

(*a*) 1.07×10^{21} (*b*) 1.52×10^{21}

(*c*) 3.04×10^{21} (*d*) 4.30×10^{21}

151. The effecitve density of states at the conduction band edge of Ge is 1.04×10^{19} cm^{-3} at room temperature (300 K). Ge has an optical bandgap of 0.66 eV. The intrinsic carrier concentration (in cm^{-3}) in Ge at room temperature (300 K) is approximately

(*a*) 3×10^{10} (*b*) 3×10^{13}

(*c*) 3×10^{16} (*d*) 6×10^{16}

152. For a conventional superconductor, which of the following statements is NOT true ?
(a) Specific heat is discontinuous at transition temperature T_c
(b) The resistivity falls sharply at T_c
(c) It is diamagnetic below T_c
(d) It is paramagnetic below T_c

2004

153. The c/a ratio for an ideal hexagonal closed packed structure is
(a) $\frac{2}{\sqrt{3}}$ (b) $\sqrt{8}$
(c) $\sqrt{5}$ (d) $\sqrt{\frac{8}{3}}$

154. The number of independent elastic constants in an isotropic cubic solid is
(a) 1 (b) 2
(c) 3 (d) 4

155. The effective mass of an electron in a semiconductor
(a) can never be positive
(b) can never be negative
(c) can be positive or negative
(d) depends on its spin

156. The critical magnetic field for a solid in superconducting state
(a) does not depend upon temperature
(b) increases if the temperature increases
(c) increases if the temperature decreases
(d) does not depend on the transition temperature

157. The equation of state of a dilute gas at very high temperatures is described by $\frac{pv}{k_B T} \approx 1 + \frac{B(T)}{v}$, where v is the volume per particle and B(T) is a negative quantity. One can conclude that this is a property of
(a) a van der Waals gas
(b) an ideal Fermi gas
(c) an ideal Bose gas
(d) an ideal inert gas

158. A doped Germanium crystal of length 2 cm, breadth 1 cm and width 1 cm, carries a current of 1 mA along its length parallel to $+x$ axis. A magnetic field of 0.5 T is applied along $+z$ axis. Hall voltage of 6 mV is measured with negative polarity at $y = 0$ plane. The sign and concentration of the majority charge carrier are, respectively.
[Given : $e = 1.6 \times 10^{-19}$C]

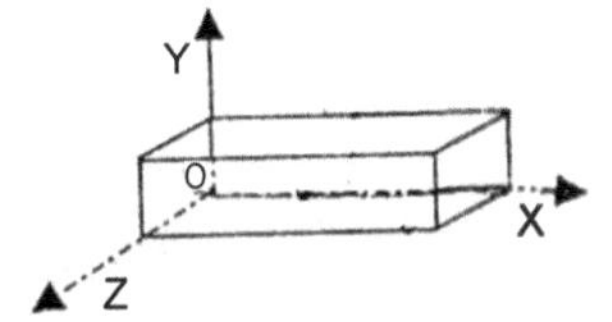

(a) positive and 5.2×10^{19} m^{-3}
(b) negative and 5.2×10^{19} m^{-3}
(c) positive and 10.4×10^{19} m^{-3}
(d) negative and 10.4×10^{19} m^{-3}

159. The temperature dependence of the electrical conductivity σ of two intrinsic semiconductors A and B is shown in the figure. If E_A and E_B are the band gaps of A and B respectively, which one of the following is TRUE ?

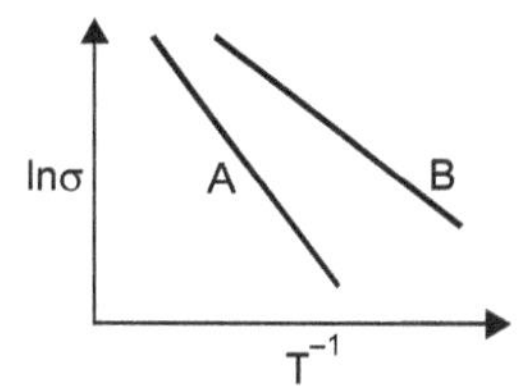

(a) $E_A > E_B$
(b) $E_A < E_B$
(c) $E_A = E_B$
(d) E_A and E_B both depend on temperature

160. If the static dielectric constant of NaCl crystal is 5.6 and its optical refractive index is 1.5, the ratio of its electric polarizability to its total polarizability is
(a) 0.5 (b) 0.7
(c) 0.8 (d) 0.9

2003

161. Consider the Fermi-Dirac distribution function f(E) at room temperature (300 K) where E refers to energy. If E_F is the Fermi energy, which of the following is true?
(a) f(E) is a step function
(b) $f(E_F)$ has a value of 1/2
(c) States with $E < E_F$ are filled completely
(d) f(E) is large and tends to infinity as E decreases much below E_F

162. If the ionic radii of Mn and S are 0.80 and 0.184 nm respectively, the structure of MnS will be
(a) cubic closed packed
(b) body centered cubic
(c) NaCl type
(d) primitive cubic cell

163. A cubic cell consists of two atoms of masses m_1 and m_2 ($m_1 > m_2$) with m_1 and m_2 atoms situated on alternate planes. Assuming only nearest neighbor interactions, the center of mass of the two atoms

(*a*) moves with the atoms in the optical mode and remains fixed in the acoustic mode

(*b*) remains fixed in the optical mode and moves with the atoms in the acoustic mode

(*c*) remains fixed in both optical and acoustic modes

(*d*) moves with the atoms in both optical and acoustic modes

164. In simple metals the phonon contribution to the electrical resistivity at temperature T is

(*a*) directly proportional to T above Debye temperature and to T^3 below it

(*b*) inversely proportional to T for all temperatures

(*c*) independent of T for all temperatures

(*d*) directly proportional to T above Debye temperature and to T^5 below it

Data for Q. 165 – 166 :

A crystal belongs to a face centered cubic lattice with four atoms in the unit cell. The size of the crystal is 1 cm and its unit cell dimension is 1 nm. f is the scattering factor of the atom

165. The number of atoms in the crystal is

(*a*) 2×10^{21} (*b*) 4×10^{21}

(*c*) 2×10^{23} (*d*) 4×10^{24}

166. The structure factors for (0 1 0) and (2 0 0) reflections respectively are

(*a*) $2f$ and zero (*b*) zero and $4f$

(*c*) $2f$ and $2f$ (*d*) zero and zero

2002

167. In a one-dimensional Kronig Penny model, the total number of possible wave functions is equal to

(*a*) twice the number of unit cells

(*b*) number of unit cells

(*c*) half the number of unit cells

(*d*) independent of the number of unit cells

168. The potential in a divalent solid at a particular temperature is represented by a one-dimensional periodic model. The solid should behave electrically as

(*a*) a semiconductor (*b*) a conductor

(*c*) an insulator (*d*) a superconductor

169. In a cubic system with cell edge a, two phonons with wave vectors $\vec{q_1}$ and $\vec{q_2}$ collide and produce a third phonon with a wave vector $\vec{q_3}$ such that $\vec{q_1} + \vec{q_2} = \vec{q_3} + \vec{R}$, where $\vec{R}$ is a lattice vector. Such a collision process will lead to

(*a*) finite thermal resistance

(*b*) zero thermal resistance

(*c*) an infinite thermal resistance

(*d*) a finite thermal resistance for certain $|\vec{R}|$ only

170. For a prefect free-electron gas in a metal, the magnitudes of phase velocity (v_p) and group velocity (v_g) are such that

(*a*) $v_p = v_g$ (*b*) $v_p = \frac{1}{2} v_g$

(*c*) $v_p = \sqrt{2}\, v_g$ (*d*) $v_p = 2\, v_g$

171. A metal has free-electron density $n = 10^{29}$ m^{-3}. Which of the following wave-lengths will excite plasma oscillations ?

(*a*) 0.033 μm (*b*) 0.330 μm

(*c*) 3.300 μm (*d*) 33.000 μm

172. Silicon has diamond structure with unit-cell-edge a = 0.542 nm. The interatomic separation is

(*a*) 0.122 nm

(*b*) 0.234 nm

(*c*) 0.383 nm

(*d*) 0.542 nm

2001

173. For an energy state E of a photon gas, the density of states is proportional to

(*a*) $\sqrt{E}$ (*b*) E

(*c*) $E^{3/2}$ (*d*) E^2

174. The probability that a state which is 0.2 eV above the Fermi energy in a metal atom at 700 K is

(*a*) 96.2% (*b*) 62.3%

(*c*) 3.5% (*d*) 37.7%

175. A piece of semiconducting material is introduced into a circuit. If the temperature of the material is raised, the circuit current will

(*a*) increase

(*b*) remain the same

(*c*) decrease

(*d*) cease to flow

176. A copper wire of uniform cross-sectional area 1.0×10^{-6} m^2 carries a current of 1 A. Assuming that each copper atom contributes one electron to the electron gas, the drift velocity of the free electrons (density of copper is 8.94×10^3 kg/m^3 and its atomic mass is 1.05×10^{-25} kg) is

(*a*) 7.4×10^{-4} m/s (*b*) 74×10^{-4} m/s

(*c*) 74×10^{-3} m/s (*d*) 7.4×10^{-5} m/s

2000

177. Which of the following is an example of a first order phase transition?

(*a*) A liquid - gas phase transition at the critical point.

(*b*) A paramagnet - ferromagnet phase transition.

(*c*) A normal metal - superconductor phase transition.

(*d*) A liquid - gas phase transition away from the critical point.

178. The Fermi energy of a free electron gas depends on the electron density ρ as,

(*a*) $\rho^{1/3}$ (*b*) $\rho^{2/3}$

(*c*) $\rho^{-1/3}$ (*d*) $\rho^{-2/3}$

179. If the output of the logic circuit shown in the figure is 1, the input could be

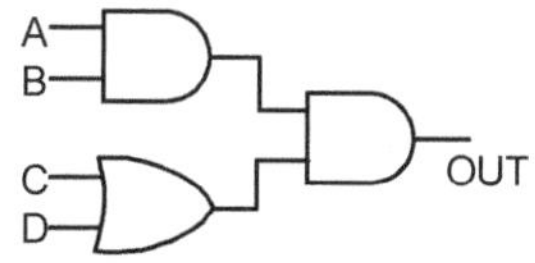

(*a*) A = 1, B = 1, C = 1, D = 0

(*b*) A = 1, B = 1, C = 0, D = 0

(*c*) A = 1, B = 0, C = 1, D = 1

(*d*) A = 0, B = 1, C = 1, D = 1

180. What is the second nearest-neighbour distance in a face centred cubic lattice whose conventional unit cell parameter is a?

(*a*) $a/\sqrt{2}$ (*b*) a/2

(*c*) a (*d*) $\sqrt{2}/a$

ANSWERS

1. (a) **2.** (a) **3.** (b) **4.** (b,c,d) **5.** (115 to 116) **6.** (b) **7.** (d) **8.** (a)
9. (14.0 to 15.0) **10.** (5.40 to 5.50) **11.** (d) **12.** (b) **13.** (5 to 5) **14.** (c) **15.** (b) **16.** (d)
17. (c) **18.** (8 to 8) **19.** (2 to 2) **20.** (a) **21.** (c) **22.** (b) **23.** (1.07 to 1.11)
24. (1 to 1) **25.** (0.244 to 0.248) **26.** (d) **27.** (c) **28.** (d) **29.** (a) **30.** (22.0 to 22.2)
31. (0.333 to 0.333) **32.** (c) **33.** (a) **34.** (a) **35.** (a) **36.** (c) **37.** (– 3.7 to – 3.5)
38. (1 to 1) **39.** (a) **40.** (b) **41.** (1.55 to 1.58) **42.** (0.70 to 0.73) **43.** 39.30 to 39.50 **44.** (a)
45. 2.50 to 2.55 **46.** (a) **47.** (d) **48.** (d) **49.** 2.32 to 2.40 **50.** 0.90 to 1.10 **51.** 2.35 to 2.45
52. 1.30 to 1.36 **53.** (a) **54.** 4.75 to 5.01 **55.** 9.80 to 10.20 **56.** (400 : 400) **57.** (c) **58.** (0.40:0.42)
59. (0.20 : 0.24) **60.** (0.445 to 0.450) **61.** (d) **62.** (0) **63.** (c) **64.** (0.394 to 0.395) **65.** (b) **66.** (c)
67. (114 to 117) **68.** (3.91 to 4.15) **69.** (9.99 to 10.01) **70.** (c) **71.** (c) **72.** (a) **73.** (a) **74.** (c)
75. (b) **76.** (d) **77.** (c) **78.** (c) **79.** (a) **80.** (d) **81.** (a) **82.** (b) **83.** (c) **84.** (c) **85.** (c)
86. (c) **87.** (b) **87.** (d) **89.** (c) **90.** (d) **91.** (d) **92.** (a) **93.** (c) **94.** (c) **95.** (b) **96.** (c)
97. (a) **98.** (a) **99.** (c) **100.** (a) **101.** (a) **102.** (c) **103.** (c) **104.** (c) **105.** (c) **106.** (b) **107.** (b)
108. (d) **109.** (b) **110.** (a) **111.** (d) **112.** (c) **113.** (c) **114.** (c) **115.** (c) **116.** (c) **117.** (d) **118.** (a)
119. (b) **120.** (a) **121.** (a) **122.** (b) **123.** (b) **124.** (b) **125.** (b) **126.** (d) **127.** (c) **128.** (d) **129.** (a)
130. (a) **131.** (d) **132.** (d) **133.** (b) **134.** (c) **135.** (c) **136.** (d) **137.** (b) **138.** (a) **139.** (b) **140.** (d)
141. (b) **142.** (b) **143.** (b) **144.** (a) **145.** (a) **146.** (d) **147.** (b) **148.** (a) **149.** (a) **150.** (a) **151.** (b)
152. (d) **153.** (d) **154.** (c) **155.** (c) **156.** (c) **157.** (a) **158.** (b) **159.** (a) **160.** (a) **161.** (b) **162.** (b)
163. (d) **164.** (d) **165.** (b) **166.** (a) **167.** (b) **168.** (b) **169.** (a) **170.** (a) **171.** (a) **172.** (c) **173.** (a)
174. (c) **175.** (a) **176.** (d) **177.** (d) **178.** (b) **179.** (a) **180.** (c)

EXPLANATIONS

1. Dielectric constant of a ferromagnetic material increases exponentially upto Tc and then decreases exponentially. So option (a) is correct.

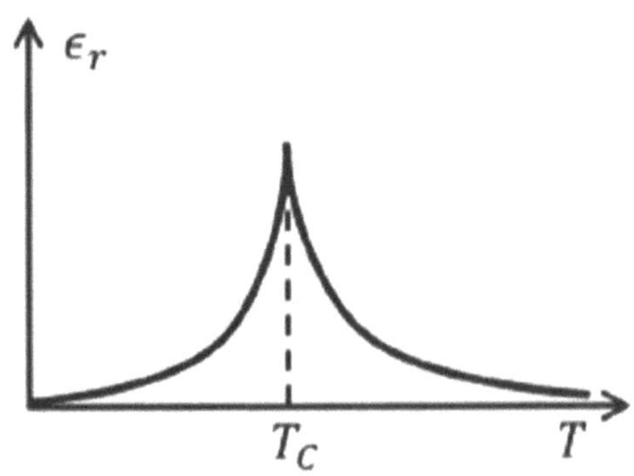

2. Given

φ_M (work function of Metal) > φ_s (work function of semicircles)

The given arrangement is act as two

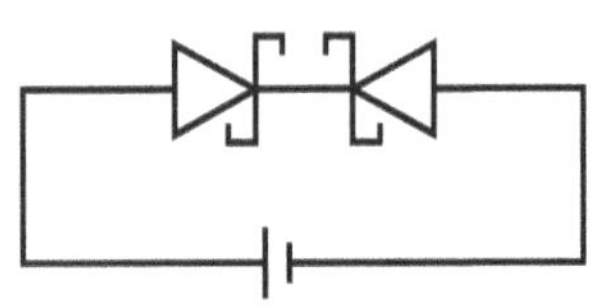

So for I-V characteristics curve is same for forward and Reverse Bias.

So, option (a) is correct.

3. Since, metals have positive coefficient of resistance,

So, resistance increase with increase in temperature

So, option (b) is correct.

4. Bond gap can occurs at the Brillown zone boundary

For first brillouin boundries at

$\left(\pm\frac{\pi}{a}, \pm\frac{\pi}{a}\right)$

So, option (b) is correct.

5. Fermi Energy is

$$E_f = E_c - K_B T\, ln\left(\frac{N_a}{N_d}\right) \quad ...(i)$$

when doping is increased by factor 100, fermi energy is

$$E_F = E_c - k_B T\, ln\left(\frac{Na}{100Nd}\right) \quad ...(ii)$$

eq (ii) – eq(i)

$$E_f - E_f = E_c - k_B T\, ln\left(\frac{N_a}{100N_d}\right) - E_c + k_B T\, ln\left(\frac{N_a}{N_d}\right)$$

$$E_f - E_f = -k_B T\, ln\left(\frac{N_a}{N_d}\right) - k_B T\, ln(100) + k_B T$$

$$ln\left(\frac{N_a}{N_d}\right)$$

$E_f - E_f = k_B T\, ln(100)$

$= 25\ y\ ln(100)$

$= 115.129$

So, shift in fermi level is 115.129.

6.

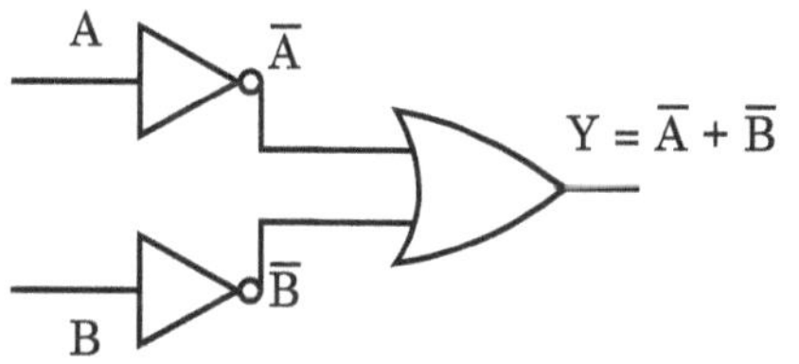

Output $y = \overline{A} + \overline{B}$

A	B	$\overline{A}$	$\overline{B}$	$Y = \overline{A} + \overline{B}$
0	0	1	1	1
0	1	1	0	1
1	0	0	1	1
1	1	0	0	0

So, Combination represents NAND gates

Hence, option (b) is the correct answer.

7. We know that, mobility $\mu \propto \frac{\tau}{m}$

$$\frac{\mu_h}{\mu_e} = \frac{\tau_h}{\tau_e} \times \frac{m_e}{m_h} \Rightarrow \frac{1}{2} \times \frac{11}{2} \Rightarrow \frac{11}{4}$$

$$\frac{\mu_h}{\mu_e} = \frac{11}{4}$$

So, option (d) is the correct answer.

8. For Antiferromagnetic substances

$$\chi = \frac{C}{T + T_N} \qquad \text{Line (1)}$$

For ferromagnetic substance

$$\chi = \frac{C}{T - T_C} \qquad \text{Line (3)}$$

For paramagnetic substance

$$\chi = \frac{C}{T} \qquad \text{Line (2)}$$

So, option (a) is the correct answer.

9. Given frequency F = 10MHz

$F = 10^6$ Hz

Time period $T = \frac{1}{F} = 10^{-6}$ sec

Velocity $V = 2.9 \times 10^8$ m/s

Minimum length $= V \times \frac{T}{2}$ ($\frac{T}{2}$ half AC cycle)

$= 2.9 \times 10^8 \times \frac{10^{-6}}{2}$

$= 14.5$ m

10. Paramagnetic susceptibility

$$\chi = \frac{n\mu_0 (\mu_\beta)^2 g^2 J(J+1)}{3K_B T}$$

Here $l = 0$, $S = \frac{7}{2}$, $J = \frac{7}{2}$

$$g = 1 + \frac{J(J+1) + S(S+1) - l(l+1)}{2J(J+1)}$$

By putting values

$g = 1 + 1$

$g = 2$

and $g^2 = 4$

Putting Values

$$\chi = \frac{10^{20} \times 4\pi \times 10^{-7} \times (9.274 \times 10^{-24})^2 \times 4 \times \frac{7}{2} \times \frac{9}{2}}{3 \times 1.3807 \times 10^{-23} \times 300}$$

$\chi = 5.47 \times 10^{-11}$

So $\boxed{m = 5.47}$

11. NAND is a universal logic gate. A universal gate is a gate which can implement any Boolean function without need to used any other gate type.

12. F6C

$$\begin{matrix} 15 & 6 & C \\ \downarrow & \downarrow & \downarrow \\ (1111 & 0110 & 1100)_2 \end{matrix} = (111101101100)_2$$

A – 10
B – 11
C – 12
D – 13
E – 14
F – 15

13. Can be constructed in 5 ways.

14.
$$\omega = \frac{1}{RC}$$
$$V_i = 2\cos(100t)$$
$$V_o = 2\cos\left(100t - \frac{\pi}{2}\right)$$
$$C = \frac{1}{\omega R}$$
$$C = \frac{1}{100 \times 1 \times 10^3} = 10\ \mu F$$

17. For 2D, Acoustic branch = 2 m

For 2D, optical branch = 2m(p – 1)

p = number of basis

p = 3

Optical branch (per unit cell)

= (3 – 1)2

= 4

18.
$$I = \frac{10.1 - 3 \times 0.7}{1}$$
$$\boxed{I = 8\ mA}$$

19.
$$S = \sum_{i=1}^{n} e^{2\pi i \vec{K}.\vec{G}}$$
$$= e^{2\pi i(0+0+0)} + e^{2\pi i\left(\frac{9}{4}n_1 + \frac{9}{4}n_2 + \frac{9}{4}n_3\right)}$$
$$S = 1 + e^{4\pi i} = 2$$

23. At r = R, U(R) → min

$$\frac{dU}{dR} = 0$$
$$2N \in \left[-p\sigma^{12}(12)\frac{1}{R^{13}} + 7q\sigma^6 \frac{1}{R^7}\right] = 0$$
$$2N \in \frac{\sigma^6}{R^7}\left[-12\frac{\sigma^6}{R^6} + 6q\right] = 0$$
$$R \neq 0,\ \sigma q = 12p\frac{\sigma^6}{R^6}$$
$$R^6 = \frac{2p}{q}\sigma^6 \Rightarrow R = \left(\frac{2p}{q}\right)^{\frac{1}{6}}\sigma$$
$$R = \left(\frac{2 \times 12.13}{14.45}\right)^{\frac{1}{6}}\sigma$$
$$R = 1.09\sigma.$$

24. E = CK

$$D(E) \propto E^p$$
$$D(E) \propto E^{\frac{d}{s}-1}$$
$$d = 2,\ s = 1$$
$$D(E) \propto E^{\frac{2}{1}-1}$$
$$D(E) \propto E^{\frac{2}{1}-1} = E$$
$$D(E) \propto E$$
$$\therefore\ E = 1$$

26.
$$Y = (\overline{A} + \overline{B})\left[\overline{A(B+C)}\right] + A(\overline{B} + \overline{C})$$
$$Y = (\overline{A} + \overline{B})[\overline{A} + \overline{B}.\overline{C}] + A(\overline{B} + \overline{C})$$
$$Y = \overline{A}.\overline{A} + \overline{A}\,\overline{B}\,\overline{C} + \overline{B}\,\overline{A} + \overline{B}\,\overline{B}\,\overline{C} + A\overline{B} + A\overline{C}$$
$$Y = \overline{A} + \overline{B}\,\overline{C} + (\overline{A} + 1) + \overline{B}(\overline{A} + A) + A\overline{C}$$
$$Y = (\overline{A} + A\overline{C}) + \overline{B}\,\overline{C} + \overline{B}$$
$$Y = \overline{A} + \overline{C} + \overline{B}(\overline{C} + 1)$$
$$Y = \overline{A} + \overline{C} + \overline{B} = \overline{A} + \overline{B} + \overline{C}$$
$$= \overline{A.\,B\,C} - \text{NAND GATE}$$

27. $V_{net} = 10v,\ n = 3$

$$\text{Resultant} = R = \frac{V_{net}}{e^n - 1} = \frac{10}{2^3 - 1} = \frac{10}{7}$$
$$= 1.428\ V$$
$$\frac{6}{1.428} = 4.2$$

0 → 0 0 0 0 – 1.25 V

1 → 0 0 1 → 125 – 2.5 V

2 → 0 1 0 → 2.5 – 3.75 V

3 → 0 1 1 → 3.75 – 5.00 V

$$\boxed{4 \rightarrow 1\,0\,0 \rightarrow 5.00 - 6.25V}$$

28. $\Delta E\, \Delta t \approx h$

$$\Delta t = T$$
$$\Delta E = \int_0^d F\, dx = F.d$$

$$\Delta t = \frac{h}{\Delta E}$$

$$\boxed{T = \frac{h}{Fd}}$$

29. R – S Flip Flop

X	Y	$\overline{X+Y}$
0	0	1
0	1	0
1	0	0
1	1	0

$X = 1 \,\&\, Y = 0$

32. Lattices can be classified into 'systems' each system being characterized by the shape of its associated unit cell. In three dimensions, the lattices are categorized into seven crystal lattice 'systems'. Within several of these lattices supporting non-primitive unit cell can be defined. The classification scheme yields a total of 14 possible lattices i.e. called Bravis lattices.

33. $\overline{A}BC + \overline{A}\overline{B}C + AB\overline{C} + A\overline{B}\overline{C}$

$= \overline{A}C(B + \overline{B}) + A\overline{C}(B + \overline{B})$

$= \overline{A}C + A\overline{C}$ $\quad (\because B + \overline{B} = 1)$

$= A \oplus C$

34. At low temperature, the specific heat of common metals is described by

$$E = \alpha T + \beta T^3.$$

35. Let C represent control line, A_0 and A_1 the inputs, and X be the output, then equation to describe the desired operation is

$$X = \overline{C}A_0 + CA_1$$

(*i*) If C = 0 then CB = 0 and $\overline{C}A = A_0$ so that input A_0 is selected

(*ii*) If C = 1 then CB = A_1 and $\overline{C}A = 0$ so that input A_1 is selected.

We note that only one input is selected at a time with the help of control line. So, the correct implementation of this multiplexer is shown below

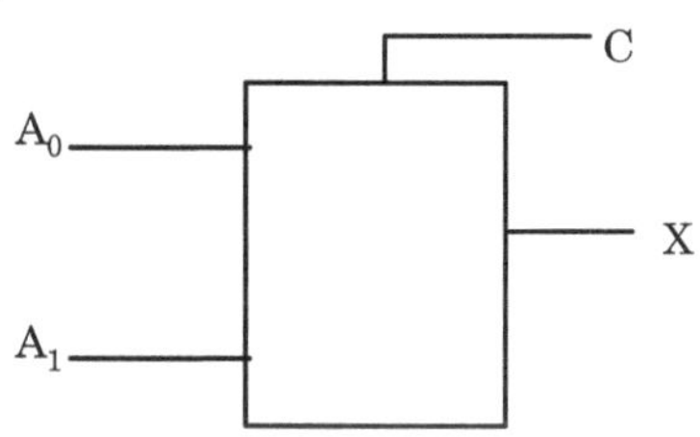

Fig (a)

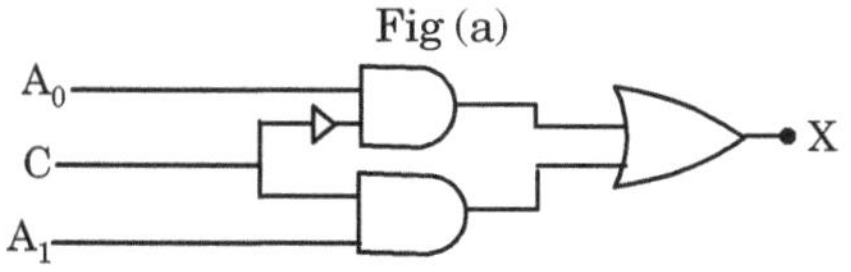

Fig (b)

Hence, option (*a*) is correct choice.

36. Magnetic susceptibility $\chi \propto \frac{1}{T+\theta}$

$$\therefore \quad \chi = \frac{C}{T+\theta}$$

The three behaviours:

For paramagnetic; $\theta = 0$

For ferromagnetic; $\theta < 0$

For anti-ferromagnetic; $\theta > 0$

37. For ideal amplifier input current is zero.

Then,

$$V_0 = -\left[\frac{R_2}{r_1} \times V_1 + \frac{R_2}{r_2} V_2\right]$$

Given

$R_2 = 4k\Omega$, $r_1 = 2K\Omega$, $r_2 = 5K\Omega$, $V_1 = 1V$ and $V_2 = 2V$

$$\therefore \quad V_0 = -\left[\frac{4}{2} \times 1 + \frac{4}{5} \times 2\right]$$

$$= -\left(2 + \frac{8}{5}\right) = -\left[\frac{10+8}{5}\right]$$

$$= -\frac{18}{5} = -3.6V$$

38. If X is the dimensionality of free electron gas the value of x = 2, then

$$y = \frac{1}{2} \times 2 = 1$$

39. The effective mass of electron

$$m_{eff} = \frac{h^2}{\left(\frac{d^2E}{dk^2}\right)}$$

$$\therefore \quad E(k) = E_o - \frac{1}{2} W \cos ka$$

$$\frac{dE(k)}{dk} = 0 - \frac{1}{2} Wa \sin ka$$

$$\Rightarrow \quad \frac{d^2E(k)}{dk^2} = \frac{1}{2} Wa^2 \cos ka$$

($\therefore$ cos *ka* = 1 because electron near the bottom of the band.)

$$= \frac{1}{2} Wa^2$$

$$\therefore \quad m_{eff} = \frac{h^2}{\frac{1}{2} Wa^2} = \frac{2h^2}{Wa^2}$$

40. Electrical Resistivity (ρ) : At a certain temperature and often within a very narrow temperature range, resistivity of certain metals becomes zero.

Specific Heat (C) : The electronic specific heat C is not linear with temperature. It instead fits an exponential form.

$$C = A\,e^{-bT_0/T}$$

Magnetic Susceptibility (χ) : This quantity shows a very sharp change at the superconducting transition temperature.

43. Formula for the Resolution of 7 bit A/D convertor is given by

$$R = \left(\frac{1}{2^7}\times 5\right)V = \left(\frac{1}{128}\times 5\right)V$$

$= 0.039 = 39 \times 10^{-3}V$

$= 39$ mV.

44. As transformer is a voltage varying with time.

The plot is :-

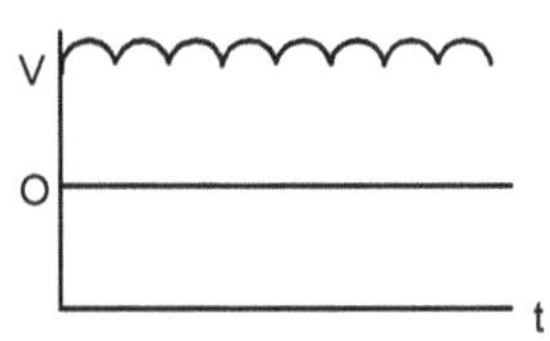

45. The atomic mass and mass density are 23 and 0.968 g/cm^3.

The number density of valence electron is given by

$$n = \frac{N_A \times (\text{density of sodium})}{\text{Atomic mass}}$$

$$= \frac{6.023\times 10^{23}\times 0.968}{23}$$

$$= 0.2534\times 10^{23}$$

$$= 2.54\times 10^{22}\ \text{cm}^{-3}$$

46. Weak periodic potential is given by

$U(x) = U_0$ as $\left(\frac{2\pi x}{a}\right)$. The gap at the edge of the Brillonin Zone $\left(k = \frac{\pi}{a}\right)$ is U_0.

47. Heat capacity of triatomic molecule at higher temperature is given by $6k_B$.

As no. of degree of freedom for non-linear triatomic molecule is $(3N - 1) = 6$.

Heat capacity is given by $fk_B = 6k_B$. Where k_B is boltzman constant.

48. Minimum number of NAND gates required to construct 'OR Gate' is 3.

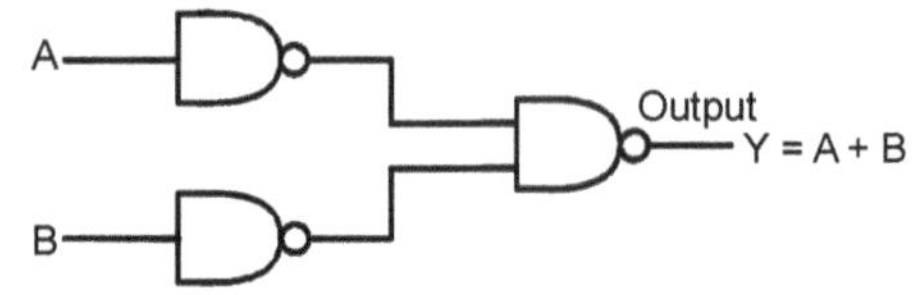

49. Given : The density of two dimensional electron gas $= 10^{19}\text{m}^{-2}$

Fermi energy is given by

$$\varepsilon_F = \frac{\eta a h^2}{m}$$

$$\eta = \frac{N}{A} = 10^{19}\ \text{m}^{-2}$$

$$= 10^{19}\times\frac{3.14\times(6.626\times 10^{-34})^2}{9.31\times 10^{-31}}$$

$= 14.80753713 \times 10^{-18} \times 1.6 \times 10^{-19}$

$= 2.36 \times 10^{-36}$ eV.

50. The total energy of an insert crystal is given by

$$E(R) = \frac{0.5}{R^{12}} - \frac{1}{R^6}(\text{in ev})$$

The equilibrium separation between the atoms is-

$$E(R) = \frac{0.5}{R^{12}} - \frac{1}{R^6}$$

$$\Rightarrow \frac{0.5}{R^{12}} - \frac{1}{R^6} = ECR$$

$$\Rightarrow 0.5 = \frac{R^{12}}{R^6} \Rightarrow 0.5 = R^6 \Rightarrow R = (0.5)^{1/6}$$

$= 0.91$ Å.

51. We know the relation

$$W_P^2 = \frac{ne^2}{m\varepsilon_0}$$

$$\left(\begin{array}{l}\text{given } n = 1.8\times 10^{23}\\ m_e = 9.1\times 10^{-31}\text{kg}, \varepsilon_0 = 8.854\times 10^{-12}\text{F/m}\end{array}\right)$$

$$= \frac{1.8\times 10^{23}\times(1.6\times 10^{-19})^2}{9.1\times 10^{-31}\times 8.854\times 10^{-12}}$$

$$= \frac{4.608}{80.5714}\times 10^{28} = 0.057191509\times 10^{28}$$

$$W_P = \sqrt{0.05719} = \sqrt{2.38\times 10^{27}}$$

$$= \sqrt{0.238\times 10^{28}}$$

$$= 2.38\times 10^{14}\text{Hz}.$$

52. Given u = aT^4 and P = u/3

The entropy per unit volume is given by $\alpha \propto T^3$.

dV = TdS

$$dS = \frac{1}{T}dU = 4aVT^2dT$$

integrating $S = \frac{4}{3}aVT^3$ + constant

have $\alpha = \frac{4}{3} = 1.33.$

53. $\overrightarrow{a_1} = a\hat{x},\ \overrightarrow{a_2} = \frac{a}{2}(\hat{x} + \sqrt{3}\,\hat{y})$

The reciprocal space unit vectors $\overrightarrow{b_1}$ and $\overrightarrow{b_2}$

$$\overrightarrow{b_1} = 2\pi\frac{-Ra_2}{-a_1 \cdot Ra_2}$$

$$\overrightarrow{b_2} = 2\pi\frac{2a\hat{x}}{\frac{a}{2}(\hat{x} + \sqrt{3}\hat{y}) - 2a\hat{x}}$$

$$\overrightarrow{b_2} = 2\pi\frac{Ra_1}{a_2 \cdot Ra_1}$$

$$\Rightarrow \frac{4\pi a\hat{x}}{\frac{a}{2}\hat{x} + \frac{\sqrt{3}}{2}a\hat{y} - 2a\hat{x}} = \frac{4\pi}{a\sqrt{3}}\hat{y}$$

$$\overrightarrow{b_1} = 2\pi\frac{-Ra_2}{-a_1 \cdot Ra_2}$$

$$= \frac{2\pi - 2\times\frac{a}{2}(\hat{x} + \sqrt{3}\,\hat{y})}{-a\hat{x}\cdot 2\frac{a}{2}(\hat{x} + \sqrt{3}\,\hat{y})}.$$

$$= \frac{2\pi a\hat{x} - \sqrt{3}a\hat{y}}{-a\hat{x}\cdot\left(a\hat{x} + a\sqrt{3}\hat{y}\right)}$$

$$= \frac{2\pi a\hat{x} - \sqrt{3}a\hat{y}}{-a^2\hat{x} - a^2\sqrt{3}\hat{y}} = \frac{2\pi}{a}\left(\hat{x} - \frac{\hat{y}}{\sqrt{3}}\right)$$

Hence, solution is

$$= \frac{2\pi}{a}\left(\hat{x} - \frac{\hat{y}}{\sqrt{3}}\right) \text{ and } \frac{4\pi}{a\sqrt{3}}\hat{y}$$

54. $\frac{1}{R_{in}} = \frac{1}{R_1} + \frac{1}{R_2} + \frac{1}{\beta R_E}$

$$= \frac{1}{10} + \frac{1}{10} + \frac{1}{99(1)}$$

$$= \frac{1}{10} + \frac{1}{10} + \frac{1}{99} = 0.1 + 0.1 + 0.010$$

$= 0.21010$

$R_{in} = 4.7596 = 4.75.$

55. Given V = 10 mV

I = ?

$R_2 = 1 \times 10^3\,\Omega$

V = IR

$$\Rightarrow I = \frac{V}{R} = \frac{10\times10^{-3}}{1\times10^3} = 10\times10^{-6}\,A$$

56. In intrinsic semiconductor

$n_i = n_e nh$

$n_e = n_h$

$n_i = n_e^2 = (2 \times 10^{19})^2$

$\Rightarrow \quad n_i = 4 \times 10^{38}$

For doping with donor impurity, in *n* type semiconductor

$n_i = n_e n_h$

$n_i = n_e^2 \quad n_h$ is neglected

$n_i = (4 \times 10^{20})^2$

$= 16 \times 10^{40}$

Ratio of majority to minority

$$= \frac{16\times10^{40}}{4\times10^{38}} = 400$$

57. Susceptibility is defined as $X = \frac{\mu}{\mu_0} - 1$, where μ and μ_0 are permeability of medium and vaccum respectively.

(i) For Diamagnet ; $\mu < \mu_0$, Thus X < 0 i.e, X is negative

(ii) For Ferromagnet ; $\mu >> \mu_0$, thus X >> 0 i.e, X is positive.

(iii)For Superconductor, μ = 0, thus X = –1.

58. The new location of atom is $\left(0, \frac{a}{2}, 0\right)$ ie on middle of *y*-axis,

If new atom of radius r fits on it then,

$$\frac{a}{2} = r + R \quad ...(i)$$

For FCC, $\sqrt{2}a = 4R$

$$\Rightarrow \quad a = \frac{4}{\sqrt{2}}R = 2\sqrt{2}R \quad ...(ii)$$

From (*i*) and (*ii*)

$$\frac{2\sqrt{2}}{2}R = R + r$$

or $(\sqrt{2}-1)R = r$

$$\therefore \quad \frac{r}{R} = \sqrt{2}-1 = 0.414$$

59. $E = A(ka)^2 + B(ka)^4$

$$\frac{\partial E}{\partial k} = 2Ak\, a^2 + 4Bk^3a^4$$

$$\frac{\partial^2 E}{\partial k^2} = 2\, Aa^2 + 12Bk^2a^4$$

At bottom of band $k = 0$

Effective Mass

$$m_e = \frac{\hbar^2}{\frac{\partial^2 E}{\partial k^2}} = \frac{\hbar^2}{2Aa^2}$$

$$\Rightarrow \quad m_e = \frac{(1.05\times10^{-34})^2}{2\times6.3\times10^{-19}\times(2.1\times10^{-10})^2}$$

$$= \frac{1.1025\times10^{-68}}{55.57\times10^{-39}}$$

$$= 0.01984 \times 10^{-29}$$

$$= 19.84 \times 10^{-32}\text{ kg}$$

$$\therefore \quad \frac{m_e}{m} = \frac{19.84\times10^{-32}}{9.1\times10^{-31}}$$

$$= 2.18 \times 10^{-1}$$

$$\approx 0.218 = 0.22$$

60. $$d_{hkl} = \frac{1}{\sqrt{\frac{h^2}{a^2}+\frac{k^2}{b^2}+\frac{l^2}{c^2}}}$$

$$d_{110} = \frac{1}{\sqrt{\frac{1}{a^2}+\frac{1}{\left(\frac{a}{2}\right)^2}+0}} = \frac{a}{\sqrt{5}}$$

$= 0.447$

$\because \; a = 2b = 3c$

61. Exclusive OR

if A & B is input then

truth table is written as for exclusive OR

A	B	$Y = \overline{A}B + A\overline{B}$
0	0	0
0	1	1
1	0	1
1	1	0

taking (*a*) option

$$= (A+B)\,\overline{AB}$$

$$= (A+B)\,(\overline{A}+\overline{B}) \text{ by Demorgan's Theorem}$$

$$\therefore \quad \overline{AB} = \overline{A}+\overline{B}$$

Which satisfy truth table

Similarly (*b*) and (*c*) also satisfy the truth table for exclusive OR

Hence (*d*) will be the right.

62. $$\omega(k) = \sqrt{\frac{2C}{M}[1-\cos(ka)]}$$

For long wavelength modes ($\lambda >> a$)

$$\because \; \cos(Ka) \cong 1 - \frac{(Ka)^2}{2}$$

$$\Rightarrow \omega(K) = \sqrt{\frac{2C}{M}\left[1-1+\frac{(Ka)^2}{2}\right]}$$

$$= a\sqrt{\frac{C}{M}}K.$$

Phase velocity $v_p = \frac{\omega}{K} = a\sqrt{\frac{C}{M}}$

and Group velocity $V_g = \frac{d\omega}{dK} = a\sqrt{\frac{C}{M}}$

$$\Rightarrow \frac{V_p}{V_g} = 1$$

63. In Hall effect experiment polarity of Hall voltage determine weather the semiconductor specimen is N-type or P-type.

64. For intrinsic semiconductor,

$$E_F = \left(\frac{E_C+E_V}{2}\right) + \frac{3}{4}KT\,\ln\left(\frac{m_n}{m_e}\right)$$

$$= 0.72\ ev + \frac{3}{4}\text{KT}\ \ln\left(\frac{6m_e}{m_e}\right)$$

and $$\text{K} = 1.38\times10^{-23}\ \text{J/K}$$

$$= \frac{1.38\times10^{-23}}{1.6\times10^{-19}}\ ev/\text{K}$$

$$= 8.62\times10^{-5}\ ev/\text{K}$$

$$\text{K T} = 8.62\times10^{-5}\times300$$

$$= 0.02586\ \text{ev}$$

$$\text{E}_\text{F} = 0.72 + \frac{3}{4}\times0.02586\ \ln(6)$$

$$\text{E}_\text{F} = 0.72 + 0.0347511$$

$$\text{E}_\text{F} = 0.75475$$

And Fermi level with respect to the edge of valence band will be

$$\text{E}_\text{F} = 0.3944$$

65. In normal slide, some states below Fermi levels are empty and equal number of states above fermi levels are filled. If materials converts into a super conductor, electron above the fermi level makes copper pair and they fall back below level fermi level as same energy released during copper pair information. Therefore, correct option is (d).

66. The equation of plane is given by

$x f_x + y f_y + z f_z = 1$

For point (0, 0, 1)

$0 + 0 + f_z = 1$

$\therefore f_z = 1$

For point (1, 0, 0)

$1\times f_x + 0 + 0 = 1$

$\therefore f_x = 1$

For point $\left(\frac{1}{2}, \frac{1}{2}, \frac{1}{4}\right)$

$\frac{1}{2}f_x + \frac{1}{2}f_y + \frac{1}{4}f_z = 1$

$\therefore \frac{1}{2} + \frac{f_y}{2} + \frac{1}{4} = 1 \qquad (\because f_x = f_z = 1)$

$\therefore \frac{f_y}{2} = \frac{1}{4}$

$\therefore f_y = \frac{1}{2}$

we got the point $\Rightarrow \left(1, \frac{1}{2}, 1\right)$ then

Miller indices $= \left(\frac{1}{1}, \frac{1}{\frac{1}{2}}, \frac{1}{1}\right) = (1, 2, 1)$

67. $\text{E}_\text{c} - \text{E}_\text{F} = \text{kT In}\left(\frac{\text{N}_\text{c}}{\text{N}_\text{d}}\right)$ and

$$\text{E}_\text{c} - \text{E}_\text{F}^1 = \text{kT In}\left(\frac{\text{N}_\text{c}}{100\text{N}_\text{d}}\right)$$

$$= \text{kT In}\left(\frac{\text{N}_\text{c}}{\text{N}_\text{d}}\right) - \text{kT In}(100)$$

Thus, Shift is $\Delta\text{E} = \text{kT In}(100)$

$= 25\ \ln(100)$ MeV

$= 115.15$ MeV

68. $\text{P} = \frac{h}{\lambda} \Rightarrow \lambda = \frac{h}{m_n}$

$$\therefore \lambda = \frac{6.626\times10^{-34}\ \text{J.s}}{1.675\times10^{-27}\text{kg}\times10^{3}\ \text{m/sec}}$$

$\lambda = 3.9558$ Å

Now, according to Bragg's Law $2d \sin\theta = \lambda$

$$\therefore \quad d = \frac{\lambda}{2\sin\theta} = \frac{3.9558}{2\times\sin30°}$$

$$= \frac{3.9558}{2\times\frac{1}{2}}$$

$d = 3.9558$ Å

69. Resolution $= \frac{\text{maximum voltage}}{2n}$

$$10^{-3} = \frac{1}{2^n}$$

$$2^n = 1000$$

$$n \log 2 = 3 \log 10$$

$$n = \frac{3}{0.3010} = 9.99$$

$$n = 10$$

71. Quantization of magnetic flux in super conduction ring in the unit of $\left(\frac{\text{h}}{\text{e}}\right)$.

72. The reciprocal lattice vectors are

$$\vec{a}^* = 2\pi\frac{\vec{b}\times\vec{c}}{\vec{a}.\left(\vec{b}\times\vec{c}\right)} = \frac{\pi}{2}\left(-\hat{i}+\hat{j}+\hat{k}\right)\text{Å}^{-1}$$

$$\vec{b}^* = 2\pi\frac{\vec{c}\times\vec{a}}{\vec{a}.\left(\vec{b}\times\vec{c}\right)} = \frac{\pi}{2}\left(\hat{i}-\hat{j}+\hat{k}\right)\text{Å}^{-1}$$

$$\vec{c}^* = 2\pi\frac{\vec{a}\times\vec{b}}{\vec{a}.\left(\vec{b}\times\vec{c}\right)} = \frac{\pi}{2}\left(\hat{i}+\hat{j}-\hat{k}\right)\text{Å}^{-1}$$

73. At $r = r_0$, $\left.\frac{dE}{dr}\right|_{r=r_0} = 0 = \frac{\alpha e^2}{4\pi\varepsilon_0 r_0^{\,2}} - \frac{9B}{r_0^{\,10}}$

$$\Rightarrow B = \frac{\alpha e^2 r_0^{\,8}}{36\pi\varepsilon_0}$$

74. $\frac{3\pi^2\hbar^2}{2ma^2} = 9.9\times10^{-17}$

75. Silicon doped with Arsenic is an n-type semiconductor so, the correct energy band diagram is in option (*b*).

76. Thermal expansion in solid can only be explained if solid behave as a anharmonic oscillator.

77. The susceptibility

$$\chi = \frac{C}{T-T_C} = \frac{C}{T-C\lambda}$$

$$\Rightarrow \quad \frac{1}{\chi} = \frac{T}{C} - \lambda \quad \dots (i)$$

At T = 0, $\frac{1}{\chi} = -2\times10^4$

Putting it in eq. (*i*), we get

$\lambda = 2\times10^4$

Now, $T_C = C\lambda$

$$C = \frac{T_C}{\lambda} = \frac{600}{2\times10^4} = 3\times10^{-2}$$

78. For simple cubic crystal $a = b = c$

$$dsc = \frac{1}{\sqrt{\frac{h^2}{a^2}+\frac{x^2}{b^2}+\frac{2}{c}}}$$

$$dsc|_{(101)} = \frac{1}{\sqrt{\frac{1}{a^2}+\frac{0}{a^2}+\frac{1}{a^2}}} = \frac{a}{\sqrt{2}}$$

For tetragonal structure $a_t = b_t = \sqrt{2a}$ and $c_t = 2a$

$$d\,TS|_{(101)} = \frac{1}{\sqrt{\frac{1}{(\sqrt{2}a)^2}+\frac{0}{(\sqrt{2}a)^2}+\frac{1}{(2a)^2}}}$$

$$= \frac{2}{\sqrt{3}}a$$

$$\left.\frac{dsc}{dTS}\right|_{(101)} = \frac{a}{\sqrt{2}}\times\frac{\sqrt{3}}{2a} = \sqrt{\frac{3}{8}}$$

79. $V_{CC} - I_CR_C - V_{CE} - I_CR_E = 0$

(where $I_C \approx I_E$)

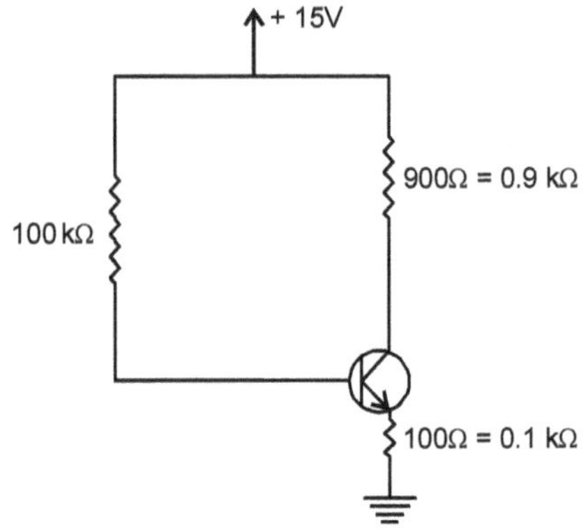

$$I_C = \frac{V_{CC}-V_{CE}}{R_C+R_C^S}$$

$$= \frac{15-2}{0.1+0.9} = 13 \text{ mA (Putting } V_{CE} = 2V)$$

Hence Q = 2V, 13 mA

80. At the first Brillouin zone the frequency is maximum and the group velocity which is the derivative of the angular frequency is zero.

81. $\because \omega = \sqrt{\frac{4\beta}{M}}\sin\left(\frac{k\alpha}{2}\right)$

$$\Rightarrow A = \sqrt{\frac{4B}{M}}$$

$$\Rightarrow \beta = \frac{MA^2}{4}$$

82. The magnetic susceptibility χ of a ferromagnetic substance with curie Temperature (T_C) is given by

$$\chi = \frac{C}{T-T_C} \text{ for } T > T_C$$

83. The order of energy gap b/w valence and conduction bands of a semiconductor is ᴗ l ev.

85. The fermi level for intrinsic semi-conductor is

$$E_F = \frac{E_C+E_V}{2} + \frac{3}{4}k_BT\,\text{In}\left(\frac{m_h^*}{m_e^*}\right)$$

86. By Bragg's low $2d\sin\theta = \lambda$

$$\Rightarrow \quad d = \frac{2.1}{2\times\sin 30} = 2.1$$

Volume of unit cell = $(2.1)^3 = 9.3(\text{Å})^3$

88. For No diffraction d should be minimum

$$d = \frac{a}{\sqrt{h^2 + k^2 + l^2}}$$

and d is minimum for (3, 1, 1)

89. $\because$ $R_H = \frac{1}{ne}$ and sodium is a metal hence, R_H only depends on charge carrier density n only.

90. The Bloch theorem is

$$\psi(\vec{r}) = u(\vec{r})e^{i\vec{k}\vec{r}}$$

Where $u(\vec{r}) = u(\vec{r} + \vec{A})$

91. For a ferromagenetic substance

$$C_V = -2\lambda Ms\left(\frac{\partial Ms}{\partial T}\right)$$

$$M_s = N\mu \tan a\left(\frac{\mu_0 h\lambda\, Ms}{k_{BT}}\right)$$

N means C_V rises maximum at T_C and then falls absolutely to zero as shown

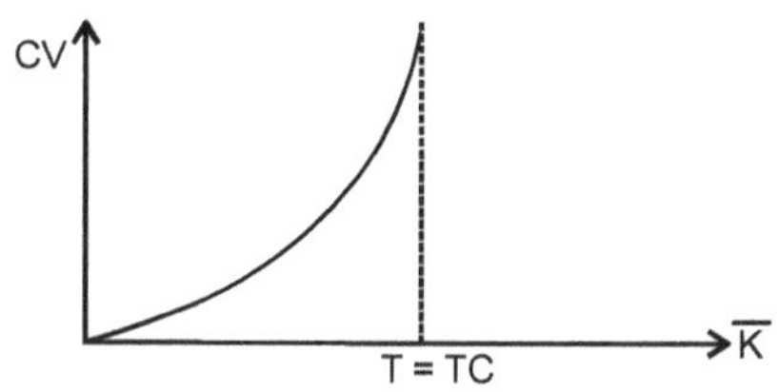

92.

93.

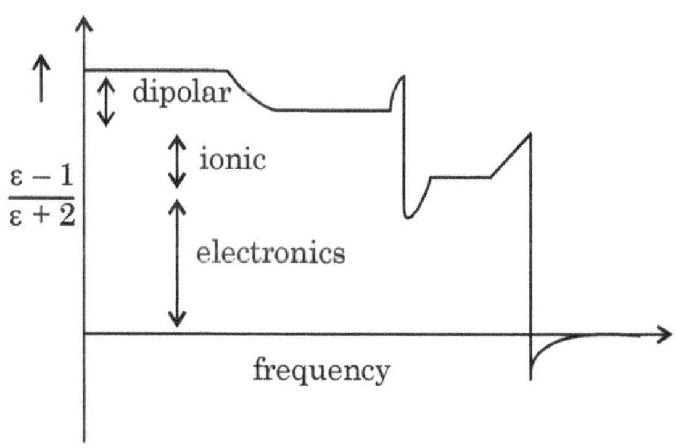

94.

$$L = \frac{1}{2}m\left(\dot{r}^2 + r^2\dot{\theta}^2 + r^2\dot{\phi}^2\sin^2\theta\right)$$

$$\Rightarrow \quad \frac{\partial L}{\partial \dot{\phi}} = 2r^2\sin^2\theta\,\dot{\phi}$$

and $\frac{\partial L}{\partial \phi} = 0$

Using Lagrang's equation

$$\frac{d}{dt}\left(\frac{\partial L}{\partial \dot{\phi}}\right) - \frac{\partial L}{\partial \phi} = 0$$

$$\Rightarrow \quad \frac{d}{dt}\left(\frac{\partial L}{\partial \dot{\phi}}\right) = 0$$

$$\Rightarrow \quad \frac{\partial L}{\partial \dot{\phi}} = \text{constant}$$

Hence, $\frac{\partial L}{\partial \dot{\phi}}$ is conserved

95. $$\text{CMRR} = \frac{Ad}{Acm}$$

where Ad = differential voltage gain
= large signal voltage gain

$$Acm = \frac{V_0 cm}{Vcm}$$

$$= \frac{\text{output common mode voltage}}{\text{input common mode voltage}}$$

The CMRR is generally expressed in db and denoted by

$$\rho = z_0 \log_{10}\frac{Ad}{Acm}$$

$$Ad = \frac{V_0}{V_{in1} - V_{in2}}$$

$$= \frac{\text{Output voltage}}{\text{Differential input}}$$

$$\because \quad = \frac{2}{200 \times 10^{-6}} = 10^4$$

Hence, $\rho = 20\log_{10}\frac{10^4}{Acm} = 100\,dB$

$$\therefore \quad \log\frac{10^4}{Acm} = 5$$

or $\frac{10^4}{Acm} = 10^5$

$$\Rightarrow \quad Acm = 0.1$$

96.

$$k_1\frac{dx}{dt} + k_2 x = k_3$$

$$\frac{dx}{dt} + \frac{k_2}{k_1}x = \frac{k_3}{k_1}$$

Differentiating w.r.t. to t, we get

$$\frac{d^2x}{dx^2}+\frac{k_2}{k_1}\frac{dx}{dt} = 0$$

Putting, $\frac{dx}{dt} = m$

$$m^2+\frac{k_2}{k_1}m = 0$$

$$m = 0, \quad -\frac{k_2}{k_1}$$

Hence, solution is given by

$$x = C_1 + C_2e^{-k_2/k_1 t}$$

By given condition $x = 0$ when $t = 0$

$$0 = C_1 + C_2$$

$$\Rightarrow \quad C_2 = -C_1$$

$$x = C_1\left[1-e^{-k_2/k_1 t}\right]$$

The correct variation is given by curve (b)

99. ∴ For parameter

$$\chi = \frac{C}{T}$$

Which is represent by curve Q

For anti ferromagnet

$$\chi = \frac{C}{T+T_C}$$

At $T = 0$

$$\chi_0 = \frac{C}{T_C}$$

102. Here $\hat{a}_1 = \frac{a}{2}(\hat{j}+\hat{k})$,

$$\hat{a}_2 = \frac{a}{2}(\hat{i}+\hat{k})$$

and $\hat{a}_3 = \frac{a}{2}(\hat{i}+\hat{j})$

$$\hat{a}_1-(\hat{a}_2\times\hat{a}_3)\begin{vmatrix} 0 & \frac{a}{2} & \frac{a}{2} \\ \frac{a}{2} & 0 & \frac{a}{2} \\ \frac{a}{2} & \frac{a}{2} & 0 \end{vmatrix}$$

$$\Rightarrow \left(\frac{a}{2}\right)^3\begin{vmatrix} 0 & 1 & 1 \\ 1 & 0 & 1 \\ 1 & 1 & 0 \end{vmatrix}$$

$$\left(\frac{a}{2}\right)^3[1+1] = \frac{a^3}{8}\times 2 = \frac{a^3}{4}$$

$$b\hat{i} = 2\pi\frac{\hat{a}_2\times\hat{a}_3}{\hat{a}_1(\hat{a}_2+\hat{a}_3)}$$

$$= \frac{2\pi}{\frac{a^3}{4}}\begin{vmatrix} \hat{i} & j & k \\ \frac{a}{2} & 0 & \frac{a}{2} \\ \frac{a}{2} & \frac{a}{2} & 0 \end{vmatrix}$$

$$\hat{b}_1 = \frac{8\pi}{a^3}\left[\hat{i}\left(-\frac{a^2}{4}\right)+\hat{j}\left(\frac{a^2}{4}\right)+\hat{k}\left(\frac{a^2}{4}\right)\right]$$

$$= \frac{2\pi}{a}\left(-\hat{i}+\hat{j}+\hat{k}\right)$$

$$\hat{b}_2 = 2\pi\frac{\hat{a}_3\times\hat{a}_1}{\hat{a}_1(\hat{a}_2\times\hat{a}_3)}$$

$$= \frac{2\pi}{\frac{a^3}{4}}\begin{vmatrix} \hat{i} & j & \hat{k} \\ \frac{a}{2} & \frac{a}{2} & 0 \\ 0 & \frac{a}{2} & \frac{a}{2} \end{vmatrix}$$

$$= \frac{8\pi}{a^3}\left[\hat{i}\left(\frac{a^2}{4}\right)-\hat{j}\left(\frac{a^2}{4}\right)+\hat{k}\left(\frac{a^2}{4}\right)\right]$$

$$= \frac{2\pi}{a}\left(\hat{i}-j+\hat{k}\right),$$

$$\hat{b}_3 = \frac{2\pi}{a}\left(\hat{i}+\hat{j}-\hat{k}\right)$$

103. Volume of primitive cells is

$$V = \frac{(2\pi)^3}{\hat{a}_1\cdot(\hat{a}_2\times\hat{a}_3)}$$

$$= \frac{(2\pi)^3}{\frac{a^3}{4}}$$

$$= 4\left(\frac{2\pi}{a}\right)^3$$

105. $V_{\text{out 2}} = V_{\text{out 1}}\left[\frac{R}{R}V_a+\frac{R}{R}V_b+\frac{R}{R}V_C\right]$

$$= \frac{1}{3}(V_a+V_b+V_c)+(V_a+V_b+V_c)$$

$$= \frac{4}{3}(V_a+V_b+V_c)$$

106. For b_{CC} structure number of atoms Per unit cell = 2

$$\frac{2}{L^3} = \frac{2}{(4\times10^{-8})^3} = \frac{2}{64\times10^{-29}}$$

$$\frac{100}{32}\times10^{22} = 3.125\times10^{22}\ \text{cm}^3$$

107. For anti-ferromagnetic substance

$$\frac{1}{\chi} = \left(\frac{T+T_C}{C}\right)$$

$\Rightarrow$ at $T = -T_C$

$$\frac{1}{\chi} = 0$$

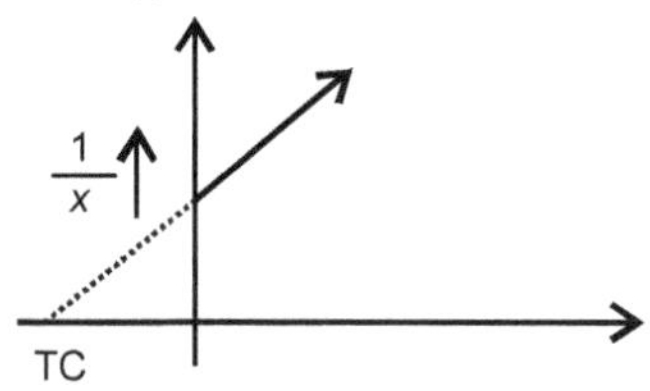

111.

$$\vec{a} = \frac{a}{2}[\hat{x}+\hat{y}-\hat{z}]$$

$$\vec{b} = \frac{a}{2}[-\hat{x}+\hat{y}+\hat{z}]$$

$$\vec{c} = \frac{a}{2}[\hat{x}-\hat{y}+\hat{z}]$$

$$\vec{a}-\vec{b}\times\vec{c} = \begin{vmatrix} \frac{a}{2} & \frac{a}{2} & \frac{-a}{2} \\ \frac{-a}{2} & \frac{a}{2} & \frac{a}{2} \\ \frac{a}{2} & \frac{-a}{2} & \frac{a}{2} \end{vmatrix}$$

$$= \left(\frac{a}{2}\right)^3 \begin{vmatrix} 1 & 1 & -1 \\ -1 & 1 & 1 \\ 1 & -1 & 1 \end{vmatrix}$$

$$= \left(\frac{a}{2}\right)^3 ((1+1)+(1+1))$$

$$= \frac{a^3}{2}$$

112.

$$S_{hkl} = f\sum_i \exp\left(-i2\pi\left(x_j h + y_j k + z_j l\right)\right)$$

$$S = f\exp\left(\vec{i}_g, \vec{r}_j\right)$$

$$\vec{r}_L = \frac{a}{2}(\hat{x}+\hat{y}+\hat{z})$$

$$S_B = f\left(1+\exp\left(i\pi(h+k+l)\right)\right)$$

$$= f(1 + (-1)^{h+k+l})$$

$$= \frac{2\pi}{a}(h\hat{x}+\hat{k}+l\hat{z})$$

$= 0$ if $h + k + l = 0$

$= 2f$

$Sy = f(1\pm i)$ (if $h+k+l =$ an odd number)

$= 0$ (if $h + k + l = 2 \times$ odd number)

$= 2f$ (if $h + k + l = 2 \times$ even number)

113. $C \approx 3R = \text{constent}$

$$C_V = 3R\left(\frac{hv}{k_B T}\right)\exp\left(\frac{-hv}{k_B T}\right)$$

$$C_V = 3R\left(\frac{a}{T}\right)^2 \exp\left(\frac{-q}{T}\right)$$

$$C_V \propto \frac{\exp(-q/T)}{T^2}$$

By $D_\theta\ by_e\ C_V \propto (T/A)^3$

115. The kinetic energy for a free electron

$$E = \frac{h^2}{2m}k^2$$

$$k_x = \pm\frac{\pi}{a}$$

$$k_y = \pm\frac{\pi}{a}$$

$$k_x + k_y = \frac{2\pi}{a} \text{ (at constant)}$$

$$E_{cor} = \frac{h^2}{2m}\left(\frac{2\pi}{a}\right)^2$$

$$E \text{ centre} = \frac{h^2}{2\pi}\left(\frac{\pi}{a}\right)^2$$

$$E_{cor} = 4\,E_{centre}$$

116.

$$E_F = \frac{E_V + E_C}{2} + \frac{3k_B T}{4}\log\left(\frac{mR}{me}\right)$$

117. For step down transformer $E_2 < E_1$

$$E_1 = -N_1\frac{d\phi}{dt}$$

$$E_2 = N_2\frac{d\phi}{dt}$$

$$\therefore \quad \frac{E_2}{E_1} = \frac{N_2}{N_1} = \frac{1}{10}$$

$$\because \quad E_1 = 220\,V_{(rms)}$$

$$\Rightarrow \quad E_2 = 22\,V_{(rms)}$$

For full wave rectifier the DC current

$$I_{DC} = \frac{1}{2\pi}\int_0^{2\pi} I d\,(wt)$$

$$= \frac{1}{2\pi}\int_0^{\pi}\frac{E_m}{R}\sin w\,(wt)$$

$$+\frac{1}{2\pi}\int_0^{2\pi}\frac{E_m}{R}\sin wt\,d(wt)$$

$$= 2\frac{E_m}{\pi RL}$$

$$E_m = E_2 = 22\,V_{(rms)}$$

So output DC voltage

$$E_{DC} = I_{DC} \times R_L$$

$$= \frac{2E_m}{\pi}$$

$$= \frac{44}{\pi}$$

118.

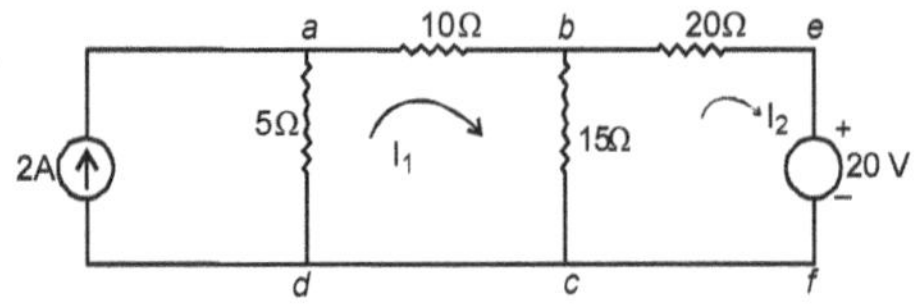

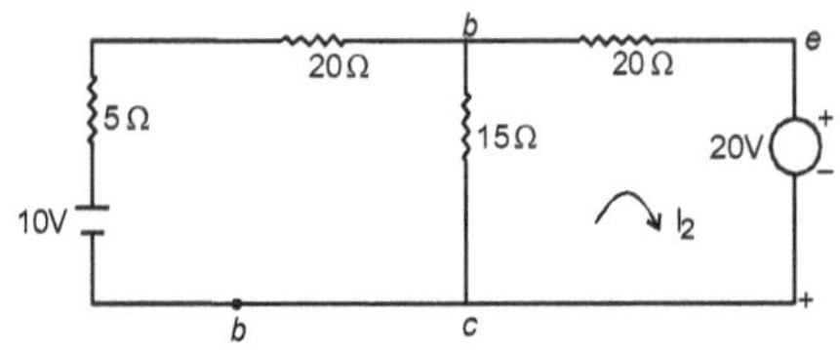

Applying KVL for the loop

$$-5\,l_1 - 10\,l_1 - (l_1 - l_2)\,15 + 10 = 0$$

$$-30\,l_1 + 15\,l_2 + 10 = 0$$

or $\quad 30\,l_1 - 15\,l_2 = 10$

119. $y = P + PQ$

$$= P(1 + Q) = P \quad (\because\ 1 + Q = 1)$$

The output of circuite (*b*) is

$$y = P + \bar{P} + Q = 1 + Q = 1 \quad (\because P + \bar{P} = 1)$$

and (1 + Q = 1)

Hence logic gate for $y = P + PQ$ is (*b*)

120. The number of comparators is

$$2N - 1 = 2^2 - 1$$

$$= 4 - 1$$

$$= 3$$

The voltage are

$$\frac{V}{2N}\frac{2V}{2N} \text{ and } \frac{3V}{2N}$$

$$\frac{V}{2^2}, \frac{2V}{2^2} \text{ and } \frac{3V}{2^2}$$

$$\frac{V}{4}, \frac{2V}{2} \text{ and } \frac{3V}{4}$$

123. $\mu = \frac{eJ}{m} = \frac{e}{m} \times \frac{m}{ne^2P} = \frac{1}{ne\rho}$

$$= \frac{1}{5.85 \times 10^{28} \times 1.6 \times 10^{-19} \times 1.6 \times 10^{-8}}$$

$$= 6.67 \times 10^{-3}$$

124. $\tau = \frac{\mu m_e}{e}$

m_e = mass of electron

$$\tau = \frac{6.67 \times 10^{-3} \times 9.1 \times 10^{-31}}{1.6 \times 10^{-19}}$$

$$\tau = 3.79 \times 10^{-14}$$

126. In super conductor below T_c Magnetic flux expels out since, surface current induces a field opposite to the external field.

127. Since the given potential is for the harmonic oscillator.

Hence, energy levels are described by

$$E_n = \left(n + \frac{1}{2}\right)^{+} \hbar\omega$$

129. Since O_2 is diamagnetic and diamagnetic substances. magnetized oppsite to external magnetic field in the presence of external magnetic field.

130. $V(r) = \frac{-P^2}{r^6} + \frac{Q}{r^{12}}$

$$\frac{dv}{dr} = \frac{6P}{r^6} - \frac{12Q}{r^{13}}$$

for bond length *i.e.* in equilibrium

$$\left(\frac{dv}{dr}\right)_{r=r_0} = 0 \Rightarrow \frac{6P}{r_0^7} - \frac{12Q}{r_0^{13}} = 0$$

$$r_0 = \left(\frac{2Q}{P}\right)^{1/6} = \left(\frac{2Q}{P}\right)^{-6}$$

131. The APF for f_{cc} and hexagonal closed packed (*ncp*) lattice are 74% or 0.74 both.

132. By Bragg's Law

$$2d \sin\theta = \lambda$$

First reflection by plane (1, 1, 1)

$$\text{Now } d(1, 1, 1) = {}^{9}\!/\!_{\sqrt{3}}$$

$$2d_{111} \sin\theta_1 = \lambda$$

Second peak by plane (001 or 100)

$$d(001) = d\,(100) = a$$

$$2d_{001} \sin\theta_2 = 2\lambda\ (n = 2)$$

$$\sin\theta_2 = \frac{2d_{111}}{d_{00}} \sin\theta_1$$

$$= 2 \times \frac{a}{\sqrt{3}a} r \sin 30° = {}^{1}\!/\!_{\sqrt{3}}$$

$$\theta_2 = \sin^{-1}\left(\frac{1}{\sqrt{3}}\right) = 35.3°$$

133. For a super conductor $\rho = 0$

At $T = T_c$ (critical temperature)

Hence, Q for super conductor and ρ becomes constant at lower temperature for a conducter, hence P for conductor.

134. $N_D(x) = N_0 \exp\left(-\frac{x}{L}\right)$

$N_D(0) = N_0 \exp(0) = N_0$

$N_D(L) = N_0 \exp(-1)$

$$R = \frac{\rho L}{A} = \frac{L}{A}\rho$$

Now, $\rho = \frac{[NL - N_0]}{\mu e N_0^2}$

$$R = \frac{L}{A} \times \frac{N_0}{\mu e N_0^2}(\exp(-1) - 1)$$

$$R = \frac{L}{A\mu e N_0}[\exp(-1) - 1]$$

135. Silicon is an indirect-Band gap seniconductor.

136. Five-fold rotational symmetry is not possible for crystall in 3-dimensions.

137. Magnetic moment for spin $\frac{1}{2}$ system

$$\mu = I_s m_s \mu_B = \pm 2.\frac{1}{2}\mu_B = \pm \mu_B$$

∴ Partition function,

$$Z = \sum_i e^{-\beta\epsilon_i} \quad \left[\varepsilon = -\bar{\mu}.\bar{\beta} = \pm\mu_B B\right]$$

$$Z = e^{\mu_B B/kT} + e^{-\frac{\mu_B B}{kT}}$$

thus $Z = 2\cos h\left(\frac{\mu_B B}{kT}\right)$

For N spins, $Z_N = \left[2\cos h\left(\frac{\mu_B B}{kT}\right)\right]^N$

Helmhotz free energy,

$$F = -k_B T \ln Z$$

$$= -N k_B T \ln\left[2\cos h\frac{\mu_B B}{kT}\right]$$

138. For free particle momentum p_x is doubly degenerate, for each energy E,

$$\frac{p_x^2}{2m} = E \quad \Rightarrow \quad p_x = \pm\sqrt{2mE}$$

which is independent of X.

but X can only be +ve within 0 to L.

So picture (*a*) is correct.

139. Brillouin zones are so constructed so that all of them have equal areas.

140. Number of wave vectors k = Number of primitive cells N

Since each R level can accomodate 2 monovalent atoms highest occupied energy band is completely filled.

141. Plasma frequency,

$$\omega_p^2 = \frac{ne^2}{\varepsilon_0 m}$$

$$\frac{\omega_1}{\omega_2} = \sqrt{\frac{n_1}{n_2}}$$

142. Meissner effect as well as normal metallic behaviour exists (mixed state Meissner effect or vortex state) is present in type II superconductors, below H_{C_2}.

144. $\overrightarrow{OP} = \left[1, 1, \frac{1}{2}\right]$

$= [2, 2, 1]$

148. Since, $4R = \sqrt{3}a$

∴ $a = \frac{4}{\sqrt{3}}R$

149. $\Omega = \frac{n_e \times e \times V_d}{E}$

$$= \frac{3 \times 10^{20} \times 1.6 \times 10^{-19} \times 100}{200}$$

$= 24$

150. Density of states $\propto \sqrt{E}$

153. $\left(\frac{c}{a}\right)^{1/2} = \frac{8}{3}$

$\Rightarrow \left(\frac{c}{a}\right) = \sqrt{\frac{8}{3}}$.

154. Cubic solid symmetry $\sigma_{xx}, \sigma_{yy}, \sigma_{zz}$.

155. $m^* = \frac{\hbar^2}{\frac{d^2E}{dk^2}}$

so m^* can be positive or negative by the curve.

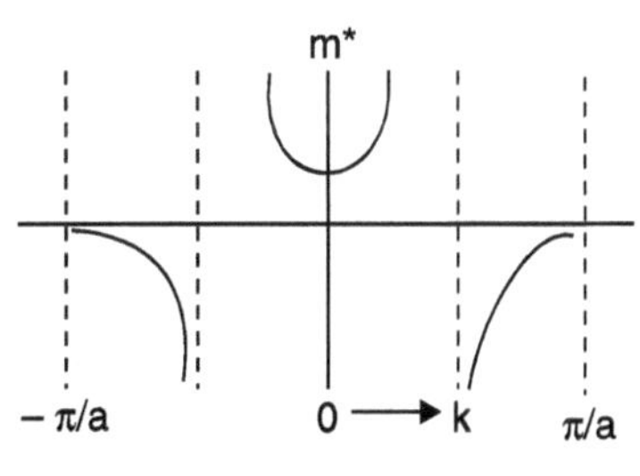

156. So the critical magnetic field for a solid in superconducting state increase if the temperature decreases.

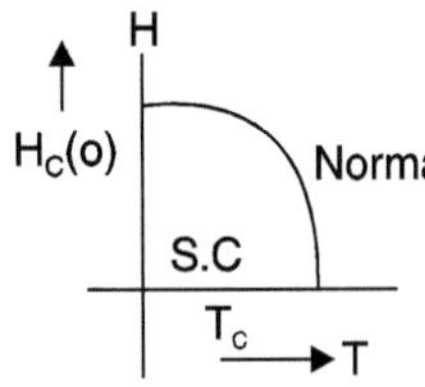

157. A Van der Waal's gas equation is

$$\left(P+\frac{a}{V^2}\right)(V-b) = RT$$

$$\Rightarrow \quad P = -\frac{a}{V^2} + \frac{RT}{V-b}$$

If multiply by volume in both sides, we get

$$PV = RT - \frac{a}{V} \quad [\because \ R = Nk_B; N = 1]$$

$$\Rightarrow \quad PV = k_BT - \frac{a}{V}$$

$$\frac{PV}{k_BT} = 1 - \frac{a}{k_BTV}$$

$$\Rightarrow \quad \frac{PV}{k_BT} \approx 1 + \frac{B(T)}{V}$$

$$\left[\because \ -\frac{a}{k_BT} = B(T)\right]$$

158. $l = 2$ cm, $d = 1$ cm $= 1 \times 10^{-2}$ m

$I_x = 1\times10^{-3}$ A, $V_H = -6\times10^{-3}$ V, $B_z = 0.5$ T

$$R_H = \frac{V_H\, d}{I_x\, B_z} = \frac{-6\times10^{-3}\times1\times10^{-2}}{1\times10^{-3}\times0.5} = -0.12$$

So the negative sign of the R_H value show the majority carrier are electrons.

$$R_H = -\frac{1}{ne}$$

$$\Rightarrow \quad -0.12 = -\frac{1}{n\times1.6\times10^{-19}}$$

$$\Rightarrow \quad n = 5.2\times10^{19}\ \text{m}^{-3}, \text{negative.}$$

159. For intrinsic semiconductors

$$n_i = \text{const}\exp\left(\frac{-E_g}{2k_BT}\right)$$

$$\sigma = n_i e(\mu_n + \mu_p)$$

$$\propto e(\mu_n + \mu_p)\exp\left(\frac{-E_g}{2k_BT}\right)$$

$$\Rightarrow \quad \ln\sigma = \log k - \frac{E_g}{2k_BT}$$

Plot ln σ vs $\frac{1}{T}$. The slope of line of this plot $= -\frac{E_g}{2k_B}$ will provide the band gap E_2. Here slope of A curve is large compared to curve B

$$\therefore \quad \frac{E_A}{2K_B}\ \text{term} > \frac{E_B}{2K_B}\ \text{term}$$

$$\Rightarrow \quad E_A > E_B.$$

160. By CM relation, $\left(\frac{\varepsilon_s - 1}{\varepsilon_s + 2}\right) = \frac{N\alpha}{3\varepsilon_0}$

where, α total polarizability

$$\frac{N\alpha}{3\varepsilon_0} = \left(\frac{5.6-1}{5.6+2}\right) = 0.60\ [\because \varepsilon_s = 5.6] \quad \ldots(i)$$

Given, Optical refractive index, $n = 1.5$,

We know, $n^2 = \varepsilon_r$

$$n = \sqrt{\varepsilon_r}$$

$$\Rightarrow \quad \varepsilon_r = (1.5)^2 = 2.25$$

$$\Rightarrow \quad \left(\frac{\varepsilon_r - 1}{\varepsilon_r + 2}\right) = \frac{N\alpha}{3\varepsilon_0}$$

$$\Rightarrow \quad \frac{N\alpha_e}{3\varepsilon_0} = \left(\frac{2.25-1}{2.25+2}\right) = 0.29 \quad \ldots(ii)$$

By equations (*i*) and (*ii*), $\dfrac{\frac{N\alpha_e}{3\varepsilon_0}}{\frac{N\alpha}{3\varepsilon_0}} = \frac{0.29}{0.60}$

$$\Rightarrow \quad \frac{\alpha_e}{\alpha} = 0.483 \approx \mathbf{0.5.}$$

165. Number of atoms $= 4\times\left(\frac{1\ \text{cm}}{1\ \text{nm}}\right)^3 = 4\times10^{21}$

168. Example of even conductors are Alkaline earth metal.

171. If (h, k, l) be the miller indicaes then distance between atomic planes,

$$d = \frac{a}{\sqrt{h^2+k^2+l^2}} \text{ for cubic crystal.}$$

For Bragg refelection,

$$2d\sin\theta = n\lambda$$

$$\Rightarrow \quad \sin\theta = \frac{n\lambda}{2d}$$

For minimum θ, $n = 1$ and d must be maximum.

For maximum $\sqrt{h^2+k^2+l^2}$ must be minimum.

Choice (A) gives minimum $\sqrt{h^2+k^2+l^2}$

172. Diamond has *fcc* structure

∴ Interatomic separation

$$r = \frac{a\sqrt{2}}{2} = \frac{a}{\sqrt{2}} = \frac{0.542}{1.4} nm$$

= 0.383 nm

174. Probability of occupancy of a given state with energy E is given by th Fermi Dirac distribution formula, which is

$$f(E) = \frac{1}{1 + \exp\{(E - E_F)/kT\}}$$

$$= \frac{1}{1 + \exp\left(\frac{0.2 \times 1.6 \times 10^{-19}}{1.38 \times 10^{-23} \times 700}\right)}$$

= 3.5%

176. Drift Velocity,

$$V_d = \frac{J}{n\,e} - \frac{I}{A\,ne}$$

$$n = \frac{8.94 \times 10^3}{1.05 \times 10^{-25}} = \frac{\rho_{cu}}{A_{cu}}$$

$$V_d = \frac{1.05 \times 10^{-25}}{8.94 \times 10^3 \times 10^{-6} \times 1.6 \times 10^{-19}}$$

$= 7.4 \times 10^{-5}$ m/s

179.

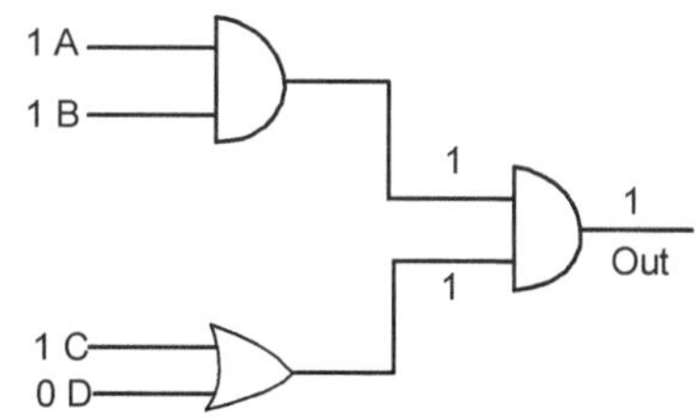

8 CHAPTER Nuclear and Particle Physics

2021

1. In an experiment, it is seen that an electric-dipole ($E1$) transition can connect an initial nuclear state of spin-parity $J_i^\pi = 2^+$ to a final state J_f^π. All possible values of J_f^π are

(a) $1^+, 2^+$ (b) $1^+, 2^+, 3^+$

(c) $1^-, 2^-$ (d) $1^-, 2^-, 3^-$

2.

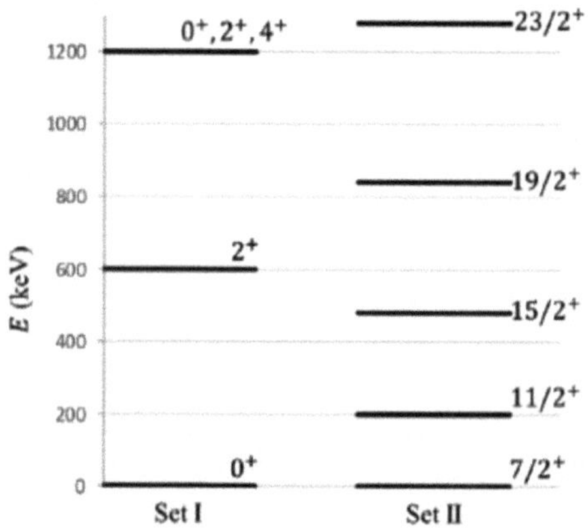

For the given sets of energy levels of nuclei X and Y whose mass numbers are odd and even, respectively, choose the best suited interpretation.

(a) Set I: Rotational band of X
Set II: Vibrational band of Y

(b) Set I: Rotational band of Y
Set II: Vibrational band of X

(c) Set I: Vibrational band of X
Set II: Rotational band of Y

(d) Set I: Vibrational band of Y
Set II: Rotational band of X

3. The Gell-Mann - Okuba mass formula defines the mass of baryons as $M = M_0 + aY + b\,[I(I+1) - \frac{1}{4}Y^2]$, where M_0, a and b are constants, I represents the isospin and Y represents the hypercharge. If the mass of Σ hyperons is same as that of a hyperons, then the correct option(s) is(are)

(a) $M \propto I(I+1)$

(b) $M \propto Y$

(c) M does not depend on I

(d) M does not depend on Y

2020

4. A particle X is produced in the process $\pi^+ + p \to K^+ + X$ via the strong interaction. If the quark content of the K^+ is $u\bar{s}$, the quark content of X is

(a) $c\bar{s}$ (b) uud

(c) uus (d) $u\bar{d}$

5. If $\vec{E}$ and $\vec{B}$ are the electric and magnetic fields respectively, the $\vec{E}\cdot\vec{B}$ is

(a) odd under parity and even under time reversal

(b) even under parity and odd under time reversal

(c) odd under parity and odd under time reversal

(d) even under parity and even under time reversal

6. A particle Y undergoes strong decay $Y \to \pi^- + \pi^-$Y the isospin of is ______.

7. According to the Fermi gas model of the nucleus, the nucleons move in a spherical volume of radius $R (= R_0 A^{\frac{1}{3}}$, where A is the mass number and R_0 is an empirical constant with the dimensions of length). The Fermi energy of the nucleus E_F is proportional to

(a) R_0^2 (b) $\frac{1}{R_0}$

(c) $\frac{1}{R_0^2}$ (d) $\frac{1}{R_0^3}$

8. In the center of mass frame, two protons each having energy 7000 GeV, collide to produce protons and anti-protons. The maximum number of anti-protons produced is ______.

(Assume the proton mass to be 1 GeV/c²)

2019

9. Considering baryon number and lepton number conservation laws, which of the following

(i) $p \to \pi^0 + e^+ + \nu_e$

(ii) $e^+ + \nu_e \to \mu^+ + \nu_\mu$

(*a*) both (i) and (ii) (*b*) only (i)

(*c*) only (ii) (*d*) neither (i) nor (ii)

10. A massive particle X in free space decays spontaneously into two photons. Which of th following statements is true for X ?

(*a*) X is charged

(*b*) Spin of X must be greater than of equal to 2

(*c*) X is a boson

(*d*) X must be a baryon

11. The nucleare spin and parity of $_{20}^{40}\text{Ca}$ in its ground state is

(*a*) 0^+ (*b*) 0^-

(*c*) 1^+ (*d*) 1^-

12. Low energy collision (s–wave scattering) of pion $\left(\pi^+\right)$ with deuteron (*d*) results in the production of two protons ($\pi^+ + d \to p + p$). The relative orbital angular momentum (in units of $\hbar$) of the resulting two–proton system for this reaction is

(*a*) 0 (*b*) 1

(*c*) 2 (*d*) 3

13. A radioactive element X has a half-life of 30 hours. It decays via alpha, beta and gamma emissions with the branching ratio for beta decay being 0.75. The partial half-life for beta decay in unit of hours is________

2018

14. Among electric field $(\vec{E})$, magnetic field $(\vec{B})$, angular momentum $(\vec{L})$, and vector potential $(\vec{A})$, which is/are **odd** under parity (space inversion) operation?

(*a*) $\vec{E}$ only (*b*) $\vec{E}$ & $\vec{A}$ only

(*c*) $\vec{E}$ & $\vec{B}$ only (*d*) $\vec{B}$ & $\vec{L}$ only

15. The elementary particle Ξ^0 is placed in the baryon decuplet, shown below, at

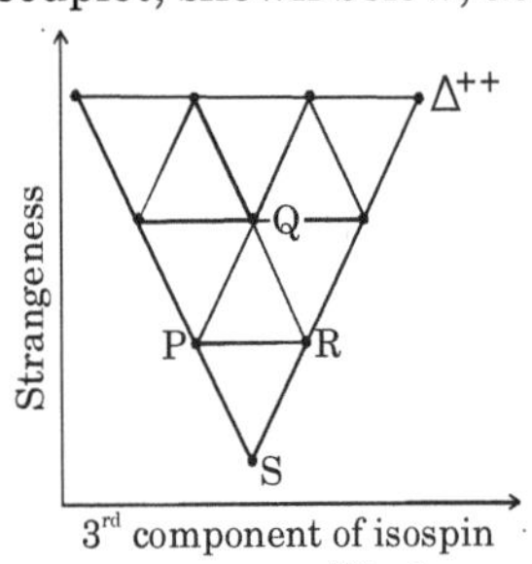

(*a*) P (*b*) Q

(*c*) R (*d*) S

16. In the decay, $\mu^+ \to e^+ + \upsilon_e + X$, what is *X*?

(*a*) γ (*b*) $\bar{\upsilon}_e$

(*c*) υ_μ (*d*) $\bar{\upsilon}_\mu$

17. For nucleus ^{164}Er, a $J^\pi = 2^+$ state is at 90 keV. Assuming ^{164}Er to be a rigid rotor, the energy of its 4^+ state is______keV (up to one decimal place).

18. Inside a large nucleus, a nucleon with mass 939 MeVc^{-2} has Fermi momentum 1.40 fm^{-1} at absolute zero temperature. Its velocity is *Xc*, where the value of *X* is_____(up to two decimal places).

$(\hbar c = 197\,\text{MeV-fm})$

19. 4 MeV γ-rays emitted by the de-excitation of ^{19}F are attributed, assuming spherical symmetry, to the transition of protons from $^1d_{3/2}$ state to $^1d_{5/2}$ state. If the contribution of spin-orbit term to the total energy is written as $C\langle \vec{\imath} \cdot \vec{s} \rangle$, the magnitude of *C* is_______ MeV (up to one decimal place).

20. An α particle is emitted by a $_{90}^{230}\text{Th}$ nucleus. Assuming the potential to be purely Coulombic beyond the point of separation, the height of the Coulomb barrier is _______ MeV (up to two decimal places).

$\left(\frac{e^2}{4\pi\varepsilon_0} = 1.44\,\text{MeV-fm}, r_0 = 1.30\,\text{fm}\right)$

2017

21. Which one of the following conservation laws is violated in the decay $\tau^+ \to \mu^+ \mu^+ \mu^-$

(*a*) Angular momentum

(*b*) Total Lepton number

(*c*) Electric charge

(*d*) Tau number

22. Electromagnetic interactions are :

(*a*) C conserving

(*b*) C non-conserving but CP conserving

(*c*) CP non-conserving but CPT conserving

(*d*) CPT non-conserving

23. In the nuclear reaction $^{13}C_6 + v_e \to {}^{13}N_7 + X$, the particle X is

(*a*) an electron

(*b*) an anti-electron

(*c*) a muon

(*d*) a pion

24. The π^+ decays at rest to μ^+ and υ^μ. Assuming the neutrino to be massless, the momentum of the neutrino is _______ MeV/c. (up to two decimal places) (m_π = 139 MeV/c^2, m_μ = 105 MeV/c^2).

2016

25. There are four electrons in the 3d shell of an isolated atom. The total magnetic moment of the atom in units of Bohr magneton is _______.

26. In the SU(3) quark model, the triplet of mesons (π^+, π^0, π^-) has

(*a*) Isospin = 0 , Strangeness = 0
(*b*) Isospin = 1 , Strangeness = 0
(*c*) Isospin = 1/2, Strangeness = +1
(*d*) Isospin = 1/2, Strangeness = –1

27. The magnitude of the magnetic dipole moment associated with a square shaped loop carrying a steady current I is m. If this loop is changed to a circular shape with the same current I passing through it, the magnetic dipole moment becomes $\frac{pm}{\pi}$. The value of p is ______.

28. The kinetic energy of a particle of rest mass m_0 is equal to its rest mass energy. Its momentum in units of m_0c, where c is the speed of light in vacuum, is _______. (Give your answer upto two decimal places)

29. Consider the reaction ${}^{54}_{25}\text{Mn} + e^- \rightarrow {}^{54}_{24}\text{Cr} + X$. The particle X is

(*a*) γ (*b*) V_e
(*c*) n (*d*) π^0

30. The scattering of particles by a potential can be analyzed by Born approximation. In particular, if the scattered wave is replaced by an appropriate plane wave, the corresponding Born approximation is known as the first Born approximation. Such an approximation is valid for

(*a*) large incident energies and weak scattering potentials.
(*b*) large incident energies and strong scattering potentials.
(*c*) small incident energies and weak scattering potentials.
(*d*) small incident energies and strong scattering potentials.

31. Consider an elastic scattering of particles in l = 0 states. If the corresponding phase shift δ_0 is 90° and the magnitude of the incident wave vector is equal to $\sqrt{2\pi}$ fm^{-1} then the total scattering cross section in units of fm^2 is _______.

32. The number of spectroscopic terms resulting from the L.S coupling of a 3p electron and a 3d electron is _______.

33. Which of the following statements is NOT correct?

(*a*) A deuteron can be disintegrated by irradiating it with gamma rays of energy 4 MeV.
(*b*) A deuteron has no excited states.
(*c*) A deuteron has no electric quadrupole moment.
(*d*) The ${}^1S_0$ state of deuteron cannot be formed.

34. An alpha particle is accelerated in a cyclotron. It leaves the cyclotron with a kinetic energy of 16 MeV. The potential difference between the D electrodes is 50 kilovolts. The number of revolutions the alpha particle makes in its spiral path before it leaves the cyclotron is ______.

35. Protons and α-particles of equal initial momenta are scattered off a gold foil in a Rutherford scattering experiment. The scattering cross sections for proton on gold and α-particle on gold are σ_p and σ_α respectively. The ratio σ_α / σ_p is ________.

36. A two-level system has energies zero and E. The level with zero energy is non-degenerate, while the level with energy E is triply degenerate. The mean energy of a classical particle in this system at a temperature T is

(*a*) $\dfrac{Ee^{-E/k_BT}}{1+3e^{-E/k_BT}}$ (*b*) $\dfrac{Ee^{-E/k_BT}}{1+e^{-E/k_BT}}$

(*c*) $\dfrac{3Ee^{-E/k_BT}}{1+e^{-E/k_BT}}$ (*d*) $\dfrac{3Ee^{-E/k_BT}}{1+3e^{-E/k_BT}}$

37. According to the nuclear shell model, the respective ground state spin-parity values of ${}^{15}_{8}\text{O}$ and ${}^{17}_{8}\text{O}$ nuclei are

(*a*) $\frac{1^+}{2}, \frac{1^-}{2}$ (*b*) $\frac{1^-}{2}, \frac{5^+}{2}$

(*c*) $\frac{3^-}{2}, \frac{5^+}{2}$ (*d*) $\frac{3^-}{2}, \frac{1^-}{2}$

2015

38. The decay $\mu^+ \rightarrow e^+ + \gamma$ is forbidden, because it violates

(*a*) momentum and lepton number conservations
(*b*) baryon and lepton number conservations
(*c*) angular momentum conservation
(*d*) lepton number conservation

39. The mean kinetic energy of a nucleon in a nucleus of atomic weight A varies as A^n, where n is __________ (upto two decimal places)

40. The binding energy per molecule of NaCl (lattice parameter is 0.563 nm) is 7.95 eV. The repulsive term of the potential is of the form $\frac{K}{r^9}$, where K is a constant. The value of the Madelung constant is ______ (upto three decimal places)

(Electron charge e = –1.6 × 10^{-19} C; ε_0 = 8.854 × 10^{-12} $C^2N^{-1}m^{-2}$)

41. In the nuclear shell model, the potential is modeled as

$$V(r) = \frac{1}{2}m\omega^2 r^2 - \lambda\,\vec{L}\cdot\vec{S}, \quad \lambda > 0$$

The correct spin – parity and isospin assignments for the ground state of ^{13}C is

(*a*) $\frac{1^-}{2}; \frac{-1}{2}$ (*b*) $\frac{1^+}{2}; \frac{-1}{2}$

(*c*) $\frac{3^+}{2}; \frac{1}{2}$ (*d*) $\frac{3^-}{2}; \frac{-1}{2}$

42. The atomic masses of $^{152}_{63}Eu$, $^{152}_{62}Sm$, $^{1}_{1}H$ and neutron are 151.921749, 151.919756, 1.007825 and 1.008665 in atomic mass units (amu), respectively. Using the above information, the Q-value of reaction $^{152}_{63}Eu + n \rightarrow ^{152}_{62}Sm + p$ is ________ $\times 10^{-3}$ amu (upto three decimal places)

2014

43. Which one of the following high energy processes is allowed by conservation laws?

(*a*) $p + \bar{p} \rightarrow \Lambda^0 + \Lambda^0$ (*b*) $\pi^- + p \rightarrow \pi^0 + n$

(*c*) $n \rightarrow p + e^- + v_e$ (*d*) $\mu^+ \rightarrow e^+ + \gamma$

44. Which one of the following is a fermion?

(*a*) α particle (*b*) $_4Be^7$ nucleus

(*c*) Hydrogen atom (*d*) Deuteron

45. Which one of the following three-quark states (*qqq*), denoted by X, CANNOT be a possible baryon? The corresponding electric charge is indicated in the superscript.

(*a*) X^{++} (*b*) X^+

(*c*) X^- (*d*) X^{--}

46. Consider the process $\mu^+ + \mu^- \rightarrow \pi^+ + \pi^-$. The minimum kinetic energy of the muons (μ) in the centre of mass frame required to produce the pion (π) pairs at rest is _______ MeV.

(Given: $m_\mu = 105$ MeV/c^2,
$m_\pi = 140$ MeV/c^2).

47. A nucleus X undergoes a first forbidden β-decay to a nucleus Y. If the angular momentum (I) and parity (P), denoted by I^P as $\frac{7^-}{2}$ for X, which of the following is a possible I^P value for Y?

(*a*) $\frac{1^+}{2}$ (*b*) $\frac{1^-}{2}$

(*c*) $\frac{3^+}{2}$ (*d*) $\frac{3^-}{2}$

2013

48. The decay process $n \rightarrow p^+ + e^- + \bar{v}_e$ violates

(*a*) baryon number (*b*) lepton number

(*c*) isospin (*d*) strangeness

49. The isospin (I) and baryon number (B) of the up quark is

(*a*) I = 1, B = 1 (*b*) I = 1, B = 1/3

(*c*) I = 1/2, B = 1 (*d*) I = 1/2, B = 1/3

50. Consider the scattering of neutrons by protons at very low energy due to a nuclear potential of range r_0. Given that,

$$\cot(kr_0 + \delta) \approx -\frac{y}{k}$$

where δ is the phase shift, k the wave number and $(-\gamma)$ the logarithmic derivative of the deuteron ground state wave function, the phase shift is

(*a*) $\delta \approx -\frac{k}{\gamma} - kr_0$ (*b*) $\delta \approx -\frac{\gamma}{k} - kr_0$

(*c*) $\delta \approx \frac{\pi}{2} - kr_0$ (*d*) $\delta \approx -\frac{\pi}{2} - kr_0$

51. In the β decay process, the transition $2^+ \rightarrow 3^+$, is

(*a*) allowed both by Fermi and Gamow-Teller selection rule

(*b*) allowed by Fermi and but not by Gamow-Teller selection rule

(*c*) not allowed by Fermi but allowed by Gamow-Teller selection rule

(*d*) not allowed both by Fermi and Gamow-Teller selection rule

52. Group I contains elementary excitations in solids. Group II gives the associated fields with these excitations. MATCH the excitations with their associated field and select your answer as per codes given below.

Group I	Group II
P. phonon	(i) photon + lattice vibration
Q. plasmon	(ii) electron + elastic deformation
R. polaron	(iii) collective electron oscillations
S. polariton	(iv) elastic wave

Codes

(a) (P-iv), (Q-iii), (R-i), (S-ii)

(b) (P-iv), (Q-iii), (R-ii), (S-i)

(c) (P-i), (Q-iii), (R-ii), (S-iv)

(d) (P-iii), (Q-iv), (R-ii), (S-i)

53. The electromagnetic form factor $F(q^2)$ of a nucleus is given by,

$$F(q^2) = exp\left[-\frac{q^2}{2Q^2}\right]$$

where Q is a constant. Given that

$$F(q^2) = \frac{4\pi}{q}\int_0^{\infty} rdr\ \rho(r)\sin qr$$

$$\int d^3r\ \rho(r) = 1$$

where $\rho(r)$ is the charge density, the root mean square radius of the nucleus is given by,

(a) 1/Q (b) $\sqrt{2}/Q$

(c) $\sqrt{3}/Q$ (d) $\sqrt{6}/Q$

54. Consider the decay of a pion into a muon and an anti-neutrino $\pi^- \rightarrow \mu^- + \bar{\upsilon}_\mu$ in the pion rest frame.

$m_\pi = 139.6$ MeV/c², $m_\mu = 105.7$ MeV/c², $m_\upsilon \approx 0$

The energy (in MeV) of the emitted neutrino, to the nearest integer is ____

Common Data for Questions 55 and 56:

There are four energy levels E, $2E$, $3E$ and $4E$ (where $E > 0$). The canonical partition function of two particles is , if these particles are

55. two identical fermions

(a) $e^{-2\beta E} + e^{-4\beta E} + e^{-6\beta E} + e^{-8\beta E}$

(b) $e^{-3\beta E} + e^{-4\beta E} + 2e^{-5\beta E} + e^{-6\beta E} + e^{-7\beta E}$

(c) $(e^{-\beta E} + e^{-2\beta E} + e^{-3\beta E} + e^{-4\beta E})^2$

(d) $e^{-2\beta E} - e^{-4\beta E} + e^{-6\beta E} - e^{-8\beta E}$

56. two distinguishable particles

(a) $e^{-2\beta E} + e^{-4\beta E} + e^{-6\beta E} + e^{-8\beta E}$

(b) $e^{-3\beta E} + e^{-4\beta E} + 2e^{-5\beta E} + e^{-6\beta E} + e^{-7\beta E}$

(c) $(e^{-\beta E} + e^{-2\beta E} + e^{-3\beta E} + e^{-4\beta E})^2$

(d) $e^{-2\beta E} - e^{-4\beta E} + e^{-6\beta E} - e^{-8\beta E}$

2012

57. Which one of the following sets corresponds to fundamental particles?

(a) proton, electron and neutron

(b) proton, electron and photon

(c) electron, photon and neutron

(d) quark, electron and meson

58. In case of a Geiger-Muller (GM) counter, which one of the following statements is CORRECT?

(a) Multiplication factor of the detector is of the order of 10^{10}

(b) Type of the particles detected can be identified

(c) Energy of the particles detected can be distinguished

(d) Operating voltage of the detector is few tens of Volts

59. Deuteron has only one bound state with spin parity 1^+, isospin 0 and electric quadrupole moment 0.286 efm². These data suggest that the nuclear forces are having

(a) only spin and isospin dependence

(b) no spin dependence and no tensor components

(c) spin dependence but no tensor components

(d) spin dependence along with tensor components

60. Choose the CORRECT statement from the following.

(a) Neutron interacts through electromag-netic interaction

(b) Electron does not interact through weak interaction

(c) Neutrino interacts through weak and electromagnetic interaction

(d) Quark interacts through strong interac-tion but not through weak interaction

61. Total binding energies of O^{15}, O^{16} and O^{17} are 111.96 MeV,127.62 MeV and 131.76 MeV, respectively. The energy gap between $1p_{1/2}$ and $1d_{5/2}$ neutron shells for the nuclei whose mass number is close to 16, is

(a) 4.1 MeV (b) 11.5 MeV

(c) 15.7 MeV (d) 19.8 MeV

2011

62. The semi-empirical mass formula for the bindng energy of nucleus contains a surface correction term. This term depends on the mass number A of the nucleus as

(a) $A^{-1/3}$ (b) $A^{1/3}$

(c) $A^{2/3}$ (d) A

63. A neutron passing through a detector is detected because of

(a) the ionization it produces

(b) the scintillation light it produces

(c) the electron-hole pairs it produces

(d) the secondary particles produced in a nuclear reaction in the detector medium

64. According to the single particle nuclear shell model, the spin-parity of the ground state of $^{17}{}_{8}O$ is

(*a*) $\frac{1}{2}^{-}$ (*b*) $\frac{3}{2}^{-}$

(*c*) $\frac{3}{2}^{+}$ (*d*) $\frac{5}{2}^{+}$

65. In the β-decay of neutron $n \rightarrow p + e^{-} + \bar{\nu}_e$, the anti-neutrino $\bar{\nu}_e$ escapes detection. Its existence is inferred from the measurement of

(*a*) energy distribution of electrons

(*b*) angular distribution of electrons

(*c*) helicity distribution of electrons

(*d*) forward-backward asymmetry of electrons

66. The isospin and the strangeness of Ω^{-} baryon are

(*a*) 1, –3 (*b*) 0, – 3

(*c*) 1, 3 (*d*) 0, 3

2010

67. The basic process underlying the neutron β-decay is

(*a*) $d \rightarrow u + e^{-} + \overline{V_e}$ (*b*) $d \rightarrow u + e^{-}$

(*c*) $s \rightarrow u + e^{-} + \overline{V_e}$ (*d*) $u \rightarrow d + e^{-} + \overline{V_e}$

68. In the nuclear shell model the spin parity of ^{15}N is given by

(*a*) $\frac{1}{2}^{-}$ (*b*) $\frac{1}{2}^{+}$

(*c*) $\frac{3}{2}^{-}$ (*d*) $\frac{3}{2}^{+}$

69. Match the reactions on the left with the associated interactions on the right.

(1) $\pi^{+} \rightarrow \mu^{+} + \nu_{\mu}$ (*i*) Strong

(2) $\pi^{0} \rightarrow \gamma + \gamma$ (*ii*) Electromagnetic

(3) $\pi^{0} \rightarrow n \rightarrow \pi^{-} + p$ (*iii*) Weak

(*a*) (1, *iii*), (2, *ii*), (3, *i*)

(*b*) (1, *i*), (2, *ii*), (3, *iii*)

(*c*) (1, *ii*), (2, *i*), (3, *iii*)

(*d*) (1, *iii*), (2, *i*), (3, *ii*)

70. The first three energy levels of $^{228}Th_{90}$ are shown below

4^{+} ______ 187 keV

2^{+} ______ 57.5 keV

0^{+} ______ 0 keV

The expected spin-parity and energy of the next level are given by

(*a*) (6^{+}; 400 keV) (*b*) (6^{+}; 300 keV)

(*c*) (2^{+}; 400 keV) (*d*) (4^{+}; 300 keV)

71. The quark content of Σ^{+}, K^{-}, π^{-} and p is indicated:

$|\Sigma^{+}\rangle = |uus\rangle; |K^{+}\rangle = |s\bar{u}\rangle; |\pi^{-}\rangle = |\bar{u}d\rangle;$

$|p\rangle = |uud\rangle$

In the process, $\pi^{-} + p \rightarrow K^{-} + \Sigma^{+}$, consid-ering strong interactions only, which of the following statements is true?

(*a*) The process is allowed because $\Delta S = 0$

(*b*) The process is allowed because $\Delta I_3 = 0$

(*c*) The process is not allowed because $\Delta S \neq 0$ and $\Delta I_3 \neq 0$

(*d*) The process is not allowed because the baryon number is violated

72. Consider the operations $P : \vec{r} \rightarrow -\vec{r}$ (parity) and $T : t \rightarrow -t$ (time-reversal). For the electric and magnetic fields $\vec{E}$ and $\vec{B}$, which of the following set of transformations is correct?

(*a*) $P : \vec{E} \rightarrow -\vec{E}, \vec{B} \rightarrow \vec{B};$ $T : \vec{E} \rightarrow \vec{E}, \vec{B} \rightarrow -\vec{B}$

(*b*) $P : \vec{E} \rightarrow \vec{E}, \vec{B} \rightarrow \vec{B};$ $T : \vec{E} \rightarrow \vec{E}, \vec{B} \rightarrow \vec{B}$

(*c*) $P : \vec{E} \rightarrow -\vec{E}, \vec{B} \rightarrow \vec{B};$ $T : \vec{E} \rightarrow -\vec{E}, \vec{B} \rightarrow -\vec{B}$

(*d*) $P : \vec{E} \rightarrow \vec{E}, \vec{B} \rightarrow -\vec{B};$ $T : \vec{E} \rightarrow -\vec{E}, \vec{B} \rightarrow \vec{B}$

73. A π^{0} meson at rest decays into two photons, which move along the x-axis. They are both detected simultaneously after a time, t = 10 s. In an inertial frame moving with a velocity V = 0.6c in the direction of one of the photons, the time interval between the two detections is

(*a*) 15*s* (*b*) 0*s*

(*c*) 10*s* (*d*) 20*s*

2009

74. Pick the WRONG statement.

(*a*) The nuclear force is independent of electric charge

(*b*) The Yukawa potential is proportional to r^{-1} $\exp\left(\frac{mc}{\hbar}r\right)$, where r is the separation between two nucleons

(c) The range of nuclear force is of the order of $10^{-15}\,m - 10^{-14}\,m$

(d) The nucleons interact among each other by the exchange of mesons

75. In the quark model which one of the following represents a proton?

(a) udd (b) uud

(c) $u\bar{b}$ (d) $c\bar{c}$

76. Consider a nucleus with N neutrons and Z protons. If m_p, m_n and BE represent the mass of the proton, the mass of the neutron and the binding energy of the nucleus respectively and c is the velocity of light in free space, the mass of the nucleus is given by

(a) $Nm_n + Zm_p$

(b) $Nm_p + Zm_n$

(c) $Nm_n + Zm_p + \frac{BE}{c^2}$

(d) $Nm_p + Zm_n + \frac{BE}{c^2}$

77. Consider the following expression for the mass of a nucleus with Z protons and A nucleons :

$M(A, Z) = \frac{1}{c^2}\ (f(A) + yZ + zZ^2)$. Here $f(A)$ is function of A,

$y = -4a_A$,

$z = a_c A^{-1/3} + 4a_A A^{-1}$,

a_A and a_c are constants of suitable dimensions. For a fixed A, the expression of Z for the most stable nucleus is

(a) $Z = \dfrac{A/2}{1+\left(\dfrac{a_c}{a_A}\right)A^{2/3}}$ (b) $Z = \dfrac{A/2}{1+\left(\dfrac{a_c}{4a_A}\right)A^{2/3}}$

(c) $Z = \dfrac{A/2}{1+\left(\dfrac{a_c}{4a_A}\right)A^{2/3}}$ (d) $Z = \dfrac{A}{1+A^{2/3}}$

78. The disintegration energy is defined to be the difference in the rest energy between the initial and final states. Consider the following process :

$$^{240}_{94}\text{Pu} \rightarrow {}^{236}_{92}\text{U} + {}^{4}_{2}\text{He}.$$

The emitted α particle has a kinetic energy 5.17 MeV. The value of the disintegration energy is

(a) 5.26 MeV (b) 5.17 MeV

(c) 5.08 MeV (d) 2.59 MeV

2008

79. According to the quark model, the K^+ meson is composed of the following quarks:

(a) $u\,u\,d$ (b) $u\bar{c}$

(c) $u\bar{s}$ (d) $s\bar{u}$

80. An O^{16} nucleus is spherical and has a charge radius R and a volume $V \equiv \frac{4}{3}\pi R^3$. According the empirical observations of the charge radii, the volume of the $_{54}Xe^{128}$ nucleus, assumed to be spherical is

(a) 8V (b) 2V

(c) 6.75V (d) 1.89V

81. Choose the correct statement from the following

(a) The reaction $K^+K^- \rightarrow p\bar{p}$ can proceed irrespective of the kinetic energies of K^+ and K^-.

(b) The reaction $K^+K^- \rightarrow p\bar{p}$ is forbidden by the baryon number conservation.

(c) The reaction $K^+K^- \rightarrow 2\gamma$ is forbidden by strangeness conservation.

(d) The decay $K^0 \rightarrow \pi^+\pi^-$ proceeds via weak interactions.

82. The following gives a list of pairs containing (i) a nucleus (ii) one of its properties. Find the pair which is **inappropriate.**

(a) (i) $_{10}Ne^{20}$ nucleus; (ii) stable nucleus

(b) (i) A spheroidal nucleus; (ii) an electric quadrupole moment

(c) (i) $_8O^{16}$ nucleus; (ii) nuclear spin $J = \frac{1}{2}$

(d) (i) U^{238} nucleus; (ii) Binding energy = 1785 MeV (approximately)

83. The four possible configurations of neurtorns in the ground state of $_4Be^9$ nucleus, according to the shell model, and the associated nuclear spin are listed below. Choose the correct one

(a) $(1s_{1/2})^2(1p_{3/2})^3; J = \frac{3}{2}$

(b) $(1s_{1/2})^2(1p_{1/2})^2(1p_{3/2})^1; J = \frac{3}{2}$

(c) $(1s_{1/2})^1(1p_{3/2})^4; J = \frac{1}{2}$

(d) $(1s_{1/2})^2(1p_{3/2})^2(1p_{1/2})^1; J = \frac{1}{2}$

84. The mass difference between the pair of mirror nuclei $_6C^{11}$ and $_5B^{11}$ is given to be $\Delta MeV/c^2$. According to the semi-empirical mass formula, the mass difference between the pir of mirror nuclei $_9F^{17}$ and $_8O^{17}$ will approximately be (rest mass of proton $m_p = 938.27\ MeV/c^2$ and rest mass of neutron $m_n = 939.57\ MeV/c^2$)

(*a*) $1.39\ \Delta\ MeV/c^2$

(*b*) $(1.39\ \Delta + 0.5)\ MeV/c^2$

(*c*) $0.86\ \Delta\ MeV/c^2$

(*d*) $(1.6\Delta + 0.78)\ MeV/c^2$

85. A heavy nucleus is found to contain more nuetrons than protons. This fact is related to which one of the following statements.

(*a*) The nuclear force between neutrons is stronger than that between protons.

(*b*) The nuclear force between protons is of a shorter range than those between neutrons, so that a smaller number of protons are held together by the nuclear force.

(*c*) Protons are unstable, so their number in a nucleus diminishes.

(*d*) It cots more energy to add a proton to a (heavy) nucleus than a neutron because of the Coulomb repulsion between protons.

86. A neutral *pi* meason (π^0) has a rest-mass of approximately 140 MeV / c^2 and a lifetime of τ sec. A π^0 produced in the laboratory is found to decay after 1.25 τ sec into two photons. Which of the following sets represents a possible set of energies of the two photons as seen in the laboratory?

(*a*) 70 MeV and 70 MeV

(*b*) 35 MeV and 100 MeV

(*c*) 75 MeV and 100 MeV

(*d*) 25 MeV and 150 MeV

Statement for Linked Answer Questions 87 and 88:

Consider the β–decay of a free neutron at rest in the laboratory.

87. Which of the following configurations of the decay products correspond to the largest energy of the anti-neutrino $\bar{\upsilon}$? (rest mass of electron $m_e = 0.51 MeV/c^2$, rest mass of proton $m_p = 938.27 MeV/c^2$ and rest mass of neutron $m_n = 939.57 MeV/c^2$)

(*a*) In the laboratory, proton is produced at rest.

(*b*) In the laboratory, momenta of proton, electron and the anti-neutrino all have the same magnitude.

(*c*) In the laboratory, proton and electron fly-off with (nearly) equal and opposite momenta.

(*d*) In the laboratory, electron is produced at rest.

88. Using the result of the above problem, answer the following. Which of the following represents approximately the maximum allowed energy of the anti-neutrino $\bar{\upsilon}$?

(*a*) 1.3 MeV (*b*) 0.8 MeV

(*c*) 0.5 MeV (*d*) 2.0 MeV

2007

89. Fission fragments are generally radioactive as

(*a*) they have excess of neutrons.

(*b*) they have excess of protons.

(*c*) they are products of radioactive nuclides

(*d*) their total kinetic energy is of the order of 200 MeV.

90. The strangeness quantum number is conserved in

(*a*) strong, weak and electromagnetic interactions.

(*b*) weak and electromagnetic interactions only.

(*c*) strong and weak interactions only.

(*d*) strong and electromagnetic interactions only.

91. A ferromagnetic mixture of iron and copper having 75% atoms of Fe exhibits a saturation magnetization of $1.3 \times 10^6\ A\ m^{-1}$. Assume that the total number of atoms per unit volume is $8 \times 10^{28}\ m^{-3}$. The magnetic moment of an iron atom, in terms of the Bohr Magneton, is

(*a*) 1.7 (*b*) 2.3

(*c*) 2.9 (*d*) 3.8

92. Half life of a radio-isotope is 4×10^8 years. If there are 10^3 radioactive nuclei in a sample today, the number of such nuclei in the sample 4×10^9 year ago were

(*a*) 128×10^3 (*b*) 256×10^3

(*c*) 512×10^3 (*d*) 1024×10^3

93. In the deuterium + tritium ($d + t$) fusion more enrgy is relased as compared to deuterium + deuterium ($d + d$) fusion because

(*a*) tritium is radioactive

(*b*) more nucleons participate in fusion

(*c*) the Coulomb barrier is lower for the $d + t$ system than $d + d$ system.

(*d*) the reaction product ^4He is more tightly bound.

94. According to the shell model the ground state spin of the ^{17}O nucleus is

(a) $\frac{3^+}{2}$ (b) $\frac{5^+}{2}$

(c) $\frac{3^-}{2}$ (d) $\frac{5^-}{2}$

95. A relativistic particle travels a length of 3×10^{-3} m in air before decaying. The decay process of the particle is dominated by

(a) strong interactions.

(b) electromagnetic interactions.

(c) weak interactions.

(d) gravitational interactions.

96. The strange baryon Σ^+ has the quark structure

(a) uds (b) uud

(c) uus (d) $u\bar{s}$

97. A neutron scatters elastically from a heavy nucleus. The initial and final states of the neutron have the

(a) same energy.

(b) same energy and linear momentum.

(c) same energy and angular momentum.

(d) same linear and angular momenta.

Statement for Linked Answer Questions 98 & 99:

A 16μA beam of alpha particles, having cross-sectional area $10^{-4}\,m^2$, is incident on a rhodium target of thickness 1 μm. This produces neutrons through the reaction

$$\alpha + {}^{100}Rh \rightarrow {}^{100}Pd + 3n.$$

98. The number of alpha particles hitting the target per second is

(a) 0.5×10^{14} (b) 1.0×10^{14}

(c) 2.0×10^{20} (d) 4.0×10^{20}

99. The neutrons are observed at the rate of $1.806 \times 10^8\ s^{-1}$. If the density of rhodium is approximated as 10^4 kg m^{-3} the cross-section for the reaction (in barns) is

(a) 0.1 (b) 0.2

(c) 0.4 (d) 0.8

2006

100. Weak nuclear forces at on

(a) both hadrons and leptons

(b) hadrons only

(c) all particles

(d) all charged particle

101. Which one of the following distintegration series of the heavy elements will give ^{209}Bi as a stable nucleus?

(a) Thorium series (b) Neptunium series

(c) Uranium series (d) Actinium series

102. The order of magnitude of the binding energy per nucleon in a nucleus is

(a) 10^{-5} MeV (b) 10^{-3} MeV

(c) 0.1 MeV (d) 10 MeV

103. The interaction potential between two qarks, separated by a distance r inside a nucleon, can be described by (a, b and β are positive constants)

(a) $ae^{-\beta r}$ (b) $\frac{a}{r}+br$

(c) $-\frac{a}{r}+br$ (d) $\frac{a}{r}$

104. The experimentally measured spiun g factors of proton and a neutron indicate that

(a) both proton and neutron are elementary point particles

(b) both proton and neutron are not elementary point particles

(c) while proton is an elementary point particle, neutron is not

(d) while neutron is an elementary point particle, proton is not

105. By capturing an electron, ${}^{54}_{25}Mn_{29}$ transforms into ${}^{54}_{25}Cr_{30}$ releasing

(a) a neutrino (b) an antineutrino

(c) an α-particle (d) a positron

106. Which one of the following nuclear processes is forbidden?

(a) $\bar{v} + p \rightarrow n + e^+$

(b) $\pi^- \rightarrow e^- + v_e + \pi^0$

(c) $\pi^- + p^- \rightarrow n + K^+ + K^-$

(d) $\mu^- \rightarrow e^- + \bar{v}_e + v_\mu$

107. To explain the observed magnetic moment of deuteron (0.8574 μ_N), its ground state wavefunction is taken to be an admixture of S and D states. The expectation values fo the z-component of the magnetic moment in pure S and pure D states are 0.8797μ_N and 0.3101μ_N respectively. The contribution of the D state to the mixed ground state is approximately

(a) 40% (b) 4%

(c) 0.4% (d) 0.04%

Statement for Linked Answer Questions 108 and 109:

Consider Fermi theory of β-decay.

108. The number of final states of electrons corresponding to momenta between p and $p + dp$ is

(*a*) independent of p

(*b*) proportional to pdp

(*c*) proportional to $p^2\,dp$

(*d*) proportinal to p^3dp

109. The number of emitted electrons with momenmtum p and energy E, in the allowed approximation, is proportional to (E_0 is the total energy given up by the nucleus)

(*a*) $(E_0 - E)$ (*b*) $p\,(E_0 - E)$

(*c*) $p^2\,(E_0 - E)^2$ (*d*) $p\,(E_0 - E)^2$

Statement for Linked Answer Questions 110 and 111:

Consider a radiation cavity of volume V at temperature T.

110. The density of states at energy E of the quantized radiation (photons) is

(*a*) $\frac{8\pi V}{h^3c^3}E^2$ (*b*) $\frac{8\pi V}{h^3c^3}E^{3/2}$

(*c*) $\frac{8\pi V}{h^3c^3}E$ (*d*) $\frac{8\pi V}{h^3c^3}E^{1/2}$

111. The average number of photons in equilibrium inside the cavity is proportinal to

(*a*) T (*b*) T^2

(*c*) T^3 (*d*) T^4

2005

112. The evidence for the nonconservation of parity in β decay has been obtained from the observation that the β intensity

(*a*) antiparallel to the nuclear spin directions is same as that along the nuclear spin direction

(*b*) antiparallel to the nuclear spin direction is not the same as that along the nuclear spin direction

(*c*) shows a continuous distribution as a function of momentum

(*d*) is independent of the nuclear spin direction

113. Which of the following expressions for total binding energy B of a nucleus is correct (a_1, a_2, a_3, $a_4 > 0$) ?

(*a*) $B = a_1A - a_2A^{2/3} - a_3\frac{Z(Z-1)}{A^{1/3}} - a_4\frac{(A-2Z)^2}{A} + \delta$

(*b*) $B = a_1A + a_2A^{2/3} - a_3\frac{Z(Z-1)}{A^{1/3}} - a_4\frac{(A-2Z)^2}{A} + \delta$

(*c*) $B = a_1A + a_2A^{1/3} - a_3\frac{Z(Z-1)}{A^{1/3}} - a_4\frac{(A-2Z)^2}{A} + \delta$

(*d*) $B = a_1A - a_2A^{1/3} - a_3\frac{Z(Z-1)}{A^{1/3}} - a_4\frac{(A-2Z)^2}{A} + \delta$

114. Which of the following decay is forbidden ?

(*a*) $\mu^- \rightarrow e^- + v_\mu + \bar{v}_c$

(*b*) $\pi^+ \rightarrow \mu^+ + v_\mu$

(*c*) $\pi^+ \rightarrow e^+ + v_e$

(*d*) $\mu^- \rightarrow e^+ + e^- + e^-$

115. With reference to nuclear forces which of the following statements is NOT true ?

The nuclear forces are

(*a*) short range

(*b*) charge independent

(*c*) velocity dependent

(*d*) spin independent

116. A nucleus having mass number 240 decays by α emission to the ground state of its daughter nucleus. The Q value of the process is 5.26 MeV. The energy (in MeV) of the α particle is

(*a*) 5.26 (*b*) 5.17

(*c*) 5.13 (*d*) 5.09

117. The threshold temperature above which the thermonuclear reaction

$${}^3_2He + {}^3_2He \rightarrow {}^4_2He + 2\,{}^1_1H + 12.86\text{ MeV}$$

can occur is (use $e^2/4\pi\varepsilon_0 = 1.44 \times 10^{-15}$ MeVm)

(*a*) 1.28×10^{10} K

(*b*) 1.28×10^{9} K

(*c*) 1.28×10^{8} K

(*d*) 1.28×10^{7} K

118. According to the shell model, the ground state of ${}^{15}_8O$ nucleus is

(*a*) $\frac{3}{2}^+$ (*b*) $\frac{1}{2}^+$

(*c*) $\frac{3}{2}^-$ (*d*) $\frac{1}{2}^-$

119. The plot of log A vs. time t, where A is activity, as shown in the figure, corresponds to decay

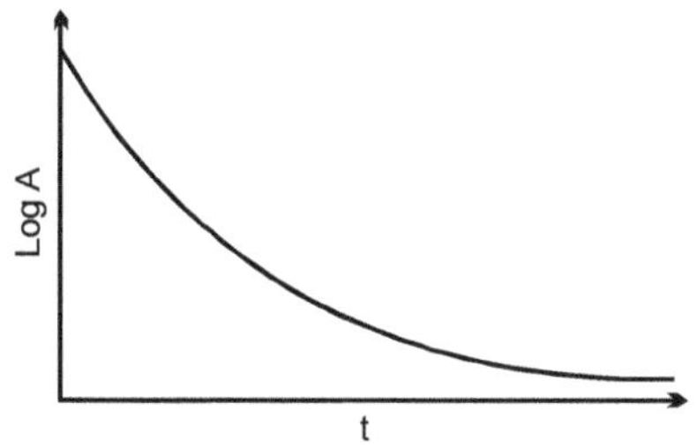

(*a*) from only one kind of radioactive nuclei having same half life

(*b*) from only neutron activated nuclei

(*c*) from a mixture of radioactive nuclei having different half lives

(*d*) which is unphysical

2004

120. The resonance widths Γ of ρ, ω and φ particle resonances satisfy the relation $\Gamma_\rho > \Gamma_\omega > \Gamma_\varphi$. Their life-times τ satisfy the relation

(*a*) $\tau_\rho > \tau_\omega > \tau_\varphi$ (*b*) $\tau_\rho < \tau_\omega < \tau_\varphi$

(*c*) $\tau_\rho < \tau_\omega < \tau_\varphi$ (*d*) $\tau_\rho > t_w < \tau_\varphi$

121. The volume of a nucleus in an atom is proportional to the

(*a*) mass number (*b*) proton number

(*c*) neutron number (*d*) electron number

122. As one moves along the line of stability from ^{56}Fe to ^{235}U nucleus, the nuclear binding energy per particle decreases from about 8.8 MeV to 7.6 MeV. This trend is mainly due to the

(*a*) short range nature of the nuclear forces

(*b*) long range nature of the Coulomb forces

(*c*) tensor nature of the nuclear forces

(*d*) spin dependence of the nuclear forces

123. A thermal neutron having speed v impinges on a ^{235} U nucleus. The reaction cross-section is proportional to

(*a*) v^{-1} (*b*) v

(*c*) $v^{1/2}$ (*d*) $v^{-1/2}$

124. Choose the particle with zero Baryon number from the list given below.

(*a*) pion (*b*) neutron

(*c*) proton (*d*) Δ^+

125. An electron gains energy so that its mass becomes $2m_0$. Its speed is

(*a*) $\frac{\sqrt{3}}{2}c$ (*b*) $\frac{3}{4}c$

(*c*) $\frac{3}{2}c$ (*d*) $\sqrt{\frac{3}{2}}c$

126. Deuteron in its ground state has a total angular momentum J = 1 and a positive parity. The corresponding orbital angular momentum L and spin S combinations are

(*a*) L = 0, S = 1 and L = 2, S = 0

(*b*) L = 0, S = 1 and L = 1, S = 1

(*c*) L = 0, S = 1 and L = 2, S = 1

(*d*) L = 1, S = 1 and L = 2, S = 1

127. Which one of the following reaction is allowed ?

(*a*) $p \to n + e^+$

(*b*) $p \to e^+ + \upsilon_e$

(*c*) $p \to \pi^+ + \gamma$

(*d*) $\overline{p} + n \to \pi^- + \pi^0$

Data for Q. 128-129

The nucleus ^{41}Ca can be described by the single particle shell model.

128. The single particle states occupied by the last proton and the last neutron, respectively, are given by

(*a*) $d_{5/2}$ and $f_{7/2}$

(*b*) $d_{3/2}$ and $f_{5/2}$

(*c*) $d_{5/2}$ and $f_{5/2}$

(*d*) $d_{3/2}$ and $f_{7/2}$

129. The ground state angular momentum and parity of ^{41}Ca are

(*a*) $\frac{7}{2^-}$ (*b*) $\frac{3}{2^+}$

(*c*) $\frac{5}{2^+}$ (*d*) $\frac{5}{2^-}$

2003

130. The effective mass of an electron in a semiconductor can be

(*a*) negative near the bottom of the band

(*b*) a scalar quantity with a small magnitude

(*c*) zero at the center of the band

(*d*) negative near the top of the band

131. The nucleus of the atom $^9Bc_4$ consists of

(*a*) 13 up quarks and 13 down quarks

(*b*) 13 up quarks and 14 down quarks

(*c*) 14 up quarks and 13 down quarks

(*d*) 14 up quarks and 14 down quarks

132. Which one of the following nuclear reactions is possible ?

(*a*) $^{14}N_7 \to {}^{13}C_6 + \beta^+ + \nu_c$

(*b*) $^{13}N_7 \to {}^{13}C_6 + \beta^+ + \nu_c$

(*c*) $^{13}N_7 \to {}^{13}C_6 + \beta^+$

(*d*) $^{13}N_7 \to {}^{13}C_7 + \beta^+ + \nu_c$

133. Suppose that a neutron at rest in free space decays into a proton and an electron. This process would violate

(*a*) conservation of charge

(*b*) conservation of energy

(*c*) conservation of linear momentum

(*d*) conservation of angular momentum

Data for Q. 134 – 135 :

An atomic bomb consisting of ^{235}U explodes and releases an energy of 10^{14} J. It is known that each ^{235}U which undergoes fission releases 3 neutrons and about 200 MeV of energy. Further, only 20% of the ^{235}U atoms in the bomb undergo fission

134. The total number of neutrons released is about

(*a*) 4.7×10^{24}

(*b*) 9.7×10^{24}

(*c*) 1.9×10^{25}

(*d*) 3.7×10^{25}

135. The mass of ^{235}U in the bomb is about

(*a*) 1.5 kg

(*b*) 3.0 kg

(*c*) 6.1 kg

(*d*) 12 kg

2002

136. The baryon number of proton, the lepton number of proton, the baryon number of electron and the lepton number of electron are respectively

(*a*) zero, zero, one and zero

(*b*) one, one, zero and one

(*c*) one, zero, zero and one

(*d*) zero, one, one and zero

137. Typical energies released in a nuclear fission and a nuclear fusion reaction are respectively

(*a*) 50 MeV and 1000 MeV

(*b*) 200 MeV and 1000 MeV

(*c*) 1000 MeV and 50 MeV

(*d*) 200 MeV and 10 MeV

138. Nuclear forces are

(*a*) spin dependent and have no non-central part

(*b*) spin dependent and have a non-central part

(*c*) spin independent and have no non-central part

(*d*) spin independent and have a non-central part

139. The nuclear spins of $_6C^{14}$ and $_{12}Mg^{25}$ nuclei are respectively

(*a*) zero and half-integer

(*b*) half-integer and zero

(*c*) an integer and half-integer

(*d*) both half-integers

140. The spin and parity of $_4Be^9$ nucleus, as predicted by the shell model, are respectively

(*a*) 3/2 and odd (*b*) 1/2 and odd

(*c*) 3/2 and even (*d*) 1/2 and even

2001

141. A thin conducting wire is bent into a circular loop of radius r and placed in a time dependent magnetic field of magnetic induction. $\vec{B}(t) = B_0 e^{-\alpha t} \hat{e}_z$, ($B_0 > 0$ & $\alpha > 0$), such that, the plane of the loop is perpendicular to $\vec{B}(t)$. Then the induced emf in the loop is

(*a*) $\pi r^2 \alpha B_0 e^{-\alpha t}$

(*b*) $\pi r^2 B_0 e^{-\alpha t}$

(*c*) $-\pi r^2 \alpha B_0 e^{-\alpha t}$

(*d*) $-\pi r^2 B_0 e^{-\alpha t}$

142. Infrared absorption can be observed in which of the following molecules ?

(*a*) N_2 (*b*) O_2

(*c*) HCl (*d*) C_2

143. The cross-sections of the reactions

$p + \Pi^- \rightarrow \Sigma^- + K^+$ and $p^- + \Pi^+ \rightarrow \overline{\Sigma}^- + K^-$

at a given energy are the same due to

(*a*) baryon number conservation

(*b*) time-reversal invariance

(*c*) charge conjugation

(*d*) parity conservation

144. Which of the following functions describes the nature of interaction potential V(r) between two quarks inside a nucleon ? (r is the distance between the quarks and a and b positive constants of suitable dimensions).

(*a*) $V(r) = \frac{a}{r} + br$ (*b*) $V(r) = -\frac{a}{r} + br$

(*c*) $V(r) = \frac{a}{r} - br$ (*d*) $V(r) = -\frac{a}{r} - br$

2000

145. The mean momentum of a nucleon in a nucleus with mass number A varies as

(*a*) A (*b*) A^2

(*c*) $A^{-2/3}$ (*d*) $A^{-1/3}$

146. Magnons in ferromagnets

(*a*) decrease the magnetization

(*b*) increase the magnetization

(*c*) stabilize the magnetization

(*d*) cause critical magnetic fluctuations

147. The deutron is known to be in a state with S = 1, J = 1, I = 0 where S, J, I refer to spin, total angular momentum and isospin quantum numbers respectively. The allowed values of the orbital angular momentum quantum number L are

(*a*) 0 (*b*) 0, 1, 2

(*c*) 1 (*d*) 0, 2

148. Identify the reaction which has the same transition probability as $\pi^+ + p \to \pi^+ + p$

(*a*) $\pi^+ + n \to \pi^+ + n$

(*b*) $\pi^- + p \to \pi^0 + n$

(*c*) $\pi + n \to \pi + n$

(*d*) $\pi^0 + p \to \pi^+ + n$

ANSWERS

1. (d) **2.** (d) **3.** (b,c) **4.** (*c*) **5.** (*c*) **6.** (2 to 2) **7.** (*c*) **8.** (6999 to 6999) **9.** (*c*)
10. (*c*) **11.** (*a*) **12.** (*b*) **13.** (40 to 40) **14.** (*b*) **15.** (*c*) **16.** (*d*) **17.** (299.9 to 300.1)
18. (0.28 to 0.31) **19.** (1.6 to 1.6) **20.** (24.4 to 27.6) **21.** (*d*) **22.** (*a*) **23.** (*a*) **24.** 29.50 to 30.10
25. (0 : 0) **26.** (*b*) **27.** (4 : 4) **28.** (1.72 : 1.74) **29.** (*b*) **30.** (*a*) **31.** (2 : 2)
32. (12 : 12, 6 : 6) **33.** (*c*) **34.** (80 : 80) **35.** (4 : 4) **36.** (*d*) **37.** (*b*) **38.** (*d*)
39. (–0.67 to – 0.66) **40.** (1.745 to 1.751) **41.** (*a*) **42.** (2.830 to 2.835)
43. (*b*) **44.** (*b*) **45.** (*d*) **46.** (34.9 to 35.1) **47.** (*c*) **48.** (*c*) **49.** (*d*) **50.** (*a*) **51.** (*c*) **52.** (*b*)
53. (*c*) **54.** (30) **55.** (*b*) **56.** (*c*) **57.** (*a*) **58.** (*a*) **59.** (*d*) **60.** (*a*) **61.** (*a*) **62.** (*c*) **63.** (*d*)
64. (*d*) **65.** (*a*) **66.** (*b*) **67.** (*a*) **68.** (*a*) **69.** (*a*) **70.** (*a*) **71.** (*c*) **72.** (*d*) **73.** (*a*) **74.** (*b*)
75. (*b*) **76.** (*c*) **77.** (*c*) **78.** (*a*) **79.** (*c*) **80.** (*a*) **81.** (*a,d*) **82.** (*c*) **83.** (*a*) **84.** (*a*) **85.** (*d*)
86. (*c*) **87.** (*c*) **88.** (*a*) **89.** (*a*) **90.** (*d*) **91.** (*b*) **92.** (*d*) **93.** (*a*) **94.** (*b*) **95.** (*c*) **96.** (*c*)
97. (*a*) **98.** (*a*) **99.** (*a*) **100.** (*c*) **101.** (*b*) **102.** (*d*) **103.** (*c*) **104.** (*b*) **105.** (*a*) **106.** (*b*) **107.** (*b*)
108. (*c*) **109.** (*b*) **110.** (*d*) **111.** (*c*) **112.** (*a*) **113.** (*a*) **114.** (*d*) **115.** (*d*) **116.** (*b*) **117.** (*d*) **118.** (*b*)
119. (*a*) **120.** (*c*) **121.** (*a*) **122.** (*b*) **123.** (*a*) **124.** (*a*) **125.** (*a*) **126.** (*c*) **127.** (*d*) **128.** (*d*) **129.** (*a*)
130. (*d*) **131.** (*b*) **132.** (*b*) **133.** (*d*) **134.** (*b*) **135.** (*c*) **136.** (*c*) **137.** (*d*) **138.** (*b*) **139.** (*a*) **140.** (*a*)
141. (*a*) **142.** (*c*) **143.** (*c*) **144.** (*b*) **145.** (*d*) **146.** (*d*) **147.** (*b*) **148.** (*b*)

EXPLANATIONS

1. We know that

If $\Delta\pi \neq 0$ E is odd

and If $\Delta\pi \neq 0$, E is even

Here $l = 1$, $J_i = 2$

$J_f = |J_i - l|$ to $|J_i + l|$

$J_f = 1, 2, 3$

Since parity changes, than All 'possible values of J_f^π are $1^-, 2^-, 3^-$

So, option (d) is correct.

2. As me know that from Collective mode of Nucleus in Set-I, energy levels are equal spaced then it is Vibrational bond for even Nuclei for (O^+), (2^+), $(O^+, 2^+, 4^+,)$......

In Set-II difference between energy levels are increasing then it is rotational bond for odd nuclei

for $\frac{7^+}{2}, \frac{11^+}{2}, \frac{15^+}{2}$

So, option (d) is the correct answer.

3. Given

$$M = M_0 + aY + b\left[I(I+1) - \frac{1}{4}Y^2\right]$$

$M_\Sigma = M_{\AA}$

$Y_\Sigma = Y_{\AA} = 0$

$I_\Sigma = 1, I_{\AA} = 0$

Now,

$$M_\Sigma = M_0 + aY_\Sigma + b\left[I_\Sigma(I_\Sigma+1) - \frac{1}{4}Y_\Sigma^2\right]$$

$M_\Sigma = M_0 + o + b\,[1(1 + 1) – 0)]$

$M_\Sigma = M_0 + 2b$ ... (i)

$$\& \;\; M_{\AA} = M_0 + aY_{\AA} + b\left[I_{\AA}(I_{\AA}+1) - \frac{1}{4}Y_{\AA}^2\right]$$

$M_{\AA} = M_0 + o + b[0 – 0]$

$M_{\AA} = M_0$ ... (ii)

According to question

$M_\Sigma = M_{\AA}$ From (i) & (ii)

$M_0 + 2b = M$

$\boxed{b = 0}$

Now,

$M = M_0 + aY$

$M \propto Y$

Hence, M does Not depend on I

4. $P + \pi^+ \rightarrow K^+ + Y$

The quark content

$\pi^+ + P \rightarrow Y + K^+$

$\underbrace{u\bar{d} + uud}_{uuu} \rightarrow \underbrace{Y + u\bar{s}}_{uuu}$

$uus + u\bar{s} \rightarrow uuu$

Answer is uus

5. $\vec{E} \cdot \vec{B}$ is odd under parity and odd under time reversal

6. $Y \rightarrow \pi^- + \pi^-$

	π^-	π^0	π^+
Isospin	1	1	1
I_z	–1	0	+1

$I = 1 + 1$

$= 2$

$\Delta I = 0$

Isospin of $\pi^+ = +1$

7. $E_F \propto \frac{1}{R_0^2}$

For 3D

$$K_F = \left(\frac{3\pi^2 N}{V}\right)^{1/3}$$

$$E_F = \frac{K_F^2 \hbar^2}{2m} \propto K_F^2$$

$$E_F \propto \frac{1}{V^{\frac{2}{3}}}$$

$$V = \frac{4}{3}\pi R^3$$

$$E_F \propto \frac{1}{R^2}$$

$$R = R_0 A^{\frac{1}{3}}$$

$$E_F \propto \frac{1}{\left(R_0 A^{\frac{1}{3}}\right)^2}$$

$$E_F \propto \frac{1}{R_0^2}$$

10. $X \longrightarrow Y + Y$

Boson Spiral Spiral

Boson Boson

X is a Boson

12. $\pi^+ + d \longrightarrow p + q$

$0 + 1 \longrightarrow 1 + 1$

Deutron (Ground state)

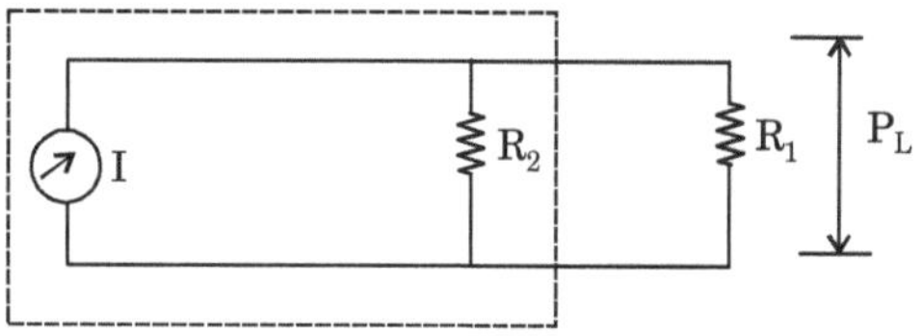

D.C. Current source

$$P = VI = \frac{V^2}{R} = I^2 R$$

$$\rho_{Total} = I^2 R_{Total}$$

$$R_{Total} = \frac{R_L R_2}{R_L + R_2}$$

$$\rho_{Total} = \frac{I^2 R_2 R_L}{R_2 + R_L}$$

$$\eta_2 = \frac{\rho_{load}}{\rho_{Total}} = \frac{R_2 + R_L}{R_2}$$

D.C. voltage source

$$\rho_{load} = \frac{V^2}{R_L}$$

$$\rho_{Total} = \frac{V^2}{R_1 + R_L}$$

$$\eta_1 = \frac{\rho_{load}}{\rho_{Total}} = \frac{R_1 + R_L}{R_L}$$

$\eta_1 = \eta_2.$

$$\frac{R_1 + R_L}{R_L} = \frac{R_2 + R_L}{R_2}$$

$$\Rightarrow R_1 R_2 = R_L^2 = (4.7)^2 = 22.1\Omega^2$$

13. Branching ratio

$$= \frac{\lambda_\beta}{\lambda} = \frac{\lambda_\beta}{\lambda_\propto + \lambda_\beta} = 0.75 = \frac{4}{3}$$

$$\Rightarrow 4\lambda_\beta = 3\lambda_\propto + 3\lambda_\beta$$

$$\Rightarrow \lambda_\propto = \frac{\lambda_\beta}{3}$$

$$t_{\frac{1}{2}} = \frac{\ln 2}{\lambda} = \frac{0.693}{\lambda}$$

$$\lambda = \frac{.693}{t_{\frac{1}{2}}}$$

$$\frac{.693}{t_{\frac{1}{2}}} = \frac{4}{3} \times \frac{.693}{t^{\beta}_{\frac{1}{2}}}$$

$$t^{\beta}_{\frac{1}{2}} = \frac{4}{3} \times t_{\frac{1}{2}} = \frac{4}{3} \times 30 = 40\text{hr}$$

14. The various field and source quantities have various symmetry properties under symmetry operations.

Such as

(1) P = Parity (space-Inversion, $\vec{r} \to -\vec{r}$ → Reflection in a mirror.

(2) E = Electric charge conjugation

(3) M = Magnetic charge conjugation

(Magnetically charged particle → magnetically charged anti-particle i.e, $g_N \to g_S$)

So, in the given options, choice (b) is odd under parity operation.

15. The baryon decuplet for intrinsic spin $\frac{3}{2}$ is shown in given figure, so,

$Q = Y^0_1$

$P = \Xi^-$

$R = \Xi^0$

$S = \Omega^-$

The elementary particle Ξ^0 is placed in the baryon decuplet at R.

16. In decay $\mu^+ \longrightarrow e^+ + v_e + X$

$$\mu^+ \longrightarrow e^+ + v_e + \bar{v}_\mu$$

Here $X = \bar{v}_\mu$ = anti-neutrinos.

17. As we know, the rotational energy of a molecule is

$$E = \frac{h^2}{8\pi^2 I} J(J+1)$$

or $E \propto J(J + 1)$

or $$\frac{E_1}{E_2} = \frac{J_1(J_1+1)}{J_2(J_2+1)}$$

$$\therefore \frac{90}{E_2} = \frac{2(2+1)}{4(4+1)} = \frac{6}{20}$$

$$\Rightarrow E_2 = \frac{90 \times 20}{6} = 300.0 \text{ KeV}$$

21. $\tau^+ \to \mu^+\mu^+\mu^-$

conservation of Tau number.

22. Electromagnetic interactions are charge conserving.

23. Nuclear Reaction :

$^{13}C_6 + V_e \to {}^{13}N_7 + X$ the particle is an electron.

β decay transformation

$$\begin{pmatrix} n \to P + \bar{\beta} + \bar{\upsilon} \\ n \to P + \bar{e} + \bar{\upsilon} \end{pmatrix}$$

$$^{13}C_6 + \upsilon^{-1} \to {}^{13}N_7 + 0^{e^{-1}}$$

Hence X is electron.

24. $E_\pi = m_\pi C^2 = E_\mu + E_\upsilon$

$P_\pi = 0 = P_\mu + P_\upsilon$

But $P_\upsilon = -P_\mu$

$$m_\pi C = \sqrt{P_\mu^2 + m_\mu^2 C^2 - P_\mu}$$

$$\sqrt{P_\mu^2 + m_\mu^2 C^2} = m_\pi C + P_\mu$$

$$P_\mu = \frac{m_\mu^2 - m_\pi^2 C}{2m_\pi}$$

$$= \frac{(139)^2 - (105)^2}{2 \times 139} \times 3 \times 10^8$$

$$= \frac{19321 - 1102}{278} \frac{MeV}{C^2} \times C = 29.8.41 \frac{MeV}{C}$$

$= 29.80\ MeV/C.$

25. For an isolated atom,

Since it is case of less than half filled subshell, according to Hund's Rule, lower J will be in ground state.

$S = 2, \quad L = 2$

$J = L - S = 0$

Thus $\vec{\mu} = 0$

26. According to SU(3) Quark model, the isospin of π mesons (π^+, π^0, π^-) is 1 and strangeness is zero.

27. Mag. dip. moment $m_1 = Ia^2$

For circle $m_2 = I\pi r^2$

Here, $4a = 2\pi r$

$$r = \frac{2a}{\pi}$$

$\therefore$ Magnetic dipole moment

$$= I\pi\left(\frac{2a}{\pi}\right)^2$$

$$= \frac{4Ia^2}{\pi} = \frac{4m_1}{\pi}$$

Comparing with $\frac{pm}{\pi}$ we get p = 4

28. $(m - m_0)C^2 = m_0C^2$

$mC^2 - m_0C^2 = m_0C^2$

$$m = 2m_0$$

$$m = \frac{m_0}{\sqrt{1-\frac{v^2}{C^2}}}$$

$$2m_0 = \frac{m_0}{\sqrt{1-\frac{v^2}{C^2}}}$$

$$\frac{1}{2} = \sqrt{1-\frac{v^2}{C^2}}$$

$$1-\frac{v^2}{C^2} = \frac{1}{4}$$

$$\Rightarrow \quad \frac{v^2}{C^2} = 1-\frac{1}{4} = \frac{3}{4}$$

$$v = \frac{\sqrt{3}}{2}C$$

$$p = mV = 2m_0 \times \frac{\sqrt{3}}{2}C$$

$$p = \sqrt{3}\ C$$

$$p = 1.74$$

31. Total scattering cross section

$$\sigma_t = \sum_l (2l+1)\sin^2\delta_l \times \frac{4\pi}{k^2}$$

$l = 0 \quad k \rightarrow$ wave vector

$$= \sqrt{2}\pi\ fm^{-1}$$

$$\sigma_t = (0+1)\sin^2\delta_l \times \frac{4\pi}{k^2}$$

$$= \sin^2 90 \times \frac{4\pi}{(\sqrt{2}\pi)^2}$$

$$\sigma_t = 1 \times \frac{4\pi}{2\pi} = 2\ fm^2$$

$$\Rightarrow \quad \sigma = 2\ fm^2$$

32. 3p 3d electrons

$l_1 = 1 \quad l_2 = 2; \quad L = 3, 2, 1$

$s_1 = \frac{1}{2} \quad s_2 = \frac{1}{2}; \quad s = 1, 0$

Spectroscopic terms ${}^{2s+1}L_J$

Total terms are

${}^3F_{4,3,2} \quad {}^3D_{3,2,1} \quad {}^3P_{2,1,0}$

${}^1F_3 \quad {}^1D_3 \quad {}^1P_3$

Total terms = 12

34. K.E. = $2n$ qV.

Charge = $2e$

16×10^6 eV = $2 \times n \times 2e \times 50 \times 10^3$ V

$$\Rightarrow \quad n = \frac{16\times10^6}{2\times2\times50\times10^3}$$

$$= \frac{4\times10^6}{50\times10^3}$$

$$= 0.08 \times 10^3$$

$$= 80$$

35. Total scattering cross section

$$\sigma_p = 2\pi\int_0^{15}\sigma(\theta)\sin\theta\, d\theta$$

diff scattering Cross Section

$$\sigma(\theta) = \frac{1}{4}\left(\frac{zz'e^2}{2E}\right)^2 \text{cosec}\frac{4\theta}{2}$$

$$\sigma_p \propto \left(\frac{zz'e^2}{2E}\right)^2$$

and $E \propto z^2$

$z = 1$ for p

$z' = 79$ for gold

$E \propto z^2$

For α atom $z = 2$ and $z' = 79$

$$\sigma_p \propto \left(\frac{1\times79e^2}{2\times1}\right)^2$$

$$\sigma_\alpha \propto \left(\frac{79\times2e^2}{4\times1}\right)^2$$

$$\Rightarrow \quad \frac{\sigma_\alpha}{\sigma_p} = 4$$

36.
$$\langle E\rangle = \frac{\Sigma g_i E_i e^{-E_i/k_BT}}{\Sigma g_i\ e^{-E_i/k_BT}}$$

$$= \frac{0\times e^{-0/k_BT} + 3\times E\times e^{-E/k_BT}}{e^{-0/k_BT} + 3\times e^{-E/k_BT}}$$

$$= \frac{3E e^{-E/k_BT}}{1+3e^{-E/k_BT}}$$

37. $_8O^{17} : (^1S_{1/2})^2\,(^1P_{3/2})^4\,(^1P_{1/2})^2\,(^1d_{5/2})^1$

Hence $I = +\frac{5}{2}$ $l = 2$, even parity

$_8O^{15} : (^1S_{1/2})^2\,(^1P_{3/2})^4\,(^1P_{1/2})$

$I = \frac{1}{2}$ $l = 1$ odd parity $I = -\frac{1}{2}$

38. $\mu^+ \longrightarrow e^+ + r$. In this decay lepton number is not conserved.

39. $$\langle T\rangle = \frac{\int_0^R \frac{-\hbar^2}{2m}\left(\frac{d^2}{dr^2} + \frac{1}{r}\frac{d}{dr}\right)4\pi r^2 dr}{\int_0^R 4\pi r^2 dr}$$

$$= \frac{-\frac{\hbar^2}{2m}4\pi\int_0^R (2+2)dr}{\int_0^R 4\pi r^2 dr}$$

$$= \frac{-\frac{\hbar^2}{2m}4\pi \times 4R}{\frac{4\pi R^3}{3}}$$

$$\Rightarrow \langle T\rangle \propto \frac{1}{R^2} = \frac{1}{\left(R_0 A^{\frac{1}{3}}\right)^2} = A^{\frac{-2}{3}}$$

$$\Rightarrow n = \frac{-2}{3} = -0.667$$

$= -0.67$

41. $^{13}C_6$, N = 7, Z = 6 for N = 7 ;

$$\left(1S_{\frac{1}{2}}\right)^2\left(1P_{\frac{3}{2}}\right)^4\left(P_{\frac{1}{2}}\right)^1$$

$$\Rightarrow j = \frac{1}{2} \text{ and } l = 1$$

Thus spin-parity is $\left(\frac{1}{2}\right)^-$

42. $Q = 152.930414 - (152.927581)$

$= 2.833 \times 10^{-3}$ a.m.u

44. Total spin of $_4Be^7$ is

$$I_{total} = \frac{1}{2}+\frac{1}{2}+\frac{1}{2}+\frac{1}{2}+\frac{1}{2}+\frac{1}{2}+\frac{1}{2} = \frac{7}{2}$$

Hence it shows fermion

45.

quark	symbol	charge
up	u	+2/3
down	d	– 1/3
chorm	c	+ 2/3
strange	s	– 1/3
Top	t	+ 2/3
bottom	b	– 1/3

Hence X^{--} cannot be a possible baryon.

46. Use conservation of energy and momentum in relativistic form.

$$E_\mu = \frac{\left(m_{\pi^+} + m_{\pi^-}\right)c^2 - \left(m_{\mu^+} + m_\mu\right)^2 c^2}{2m\mu}$$

$$\Rightarrow E_\mu = \frac{(280)^2\text{ MeV} - (210)^2\text{ MeV}}{2\times 10^5}$$

= 163.3 MeV

For pair it will be $\frac{163.3}{2}$MeV

= 81.7 MeV

47. For first Forbidden β-decay; $\Delta I = 0$, 1 or 2 and parity does change.

61. Nuclear configuration

$$O^{17} \equiv \left(^1S_{1/2}\right)^2\left(^1P_{3/2}\right)^4\left(^1P_{1/2}\right)^2\left(^1d_{5/2}\right)^1 \equiv E^{17}$$

= 131.76

$$O^{16} \equiv \left(^1S_{1/2}\right)^2\left(^1P_{3/2}\right)^4\left(^1P_{1/2}\right)^2\left(^1d_{5/2}\right)^0 \equiv E^{16}$$

= 127.62

Energy gap between $^1P_{1/2}$ and $'d_{5/2}$

= 131.76 – 127.62

= 4.10 MeV

62. A nucleon near the surface of the nucleus like an atom

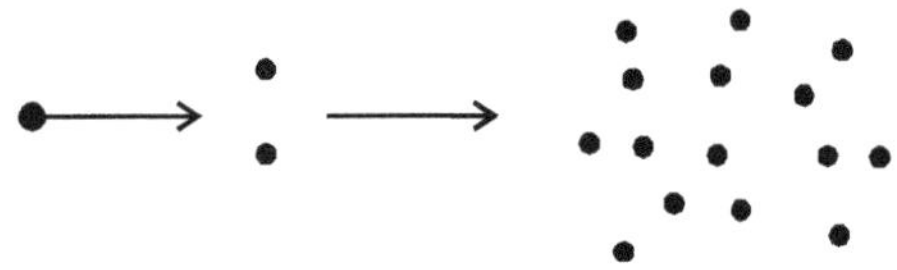

(i) (ii)

(iii)

(i) The nucleon adds the binding to two neighbours

(ii) The nucleon adds the binding to as many neighbours as it can have, the binding now has the saturation value.

(iii) A is much larger here than in (ii) but the added nucleon still has the same number of neighbours and adds only the same saturation binding energy in (i)

64. For $^{17}_{8}O$

The No. of protons = 8 = z

The No. of neutrons = 9 = N

The proton configuration is

$1S_{1/2}$ $1P_{3/2}$ $1P_{1/2}$,

Hence the all subshells are compeletely filled.

The neutron configuration is $1S_{1/2}$, $1P_{3/2}$, $1P_{1/2}$ and last unparied neutron will go in $1d_{5/2}$ subshell which is partially filled. Hence, the ground state spin of $^{17}_{8}O$ should be $l = \frac{5}{2}$.

Since $l = 2$ for the d state for which the parity is even. Therefore, the ground state should be $\frac{5t}{2}$.

65. $z = 7$; $(S_{1/2})^2\ (P_{3/2})^4\ (P_{1/2})^1$ and N = 8

$$l = 1,\ J = \frac{1}{2} \Rightarrow \text{Parity} = (-1)^1 = -1$$

$$\text{Spin-parity} = \left(\frac{1}{2}\right)^-$$

70. $$\frac{E_2}{E_1} = \frac{J_2(J_2+1)}{J_1(J_1+1)}$$

$$\Rightarrow \frac{E_6}{E_4} = \frac{6(6+1)}{4(4+1)}$$

$$\Rightarrow E_6 = 393 \text{ kev}$$

72. Since $\vec{E} \propto \frac{1}{r^2}$ and $\vec{B} \propto \frac{1}{r}$

Hence, $P, \vec{E} \to E, \vec{B} \to -\vec{B}$

Also $\vec{E}$ and $\vec{B}$ are not time invariant

i.e., $T : \vec{E} \to -E, \vec{B} \to \vec{B}$

79. kt meson has

(Q) change = +1

(B) Baryon number = 0

Strangeres number (s) = +1

$$l = \frac{1}{2} \quad J = 0 \text{ party } (-)$$

80. $$N = \frac{4}{3}\pi R^3 \alpha A$$

$$R \propto A^{1/3}$$

$$R = r_0 A^{1/3}$$

For $^{16}_{8}O$ $V = \frac{4}{3}\pi R_0^3$ ... (ii)

(i) and (ii), we have

$$V_0 = \frac{4}{3}\pi(r_0^3 A)$$

$$R_0 = r_0 (16)^{1/3} \quad \text{... (iii)}$$

$$Rx_e = r_0(128)^{1/3}$$

$$\frac{Rx_e}{R_0} = \left(\frac{128}{16}\right)^{1/3} = 2$$

$$\Rightarrow \quad Rx_e = 2R_0$$

$$Vx_e = \frac{4}{3}\pi(2R_0)^3$$

$$8\left(\frac{4}{3}\pi R_0^3\right) = 8V$$

84. $E = mc^2$ for $^{11}_{6}C$

$$E_1 = [6\,m_P + 5\,m_n - 11]\,c^2$$

$$E_2 = [5\,m_P + 6\,m_n - 11]\,c^2 \text{ for } ^{11}_{5}B$$

$$E_1 - E_2 = (m_P - m_n)\,\Delta\text{MeV}$$

Now, $$E_0 = [8\,m_P + 9\,m_n - 17]\,c^2$$

$$E_F = [9\,m_P + 9\,m_n - 17]\,c^2$$

$$E_F - E_0 = [m_P - m_n]\,c^2$$

$$= E_0 + \Delta\text{MeV}$$

$$E_F - E_0 = [938.27 - 939.57]$$

$$E_F - E_0 = -1.30$$

$$E_F = E_0 - 1.30$$

$$\approx 1.39 \text{ MeV}$$

89. We know that the neutron excess (N-Z) increases with the mass number A. For heavist element uranium, two stable isotopes ^{235}U and ^{238}U have (N-Z) values 51 and 54 respectively. On the other hand, the sum of the neutron excesses for the heavist stable isotopes of B*a* and K*r* given below is 40. Thus when a Uranium nucleus undergo fisson both the fission fragments generally contain many neutron is excess of the numbers required for their stability which makes them β^- active. Even, it one of the fragments has the proper neutron excess to make it stable, the other must have so great a neutron. Excess that it will definitely be β^--active.

$$^{141}B_a \xrightarrow[18\text{min}]{} {}^{141}L_a \xrightarrow[3.7h]{} {}^{141}C_e \xrightarrow[33d]{} {}^{141}P_r \text{ (stable)}$$

$$^{92}K_r \xrightarrow[35]{} {}^{92}R_b \xrightarrow[805]{} {}^{92}S_r \xrightarrow[2.7h]{} {}^{92}Y \xrightarrow[3.5h]{} {}^{92}Z_r \text{ (stable)}$$

90. The strangeness quantum number is conserved in strong and electro magnetic interactions.

91. Magnetization = 1.3×10^6 Am^{-1}

Number of atoms per unit volume $= 8 \times 10^{28}\, m^{-3}$

75% atom of F*e* has saturation magnetization. Hence the number of atoms of saturation magnetization.

$$= 75\% \times 8 \times 10^{28}$$

$$= \frac{3}{4} \times 8 \times 10^{28} = 6 \times 10^{28}$$

$$\text{Magnetic moment} = \frac{1.3 \times 10^6}{6 \times 10^{28}}$$

$$= \frac{13}{6} \times 10^{-22} \approx 2.3 \times 10^{-22}$$

94. For $^{17}_{8}\text{O}$

The number of protons = 8

The number of neutrons = 9

Proton configuration

$\Rightarrow (^1S_{1/2})^2\, (^1P_{3/2})^4\, (^1P_{1/2})^2$

Hence, all sub shells for the proton are completely filled.

Neutron configuration

$\Rightarrow (^1S_{1/2})^2\, (^1P_{3/2})^4\, (^1P_{1/2})^2\, (^1d_{s/2})^1$

Hence, last subshell for the neutron is partially filled.

Therefore, the ground state spin of the $^{17}_{8}\text{O}$ nucleus is $\left(\frac{5^+}{2}\right)$

96. The quark structure of Σ^t is 44s.

100. All matter is acted upon by weak nuclear force.

Strong nuclear force act on hadrons (not on Lep tons)

101. (*i*) Th-series $\longrightarrow Pb^{208} (A = 4n + 1)$

(*ii*) Actinium-serires $\rightarrow Pb^{207}$

(*iii*) Neptunium- series $\rightarrow Bi^{209}$ (A = 4*n* +2)

102. $\frac{E_B}{A} \simeq 10$ MeV for average nucleon.

103.

$$V = -\frac{a}{r} + br$$

↓ short range attraction term; ↓ Long range repulsion term

105. $^{0}_{-1}e + {}^{54}_{25}Mn_{29} \rightarrow {}^{54}_{24}Cr_{30} + \gamma_e$

To conserve Lepton number, a neutrino is emitted

Lepton Number L for electron = 1

For neutrino L = 1

106. Second reaction does not conserve Lepton number.

107. Observed value is 4%.

110. Density of states,

$$g(E).dE = \frac{1}{(2\pi)^3}\frac{4\pi}{3}\left(\frac{2m}{\hbar}\right)^{3/2}\frac{3}{2}E^{1/2}dE$$

$$= \frac{1}{4\pi^2}\left(\frac{2m}{\hbar}\right)\sqrt{\frac{2mE}{\hbar^2}}\,dE$$

$$g(E) \sim E^{1/2}$$

111. T^3.

116. We know, $Q = E_y\left[1 + \frac{m_x}{m_y}\right]$

Here, $m_x = 4$ amu

and $m_y = 240 - 4 = 236$ amu

$$\therefore \quad E_y = \frac{Q}{1 + \frac{4}{236}} = \frac{5.26}{1 + \frac{1}{59}}$$

$$= \frac{59}{60} \times 5.26 = 5.17$$

119. $A = A_0 e^{-\lambda t}$

Taking log, $\log A = \log A_0 - \lambda t\,(2.303)$

Hence,graph of log A v_s t is straight line with negative slope.

120. Here $\Gamma = \hbar\lambda$

Since, $\lambda(\text{radioactivity}) = \dfrac{1}{\tau(\text{life time})}$

$$\Rightarrow \quad \Gamma = \frac{\hbar}{\tau} \quad \text{or} \quad \Gamma \alpha \frac{1}{\tau}$$

If $\Gamma_p > \Gamma_\omega > \Gamma_\phi$, then $\tau_p < \tau_\omega < \tau_\phi$.

121. $V = \dfrac{4}{3}\pi R^3$

where, $R = r \,.\, A^{1/3}$

$$\therefore \quad V = \frac{4}{3}\pi r^3 \,.\, A$$

$\Rightarrow$ $V \propto A$, mass number.

122. The nuclear binding energy per particle decrease from 8.8 to 7.6 MeV because as increase ^{56}Fe to ^{235}U, number, the coulomb forces increase at large range.

123. By Breit-Wignes formula

$$\sigma(n,r) \propto \frac{1}{v} \propto v^{-1}.$$

124. Pion has meson which do not have B no. but n, p and Δ^+ are Baryon.

125. Total energy, $mc^2 = 2\, m_0\, c^2$

$$\frac{m_0 c^2}{\sqrt{1-\frac{v^2}{c^2}}} = 2\, m_0\, c^2 \Rightarrow \frac{1}{\sqrt{1-\frac{v^2}{c^2}}} = 2$$

$$\Rightarrow \sqrt{1-\frac{v^2}{c^2}} = \frac{1}{2} \qquad \Rightarrow 1-\frac{v^2}{c^2} = \frac{1}{4}$$

$$\Rightarrow \frac{v^2}{c^2} = 1 - \frac{1}{4} = \frac{3}{4}$$

$$\Rightarrow \quad \boldsymbol{v = \frac{\sqrt{3}}{2}c.}$$

126. $J = 1$ and positive parity

Mean Parity= $(-1)^L = +\text{ve}$

If $L = 0$ and $L = 2$

so $J = (L + S) \ldots |L - S)|$

If $L = 0$, $J = 1$ then $S = 1$ by $J = L + S$

and $L = 2$, $J = 1$ then by $J = L - S$

$S = L - J = (2 - 1) = 1$

$\therefore$ $L = 0, S = 1$ and $L = 2, S = 1$.

127. (*a*) $p \to n + e^+$. α is not conserved

(*b*) $p \to e^+ + v_e$ β is not conserved

(*c*) $p = \pi^+ + \gamma$, β is not conserved

(*d*) $p + n \to \pi^- + \pi^0$, change, β, T_3, S all are conserved, so this reaction is allowed.

128. Nucleus ^{41}Ca, $Z = 20$, $N = 21$

so, Proton $Z = 20$ $\to 1d_{3/2}$ complete

Neutron $N = 21 \to 1\,f_{7/2}$ uncomplete

so last proton in $d_{3/2}$ and last neutron in $f_{7/2}$.

129. Ground state $1\,f_{7/2}$

for f shell l value is 3

Parity $= (-1)^l = (-1)^3 = -1$ odd

spin $= \dfrac{7}{2}$

Ground state angular momentum and parity of ^{41}Ca are $\dfrac{7}{2}$.

131. up quark $\to \dfrac{2}{3}e$

down quark $\to -\dfrac{1}{3}e$

134. $\left(\dfrac{10^{14}\,J}{200\text{ MeV}}\right) \times 3 = 9.4 \times 10^{24}$

139. $_6C^{14}$ has even number of proton and neutron.

$_{12}Mg^{25}$ has even no. of proton but odd number of neutron.

140. Numbeer of proton in $_4Be^9 = 4$ (even)

$\therefore$ Spin due to proton = 0

Number of neutron in $_4Be^9 = 5$

Shell mode term are $|S^2_{1/2}|\ |P^2_{3/2}|$

$\therefore$ spoin of $_4Be^9 = \dfrac{3}{2}$

Parity $= (-1)^l = (-1)^l = -1$

$\therefore$ $_4Be^9$ has odd parity

141. $E = -\dfrac{d\phi}{dt} = -\pi r^2 \dfrac{d\bar{B}}{dt} = \pi r^2 \alpha B_0 e^{-\alpha t}$

142. HCI molecule has permanent dipole moment.

143. Both the nuclear reaction is out cone of strong interaction. Because K mesons are one of the products.

145. Size of nucleus, $r = r_0 A^{1/3}$

By Heisenberg uncertainity principle

$$\Delta x . \delta p \approx \hbar$$

$$\Delta p \propto A^{-1/3}$$

147. For S = 1, possible values of L = J – 1, J, J + 1

$\therefore$ Possible values of L = 0, 1, 2

148. In the scattering process $\pi^+ + p \rightarrow \pi^+ + p$

π^+ measons are scattered more strongly in the backward direction than in the forward direction. In the scattering

process $\pi^- + n \rightarrow \pi^- + n$

π^- measons are scattered more strongly in forward direction than in the backward direction.

In the scattering process $\pi^- + p \rightarrow \pi^\circ + n$

π^- converts itself into π^0 and gets scattered more stronlgy in the backward direction than in the forward direction.

■■

CHAPTER 9 Electronics

2016

1. For the digital circuit given below, the output X is

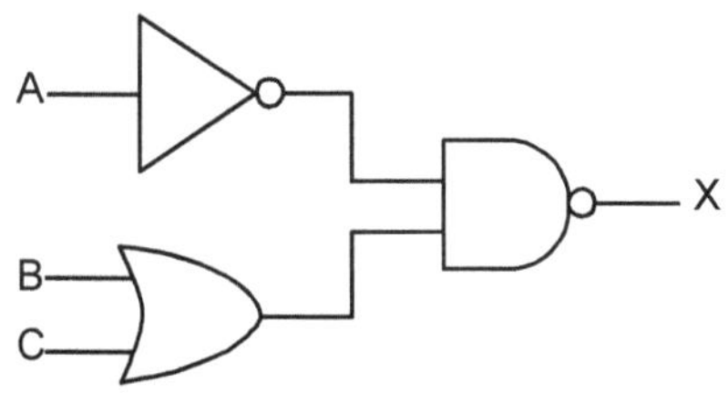

(*a*) $\overline{\overline{A}+B\cdot C}$ (*b*) $\overline{\overline{A}\cdot(B+C)}$

(*c*) $\overline{A}\cdot(B+C)$ (*d*) $A+\overline{(B\cdot C)}$

2. For the transistor shown in the figure, assume V_{BE} = 0.7 V and β_{dc} = 100. If V_{in} = 5V, V_{out} (in Volts) is ________. (Give your answer upto one decimal place)

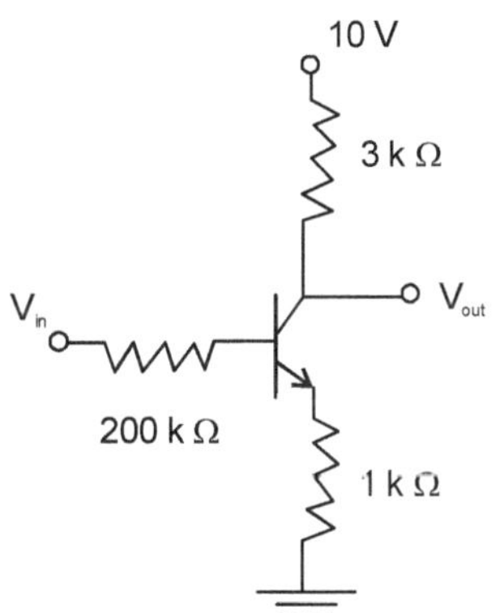

2015

3. In the given circuit, the voltage across the source resistor is 1 V. The drain voltage (in V) is ______

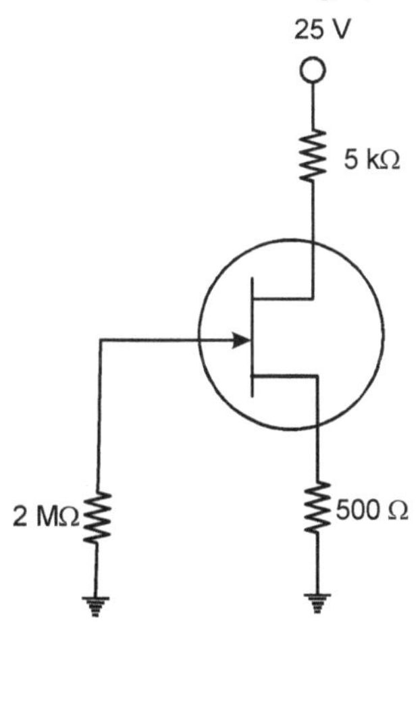

4. In the given circuit, if the open loop gain A = 10^5, the feedback configuration and the closed loop gain A_f are

(*a*) series-shunt, A_f = 9

(*b*) series-series, A_f = 10

(*c*) series-shunt, A_f = 10

(*d*) shunt-shunt, A_f = 10

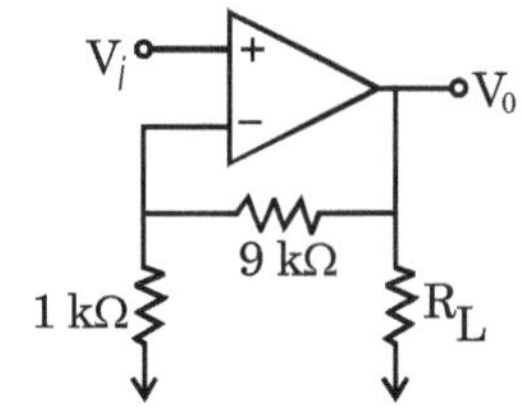

5. In the *simple current source* shown in the figure, Q_1 and Q_2 are identical transistors with current gain β = 100 and V_{BE} = 0.7 V

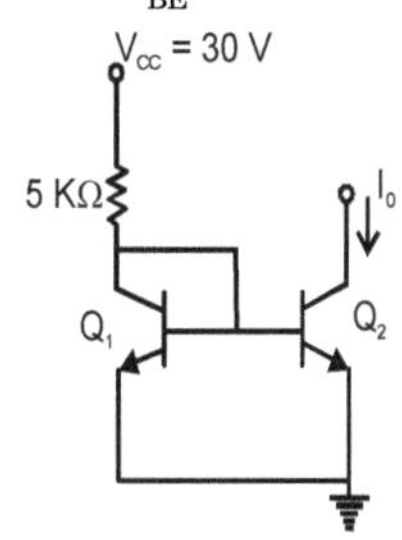

The current I_0 (in mA) is____(upto two decimal places)

6. Consider the circuit shown in the figure, where RC = 1. For an input signal V_i shown below, choose the correct V_o from the options :

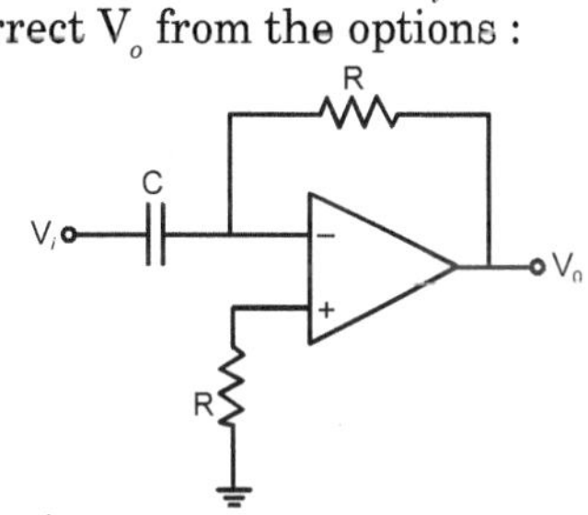

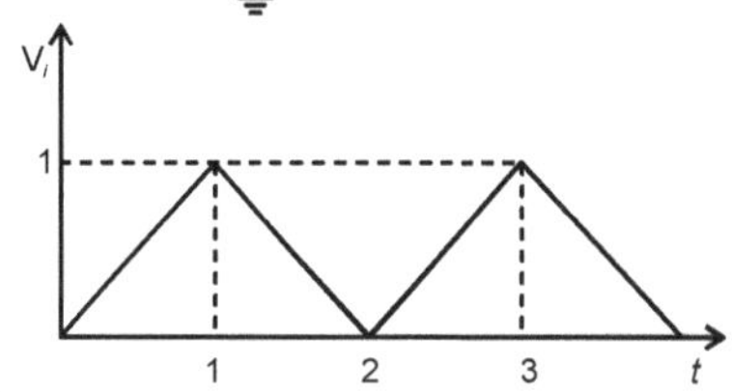

(*a*)

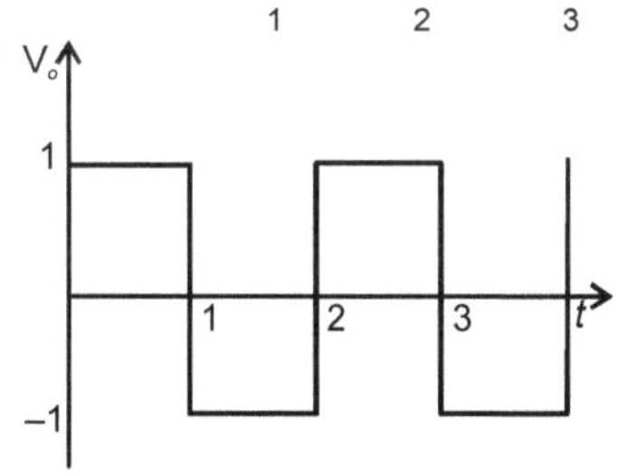

(b) 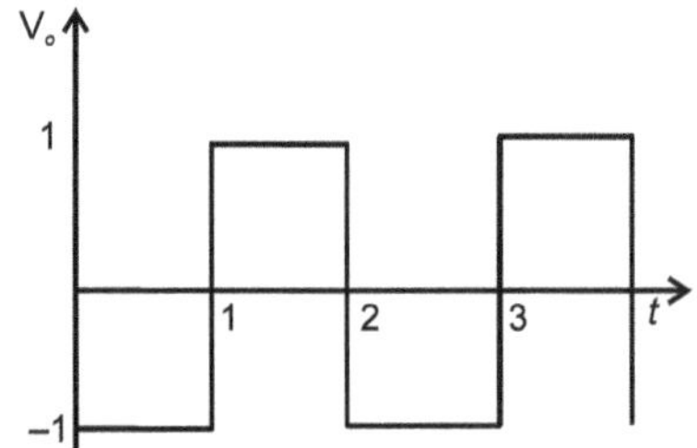

(c) 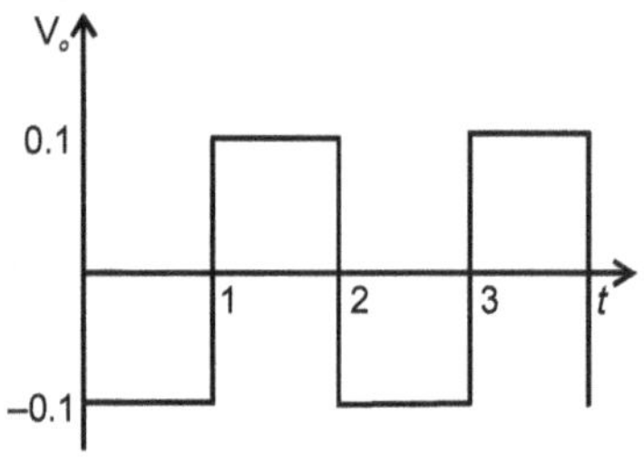

(d) 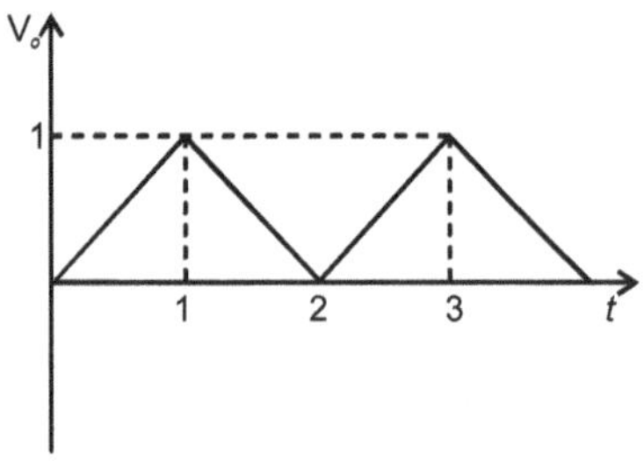

2014

7. The minimum number of flip-flops required to construct a mod-75 counter is _______.

8. The current gain of the transistor in the following circuit is $\beta_{dc} = 100$. The value of collector current I_c is ______ mA.

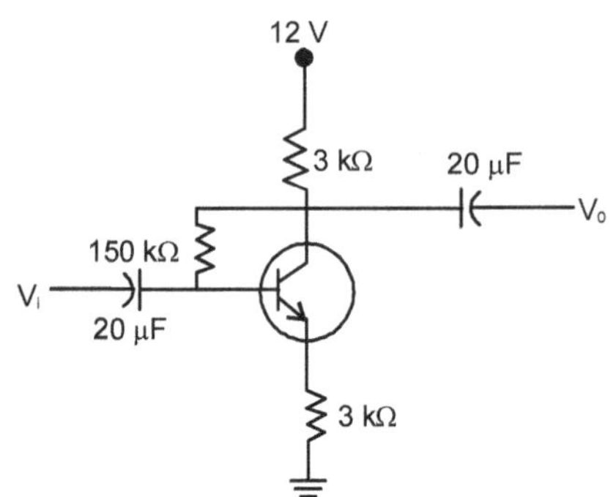

9. A low pass filter is formed by a resistance R and a capacitance C. At the cut-off angular frequency $\omega_c = \frac{1}{RC}$, the voltage gain and the phase of the output voltage relative to the input voltage respectively, are

(a) 0.71 and 45°

(b) 0.71 and – 45°

(c) 0.5 and –90°

(d) 0.5 and 90°

10. The input given to an ideal OP-AMP integrator circuit is

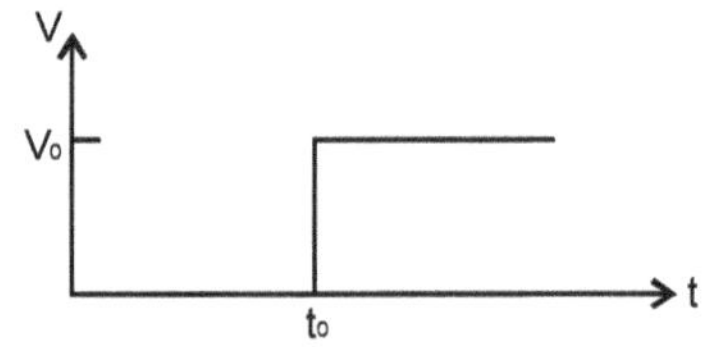

The correct output of the integrator circuit is

(a) 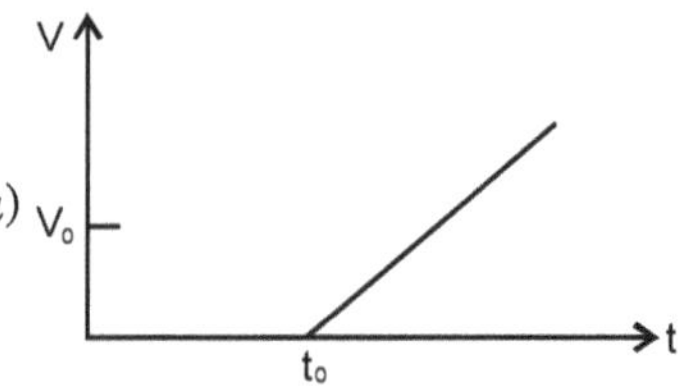

(b) 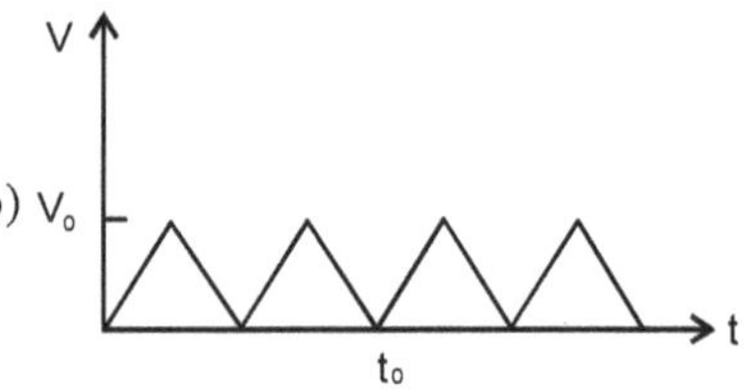

(c) 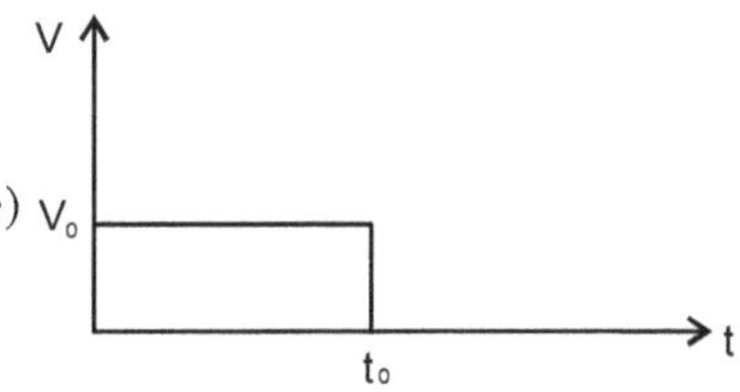

(d) 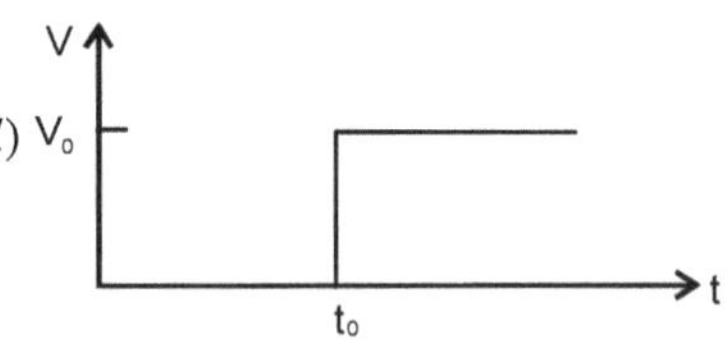

2013

11. What should be the clock frequency of a 6-bit A/D converter so that its maximum conversion time is 32 μs?

(a) 1 MHz

(b) 2 MHz

(c) 0.5 MHz

(d) 4 MHz

12. A voltage regulator has ripple rejection of –50dB. If input ripple is 1 mV, what is the output ripple voltage in μV? The answer should be up to two decimal places. __________

13. An n-channel junction field effect transistor has $5mA$ source to drain current at shorted gate (I_{DSS}) and 5V pinch off voltage (V_P). Calculate the drain current in mA for a gate-source voltage (V_{GS}) of –2.5V. The answer should be up to two decimal places. _______

Statement for Linked Answer Q. 14 and 15:

Consider the following circuit

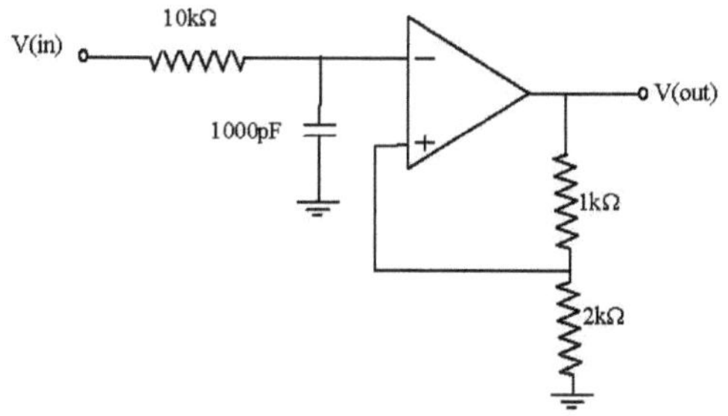

14. For this circuit the frequency above which the gain will decrease by 20 *d*B per decade is

(*a*) 15.9 kHz (*b*) 1.2 kHz

(*c*) 5.6 kHz (*d*) 22.5 kHz

15. At 1.2 kHz the closed loop gain is

(*a*) 1 (*b*) 1.5

(*c*) 3 (*d*) 0.5

2012

16. If the peak output voltage of a full wave rectifier is 10 V, its d.c. voltage is

(*a*) 10.0 V (*b*) 7.07 V

(*c*) 6.36 V (*d*) 3.18 V

17. A Ge semiconductor is doped with acceptor impurity concentration of 10^{15} atoms/cm^3. For the given hole mobility of 1800 cm^2/V-s, the resistivity of this material is

(*a*) 0.288 Ω cm (*b*) 0.694 Ω cm

(*c*) 3.472 Ω cm (*d*) 6.944 Ω cm

18. Consider the following OP-AMP circuit.

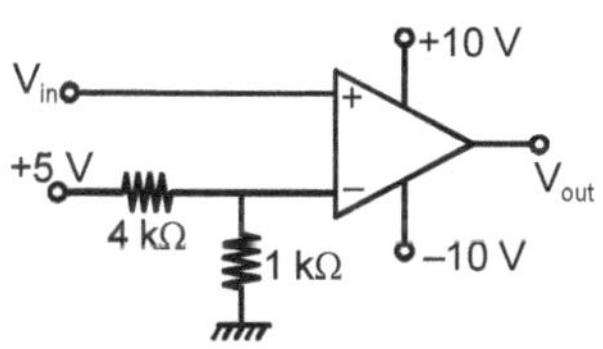

Which one of the following correctly represents the output V_{out} coresponding to the input V_{in}?

(*a*)

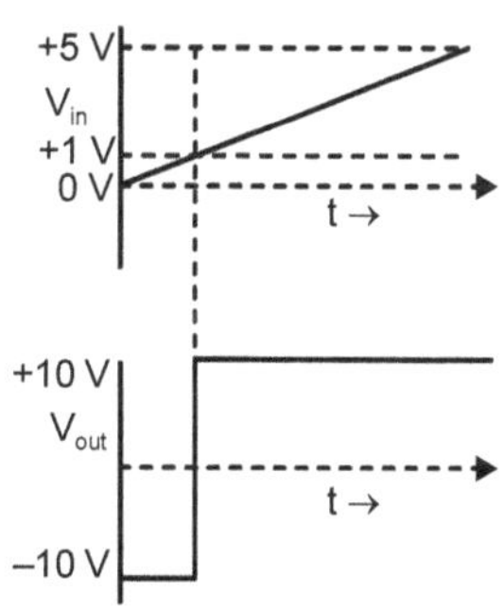

(*b*)

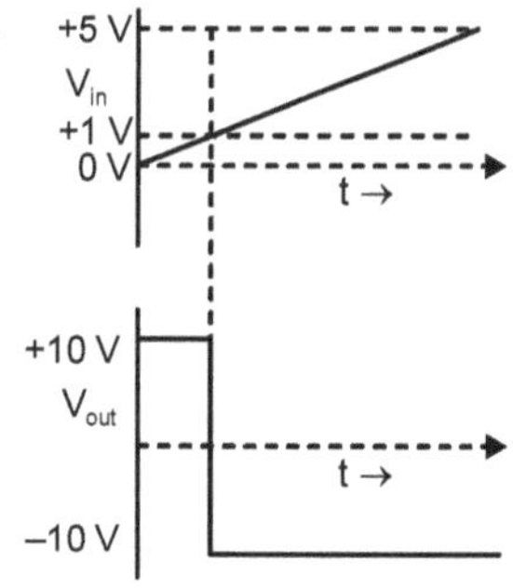

(*c*)

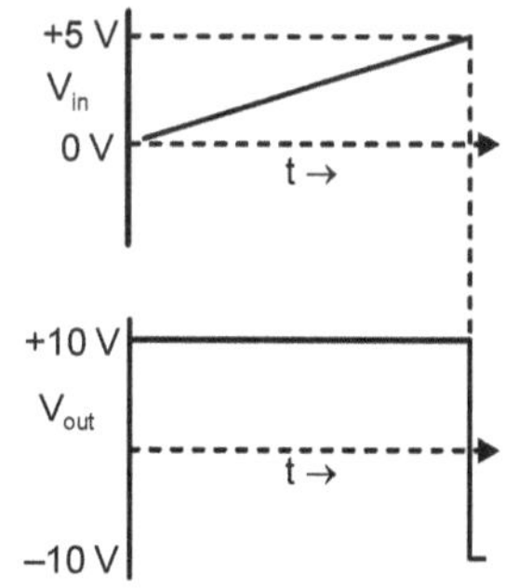

(*d*)

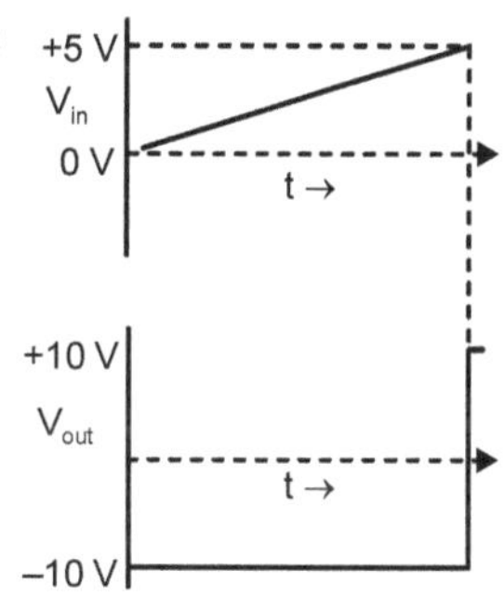

19. In the following circuit, for the output voltage to be $V_0 = (-V_1 + V_2/2)$, the ratio R_1/R_2 is

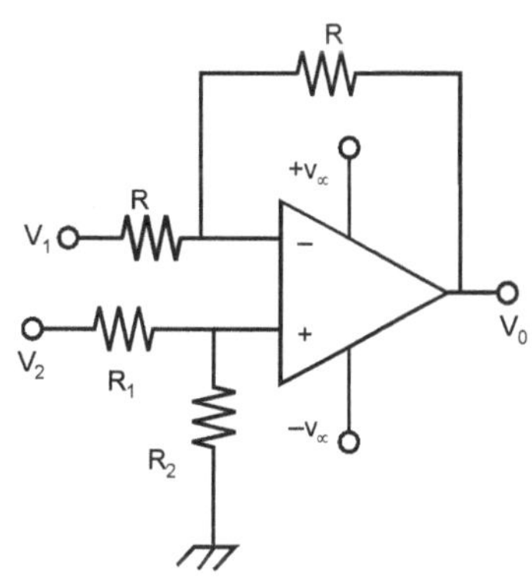

(*a*) 1/2 (*b*) 1

(*c*) 2 (*d*) 3

20. In the following circuit, the voltage drop across the ideal diode in forward bias condition is 0.7 V.

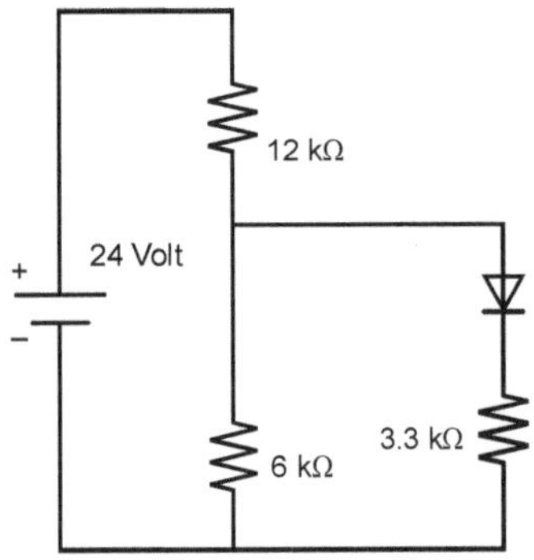

The current passing through the diode is

(*a*) 0.5 mA (*b*) 1.0 mA

(*c*) 1.5 mA (*d*) 2.0 mA

2011

21. Which of the following statements is CORRECT for a common emitter amplifier circuit?

(*a*) The output is taken from the emitter

(*b*) There is 180° phase shift between input and output voltages

(*c*) There is no phase shift between input and output voltages

(*d*) Both *p-n* junctions are forward biased

22. In the following circuit, the voltage across and the current through the 2kΩ resistance are

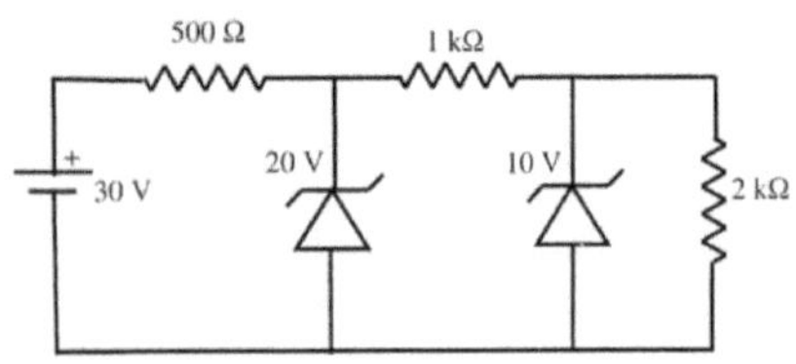

(*a*) 20 V, 10 mA (*b*) 20 V, 5 mA

(*c*) 10 V, 10 mA (*d*) 10 V, 5 mA

23. In the following circuit, Tr1 and Tr2 are identical transistors having V_{BE} = 0.7 V. The current passing through the transistor Tr2 is

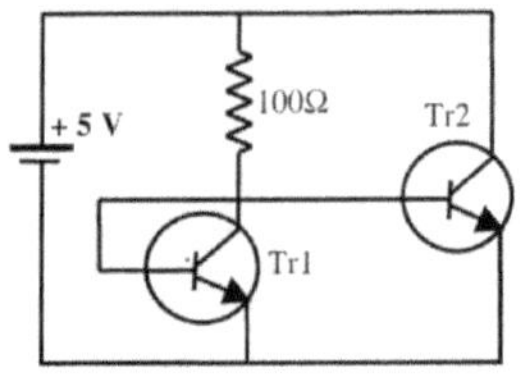

(*a*) 57 mA

(*b*) 50 mA

(*c*) 48 mA

(*d*) 43 mA

24. The following Boolean expression

$$Y = A\bullet\overline{B}\bullet\overline{C}\bullet\overline{D} + \overline{A}\bullet B\bullet\overline{C}\bullet D + \overline{A}\bullet\overline{B}\bullet\overline{C}\bullet D + \overline{A}\bullet\overline{B}\bullet C\bullet D + \overline{A}\bullet B\bullet C\bullet D + A\bullet\overline{B}\bullet\overline{C}\bullet D$$

can be simplified to

(*a*) $\overline{A}\bullet\overline{B}\bullet C + A\bullet\overline{D}$

(*b*) $\overline{A}\bullet B\bullet\overline{C} + A\bullet\overline{D}$

(*c*) $A\bullet\overline{B}\bullet\overline{C} + \overline{A}\bullet D$

(*d*) $A\bullet\overline{B}\bullet C + \overline{A}\bullet D$

25. Consider the following circuit.

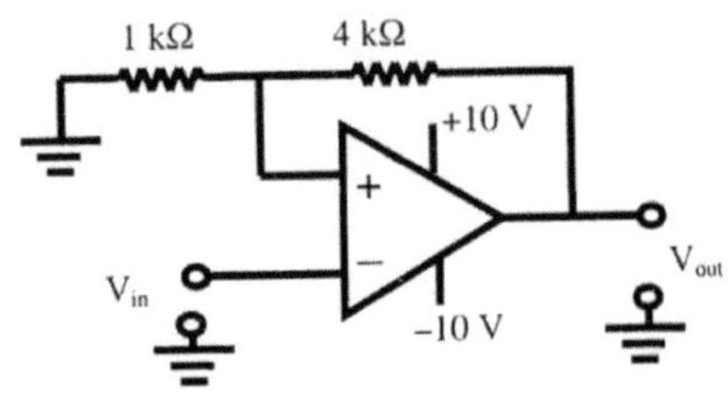

Which of the following correctly represents the output V_{out} corresponding to the input V_{in}?

(*a*)

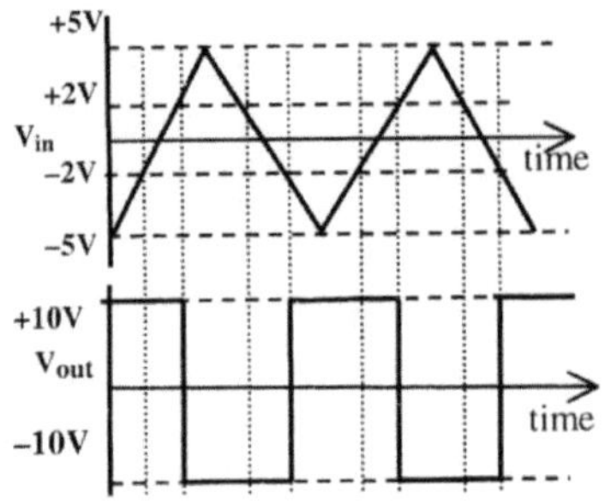

(*b*)

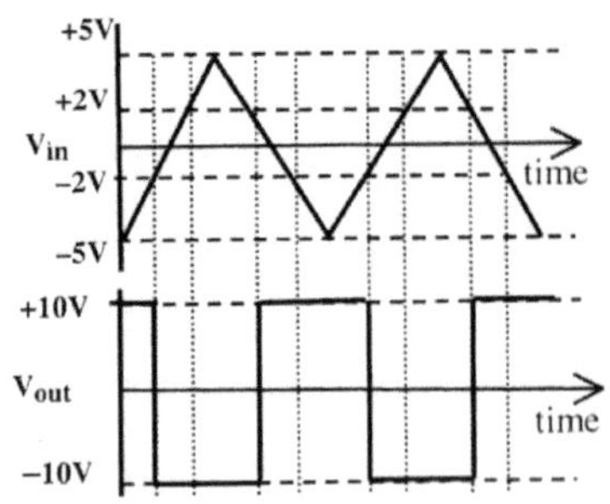

(*c*)

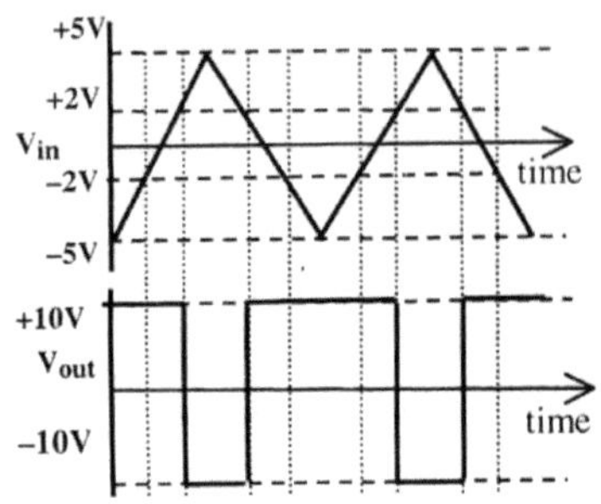

(*d*)

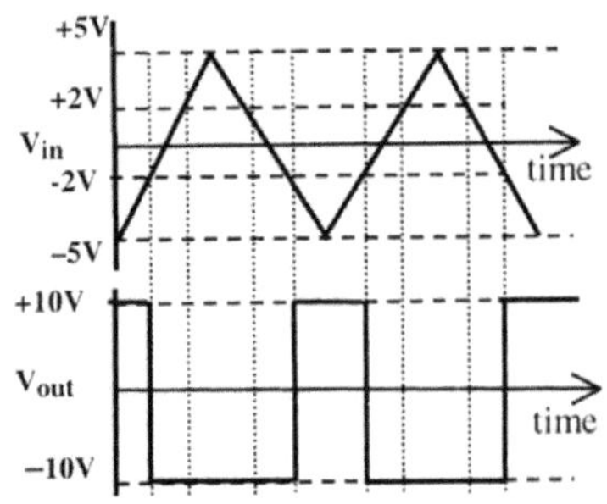

Common Data for Questions 26 and 27:

The tight binding energy dispersion (E-k) relation for electrons in a one-dimensional array of atoms having lattice constant a and total length L is

$$E = E_0 - \beta - 2\gamma \cos(ka),$$

where E_0, β and γ are constants and k is the wave-vector.

26. The density of states of electrons (including spin degeneracy) in the band is given by

(a) $\dfrac{L}{\pi\gamma a \sin(ka)}$ (b) $\dfrac{L}{2\pi\gamma a \sin(ka)}$

(c) $\dfrac{L}{2\pi\gamma a \cos(ka)}$ (d) $\dfrac{L}{\pi\gamma a \cos(ka)}$

27. The effective mass of electrons in the band is given by

(a) $\dfrac{\hbar^2}{\gamma a^2 \cos(ka)}$ (b) $\dfrac{\hbar^2}{2\gamma a^2 \cos(ka)}$

(c) $\dfrac{\hbar^2}{\gamma a^2 \sin(ka)}$ (d) $\dfrac{\hbar^2}{2\gamma a^2 \sin(ka)}$

2010

28. The voltage resolution of a 12 – bit digital to analog converter (DAC), whose output varies from – 10 V to + 10 V is, approximately

(a) 1 mV (b) 5 mV

(c) 20 mV (d) 100 mV

29. In one of the following circuits, negative feedback does not operate for a negative input. Which one is it? The opamps are running from ± 15V supplies.

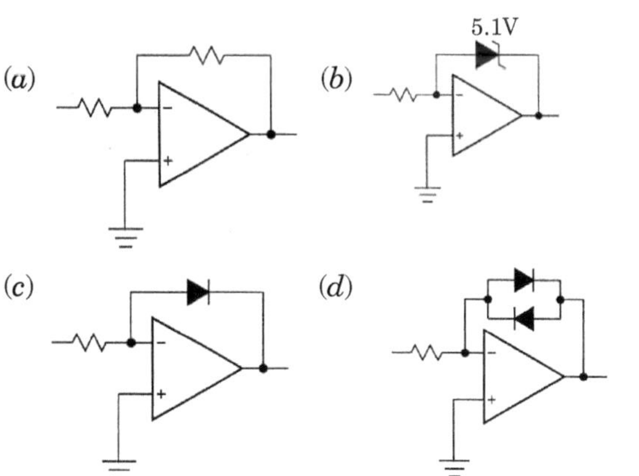

30. Consider a conducting loop of radius a and total loop resistance R placed in a region with a magnetic field B thereby enclosing a flux ϕ_0. The loop is connected to an electronic circuit as shown, the capacitor being initially uncharged.

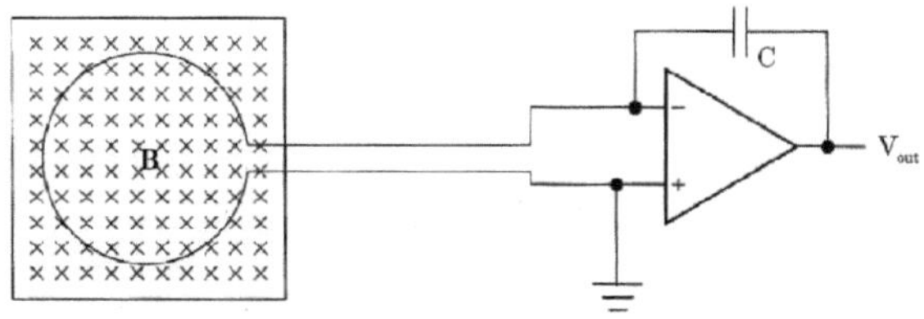

If the loop is pulled out of the region of the magnetic field at a constant speed u, the final output voltage V_{out} is independent of

(a) ϕ_0 (b) u

(c) R (d) C

31. The figure shows a constant current source charging a capacitor that is initially uncharged.

If the switch is closed at t = 0, which of the following plots depicts correctly the output voltage of the circuit as a function of time?

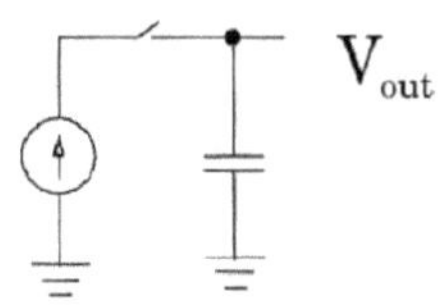

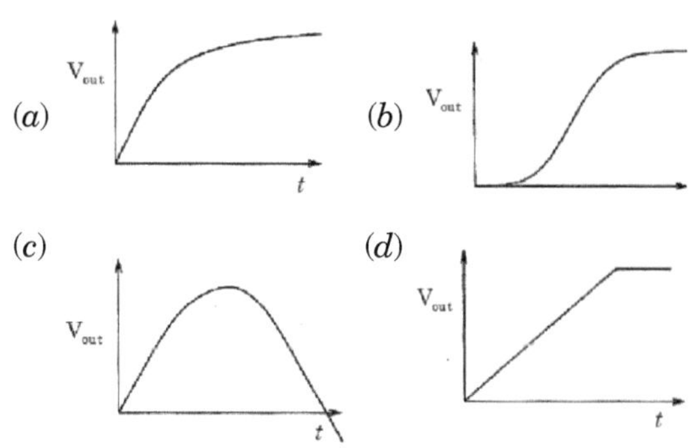

32. For any set of inputs, A and B, the following circuits give the same output, Q, except one. Which one is it?

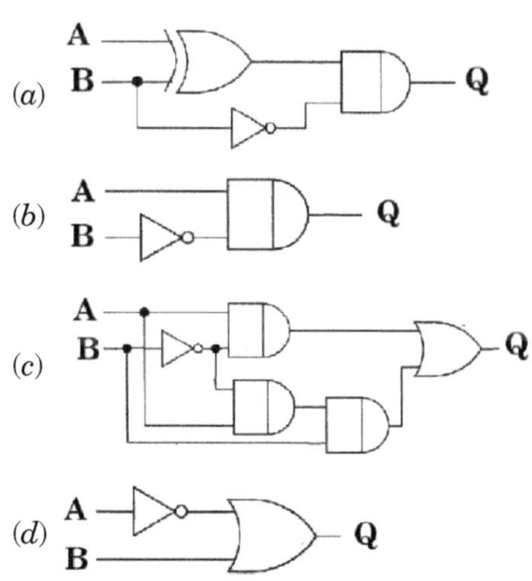

2009

33.

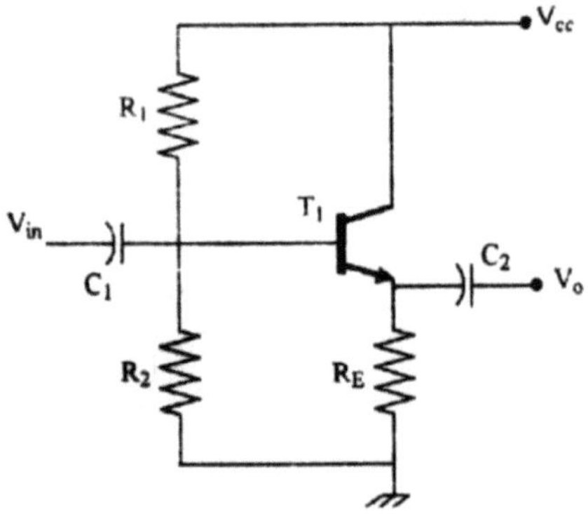

The circuit shown above

(*a*) is a common-emitter amplifier

(*b*) uses a *pnp* transistor

(*c*) is an oscillator

(*d*) has a voltage gain less than one

34.

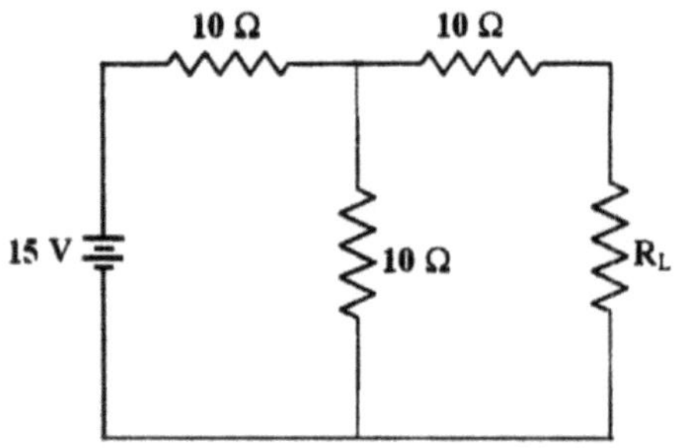

Assuming an ideal voltage source, Thevenin's resistance and Thevenin's voltage respectively for the above circuit are

(*a*) 15 Ω and 7.5 V (*b*) 20 Ω and 5V

(*c*) 10 Ω and 10 V (*d*) 30 Ω and 15 V

35. An amplifier of gain 1000 is made into a feedback amplifier by feeding 9.9% of its output voltage in series with the input opposing. If f_L = 20 Hz and f_H = 200 kHz for the amplifier without feedback, then due to the feedback

(*a*) the gain decreases by 10 times

(*b*) the output resistance increases by 10 times

(*c*) the f_H increases by 100 times

(*d*) the input resistance decreases by 100 times

35.

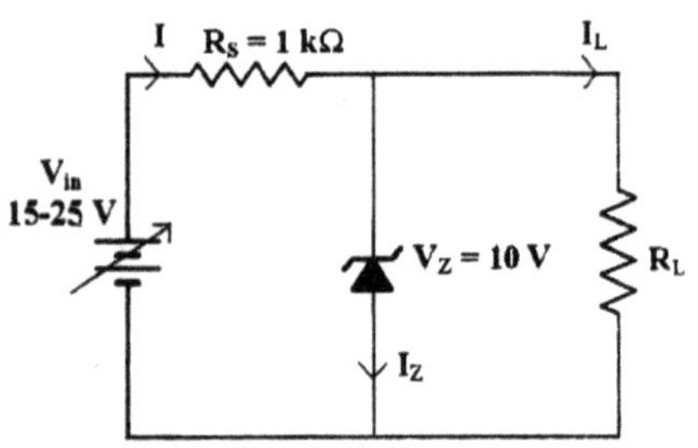

Pick the correct statement based on the above circuit.

(*a*) The maximum Zener current, $I_{Z(max)}$, when R_L = 10 kW is 15 mA

(*b*) The maximum Zener current, $I_{Z(min)}$, when R_L = 10 kΩ is 5 mA

(*c*) With V_{in} = 20V, $I_L = I_Z$, when R_L = 2 kΩ

(*d*) The power dissipated across the Zener when R_L = 10 kΩ and V_{in} = 20 V is 100 mW

Statement for Linked Answer Questions 37 and 38:

The Karnaugh map of a logic circuit is shown below :

37. The minimized logic expression for the above map is

(*a*) $Y=\overline{PR}+\overline{Q}$ (*b*) $Y=\overline{Q}\cdot PR$

(*c*) $Y=\overline{Q}+PR$ (*d*) $Y=Q\cdot\overline{PR}$

38. The corresponding logic implementation using gates is given as :

(*a*)

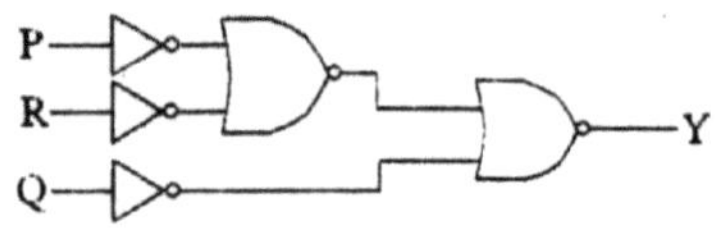

(*b*)

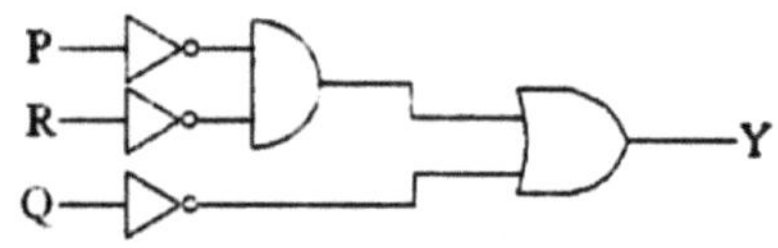

(*c*)

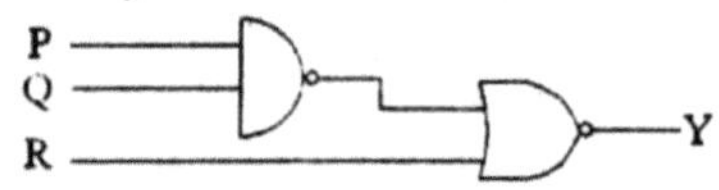

(*d*)

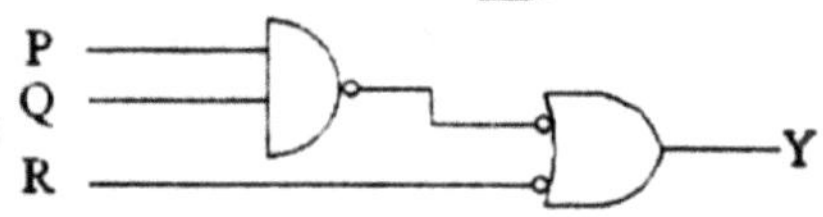

2007

39. A power amplifier gives 150 W output for an input of 1.5W. The gain, in dB, is

(*a*) 10 (*b*) 20

(*c*) 40 (*d*) 100

40. In a typical *npn* transitor the doping concentrations in emitter, base and collector regions are C_E, C_B and C_C respectively. These satisfy the relation,

(*a*) $C_E > C_C > C_B$ (*b*) $C_E > C_B > C_C$

(*c*) $C_C > C_B > C_E$ (*d*) $C_E = C_C > C_B$

41. The circuit shown based on ideal operational amplifiers. It acts as a

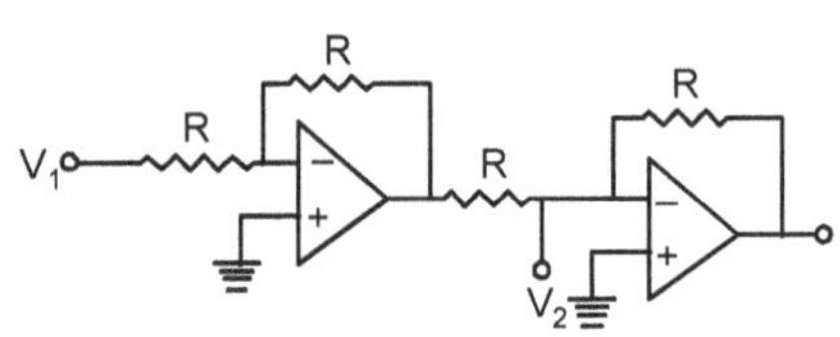

(*a*) subtractor. (*b*) buffer amplifier.

(*c*) adder. (*d*) divider.

42. Identify the function F generated by the logic network shown

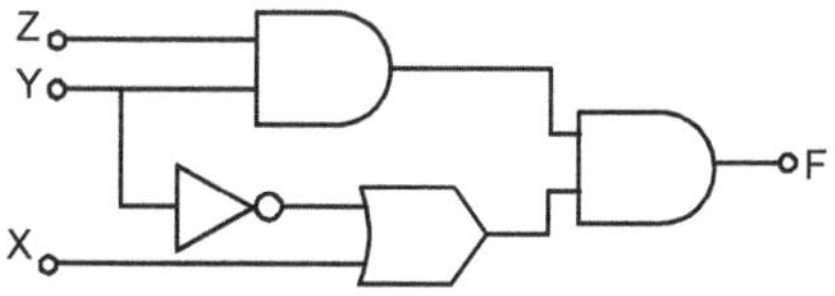

(*a*) $F = (X + Y)Z$ (*b*) $F = Z + Y + \bar{Y}X$

(*c*) $F = ZY(Y + X)$ (*d*) $F = XYX$

43. In the circuit shown, the port Q_1 and Q_2 are in state $Q_1 = 1, Q_2 = 0$. The circuit is now subjected to two complete clock pulses. The state of these ports now becomes.

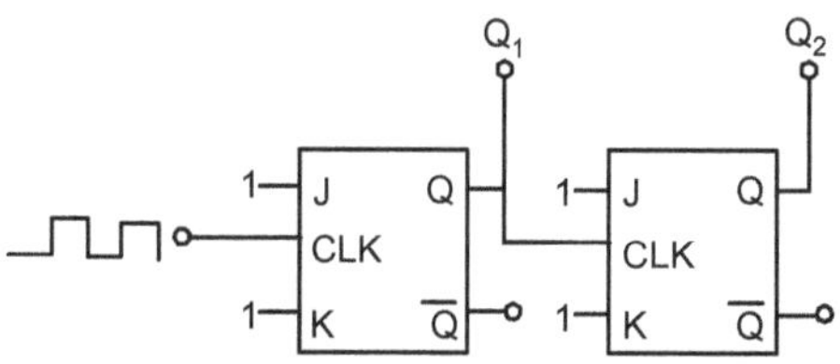

(*a*) $Q_2 = 1, Q_1 = 0$ (*b*) $Q_2 = 0, Q_1 = 1$

(*c*) $Q_2 = 1, Q_1 = 1$ (*d*) $Q_2 = 0, Q_1 = 0$

44. The registers Q_D, Q_C, Q_B and Q_A shown in the figure are initially in the state 1010 respectivelyh. An input sequence SI = 0101 is applied. After two clock pulses, the state of the shift registers (in the same sequence $Q_DQ_CQ_BQ_A$) is

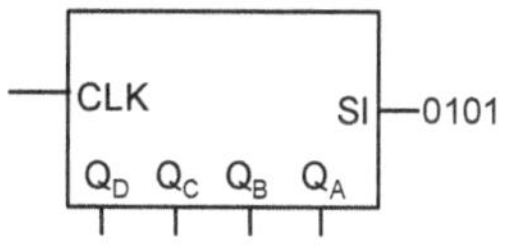

(*a*) 1001 (*b*) 0100

(*c*) 0110 (*d*) 1010

45. For the circuit shown, the potential difference (in Volts) across R_L is

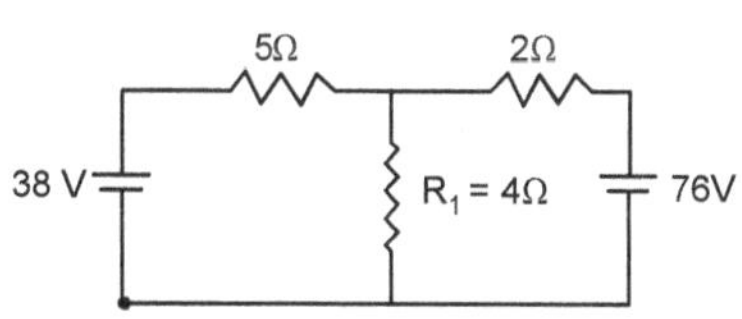

(*a*) 48 (*b*) 52

(*c*) 56 (*d*) 65

46. In the circuit shown, the voltage at test point P is 12 V and the voltage between gate and source is –2V. The value of R (in kΩ) is

(*a*) 42

(*b*) 48

(*c*) 56

(*d*) 70

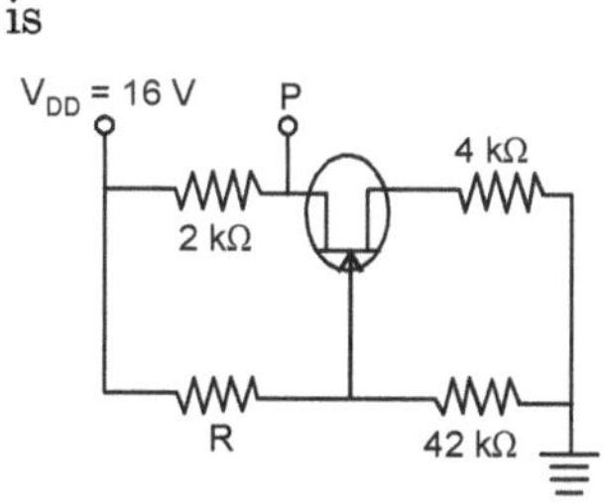

47. When an input voltage $V_{i,}$ of the form shown, is applied to the circuit given below, tghe output voltage V_0 is the form

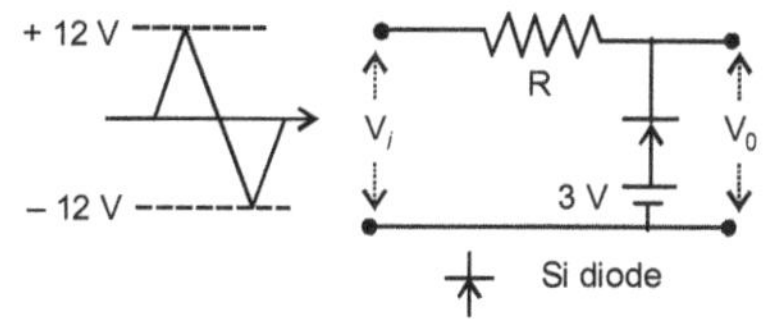

(*a*)
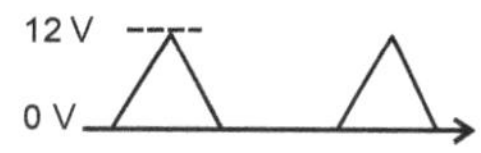

(*b*)

(*c*)

(*d*)
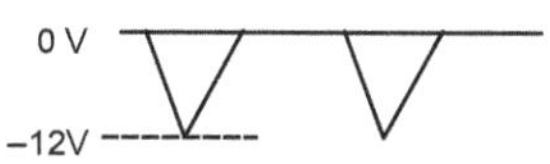

2006

48. The high input impendance of field effect transistor (FET) amplifier is due to

(*a*) the pinch-off voltage

(*b*) its very low gate current

(*c*) the source and drain being far apart

(*d*) the geometry of the FET

49. The circuit shown in the figure function as

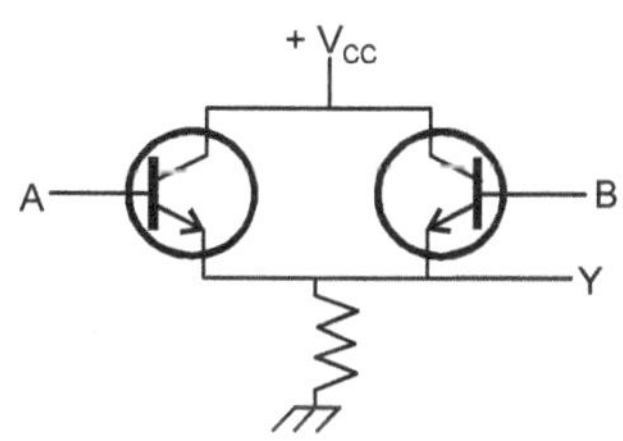

(*a*) an **OR** gate (*b*) an **AND** gate

(*c*) a **NOR** gate (*d*) a **NAND** gate

50. A sinusoidal input voltage v_{in} of frequency ω is fed to the circuit shown in the figure, where $C_1 >> C_2$. If v_m is the peak value of the input voltage, then output voltage (v_{out}) is

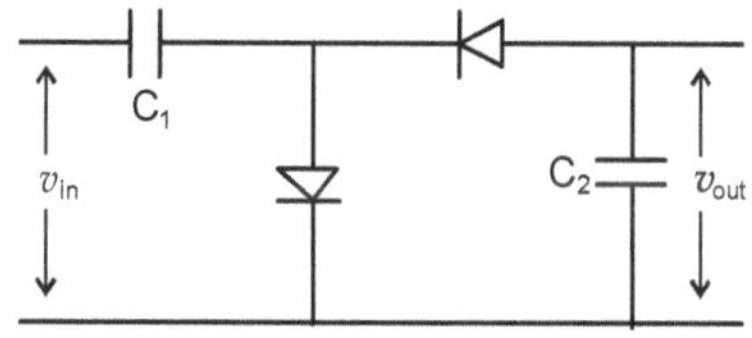

(*a*) $2v_m$ (*b*) $2v_0 \sin \omega t$

(*c*) $\sqrt{2v}_m$ (*d*) $\frac{v_m}{2} \sin \omega t$

51. The low-pass active filter shown in the figure has a cut-off frequency of 2kHz and a pass band gain of 1.5. The values of the resistors are

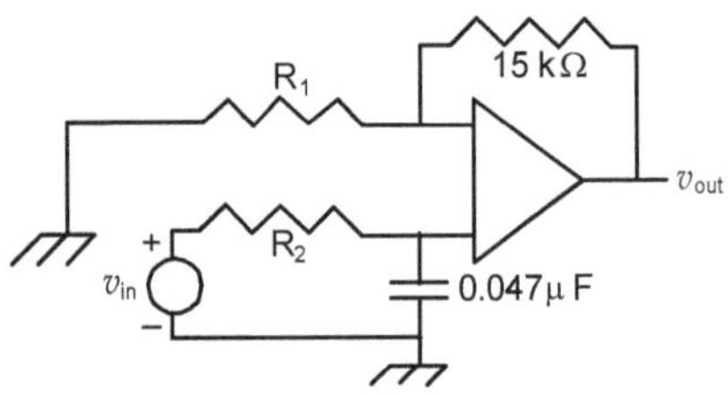

(a) $R_1 = 10k\Omega; R_2 = 1.3\Omega$

(b) $R_1 = 30k\Omega; R_2 = 1.3\Omega$

(c) $R_1 = 10k\Omega; R_2 = 1.7\Omega$

(d) $R_1 = 30k\Omega; R_2 = 1.7\Omega$

52. In order to obtain a solution of the differential equation $\frac{d^2v}{dt^2} - 2\frac{dv}{dt} + v_1 = 0$, involving voltages $v(t)$ and v_1, an operational amplifier (Op-Amp) circuit would require at least

(a) two Op-Amp integrators and one Op-Amp adder

(b) two Op-Amp differentiators and one Op-Amp adder

(c) one Op-Amp integrator and one Op-Amp adder

(d) one Op-Amp integrator, one Op-Amp differentiator and one Op-Amp adder

53. In the given digital logic circuit, A and B form the input. The output Y is

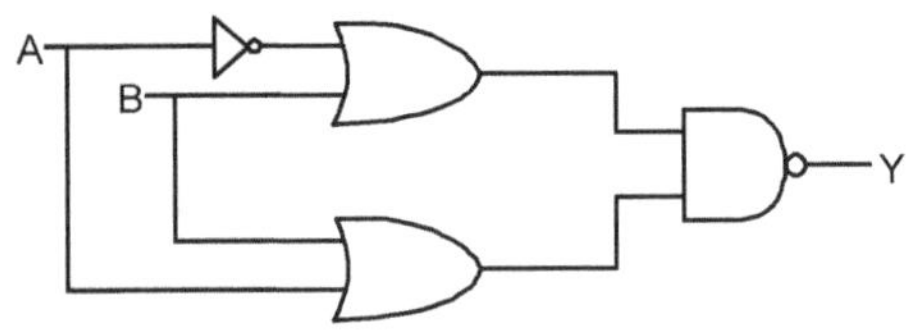

(a) $Y = \bar{A}$ (b) $Y = A\bar{B}$

(c) $Y = A \oplus B$ (d) $Y = \bar{B}$

54. The largest analog output voltage from a 6-bit digital to analog converter (DAC) which produces 1.0 V output for a digital input of 010100, is

(a) 1.6 V (b) 2.9 V

(c) 3.15 V (d) 5.0 V

55. A ripple counter designed with JK flip-flops provided with CLEAR (CL) input is shown in the figure. In order that this circuit functions as a MOD-12 counter, the NAND gate input (X_1 and X_2) should be

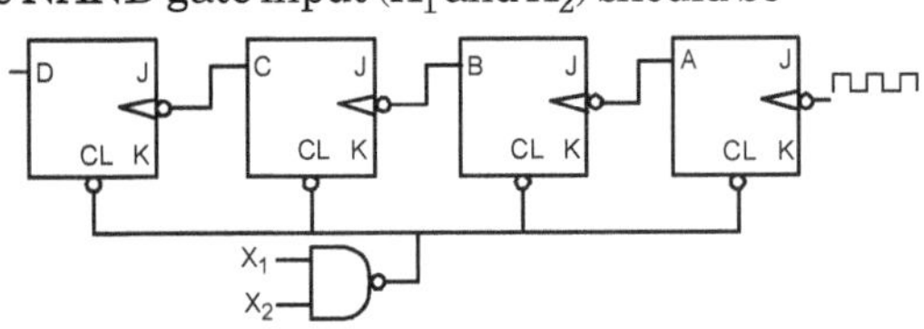

(a) A and C (b) A and D

(c) B and D (d) C and D

56. The tank circuit of a Hartley oscillator is shown in the figure. If M is the mutual inductance between the inductors, the oscillation frequency is

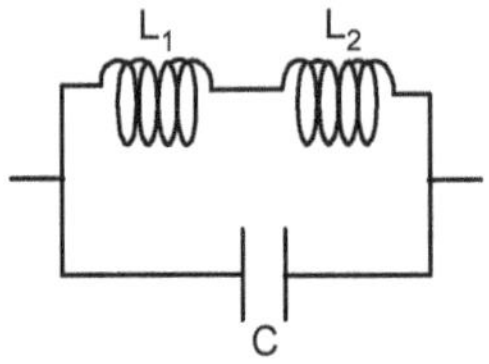

(a) $\frac{1}{2\pi\sqrt{(L_1 + L_2 + 2M)C}}$

(b) $\frac{1}{2\pi\sqrt{(L_1 + L_2 - 2M)C}}$

(c) $\frac{1}{2\pi\sqrt{(L_1 + L_2 + M)C}}$

(d) $\frac{1}{2\pi\sqrt{(L_1 + L_2 - M)C}}$

2005

57. For the rectifier circuit shown in the figure, the sinusoidal voltage (V_1 or V_2) at the output of the transformer has a maximum value of 10 V. The load resistance R_L is 1 kΩ. If I_{av} is the average current through the resistor R_L the circuit corresponds to a

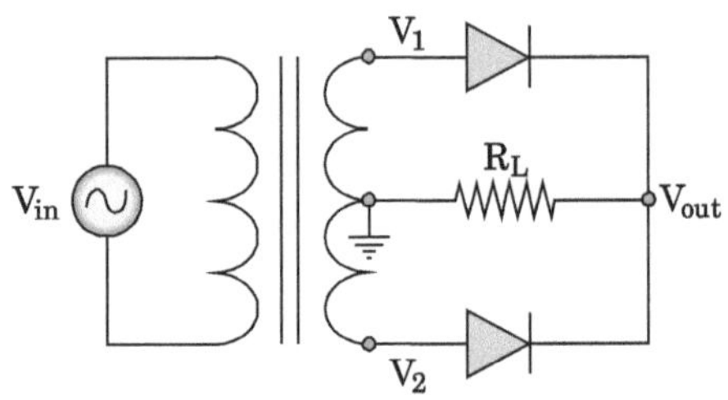

(a) full wave rectifier with $I_{av} = 20/\pi$ mA

(b) half wave rectifier with $I_{av} = 20/\pi$ mA

(c) half wave rectifier with $I_{av} = 10/\pi$ mA

(d) full wave rectifier with $I_{av} = 10/\pi$ mA

58. The Boolean expression :

$B(A + B) + A.(\bar{B} + A)$ can be realized using minimum number of

(a) 1 AND gate (b) 2 AND gates

(c) 1 OR gate (d) 2 OR gates

59. The output V_0 of the ideal opamp circuit shown in the figure is

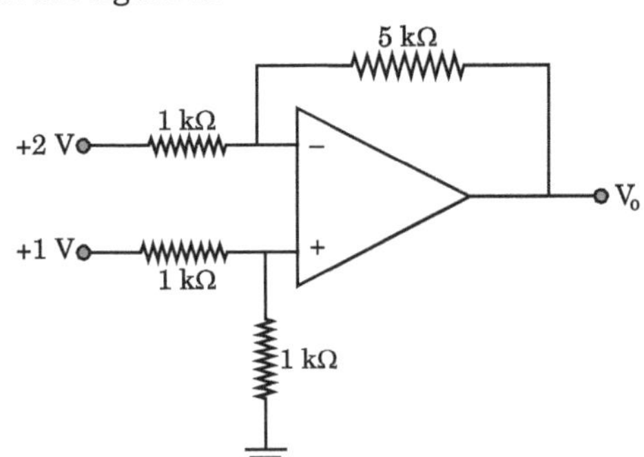

(a) –7 V (b) – 5 V

(c) 5 V (d) 7 V

60. The circuit shown in the figure can be used as a

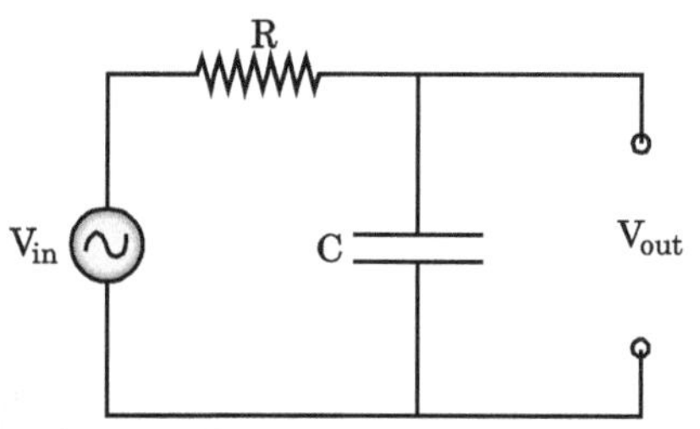

(*a*) high pass filter or a differentiator
(*b*) high pass filter or an integrator
(*c*) low pass filter or a differentiator
(*d*) low pass filter or an integrator

61. In the circuit shown in the figure the Thevenin voltage V_{Th} and Thevenin resistance R_{Th} as seen by the load resistance R_L (= 1 kΩ) are respectively

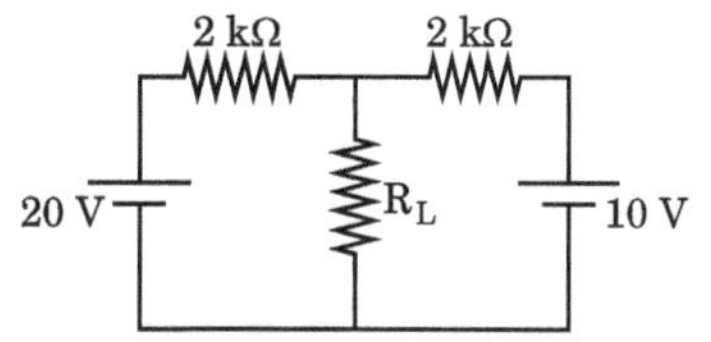

(*a*) 15 V, 1 kΩ (*b*) 30 V, 4 kΩ
(*c*) 20 V, 2 kΩ (*d*) 10 V, 5 kΩ

2004

62. A bipolar junction transistor with one junction forward biased and either the collector or emitter open, operates in the

(*a*) cut-off region (*b*) saturation region
(*c*) pinch-off region (*d*) active region

63. A field effect transistor is a

(*a*) unipolar device
(*b*) special type of bipolar junction transistor
(*c*) unijunction device
(*d*) device with low input impedance

64. The inverting input terminal of an operational amplifier (op-amp) is shorted with the output terminal apart from being grounded. A voltage signal v_i is applied to the non-inverting input terminal of the op-amp. Under this configuration, the op-amp functions as

(*a*) an open loop inverter
(*b*) a voltage to current converter
(*c*) a voltage follower
(*d*) an oscillator

65. A half-adder is a digital circuit with

(*a*) three inputs and one output
(*b*) three inputs and two outputs
(*c*) two inputs and one output
(*d*) two inputs and two outputs

66. What should be the values of the components R and R_2 such that the frequency of the Wien Bride oscillator is 300 Hz ?

[Given : C = 0.01 μF and R_1 12 **kΩ**]

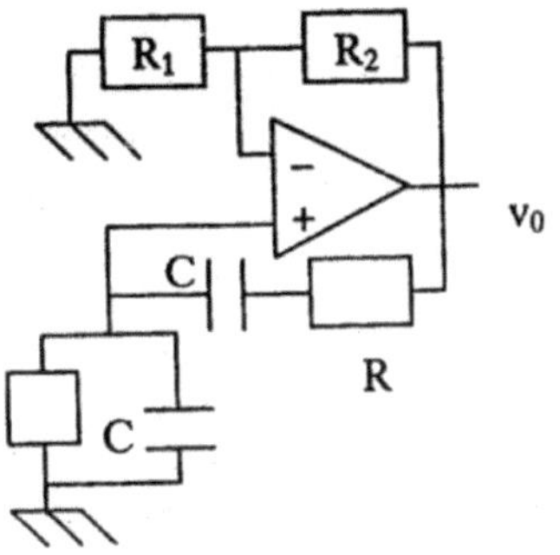

(*a*) R = 48 kΩ and R_2 = 12 kΩ
(*b*) R = 26 kΩ and R_2 = 24 kΩ
(*c*) R = 530 Ω and R_2 = 1 MΩ
(*d*) R = 53 kΩ and R_2 = 24 kΩ

67. Figure shows a common emitter amplifier with β = 100. What is the maximum peak to peak input signal (v_s) for which is distortion-free output may be obtained ? [Assume V_{BE} = 0 and r_e = 20 Ω]

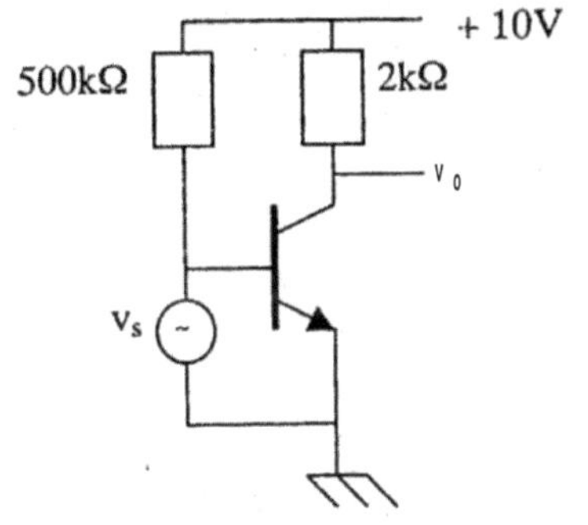

(*a*) 40 mV (*b*) 60 mV
(*c*) 80 mV (*d*) 100 mV

68. Calculate the collector voltage (v_c) of the transistor circuit shown in the figure.

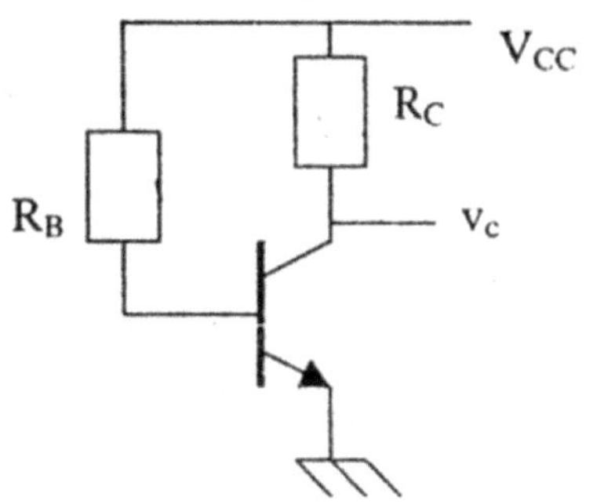

[Given : α = 0.96, I_{CB0} = 20 μA, V_{BE} = 0.3V, R_B = 100 kΩ, V_{CC} = + 10V and R_C = 2.2 kΩ]

(*a*) 3.78 V (*b*) 3.82 V
(*c*) 4.72 V (*d*) 9.7 V

69. Figure shows a practical integrator with $R_S = 30$ MΩ, $R_F = 20$ MΩ and $C_F = 0.1$ μF. If a step (*dc*) voltage of + 3V is applied as input for $0 \le t \le 4$(t is in seconds), the output voltage is

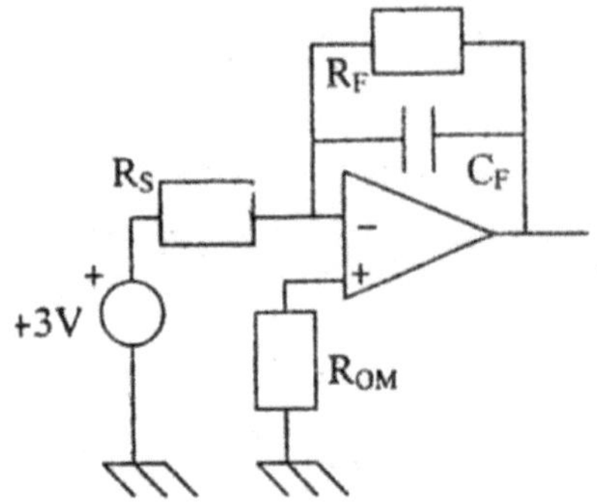

(*a*) a ramp function of – 6 V
(*b*) a step function of – 12 V
(*c*) a ramp function of – 15 V
(*d*) a ramp function of – 4 V

70. The Boolean expression

$$Y = \overline{A}\,\overline{B}\,\overline{C}D + \overline{A}\,\overline{B}CD + \overline{A}B\overline{C}D + \overline{A}BCD$$

reduces to

(*a*) $\overline{A}B$ (*b*) D
(*c*) $\overline{A}$ (*d*) $\overline{A}D$

2003

71. Which one of the following is true for a semiconductor p – n junction with no external bias ?
(*a*) The total charge in the junction is not conserved
(*b*) The *p* side of the juction is positively charged
(*c*) The *p* side of the junction is negatively charged
(*d*) No charge develops anywhere in the junction

72. Which one of the set of values given below does NOT satisfy the Boolean relation R = PQ' (where Q' denotes NOT Q) ?
(*a*) P = 1, Q = 1, R = 0
(*b*) P = 1, Q = 1, R = 1
(*c*) P = 0, Q = 0, R = 0
(*d*) P = 0, Q = 1, R = 1

73. An avalanche effect is observed in a diode when
(*a*) the forward voltage is less than the breakdown voltage
(*b*) the forward voltage exceeds the breakdown voltage
(*c*) the reverse voltage exceeds the breakdown voltage
(*d*) the diode is heavily doped and forward biased

74. Which of the given relations between the Boolean variables P and Q is NOT correct ? (In the notation used here, P' denotes NOT P and Q' denotes NOT Q)
(*a*) PQ' + PQ = P (*b*) (PQ)' = P' + Q'
(*c*) PQ' = (P' + Q)' (*d*) PQ' + Q = P

Data for Questions 75 – 76 :

The circuit below represents a non-inverting integrator

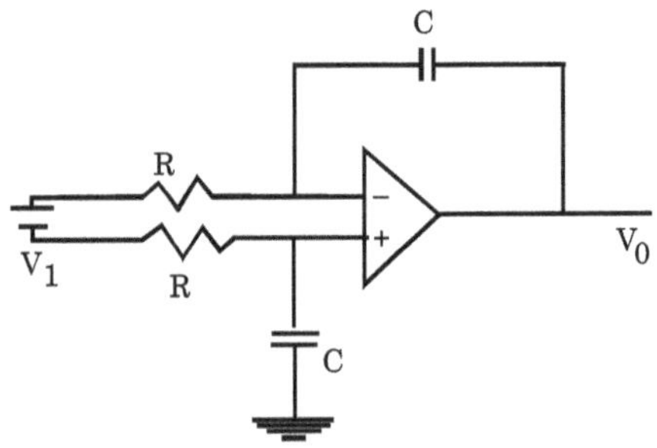

75. For high frequencies ($\omega \to \infty$) the input impedance is
(*a*) 0 (*b*) R
(*c*) R/(1 + ωRC) (*d*) ∞

76. Which of the following is NOT correct
(*a*) $V_0 = (1/RC) \int V_1 dt$
(*b*) The voltages at the inverting and non-inverting terminals of the op-amp are nearly equal
(*c*) The voltage at the non-inverting terminal of the op-amp and the current in the resistor attached to it are π/2 out of phase
(*d*) The current in the two resistors are in phase

2002

77. A large circular coil of N turns and radius R carries a time varying current $i = i_0 \sin(\omega t)$. A small circular coil of n turns and radius r ($r << R$) is placed at the center of the large coil such that the coils are concentric and coplanar. The induced emf in the small coil
(*a*) leads the current in the large coil by π/2
(*b*) lags the current in the large coil by π.
(*c*) is in phase with the current inthe large coil.
(*d*) lags the current in the large coil by π/2.

78. The asymmetry terms in the Weizsacker semi-empirical mass formula is because of
(*a*) non-spherical shape of the nucleus
(*b*) non-zero spin of nucleus
(*c*) unequal number of protons and neutrons inside the nucleus
(*d*) odd number of protons inside the nucleus

79. Which of the following options is true for a two input XOR gate ?

	Input	**Output**	
	A	**B**	
(*a*)	0	1	1
(*b*)	1	0	0
(*c*)	0	0	1
(*d*)	1	1	1

2001

80. RAM and ROM are

(*a*) charge coupled devices used in computers

(*b*) computer memories

(*c*) logic gates

(*d*) binary counters used in computers

81. In an *n-p-n* transistor, the leakage current consists of

(*a*) electrons moving from the base to the emitter

(*b*) electrons moving from the collector to the base

(*c*) electrons moving from the collector to the emitter

(*d*) electrons moving from the base to the collector

2000

82. In the circuit on the right, R_B = 1 kQ and R_C = 100Ω. If the transistor $\beta(h_{FE})$ is 100, the current through R_C will be

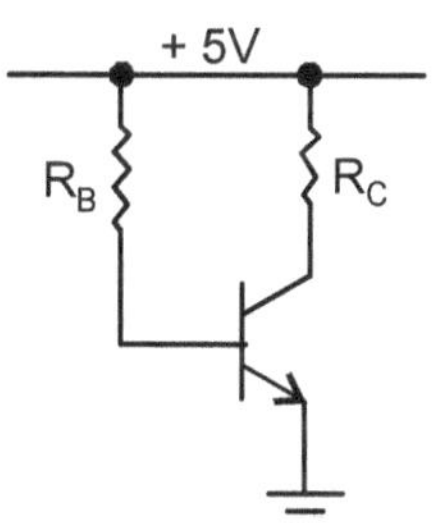

(*a*) ≈ 0.43 A

(*b*) ≈ 50 mA

(*c*) zero

(*d*) oscillating between 0 and 50 mA

83. The output of the circuit on the right will be

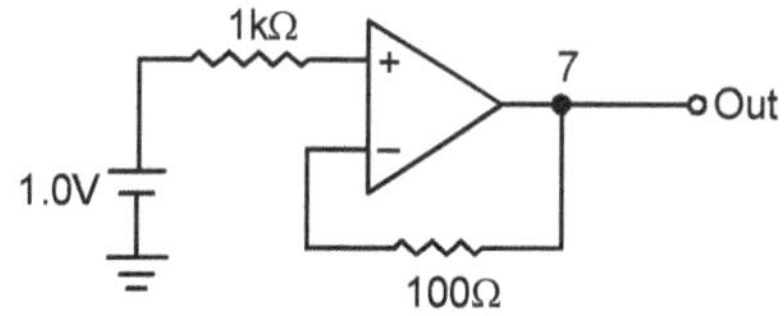

(*a*) 1 V (*b*) 11 V

(*c*) –10 V (*d*) 0V

84. The circuit in the figure on the right shows a constant current source charging a capacitor. The initial voltage across the capacitor is V_0. The switch is closed at time t = 0. The voltage across the capacitor is best described by

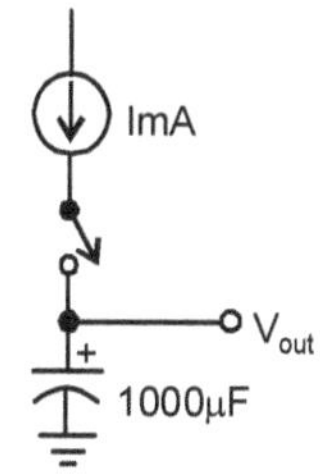

(*a*)

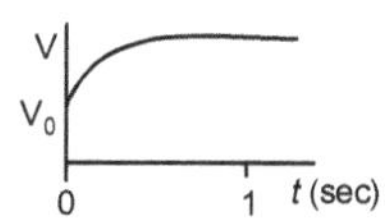

(*b*)

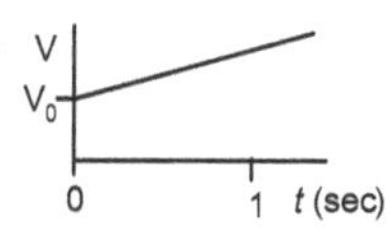

(*c*)

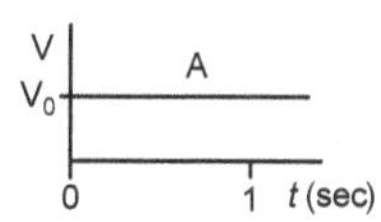

(*d*)

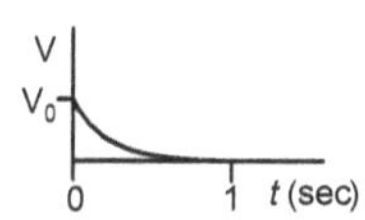

ANSWERS

1. (*b*)	**2.** (5.5 : 5.9)		**3.** (15)	**4.** (*c*)	**5.** (5.74 to 5.75)		**6.** (*b*)	**7.** (6.99 to 7.01)	
8. (1.4 to 1.7)		**9.** (*b*)	**10.** (*a*)	**11.** (*b*)	**12.** (3.16)	**13.** (1.25)	**14.** (*a*)	**15.** (*b*)	**16.** (*c*)
17. (*c*)	**18.** (*a*)	**19.** (*d*)	**20.** (*b*)	**21.** (*b*)	**22.** (*d*)	**23.** (*d*)	**24.** (*c*)	**25.** (*c*)	**26.** (*b*)
27. (*b*)	**28.** (*c*)	**29.** (*c*)	**30.** (*a*)	**31.** (*a*)	**32.** (*c*)	**33.** (*a*)	**34.** (*a*)	**35.** (*c*)	**36.** (*d*)
37. (*a*)	**38.** (*b*)	**39.** (*c*)	**40.** (*a*)	**41.** (*b*)	**42.** (*d*)	**43.** (*c*)	**44.** (*d*)	**45.** (*b*)	**46.** (*c*)
47. (*b*)	**48.** (*b*)	**49.** (*c*)	**50.** (*c*)	**51.** (*c*)	**52.** (*a*)	**53.** (*d*)	**54.** (*c*)	**55.** (*d*)	**56.** (*a*)
57. (*a*)	**58.** (*c*)	**59.** (*a*)	**60.** (*d*)	**61.** (*a*)	**62.** (*c*)	**63.** (*a*)	**64.** (*c*)	**65.** (*d*)	**66.** (*d*)
67. (*c*)	**68.** (*c*)	**69.** (*d*)	**70.** (*d*)	**71.** (*c*)	**72.** (*d*)	**73.** (*c*)	**74.** (*d*)	**75.** (*a*)	**76.** (*c*)
77. (*d*)	**78.** (*d*)	**79.** (*a*)	**80.** (*b*)	**81.** (*d*)	**82.** (*a*)	**83.** (*a*)	**84.** (*a*)		

EXPLANATIONS

1.

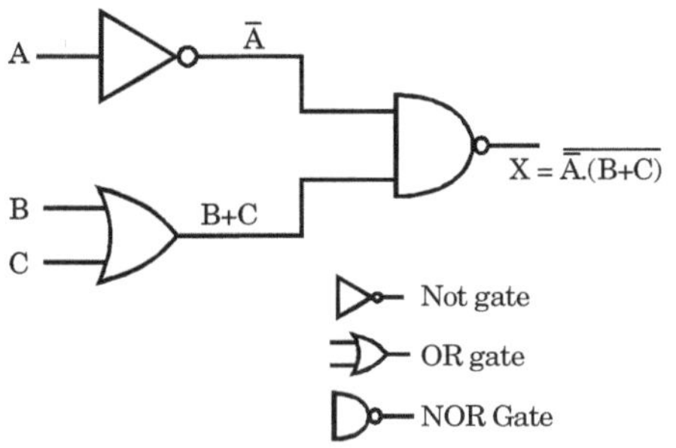

2. $$I_C = \frac{V_{EE} - V_{BE}}{R_E + R_B}$$

$$= \frac{5 - 0.7}{1 + \frac{200}{100}} = \frac{5 - 0.7}{3} = 1.43$$

$$V_C = V_{CC} - I_C R_C$$
$$= 10 - 1.43 \times 3$$
$$= 10 - 4.29$$
$$= 5.71 \text{ Volt}$$

3. $V_S = I_D R_S \Rightarrow I_D = \frac{1}{500} A$

$\Rightarrow V_D = V_{DD} - I_D R_D$

$= 25 - \frac{1}{500} \times 5000$

$\Rightarrow \boxed{V_D = 15 \text{ V}}$

4. $A_F = \left(1 + \frac{R_F}{R_L}\right)$

$= 1 + 9 = 10$

5. $-V_{CC} + I_C R_C + V_{BE} = 0$

$I_C = \frac{30 - 0.7}{5} = \frac{29.3}{5}$

$= 5.86$ mA.

6. $c\frac{dV_i}{dt} = \frac{0 - V_0}{R}$

$\Rightarrow V_0 = -RC\frac{dV_{in}}{dt}$

$= -\frac{dV_{in}}{dt}$

$\Rightarrow V_{in} = -V_0 t$

$V_{in} = +t \Rightarrow V_0 = -1V$ and $V_{in} = -t$

$\Rightarrow V_0 = +1$ V

7. $2^6 = 64 < 75$

$2^7 = 128$

Hence total 7 flip flops are required to construct mod-75 counter.

8. In order to find the collector current, given circuit can be redrawn as

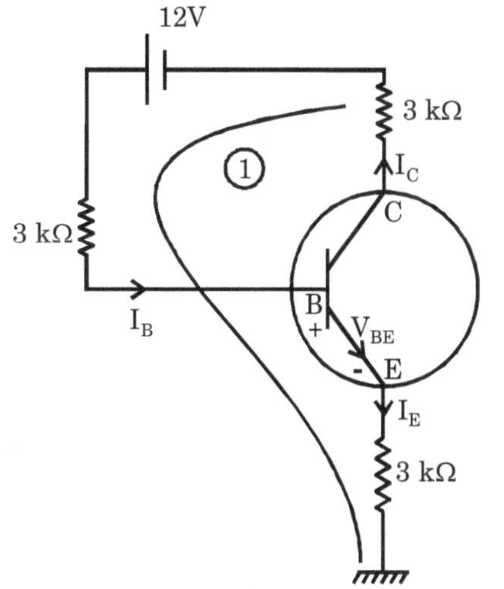

First Loop 1

$I_C(3K) - 12 + I_B(3K) + V_{BE} + 3K\,I_E = 0$

$\therefore\ (3K)I_C - 12 + \frac{I_C}{\beta}(3K) + 0.7 + (3K)\,I_E = 0$

$(\because I_C = \beta\, I_B)$

$\therefore\ (3K)I_C - 12 + \frac{I_C}{100}(3K) + 0.7 + (3K) \quad (B_{dc} + 1)\frac{I_C}{\beta} = 0$

$[\because I_E = (B_{dc} + 1)\,I_B]$

$\therefore\ (3K)(I_C - 12 + 30\,I_C + 0.7 + (3K)(100 + 1)\frac{I_C}{100} = 0$

$\therefore\ I_C\left[3K + 30 + \frac{3K(101)}{100}\right] = 12 - 0.7$

$\therefore\ I_C[3000 + 30 + 30(101)] = 11.3$

$\therefore\ I_C[3030 + 3030] = 11.3$

$\therefore\ I_C = \frac{11.3}{6060} = 1.6$ mA

9. At the cut off frequency voltage gain $A_V = \frac{1}{\sqrt{2}} = 0.707$

phase of the low pass filter at cutt off frequency

$$\phi = -\tan^{-1}(\omega RC)$$
$$= -\tan^{-1}(\tan 45°)$$
$$= -45°$$
$$= 0.71 \text{ and } -45°$$

10.

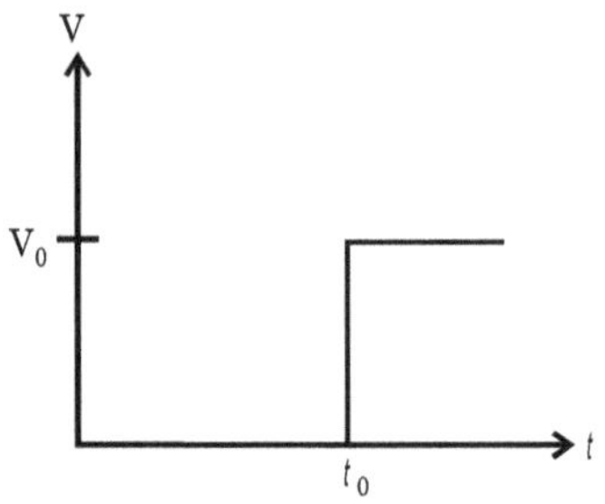

The output of the integrator circuit is

$$V_0 = -\frac{1}{RC}\int_0^t V_i(t)dt$$

$$= -\frac{1}{RC}\left[\int_0^{t_0} 0\,dt \int_{t_0}^{t} V_0\,dt\right]$$

$$= -\frac{V_0}{RC}[t - t_0]$$

Hence its output wave form is

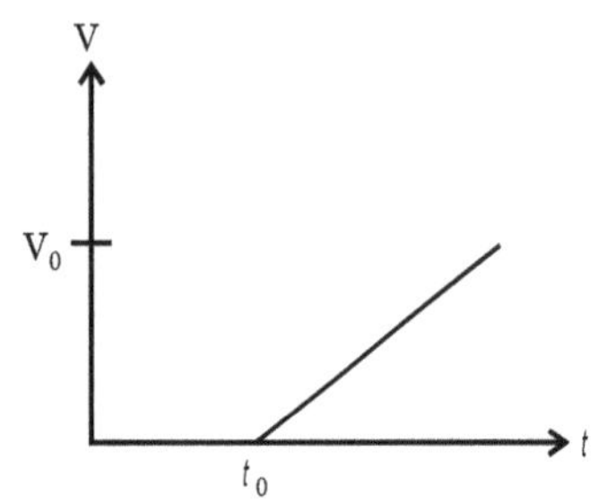

14. $f_H = \frac{1}{2\pi RC} = 16 \text{ kHz}$

15. $\left|\frac{V_0}{V_{in}}\right| = \frac{\left(1+\frac{R_F}{R_1}\right)}{\sqrt{1+\left(\frac{f}{f_H}\right)^2}} = 1.5$

16. $V_{dC} = \frac{2Vm}{\pi} = \frac{2\times 10}{\frac{22}{7}} = \frac{14\times 10}{22} = \frac{70}{11} = 6.36 \text{ V}$

17. Resistivity of material

$$f = \frac{1}{n_e\mu} = \frac{1}{10^{15}\times 1.6\times 10^{-19}\times 1800}$$

$= 3.472\ \Omega$ cm

18. Output voltages will vary between –10V to +10V and at +10V it will be constant because op-amp have very high input impedance also there is no change in polarity.

19. $V_A = \frac{V_2R_2}{R_1 + R_2} = \frac{V_2}{1 + R_1/R_2}$

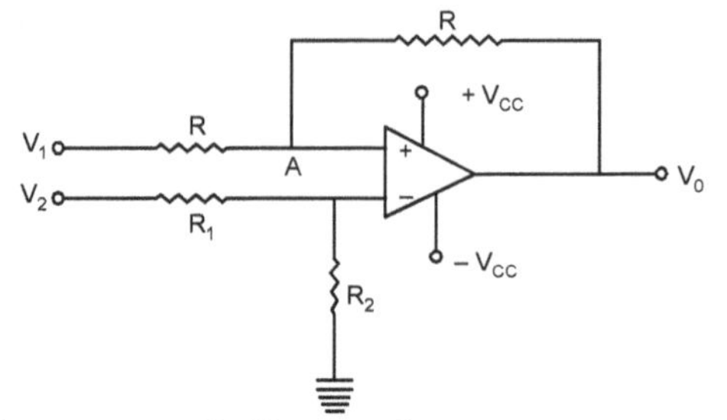

Let, $R_1/R_2 = x$, then

$$V_A = V_2/\ 1 + x$$

Now,

$$\Rightarrow \quad \frac{V_A - V_1}{R} = \frac{V_0 - V_A}{R}$$

$$\Rightarrow \quad V_0 = 2V_A - V_1 = \frac{2V_2}{1+x} - V_1$$

From question

$$V_0 = -V_1 + \frac{V_2}{2}$$

$$\frac{2V_2}{1+x} - V_1 = -V_1 + \frac{V_2}{2}$$

$$1 + x + 4$$

or $\quad x = 3$

$\therefore \quad \frac{R_1}{R_2} = 3$

20. Equivalent circuit

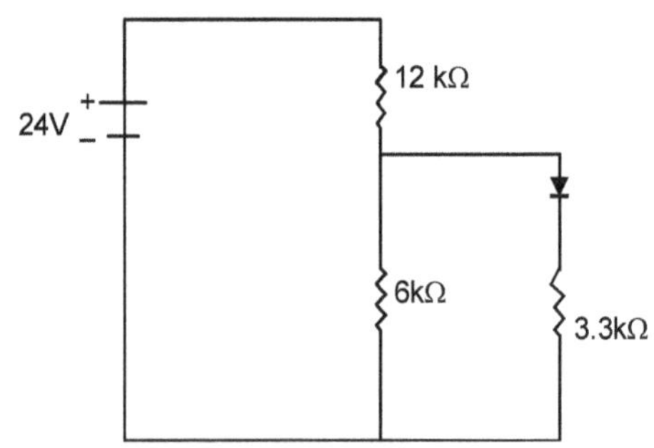

Voltage across 6kΩ

$$\frac{24\times 6}{(12+6)} = \frac{24\times 6}{18} = 8 \text{ V}$$

Equivalent resistance $= \frac{12\times 6}{12+6} = 4$

Total resistance = 4 + 3.3 = 7.3 kΩ

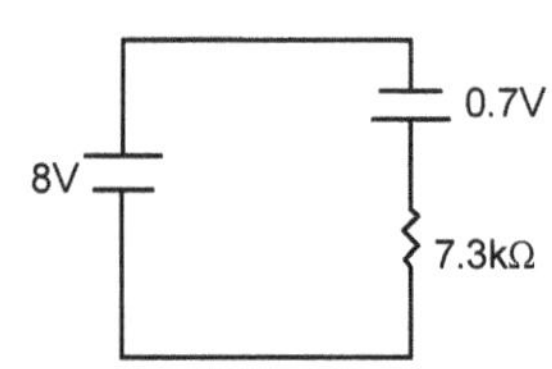

Current flowing through diode

$$= \frac{(8-0.7)}{7.3\text{k}\Omega}\text{V}$$

$$= \frac{7.3}{7.3} = 1 \text{ mA}$$

22. We know that zener diode acts as a DC voltage stablizer, hence the voltage developed across 2 kΩ resistence is 10 V. So the correct flowing through in 2kΩ is

$$\frac{10\text{V}}{2\text{k}\Omega} = 5\text{mA}$$

23. Applying kischhoff's voltage law (KVL), we get

$$5 = 100 \times l + 0.7$$

$$\Rightarrow \quad l = \frac{4.3}{100} = 43\ mA$$

24. $y = \text{A}.\overline{\text{B}}.\overline{\text{C}}.\overline{\text{D}} + \overline{\text{A}}.\text{B}.\overline{\text{C}}.\text{D} + \overline{\text{A}}\cdot\overline{\text{B}}\cdot\overline{\text{C}}\cdot\text{D}$
$+\overline{\text{A}}\cdot\overline{\text{B}}\cdot\text{C} + 0 + \overline{\text{A}}\cdot\text{B}\cdot\text{C}\cdot\text{D} + \text{A}\cdot\overline{\text{B}}\cdot\overline{\text{C}}\cdot\text{D}$

$$\Rightarrow = \text{A}\cdot\overline{\text{BC}} + \overline{\text{AC}}\text{D} + \overline{\text{A}}\cdot\text{C}\cdot\text{D} \quad (\because \text{A} + \overline{\text{A}} = 1)$$

$$= \text{A}\cdot\overline{\text{B}}\cdot\overline{\text{C}} + \overline{\text{A}}\text{D}\,(\overline{\text{C}} + \text{C})$$

$$= \text{A}\overline{\text{BC}} + \overline{\text{A}}\text{D}$$

25. The given circuit is an inverting amplifier

$$\text{V}_0 = -\left[1 + \frac{\text{R}f}{\text{R}_1}\right]\text{V}_{in}$$

$$\Rightarrow \quad \text{V}_0 = -\left[1 + \frac{4}{1}\right]\text{V}_{in}$$

$$\Rightarrow \quad \text{V}_0 = -5\text{V}_{in}$$

26.

$$\text{E} = \text{E}_0 - \text{B} - 2\gamma\cos ka$$

$$\frac{d\text{E}}{dk} = 2\gamma a \sin ka$$

$$\text{Energy density} = \frac{\text{L}}{\pi\left(\frac{d\text{E}}{dk}\right)}$$

$$= \frac{\text{L}}{(2\pi\gamma a \sin ka)}$$

27. Effective mass $m^* = \dfrac{\hbar^2}{d^2\dfrac{\text{E}}{dk^2}}$

$$\text{Now,} \quad \frac{d\text{E}}{dk} = 2\gamma a \sin ka$$

$$\frac{d^2\text{E}}{dk^2} = 2\gamma a^2 \cos ka$$

$$\Rightarrow \quad m^* = \frac{\hbar^2}{2\gamma a^2 coska}$$

28. Voltage resolution

$$= \frac{10}{z^{\text{N}-1}} = \frac{10}{2^{12-1}} = \frac{10}{2^{11}} \cong 20\,\text{mV}$$

29. Since diode does not conduct in reverse bias hence the circuit given in option (*a*) does not operate for a negative feedback.

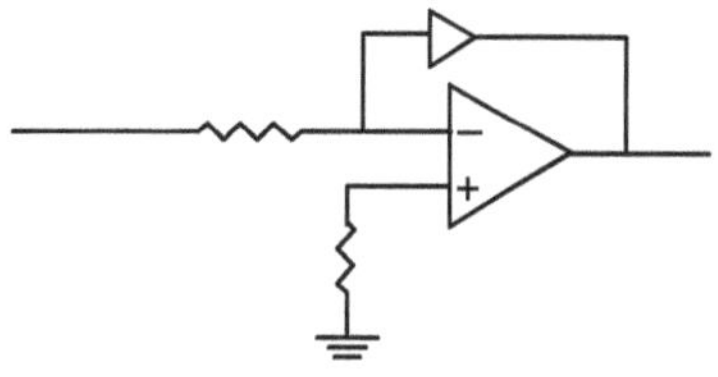

30.

$$\text{V}_{\text{emf}} = -\frac{d\phi}{dt}$$

At $t = 0, = \phi_0, \phi_0 =$ constant

Hence output voltage is independent of ϕ_0

Complete solution is

$$z = \text{CF} + \text{PI}$$

$$= \text{A}\cos ht + \text{B}\sin ht \neq \frac{e(et - e^{-t})}{2}$$

$$= \text{A}\cos ht + \text{B}\sin ht + t\sin ht$$

$$\frac{dy}{dt} = \text{A}\sin ht + \text{B}\cos ht \qquad \ldots (i)$$

$$= +\sin ht + \cos ht \qquad \ldots (ii)$$

$$y(0) = 0, \left.\frac{dy}{dt}\right|_{t=5} = 0$$

From equation (*i*) and (*ii*), we have

$$0 = \text{A} + \text{B}.0 + 0$$

$$\Rightarrow \quad \text{A} = 0$$

$$0 = \text{A}.0 + \text{B}.1 \times \text{D} + 0 = 0$$

$$y = t\sin h.t$$

31.

$$l = \frac{d\text{Q}}{dt} \quad (\text{Here } l = \text{constant})$$

$$\text{Q} = lt + \text{C}$$

At $t = 0, \text{Q} = 0$

$\Rightarrow \quad \text{C} = 0$

$\Rightarrow \quad \text{Q} = lt$

Potential across the capacitor

$$v = \frac{\text{Q}^2}{2\text{C}} = \frac{l^2t^2}{2\text{C}}$$

($\because$ l and C are constants)

$v = kdt^2$, (where k = 2C)

Hence graph between v and t is a parabolic, when capacitor becomes complete charge, V becomes constant.

34.

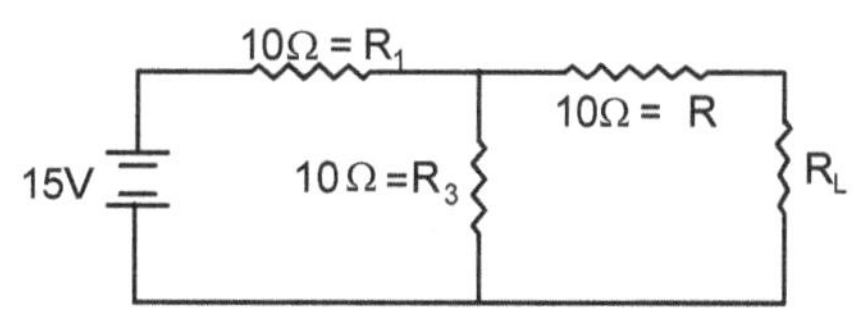

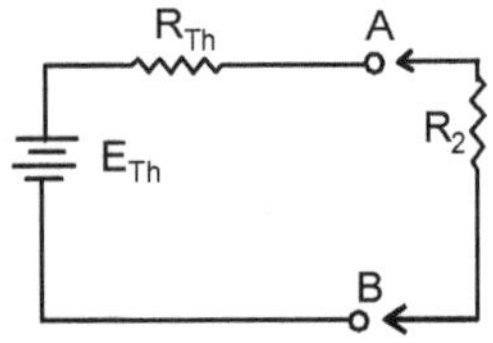

$$R_Th = R_L + \frac{R_1R_3}{R_1+R_3}$$

$$= 10 + \frac{10\times10}{10+10}$$

$$= 15\ \Omega$$

$$I = \frac{E_{Th}}{R_{Th} + R_L}$$

E_{Th} = Voltage across $R_3(10\Omega)$

= current through $R_3 \times$ resistance

$$= \left(\frac{V}{R_1+R_2}\right)\times R_3$$

$$= \frac{15}{10+10}\times 10 = 7.5\ V$$

35. $$A_l = \frac{Am}{1-j\left(\frac{f_L}{f}\right)} \quad ...(i)$$

$$A_h = \frac{Am}{1+j\left(\frac{f}{f_h}\right)} \quad ...(ii)$$

$$A_lf = \frac{A_l}{1+A_l\beta} \quad ...(iii)$$

$$A_hf = \frac{A_h}{1+A_h\beta} \quad ...(iv)$$

A_m = mid frequency,

f_l = lower half power frequency,

A_h = frequency gain,

f_h = upper half power frequency

$$A_lf = \frac{A_m}{1+A_m\beta - j\left(\frac{f_l}{f}\right)}$$

$$= \frac{A_mf}{1+j\left(\frac{f_lf}{f}\right)} \quad ...(v)$$

$$A_mf = \frac{A_m}{(1+A_m\beta)}$$

$A_lf = f_l(1 + A_m\beta)$ is lower half power

$(1 + A_m\beta) > 1$ so that $f_lf < f_l$

$$A_hf = \frac{A_mf}{1+j\left(\frac{f}{f_hf}\right)}$$

$f_hf = f_h\ (1 + A_mB)$ is upper half power frequency,

$f_hf > f_h$

$f_hf = f_h(1 + A_mB)$

$= 200(1 + 1000 \times .99)$

$= 200 \times 100$ Hz

37. k-map is given as

$$y = \bar{Q} + \bar{P}\bar{R}$$

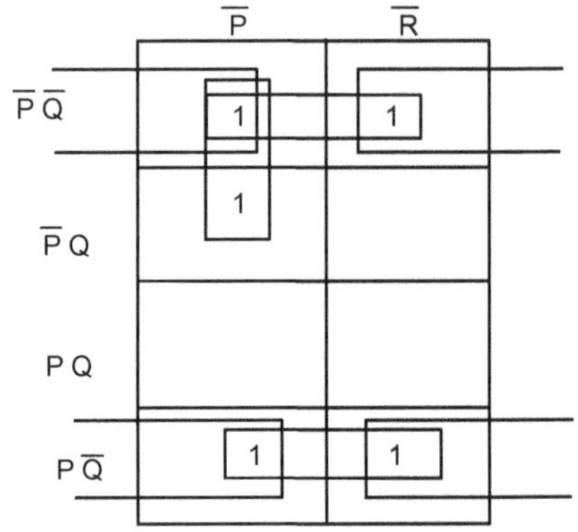

39. Gain $= 20\log_{10}\left|\frac{P_{out}}{P_{in}}\right|$

$$= 20\log_{10}\frac{150}{1.5}\times 20\times 2$$

$= 40$ dB

40. In JFET, the gate source diode always reverse biased. Because of the reverse bias, only a very small reverese current can exist in the gate load. As an approximation, the gate current is Zero. In symbols.

IG = 0

i.e. no input current. It means that the divice has an infinite input resistance.

41. Given circuit is a buffer amplifier.

42. $$F = Z\cdot y(\bar{y}+x)$$

$$= Z\cdot 4\bar{y} + Zyx$$

$$F = xyZ$$

43. For $J-k$ flip-flop

For Q_1 when J = 1, K = 1, CLK = 1 $\Rightarrow Q_1 = 1$

For Q_2 when J = 1, K = 1, CLK = Q_1 = 1 $\Rightarrow Q_2 = 1$

44. Here $Q_D\ Q_C\ Q_B\ Q_A$ = 1010

SI = 0101

After 1st pulse, output = 0010

After 2nd pulse, output = 0110

45. Applying KVL, we get

$$38 = 5I_1 + 4(I_1 + I_2)$$
$$76 = 2I_2 + 4(I_1 + I_2)$$
$$9I_1 + 4I_2 = 38$$
$$4I_1 + 6I_2 = 76$$
$$I_1 = -38/21$$
$$I_1 + I_2 = \frac{38}{21} \times \frac{38}{6}$$
$$V = (I_1 + I_2)R$$
$$= \frac{38 \times 38}{21 \times 6} \times 4 = 52V$$

47. During the positive half cycle, the diode is reverse biased and hence no current flows across the diode therefore output voltage V_0 is 3V. for negative half cycle, the diode is forward biased.

$$\psi(x) = \psi_0 \cos \frac{\pi x}{4L}$$

where $x = -2L + 0\ x = +2L$

ψ is normalized, if

$$\int_{-2L}^{2L} \psi^* \psi dx = 1$$

$$\psi_0^2 \int_{-2L}^{2L} \cos^2 \frac{2\pi x}{4L} dx = 1$$

$$\psi_0 = 1/\sqrt{2L}$$

48. Its very low gate current.

49. a NOR gate

50. Current, $i_1 = V_{in} j\omega C_1 \quad V_i = \sin \omega t$.

51. $$R_2 = \frac{1}{\omega C} = \frac{1}{2\pi f C} = 1.69\,k\Omega$$

$$A_V = \frac{15}{R_1}$$

$$R_1 = 10\ k\Omega$$

52. Since it is a 2nd order linear differential equation, two integrators are required and an OP-AMP adder to sum the components.

53. $Y = \overline{B}$

54. 010100 $\longrightarrow 2^4 \times 1 + 2^2 \times 1 = 20$

543210

$$V_0 = \frac{2^6 - 1}{20} = \frac{63}{20} = 3.15\,V.$$

56. Effective inductance $L_{eq} = L_1 + L_2 + 2M$

When two inductor L_1 and L_2 are in series; $M > 0$ for coupling in same sense.

58. $B.(A + B) + A.(\overline{B} + A)$

$$= B.A + B.B + A.\overline{B} + A.A$$
$$= A + B.B + A.A$$
$$= A + BB = A + B$$

59. We know,

$$V_o = \left(\frac{R_3 + R_4}{R_3}\right)\left(\frac{R_2}{R_1 + R_2}\right)V_1 - \frac{R_4}{R_3}V_2$$

where, $R_1 = 1k\Omega, R_2 = 1k\Omega, R_3 = 1k\Omega,$
$R_4 = 1k\Omega; V_1 = 1V, V_2 = 2V$

$$\therefore \quad = \left(\frac{1+5}{1}\right)\left(\frac{1}{1+1}\right)1 - \left(\frac{5}{1} \times 2\right)$$

$$= 3 - 10 = -\mathbf{7V}$$

61. Resistances are in parallel giving

$$R_{th} = \frac{2 \times 2}{2 + 2} = 1\ k\Omega$$

62. 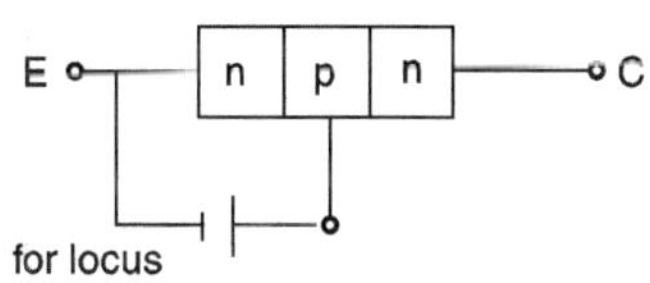

Pinch off region

63. FET have only electron carrier, so FET is the unipolar device.

64. 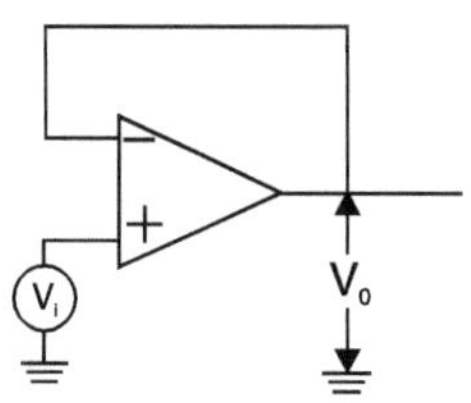

a voltage follows

65. 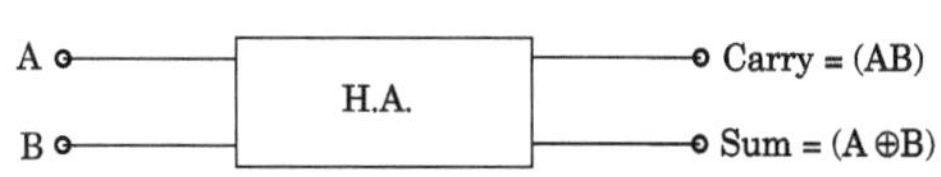

So two input and two output.

66. For zero phase shift in Wien Bridge

$$R = \frac{1}{\omega C} \quad \Rightarrow \quad \omega = \frac{1}{RC}$$

Since $\omega = 2\pi f \quad \Rightarrow f = \dfrac{1}{2\pi RC}$

$\therefore \quad R = \dfrac{1}{2\pi Cf}$

$= \dfrac{1}{2 \times 3.14 \times 0.01 \times 10^{-6} \times 300}$

$= 53\ k\Omega$

Now $\dfrac{R_2}{R_1} = 2$

$\Rightarrow \quad R_2 = 2R_1 = 2 \times 12\ k\Omega = 24\ k\Omega$

$\therefore \quad R_2 = \mathbf{24\ k\Omega}, \quad R = \mathbf{53\ k\Omega.}$

67. $I_b = \left(\dfrac{10}{500k}\right) \quad \Rightarrow \beta = \dfrac{I_c}{I_b}$

$\Rightarrow I_c = \beta\, I_b = 100 \times \left(\dfrac{10}{500\,K}\right) = 2\ mA$

$r_e = 20\,\Omega$ and $V_s = r_e = I_c \,.\, 20 \times 2 = 40\ mV$

But for maximum peak to peak input signal is

$V_s = 2 \times 40 = 80$ mV.

68. Since $\beta = \left(\dfrac{\alpha}{1-\alpha}\right) = \left(\dfrac{0.96}{0.04}\right) = 24$

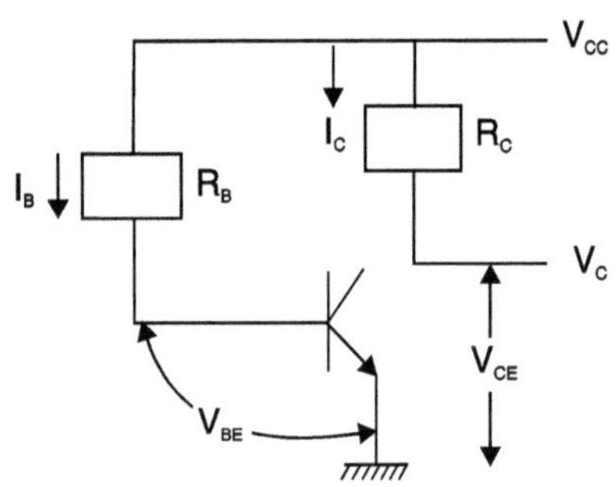

Now, $\beta = \dfrac{I_C}{I_B}, \quad \Rightarrow \quad I_B = \dfrac{I_C}{\beta}$

$V_{CC} = I_B R_B + V_{BE}$

$= \dfrac{I_C}{\beta} R_B + V_{BE}$

$\Rightarrow \quad I_C = \dfrac{(V_{CC} - V_{BE})}{\left(\dfrac{R_B}{\beta}\right)}$

But the value of $V_{BE} << V_{CC}$, so V_{BE} is neglible compared to V_{CC}

$I_C \approx \dfrac{V_{CC}\beta}{R_B} \approx \dfrac{10 \times 24}{100 \times 10^3} = 2.4\ mA$

$V_C = V_{CC} - I_C R_C$

$= 10 - 2.4 \times 10^{-3} \times 2.2 \times 10^3$

$= 10 - 5.28 = \mathbf{4.72\ volts.}$

69. $V_0 = -\dfrac{1}{C_F}\displaystyle\int_0^t \dfrac{V_s(t)}{R_s}\,dt$

$V_0 = -\dfrac{V_s \,.\, t}{R_s C_F}$

$= \dfrac{-3 \times 4}{30 \times 10^6 \times 0.1 \times 10^{-6}}$

$= -4$ volt

the output voltage is – 4 volt.

70. $Y = \overline{A}\,\overline{B}\,\overline{C}D + \overline{A}\,\overline{B}CD + \overline{A}\,B\,\overline{C}D + \overline{A}\,BCD$

$= \overline{A}\,\overline{B}D(\overline{C} + C) + \overline{A}\,BD(\overline{C} + C) \qquad [\because \overline{C} + C = 1]$

$= \overline{A}\,\overline{B}D + \overline{A}BD$

$= \overline{A}D(\overline{B} + B) \qquad [\because \overline{B} + B = 1]$

$Y = \mathbf{\overline{A}D}$

72. For $\quad P = 0, Q = 1, R = 1$

$PQ' = 0 \neq 1 = R$

74. $PQ' + Q = P$

For $\quad P = 0, Q = 1$, not correct

82. $V_{BE} = 0.7\ V$

$\therefore$ Potential across, $R_B = 4.3\ V$

$\Rightarrow \quad I_B R_B = 4.3\ V$

$\Rightarrow \quad I_B = \dfrac{4.3}{10^3} = 4.3\ mA$

Now $\quad \beta = \dfrac{I_C}{I_B}$

$\therefore \quad I_C = 100 \times 4.3\ mF$

$= \mathbf{0.43\ A}$

Printed by Libri Plureos GmbH in Hamburg,
Germany

9 789391 061265